Foreword

For four years the *World Development Indicators* has reported on progress toward the international development goals. While the challenge is immense, the prospects for success in some areas are improving. Between 1990 and 1998 the proportion of people living in extreme poverty fell from 29 to 23 percent— and in China the number in extreme poverty fell by almost 150 million. In Liberia the rate of infant deaths dropped from 155 per 1,000 live births in 1990 to 113 in 1998, and at least 25 other developing countries lowered infant mortality rates fast enough to reach the goal for 2015. This is some of the good news.

But other data are more sobering. Despite remarkable success in some countries, none of the international development goals for health and education is likely, on present trends, to be achieved at the global level. We are not likely to achieve a two-thirds decline in infant and under-five mortality or a three-fourths decline in maternal mortality. And we are not likely to have universal primary education by 2015.

With less than 15 years to reach the goals, it is time for renewed and vigorous efforts to make good on our commitment to free our fellow men, women, and children from the cruel grip of poverty. That means action by rich and poor alike.

At a time of unprecedented prosperity in developed countries, it is simply unacceptable that more than one child of seven in Africa will die before his or her fifth birthday. Now is the time for a concerted appeal to the heads of government of major aid donors to make clear, once and for all, that development assistance is a vital investment in global peace and security. We must remind them that their current levels of aid, at some 0.24 percent of gross national income, fall far short of the 0.7 percent target many promised to meet. That shortfall is $100 billion a year—for millions of children the difference between life and death.

Rich countries must also open their markets and reduce their agricultural subsidies. In 1995 tariffs in high-income countries cost developing countries more than $40 billion. Nontariff barriers and other forms of trade protection at least double that cost—far in excess of the value of official development assistance at $59 billion. Working with developing countries, we must also push ahead with debt relief. Last year we reached agreements that will bring $34 billion in debt service relief to 22 poor countries. That relief, with its focus on poverty reduction, gives some of the poorest countries the means and incentives to make real progress in their fight to reduce poverty.

In all these issues we must be judged by results. That is why the statistics in the *World Development Indicators* are so important. They remind us that in our work with governments we must widen the focus beyond growth. We must reduce infant, child, and maternal mortality and increase access to education, nutrition, and health care. And we must sustain the benefits that flow from the environment for future generations. That is what development is truly about. Knowledge and hard facts can help us do this, and that is why the *World Development Indicators* is such an important tool.

James D. Wolfensohn
President
The World Bank Group

Acknowledgments

This book and its companion volumes, the *World Bank Atlas* and the *Little Data Book,* were prepared by a team led by Eric Swanson. The team consisted of Mehdi Akhlaghi, David Cieslikowski, Richard Fix, Amy Heyman, Masako Hiraga, M. H. Saeed Ordoubadi, Sulekha Patel, K. M. Vijayalakshmi, Amy Wilson, and Estela Zamora, working closely with other teams in the Development Economics Vice Presidency's Development Data Group. The CD-ROM development team included Azita Amjadi, Elizabeth Crayford, Reza Farivari, Anat Lewin, and William Prince. The work was carried out under the management of Shaida Badiee.

The choice of indicators and textual content was shaped through close consultation with and substantial contributions from staff in the World Bank's four thematic networks—Environmentally and Socially Sustainable Development; Private Sector Development and Infrastructure; Human Development; and Poverty Reduction and Economic Management—and staff of the International Finance Corporation and the Multilateral Investment Guarantee Agency. Most important, we received substantial help, guidance, and data from our external partners. For individual acknowledgments of contributions to the book's content, please see the *Credits* section. For a listing of our key partners, see the *Partners* section.

Bruce Ross-Larson was the principal editor, and Peter Grundy the art director. The cover and page design and the layout were done by Communications Development Incorporated with Grundy & Northedge of London. Staff from External Affairs oversaw publication and dissemination of the book.

This is the 5th edition of the *World Development Indicators* in its current format, the 24th since the World Bank began publishing a comprehensive set of development indicators. We believe that statistics matter. We also believe that by publishing objective information on the condition of the world and its people, we can contribute to better decisionmaking and hasten the achievement of our goals.

With its focus on the international development goals in *World view,* the *World Development Indicators* tries to capture the most important issues facing developing countries: the health, education, and welfare of people; sustainable use of natural resources and the reversal of environmental losses; economic growth and management of the economy; the complementary roles of governments and the private sector; and the continuing integration with the world economy.

Such statistics are a classic example of a "public good." They are costly to produce, but once produced they can and should be widely distributed and used. But as valuable as good statistics may be, the public agencies that produce them are often understaffed, underfunded, and poorly equipped. Improving the quality of statistics—and the quality of the decisions they support— requires building the capacity of statistical offices. How? Through investments in people and equipment and in the knowledge to sustain a consistent level and quality of statistical output.

The World Bank is supporting these investments through a wide range of efforts. Four examples of our ongoing work:

First, as a member of the Partnership in Statistics for the 21st Century—or PARIS21, a consortium of donor agencies and developing countries formed in 1999—the World Bank is working closely with its development partners to raise awareness of the need for and value of good statistics—and to increase the resources available for statistical capacity building in developing countries.

Second, the World Bank is devoting special efforts to statistical capacity building in the countries preparing poverty reduction strategies—offering training programs through the World Bank Institute and bringing together poverty specialists, analysts, and statisticians. And as part of the International Monetary Fund's General Data Dissemination System (GDDS) initiative, the World Bank is supporting work by countries to document their current practices in producing and disseminating statistical information and to develop plans for improvement.

Third, through the International Comparison Programme (ICP), the World Bank is working to improve the quality of data for comparing standards of living across countries. It is collaborating with other agencies to prepare and launch the next global ICP round.

And fourth, thanks to the generous support of several donors, the World Bank has established a trust fund for building statistical capacity. The first grants from the trust fund to national statistical agencies were recently approved. And we hope that by working together we can continue to improve the quality of data found here.

Meanwhile, we appreciate your comments and responses. Please keep sending them to us at data@worldbank.org. And for more information on the World Bank's statistical publications, please visit our Web site at www.worldbank.org and select *data* on the menu.

Shaida Badiee
Director
Development Data Group

Contents

1   WORLD VIEW

6 GLOBAL LINKS

Partners

Defining, gathering, and disseminating international statistics is a collective effort of many people and organizations. The indicators presented in the *World Development Indicators* are the fruit of decades of work at many levels, from the field workers who administer censuses and household surveys to the committees and working parties of the national and international statistical agencies that develop the nomenclature, classifications, and standards fundamental to an international statistical system. Nongovernmental organizations and the private sector have also made important contributions, both in gathering primary data and in organizing and publishing their results. And academic researchers have played a crucial role in developing statistical methods and carrying on a continuing dialogue about the quality and interpretation of statistical indicators. All these contributors have a strong belief that available, accurate data will improve the quality of public and private decisionmaking.

The organizations listed here have made the *World Development Indicators* possible by sharing their data and their expertise with us. More important, their collaboration contributes to the World Bank's efforts, and to those of many others, to improve the quality of life of the world's people. We acknowledge our debt and gratitude to all who have helped to build a base of comprehensive, quantitative information about the world and its people.

For your easy reference we have included URLs (Web addresses) for organizations that maintain Web sites. The addresses shown were active on 1 March 2001. Information about the World Bank is also provided.

International and government agencies

Bureau of Verification and Compliance, U.S. Department of State

The Bureau of Verification and Compliance, U.S. Department of State, is responsible for international agreements on conventional, chemical, and biological weapons and on strategic forces; treaty verification and compliance; and support to ongoing negotiations, policymaking, and interagency implementation efforts.

For information contact the Public Affairs Officer, Bureau of Verification and Compliance, U.S. Department of State, 2201 C Street NW, Washington, DC 20520, USA; telephone: 202 647 6946; Web site: www.state.gov/www/global/arms/bureauvc.html.

Carbon Dioxide Information Analysis Center

The Carbon Dioxide Information Analysis Center (CDIAC) is the primary global change data and information analysis center of the U.S. Department of Energy. The CDIAC's scope includes potentially anything that would be of value to those concerned with the greenhouse effect and global climate change, including concentrations of carbon dioxide and other radiatively active gases in the atmosphere; the role of the terrestrial biosphere and the oceans in the biogeochemical cycles of greenhouse gases; emissions of carbon dioxide to the atmosphere; long-term climate trends; the effects of elevated carbon dioxide on vegetation; and the vulnerability of coastal areas to rising sea levels.

For information contact the CDIAC, Oak Ridge National Laboratory, PO Box 2008, Oak Ridge, TN 37831-6335, USA; telephone: 865 574 0390; fax: 865 574 2232; email: cdiac@ornl.gov; Web site: cdiac.esd.ornl.gov.

Food and Agriculture Organization

The Food and Agriculture Organization (FAO), a specialized agency of the United Nations, was founded in October 1945 with a mandate to raise nutrition levels and living standards, to increase agricultural productivity, and to better the condition of rural populations. The organization provides direct development assistance; collects, analyzes, and disseminates information; offers policy and planning advice to governments; and serves as an international forum for debate on food and agricultural issues.

Statistical publications of the FAO include the *Production Yearbook, Trade Yearbook,* and *Fertilizer Yearbook.* The FAO makes much of its data available on diskette through its Agrostat PC system.

FAO publications can be ordered from national sales agents or directly from the FAO Sales and Marketing Group, Viale delle Terme di Caracalla, 00100 Rome, Italy; telephone: 39 06 57051; fax: 39 06 5705/3152; email: Publications-sales@fao.org; Web site: www.fao.org.

International Civil Aviation Organization

The International Civil Aviation Organization (ICAO), a specialized agency of the United Nations, was founded on 7 December 1944. It is responsible for establishing international standards and recommended practices and procedures for the technical, economic, and legal aspects of international civil aviation operations. The ICAO works to achieve the highest practicable degree of uniformity worldwide in civil aviation issues whenever this will facilitate and improve air safety, efficiency, and regularity.

To obtain ICAO publications contact the ICAO, Document Sales Unit, 999 University Street, Montreal, Quebec H3C 5H7, Canada; telephone: 514 954 8022; fax: 514 954 6769; email: sales_unit@icao.int; Web site: www.icao.int.

International Labour Organization

The International Labour Organization (ILO), a specialized agency of the United Nations, seeks the promotion of social justice and internationally recognized human and labor rights. Founded in 1919, it is the only surviving major creation of the Treaty of Versailles, which brought the League of Nations into being. It became the first specialized agency of the United Nations in 1946. Unique within the United Nations system, the ILO's tripartite structure has workers and employers participating as equal partners with governments in the work of its governing organs.

As part of its mandate, the ILO maintains an extensive statistical publication program. The *Yearbook of Labour Statistics* is its most comprehensive collection of labor force data.

Publications can be ordered from the International Labour Office, 4 route des Morillons, CH-1211 Geneva 22, Switzerland, or from sales agents and major booksellers throughout the world and ILO offices in many countries. Telephone: 41 22 799 78 66; fax: 41 22 799 61 17; email: publns@ilo.org; Web site: www.ilo.org.

International Monetary Fund

The International Monetary Fund (IMF) was established at a conference in Bretton Woods, New Hampshire, United States, on 1–22 July 1944. (The conference also established the World Bank.) The IMF came into official existence on 27 December 1945 and commenced financial operations on 1 March 1947. It currently has 183 member countries.

The statutory purposes of the IMF are to promote international monetary cooperation, facilitate the expansion and balanced growth of international trade, promote exchange rate stability, help to establish a multilateral payments system, make the general resources of the IMF temporarily available to its members under adequate safeguards, and shorten the duration and lessen the degree of disequilibrium in the international balances of payments of members.

The IMF maintains an extensive program for the development and compilation of international statistics and is responsible for collecting and reporting statistics on international financial transactions and the balance of payments. In April 1996 it undertook an important initiative to improve the quality of international statistics, establishing the Special Data Dissemination Standard (SDDS) to guide members that have, or seek, access to international capital markets in providing economic and financial data to the public. In 1997 the IMF established the General Data Dissemination System (GDDS) to guide countries in providing the public with comprehensive, timely, accessible, and reliable economic, financial, and sociodemographic data.

The IMF's major statistical publications include *International Financial Statistics, Balance of Payments Statistics Yearbook, Government Finance Statistics Yearbook,* and *Direction of Trade Statistics Yearbook.*

For more information on IMF statistical publications contact the International Monetary Fund, Publications Services, Catalog Orders, 700 19th Street NW, Washington, DC 20431, USA; telephone: 202 623 7430; fax: 202 623 7201; telex: RCA 248331 IMF UR; email: pub-web@imf.org; Web site: www.imf.org; SDDS and GDDS bulletin board: dsbb.imf.org.

International Telecommunication Union

Founded in Paris in 1865 as the International Telegraph Union, the International Telecommunication Union (ITU) took its current name in 1934 and became a specialized agency of the United Nations in 1947. The ITU is an intergovernmental organization in which the public and private sectors cooperate for the development of telecommunications. The ITU adopts international regulations and treaties governing all terrestrial and space uses of the frequency spectrum and the use of the geostationary satellite orbit. It also develops standards for the interconnection of telecommunications systems worldwide.

The ITU fosters the development of telecommunications in developing countries by establishing medium-term development policies and strategies in consultation with other partners in the sector and providing specialized technical assistance in management, telecommunications policy, human resource management, research and development, technology choice and transfer, network installation and maintenance, and investment financing and resource mobilization.

The ITU's main statistical publication is the *Telecommunications Yearbook.*

Publications can be ordered from ITU Sales and Marketing Service, Place des Nations, CH-1211 Geneva 20, Switzerland; telephone: 41 22 730 6141 (English), 41 22 730 6142 (French), and 41 22 730 6143 (Spanish); fax: 41 22 730 5194; email: sales@itu.int; telex: 421 000 uit ch; telegram: ITU GENEVE; Web site: www.itu.int.

National Science Foundation

The National Science Foundation (NSF) is an independent U.S. government agency whose mission is to promote the progress of science; to advance the national health, prosperity, and welfare; and to secure the national defense. It is responsible for promoting science and engineering through almost 20,000 research and education projects. In addition, the NSF fosters the exchange of scientific information among scientists and engineers in the United States and other countries, supports programs to strengthen scientific and engineering research potential, and evaluates the impact of research on industrial development and general welfare.

As part of its mandate, the NSF biennially publishes *Science and Engineering Indicators,* which tracks national and international trends in science and engineering research and education.

Electronic copies of NSF documents can be obtained from the NSF's online document system (www.nsf.gov/pubsys/index.htm) or requested by email from its automated mailserver (getpub@nsf.gov). Documents can also be requested from the NSF Publications Clearinghouse by mail, at PO Box 218, Jessup, MD 20794-0218, or by telephone, at 301 947 2722.

For more information contact the National Science Foundation, 4201 Wilson Boulevard, Arlington, VA 22230, USA; telephone: 703 292 5111; Web site: www.nsf.gov.

Organisation for Economic Co-operation and Development

The Organisation for Economic Co-operation and Development (OECD) was set up in 1948 as the Organisation for European Economic Co-operation (OEEC) to administer Marshall Plan funding in Europe. In 1960, when the Marshall Plan had completed its task, the OEEC's member countries agreed to bring in Canada and the United States to form an organization to coordinate policy among industrial countries. The OECD is the international organization of the industrialized, market economy countries.

Representatives of member countries meet at the OECD to exchange information and harmonize policy with a view to maximizing economic growth in member countries and helping non-member countries develop more rapidly. The OECD has set up a number of specialized committees to further its aims. One of these is the Development Assistance Committee (DAC), whose members have agreed to coordinate their policies on assistance to developing and transition economies.

Also associated with the OECD are several agencies or bodies that have their own governing statutes, including the International Energy Agency and the Centre for Co-operation with Economies in Transition.

The OECD's main statistical publications include *Geographical Distribution of Financial Flows to Aid Recipients, National Accounts of OECD Countries, Labour Force Statistics, Revenue Statistics of OECD Member Countries, International Direct Investment Statistics Yearbook, Basic Science* and Technology Statistics, Industrial Structure Statistics, and Services: Statistics on International Transactions.

For information on OECD publications contact the OECD, 2, rue André-Pascal, 75775 Paris Cedex 16, France; telephone: 33 1 45 24 82 00; fax: 33 1 49 10 42 76; email: sales@oecd.org; Web sites: www.oecd.org and www.oecdwash.org.

United Nations

The United Nations and its specialized agencies maintain a number of programs for the collection of international statistics, some of which are described elsewhere in this book. At United Nations headquarters the Statistics Division provides a wide range of statistical outputs and services for producers and users of statistics worldwide.

The Statistics Division publishes statistics on international trade, national accounts, demography and population, gender, industry, energy, environment, human settlements, and disability. Its major statistical publications include the International Trade Statistics Yearbook, Yearbook of National Accounts, and Monthly Bulletin of Statistics, along with general statistics compendiums such as the Statistical Yearbook and World Statistics Pocketbook.

For publications contact United Nations Publications, Room DC2 853, 2 UN Plaza, New York, NY 10017, USA; telephone: 212 963 8302 or 800 253 9646 (toll free); fax: 212 963 3489; email: publications@un.org; Web site: www.un.org.

United Nations Centre for Human Settlements (Habitat), Global Urban Observatory

The Urban Indicators Programme of the United Nations Centre for Human Settlements (Habitat) was established to address the urgent global need to improve the urban knowledge base by helping countries and cities design, collect, and apply policy-oriented indicators related to urban development at the city level. In 1997 the Urban Indicators Programme was integrated into the Global Urban Observatory, the principal United Nations program for monitoring urban conditions and trends and for tracking progress in implementing the goals of the Habitat Agenda. With the Urban Indicators and Best Practices programs, the Global Urban Observatory is establishing a worldwide information, assessment, and capacity building network to help governments, local authorities, the private sector, and nongovernmental and other civil society organizations.

Contact Christine Auclair (guo@unchs.org), Urban Indicators Programme, Global Urban Observatory, UNCHS (Habitat), PO Box 30030, Nairobi, Kenya; telephone: 2542 623694; fax: 2542 624266/7; Web site: www.unchs.org.

United Nations Children's Fund

The United Nations Children's Fund (UNICEF), the only organization of the United Nations dedicated exclusively to children, works with other United Nations bodies and with governments and nongovernmental organizations to improve children's lives in more than 140 developing countries through community-based services in primary health care, basic education, and safe water and sanitation.

UNICEF's major publications include The State of the World's Children and The Progress of Nations.

For information on UNICEF publications contact UNICEF House, 3 United Nations Plaza, New York, NY 10017, USA; telephone: 212 326 7000; fax: 212 888 7465 or 7454; telex: RCA-239521; email: publications@un.org; Web site: www.unicef.org.

United Nations Conference on Trade and Development

The United Nations Conference on Trade and Development (UNCTAD) is the principal organ of the United Nations General Assembly in the field of trade and development. It was established as a permanent intergovernmental body in 1964 in Geneva with a view to accelerating economic growth and development, particularly in developing countries. UNCTAD discharges its mandate through policy analysis; intergovernmental deliberations, consensus building, and negotiation; monitoring, implementation, and follow-up; and technical cooperation.

UNCTAD produces a number of publications containing trade and economic statistics, including the *Handbook of International Trade and Development Statistics*.

For information contact UNCTAD, Palais des Nations, CH-1211 Geneva 10, Switzerland; telephone: 41 22 907 12 34 or 917 12 34; fax: 41 22 907 00 43; telex: 42962; email: reference.service@unctad.org; Web site: www.unctad.org.

United Nations Educational, Scientific, and Cultural Organization

The United Nations Educational, Scientific, and Cultural Organization (UNESCO) is a specialized agency of the United Nations established in 1945 to promote "collaboration among nations through education, science, and culture in order to further universal respect for justice, for the rule of law, and for the human rights and fundamental freedoms . . . for the peoples of the world, without distinction of race, sex, language, or religion."

UNESCO's principal statistical publications are the *Statistical Yearbook, World Education Report* (biennial), and *Basic Education and Literacy: World Statistical Indicators*.

For publications contact UNESCO Publishing, Promotion, and Sales Division, 1, rue Miollis F, 75732 Paris Cedex 15, France; fax: 33 1 45 68 57 41; email: publishing.promotion@unesco.org; Web site: www.unesco.org.

United Nations Environment Programme

The mandate of the United Nations Environment Programme (UNEP) is to provide leadership and encourage partnership in caring for the environment by inspiring, informing, and enabling nations and people to improve their quality of life without compromising that of future generations.

UNEP publications include *Global Environment Outlook* and *Our Planet* (a bimonthly magazine).

For information contact the UNEP, PO Box 30552, Nairobi, Kenya; telephone: 254 2 62 1234 or 3292; fax: 254 2 22 6886 or 62 2615; email: oedinfo@unep.org; Web site: www.unep.org.

United Nations Industrial Development Organization

The United Nations Industrial Development Organization (UNIDO) was established in 1966 to act as the central coordinating body for industrial activities and to promote industrial development and cooperation at the global, regional, national, and sectoral levels. In 1985 UNIDO became the 16th

Partners

specialized agency of the United Nations, with a mandate to help develop scientific and technological plans and programs for industrialization in the public, cooperative, and private sectors.

UNIDO's databases and information services include the Industrial Statistics Database (INDSTAT), Commodity Balance Statistics Database (COMBAL), Industrial Development Abstracts (IDA), and the International Referral System on Sources of Information. Among its publications is the *International Yearbook of Industrial Statistics.*

For information contact UNIDO Public Information Section, Vienna International Centre, PO Box 300, A-1400 Vienna, Austria; telephone: 43 1 260 26 5031; fax: 43 1 213 46 5031 or 260 26 6843; email: publications@unido.org; Web site: www.unido.org.

World Bank Group

The World Bank Group is made up of five organizations: the International Bank for Reconstruction and Development (IBRD), the International Development Association (IDA), the International Finance Corporation (IFC), the Multilateral Investment Guarantee Agency (MIGA), and the International Centre for Settlement of Investment Disputes (ICSID).

Established in 1944 at a conference of world leaders in Bretton Woods, New Hampshire, United States, the World Bank is the world's largest source of development assistance, providing nearly $16 billion in loans annually to its client countries. It uses its financial resources, trained staff, and extensive knowledge base to help each developing country onto a path of stable, sustainable, and equitable growth in the fight against poverty. The World Bank Group has 182 member countries.

For information about the World Bank visit its Web site at www.worldbank.org. For more information about development data contact the Development Data Group, World Bank, 1818 H Street NW, Washington, DC 20433, USA; telephone: 800 590 1906 or 202 473 7824; fax: 202 522 1498; email: data@worldbank.org; Web site: www.worldbank.org/data.

World Health Organization

The constitution of the World Health Organization (WHO) was adopted on 22 July 1946 by the International Health Conference, convened in New York by the Economic and Social Council. The objective of the WHO, a specialized agency of the United Nations, is the attainment by all people of the highest possible level of health.

The WHO carries out a wide range of functions, including coordinating international health work; helping governments strengthen health services; providing technical assistance and emergency aid; working for the prevention and control of disease; promoting improved nutrition, housing, sanitation, recreation, and economic and working conditions; promoting and coordinating biomedical and health services research; promoting improved standards of teaching and training in health and medical professions; establishing international standards for biological, pharmaceutical, and similar products; and standardizing diagnostic procedures.

The WHO publishes the *World Health Statistics Annual* and many other technical and statistical publications.

For publications contact Distribution and Sales, Division of Publishing, Language, and Library Services, World Health Organization Headquarters, CH-1211 Geneva 27, Switzerland; telephone:

41 22 791 2476 or 2477; fax: 41 22 791 4857; email: publications@who.ch; Web site: www.who.ch.

World Intellectual Property Organization

The World Intellectual Property Organization (WIPO) is a specialized agency of the United Nations based in Geneva, Switzerland. The objectives of WIPO are to promote the protection of intellectual property throughout the world through cooperation among states and, where appropriate, in collaboration with other international organizations and to ensure administrative cooperation among the intellectual property unions—that is, the "unions" created by the Paris and Berne Conventions and several subtreaties concluded by members of the Paris Union. WIPO is responsible for administering various multilateral treaties dealing with the legal and administrative aspects of intellectual property. A substantial part of its activities and resources is devoted to development cooperation with developing countries.

For information contact the World Intellectual Property Organization, 34, chemin des Colombettes, Geneva, Switzerland; mailing address: PO Box 18, CH-1211 Geneva 20, Switzerland; telephone: 41 22 338 9111; fax: 41 22 733 5428; telex: 412912 ompi ch; email: publications.mail@wipo.int; Web site: www.wipo.int.

World Tourism Organization

The World Tourism Organization is an intergovernmental body charged by the United Nations with promoting and developing tourism. It serves as a global forum for tourism policy issues and a source of tourism know-how. The organization began as the International Union of Official Tourist Publicity Organizations, set up in 1925 in The Hague. Renamed the World Tourism Organization, it held its first general assembly in Madrid in May 1975. Its membership includes 132 countries and territories and more than 350 affiliate members representing local governments, tourism associations, and private companies, including airlines, hotel groups, and tour operators.

The World Tourism Organization publishes the *Yearbook of Tourism Statistics, Compendium of Tourism Statistics,* and *Travel and Tourism Barometer* (triannual).

For information contact the World Tourism Organization, Capitán Haya, 42, 28020 Madrid, Spain; telephone: 34 91 567 81 00; fax: 34 91 567 82 18; email: omt@world-tourism.org; Web site: www.world-tourism.org.

World Trade Organization

The World Trade Organization (WTO), established on 1 January 1995, is the successor to the General Agreement on Tariffs and Trade (GATT). The WTO provides the legal and institutional foundation of the multilateral trading system and embodies the results of the Uruguay Round of trade negotiations, which ended with the Marrakesh Declaration of 15 April 1994. The WTO is mandated with administering and implementing multilateral trade agreements, serving as a forum for multilateral trade negotiations, seeking to resolve trade disputes, overseeing national trade policies, and cooperating with other international institutions involved in global economic policymaking.

The WTO's Statistics and Information Systems Divisions compile statistics on world trade and maintain the Integrated Database, which contains the basic records of the outcome of the Uruguay Round. Its *Annual Report* includes a statistical appendix.

Partners

For publications contact the World Trade Organization, Publications Services, Centre William Rappard, 154 rue de Lausanne, CH-1211, Geneva, Switzerland; telephone: 41 22 739 5208 or 5308; fax: 41 22 739 5792; email: publications@wto.org; Web site: www.wto.org.

Private and nongovernmental organizations

Euromoney Publications PLC

Euromoney Publications PLC provides a wide range of financial, legal, and general business information. The monthly magazine *Euromoney* carries a semiannual rating of country creditworthiness.

For information contact Euromoney Publications PLC, Nestor House, Playhouse Yard, London EC4V 5EX, UK; telephone: 44 20 7779 8999; fax: 44 20 7779 8602; telex: 2907002; email: hotline@euromoneyplc.com; Web site: www.euromoney.com.

Institutional Investor, Inc.

Institutional Investor, Inc., develops country credit ratings every six months based on information provided by leading international banks. It publishes the magazine *Institutional Investor* monthly.

For information contact Institutional Investor, Inc., 488 Madison Avenue, New York, NY 10022, USA; telephone: 212 224 3300; email: info@iimagazine.com; Web site: www.iimagazine.com.

International Road Federation

The International Road Federation (IRF) is a not-for-profit, nonpolitical service organization. Its purpose is to encourage better road and transport systems worldwide and to help apply technology and management practices that will maximize economic and social returns from national road investments. The IRF has led global road infrastructure developments and is the international point of affiliation for about 600 member companies, associations, and governments.

The IRF's mission is to promote road development as a key factor in social and economic growth, to provide governments and financial institutions with professional ideas and expertise, to facilitate business exchange among members, to establish links between members and external institutions and agencies, to support national road federations, and to give information to professional groups.

The IRF publishes *World Road Statistics*.

Contact the Geneva office at 2 chemin de Blandonnet, CH-1214 Vernier, Geneva, Switzerland; telephone: 41 22 306 0260; fax: 41 22 306 0270; or the Washington, DC, office at 1010 Massachusetts Avenue NW, Suite 410, Washington, DC 20001, USA; telephone: 202 371 5544; fax: 202 371 5565; email: info@irfnet.com; Web site: www.irfnet.org.

Monetary Research Institute

The Monetary Research Institute (MRI) was founded in 1990 to collect information about the current means of payment in the world. Its flagship publication, the quarterly *MRI Bankers' Guide to Foreign Currency*, is designed for use by banks, foreign exchange bureaus, libraries, universities, coin dealers, travel agents, and those relying on international trade. It features information on and images of all currencies and banknotes in circulation, information on travelers checks, and currency histories, news, and approaching expiration dates. It also lists tourist and parallel exchange rates for every country. The MRI maintains relationships with all currency issuing authorities.

For information contact the Monetary Research Institute, 1014 Wirt Road, Suite 200, Houston, TX 77055, USA; telephone: 713 827 1796; fax: 713 827 8665; email: info@mriguide.com; Web site: www.mriguide.com.

Moody's Investors Service

Moody's Investors Service is a global credit analysis and financial opinion firm. It provides the international investment community with globally consistent credit ratings on debt and other securities issued by North American state and regional government entities, by corporations worldwide, and by some sovereign issuers. It also publishes extensive financial data in both print and electronic form. Its clients include investment banks, brokerage firms, insurance companies, public utilities, research libraries, manufacturers, and government agencies and departments.

Moody's publishes *Sovereign, Subnational and Sovereign-Guaranteed Issuers.*

For information contact Moody's Investors Service, 99 Church Street, New York, NY 10007, USA; telephone: 212 553 1658; fax: 212 553 0882; Web site: www.moodys.com.

Netcraft

Netcraft is an Internet consultancy based in Bath, England. Most of its work relates to the development of Internet services for its clients or for itself acting as principal.

For information visit its Web site: www.netcraft.com.

PricewaterhouseCoopers

Drawing on the talents of 150,000 people in more than 150 countries, PricewaterhouseCoopers provides a full range of business advisory services to leading global, national, and local companies and public institutions. Its service offerings have been organized into six lines of service, each staffed with highly qualified, experienced professionals and leaders. These services are audit, assurance, and business advisory services; business process outsourcing; financial advisory services; global human resource solutions; management consulting services; and global tax services.

PricewaterhouseCoopers publishes *Corporate Taxes: Worldwide Summaries* and *Individual Taxes: Worldwide Summaries.*

For information contact PricewaterhouseCoopers, 1301 Avenue of the Americas, New York, NY 10019, USA; telephone: 212 596 8000; fax: 212 259 1301; Web site: www.pwcglobal.com.

The PRS Group

PRS Group is a global leader in political and economic risk forecasting and market analysis and has served international companies large and small for about 20 years. The data it contributed to this year's *World Development Indicators* come from the *International Country Risk Guide,* a monthly publication that monitors and rates political, financial, and economic risk in 140 countries. The guide's data series and commitment to independent and unbiased analysis make it the standard for any organization practicing effective risk management.

For information contact the PRS Group, 6320 Fly Road, Suite 102, PO Box 248, East Syracuse, NY 13057-0248, USA; telephone: 315 431 0511; fax: 315 431 0200; email: custserv@PRSgroup.com; Web site: www.prsgroup.com.

Standard & Poor's Equity Indexes and Rating Services

Standard & Poor's, a division of the McGraw-Hill Companies, has provided independent and objective financial information, analysis, and research for nearly 140 years. The S&P 500 index, one of its most popular products, is calculated and maintained by Standard & Poor's Index Services, a leading provider of equity indexes. Standard & Poor's indexes are used by investors around the world for measuring investment performance and as the basis for a wide range of financial instruments.

Standard & Poor's *Sovereign Ratings* provides issuer and local and foreign currency debt ratings for sovereign governments and for sovereign-supported and supranational issuers worldwide. Standard & Poor's Rating Services monitors the credit quality of $1.5 trillion worth of bonds and other financial instruments and offers investors global coverage of debt issuers. Standard & Poor's also has ratings on commercial paper, mutual funds, and the financial condition of insurance companies worldwide.

For information on equity indexes contact Standard & Poor's Index Services, 22 Water Street, New York, NY 10041, USA; telephone: 212 438 2046; fax: 212 438 3523; email: index_services@sandp.com; Web site: www.spglobal.com.

For information on ratings contact the McGraw-Hill Companies, Inc., Executive Offices, 1221 Avenue of the Americas, New York, NY 10020, USA; telephone: 212 512 4105 or 800 352 3566 (toll free); fax: 212 512 4105; email: ratings@mcgraw-hill.com; Web site: www.standardand poor.com/ratingsactions/ratingslists/.

World Conservation Monitoring Centre

WORLD CONSERVATION
MONITORING CENTRE

The World Conservation Monitoring Centre (WCMC) provides information on the conservation and sustainable use of the world's living resources and helps others to develop information systems of their own. It works in close collaboration with a wide range of people and organizations to increase access to the information needed for wise management of the world's living resources. Committed to the principle of data exchange with other centers and noncommercial users, the WCMC, whenever possible, places the data it manages in the public domain.

For information contact the World Conservation Monitoring Centre, 219 Huntingdon Road, Cambridge CB3 0DL, UK; telephone: 44 12 2327 7314; fax: 44 12 2327 7136; email: info@wcmc.org.uk; Web site: www.unep-wcmc.org.

World Information Technology and Services Alliance

The World Information Technology and Services Alliance (WITSA) is a consortium of 41 information technology (IT) industry associations from around the world. WITSA members represent more than 97 percent of the world IT market. As the global voice of the IT industry, WITSA is dedicated to advocating policies that advance the industry's growth and development; facilitating international trade and investment in IT products and services; strengthening WITSA's national industry associations by sharing knowledge, experience, and information; providing members with a network of contacts in nearly every region; and hosting the World Congress on Information Technology.

WITSA's publication, *Digital Planet 2000: The Global Information Economy,* uses data provided by the International Data Corporation.

For information contact WITSA, 8300 Boone Boulevard, Suite 450, Vienna, VA 22182, USA; telephone: 703 284 5329; Web site: www.witsa.org.

World Resources Institute

The World Resources Institute is an independent center for policy research and technical assistance on global environmental and development issues. The institute provides—and helps other institutions provide—objective information and practical proposals for policy and institutional change that will foster environmentally sound, socially equitable development. The institute's current areas of work include trade, forests, energy, economics, technology, biodiversity, human health, climate change, sustainable agriculture, resource and environmental information, and national strategies for environmental and resource management.

For information contact the World Resources Institute, Suite 800, 10 G Street NE, Washington, DC 20002, USA; telephone: 202 729 7600; fax: 202 729 7610; telex 64414 WRIWASH; email: lauralee@wri.org; Web site: www.wri.org.

Users guide

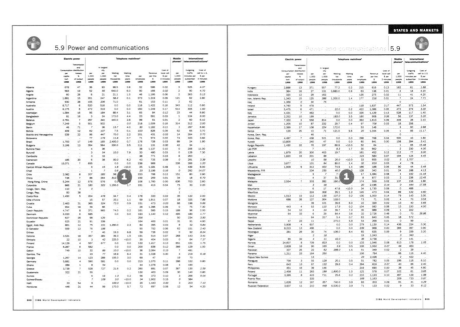

Tables

The tables are numbered by section and display the identifying icon of the section. Countries and economies are listed alphabetically (except for Hong Kong, China, which appears after China). Data are shown for 148 economies with populations of more than 1 million and for which data are regularly reported by the relevant authority, as well as for Taiwan, China, in selected tables. Selected indicators for 59 other economies—small economies with populations between 30,000 and 1 million, smaller economies if they are members of the International Bank for Reconstruction and Development (IBRD, or, as it is commonly known, the World Bank), and larger economies for which data are not regularly reported—are shown in table 1.6. The term *country,* used interchangeably with *economy,* does not imply political independence, but refers to any territory for which authorities report separate social or economic statistics. When available, aggregate measures for income and regional groups appear at the end of each table.

1 Indicators

Indicators are shown for the most recent year or period for which data are available and, in most tables, for an earlier year or period (usually 1990 in this edition). Time-series data are available on the *World Development Indicators* CD-ROM.

2 Aggregate measures for income groups

The aggregate measures for income groups include 207 economies (the economies listed in the main tables plus those in table 1.6) wherever data are available. Note that in this edition, as in the previous one, table 1.6 does not include France's overseas departments—French Guiana, Guadeloupe, Martinique, and Réunion—which are now included in the national accounts (gross national income and other economic measures) of France. To maintain consistency in the aggregate measures over time and between tables, missing data are imputed where possible. The aggregates are totals (designated by a *t* if the aggregates include gap-filled estimates for missing data, and by an *s,* for simple totals, where they do not), median values (*m*), or weighted averages (*w*). Gap filling of amounts not allocated to countries may result in discrepancies between subgroup aggregates and overall totals. For further discussion of aggregation methods see *Statistical methods.*

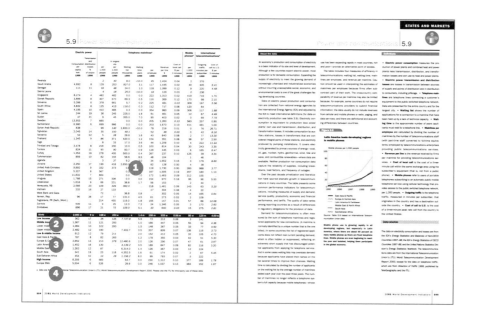

③ Aggregate measures for regions

The aggregate measures for regions include only low- and middle-income economies (note that these measures include developing economies with populations of less than 1 million, including those listed in table 1.6).

The country composition of regions is based on the World Bank's analytical regions and may differ from common geographic usage. For regional classifications see the map on the inside back cover and the list on the back cover flap. For further discussion of aggregation methods see *Statistical methods*.

④ Footnotes

Known deviations from standard definitions or breaks in comparability over time or across countries are either footnoted in the tables or noted in *About the data*. When available data are deemed to be too weak to provide reliable measures of levels and trends or do not adequately adhere to international standards, the data are not shown.

Statistics

Data are shown for economies as they were constituted in 1999, and historical data are revised to reflect current political arrangements. Exceptions are noted throughout the tables.

Additional information about the data is provided in *Primary data documentation.* That section summarizes national and international efforts to improve basic data collection and gives information on primary sources, census years, fiscal years, and other background. *Statistical methods* provides technical information on some of the general calculations and formulas used throughout the book.

Discrepancies in data presented in different editions of the *World Development Indicators* reflect updates by countries as well as revisions to historical series and changes in methodology. Thus readers are advised not to compare data series between editions of the *World Development Indicators* or between different World Bank publications. Consistent time-series data for 1960–99 are available on the *World Development Indicators* CD-ROM.

Except where noted, growth rates are in real terms. (See *Statistical methods* for information on the methods used to calculate growth rates.) Data for some economic indicators for some economies are presented in fiscal years rather than calendar years; see *Primary data documentation.* All dollar figures are current U.S. dollars unless otherwise stated. The methods used for converting national currencies are described in *Statistical methods.*

China
On 1 July 1997 China resumed its exercise of sovereignty over Hong Kong, and on 20 December 1999 it resumed its exercise of sovereignty over Macao. Unless otherwise noted, data for China do not include data for Hong Kong, China; Taiwan, China; or Macao, China.

Democratic Republic of the Congo
Data for the Democratic Republic of the Congo (Congo, Dem. Rep., in the table listings) refer to the former Zaire. The Republic of Congo is referred to as Congo, Rep., in the table listings.

Czech Republic and Slovak Republic
Data are shown whenever possible for the individual countries formed from the former Czechoslovakia—the Czech Republic and the Slovak Republic.

Jordan
Data for Jordan refer to the East Bank only unless otherwise noted.

East Timor
On 25 October 1999 the United Nations Transitional Administration for East Timor (UNTAET) assumed responsibility for the administration of East Timor. Data for Indonesia include East Timor.

Eritrea
Data are shown for Eritrea whenever possible, but in most cases before 1992 Eritrea is included in the data for Ethiopia.

Germany
Data for Germany refer to the unified Germany unless otherwise noted.

Union of Soviet Socialist Republics
In 1991 the Union of Soviet Socialist Republics came to an end. Whenever possible, data are shown for the individual countries now existing on its former territory (Armenia, Azerbaijan, Belarus, Estonia, Georgia, Kazakhstan, Kyrgyz Republic, Latvia, Lithuania, Moldova, the Russian Federation, Tajikistan, Turkmenistan, Ukraine, and Uzbekistan).

República Bolivariana de Venezuela
In December 1999 the official name of Venezuela was changed to República Bolivariana de Venezuela (Venezuela, RB, in the table listings).

Republic of Yemen
Data for the Republic of Yemen refer to that country from 1990 onward; data for previous years refer to aggregated data for the former People's Democratic Republic of Yemen and the former Yemen Arab Republic unless otherwise noted.

Former Socialist Federal Republic of Yugoslavia
Whenever possible, data are shown for the individual countries formed from the former Socialist Federal Republic of Yugoslavia—Bosnia and Herzegovina, Croatia, the former Yugoslav Republic of Macedonia, Slovenia, and the Federal Republic of Yugoslavia. All references to the Federal Republic of Yugoslavia in the tables are to the Federal Republic of Yugoslavia (Serbia/Montenegro) unless otherwise noted.

Changes in the System of National Accounts
For the first time, this edition of the *World Development Indicators* uses terminology in line with the 1993 System of National Accounts (SNA). For example, in the 1993 SNA *gross national income* replaces *gross national product.* See *About the data* for tables 1.1 and 4.9.

Most countries continue to compile their national accounts according to the 1968 SNA, but more and more are adopting the 1993 SNA. Countries that use the 1993 SNA are identified in *Primary data documentation.* A few low-income countries still use concepts from older SNA guidelines, including valuations such as factor cost, in describing major economic aggregates.

Classification of economies

For operational and analytical purposes the World Bank's main criterion for classifying economies is gross national income (GNI) per capita. Every economy is classified as low income, middle income (subdivided into lower middle and upper middle), or high income. For income classifications see the map on the inside front cover and the list on the front cover flap. Note that classification by income does not necessarily reflect development status. Because GNI per capita changes over time, the country composition of income groups may change from one edition of the *World Development Indicators* to the next. Once the classification is fixed for an edition, based on GNI per capita in the most recent year for which data are available (1999 in this edition), all historical data presented are based on the same country grouping.

Low-income economies are those with a GNI per capita of $755 or less in 1999. Middle-income economies are those with a GNI per capita of more than $755 but less than $9,266. Lower-middle-income and upper-middle-income economies are separated at a GNI per capita of $2,995. High-income economies are those with a GNI per capita of $9,266 or more. The 11 participating member countries of the European Monetary Union (EMU) are presented as a subgroup under high-income economies.

Recent revisions of 1999 GNI per capita for Costa Rica, from $2,740 to $3,570, and for Ukraine, from $750 to $840, would place these countries in higher income categories. However, since the official analytical classifications are fixed during the World Bank's fiscal year (ending on 30 June), these countries remain in the income categories in which they were classified before these revisions: Costa Rica in the lower-middle-income category and Ukraine in the low-income category.

Symbols

..

means that data are not available or that aggregates cannot be calculated because of missing data in the years shown.

0 or 0.0

means zero or less than half the unit shown.

/

in dates, as in 1990/91, means that the period of time, usually 12 months, straddles two calendar years and refers to a crop year, a survey year, or a fiscal year.

$

means current U.S. dollars unless otherwise noted.

>

means more than.

<

means less than.

Data presentation conventions

- A blank means not applicable or, for an aggregate, not analytically meaningful.
- A billion is 1,000 million.
- A trillion is 1,000 billion.
- Figures in italics refer to years or periods other than those specified.
- Data for years that are more than three years from the range shown are footnoted.

The cutoff date for data is 1 February 2001.

WORLD VIEW

International development goals

1 Halve the proportion of people living in extreme poverty between 1990 and 2015

2 Enroll all children in primary school by 2015

3 Empower women by eliminating gender disparities in primary and secondary education by 2005

4 Reduce infant and child mortality rates by two-thirds between 1990 and 2015

5 Reduce maternal mortality ratios by three-quarters between 1990 and 2015

6 Provide access to all who need reproductive health services by 2015

7 Implement national strategies for sustainable development by 2005 so as to reverse the loss of environmental resources by 2015

Progress toward the international development goals

Of the world's 6 billion people, 1.2 billion live on less than $1 a day. About 10 million children under the age of five died in 1999, most from preventable diseases. More than 113 million primary school age children do not attend school—more of them girls than boys. More than 500,000 women die each year during pregnancy and childbirth—unnecessarily. And more than 14 million adolescents give birth each year. Cause for despair? Or hope? In 1990 there were 1.3 billion living on less than $1 a day. There were more than 11 million deaths among children under five. There were fewer children out of school, but enrollment rates were also lower. So there has been progress. But is this the best we can do?

The international development goals provide a standard for measuring progress. They come from the agreements and resolutions of the world conferences organized by the United Nations in the first half of the 1990s. In September 2000 many of them were incorporated into the resolutions of the Millennium Summit, attended by 149 heads of state. The World Bank, the International Monetary Fund, and the Organisation for Economic Co-operation and Development have adopted them as well.

Reaching the goals will not be easy. It will take commitment and concerted action by citizens, governments, and international agencies to turn pledges into reality.

Global poverty rates—down 20 percent since 1990

The overall decline in poverty rates throughout the 1990s was driven by high rates of growth in countries with large numbers of poor people. In China, which had a fourth of the world's poor in 1990, GDP per capita grew by 9.5 percent a year.

Poverty rates will continue to fall if growth continues. But the gains will be offset somewhat by increasing inequality in household consumption, particularly in China and India. In the base case projection developing countries are expected to

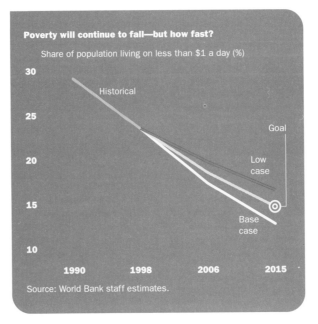

Poverty will continue to fall—but how fast?

Share of population living on less than $1 a day (%)

Historical

Goal

Low case

Base case

1990   1998   2006   2015

Source: World Bank staff estimates.

maintain average GDP per capita growth of 3.7 percent a year after the recovery from the financial crisis. If they do, the goal of reducing poverty rates to half the 1990 level will be achieved in all regions except Sub-Saharan Africa. But if their GDP per capita growth averages 2.3 percent a year (the low case), the world will fall short of the goal—and only East Asia will cut its poverty rate by more than half. And if growth falls back to the average of the 1990s (1.7 percent a year), the poverty rate will decline even more slowly, reaching 18.7 percent in 2015.

Poverty

Poor on $1 a day Millions	1990	1998	2015 (low case)	2015 (base case)
East Asia and Pacific	452	267	101	65
Excluding China	92	54	20	9
Europe and Central Asia	7	18	9	6
Latin America and the Caribbean	74	61	58	43
Middle East and North Africa	6	6	6	5
South Asia	495	522	411	297
Sub-Saharan Africa	242	302	426	361
Total	1,276	1,175	1,011	777
Excluding China	916	961	931	721

Source: World Bank, *Global Economic Prospects and the Developing Countries 2001.*

Still poor on $2 a day Millions	1990	1998	2015 (low case)	2015 (base case)
East Asia and Pacific	1,084	885	472	323
Excluding China	285	252	187	115
Europe and Central Asia	44	98	58	47
Latin America and the Caribbean	167	159	162	133
Middle East and North Africa	59	85	80	58
South Asia	976	1,095	1,214	1,078
Sub-Saharan Africa	388	489	690	637
Total	2,718	2,812	2,675	2,275
Excluding China	1,919	2,179	2,390	2,067

Source: World Bank, *Global Economic Prospects and the Developing Countries 2001.*

Even if we achieve the goal of cutting global poverty rates in half, the number of people living in extreme poverty will fall by only a third. China and India will see the largest improvements, but in Sub-Saharan Africa the number will rise. Europe and Central Asia, where the number of extremely poor people rose during the transition period, should return to 1990 levels of poverty.

Even under the most optimistic assumptions, in 2015 there are likely to be 2.3 billion people living on $2 a day or less, a limit that represents extreme poverty in many middle-income economies.

The pledge to eradicate poverty will have to be renewed by the next generation.

Rising enrollments—but too many children still out of school

UNESCO estimates that there were 113 million children out of school in 1998. About 97 percent lived in developing countries, and nearly 60 percent were girls. Sub-Saharan Africa has the largest proportion of children out of school—40 percent.

Enrollment rates tell only part of the story. The quality of the education provided is also important. Better teachers, improved school facilities, and a curriculum that attracts students and keeps them in school are needed if the goal is to be met.

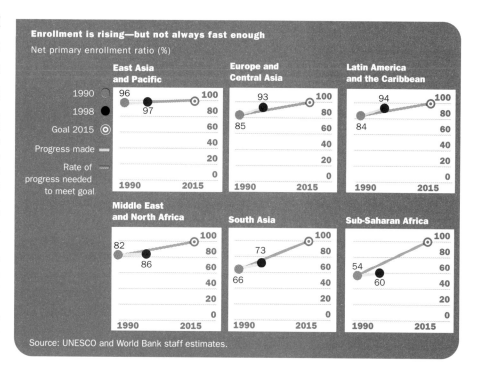

Enrollment is rising—but not always fast enough

Net primary enrollment ratio (%)

1990
1998
Goal 2015
Progress made
Rate of progress needed to meet goal

East Asia and Pacific
Europe and Central Asia
Latin America and the Caribbean
Middle East and North Africa
South Asia
Sub-Saharan Africa

Source: UNESCO and World Bank staff estimates.

Education

Demographic challenges and opportunities

Because of declining birth rates, the number of children of primary school age in developing regions will increase by only 14 million in the next 15 years. Over the same period the working-age population in most regions will grow, providing an opportunity to invest in education.

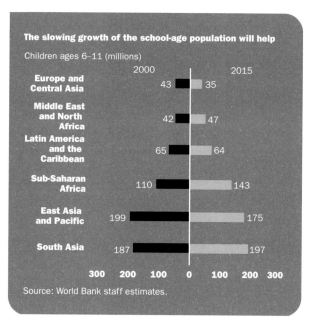

The slowing growth of the school-age population will help

Children ages 6–11 (millions)

	2000	2015
Europe and Central Asia	43	35
Middle East and North Africa	42	47
Latin America and the Caribbean	65	64
Sub-Saharan Africa	110	143
East Asia and Pacific	199	175
South Asia	187	197

Source: World Bank staff estimates.

In East Asia and Pacific the number of school-age children will drop by 24 million, making it easier to provide primary education for all. But in Sub-Saharan Africa there will be 33 million more. With 46 million already not in school, almost 80 million new places will have to be created to accommodate all children. New teachers, new buildings, and new books will be needed. But the loss of teachers to HIV/AIDS is already creating a teacher shortage in some places.

Closing the gender gap in education

The gaps between girls' and boys' enrollments have narrowed. East Asia should be close to achieving the goal by 2005. Progress has also been good in the Middle East and North Africa and in South Asia. But in Sub-Saharan Africa, where barriers to girls' schooling have traditionally been lower than in many other places, progress has been disappointing.

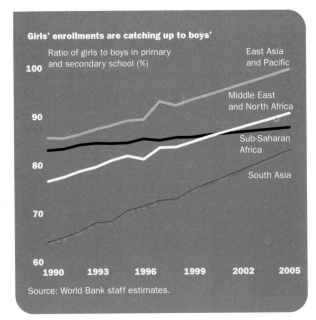

Girls' enrollments are catching up to boys'

Ratio of girls to boys in primary and secondary school (%)

East Asia and Pacific
Middle East and North Africa
Sub-Saharan Africa
South Asia

Source: World Bank staff estimates.

Many girls begin school but then drop out. To encourage girls to attend, schools need to address their needs—providing separate toilet facilities, ensuring their safety at school and between school and home, and hiring more female teachers. They must also convince parents of the value of educating girls.

Gender equality

Empowering women brings other benefits

Greater women's rights and greater participation by women in public life are associated with cleaner government and better governance. Good governance is associated in turn with better law enforcement, greater stability, and enhanced prospects for development. And economic growth reduces constraints on women's access to resources, especially education, health care, and credit, and leads to more equal opportunities for employment and income generation.

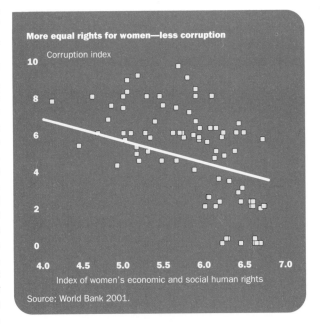

More equal rights for women—less corruption

Corruption index

Index of women's economic and social human rights

Source: World Bank 2001.

The corruption index in the figure uses data from the PRS Group's *International Country Risk Guide* and transforms them. The lower the value of the index for a country, the less corruption it is thought to have. The index of women's economic and social human rights was developed by Purdue University's Global Studies Program. The higher the value of this index, the greater the rights enjoyed by women. The regression line in the figure controls for per capita GDP in each country.

Infant and child mortality falling

There is evidence of progress everywhere. In the past decade 26 developing countries reduced infant mortality at a pace fast enough to reach or exceed the goal in 2015. But mortality rates increased in 11 countries, most of them poor and most of them in Sub-Saharan Africa. And for the rest, faster progress will be needed to reach the goal in 2015.

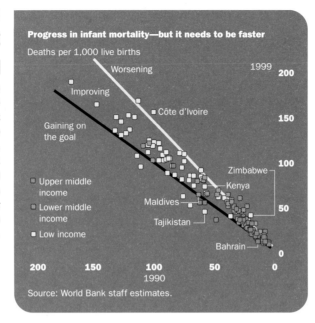

Progress in infant mortality—but it needs to be faster

Deaths per 1,000 live births

Worsening   1999

Improving

Côte d'Ivoire

Gaining on the goal

☐ Upper middle income

☐ Lower middle income

☐ Low income

Zimbabwe
Kenya
Maldives
Tajikistan
Bahrain

1990

Source: World Bank staff estimates.

We know what needs to be done to reduce infant and child deaths. Malnutrition, unsafe water, war and civil conflict, and the spread of HIV/AIDS all contribute to the annual toll. And immunization, disease prevention, and campaigns to teach treatment of diarrhea can all help to reduce deaths.

Infant and child mortality

More than 150 million underweight children in developing countries

The 1996 World Food Summit set a goal of halving the number of malnourished people in the world by 2015. Although there has been progress everywhere except in Sub-Saharan Africa, greater effort will be needed to reach the goal.

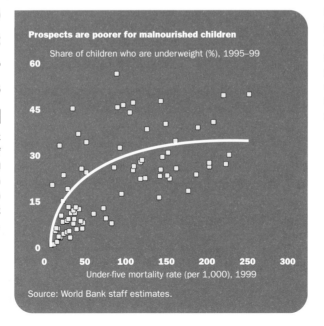

Prospects are poorer for malnourished children

Share of children who are underweight (%), 1995–99

Under-five mortality rate (per 1,000), 1999

Source: World Bank staff estimates.

While being underweight is rarely the cause of death in children, it increases the risk of disease and death and inhibits mental and physical development. The cycle of deprivation repeats itself when malnourished mothers give birth to underweight infants.

More than 500,000 women died in pregnancy and childbirth in 1995

In high-income countries maternal deaths average about 21 per 100,000 live births. In developing countries the average is 440, and in some it may be as high as 1,000.

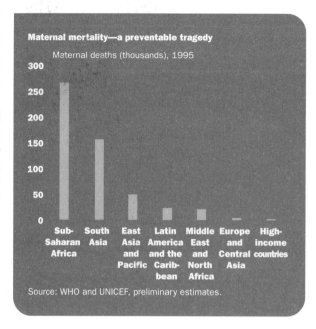

Maternal mortality—a preventable tragedy

Maternal deaths (thousands), 1995

Source: WHO and UNICEF, preliminary estimates.

The tragedy of maternal deaths is that most are preventable with proper care. Infections, blood loss, and unsafe abortions account for the majority of deaths. Better health care is the key to reducing these losses. And it requires neither high technology nor expensive drugs. Trained health workers with midwifery skills greatly increase the safety of childbirth. And by providing information about family planning, they help women care for themselves and their children.

Maternal mortality

Poor mothers have less help at birth

Just as there are large differences between rich and poor countries, there are large differences between rich and poor within a country. In Bangladesh a woman in the wealthiest fifth of the population is 16 times as likely to have trained assistance in childbirth as a woman in the poorest fifth. But even in the wealthiest quintile less than a third of births are attended by a doctor, nurse, or trained midwife.

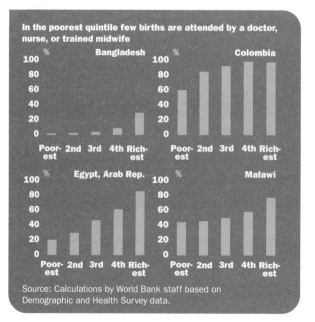

In the poorest quintile few births are attended by a doctor, nurse, or trained midwife

Source: Calculations by World Bank staff based on Demographic and Health Survey data.

The proportion of births attended by skilled personnel is a key indicator for tracking progress in reducing maternal mortality. Globally, just over half of all deliveries are attended by a skilled birth assistant. For many of the rest, mothers have the help of a relative or a traditional birth attendant. But millions deliver on their own. Much less likely than rich mothers to have access to a skilled birth attendant, poor mothers are also more likely to die (AbouZhar 2000).

The use of contraception has risen in most countries

In Sub-Saharan Africa only 26 percent of married women practice contraception. In East Asia more than 75 percent do. Reproductive health services will need to expand rapidly over the next two decades as the number of women and men of reproductive age increases by more than 300 million. The need for reproductive health services is particularly great among adolescents, who account for more than 14 million births each year and 4.4 million abortions. And 15- to 24-year-olds account for more than half of all new HIV infections.

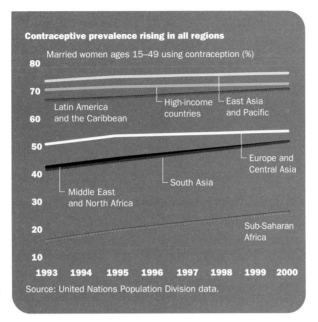

Contraceptive prevalence rising in all regions

Married women ages 15–49 using contraception (%)

Latin America and the Caribbean

High-income countries

East Asia and Pacific

Europe and Central Asia

South Asia

Middle East and North Africa

Sub-Saharan Africa

1993 1994 1995 1996 1997 1998 1999 2000

Source: United Nations Population Division data.

Reproductive health services provide women and men with the knowledge they need to protect their health and that of their families. The services include family planning, prenatal and postnatal care, prevention and treatment of sexually transmitted diseases (including HIV/AIDS), and activities to discourage harmful practices against women, such as female genital mutilation (Leete 2000).

Reproductive health

The rising toll of HIV/AIDS

In 2000 the number of new HIV infections declined slightly, from 5.6 million to 5.3 million, while the number living with HIV/AIDS rose by 2.5 million to 36.1 million. In Sub-Saharan Africa the number of new infections also declined, from 4.0 million in 1999 to 3.8 million. These numbers reflect the success of prevention programs in a limited number of countries. They also reflect the maturing of an epidemic that has already affected much of the sexually active population in countries with high prevalence rates. In countries with relatively low rates of infection, the rates will rise

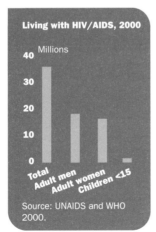

Living with HIV/AIDS, 2000

Millions

Total | Adult men | Adult women | Children <15

Source: UNAIDS and WHO 2000.

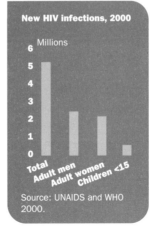

New HIV infections, 2000

Millions

Total | Adult men | Adult women | Children <15

Source: UNAIDS and WHO 2000.

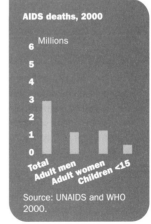

AIDS deaths, 2000

Millions

Total | Adult men | Adult women | Children <15

Source: UNAIDS and WHO 2000.

if adequate preventive measures are not adopted.

HIV/AIDS presents a formidable obstacle to reaching the international development goals. By killing

adults in their most productive years and slowing economic growth, HIV/AIDS exposes millions to the risk of prolonged destitution. In some countries the deaths of schoolteachers have left

classrooms empty. And the cost of preventing and treating the disease threatens to bankrupt health systems.

Greenhouse gas emissions still rising

In 1997 the world released 23.8 billion metric tons of carbon dioxide (CO_2), almost half of it from high-income economies. Emissions per capita dropped slightly thanks to the continuing decline in emissions in Europe and Central Asia, where old, inefficient, and highly polluting industrial plants have been closed.

Human impacts on the environment reach across borders to affect both rich and poor. What are the driving forces of environmental change?

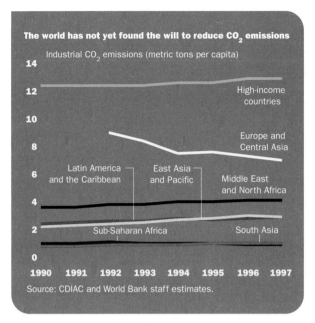

The world has not yet found the will to reduce CO₂ emissions

Industrial CO_2 emissions (metric tons per capita)

High-income countries

Europe and Central Asia

Latin America and the Caribbean — East Asia and Pacific — Middle East and North Africa

Sub-Saharan Africa   South Asia

1990 1991 1992 1993 1994 1995 1996 1997

Source: CDIAC and World Bank staff estimates.

• The size of the human population.
• The per capita consumption of resources.
• The technologies used to produce and consume those resources.

Population growth, increasing consumption, and reliance on fossil fuels all combine to drive up the release of greenhouse gases.

Despite increasing evidence of global warming, the world has not yet found the will to reduce emissions of carbon dioxide. For high-income as well as developing economies the short-run tradeoff between economic growth and safeguarding the environment appears too costly.

Environment

Water is a basic need

The largest shortfalls in the provision of such basic services as water and sanitation are in rural areas, but fast-growing urban areas will face great challenges in the next 25 years.

The goal adopted in 2000 by the Second Water Forum and endorsed in the United Nations Millennium Declaration calls for halving the proportion of people without sustainable access to an adequate amount of safe and affordable water by 2015, and providing water, sanitation, and hygiene for all by 2025.

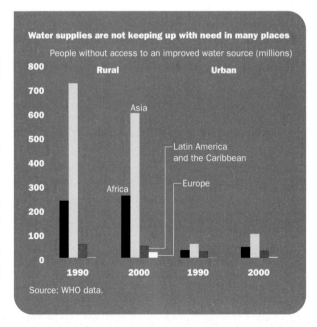

Water supplies are not keeping up with need in many places

People without access to an improved water source (millions)

Rural   Urban

Asia

Latin America and the Caribbean

Africa   Europe

1990   2000   1990   2000

Source: WHO data.

To achieve the 2015 target in Africa, Asia, and Latin America and the Caribbean will require providing an additional 1.5 billion people with access to an improved water supply. That means providing water supply services to an additional 280,000 people every day for the next 15 years. To reach universal coverage by 2025, services will need to be extended to almost 3 billion additional people.

> We are at the service of the world's peoples, and we must listen to them. They are telling us that our past achievements are not enough. They are telling us we must do more, and do it better.

Kofi A. Annan,
Secretary-General of the United Nations
We the Peoples (United Nations 2000a)

Seven steps toward achieving the goals

1 Promote fast, sustainable growth that benefits the poor and reduces inequality.

2 Strengthen the participation of poor people in political processes and local decisionmaking.

3 Reduce vulnerability to economic shocks, natural disasters, ill health, and violence.

4 Invest in people through education, health care, and basic social services.

5 Promote gender equity and eliminate other forms of social exclusion.

6 Forge effective partnerships between civil society, governments, and international agencies.

7 Encourage public discussion of the goals and the means of achieving them.

1.1 Size of the economy

	Population	Surface area	Population density	Gross national income		Gross national income per capita		PPP gross national income[a]			Gross domestic product	
	millions 1999	thousand sq. km 1999	people per sq. km 1999	$ billions 1999[b]	Rank 1999	$ 1999[b]	Rank 1999	$ billions 1999	Per capita $ 1999	Rank 1999	% growth 1998–99	Per capita % growth 1998–99
Albania	3	29	123	3.1	135	930	136	11	3,240	137	7.2	6.1
Algeria	30	2,382	13	46.5	52	1,550	117	145 ᶜ	4,840 ᶜ	105	3.3	1.8
Angola	12	1,247	10	3.3	130	270	185	14 ᶜ	1,100 ᶜ	183	2.7	−0.2
Argentina	37	2,780	13	276.1	17	7,550	58	437	11,940	57	−3.2	−4.4
Armenia	4	30	135	1.9	150	490	157	9	2,360	..	3.3	2.9
Australia	19	7,741	2	397.3	15	20,950	27	452	23,850	20	4.4	3.2
Austria	8	84	98	205.7	21	25,430	14	199	24,600	16	2.1	1.9
Azerbaijan	8	87	92	3.7	124	460	161	20	2,450	145	7.4	6.4
Bangladesh	128	144	981	47.1	50	370	170	196	1,530	167	4.9	3.2
Belarus	10	208	48	26.3	61	2,620	94	69	6,880	83	3.4	4.4
Belgium	10	33	312	252.1	19	24,650	18	263	25,710	12	2.5	2.3
Benin	6	113	55	2.3	141	380	169	6	920	189	5.0	2.1
Bolivia	8	1,099	8	8.1	94	990	134	19	2,300	150	0.6	−1.7
Bosnia and Herzegovina	4	51	76	4.7	114	1,210	127	..	..	..	12.8	9.5
Botswana	2	582	3	5.1	109	3,240	87	10	6,540	85	4.5	2.8
Brazil	168	8,547	20	730.4	8	4,350	73	1,148	6,840	84	0.8	−0.5
Bulgaria	8	111	74	11.6	81	1,410	121	42	5,070	102	2.4	3.0
Burkina Faso	11	274	40	2.6	138	240	193	11 ᶜ	960 ᶜ	187	5.8	3.2
Burundi	7	28	260	0.8	174	120	205	4 ᶜ	570 ᶜ	203	−1.0	−2.9
Cambodia	12	181	67	3.0	136	260	187	16	1,350	174	4.5	2.2
Cameroon	15	475	32	8.8	88	600	150	22	1,490	168	4.4	1.6
Canada	30	9,971	3	614.0	9	20,140	30	776	25,440	14	4.6	3.8
Central African Republic	4	623	6	1.0	168	290	181	4 ᶜ	1,150 ᶜ	181	3.4	1.7
Chad	7	1,284	6	1.6	155	210	197	6 ᶜ	840 ᶜ	191	−0.7	−3.4
Chile	15	757	20	69.6	43	4,630	70	126	8,410	72	−1.1	−2.4
China	1,254	9,598 ᵈ	134	979.9	7	780	142	4,452	3,550	127	7.1	6.1
Hong Kong, China	7	..		165.1	24	24,570	19	152	22,570	26	2.9	1.8
Colombia	42	1,139	40	90.0	37	2,170	100	232	5,580	93	−4.3	−6.0
Congo, Dem. Rep.	50	2,345	22	..	..	.. ᵉ	..	..	..	..	..	..
Congo, Rep.	3	342	8	1.6	154	550	152	2	540	205	−3.0	−5.6
Costa Rica	4	51	70	12.8	79	3,570 ᶠ	80	28	7,880	76	8.0	6.1
Côte d'Ivoire	16	322	49	10.4	84	670	147	24	1,540	166	2.8	0.1
Croatia	4	57	80	20.2	64	4,530	71	32	7,260	80	−0.3	0.5
Cuba	11	111	102	..	..	.. ᵍ	..	..	..	..	..	..
Czech Republic	10	79	133	51.6	48	5,020	66	132	12,840	54	−0.2	−0.1
Denmark	5	43	126	170.7	23	32,050	6	136	25,600	13	1.7	1.2
Dominican Republic	8	49	174	16.1	74	1,920	105	44	5,210	100	8.3	6.4
Ecuador	12	284	45	16.8	72	1,360	123	35	2,820	140	−7.3	−9.0
Egypt, Arab Rep.	63	1,001	63	86.5	39	1,380	122	217	3,460	128	6.0	4.1
El Salvador	6	21	297	11.8	80	1,920	105	26	4,260	117	3.4	1.4
Eritrea	4	118	40	0.8	177	200	198	4	1,040	185	0.8	−2.0
Estonia	1	45	34	4.9	112	3,400	83	12	8,190	74	−1.1	−0.6
Ethiopia	63	1,104	63	6.5	99	100	207	39	620	201	6.2	3.6
Finland	5	338	17	127.8	29	24,730	17	117	22,600	25	4.0	3.8
France	59	552	107	1,453.2 ʰ	4	24,170 ʰ	21	1,349	23,020	23	2.9	2.5
Gabon	1	268	5	4.0	118	3,300	85	6	5,280	98	−6.2	−8.4
Gambia, The	1	11	125	0.4	190	330	175	2 ᶜ	1,550 ᶜ	164	6.4	3.4
Georgia	5	70	78	3.4	128	620	149	14	2,540	144	3.3	3.1
Germany	82	357	235	2,103.8	3	25,620	13	1,930	23,510	21	1.5	1.4
Ghana	19	239	83	7.5	97	400	166	35 ᶜ	1,850 ᶜ	161	4.4	2.1
Greece	11	132	82	127.6	30	12,110	46	166	15,800	48	3.4	3.1
Guatemala	11	109	102	18.6	68	1,680	112	40	3,630	126	3.6	0.9
Guinea	7	246	30	3.6	126	490	157	14	1,870	158	3.3	1.0
Guinea-Bissau	1	36	42	0.2	200	160	202	1	630	200	7.8	5.7
Haiti	8	28	283	3.6	125	460	161	11 ᶜ	1,470 ᶜ	169	2.2	0.2
Honduras	6	112	56	4.8	113	760	143	14	2,270	151	−1.9	−4.5

Size of the economy | 1.1

	Population	Surface area	Population density	Gross national income		Gross national income per capita		PPP gross national income[a]			Gross domestic product	
									Per capita			Per capita
	millions 1999	thousand sq. km 1999	people per sq. km 1999	$ billions 1999[b]	Rank 1999	$ 1999[b]	Rank 1999	$ billions 1999	$ 1999	Rank 1999	% growth 1998–99	% growth 1998–99
Hungary	10	93	109	46.8	51	4,640	69	111	11,050	60	4.5	5.0
India	998	3,288	336	441.8	11	440	163	2,226	2,230	153	6.5	4.6
Indonesia	207	1,905	114	125.0	31	600	150	550	2,660	143	0.3	–1.3
Iran, Islamic Rep.	63	1,633	39	113.7	33	1,810	109	347	5,520	95	2.5	0.8
Iraq	23	438	52	..	..	.. g	..	..	..	..	..	..
Ireland	4	70	54	80.6	40	21,470	24	84	22,460	27	9.8	8.7
Israel	6	21	296	99.6	35	16,310	36	110	18,070	40	2.2	–0.2
Italy	58	301	196	1,162.9	6	20,170	29	1,268	22,000	32	1.4	1.3
Jamaica	3	11	240	6.3	101	2,430	95	9	3,390	130	–0.4	–1.2
Japan	127	378	336	4,054.5	2	32,030	7	3,186	25,170	15	0.2	0.1
Jordan	5	89	53	7.7	96	1,630	115	18	3,880	122	3.1	0.0
Kazakhstan	15	2,717	6	18.7	67	1,250	126	71	4,790	106	1.7	2.7
Kenya	29	580	52	10.7	83	360	172	30	1,010	186	1.3	–0.8
Korea, Dem. Rep.	23	121	194	..	..	.. e	..	..	..	..	..	..
Korea, Rep.	47	99	475	397.9	13	8,490	54	728	15,530	49	10.7	9.7
Kuwait	2	18	108	..	..	.. i	..	..	..	..	..	..
Kyrgyz Republic	5	199	25	1.5	158	300	180	12	2,420	147	3.7	2.2
Lao PDR	5	237	22	1.5	157	290	181	7 c	1,430 c	170	7.4	4.8
Latvia	2	65	39	5.9	106	2,430	95	15	6,220	89	0.1	0.8
Lebanon	4	10	418	15.8	75	3,700	78	..	..	..	..	..
Lesotho	2	30	69	1.2	165	550	152	5 c	2,350 c	149	2.5	0.2
Libya	5	1,760	3	..	..	.. j	..	..	..	..	..	..
Lithuania	4	65	57	9.8	85	2,640	92	24	6,490	86	–4.2	–4.1
Macedonia, FYR	2	26	79	3.3	129	1,660	113	9	4,590	109	2.7	2.1
Madagascar	15	587	26	3.7	123	250	190	12	790	193	4.7	1.5
Malawi	11	118	115	2.0	149	180	201	6	570	203	4.0	1.5
Malaysia	23	330	69	76.9	42	3,390	84	173	7,640	78	5.8	3.3
Mali	11	1,240	9	2.6	139	240	193	8	740	195	5.5	3.0
Mauritania	3	1,026	3	1.0	169	390	167	4	1,550	164	4.1	1.3
Mauritius	1	2	579	4.2	117	3,540	81	11	8,950	68	3.4	2.1
Mexico	97	1,958	51	428.9	12	4,440	72	780	8,070	75	3.5	2.1
Moldova	4	34	130	1.5	156	410	164	9	2,100	155	–4.4	–4.0
Mongolia	2	1,567	2	0.9	172	390	167	4	1,610	163	3.0	2.1
Morocco	28	447	63	33.7	55	1,190	129	94	3,320	134	–0.7	–2.3
Mozambique	17	802	22	3.8	122	220	195	14 c	810 c	192	7.3	5.2
Myanmar	45	677	68	..	..	.. e	..	..	..	..	..	..
Namibia	2	824	2	3.2	132	1,890	107	9 c	5,580 c	93	3.1	0.7
Nepal	23	147	164	5.2	108	220	195	30	1,280	176	3.9	1.6
Netherlands	16	41	466	397.4	14	25,140	16	386	24,410	17	3.6	2.9
New Zealand	4	271	14	53.3	47	13,990	43	67	17,630	43	4.4	3.9
Nicaragua	5	130	41	2.0	147	410	164	10 c	2,060 c	156	7.0	4.3
Niger	10	1,267	8	2.0	148	190	200	8 c	740 c	195	–0.6	–3.9
Nigeria	124	924	136	31.6	57	260	187	95	770	194	1.0	–1.5
Norway	4	324	15	149.3	26	33,470	5	126	28,140	8	0.9	0.2
Oman	2	212	11	..	..	.. j	..	..	..	..	..	..
Pakistan	135	796	175	62.9	44	470	160	250	1,860	159	4.0	1.5
Panama	3	76	38	8.7	89	3,080	89	15 c	5,450 c	96	3.0	1.2
Papua New Guinea	5	463	10	3.8	121	810	140	11 c	2,260 c	152	3.2	0.9
Paraguay	5	407	13	8.4	91	1,560	116	23 c	4,380 c	113	–0.8	–3.4
Peru	25	1,285	20	53.7	46	2,130	101	113	4,480	111	1.4	–0.3
Philippines	74	300	249	78.0	41	1,050	133	296	3,990	120	3.2	1.2
Poland	39	323	127	157.4	25	4,070	74	324	8,390	73	4.1	4.1
Portugal	10	92	109	110.2	34	11,030	49	158	15,860	47	3.0	2.8
Puerto Rico	4	9	439	..	..	.. j	..	..	..	..	..	..
Romania	22	238	97	33.0	56	1,470	120	134	5,970	90	–3.2	–3.0
Russian Federation	146	17,075	9	329.0	16	2,250	99	1,022	6,990	82	3.2	3.6

1.1 Size of the economy

	Population	Surface area	Population density	Gross national income		Gross national income per capita		PPP gross national income[a]			Gross domestic product	
	millions 1999	thousand sq. km 1999	people per sq. km 1999	$ billions 1999[b]	Rank 1999	$ 1999[b]	Rank 1999	$ billions 1999	Per capita $ 1999	Rank 1999	% growth 1998–99	Per capita % growth 1998–99
Rwanda	8	26	337	2.0	146	250	190	7	880	190	6.1	3.5
Saudi Arabia	20	2,150	9	139.4	27	6,900	60	223	11,050	60	0.4	−2.1
Senegal	9	197	48	4.7	115	500	156	13	1,400	172	5.1	2.3
Sierra Leone	5	72	69	0.7	180	130	204	2	440	207	−8.1	−9.9
Singapore	4	1	6,384	95.4	36	24,150	22	88	22,310	28	5.4	4.6
Slovak Republic	5	49	112	20.3	63	3,770	77	56	10,430	64	1.9	1.8
Slovenia	2	20	99	19.9	65	10,000	50	32	16,050	46	4.9	4.7
South Africa	42	1,221	34	133.6	28	3,170	88	367 [c]	8,710 [c]	70	1.2	−0.5
Spain	39	506	79	583.1	10	14,800	39	704	17,850	42	3.7	3.6
Sri Lanka	19	66	294	15.6	76	820	139	61	3,230	138	4.3	3.2
Sudan	29	2,506	12	9.4	86	330	175	..	..	..	5.2	2.9
Sweden	9	450	22	236.9	20	26,750	12	196	22,150	31	3.8	3.7
Switzerland	7	41	180	273.9	18	38,380	3	205	28,760	7	1.5	1.1
Syrian Arab Republic	16	185	85	15.2	77	970	135	54	3,450	129	5.2	2.6
Tajikistan	6	143	44	1.7	153	280	184	..	..	..	..	..
Tanzania	33	945	37	8.5 [k]	90	260 [k]	187	16	500	206	4.7	2.2
Thailand	60	513	118	121.1	32	2,010	103	358	5,950	91	4.2	3.4
Togo	5	57	84	1.4	160	310	179	6	1,380	173	2.1	−0.3
Trinidad and Tobago	1	5	252	6.1	104	4,750	68	10	7,690	77	6.8	6.1
Tunisia	9	164	61	19.8	66	2,090	102	54	5,700	92	6.2	4.9
Turkey	64	775	84	186.5	22	2,900	90	415	6,440	87	−5.1	−6.6
Turkmenistan	5	488	10	3.2	133	670	147	16	3,340	132	16.0	14.5
Uganda	21	241	108	6.8	98	320	178	25 [c]	1,160 [c]	180	7.4	4.5
Ukraine	50	604	86	42.0	53	840 [l]	138	168	3,360	131	−0.4	0.3
United Arab Emirates	3	84	34	..	..	.. [i]	..	..	..	..	..	..
United Kingdom	60	245	246	1,403.8	5	23,590	23	1,322	22,220	29	2.1	1.7
United States	278	9,364	30	8,879.5	1	31,910	8	8,878	31,910	4	3.6	2.4
Uruguay	3	177	19	20.6	62	6,220	64	29	8,750	69	−3.2	−3.9
Uzbekistan	24	447	59	17.6	70	720	146	54	2,230	153	4.4	2.6
Venezuela, RB	24	912	27	87.3	38	3,680	79	129	5,420	97	−7.2	−9.0
Vietnam	78	332	238	28.7	60	370	170	144	1,860	159	4.8	3.5
West Bank and Gaza	3	..	..	5.1	110	1,780	110	..	..	..	6.9	2.9
Yemen, Rep.	17	528	32	6.1	105	360	172	12	730	197	3.8	1.1
Yugoslavia, FR (Serb./Mont.)	11	102	..	..	..	.. [g]	..	..	..	..	..	..
Zambia	10	753	13	3.2	131	330	175	7	720	199	2.4	0.2
Zimbabwe	12	391	31	6.3	102	530	154	32	2,690	141	0.1	−1.7
World	5,978 s	133,567 s	46 w	29,994.6 t		5,020 w		41,053 t	6,870 w		2.6 w	1.2 w
Low income	2,417	34,227	73	1,008.4		420		4,522	1,870		4.1	2.1
Middle income	2,665	67,257	40	5,285.0		1,980		13,843	5,200		3.0	1.9
Lower middle income	2,093	44,751	48	2,508.3		1,200		8,887	4,250		3.6	2.5
Upper middle income	571	22,506	26	2,782.5		4,870		5,009	8,770		2.6	1.2
Low & middle income	5,082	101,484	51	6,292.1		1,240		18,321	3,610		3.2	1.7
East Asia & Pacific	1,837	16,385	115	1,854.5		1,010		6,876	3,740		6.8	5.6
Europe & Central Asia	474	24,208	20	1,023.9		2,160		2,836	5,980		1.0	0.9
Latin America & Carib.	508	20,461	25	1,932.9		3,800		3,364	6,620		0.0	−1.5
Middle East & N. Africa	290	11,024	26	598.4		2,060		1,452	5,000		2.6	0.6
South Asia	1,329	5,140	278	581.3		440		2,804	2,110		6.0	4.0
Sub-Saharan Africa	643	24,267	27	315.8		490		966	1,500		2.0	−0.5
High income	896	32,083	29	23,701.7		26,440		23,032	25,690		2.4	1.7
Europe EMU	293	2,499	122	6,513.1		22,250		6,494	22,180		2.4	2.2

a. PPP is purchasing power parity; see *Definitions*. b. Calculated using the World Bank Atlas method. c. The estimate is based on regression; others are extrapolated from the latest International Comparison Programme benchmark estimates. d. Includes Taiwan, China; Macao, China; and Hong Kong, China. e. Estimated to be low income ($755 or less). f. Included under lower-middle-income economies in calculating the aggregates based on earlier data. g. Estimated to be lower middle income ($756–2,995). h. GNI and GNI per capita estimates include the French overseas departments of French Guiana, Guadeloupe, Martinique, and Réunion. i. Estimated to be high income ($9,266 or more). j. Estimated to be upper middle income ($2,996–9,265). k. Data refer to mainland Tanzania only. l. Included under low-income economies in calculating the aggregates based on earlier data.

Size of the economy | 1.1

About the data

Population, land area, income, and output are basic measures of the size of an economy. They also provide a broad indication of actual and potential resources. Therefore, population, land area, income—as measured by gross national income (GNI)—and output—as measured by gross domestic product (GDP)—are used throughout the *World Development Indicators* to normalize other indicators.

Population estimates are generally based on extrapolations from the most recent national census. For further discussion of the measurement of population and population growth see *About the data* for table 2.1 and *Statistical methods*.

The surface area of a country or economy includes inland bodies of water and some coastal waterways. Surface area thus differs from land area, which excludes bodies of water, and from gross area, which may include offshore territorial waters. Land area is particularly important for understanding the agricultural capacity of an economy and the effects of human activity on the environment. (For measures of land area and data on rural population density, land use, and agricultural productivity see tables 3.1–3.3.) Recent innovations in satellite mapping techniques and computer databases have resulted in more precise measurements of land and water areas.

GNI (gross national product, or GNP, in previous editions) measures the total domestic and foreign value added claimed by residents. GNI comprises GDP plus net receipts of primary income (compensation of employees and property income) from nonresident sources.

The World Bank uses GNI per capita in U.S. dollars to classify countries for analytical purposes and to determine borrowing eligibility. See the *Users guide* for definitions of the income groups used in the *World Development Indicators*. For further discussion of the usefulness of national income as a measure of productivity or welfare see *About the data* for tables 4.1 and 4.2.

When calculating GNI in U.S. dollars from GNI reported in national currencies, the World Bank follows its Atlas conversion method. This involves using a three-year average of exchange rates to smooth the effects of transitory exchange rate fluctuations. (For further discussion of the Atlas method see *Statistical methods*.) Note that growth rates are calculated from data in constant prices and national currency units, not from the Atlas estimates.

Because exchange rates do not always reflect international differences in relative prices, this table also shows GNI and GNI per capita estimates converted into international dollars using purchasing power parity (PPP) rates. PPP rates provide a standard measure allowing comparison of real price levels between countries, just as conventional price indexes allow comparison of real values over time. The PPP conversion factors used here are derived from price surveys covering 118 countries conducted by the International Comparison Programme (ICP). For 62 countries data come from the most recent round of surveys, completed in 1996; the rest are from the 1993 round and have been extrapolated to the 1996 benchmark. Estimates for countries not included in the surveys are derived from statistical models using available data.

All economies shown in the *World Development Indicators* are ranked by size, including those that appear in table 1.6. Ranks are shown only in table 1.1. (The *World Bank Atlas* includes a table comparing the GNI per capita rankings based on the Atlas method with those based on the PPP method for all economies with available data.) No rank is shown for economies for which numerical estimates of GNI per capita are not published. Economies with missing data are included in the ranking process at their approximate level, so that the relative order of other economies remains consistent. Where available, rankings for small economies are shown in the *World Bank Atlas*. In 1999 Luxembourg and Liechtenstein were judged to have the highest GNI per capita in the world.

Growth in GDP and growth in GDP per capita are based on GDP measured in constant prices. Growth in GDP is considered a broad measure of the growth of an economy, as GDP in constant prices can be estimated by measuring the total quantity of goods and services produced in a period, valuing them at an agreed set of base year prices, and subtracting the cost of intermediate inputs, also in constant prices. For further discussion of the measurement of economic growth see *About the data* for table 4.1.

Definitions

• **Population** is based on the de facto definition of population, which counts all residents regardless of legal status or citizenship—except for refugees not permanently settled in the country of asylum, who are generally considered part of the population of their country of origin. The values shown are midyear estimates for 1999. See also table 2.1. • **Surface area** is a country's total area, including areas under inland bodies of water and some coastal waterways. • **Population density** is midyear population divided by land area in square kilometers. • **Gross national income** (GNI) is the sum of value added by all resident producers plus any product taxes (less subsidies) not included in the valuation of output plus net receipts of primary income (compensation of employees and property income) from abroad. Data are in current U.S. dollars converted using the World Bank Atlas method (see *Statistical methods*). • **GNI per capita** is gross national income divided by midyear population. GNI per capita in U.S. dollars is converted using the World Bank Atlas method. • **PPP GNI** is gross national income converted to international dollars using purchasing power parity rates. An international dollar has the same purchasing power over GNI as a U.S. dollar has in the United States. • **Gross domestic product** (GDP) is the sum of value added by all resident producers plus any product taxes (less subsidies) not included in the valuation of output. • **GDP per capita** is gross domestic product divided by midyear population. Growth is calculated from constant price GDP data in local currency.

Data sources

Population estimates are prepared by World Bank staff from a variety of sources (see *Data sources* for table 2.1). The data on surface and land area are from the Food and Agriculture Organization (see *Data sources* for table 3.1). GNI, GNI per capita, GDP growth, and GDP per capita growth are estimated by World Bank staff based on national accounts data collected by Bank staff during economic missions or reported by national statistical offices to other international organizations such as the Organisation for Economic Co-operation and Development. Purchasing power parity conversion factors are estimates by World Bank staff based on data collected by the International Comparison Programme.

	Prevalence of child malnutrition	Net primary enrollment ratio[a]				Infant mortality rate		Under-five mortality rate		Contraceptive prevalence rate	Maternal mortality ratio	Access to an improved water source
	Weight for age % of children under 5	Male % of relevant age group		Female % of relevant age group		per 1,000 live births		per 1,000		% of women 15–49	per 100,000 live births	% of population
	1993–99[b]	1990	1997	1990	1997	1990	1999	1990	1999	1990–99[b]	1990–99[b]	2000
Albania	8	..	101	..	103	28	24	42	..	..	..	..
Algeria	13	99	97	87	91	46	34	55	39	51	220	94
Angola	41	..	..	..	..	130	127	..	208	..	..	38
Argentina	2	..	..	..	..	25	18	28	22	..	38	79
Armenia	3	..	..	..	..	19	14	24	18	..	35	..
Australia	0	99	95	99	95	8	5	10	5	..	..	100
Austria	..	..	86	..	89	8	4	9	5	..	..	100
Azerbaijan	10	..	..	..	..	23	16	..	21	..	43	..
Bangladesh	56	68	..	60	..	91	61	136	89	54	440	97
Belarus	..	..	87	..	84	12	11	16	14	..	28	100
Belgium	..	96	99	98	98	8	5	9	6	..	..	..
Benin	29	..	80	..	47	104	87	..	145	16	500	63
Bolivia	8	95	..	87	..	80	59	120	83	49	390	79
Bosnia and Herzegovina	..	..	..	..	..	15	13	21	18	..	10	..
Botswana	17	90	79	97	83	55	58	62	95	..	330	..
Brazil	6	..	..	..	..	48	32	58	40	77	160	87
Bulgaria	..	86	93	86	91	15	14	19	17	..	15	100
Burkina Faso	33	33	37	21	24	111	105	..	210	12	..	..
Burundi	..	..	..	..	..	119	105	180	176	..	..	..
Cambodia	47	..	..	..	92	122	100	..	143	22	470	30
Cameroon	22	..	..	..	..	81	77	141	154	19	430	62
Canada	..	97	96	97	94	7	5	8	6	..	..	100
Central African Republic	23	64	..	42	..	102	96	..	151	15	1,100	60
Chad	39	..	59	..	33	118	101	209	189	4	830	27
Chile	1	..	91	..	88	16	10	20	12	..	20	94
China	9	99	101	95	102	33	30	47	37	85	55	75
Hong Kong, China	..	..	88	..	91	6	3	..	5	..	..	..
Colombia	8	..	..	..	..	30	23	40	28	72	80	91
Congo, Dem. Rep.	34	61	72	48	50	96	85	155	161	..	..	45
Congo, Rep.	..	..	..	..	..	88	89	..	144	..	..	51
Costa Rica	5	86	89	87	89	15	12	16	14	..	29	98
Côte d'Ivoire	24	..	63	..	47	95	111	150	180	15	600	77
Croatia	1	79	83	79	82	11	8	13	9	..	6	95
Cuba	..	92	101	92	100	11	7	13	8	..	27	95
Czech Republic	..	..	87	..	87	11	5	12	5	69	9	..
Denmark	..	98	99	98	99	8	5	9	6	..	10	100
Dominican Republic	6	..	..	..	..	51	39	..	47	64	230	79
Ecuador	..	..	97	..	97	45	28	..	35	66	160	71
Egypt, Arab Rep.	11	..	98	..	88	69	47	85	61	52	170	95
El Salvador	12	..	78	..	78	46	30	54	36	60	120	74
Eritrea	44	..	32	..	29	81	60	..	105	8	1,000	46
Estonia	..	..	87	..	86	12	10	17	12	..	50	..
Ethiopia	..	..	40	..	25	124	104	190	180	4	..	24
Finland	..	..	98	..	98	6	4	7	5	..	6	100
France	..	101	100	101	100	7	5	9	5	71	10	..
Gabon	..	..	..	..	..	96	84	164	133	..	600	70
Gambia, The	26	..	72	..	57	109	75	127	110	..	..	62
Georgia	3	..	87	..	87	16	15	..	20	41	70	76
Germany	..	..	86	..	87	7	5	9	5	..	8	..
Ghana	25	..	..	..	..	66	57	119	109	22	210	64
Greece	..	94	90	94	90	10	6	11	7	..	1	..
Guatemala	24	..	76	..	69	56	40	..	52	38	190	92
Guinea	..	..	50	..	33	121	96	..	167	6	670	48
Guinea-Bissau	..	..	..	..	..	145	127	246	214	..	910	49
Haiti	28	22	..	23	..	85	70	..	118	18	..	46
Honduras	25	..	..	..	..	50	34	65	46	50	110	90

	Prevalence of child malnutrition	Net primary enrollment ratio[a]				Infant mortality rate		Under-five mortality rate		Contra-ceptive prevalence rate	Maternal mortality ratio	Access to an improved water source
	Weight for age % of children under 5	Male % of relevant age group		Female % of relevant age group		per 1,000 live births		per 1,000		% of women 15–49	per 100,000 live births	% of population
	1993–99[b]	1990	1997	1990	1997	1990	1999	1990	1999	1990–99[b]	1990–99[b]	2000
Hungary	..	91	97	92	96	15	8	17	10	73	15	99
India	45	..	..	..	..	80	71	112	90	52	410	88
Indonesia	34	100	96	95	93	60	42	83	52	57	450	76
Iran, Islamic Rep.	11	..	91	..	88	47	26	59	33	73	37	95
Iraq	..	..	81	..	71	102	101	..	128	..	..	85
Ireland	..	90	91	91	93	8	6	9	5	60	6	..
Israel	..	..	..	..	..	10	6	12	8	..	5	..
Italy	..	..	100	..	100	8	5	10	6	..	7	..
Jamaica	4	96	..	96	..	25	20	32	24	65	120	71
Japan	..	100	103	100	103	5	4	6	4	..	8	..
Jordan	5	66	67	67	68	30	26	34	31	50	41	96
Kazakhstan	8	..	..	..	..	26	22	34	28	66	70	91
Kenya	22	..	..	..	..	62	76	97	118	39	590	49
Korea, Dem. Rep.	32	..	..	..	..	45	58	35	93	..	110	..
Korea, Rep.	..	103	92	104	93	12	8	..	9	..	20	92
Kuwait	2	..	62	..	62	14	11	16	13	..	5	..
Kyrgyz Republic	11	..	97	..	93	30	26	41	38	60	65	77
Lao PDR	40	..	76	..	69	108	93	..	143	25	650	90
Latvia	..	..	92	..	87	14	14	18	18	..	45	..
Lebanon	3	..	..	..	..	36	26	40	32	61	100	100
Lesotho	16	65	64	81	76	102	92	148	141	23	..	91
Libya	5	..	..	..	..	33	22	42	28	45	75	72
Lithuania	..	..	..	..	..	10	9	14	12	..	18	..
Macedonia, FYR	6	95	96	94	95	32	16	33	17	..	3	99
Madagascar	40	..	60	..	62	103	90	170	149	19	490	47
Malawi	30	52	102	48	104	135	132	230	227	22	620	57
Malaysia	20	..	102	..	102	16	8	21	10	..	39	..
Mali	27	27	38	16	25	136	120	..	223	7	580	65
Mauritania	23	..	61	..	53	105	88	..	142	..	550	37
Mauritius	15	95	98	95	98	20	19	25	23	75	50	100
Mexico	8	..	101	..	102	36	29	46	36	65	55	86
Moldova	..	..	..	..	..	19	17	25	22	74	42	100
Mongolia	13	..	79	..	83	73	58	102	73	60	150	60
Morocco	..	68	83	48	65	64	48	83	62	59	230	82
Mozambique	26	..	45	..	34	150	131	..	203	6	1,100	60
Myanmar	28	..	..	..	..	94	77	130	120	..	230	68
Namibia	..	..	..	..	..	64	63	84	108	29	230	77
Nepal	47	..	..	..	..	101	75	138	109	29	..	81
Netherlands	..	93	100	97	100	7	5	8	5	75	7	100
New Zealand	..	102	100	101	100	8	5	11	6	..	15	..
Nicaragua	12	71	77	73	79	51	34	63	5	60	150	79
Niger	50	32	30	18	19	150	116	335	252	8	590	59
Nigeria	39	..	..	..	..	86	83	136	151	6	700	57
Norway	..	100	100	100	100	7	4	9	4	..	6	100
Oman	23	73	70	68	68	22	17	30	24	..	19	39
Pakistan	38	..	..	..	..	111	90	138	126	24	..	88
Panama	..	91	..	92	..	26	20	..	25	..	70	87
Papua New Guinea	..	..	..	..	..	83	58	..	77	26	370	42
Paraguay	..	94	91	92	91	31	24	37	27	57	190	79
Peru	8	..	91	..	90	54	39	75	48	64	270	77
Philippines	30	..	..	..	..	37	31	62	41	47	170	87
Poland	..	97	95	97	94	19	9	22	10	..	8	..
Portugal	..	102	..	102	..	11	6	15	6	..	8	..
Puerto Rico	..	..	..	..	..	14	10	..	..	78	..	..
Romania	..	..	96	..	95	27	20	36	24	48	41	58
Russian Federation	3	..	93	..	93	17	16	21	20	34	50	99

1.2 Development progress

	Prevalence of child malnutrition	Net primary enrollment ratio[a]				Infant mortality rate		Under-five mortality rate		Contra-ceptive prevalence rate	Maternal mortality ratio	Access to an improved water source
	Weight for age % of children under 5	Male % of relevant age group		Female % of relevant age group		per 1,000 live births		per 1,000		% of women 15–49	per 100,000 live births	% of population
	1993–99[b]	1990	1997	1990	1997	1990	1999	1990	1999	1990–99[b]	1990–99[b]	2000
Rwanda	27	66	..	66	..	132	123	..	203	21	..	41
Saudi Arabia	..	65	63	53	60	32	19	45	25	21	..	95
Senegal	22	..	65	..	55	74	67	..	124	13	560	78
Sierra Leone	..	..	..	..	..	189	168	323	283	..	..	28
Singapore	..	..	93	..	92	7	3	8	4	..	6	100
Slovak Republic	..	..	..	..	..	12	8	14	10	..	9	100
Slovenia	..	..	95	..	94	8	5	10	6	..	11	100
South Africa	9	..	..	..	..	55	62	73	76	69	..	86
Spain	..	103	105	103	105	8	5	9	6	..	6	..
Sri Lanka	33	..	..	..	..	19	15	23	19	..	60	83
Sudan	34	..	..	..	..	85	67	125	109	10	500	75
Sweden	..	100	103	100	102	6	4	6	4	..	5	100
Switzerland	..	83	90	84	90	7	5	8	5	..	5	100
Syrian Arab Republic	13	103	95	93	87	39	26	..	30	45	110	80
Tajikistan	..	..	..	..	..	41	20	..	34	..	65	..
Tanzania	31	51	48	52	49	115	95	..	152	25	530	54
Thailand	19	..	..	..	..	37	28	41	33	72	44	80
Togo	25	87	93	62	69	81	77	142	143	24	480	54
Trinidad and Tobago	..	91	88	91	88	18	16	24	20	..	..	86
Tunisia	9	97	99	90	96	37	24	52	30	60	70	..
Turkey	8	..	102	..	96	58	36	67	45	64	130	83
Turkmenistan	..	..	..	..	..	45	33	..	45	..	65	58
Uganda	26	..	..	..	..	104	88	165	162	15	510	50
Ukraine	..	..	..	..	..	13	14	..	17	68	27	..
United Arab Emirates	7	95	79	93	78	20	8	..	9	..	3	..
United Kingdom	..	96	98	98	100	8	6	9	6	..	7	100
United States	1	96	94	96	95	9	7	..	8	64	8	100
Uruguay	4	..	92	..	93	21	15	24	17	..	26	98
Uzbekistan	19	..	..	..	..	35	22	..	29	56	21	85
Venezuela, RB	8	87	83	89	85	25	20	27	23	..	60	84
Vietnam	37	..	..	..	..	40	37	54	42	75	160	56
West Bank and Gaza	..	..	..	..	..	..	23	..	26	42	..	..
Yemen, Rep.	46	..	..	..	..	110	79	130	97	21	350	69
Yugoslavia, FR (Serb./Mont.)	2	69	..	70	..	23	12	26	16	..	10	..
Zambia	24	..	76	..	74	107	114	..	187	26	650	64
Zimbabwe	16	..	..	..	..	52	70	..	118	48	400	85
World	.. w	.. w	.. w	.. w	.. w	61 w	54 w	86 w	78 w	50 w		81 w
Low income	..	..	..	..	..	88	77	126	116	23		76
Middle income	14	..	99	..	98	38	31	49	39	53		81
Lower middle income	9	98	100	..	99	38	32	50	40	53		80
Upper middle income	..	..	..	..	..	35	27	46	34	65		87
Low & middle income	..	..	..	..	..	66	59	91	85	49		79
East Asia & Pacific	12	100	100	96	100	40	35	55	44	57		75
Europe & Central Asia	..	..	93	..	92	28	21	34	26	64		..
Latin America & Carib.	9	..	..	..	..	41	30	49	38	59		85
Middle East & N. Africa	..	..	90	..	83	60	44	71	56	52		89
South Asia	47	..	..	..	..	87	74	121	99	49		87
Sub-Saharan Africa	..	..	..	..	..	101	92	155	161	21		55
High income	..	98	95	98	95	8	6	9	6	75		..
Europe EMU	..	..	94	..	96	8	5	9	5	75		..

a. Net enrollment ratios exceeding 100 percent indicate discrepancies between the estimates of school-age population and reported enrollment data. b. Data are for the most recent year available.

Development progress | 1.2

The indicators in this table are part of the set of 21 social and environmental indicators selected for monitoring development progress by the Organisation for Economic Co-operation and Development, the World Bank, and the United Nations in consultation with countries that provide and those that receive development assistance. For some of the indicators specific targets for improvement have been announced (table 1.2a). The international development goals call for achieving equal enrollment of girls and boys in primary and secondary school by 2005 and universal primary enrollment by 2015. They also call for reducing the infant and under-five mortality rates by two-thirds—and the maternal mortality ratio by three-quarters—from their 1990 levels by 2015. And the 1995 World Summit for Social Development called for reducing preschool child malnutrition rates to half their 1990 level by 2000.

Not all indicators have specific targets. The contraceptive prevalence rate is included in the set of indicators to help monitor access to reproductive health care, but no target rate has been set. The World Summit for Social Development called for ensuring access to an adequate quantity of safe water for all people, but no date for achieving this goal has been specified. Moreover, there is no practical way today to determine whether all water sources are safe for human use.

The introduction to this section discusses progress toward the international development goals. For additional discussion of the indicators here see *About the data* for tables 2.18 (prevalence of child malnutrition), 2.12 (net enrollment ratio), 2.19 (infant and under-five mortality rates), 2.17 (maternal mortality ratio), and 2.16 (population with access to an improved water source).

• **Prevalence of child malnutrition** is the percentage of children under five whose weight for age is less than minus two standard deviations from the median for the international reference population ages 0–59 months. The reference population, adopted by the World Health Organization in 1983, is based on children from the United States, who are assumed to be well nourished.
• **Net primary enrollment ratio** is the ratio of the number of children of official school age (as defined by the education system) enrolled in school to the number of children of official school age in the population.
• **Infant mortality rate** is the number of infants dying before reaching the age of one year, per 1,000 live births in the year shown. • **Under-five mortality rate** is the probability that a newborn baby will die before reaching age five, if subject to current age-specific mortality rates. The probability is expressed as a rate per 1,000.
• **Contraceptive prevalence rate** is the percentage of women who are practicing, or whose partners are practicing, any form of contraception. It is usually measured for married women ages 15–49 only. • **Maternal mortality ratio** is the number of women who die during pregnancy and childbirth, per 100,000 live births.
• **Access to an improved water source** refers to the share of the population with reasonable access to an adequate amount of water (at least 20 liters a person a day) from an improved source, such as a household connection, public standpipe, borehole, protected well or spring, or rainwater collection. Unimproved sources include vendors, tanker trucks, and unprotected wells and springs.

The indicators here and where they appear throughout the rest of the book have been compiled by World Bank staff from primary and secondary sources. More information can be found in the *About the data, Definitions,* and *Data sources* entries that accompany each table in subsequent sections.

More information about the international development goals and related indicators can be found at www.oecd.org/dac/indicators. For a broader set of goals and indicators used by the United Nations in its common country assessments see www.cca-undaf.org. Data for the international development goals and related indicators are available from the World Bank at www.world bank.org/data. The International Monetary Fund provides links to national data sources and information on data quality and standards through its Dissemination Standards Bulletin Board (dsbb.imf.org).

Table 1.2a

Indicators for the international development goals

	Goal	Indicators
Economic well-being	**Reduction of extreme poverty** Reduce the proportion of people living in extreme poverty in developing countries by at least half between 1990 and 2015.	Share of population living on less than $1 a day Poverty gap ratio (mean shortfall below poverty line) Income inequality Child malnutrition rate
Social development	**Universal primary education** Achieve universal primary education in all countries by 2015.	Net primary enrollment ratio Pupils completing fourth grade of primary education Literacy rate of 15- to 24-year-olds
	Gender equality Demonstrate progress toward gender equality and the empowerment of women by eliminating gender disparity in primary and secondary education by 2005.	Ratio of girls to boys in primary and secondary education Ratio of literate females to males (ages 15–24)
	Reduction in infant and child mortality Reduce the death rates of infants and children under five in each developing country by two-thirds between 1990 and 2015.	Infant mortality rate Under-five mortality rate
	Reduction of maternal mortality Reduce maternal mortality ratios by three-quarters between 1990 and 2015.	Maternal mortality ratio Births attended by skilled health personnel
	Reproductive health Ensure access through the primary health care system to reproductive health services for all individuals of appropriate ages by 2015.	Contraceptive prevalence rate HIV prevalence in pregnant women ages 15–24
Environment	**Environmental sustainability and regeneration** Implement a national strategy for sustainable development in every country by 2005, so as to reverse the current trends in the loss of environmental resources at both global and national levels by 2015.	Countries with effective processes for sustainable development Population with access to an improved water source Forest area as a percentage of national surface area Biodiversity (protected land area) Energy efficiency (GDP per unit of energy use) Carbon dioxide emissions per capita

1.3 Women in development

	Female population	Life expectancy at birth		Prevalence of HIV		Pregnant women receiving prenatal care	Literacy gender parity index	Labor force participation ratio		Maternity leave benefits	Women in decision-making positions	
										% of wages paid in covered period	% at ministerial level	
	% of total	Male years	Female years	Male % ages 15–24	Female % ages 15–24	%	ages 15–24	female-to-male ratio				
	1999	1999	1999	1999ᵃ	1999ᵃ	1996	1995–99ᵇ	1990	1999	1998	1994	1998
Albania	48.7	69	75	..	..			0.7	0.7	..	0	11
Algeria	49.4	69	72	..	..	58	0.9	0.3	0.4	100	4	0
Angola	50.6	45	48	2.7	1.3	25		0.9	0.9	100	7	14
Argentina	50.9	70	77	0.3	0.9			0.4	0.5	100	0	8
Armenia	51.4	71	78	..	..	95	1.0	0.9	0.9	..	3	0
Australia	50.1	76	82	0.0	0.1	..	..	0.7	0.8	0	13	14
Austria	50.8	75	81	0.1	0.2	..	..	0.7	0.7	100	16	20
Azerbaijan	51.0	68	75	..	..	95	1.0	0.8	0.8	..	5	10
Bangladesh	49.5	60	61	0.0	0.0	23	0.7	0.7	0.7	100	8	5
Belarus	53.0	63	74	0.2	0.4	..	1.0	1.0	1.0	100	3	3
Belgium	51.0	75	81	0.1	0.1	..	..	0.7	0.7	82ᶜ	11	3
Benin	50.7	51	55	2.2	0.9	60	0.5	0.9	0.9	100	10	13
Bolivia	50.3	60	64	0.0	0.1	52	1.0	0.6	0.6	70ᵈ	0	6
Bosnia and Herzegovina	50.4	71	75	..	..	..	1.0	0.6	0.6	..	0	6
Botswana	50.9	40	39	34.3	15.8	92	..	0.9	0.8	25	6	14
Brazil	50.6	63	71	0.3	0.7	74	..	0.5	0.5	100	5	4
Bulgaria	51.3	68	75	..	..	..		0.9	0.9	100	0	..
Burkina Faso	50.5	44	46	5.8	2.3	59		0.9	0.9	100	7	10
Burundi	50.9	41	43	11.6	5.7	88	..	1.0	1.0	50	7	8
Cambodia	51.5	52	55	3.5	2.4	52	0.8	1.2	1.1	50	0	..
Cameroon	50.3	50	52	7.8	3.8	73	0.9	0.6	0.6	100	3	6
Canada	50.4	76	82	0.1	0.3	..	..	0.8	0.8	55ᵉ	14	..
Central African Republic	51.4	43	45	14.1	6.9	67	..	..	..	50	5	4
Chad	50.5	47	50	3.0	1.9	30	..	0.8	0.8	50	5	0
Chile	50.5	73	79	0.1	0.3	91	1.0	0.4	0.5	100	13	13
China	48.4	68	72	0.0	0.1	79	..	0.8	0.8	100	6	..
Hong Kong, China	50.0	77	82	0.0	0.1	100	..	0.6	0.6	..	..	..
Colombia	50.6	67	74	0.1	0.4	83	1.0	0.6	0.6	100	11	18
Congo, Dem. Rep.	50.5	45	47	5.1	2.5	66	..	0.8	0.8	67	6	..
Congo, Rep.	51.1	46	50	6.5	3.2	55	..	0.8	0.8	100	6	6
Costa Rica	49.3	75	79	0.3	0.6	95	..	0.4	0.4	100	10	15
Côte d'Ivoire	49.1	46	47	9.5	3.8	83	..	0.5	0.5	100	8	3
Croatia	51.6	69	77	0.0	0.0	..	..	0.7	0.8	..	4	12
Cuba	49.9	74	78	0.0	0.1	100	..	0.6	0.6	100	0	5
Czech Republic	51.3	71	78	0.0	0.1	..	..	0.9	0.9	..	0	17
Denmark	50.4	73	78	0.1	0.2	..	..	0.9	0.9	100ᶠ	29	41
Dominican Republic	49.3	69	73	2.8	2.6	97	..	0.4	0.4	100	4	10
Ecuador	49.8	68	71	0.1	0.4	75	1.0	0.3	0.4	100	6	20
Egypt, Arab Rep.	49.1	65	68	..	..	53	..	0.4	0.4	100	4	6
El Salvador	50.9	67	72	0.3	0.7	69	..	0.5	0.6	75	10	6
Eritrea	50.4	49	52	..	..	19	..	0.9	0.9	..	7	5
Estonia	53.2	65	76	..	..	..	..	1.0	1.0	..	15	12
Ethiopia	49.8	41	43	11.9	7.5	20	..	0.7	0.7	100	10	5
Finland	51.2	74	81	0.0	0.0	..	..	0.9	0.9	80	39	29
France	51.3	75	82	0.2	0.3	..	..	0.8	0.8	100	7	12
Gabon	50.6	51	54	4.7	2.3	86	..	0.8	0.8	100	7	3
Gambia, The	50.5	52	55	2.2	0.9	91	..	0.8	0.8	100	0	29
Georgia	52.3	69	77	..	..	95	1.0	0.9	0.9	..	0	4
Germany	51.0	74	80	0.0	0.1	..	..	0.7	0.7	100	16	8
Ghana	50.3	57	59	3.4	1.4	86	..	1.0	1.0	50	11	9
Greece	50.8	75	81	0.1	0.1	..	..	0.5	0.6	75	4	5
Guatemala	49.6	62	68	0.9	1.2	53	1.0	0.3	0.4	100	19	0
Guinea	49.7	46	47	1.4	0.6	59	..	0.9	0.9	100	9	8
Guinea-Bissau	50.8	43	45	2.5	1.0	50	..	0.7	0.7	100	4	18
Haiti	50.8	51	56	2.9	4.9	68	..	0.8	0.8	100ᵍ	13	0
Honduras	49.6	67	72	1.7	1.4	73	..	0.4	0.5	100ʰ	11	11

Women in development 1.3

	Female population	Life expectancy at birth		Prevalence of HIV		Pregnant women receiving prenatal care	Literacy gender parity index	Labor force participation ratio		Maternity leave benefits	Women in decision-making positions	
	% of total	Male years	Female years	Male % ages 15–24	Female % ages 15–24	%	ages 15–24	female-to-male ratio		% of wages paid in covered period	% at ministerial level	
	1999	1999	1999	1999[a]	1999[a]	1996	1995–99[b]	1990	1999	1998	1994	1998
Hungary	52.1	66	75	0.0	0.1	..	..	0.8	0.8	100	0	5
India	48.4	62	64	0.6	0.4	62	0.8	0.5	0.5	100	3	..
Indonesia	50.1	64	68	0.0	0.0	82	..	0.6	0.7	100	6	3
Iran, Islamic Rep.	49.8	70	72	..	..	62	1.0	0.3	0.4	67	0	0
Iraq	49.1	58	60	..	..	59	..	0.2	0.2	100	0	0
Ireland	50.1	74	79	0.0	0.1	..	..	0.5	0.5	70[f]	16	21
Israel	50.3	76	80	0.1	0.1	90	..	0.6	0.7	75	4	0
Italy	51.4	75	82	0.2	0.3	..	..	0.6	0.6	80	12	13
Jamaica	50.4	73	77	0.4	0.6	98	..	0.9	0.9	100[i]	5	12
Japan	51.0	77	84	0.0	0.0	..	..	0.7	0.7	60	6	0
Jordan	48.2	70	73	..	..	80	1.0	0.2	0.3	100	3	2
Kazakhstan	51.5	59	70	..	0.1	92	1.0	0.9	0.9	..	6	5
Kenya	49.9	47	48	13.0	6.4	95	..	0.8	0.9	100	0	0
Korea, Dem. Rep.	49.8	59	62	..	..	100	1.0	0.8	0.8	..	0	..
Korea, Rep.	49.6	69	77	0.0	0.0	96	1.0	0.6	0.7	100	4	..
Kuwait	47.2	74	80	..	..	99	1.0	0.3	0.5	100	0	0
Kyrgyz Republic	51.0	63	72	..	..	90	..	0.9	0.9	..	0	4
Lao PDR	50.5	53	56	0.1	0.0	25	0.8	..	..	100	0	0
Latvia	54.1	64	76	0.1	0.2	..	1.0	1.0	1.0	..	0	7
Lebanon	50.9	68	72	..	..	85	1.0	0.4	0.4	100	0	0
Lesotho	50.8	44	45	26.4	12.1	91	..	0.6	0.6	0	6	6
Libya	48.2	69	73	..	..	100	1.0	0.2	0.3	50	0	7
Lithuania	52.8	67	77	..	..	..	..	0.9	0.9	..	0	6
Macedonia, FYR	50.0	71	75	..	..	..	..	0.7	0.7	..	8	9
Madagascar	50.1	53	56	0.1	0.0	78	0.9	0.8	0.8	100[f]	0	19
Malawi	50.5	39	40	15.3	7.0	90	..	1.0	0.9	..	9	4
Malaysia	49.4	70	75	0.1	0.6	90	..	0.6	0.6	100	7	16
Mali	50.7	41	44	2.1	1.3	25	..	0.9	0.9	100	10	21
Mauritania	50.4	52	56	0.6	0.4	49	0.7	0.8	0.8	100	0	4
Mauritius	50.1	67	75	0.0	0.0	99	..	0.4	0.5	100	3	..
Mexico	50.5	69	75	0.1	0.4	71	1.0	0.4	0.5	100	5	5
Moldova	52.2	63	70	0.1	0.3	..	..	0.9	0.9	..	0	0
Mongolia	49.8	65	68	..	..	90	1.0	0.9	0.9	..	0	0
Morocco	50.0	65	69	..	..	45	..	0.5	0.5	100	0	0
Mozambique	51.5	42	44	14.7	6.7	54	..	0.9	0.9	100	4	0
Myanmar	50.2	58	61	1.7	1.0	80	..	0.8	0.8	67	0	0
Namibia	50.2	50	50	19.8	9.1	88	..	0.7	0.7	..	10	8
Nepal	49.4	58	58	0.2	0.1	15	0.7	0.7	0.7	100	0	3
Netherlands	50.5	75	81	0.1	0.2	..	..	0.6	0.7	100	31	28
New Zealand	50.8	75	80	0.0	0.1	..	..	0.8	0.8	0	8	8
Nicaragua	50.3	66	71	0.1	0.2	71	..	0.5	0.6	60	10	5
Niger	50.6	44	48	1.5	0.9	30	..	0.5	0.6	50	3	6
Nigeria	50.7	47	48	5.1	2.5	60	..	0.8	0.9	100	0	0
Norway	50.1	76	81	0.0	0.1	..	..	0.8	0.9	100	35	20
Oman	46.8	72	75	..	..	98	0.9	0.1	0.2	..	0	0
Pakistan	48.2	62	64	0.0	0.1	27	0.6	0.3	0.4	100	4	7
Panama	49.5	72	76	1.4	1.6	72	..	0.5	0.5	100	13	6
Papua New Guinea	48.5	58	59	0.2	0.1	70	..	0.7	0.7	0	0	0
Paraguay	49.6	68	72	0.0	0.1	83	..	0.4	0.4	50[j]	0	7
Peru	50.3	66	71	0.2	0.4	64	..	0.4	0.4	100	6	10
Philippines	49.6	67	71	0.1	0.0	83	..	0.6	0.6	100	8	10
Poland	51.3	69	77	..	..	..	1.0	0.8	0.9	100	17	12
Portugal	52.1	72	79	0.2	0.6	..	1.0	0.7	0.8	100	10	10
Puerto Rico	51.8	71	80	..	..	99	..	0.5	0.6	..	..	..
Romania	50.9	66	73	0.0	0.0	..	..	0.8	0.8	50–94	0	8
Russian Federation	53.2	60	72	0.1	0.3	..	..	0.9	1.0	100	0	8

1.3 | Women in development

	Female population	Life expectancy at birth		Prevalence of HIV		Pregnant women receiving prenatal care	Literacy gender parity index	Labor force participation ratio		Maternity leave benefits	Women in decision-making positions	
		Male	Female	Male % ages 15–24	Female % ages 15–24		ages	female-to-male ratio		% of wages paid in covered period	% at ministerial level	
	% of total 1999	years 1999	years 1999	1999[a]	1999[a]	% 1996	15–24 1995–99[b]	1990	1999	1998	1994	1998
Rwanda	50.5	40	41	10.6	5.2	94	..	1.0	1.0	67	9	5
Saudi Arabia	44.8	71	74	..	..	87	..	0.1	0.2	50–100	0	0
Senegal	50.2	51	54	1.6	0.7	74	..	0.7	0.7	100	7	7
Sierra Leone	50.9	36	39	2.9	1.2	30	..	0.6	0.6	..	0	10
Singapore	49.7	76	80	0.2	0.2	100	..	0.6	0.6	100	0	0
Slovak Republic	51.3	69	77	0.0	0.0	..	..	0.9	0.9	..	5	19
Slovenia	51.4	71	79	0.0	0.0	..	..	0.9	0.9	..	5	0
South Africa	51.8	47	50	24.8	11.3	89	1.1	0.6	0.6	45	6	..
Spain	51.1	75	82	0.2	0.5	..	..	0.5	0.6	100	14	18
Sri Lanka	49.2	71	76	0.1	0.0	100	..	0.5	0.6	100	3	13
Sudan	49.8	54	57	..	..	54	0.8	0.4	0.4	100	0	0
Sweden	50.5	77	82	0.0	0.1	..	..	0.9	0.9	75	30	43
Switzerland	50.4	77	83	0.3	0.4	..	..	0.6	0.7	100	17	17
Syrian Arab Republic	49.4	67	72	..	..	33	0.9	0.3	0.4	100	7	8
Tajikistan	50.2	66	72	..	..	90	..	0.7	0.8	..	3	6
Tanzania	50.5	44	46	8.1	4.0	92	..	1.0	1.0	100	13	13
Thailand	50.1	67	71	2.3	1.2	77	..	0.9	0.9	100 [k]	0	4
Togo	50.4	48	50	5.5	2.2	43	0.8	0.7	0.7	100	5	9
Trinidad and Tobago	50.2	70	75	0.6	0.8	98	..	0.5	0.5	60–100	19	14
Tunisia	49.5	71	74	..	..	71	..	0.4	0.5	67	4	3
Turkey	49.5	67	72	..	..	62	1.0	0.5	0.6	67	5	5
Turkmenistan	50.5	63	70	..	..	90	..	0.8	0.8	..	3	4
Uganda	50.1	42	42	7.8	3.8	87	..	0.9	0.9	100 [l]	10	13
Ukraine	53.5	62	73	0.8	1.3	..	..	1.0	1.0	100	0	5
United Arab Emirates	33.7	74	77	..	..	95	1.0	0.1	0.2	100	0	0
United Kingdom	50.8	75	80	0.0	0.1	..	..	0.7	0.8	90 [m]	9	24
United States	50.7	74	80	0.2	0.5	..	..	0.8	0.8	0	14	26
Uruguay	51.5	70	78	0.2	0.4	80	1.0	0.6	0.7	100	0	7
Uzbekistan	50.3	66	73	..	..	90	1.0	0.8	0.9	..	3	3
Venezuela, RB	49.7	70	76	0.1	0.7	74	1.0	0.5	0.5	100	11	3
Vietnam	51.1	66	71	0.1	0.3	78	1.0	1.0	1.0	100	5	0
West Bank and Gaza	49.2	70	74	..	..	..	..	..	..	..	..	..
Yemen, Rep.	48.9	55	57	..	..	26	..	0.4	0.4	100	0	0
Yugoslavia, FR (Serb./Mont.)	50.2	70	75	..	..	..	..	0.7	0.7	..	..	5
Zambia	50.3	38	39	17.8	8.2	92	0.9	0.8	0.8	100	5	3
Zimbabwe	50.4	41	40	24.5	11.3	93	1.0	0.8	0.8	60–75	3	12
World	**49.6 w**	**65 w**	**69 w**	**1.1 w**	**0.7 w**	**70 w**		**0.7 w**	**0.7 w**		**6 w**	**.. w**
Low income	49.4	58	60	2.0	1.1	62		0.6	0.6		4	..
Middle income	49.5	67	72	0.6	0.5	77		0.7	0.7		5	..
Lower middle income	49.3	67	72	0.2	0.2	76		0.7	0.8		5	..
Upper middle income	50.5	66	73	2.2	1.5	80		0.5	0.6		6	6
Low & middle income	49.5	63	66	1.3	0.8	70		0.7	0.7		5	..
East Asia & Pacific	48.9	67	71	0.2	0.2	80		0.8	0.8		5	..
Europe & Central Asia	51.9	64	73	..	0.4	..		0.8	0.9		3	7
Latin America & Carib.	50.5	67	73	0.3	0.7	75		0.5	0.5		6	7
Middle East & N. Africa	49.3	67	69	..	..	58		0.3	0.4		2	2
South Asia	48.5	62	63	0.5	0.3	55		0.5	0.5		4	..
Sub-Saharan Africa	50.5	46	48	9.2	4.5	65		0.7	0.7		6	7
High income	50.4	75	81	0.1	0.3	..		0.7	0.8		12	16
Europe EMU	51.2	74	81	0.2	0.3	..		0.7	0.7		14	13

a. Average of high and low estimates. b. Data are for the most recent year available. c. For 30 days; 75 percent thereafter. d. Benefit is 70 percent of wages above the minimum wage, 100 percent of the national minimum wage. e. For 15 weeks. f. Up to a ceiling. g. For 6 weeks. h. For 84 days. i. For 8 weeks. j. For 9 weeks. k. Benefit is 100 percent for the first 45 days, then 50 percent for 15 days. l. For 1 month. m. For 6 weeks; flat rate thereafter.

Women in development | 1.3

Despite considerable progress in recent decades, gender inequalities remain pervasive in many dimensions of life—worldwide. But while disparities exist throughout the world, they are most prevalent in poor developing countries. The differences in outcomes between men and women—and between boys and girls—are a consequence of differences in the opportunities and resources available to them. Inequalities in the allocation of such resources as education, health care, and nutrition matter because of the strong association of these resources with well-being, productivity, and growth. This pattern of inequality begins at an early age, with boys routinely receiving a larger share of education and health spending than girls do, for example.

Life expectancy has increased for both men and women in all regions, but female morbidity and mortality rates sometimes exceed male rates, particularly during early childhood and the reproductive years. In high-income countries women tend to outlive men by four to eight years on average, while in low-income countries the difference is narrower—about two to three years. The female disadvantage is best reflected in differences in child mortality rates (see table 2.19). Child mortality captures the effect of preferences for boys because adequate nutrition and medical interventions are particularly important for the age group 1–5. Because of the natural female biological advantage, when female child mortality is as high as or higher than male child mortality, there is good reason to believe that girls are discriminated against.

Female disadvantage in mortality is carried into adolescence and the reproductive years. Serious health risks for adolescents arise when they become sexually active, and one of the most important health concerns is the prevalence of sexually transmitted diseases, including HIV. It is estimated that half of all new HIV infections occur in the age group 15–24, and in some countries the rate of infection in this age group is higher among women than men. And while in high-income countries women have universal access to health care during pregnancy, in developing countries it is estimated that 35 percent of pregnant women—some 45 million each year—receive no care at all (United Nations 2000b). Prenatal care is essential for recognizing, diagnosing, and promptly treating complications that arise during pregnancy.

Girls in many developing countries are allowed less education by their families than boys are—a disparity reflected in lower female primary enrollment (see table 1.2) and higher female illiteracy. As a result, women have fewer employment opportunities, especially in the formal sector. A labor force par-

ticipation ratio of less than 1.0 shows that women's labor force participation in the formal sector is lower than men's (a ratio of 1.0 indicates gender equality).

Women who work outside the home continue to bear a disproportionate share of the responsibility for housework and child-rearing. They also face discriminatory practices in the workplace, especially relating to equal pay and maternity benefits. The maternity benefits data in the table relate only to legislated benefits and do not include contractual benefits negotiated through labor union contracts. The benefits generally apply only in the formal sector, leaving out the vast majority of working women in developing countries. As a result, while the situation in the United States is much better than the data indicate, the situation in Thailand is likely to be much worse.

Women are vastly underrepresented in decision-making positions in government, although there is some evidence of recent improvement. While 6 percent of the world's cabinet ministers were women in 1994, 8 percent were in 1998. Without representation at this level, it is difficult for women to influence policy.

For information on other aspects of gender, see tables 1.2 (development progress), 2.3 (employment by economic activity), 2.4 (unemployment), 2.13 (education efficiency), 2.14 (education outcomes), 2.17 (reproductive health), 2.18 (health: risk factors and future challenges), and 2.19 (mortality).

• **Female population** is the percentage of the population that is female. • **Life expectancy at birth** is the number of years a newborn infant would live if prevailing patterns of mortality at the time of its birth were to stay the same throughout its life. • **Prevalence of HIV** refers to the percentage of people ages 15–24 who are infected with HIV. • **Pregnant women receiving prenatal care** are the percentage of women attended at least once during pregnancy by skilled health personnel for reasons related to pregnancy. • **Literacy gender parity index** is the ratio of the female literacy rate to the male rate, for the age group 15–24. • **Labor force participation ratio** is the ratio of the percentage of women who are economically active to the percentage of men who are. According to the International Labour Organization (ILO) definition, the economically active population is all those who supply labor for the production of goods and services during a specified period. It includes both the employed and the unemployed. While national practices vary in the treatment of such groups as the armed forces and seasonal or part-time workers, in general the labor force includes the armed forces, the unemployed, and first-time job-seekers, but excludes homemakers and other unpaid caregivers and workers in the informal sector. • **Maternity leave benefits** refer to the compensation provided to women during maternity leave, as a share of their full wages. • **Women in decisionmaking positions** are those in ministerial or equivalent positions in the government.

Data sources

The data are from the World Bank's population database; electronic databases of the Joint United Nations Programme on HIV/AIDS (UNAIDS) and the United Nations Educational, Scientific, and Cultural Organization (UNESCO); the ILO database Estimates and Projections of the Economically Active Population, 1950–2010; and the United Nations' *World's Women: Trends and Statistics 2000*.

1.4 Trends in long-term economic development

	Gross domestic product		Population		Value added			Household final consumption expenditure	Gross fixed capital formation	Exports of goods and services
	average annual % growth		average annual % growth		average annual % growth			average annual % growth	average annual % growth	average annual % growth
	Total 1965–99	Per capita 1965–99	Total 1965–99	Labor force 1965–99	Agriculture 1965–99	Industry 1965–99	Services 1965–99	1965–99	1965–99	1965–99
Albania	–0.3	–1.4	1.8	2.2	3.3	–4.6	–0.6	..	..	..
Algeria	3.9	1.0	2.7	3.2	4.7	2.9	4.0	4.6	3.3	2.7
Angola	0.9	–2.1	2.5	2.1	..	..	..	..	..	..
Argentina	1.9	0.4	1.5	1.5	1.7	1.1	2.6	2.4	1.1	5.3
Armenia	..	..	1.6	2.3						
Australia	3.4	1.9	1.5	2.1	2.2	2.4	3.6	3.4	3.5	5.8
Austria	2.8	2.5	0.3	0.4	0.7	2.0	2.5	2.9	2.8	6.2
Azerbaijan	..	..	1.7	2.1	..	..	..	..	..	..
Bangladesh	3.8	1.3	2.3	2.3	2.1	4.4	4.3	3.6	4.7	7.8
Belarus	..	..	0.5	0.6		..		..		..
Belgium	2.4	2.2	0.2	0.5	1.8	2.0	2.1	2.5	2.0	5.0
Benin	3.1	0.2	2.8	2.3	4.1	4.0	2.5	2.7	2.9	3.1
Bolivia	2.0	–0.3	2.3	2.4	..	..	..	2.5	2.3	3.1
Bosnia and Herzegovina	..	..	0.2	0.6	..	..	..	..		..
Botswana	10.6	7.1	3.2	2.9	3.2	12.8	10.8	6.7	7.0	4.8
Brazil	4.5	2.4	2.0	2.9	3.4	4.4	4.8	4.3	1.7	8.2
Bulgaria	–0.6	–0.2	0.0	–0.1	–2.1	–1.5	1.5	0.9	–4.7	–11.2
Burkina Faso	3.4	1.1	2.3	1.7	2.6	2.6	5.5	3.0	5.8	3.2
Burundi	2.9	0.6	2.2	2.0	2.5	3.3	3.5	2.9	..	3.5
Cambodia	..	..	1.9	1.9	..	..	..	..	..	..
Cameroon	3.9	1.1	2.7	2.2	3.4	6.2	3.4	3.2	0.1	6.2
Canada	3.2	1.9	1.3	2.3	1.2	2.0	3.0	3.2	4.1	6.0
Central African Republic	1.2	–1.1	2.2	..	1.7	2.1	0.2	2.7	..	2.1
Chad	1.8	–0.6	2.4	2.2	1.7	1.6	2.3	2.6	..	1.7
Chile	4.2	2.5	1.7	2.3	3.5	3.3	4.2	3.3	4.9	8.3
China	8.1	6.4	1.7	2.1	4.1	10.9	9.3	7.4	10.0	11.2
Hong Kong, China	7.3	5.4	1.8	2.6	..	..	..	7.7	7.5	11.7
Colombia	4.3	2.1	2.2	3.2	2.5	4.3	4.7	4.0	4.5	5.7
Congo, Dem. Rep.	–0.4	–3.4	3.1	2.7	2.0	–2.9	–2.2	0.1	–0.3	2.4
Congo, Rep.	4.6	1.7	2.8	2.6	2.8	7.0	3.8	3.6	..	6.3
Costa Rica	4.2	1.4	2.7	3.5	3.4	4.9	4.1	3.5	5.0	7.2
Côte d'Ivoire	3.0	–0.7	3.7	3.4	2.2	6.1	2.7	2.5	0.4	5.1
Croatia	..	..	0.1	0.2	..	..	..	..	..	..
Cuba	..	..	1.1	2.2	..	..	..	..	..	..
Czech Republic	..	..	0.2	0.4	..	..	..	..	..	..
Denmark	1.9	1.6	0.3	0.8	2.5	1.2	2.1	1.5	1.0	4.5
Dominican Republic	4.9	2.5	2.3	3.2	2.9	5.7	5.0	4.1	5.3	5.8
Ecuador	4.6	1.9	2.6	3.1	3.4	5.9	4.4	4.1	2.8	7.0
Egypt, Arab Rep.	5.6	3.3	2.2	2.4	2.8	6.5	7.8	5.1	5.8	5.4
El Salvador	1.6	–0.3	2.1	2.9	0.6	0.9	2.2	2.0	2.7	1.6
Eritrea	..	..	2.7	2.5	..	..	..	..	..	..
Estonia	–1.4	–1.3	0.4	0.5	..	..	..	..	..	..
Ethiopia	2.4	–0.3	2.7	2.4	1.8	0.7	3.9	2.5	3.8	1.3
Finland	2.9	2.5	0.4	0.6	0.3	3.1	3.0	2.8	1.2	5.1
France	2.7	2.1	0.5	0.7	1.6	1.3	2.5	2.5	1.9	5.8
Gabon	3.8	0.8	2.6	2.0	–0.5	2.6	2.5	3.5	–2.0	5.4
Gambia, The	4.0	0.5	3.3	3.1	1.9	4.0	4.1	1.5	9.4	3.2
Georgia	..	..	0.6	0.8	..	..	..	..	..	..
Germany	..	..	0.2	0.4	..	..	..	..	..	..
Ghana	2.0	–0.7	2.6	2.6	1.4	0.4	3.4	1.6	0.8	–0.1
Greece	3.1	2.4	0.6	0.9	1.1	2.9	3.8	3.3	1.7	7.3
Guatemala	3.3	0.7	2.6	2.8	2.8	3.6	3.5	3.3	2.6	2.5
Guinea	..	..	2.1	1.8	..	..	..	..	..	..
Guinea-Bissau	2.8	0.0	2.4	2.1	1.6	2.6	4.9	1.1	..	3.4
Haiti	1.0	–0.9	1.9	1.2	0.1	1.3	1.6	1.5	2.0	0.5
Honduras	3.8	0.6	3.1	3.3	2.5	4.3	4.3	3.7	4.1	2.5

Trends in long-term economic development 1.4

	Gross domestic product		Population		Value added			Household final consumption expenditure	Gross fixed capital formation	Exports of goods and services
	average annual % growth		average annual % growth		average annual % growth			average annual % growth	average annual % growth	average annual % growth
	Total 1965–99	Per capita 1965–99	Total 1965–99	Labor force 1965–99	Agriculture 1965–99	Industry 1965–99	Services 1965–99	1965–99	1965–99	1965–99
Hungary	2.1	2.1	0.0	−0.2	−1.8	−1.2	0.7	1.0	2.2	4.5
India	4.6	2.4	2.1	2.1	2.8	5.5	5.8	4.3	5.5	7.3
Indonesia	6.9	4.8	2.0	2.7	3.8	8.7	7.5	7.0	7.6	5.6
Iran, Islamic Rep.	1.7	−1.0	2.8	2.7	4.5	0.7	2.0	3.5	−0.3	−1.1
Iraq	−0.3	−3.5	3.1	2.9	..	..	..	..	..	..
Ireland	4.4	3.5	0.8	1.0	..	..	..	3.1	..	9.1
Israel	5.0	2.4	2.6	3.0	..	..	..	5.6	3.1	7.2
Italy	2.8	2.5	0.3	0.6	1.2	1.9	2.8	3.2	1.6	5.5
Jamaica	1.0	−0.2	1.2	2.0	0.9	0.1	1.9	1.2	0.1	1.9
Japan	4.1	3.4	0.7	1.0	−0.2	4.3	4.6	4.0	4.4	7.3
Jordan	4.7	0.4	4.3	4.4	6.5	5.3	4.0	4.7	4.5	7.1
Kazakhstan	..	..	0.7	1.2	..	..	..	..	..	..
Kenya	4.7	1.2	3.3	3.3	3.4	5.4	5.3	4.0	1.5	2.9
Korea, Dem. Rep.	..	..	2.0	2.6	..	..	..	..	..	..
Korea, Rep.	8.1	6.6	1.5	2.6	2.1	10.8	7.6	7.3	11.5	15.6
Kuwait	0.0	−3.9	4.2	4.3	9.7	−4.1	6.2	7.8	8.5	−3.0
Kyrgyz Republic	..	..	1.9	2.1	..	..	..	..	..	..
Lao PDR	..	..	2.2	1.9	..	..	..	..	..	..
Latvia	1.5	1.2	0.2	0.3	−3.8	−5.6	1.1	..	..	..
Lebanon	..	..	1.9	2.5	..	..	..	..	..	..
Lesotho	5.3	2.8	2.3	2.0	−0.2	11.7	6.2	5.1	7.5	7.6
Libya	0.5	−3.6	3.6	3.2	10.3	−1.2	11.4	12.1	..	−1.2
Lithuania	..	..	0.7	0.8	..	..	..	..	..	..
Macedonia, FYR	..	..	..	..	..	..	..	..	..	..
Madagascar	0.9	−1.7	2.6	2.4	1.6	0.1	0.5	0.3	1.2	−0.3
Malawi	3.7	0.6	3.0	2.6	2.9	3.5	4.0	3.0	−3.6	3.7
Malaysia	7.0	4.3	2.6	3.1	2.9	8.1	7.7	5.9	9.9	9.9
Mali	2.3	−0.1	2.3	2.1	3.2	3.3	2.4	2.4	1.9	6.9
Mauritania	2.4	−0.2	2.5	2.2	1.5	2.6	3.2	3.7	..	2.1
Mauritius	5.2	3.9	1.3	2.4	−0.3	7.4	6.5	5.0	4.4	5.9
Mexico	4.0	1.5	2.4	3.3	2.1	4.1	4.2	3.6	3.6	10.0
Moldova	..	..	0.8	0.7	..	..	..	..	..	..
Mongolia	1.3	−0.5	2.3	2.5	1.0	0.5	2.1	..	..	..
Morocco	4.2	1.9	2.2	2.6	2.3	3.9	5.2	4.4	4.2	5.3
Mozambique	3.0	1.3	2.2	1.9	..	..	..	1.2	6.2	5.8
Myanmar	3.5	1.5	1.8	2.0	3.3	4.3	3.4	2.8	5.4	3.8
Namibia	2.7	0.0	2.6	2.2	3.6	1.4	2.6	1.3	2.2	2.9
Nepal	3.7	1.2	2.4	2.0	2.3	7.9	4.6	3.9	6.1	8.8
Netherlands	2.6	1.9	0.7	1.5	3.6	1.4	2.7	2.6	2.2	5.0
New Zealand	1.9	0.9	1.1	1.9	3.4	1.1	2.7	1.6	2.8	4.3
Nicaragua	−0.1	−2.9	3.0	3.6	0.1	0.0	−0.2	−0.4	0.8	0.6
Niger	0.7	−2.3	3.1	2.8	0.3	4.8	0.3	1.4	−4.9	−0.3
Nigeria	3.0	0.0	2.9	2.7	1.6	3.9	4.7	2.4	−0.1	2.6
Norway	3.5	3.0	0.5	1.3	1.6	3.9	2.6	2.7	1.9	5.3
Oman	9.5	5.0	3.9	3.8	..	..	..	..	..	..
Pakistan	5.6	2.7	2.8	2.9	4.1	6.5	6.1	5.1	4.2	6.2
Panama	3.4	1.1	2.3	2.9	2.1	2.7	2.6	4.3	5.0	0.0
Papua New Guinea	3.1	0.7	2.3	2.1	3.0	5.8	2.6	2.8	1.0	7.0
Paraguay	5.0	2.1	2.8	3.0	4.3	5.7	5.2	5.2	6.6	8.3
Peru	2.0	−0.3	2.3	2.9	1.9	2.3	1.9	2.2	2.8	2.6
Philippines	3.4	0.9	2.5	2.8	2.3	3.5	4.0	3.7	4.3	6.4
Poland	..	..	0.6	0.6	..	..	..	..	..	..
Portugal	3.6	3.2	0.3	1.1	..	..	..	3.4	..	5.6
Puerto Rico	3.9	2.7	1.2	2.0	1.7	4.2	3.2	2.8	0.4	4.4
Romania	−0.3	−0.5	0.5	0.0	..	..	..	..	..	..
Russian Federation	..	..	0.4	0.7	..	..	..	..	..	..

1.4 Trends in long-term economic development

	Gross domestic product		Population		Value added			Household final consumption expenditure	Gross fixed capital formation	Exports of goods and services
	average annual % growth		average annual % growth		average annual % growth			average annual % growth	average annual % growth	average annual % growth
	Total 1965–99	Per capita 1965–99	Total 1965–99	Labor force 1965–99	Agriculture 1965–99	Industry 1965–99	Services 1965–99	1965–99	1965–99	1965–99
Rwanda	2.7	−0.1	2.8	2.8	2.1	2.6	4.2	3.2	5.9	2.7
Saudi Arabia	4.6	−0.1	4.3	4.8	7.4	3.2	6.9	..	..	..
Senegal	2.4	−0.4	2.8	2.5	1.1	3.8	2.5	2.4	3.3	1.6
Sierra Leone	0.9	−1.2	2.1	1.8	3.1	−1.0	−0.9	−1.0	..	−5.4
Singapore	8.3	6.3	1.9	3.1	−1.5	8.4	8.4	6.6	9.4	..
Slovak Republic	..	..	0.6	1.3	..	..	..	..	..	..
Slovenia	..	..	0.6	0.8	..	..	..	..	..	..
South Africa	2.3	0.0	2.2	2.4	2.0	1.8	3.1	3.2	1.6	1.8
Spain	3.0	2.4	0.6	1.0	..	..	..	2.8	4.1	7.3
Sri Lanka	4.6	3.0	1.6	2.2	2.7	5.1	5.3	4.2	7.7	4.2
Sudan	3.1	0.5	2.5	2.7	3.1	3.7	3.5	3.9	..	−2.1
Sweden	1.9	1.5	0.4	1.0	0.5	1.7	2.2	1.4	0.4	4.5
Switzerland	1.6	1.1	0.6	1.0	..	..	..	1.7	1.8	3.7
Syrian Arab Republic	5.7	2.3	3.2	3.3	4.3	8.4	6.2	4.5	0.7	6.2
Tajikistan	..	..	2.7	2.7	..	..	..	..	..	..
Tanzania	..	..	3.0	2.9	..	..	..	..	..	..
Thailand	7.3	5.1	2.0	2.6	3.9	9.4	7.2	6.1	8.1	11.2
Togo	2.4	−0.5	3.1	2.8	3.6	2.8	1.4	3.4	−2.3	3.1
Trinidad and Tobago	2.9	1.8	1.1	1.9	..	0.2	..	2.9	..	3.7
Tunisia	5.0	2.7	2.1	2.8	3.9	5.9	5.0	5.6	4.3	6.7
Turkey	4.4	2.2	2.2	2.1	1.3	5.5	4.9	..	..	..
Turkmenistan	..	..	2.8	3.1	..	..	..	..	..	..
Uganda	5.5	2.5	2.9	2.6	3.3	9.1	6.1	5.0	7.6	8.1
Ukraine	..	..	0.3	0.3	..	..	..	..	..	..
United Arab Emirates	3.3	−3.9	9.5	10.5	11.4	1.1	6.4	..	..	..
United Kingdom	2.2	2.0	0.3	0.5	1.2	1.9	2.7	2.6	2.3	4.3
United States	3.0	2.0	1.1	1.7	..	..	..	3.2	3.4	6.5
Uruguay	2.0	1.4	0.6	1.0	1.5	1.3	2.5	1.8	2.6	5.7
Uzbekistan	..	..	2.6	2.8	..	..	..	..	..	..
Venezuela, RB	2.1	−0.8	2.8	3.7	2.7	1.7	2.5	2.2	1.1	2.1
Vietnam	..	..	2.1	2.1	..	..	..	..	..	..
West Bank and Gaza	..	..	..	..	..	..	..	..	..	..
Yemen, Rep.	..	..	3.2	2.8	..	..	..	..	..	..
Yugoslavia, FR (Serb./Mont.)	..	..	0.7	0.9	..	..	..	..	..	..
Zambia	1.0	−2.0	3.0	2.7	2.9	0.6	0.6	−0.1	−5.8	−0.9
Zimbabwe	3.9	0.9	2.9	2.9	2.3	1.8	4.4	4.1	2.5	7.0
World	**3.3** w	**1.6** w	**1.7** w	**2.0** w	**2.2** w	**..** w	**..** w	**3.4** w	**3.3** w	**5.9** w
Low income	4.1	1.8	2.3	2.2	2.7	4.9	5.2	4.0	3.3	3.7
Middle income	4.2	2.4	1.7	2.1	3.0	4.1	4.3	4.4	3.2	5.7
Lower middle income	4.3	2.6	1.7	2.0	3.1	5.5	4.7	..	4.4	3.9
Upper middle income	4.1	2.2	1.8	2.3	2.7	3.4	4.0	4.1	3.4	7.2
Low & middle income	4.2	2.2	2.0	2.1	2.9	4.3	4.6	4.1	3.3	5.3
East Asia & Pacific	7.4	5.6	1.8	2.2	3.6	9.6	7.8	6.7	9.7	10.1
Europe & Central Asia	..	..	0.8	0.9	..	..	..	..	..	..
Latin America & Carib.	3.5	1.4	2.1	2.8	2.7	3.3	3.8	3.5	1.9	6.0
Middle East & N. Africa	3.0	0.1	2.8	2.8	4.2	1.3	4.0	..	..	..
South Asia	4.7	2.4	2.2	2.2	2.9	5.5	5.6	4.4	5.3	7.2
Sub-Saharan Africa	2.6	−0.2	2.7	2.6	1.9	2.4	3.1	2.8	0.1	2.4
High income	3.2	2.4	0.8	1.2	..	..	..	3.2	3.3	5.9
Europe EMU	..	..	0.4	0.7	..	..	..	..	..	5.5

Trends in long-term economic development | 1.4

The long-term trends shown in this table provide a view of the relative rates of change in key social and economic indicators over the period 1965–99. In viewing these growth rates, it may be helpful to keep in mind that a quantity growing at 2.3 percent a year will double in 30 years, while a quantity growing at 7 percent a year will double in 10 years. But like all averages the rates reflect the general tendency and may disguise considerable year-to-year variation, especially for economic indicators.

Average annual growth rates of gross domestic product (GDP), value added, household final consumption expenditure (private consumption in previous editions), gross fixed capital formation, and exports of goods and services are calculated from data in 1995 constant prices using the least-squares method. For more information on the calculation of growth rates see *Statistical methods*. As noted in *About the data* for table 4.1, the growth rates of GDP and its components are calculated using constant price data in local currency. Regional and income group growth rates are calculated after converting local currencies to constant price U.S. dollars. Because the data have been rescaled to a common reference year, the weighted average of the sector growth rates generally will not equal the GDP growth rate.

All the indicators shown here appear elsewhere in the *World Development Indicators*. For more information about them see *About the data* for tables 1.1 (GDP and GDP per capita), 2.1 (population), 2.2 (labor force), 4.1 (growth of GDP), 4.2 (value added by industrial origin), 4.9 (exports of goods and services), and 4.10 (household final consumption expenditure and gross capital formation).

Definitions

• **Gross domestic product** (GDP) is the sum of value added by all resident producers plus any taxes (less subsidies) not included in the valuation of output. • **GDP per capita** is gross domestic product divided by midyear population. • **Average annual growth of total population and labor force** is calculated using the exponential endpoint method. Labor force comprises all people who meet the International Labour Organization's definition of the economically active population. • **Value added** is the net output of an industry after adding up all outputs and subtracting intermediate inputs. It is calculated without making deductions for depreciation of fabricated assets or depletion and degradation of natural resources. The industrial origin of value added is determined by the International Standard Industrial Classification (ISIC) revision 3. • **Agriculture** corresponds to ISIC divisions 1–5. • **Industry** comprises ISIC divisions 10–45. • **Services** correspond to ISIC divisions 50–99. • **Household final consumption expenditure** is the market value of all goods and services, including durable products, purchased by households. It excludes purchases of dwellings but includes imputed rent for owner-occupied dwellings. The *World Development Indicators* includes in household consumption expenditure the expenditures of nonprofit institutions serving households. • **Gross fixed capital formation** consists of outlays on additions to the fixed assets of the economy. • **Exports of goods and services** are the value of all goods and market services provided to the rest of the world.

Data sources

The indicators here and throughout the rest of the book have been compiled by World Bank staff from primary and secondary sources. More information about the indicators and their sources can be found in the *About the data*, *Definitions*, and *Data sources* entries that accompany each table in subsequent sections.

1.5 Long-term structural change

	Agriculture value added % of GDP		Employment in agriculture % of total labor force		Urban population % of total population		Trade % of GDP		Central government revenue % of GDP		Money and quasi money % of GDP	
	1970	1999	1980	1998	1970	1999	1970	1999	1970	1999	1970	1999
Albania	..	53	57	..	32	39	..	41	..	19	..	53
Algeria	11	11	36	..	40	60	51	51	..	30	51	43
Angola	..	7	76	..	15	34	..	105	..	..	..	14
Argentina	10	5	13	2	78	89	10	21	..	14	21	31
Armenia	..	29	21	..	59	70	..	71	..	..	..	10
Australia	..	3	7	5	85	85	28	40	20	24	42	65
Austria	..	2	11	7	65	65	59	91	28	38	..	..
Azerbaijan	..	23	35	29	50	57	..	84	..	20	..	12
Bangladesh	44	25	73	63	8	24	17	32	..	..	..	29
Belarus	..	13	26	..	44	71	..	127	..	30	..	13
Belgium	5	1	3	..	94	97	105	148	35	44	..	..
Benin	36	38	67	..	17	42	40	45	..	..	10	23
Bolivia	..	18	47	2	41	64	62	44	..	17	16	48
Bosnia and Herzegovina	..	15	30	..	27	43	..	90	..	..	..	..
Botswana	28	4	5	..	8	50	71	61	17	..	..	28
Brazil	12	9	..	24	56	81	14	22	..	24	17	30
Bulgaria	..	15	24	26	52	69	..	96	..	35	..	29
Burkina Faso	35	31	92	..	6	18	23	41	..	..	8	23
Burundi	71	52	93	..	2	9	22	27	..	18	9	20
Cambodia	..	51	76	..	12	16	14	78	..	..	..	11
Cameroon	31	44	73	..	20	48	51	49	..	16	14	15
Canada	4	..	5	4	76	77	43	84	19	22	36	61
Central African Republic	35	55	85	..	30	41	74	41	..	..	15	16
Chad	40	36	88	..	12	23	38	47	8	..	7	11
Chile	7	8	16	14	75	84	29	56	29	23	12	48
China	35	18	69	47	17	32	4	41	..	6	..	138
Hong Kong, China	..	0	1	0	88	100	181	261	..	..	..	214
Colombia	26	13	1	1	57	74	30	37	11	12	18	24
Congo, Dem. Rep.	15	58	72	..	30	30	34	46	11	5	8	..
Congo, Rep.	18	10	58	..	33	62	92	148	22	28	17	14
Costa Rica	25	11	27	20	40	51	63	101	12	21	15	30
Côte d'Ivoire	32	26	65	..	27	46	65	82	..	20	25	24
Croatia	..	9	..	17	40	57	..	89	..	43	..	40
Cuba	..	..	24	..	60	75	..	..	..	..	..	..
Czech Republic	..	4	13	6	52	75	..	129	..	34	..	67
Denmark	6	2	..	4	80	85	57	70	34	38	43	56
Dominican Republic	23	11	33	20	40	64	42	69	18	17	18	29
Ecuador	24	12	40	7	40	62	33	63	..	..	20	33
Egypt, Arab Rep.	29	17	42	..	42	45	33	40	..	26	34	75
El Salvador	40	10	38	26	39	46	49	62	11	14	20	46
Eritrea	..	17	83	..	11	18	..	89	..	..	..	..
Estonia	..	6	15	10	65	69	..	159	..	31	..	31
Ethiopia	..	52	89	..	9	17	..	43	..	..	..	38
Finland	11	3	13	6	50	67	51	67	25	32	..	..
France	..	3	8	..	71	75	30	50	33	41	..	..
Gabon	19	8	66	..	31	80	88	83	..	..	15	17
Gambia, The	34	31	84	..	15	32	79	117	16	..	15	31
Georgia	..	36	32	..	48	60	..	73	..	12	..	7
Germany	..	1	..	3	80	87	..	57	..	31	..	..
Ghana	47	36	62	..	29	38	44	84	15	..	18	17
Greece	12	7	30	20	53	60	24	44	21	23	33	47
Guatemala	27	23	54	..	36	40	36	46	9	..	17	22
Guinea	..	24	91	..	14	32	..	45	..	12	..	6
Guinea-Bissau	50	62	87	..	15	23	34	70	..	..	..	28
Haiti	..	29	..	..	20	35	31	40	..	..	12	30
Honduras	32	16	57	35	29	46	62	100	12	..	19	41

Long-term structural change | 1.5

	Agriculture value added		Employment in agriculture		Urban population		Trade		Central government revenue		Money and quasi money	
	% of GDP		% of total labor force		% of total population		% of GDP		% of GDP		% of GDP	
	1970	1999	1980	1998	1970	1999	1970	1999	1970	1999	1970	1999
Hungary	..	6	22	8	49	64	63	108	..	38	..	43
India	46	28	70	..	20	28	8	27	..	12	18	49
Indonesia	45	19	56	45	17	40	28	62	13	18	8	54
Iran, Islamic Rep.	..	21	39	23	42	61	..	38	..	24	..	38
Iraq	..	..	29	..	56	76	..	..	..	..	22	..
Ireland	..	5	18	9	52	59	77	161	29	32	..	..
Israel	..	..	6	2	84	91	79	81	33	41	47	91
Italy	8	3	14	7	64	67	32	49	..	42	..	..
Jamaica	7	7	37	21	42	56	71	107	..	..	30	45
Japan	6	2	10	5	71	79	20	19	11	..	69	124
Jordan	12	2	..	..	51	74	..	105	..	27	54	102
Kazakhstan	..	11	24	..	50	56	..	85	..	9	..	11
Kenya	33	23	23	..	10	32	60	56	17	..	27	43
Korea, Dem. Rep.	..	..	45	..	54	60	..	..	..	..	..	..
Korea, Rep.	26	5	34	12	41	81	37	77	15	20	29	61
Kuwait	0	..	2	..	78	97	84	84	42	34	36	85
Kyrgyz Republic	..	38	34	49	37	34	..	99	..	13	..	12
Lao PDR	..	53	80	..	10	23	..	86	..	..	..	12
Latvia	..	4	16	19	62	69	..	104	..	33	..	27
Lebanon	..	12	14	..	59	89	..	62	..	17	..	143
Lesotho	36	18	40	..	9	27	60	136	17	45	..	32
Libya	2	..	25	..	45	87	89	..	..	..	19	..
Lithuania	..	9	28	19	50	68	..	90	..	26	..	20
Macedonia, FYR	..	12	36	..	47	62	..	97	..	..	..	17
Madagascar	24	30	82	..	14	29	41	57	14	..	17	18
Malawi	44	38	87	..	6	15	63	70	16	..	18	13
Malaysia	29	11	37	19	34	57	79	218	19	23	30	98
Mali	66	47	89	..	14	29	31	61	..	..	13	23
Mauritania	29	25	71	..	14	56	66	88	..	..	9	14
Mauritius	16	6	29	..	42	41	85	133	..	21	35	76
Mexico	13	5	26	20	59	74	17	63	10	13	15	25
Moldova	..	25	43	46	32	46	..	115	..	24	..	18
Mongolia	..	32	40	45	45	58	..	105	..	21	..	21
Morocco	20	15	56	..	35	55	39	64	19	..	28	75
Mozambique	..	33	84	..	6	39	..	49	..	..	..	22
Myanmar	38	60	67	63	23	27	14	1	..	7	24	23
Namibia	..	13	47	..	19	30	..	116	..	..	..	42
Nepal	67	42	94	..	4	12	13	53	5	10	11	44
Netherlands	..	3	5	3	86	89	97	116	..	44	..	..
New Zealand	..	..	11	9	81	87	..	61	28	32	20	90
Nicaragua	25	32	3	..	47	64	56	122	12	..	14	61
Niger	65	41	6	..	9	20	29	38	..	..	5	7
Nigeria	41	39	54	..	20	43	20	79	10	..	9	19
Norway	..	2	8	5	65	75	74	72	32	42	49	55
Oman	16	..	50	..	11	82	93	..	38	25	..	36
Pakistan	37	27	53	44	25	36	22	35	..	17	41	44
Panama	..	7	..	19	48	57	..	74	..	25	22	79
Papua New Guinea	37	30	82	..	10	17	72	87	..	19	..	31
Paraguay	32	29	45	5	37	55	31	60	11	..	17	31
Peru	19	7	40	5	57	72	34	32	14	16	18	32
Philippines	30	18	52	40	33	58	43	101	13	16	23	59
Poland	..	3	30	19	52	65	..	59	..	32	..	39
Portugal	..	4	27	14	26	63	49	71	..	35	..	..
Puerto Rico	3	..	5	3	58	75	107	..	..	..	..	..
Romania	..	16	30	40	42	56	..	64	..	26	..	22
Russian Federation	..	7	16	..	63	..	..	74	..	22	..	18

	Agriculture value added % of GDP		Employment in agriculture % of total labor force		Urban population % of total population		Trade % of GDP		Central government revenue % of GDP		Money and quasi money % of GDP	
	1970	1999	1980	1998	1970	1999	1970	1999	1970	1999	1970	1999
Rwanda	66	46	93	..	3	6	27	27	..	..	11	15
Saudi Arabia	4	7	44	..	49	85	89	68	..	..	13	56
Senegal	24	18	81	..	33	47	56	72	16	..	14	23
Sierra Leone	30	43	70	..	18	36	65	34	..	10	13	14
Singapore	2	0	1	0	100	100	232	..	21	25	62	116
Slovak Republic	..	4	14	8	41	57	..	128	..	37	..	61
Slovenia	..	4	16	12	37	50	..	109	..	40	..	44
South Africa	7	4	17	..	48	55	46	48	21	28	58	55
Spain	..	4	19	8	66	77	26	56	17	29	..	..
Sri Lanka	28	21	..	42	22	23	54	78	20	18	22	30
Sudan	44	40	72	..	16	35	33	..	173	..	17	9
Sweden	..	..	6	3	81	83	48	82	28	40	..	..
Switzerland	..	..	7	5	55	68	64	77	14	25	100	155
Syrian Arab Republic	20	..	..	..	43	54	39	69	25	22	34	48
Tajikistan	..	19	45	46	37	28	..	132	..	9	..	..
Tanzania	..	45[a]	86	..	7	27	..	41[a]	..	..	..	17
Thailand	26	10	71	51	13	21	34	102	12	16	27	104
Togo	34	41	69	..	13	33	88	70	..	..	17	24
Trinidad and Tobago	5	2	10	8	63	74	84	93	..	..	27	48
Tunisia	17	13	39	..	45	65	47	86	23	29	32	48
Turkey	40	16	60	43	38	74	10	50	14	26	20	39
Turkmenistan	..	27	39	..	48	45	..	104	..	..	..	13
Uganda	54	44	87	..	8	14	43	34	14	..	17	14
Ukraine	..	13	25	26	55	68	..	104	..	..	..	15
United Arab Emirates	..	..	5	..	57	85	..	..	..	3	..	56
United Kingdom	..	1	3	2	89	89	44	53	37	36	..	..
United States	..	..	4	3	74	77	11	24	17	21	62	60
Uruguay	18	6	..	4	82	91	29	38	24	27	20	46
Uzbekistan	..	33	40	..	37	37	..	38	..	..	..	..
Venezuela, RB	6	5	15	11	72	87	38	37	17	17	19	18
Vietnam	..	25	73	71	18	..	..	95	..	17	..	27
West Bank and Gaza	..	9	..	..	..	..	..	92	..	..	..	..
Yemen, Rep.	..	17	73	..	13	24	..	84	..	26	..	33
Yugoslavia, FR (Serb./Mont.)	..	..	..	..	39	52	..	..	..	..	..	..
Zambia	12	25	76	..	30	44	90	63	22	..	25	17
Zimbabwe	17	20	32	..	17	35	..	91	..	29	..	19
World	..w	5w	51w	..w	37w	44w	28w	52w	18w	26w		
Low income	43	26	66	..	20	31	20	50	..	..	..	15
Middle income	20	10	54	40	34	46	26	55	..	..	..	19
Lower middle income	31	14	58	45	28	37	19	59	..	..	..	14
Upper middle income	14	6	24	20	55	76	30	52	7	..	..	22
Low & middle income	24	12	59	..	28	39	25	54	..	..	..	19
East Asia & Pacific	33	14	66	46	19	34	24	70	..	..	..	10
Europe & Central Asia	..	10	26	..	52	43	..	77	..	..	..	26
Latin America & Carib.	13	8	..	19	57	75	20	34	1	..	..	20
Middle East & N. Africa	13	14	42	..	41	58	..	57	..	..	..	..
South Asia	44	27	69	..	19	28	12	30	..	..	..	13
Sub-Saharan Africa	21	15	67	..	19	34	47	60	24	..	..	..
High income	..	..	8	4	73	77	29	43	19	29	..	..
Europe EMU	..	2	13	6	71	78	..	64	..	37		

a. The data for GDP and its components refer to mainland Tanzania only.

Long-term structural change | 1.5

About the data

Over a period of 30 years cumulative processes of change reshape an economy and the social order built on that economy. This table highlights some of the notable trends at work for much of the past century: the shift of production from agriculture to manufacturing and services; the reduction of the agricultural labor force and the growth of urban centers; the expansion of trade; the increasing size of the central government in most countries—and the reversal of this trend in some; and the monetization of economies that have achieved stable macroeconomic management. All the indicators shown here appear elsewhere in the *World Development Indicators*. For more information about them see tables 4.2 (agriculture value added), 2.3 (labor force employed in agriculture), 3.10 (urban population), 4.9 and 6.1 (trade), 4.11 (central government revenue), and 4.14 (money and quasi money).

Definitions

• **Agriculture value added** is the sum of outputs of the agricultural sector (International Standard Industrial Classification [ISIC] major divisions 1–5) less the cost of intermediate inputs, measured as a share of gross domestic product (GDP). • **Employment in agriculture** is the percentage of the total labor force in agriculture, hunting, forestry, and fishing, corresponding to major division 1 (ISIC revision 2) or tabulation categories A and B (ISIC revision 3). • **Urban population** is the share of the total population living in areas defined as urban in each country. • **Trade** is the sum of exports and imports of goods and services, measured as a share of GDP. • **Central government revenue** includes all revenue to the central government from taxes and nonrepayable receipts (other than grants), measured as a share of GDP. • **Money and quasi money** comprise the sum of currency outside banks, demand deposits other than those of the central government, and the time, savings, and foreign currency deposits of resident sectors other than the central government. This measure of the money supply is commonly called M2.

Data sources

The indicators here and throughout the rest of the book have been compiled by World Bank staff from primary and secondary sources. More information about the indicators and their sources can be found in the *About the data, Definitions,* and *Data sources* entries that accompany each table in subsequent sections.

	Population	Surface area	Population density	Gross national income				Gross domestic product		Life expectancy at birth	Adult illiteracy rate	Carbon dioxide emissions
						PPPª						
					Per capita		Per capita		Per capita		% of people 15	thousand
	thousands	thousand sq. km	people per sq. km	$ millions	$	$ millions	$	% growth	% growth	years	and above	metric tons
	1999	**1999**	**1999**	**1999ᵇ**	**1999ᵇ**	**1999**	**1999**	**1998–99**	**1998–99**	**1999**	**1999**	**1997**
Afghanistan	25,869	652.1	40	..	.. ᶜ	..	..	..	..	46	64	1,153
American Samoa	64	0.2	320	..	.. ᵈ	..	..	..	..	..	..	282
Andorra	66	0.5	147	..	.. ᵉ	..	..	..	..	..	..	..
Antigua and Barbuda	67	0.4	153	606	8,990	665	9,870	4.6	3.7	75	..	337
Aruba	98	0.2	516	..	.. ᵉ	..	..	..	..	..	..	1,872
Bahamas, The	298	13.9	30	..	.. ᵉ	4,620	15,500	5.5	4.1	73	4	1,740
Bahrain	666	0.7	966	..	.. ᵈ	..	..	..	..	73	13	14,932
Barbados	267	0.4	620	2,294	8,600	3,737	14,010	1.3	0.9	76	..	984
Belize	247	23.0	11	673	2,730	1,173	4,750	4.5	1.0	72	7	388
Bermuda	64	0.1	1,280	..	.. ᵉ	..	..	..	..	..	..	462
Bhutan	782	47.0	17	399	510	985	1,260	7.0	3.9	61	..	472
Brunei	322	5.8	61	..	.. ᵉ	..	..	..	..	76	9	5,454
Cape Verde	428	4.0	106	569	1,330	1,903 ᶠ	4,450 ᶠ	8.0	4.8	69	26	121
Cayman Islands	39	0.3	150	..	.. ᵉ	..	..	..	..	..	..	282
Channel Islands	149	0.3	479	..	.. ᵉ	..	..	..	..	79	..	..
Comoros	544	2.2	244	189	350	778 ᶠ	1,430 ᶠ	–1.4	–3.9	61	41	66
Cyprus	760	9.3	82	9,086	11,950	14,511 ᶠ	19,080 ᶠ	4.5	3.6	78	3	5,954
Djibouti	648	23.2	28	511	790	..	..	..	..	47	37	366
Dominica	73	0.8	97	238	3,260	368	5,040	0.1	0.1	76	..	81
Equatorial Guinea	443	28.1	16	516	1,170	1,729	3,910	15.1	12.2	51	18	612
Faeroe Islands	44	1.4	31	..	.. ᵉ	..	..	..	..	..	..	634
Fiji	801	18.3	44	1,848	2,310	3,828	4,780	7.8	6.4	73	7	797
French Polynesia	231	4.0	63	3,908	16,930	5,124	22,200	4.0	2.5	73	..	561
Greenland	56	341.7	0	..	.. ᵉ	..	..	..	..	..	..	520
Grenada	97	0.3	285	334	3,440	614	6,330	8.2	7.3	72	..	183
Guam	152	0.6	276	..	.. ᵉ	..	..	..	..	78	..	4,078
Guyana	856	215.0	4	651	760	2,850	3,330	3.0	2.2	64	2	1,022
Iceland	278	103.0	3	8,197	29,540	7,552	27,210	4.3	3.0	79	..	2,140
Isle of Man	76	0.6	129	..	.. ᵈ	..	..	..	..	..	..	..

This table shows data for 59 economies—small economies with populations between 30,000 and 1 million, smaller economies if they are members of the World Bank, and larger economies for which data are not regularly reported. Where data on gross national income (GNI) per capita are not available, the estimated range is given. For more information on the calculation of GNI (gross national product, or GNP, in previous editions), see *About the data* for table 1.1. As in last year's edition, this table excludes France's overseas departments—French Guiana, Guadeloupe, Martinique, and Réunion—for which GNI and other economic measures are now included in the French national accounts.

• **Population** is based on the de facto definition of population, which counts all residents regardless of legal status or citizenship—except for refugees not permanently settled in the country of asylum, who are generally considered part of the population of their country of origin. The values shown are midyear estimates for 1999. See also table 2.1. • **Surface area** is a country's total area, including areas under inland bodies of water and some coastal waterways. • **Population density** is midyear population divided by land area in square kilometers. • **Gross national income** (GNI) is the sum of value added by all resident producers plus any product taxes (less subsidies) not included in the valuation of output plus net receipts of primary income (compensation of employees and

Key indicators for other economies | 1.6

	Population	Surface area	Population density	Gross national income				Gross domestic product		Life expectancy at birth	Adult illiteracy rate	Carbon dioxide emissions
						PPP[a]						
					Per capita		Per capita		Per capita		% of people 15	
	thousands	thousand sq. km	people per sq. km	$ millions	$	$ millions	$	% growth	% growth	years	and above	thousand metric tons
	1999	1999	1999	1999[b]	1999[b]	1999	1999	1998–99	1998–99	1999	1999	1997
Kiribati	88	0.7	121	81	910	..	..	2.5	–0.3	61	..	22
Liberia	3,044	111.4	32	..	.. [c]	..	..	..	..	47	47	339
Liechtenstein	32	0.2	200	..	.. [e]	..	..	..	..	..	..	..
Luxembourg	432	2.6	166	18,545	42,930	17,810	41,230	7.5	6.2	77	..	8,241
Macao, China	434	..	..	6,161	14,200	7,350	16,940	–2.9	–4.7	78	..	1,473
Maldives	269	0.3	898	322	1,200	..	..	..	..	68	4	304
Malta	379	0.3	1,184	3,492	9,210	..	..	..	..	77	8	1,759
Marshall Islands	51	0.2	255	99	1,950	..	..	0.5	..	..	..	..
Mayotte	140	0.4	350	..	.. [d]	..	..	..	..	..	..	..
Micronesia, Fed. Sts.	116	0.7	166	212	1,830	..	..	0.2	–1.9	68	..	..
Monaco	32	0.0	16,410	..	.. [e]	..	..	..	..	..	..	..
Netherlands Antilles	215	0.8	268	..	.. [e]	..	..	..	..	76	4	6,760
New Caledonia	209	18.6	11	3,169	15,160	4,415	21,130	0.9	–1.0	73	..	1,801
Northern Mariana Islands	69	0.5	143	..	.. [e]	..	..	..	..	..	..	..
Palau	19	0.5	41	..	.. [d]	..	..	..	..	..	..	238
Qatar	565	11.0	51	..	.. [e]	..	..	..	..	75	19	38,264
Samoa	169	2.8	60	181	1,070	686	4,070	1.0	0.5	69	20	132
São Tomé and Principe	145	1.0	151	40	270	..	..	2.5	0.2	65	..	77
Seychelles	80	0.5	178	520	6,500	..	..	1.5	0.0	72	..	198
Solomon Islands	429	28.9	15	320	750	880 [f]	2,050 [f]	–0.5	–3.5	71	..	161
Somalia	9,388	637.7	15	..	.. [c]	..	..	..	..	48	..	30
San Marino	26	0.1	433	..	.. [e]	..	..	..	..	..	..	..
St. Kitts and Nevis	41	0.4	114	259	6,330	425	10,400	2.8	2.7	71	..	103
St. Lucia	154	0.6	253	590	3,820	801	5,200	3.1	1.6	72	..	198
St. Vincent and the Grenadines	114	0.4	293	301	2,640	569	4,990	4.0	3.2	73	..	132
Suriname	413	163.3	3	..	.. [g]	1,564	3,780	–1.0	–1.3	70	..	2,135
Swaziland	1,019	17.4	59	1,379	1,350	4,468	4,380	2.0	–0.9	46	21	399
Tonga	100	0.8	138	172	1,730	..	..	3.5	2.6	71	..	121
Vanuatu	193	12.2	16	227	1,180	556 [f]	2,880 [f]	–2.5	–6.0	65	..	62
Virgin Islands (U.S.)	120	0.3	352	..	.. [e]	..	..	..	..	77	..	11,553

a. PPP is purchasing power parity; see *Definitions*. b. Calculated using the World Bank Atlas method. c. Estimated to be low income ($755 or less). d. Estimated to be upper middle income ($2,996–9,265). e. Estimated to be high income ($9,266 or more). f. The estimate is based on regression; others are extrapolated from the latest International Comparison Programme benchmark estimates. g. Estimated to be lower middle income ($756–2,995).

property income) from abroad. Data are in current U.S. dollars converted using the World Bank Atlas method (see *Statistical methods*). • **GNI per capita** is gross national income divided by midyear population. GNI per capita in U.S. dollars is converted using the World Bank Atlas method. • **PPP GNI** is gross national income converted to international dollars using purchasing power parity rates. An international dollar has the same purchasing power over GNI as a U.S. dollar has in the United States. • **Gross domestic product** (GDP) is the sum of value added by all resident producers plus any product taxes (less subsidies) not included in the valuation of output. Growth is calculated from constant price GDP data in local currency. • **Life expectancy at birth** is the number of years a

newborn infant would live if prevailing patterns of mortality at the time of its birth were to stay the same throughout its life. • **Adult illiteracy rate** is the percentage of adults ages 15 and above who cannot, with understanding, read and write a short, simple statement about their everyday life. • **Carbon dioxide emissions** are those stemming from the burning of fossil fuels and the manufacture of cement. They include carbon dioxide produced during consumption of solid, liquid, and gas fuels and gas flaring.

Data sources

The indicators here and throughout the rest of the book have been compiled by World Bank staff from primary and secondary sources. More information about the indicators and their sources can be found in the *About the data*, *Definitions*, and *Data sources* entries that accompany each table in subsequent sections.

PEOPLE

The three key elements of the World Bank's long-term strategy to promote gender equality and strengthen countries' abilities to attack poverty:

Establish a policy and institutional environment that provides for equal rights and opportunities for women and men.

Foster economic development that strengthens incentives for more equal allocation of resources.

Take active measures to redress persistent inequalities in society. Source: World Bank 2001.

The status of women has improved considerably in most developing countries in the past quarter century. Yet in no region do women enjoy equal legal, social, and economic rights. Women have fewer resources than men, and more limited economic opportunities and political participation. Women and girls bear the most direct cost of these inequalities—but the harm ultimately extends to everyone. Because gender gaps are often largest and most costly among the poor, gender equality is a core development issue.

Gender inequalities persist because they are supported by social norms and legal institutions, by the choices and behaviors of households, and by regulations and incentives that affect the way economies function. A strategy to reduce gender inequalities must address these factors.

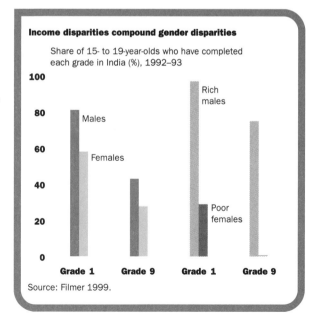

Income disparities compound gender disparities

Share of 15- to 19-year-olds who have completed each grade in India (%), 1992–93

Source: Filmer 1999.

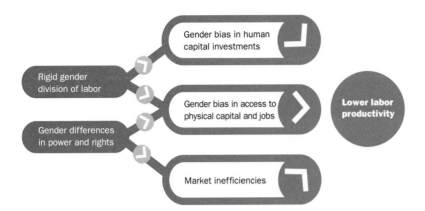

Rigid gender division of labor

Gender differences in power and rights

Gender bias in human capital investments

Gender bias in access to physical capital and jobs

Market inefficiencies

Lower labor productivity

Gender inequalities hinder productivity, efficiency, and economic progress. By hampering the accumulation of human capital in the home and the labor market and systematically excluding women from access to resources, public services, and certain productive activities, gender discrimination diminishes an economy's capacity to prosper and to provide for its people.

The great costs of gender inequality

Foremost among the costs of gender inequality is its toll on the quality of human lives. Evidence suggests that societies with large and persistent gender inequalities pay the price of more poverty, illness, malnutrition, and other deprivations, even death. This makes a compelling case for public and private action to eliminate inequality. Public action is particularly important, since many social, legal, and economic institutions that perpetuate gender inequalities are extremely difficult for individuals to change.

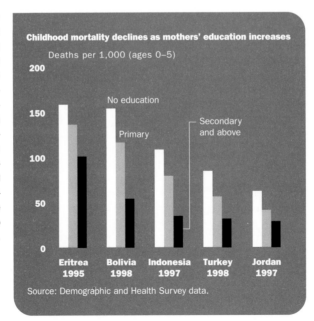

Childhood mortality declines as mothers' education increases

Deaths per 1,000 (ages 0–5)

No education

Primary

Secondary and above

Eritrea 1995
Bolivia 1998
Indonesia 1997
Turkey 1998
Jordan 1997

Source: Demographic and Health Survey data.

Public policy can:
• Establish an enabling policy and institutional environment that provides for equal rights and opportunities for women and men.
• Foster economic development that strengthens incentives for more equal allocation of resources.
• Take active measures to redress persistent inequalities in society.

These are the three key elements of the World Bank's long-term strategy to promote gender equality and strengthen countries' abilities to eliminate gender disparities and attack poverty.

Q: **Why is there a paucity of data disaggregated by gender?**

A: **Resource constraints**

Constraints of time, money, and technical expertise limit data collection to aggregate statistics.

A: **Assumptions in the design and implementation of policy**

Policymakers often believe that the benefits of economic growth will accrue equally to all people, although empirical evidence does not always bear out this altruistic model. If it is recognized, as it increasingly now is, that men and women have differing needs and benefit differently from programs, policies, and projects, the case for collecting data for both men and women separately becomes stronger.

A: **Agencies' priorities and conventional wisdom**

Data collection and indicators often reflect the policy thrust of particular agencies and the conventional wisdom that gender matters only in certain sectors. Thus indicators disaggregated by gender exist for primary education and basic health, but not for rural development and infrastructure.

Monitoring gender equality—and measuring it

How do countries know how they are doing with respect to gender equality?

The first step toward incorporating gender issues in policymaking, and determining appropriate measures for strengthening women's participation, is to obtain good information—on gender roles, existing institutions, and the constraints operating against women. This requires a combination of administrative data, qualitative social assessments, and quantitative and time use surveys. But there is no single accepted set of indicators for monitoring progress toward gender equality, and the methodological quest for specific sets of relevant indicators is often complicated by the lack of data disaggregated by gender and the lack of conceptual clarity.

Nevertheless, progress can be measured by monitoring trends in the public and private dimensions of women's lives. These dimensions are identified in the Bank's strategy to promote gender equality. But the relevance of many measures of gender inequality varies by cultural context.

REFORM INSTITUTIONS   p. 38

> Legal frameworks
> Markets and institutions

FOSTER ECONOMIC DEVELOPMENT p. 40 >

> Education
> Health and reproductive health
> Infrastructure

REDRESS INEQUALITIES   p. 42

> Employment and earnings
> Political participation
> Gender and violence

> Reform institutions

Gender equality in legal, political, and economic rights provides a supportive institutional environment in which women can be as productive in society as men. It also enhances their ability to participate in and benefit from the development process.

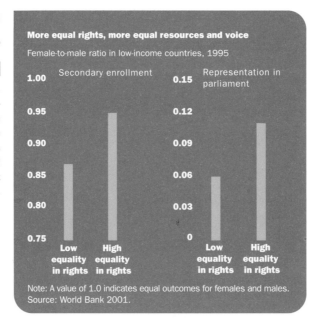

More equal rights, more equal resources and voice

Female-to-male ratio in low-income countries, 1995

Secondary enrollment

1.00	
0.95	
0.90	
0.85	
0.80	
0.75	

Low equality in rights / High equality in rights

Representation in parliament

0.15	
0.12	
0.09	
0.06	
0.03	
0	

Low equality in rights / High equality in rights

Note: A value of 1.0 indicates equal outcomes for females and males.
Source: World Bank 2001.

But discrimination against women is still widely embodied in both laws and customs. Even where supportive legislation exists, legal rights may be weakly enforced or overridden by customary law. In South Africa, for example, reproductive rights are guaranteed in the constitution, but their exercise has been restricted by appeals to customary law (UNFPA 2000). In many regions girls do not attend school because traveling without chaperones violates social norms (Narayan, Patel, Schafft, Rademacher, and Koch-Schulte 2000).

A foundation of equal rights . . .

Legal frameworks

A fundamental step is to establish equal rights for women, especially in family law, property rights, political rights, and protection against gender-related violence. Where these rights exist, judicial and administrative enforcement should be strengthened. Equally important is extending legal aid to women. Sustained efforts to provide legal assistance, legal literacy training, and improved access to justice at the local level are necessary for women to have the confidence to claim their rights or seek redress. An independent, well-trained, well-equipped, and gender-sensitive judiciary, with female judges especially at local levels, can improve the enforcement of laws.

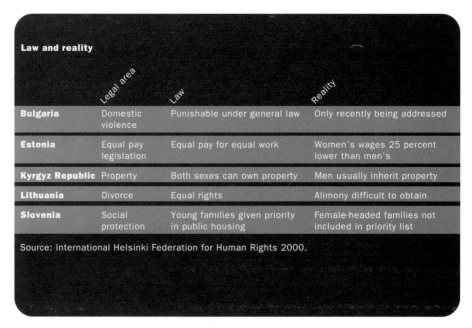

Law and reality

	Legal area	Law	Reality
Bulgaria	Domestic violence	Punishable under general law	Only recently being addressed
Estonia	Equal pay legislation	Equal pay for equal work	Women's wages 25 percent lower than men's
Kyrgyz Republic	Property	Both sexes can own property	Men usually inherit property
Lithuania	Divorce	Equal rights	Alimony difficult to obtain
Slovenia	Social protection	Young families given priority in public housing	Female-headed families not included in priority list

Source: International Helsinki Federation for Human Rights 2000.

Markets and institutions

The structure of economic institutions can promote or impede gender equality, thereby affecting a country's long-run prospects for economic growth. Markets embody incentives that influence decisions to work, save, and consume. Thus the structure of markets largely determines women's lower wages and their segregation in a narrow range of activities (Anker 1998).

Women are clustered in lower-paying, lower-status occupations, primarily in the service sectors, and in such professions as teaching and registered nursing.

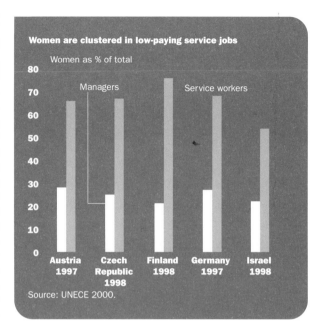

Women are clustered in low-paying service jobs

Women as % of total

Managers   Service workers

Austria 1997   Czech Republic 1998   Finland 1998   Germany 1997   Israel 1998

Source: UNECE 2000.

This clustering implies that factors other than efficiency determine labor supply and demand in an economy. Overcoming occupational segregation could be an important step in ensuring an efficient allocation of resources. And by ensuring that competent women are not overlooked, it will create a more flexible and responsive labor market.

. . . and labor market opportunities empowers women

Developing countries tend to have limited job creation in the formal sector and uneven development of local industries and of services for reducing unemployment. Thus women, especially those in rural areas, see self-employment or entrepreneurship as a way out of poverty. Access to microcredit programs that finance income generating activities can help them become self-employed. More important, an increase in assets and cash income may empower women within the household, helping to increase their consumption and that of their children and contributing to other measures of welfare.

Grameen Bank loans to women in Bangladesh tend to boost welfare more

%

Welfare change	Effect of	
	Male borrowing	Female borrowing
Increase in boys' schooling	7.2	6.1
Increase in girls' schooling	3.0	4.7
Increase in per capita spending	1.8	4.3
Reduction in recent fertility	7.4	3.5
Increase in women's labor supply in cash income earning activities	0.0	10.4
Increase in women's nonland assets	0.0	19.9

Source: World Bank 1995b.

❯ Foster economic development

In most settings economic development is associated with improved circumstances for women and girls and with greater gender equality, primarily through the investments in basic services and infrastructure that accompany development. Cross-country analysis supports the conclusion that economic development provides an enabling environment for gender equality—though its effects are not immediate or without short-term costs (World Bank 2001). For example, economic development may not benefit all women and men, and it may

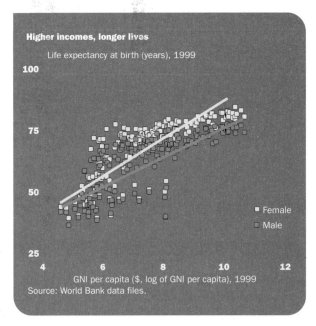

Higher incomes, longer lives

Life expectancy at birth (years), 1999

GNI per capita ($, log of GNI per capita), 1999
Source: World Bank data files.

■ Female
□ Male

actually harm some. But countries with higher per capita incomes have better health, education, and related outcomes than countries with lower per capita incomes.

Data from 127 countries over four periods show that income growth leads to greater gender equality in life expectancy, political representation, and secondary school attainment. But the relationship is nonlinear: improvement is slow at low per capita incomes, but increases rapidly once countries reach middle-income status. For many poor countries the salutary effects of income growth on gender equality may take a long time to realize (World Bank 2001).

A more vibrant economy can increase gender equality. . .

Education

Enrollment has improved more for girls than for boys, narrowing the gender gap in primary and secondary school enrollment (see table 1.2). These improvements in enrollment over the years have raised literacy rates among younger women (see table 2.14). But a gender gap in favor of boys continues to disadvantage girls in South Asia and the Middle East and North Africa (Filmer 1999). In these regions girls' access to primary and secondary education is still limited in rural areas, and girls are more likely than boys to drop out of school. Beyond improving supply, special measures may be needed to encourage girls' enrollment, including

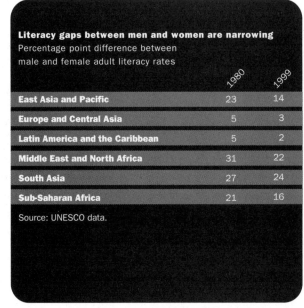

Literacy gaps between men and women are narrowing
Percentage point difference between male and female adult literacy rates

	1980	1999
East Asia and Pacific	23	14
Europe and Central Asia	5	3
Latin America and the Caribbean	5	2
Middle East and North Africa	31	22
South Asia	27	24
Sub-Saharan Africa	21	16

Source: UNESCO data.

providing subsidies and other financial inducements, improving the relevance of education (and thereby increasing returns to families), and enacting supportive national policies.

Health and reproductive health

Globally, girls have a greater chance of surviving childhood than boys—except where sex discrimination is greatest. Among young children in developing countries, female disadvantage in access to health services is small and diminishing, but the female disadvantage in morbidity and mortality is carried into adolescent and reproductive years. Serious health risks arise for adolescents when they become sexually active. One of the most important concerns is the prevalence of sexually transmitted diseases, including HIV. It is

More boys than girls get treatment for respiratory infection or fever

Share receiving treatment (%)

	Year	Boys	Girls
Cameroon	1991	53	42
	1998	31	36
Egypt, Arab Rep.	1992	66	56
	1995	68	61

Source: Demographic and Health Survey data.

Virtually all maternal deaths occur in developing countries
Maternal deaths, 1995

East Asia and Pacific	48,000
Europe and Central Asia	3,000
Latin America and the Caribbean	22,000
Middle East and North Africa	20,000
South Asia	155,000
Sub-Saharan Africa	265,000
High-income countries	1,000

Source: WHO and UNICEF, preliminary estimates.

estimated that half of all new HIV infections in 1996 occurred in the age group 15–24, and in some countries the rate of infection in this age group is higher among women than men (see table 1.3). In high-income countries women's access to reproductive health care is universal, especially during pregnancy and childbirth, but in developing countries many women receive little or no skilled prenatal or delivery care (see tables 1.3 and 2.17). This results in preventable deaths and injuries during pregnancy and childbirth: 99 percent of all maternal deaths occur in developing countries.

... through many pathways

Infrastructure

Investments in certain types of infrastructure can be important in facilitating greater gender equality in access to resources and in economic participation. While infrastructure investments generally benefit both females and males, they often benefit them differently. In poor rural areas lack of water and energy infrastructure can mean long hours for women and girls collecting water and fuel. A study of Indian villages showed that in resource-depleted areas women spend an average of four to five hours a day collecting household fuel (World Bank 1991b). And girls in several countries identified the

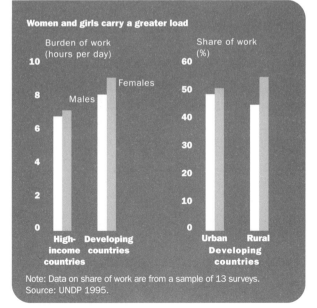

Women and girls carry a greater load

Note: Data on share of work are from a sample of 13 surveys.
Source: UNDP 1995.

need to help out at home as one of the main reasons for dropping out of school (Gardner 1998).

Women in all countries work more unvalued hours—in the home and community—than men, though this gap is smaller in high-income countries. A 1990 survey in the Republic of Korea showed that married women spent an average of more than five hours a day on household chores and childcare, compared with an average of 37 minutes for men (World Bank 2001). Investments in water, energy, and transport infrastructure can substantially reduce the time women and girls devote to household maintenance, freeing them to participate in other activities.

Redress inequalities

The combined effects of institutional reform and economic development take time to become apparent, and active measures are needed to redress disparities in command of resources and political voice. Because active measures have real resource costs, policymakers will need to be selective about which measures to undertake, focusing strategically on areas where government intervention has the largest benefits.

In almost all societies women and girls have primary responsibility for housework and childcare activities. Several types of interventions can reduce the personal costs of household roles to women and girls:
• Interventions that increase education, wages, and labor force participation, coupled with access to reproductive health services, can strengthen women's role in reproductive decisions.
• Providing support for childcare services can allow greater economic participation for women.
• Selected investments in water, fuel, and transport can reduce women's and girls' domestic workloads.

• Protective labor legislation that spreads the cost of maternity and related benefits across employers, workers, and the state can help to avoid bias against hiring women.

Sometimes active measures are needed

Employment and earnings

Women's employment in the formal sector has increased considerably in many countries in recent years (see table 2.2). But women still face formidable barriers to obtaining skilled employment in the formal sector.

A larger share of women than men are classified as contributing family workers in all regions for which data are reported (United Nations 2000b). Where women do participate in the formal sector, they face discriminatory practices relating to pay and benefits,

Female-to-male pay ratios—still too low
Female monthly wages as % of male wages

	Year	Ratio
Czech Republic	1987	66
	1992	73
	1996	81
Hungary	1986	74
	1992	81
	1997	78
Russian Federation	1989	71
	1992	69
	1996	70

Source: UNICEF 1999.

especially maternity benefits (see table 1.3). This discrimination constrains their income earning capacity and imposes a cost on economic growth. Data from Latin America and the Caribbean suggest that if female and male wages were equal, output would be 6 percent higher (World Bank 2001). Women spend considerably more time on unpaid work, but less on paid work, and their unemployment rates are higher than men's (see table 2.4). Legislating against discriminatory labor practices and enforcing these laws is likely to result in productivity gains to firms and to countries as a whole.

Political participation

Women's participation and voice in politics, government, and policymaking remain limited in all regions, making it difficult for women to influence policy. Except in Western Europe, women make up less than 15 percent of national parliaments on average, and they are largely excluded from executive branches of government.

Women are underrepresented in national parliaments
Parliamentary seats held by women (%)

	1987	1995	1999
World average	9	9	11
Northern Africa	3	4	3
Southern Africa	7	9	10
Caribbean	9	11	13
Central America	8	10	13
South America	7	9	13
Eastern Asia	18	12	13
Southern Asia	5	5	5
Central Asia	..	8	8
Western Asia	4	4	4
Eastern Europe	26	9	10
Western Europe	14	18	21

Source: United Nations 2000b.

Proactive measures to increase women's participation in politics and decisionmaking in the short run include "reservations," which guarantee a share of political positions for women. Like any form of affirmative action, political reservations are not without their detractors. Nevertheless, even critics acknowledge that female representation in electoral bodies has expanded in countries that have reservations.

It is essential to design, implement, and monitor with the full participation of women . . .

Beijing Declaration, 1995

Gender and violence

Women are more likely than men to be victims of violence perpetrated by family members, and abuse by a husband or intimate partner is the most common form of violence against women. Many cultures condone or at least tolerate a certain amount of violence against women.

In parts of Asia and Africa men are seen as having a right to discipline their wives as they see fit. In many countries laws that ostensibly protect women from gender-based violence contain biases against victims. The first goal of

Many women have been physically abused by an intimate partner
Women physically assaulted (%), various years, 1991–99

	In past 12 months	Ever (in any relationship)
Australia	3	23
Canada	3[a]	29[a]
Colombia	..	19[b,c]
Egypt, Arab Rep.	..	34
Nicaragua	12	28
Puerto Rico	..	48[c]
South Africa	6	16
Switzerland	6	13
United States	1[a]	22[a]

a. Physical or sexual contact. b. In current relationship only. c. Sample group included women who had never been in a relationship and therefore were not exposed.
Source: United Nations 2000b.

legal reform should be to correct gender biases in existing laws.

The root cause of violence against women is poorly understood, and efforts to address violence are often reactive and fragmented. Insufficient resources have been allocated for anti-violence measures, and conflicting values and beliefs about women and their place in society undermine the development and implementation of such measures. The paucity of data and statistics on the various forms of violence against women is another obstacle to its eradication.

2.1 Population dynamics

	Total population (millions)			Average annual population growth rate					Crude death rate	Crude birth rate	Age dependency ratio	
				Total %	Ages 0–14 %	Ages 15–64 %	Ages 65+ %		per 1,000 people	per 1,000 people	dependents as proportion of working-age population	
	1980	1999	2015	1980–99	1999–2015	1999–2015	1999–2015	1999–2015	1999	1999	1980	1999
Albania	2.7	3.4	3.9	1.2	1.0	−0.7	1.4	2.6	5	16	0.7	0.6
Algeria	18.7	30.0	39.1	2.5	1.7	−0.2	2.5	2.9	6	25	1.0	0.7
Angola	7.0	12.4	19.0	3.0	2.7	2.6	3.1	2.1	19	48	0.9	1.0
Argentina	28.1	36.6	42.8	1.4	1.0	−0.1	1.3	1.6	8	19	0.6	0.6
Armenia	3.1	3.8	4.1	1.1	0.4	−2.0	0.9	1.8	6	11	0.6	0.5
Australia	14.7	19.0	21.5	1.3	0.8	−0.2	0.8	2.4	7	13	0.5	0.5
Austria	7.6	8.1	8.0	0.4	−0.1	−1.7	−0.1	1.5	10	10	0.6	0.5
Azerbaijan	6.2	8.0	9.3	1.4	0.9	−0.4	1.7	1.8	6	15	0.7	0.6
Bangladesh	86.7	127.7	166.1	2.0	1.6	−0.2	2.3	2.8	9	28	1.0	0.7
Belarus	9.6	10.0	9.3	0.2	−0.4	−2.0	−0.1	−0.4	14	9	0.5	0.5
Belgium	9.8	10.2	10.3	0.2	0.0	−1.0	0.0	1.1	10	11	0.5	0.5
Benin	3.5	6.1	9.0	3.0	2.4	1.6	3.2	2.0	13	40	1.0	1.0
Bolivia	5.4	8.1	10.9	2.2	1.8	0.5	2.6	2.8	9	32	0.9	0.8
Bosnia and Herzegovina	4.1	3.9	4.3	−0.3	0.6	1.9	3.1	6.0	7	13	0.5	0.4
Botswana	0.9	1.6	1.7	3.0	0.6	−0.4	1.8	−0.6	18	33	1.0	0.8
Brazil	121.7	168.0	199.8	1.7	1.1	0.0	1.4	2.9	7	20	0.7	0.5
Bulgaria	8.9	8.2	7.3	−0.4	−0.7	−2.2	−0.6	0.5	14	8	0.5	0.5
Burkina Faso	7.0	11.0	15.5	2.4	2.1	1.8	2.8	1.0	19	44	1.0	1.0
Burundi	4.1	6.7	8.8	2.5	1.7	1.1	2.6	0.4	20	41	0.9	0.9
Cambodia	6.8	11.8	14.8	2.9	1.4	0.1	2.8	3.3	12	32	0.7	0.8
Cameroon	8.7	14.7	19.4	2.8	1.8	1.0	2.7	1.9	13	38	0.9	0.9
Canada	24.6	30.5	33.5	1.1	0.6	−0.6	0.5	2.1	7	11	0.5	0.5
Central African Republic	2.3	3.5	4.5	2.2	1.4	1.0	2.1	−0.4	19	36	0.8	0.9
Chad	4.5	7.5	11.7	2.7	2.8	2.0	3.6	0.6	16	45	0.8	1.2
Chile	11.1	15.0	17.7	1.6	1.0	−0.5	1.4	3.1	5	18	0.6	0.6
China	981.2	1,253.6	1,393.7	1.3	0.7	−0.9	1.0	2.3	7	16	0.7	0.5
Hong Kong, China	5.0	6.7	7.5	1.5	0.7	−0.4	1.0	2.4	5	8	0.5	0.4
Colombia	28.4	41.5	51.4	2.0	1.3	−0.3	2.0	2.9	6	23	0.8	0.6
Congo, Dem. Rep.	27.0	49.8	75.0	3.2	2.6	2.4	3.3	2.6	15	45	1.0	1.0
Congo, Rep.	1.7	2.9	4.3	2.8	2.6	2.2	3.0	1.1	16	43	0.9	1.0
Costa Rica	2.3	3.6	4.4	2.4	1.3	−0.8	1.9	3.8	4	21	0.7	0.6
Côte d'Ivoire	8.2	15.5	20.4	3.4	1.7	0.8	2.3	0.8	17	37	1.0	0.9
Croatia	4.6	4.5	4.3	−0.1	−0.2	−1.7	−0.7	0.8	12	10	0.5	0.5
Cuba	9.7	11.2	11.7	0.7	0.3	−1.7	0.3	2.8	7	13	0.7	0.4
Czech Republic	10.2	10.3	9.9	0.0	−0.2	−1.9	−0.3	1.7	11	9	0.6	0.4
Denmark	5.1	5.3	5.4	0.2	0.0	−0.8	−0.2	1.7	11	12	0.5	0.5
Dominican Republic	5.7	8.4	10.4	2.0	1.3	−0.5	2.0	3.2	5	24	0.8	0.6
Ecuador	8.0	12.4	15.8	2.3	1.5	−0.3	2.1	3.0	6	24	0.9	0.6
Egypt, Arab Rep.	40.9	62.7	80.0	2.2	1.5	−0.1	2.1	2.7	7	26	0.8	0.7
El Salvador	4.6	6.2	7.9	1.5	1.6	0.2	2.3	2.1	6	27	0.9	0.7
Eritrea	2.4	4.0	5.6	2.7	2.2	1.6	2.8	2.6	13	39	..	0.9
Estonia	1.5	1.4	1.3	−0.1	−0.5	−2.1	−0.3	0.7	13	9	0.5	0.5
Ethiopia	37.7	62.8	88.1	2.7	2.1	1.6	2.6	0.5	20	44	0.9	1.0
Finland	4.8	5.2	5.3	0.4	0.1	−0.8	−0.2	2.1	10	11	0.5	0.5
France	53.9	58.6	61.1	0.4	0.3	−0.4	0.2	1.2	9	13	0.6	0.5
Gabon	0.7	1.2	1.7	2.9	2.1	2.0	2.3	1.2	16	36	0.7	0.8
Gambia, The	0.6	1.3	1.8	3.5	2.2	1.6	2.6	3.0	13	41	0.8	0.8
Georgia	5.1	5.5	5.3	0.4	−0.1	−2.6	0.4	1.0	8	9	0.5	0.5
Germany	78.3	82.1	79.4	0.2	−0.2	−1.6	−0.4	1.3	10	9	0.5	0.5
Ghana	10.7	18.8	24.5	2.9	1.7	1.0	2.9	2.9	10	30	0.9	0.9
Greece	9.6	10.5	10.3	0.5	−0.1	−1.1	−0.3	1.1	10	9	0.6	0.5
Guatemala	6.8	11.1	16.4	2.6	2.4	0.6	3.1	2.1	7	34	1.0	0.9
Guinea	4.5	7.3	9.8	2.6	1.9	1.1	2.7	2.1	17	40	0.9	0.9
Guinea-Bissau	0.8	1.2	1.6	2.1	1.8	1.3	2.3	1.0	21	41	0.8	0.9
Haiti	5.4	7.8	10.0	2.0	1.6	0.3	2.4	1.8	13	31	0.9	0.8
Honduras	3.6	6.3	8.9	3.0	2.1	0.4	3.1	2.8	5	32	1.0	0.8

Population dynamics | 2.1

	Total population			Average annual population growth rate					Crude death rate	Crude birth rate	Age dependency ratio	
				Total %	Ages 0–14 %	Ages 15–64 %	Ages 65+ %		per 1,000 people	per 1,000 people	dependents as proportion of working-age population	
	millions											
	1980	**1999**	**2015**	**1980–99**	**1999–2015**	**1999–2015**	**1999–2015**	**1999–2015**	**1999**	**1999**	**1980**	**1999**
Hungary	10.7	10.1	9.4	−0.3	−0.4	−1.6	−0.4	0.7	14	9	0.5	0.5
India	687.3	997.5	1,221.9	2.0	1.3	−0.1	1.9	2.3	9	26	0.7	0.6
Indonesia	148.3	207.0	250.5	1.8	1.2	−0.2	1.6	3.0	7	22	0.8	0.6
Iran, Islamic Rep.	39.1	63.0	82.1	2.5	1.7	0.1	2.4	2.1	6	21	0.9	0.7
Iraq	13.0	22.8	31.3	3.0	2.0	0.5	2.8	3.9	10	32	0.9	0.8
Ireland	3.4	3.8	4.3	0.5	0.8	0.0	0.7	1.8	9	14	0.7	0.5
Israel	3.9	6.1	7.9	2.4	1.6	0.2	1.7	2.3	6	21	0.7	0.6
Italy	56.4	57.6	54.8	0.1	−0.3	−1.6	−0.6	1.3	10	9	0.5	0.5
Jamaica	2.1	2.6	3.0	1.0	0.9	−0.9	1.5	1.6	6	22	0.9	0.6
Japan	116.8	126.6	124.3	0.4	−0.1	−0.9	−0.7	2.4	8	10	0.5	0.5
Jordan	2.2	4.7	6.8	4.1	2.3	1.0	2.9	4.3	4	30	1.1	0.7
Kazakhstan	14.9	14.9	15.2	0.0	0.1	−0.8	0.5	1.1	10	14	0.6	0.5
Kenya	16.6	29.4	37.5	3.0	1.5	0.4	2.6	0.0	13	35	1.1	0.9
Korea, Dem. Rep.	17.7	23.4	25.6	1.5	0.6	−0.9	1.1	3.0	10	20	0.8	0.5
Korea, Rep.	38.1	46.9	51.1	1.1	0.5	−0.9	0.6	3.8	6	14	0.6	0.4
Kuwait	1.4	1.9	2.9	1.8	2.5	0.5	3.2	8.4	2	22	0.7	0.6
Kyrgyz Republic	3.6	4.9	5.8	1.5	1.1	−0.9	2.1	0.6	7	21	0.8	0.7
Lao PDR	3.2	5.1	7.2	2.4	2.2	1.3	2.8	1.7	13	37	0.8	0.9
Latvia	2.5	2.4	2.2	−0.2	−0.7	−2.8	−0.6	0.7	14	8	0.5	0.5
Lebanon	3.0	4.3	5.2	1.9	1.2	−0.6	2.0	1.3	6	21	0.8	0.6
Lesotho	1.3	2.1	2.5	2.4	0.9	0.3	2.3	2.4	13	34	0.9	0.8
Libya	3.0	5.4	7.4	3.0	2.0	0.4	2.5	4.9	4	28	1.0	0.7
Lithuania	3.4	3.7	3.6	0.4	−0.1	−1.7	0.0	1.1	11	10	0.5	0.5
Macedonia, FYR	1.9	2.0	2.2	0.4	0.4	−0.8	0.6	2.1	8	14	0.6	0.5
Madagascar	8.9	15.1	22.5	2.8	2.5	1.7	3.3	3.7	12	41	0.9	0.9
Malawi	6.2	10.8	14.3	2.9	1.8	1.6	2.7	2.1	24	46	1.0	0.9
Malaysia	13.8	22.7	29.3	2.6	1.6	0.0	2.2	4.1	4	24	0.8	0.6
Mali	6.6	10.6	15.0	2.5	2.2	2.1	3.2	1.7	19	46	1.0	1.0
Mauritania	1.6	2.6	3.7	2.7	2.2	1.3	2.9	2.4	13	39	0.9	0.9
Mauritius	1.0	1.2	1.4	1.0	0.9	−0.1	1.0	3.0	7	17	0.6	0.5
Mexico	67.6	96.6	118.3	1.9	1.3	−0.2	1.9	3.2	5	27	1.0	0.6
Moldova	4.0	4.3	4.2	0.4	−0.2	−1.5	0.2	0.4	11	12	0.5	0.5
Mongolia	1.7	2.4	3.0	1.9	1.5	−0.4	2.3	2.2	6	21	0.9	0.6
Morocco	19.4	28.2	35.3	2.0	1.4	−0.1	1.9	2.6	7	25	0.9	0.6
Mozambique	12.1	17.3	22.6	1.9	1.7	1.3	2.7	1.0	20	40	0.9	0.9
Myanmar	33.8	45.0	53.8	1.5	1.1	0.9	1.2	1.8	10	26	0.8	0.5
Namibia	1.0	1.7	2.0	2.6	1.1	0.8	2.4	1.0	15	35	0.9	0.8
Nepal	14.5	23.4	32.5	2.5	2.1	1.1	2.7	2.7	10	34	0.8	0.8
Netherlands	14.2	15.8	16.8	0.6	0.4	−0.9	0.1	1.9	9	13	0.5	0.5
New Zealand	3.1	3.8	4.2	1.1	0.6	−0.4	0.6	2.2	7	15	0.6	0.5
Nicaragua	2.9	4.9	6.9	2.7	2.1	0.4	3.2	3.1	5	30	1.0	0.8
Niger	5.6	10.5	16.8	3.3	3.0	2.7	3.3	2.1	18	51	1.0	1.1
Nigeria	71.1	123.9	169.4	2.9	2.0	2.0	2.8	3.2	16	40	1.0	0.9
Norway	4.1	4.5	4.7	0.5	0.4	−0.7	0.3	1.3	10	13	0.6	0.5
Oman	1.1	2.3	3.3	4.0	2.2	0.2	3.1	5.7	3	28	0.9	0.8
Pakistan	82.7	134.8	192.9	2.6	2.2	1.1	3.1	3.3	8	34	0.9	0.8
Panama	2.0	2.8	3.4	1.9	1.2	−0.6	1.8	3.2	5	21	0.8	0.6
Papua New Guinea	3.1	4.7	6.2	2.2	1.7	0.8	2.3	3.2	10	31	0.8	0.7
Paraguay	3.1	5.4	7.5	2.9	2.1	0.2	2.9	3.1	5	30	0.9	0.8
Peru	17.3	25.2	31.8	2.0	1.4	0.0	2.0	2.8	6	24	0.8	0.6
Philippines	48.3	74.3	96.5	2.3	1.6	0.1	2.3	3.9	6	27	0.8	0.7
Poland	35.6	38.7	38.8	0.4	0.0	−1.5	0.2	1.3	10	10	0.5	0.5
Portugal	9.8	10.0	9.9	0.1	0.0	−0.8	0.1	0.2	11	12	0.6	0.5
Puerto Rico	3.2	3.9	4.4	1.0	0.7	−0.3	0.9	2.4	8	16	0.7	0.5
Romania	22.2	22.5	21.3	0.1	−0.3	−1.8	−0.1	0.5	12	10	0.6	0.5
Russian Federation	139.0	146.2	134.5	0.3	−0.5	−1.8	−0.2	0.2	14	9	0.5	0.4

	Total population (millions)			Average annual population growth rate					Crude death rate (per 1,000 people)	Crude birth rate (per 1,000 people)	Age dependency ratio (dependents as proportion of working-age population)	
				Total %		Ages 0–14 %	Ages 15–64 %	Ages 65+ %				
	1980	1999	2015	1980–99	1999–2015	1999–2015	1999–2015	1999–2015	1999	1999	1980	1999
Rwanda	5.2	8.3	11.1	2.5	1.8	1.9	2.7	2.5	22	45	1.0	0.9
Saudi Arabia	9.4	20.2	32.1	4.0	2.9	2.2	3.0	5.5	4	34	0.9	0.8
Senegal	5.5	9.3	13.3	2.7	2.3	1.6	2.8	2.2	13	38	0.9	0.9
Sierra Leone	3.2	4.9	6.7	2.2	1.9	1.2	2.6	2.4	25	45	0.9	0.9
Singapore	2.4	4.0	4.9	2.6	1.4	−1.1	1.1	4.3	5	13	0.5	0.4
Slovak Republic	5.0	5.4	5.4	0.4	0.0	−1.3	0.3	1.2	10	10	0.6	0.5
Slovenia	1.9	2.0	1.9	0.2	−0.2	−1.6	−0.3	1.7	10	9	0.5	0.4
South Africa	27.6	42.1	45.8	2.2	0.5	0.1	1.5	1.0	14	26	0.7	0.6
Spain	37.4	39.4	38.1	0.3	−0.2	−0.9	−0.2	0.6	10	9	0.6	0.5
Sri Lanka	14.7	19.0	22.6	1.3	1.1	0.3	1.2	3.4	6	17	0.7	0.5
Sudan	18.7	29.0	40.6	2.3	2.1	1.8	2.3	2.5	11	33	0.9	0.7
Sweden	8.3	8.9	8.7	0.3	−0.1	−2.0	−0.2	1.4	11	10	0.6	0.6
Switzerland	6.3	7.1	7.1	0.6	0.0	−1.8	−0.2	2.0	9	11	0.5	0.5
Syrian Arab Republic	8.7	15.7	21.9	3.1	2.1	0.3	3.1	3.0	5	29	1.1	0.8
Tajikistan	4.0	6.2	7.9	2.4	1.5	−0.8	2.8	1.0	5	22	0.9	0.8
Tanzania	18.6	32.9	43.8	3.0	1.8	1.2	2.6	0.8	17	40	1.0	0.9
Thailand	46.7	60.2	68.7	1.3	0.8	−0.2	1.1	3.3	7	17	0.8	0.5
Togo	2.6	4.6	6.3	2.9	2.0	1.0	2.8	1.0	15	38	0.9	0.9
Trinidad and Tobago	1.1	1.3	1.5	0.9	0.7	−0.9	0.9	2.6	7	15	0.7	0.5
Tunisia	6.4	9.5	11.5	2.1	1.2	−0.5	1.8	2.0	6	17	0.8	0.6
Turkey	44.5	64.4	77.8	1.9	1.2	0.0	1.4	2.4	6	21	0.8	0.5
Turkmenistan	2.9	4.8	5.8	2.7	1.2	−0.7	2.4	1.5	6	21	0.8	0.7
Uganda	12.8	21.5	31.4	2.7	2.4	1.6	2.9	−0.3	19	46	1.0	1.0
Ukraine	50.0	50.0	44.0	0.0	−0.8	−2.1	−0.6	−0.4	15	9	0.5	0.5
United Arab Emirates	1.0	2.8	3.8	5.2	1.9	0.3	1.9	10.3	3	18	0.4	0.4
United Kingdom	56.3	59.5	59.8	0.3	0.0	−1.0	0.0	1.1	11	12	0.6	0.5
United States	227.2	278.2	316.4	1.1	0.8	−0.1	0.7	2.1	9	15	0.5	0.5
Uruguay	2.9	3.3	3.7	0.7	0.6	−0.2	0.8	0.7	10	16	0.6	0.6
Uzbekistan	16.0	24.4	30.2	2.2	1.3	−1.2	2.5	1.6	6	23	0.9	0.7
Venezuela, RB	15.1	23.7	30.2	2.4	1.5	−0.2	2.1	3.9	4	24	0.8	0.6
Vietnam	53.7	77.5	94.4	1.9	1.2	−0.8	2.1	1.0	6	20	0.9	0.6
West Bank and Gaza	..	2.8	5.0	..	3.5	2.8	4.5	2.4	4	41	..	1.0
Yemen, Rep.	8.5	17.0	26.5	3.6	2.8	2.1	3.5	1.4	12	40	1.1	1.0
Yugoslavia, FR (Serb./Mont.)	9.8	10.6	10.7	0.4	0.1	−0.8	0.2	0.6	11	12	0.5	0.5
Zambia	5.7	9.9	12.2	2.9	1.3	0.7	2.5	0.9	21	41	1.1	0.9
Zimbabwe	7.0	11.9	13.1	2.8	0.6	−0.6	2.1	0.1	16	30	1.0	0.8
World	4,430.1 s	5,978.0 s	7,084.3 s	1.6 w	1.1 w	0.1 w	1.4 w	2.1 w	9 w	22 w	0.7 w	0.6 w
Low income	1,612.9	2,417.1	3,086.1	2.1	1.5	0.6	2.1	2.2	11	29	0.8	0.7
Middle income	2,028.1	2,664.5	3,055.0	1.4	0.9	−0.5	1.2	2.2	8	18	0.7	0.5
Lower middle income	1,607.9	2,093.0	2,382.8	1.4	0.8	−0.6	1.2	2.1	8	17	0.7	0.5
Upper middle income	420.2	571.5	672.2	1.6	1.0	−0.1	1.4	2.5	7	21	0.7	0.6
Low & middle income	3,641.0	5,081.6	6,141.1	1.8	1.2	0.1	1.6	2.2	9	24	0.8	0.6
East Asia & Pacific	1,397.8	1,836.6	2,097.3	1.4	0.8	−0.6	1.2	2.5	7	18	0.7	0.5
Europe & Central Asia	425.8	474.4	478.1	0.6	0.0	−1.2	0.4	0.7	11	12	0.6	0.5
Latin America & Carib.	360.0	508.2	621.6	1.8	1.3	−0.1	1.7	2.8	7	23	0.8	0.6
Middle East & N. Africa	174.4	290.3	389.7	2.7	1.8	0.5	2.5	2.8	7	26	0.9	0.7
South Asia	902.6	1,329.3	1,676.3	2.0	1.4	0.1	2.0	2.5	9	27	0.8	0.7
Sub-Saharan Africa	380.5	642.8	878.1	2.8	1.9	1.6	2.7	1.8	16	40	0.8	0.9
High income	789.1	896.3	943.2	0.7	0.3	−0.6	0.2	1.8	9	12	0.5	0.5
Europe EMU	277.0	292.8	290.1	0.3	−0.1	−1.1	−0.2	1.2	10	10	0.5	0.5

About the data

Population estimates are usually based on national population censuses, but the frequency and quality of these vary by country. Most countries conduct a complete enumeration no more than once a decade. Pre- and postcensus estimates are interpolations or extrapolations based on demographic models. Errors and undercounting occur even in high-income countries; in developing countries such errors may be substantial because of limits in the transport, communications, and other resources required to conduct a full census. The quality and reliability of official demographic data are also affected by the public trust in the government, the government's commitment to full and accurate enumeration, the confidentiality and protection against misuse accorded to census data, and the independence of census agencies from undue political influence. Moreover, the international comparability of population indicators is limited by differences in the concepts, definitions, data collection procedures, and estimation methods used by national statistical agencies and other organizations that collect population data.

Of the 148 economies listed in the table, 117 (about 80 percent) conducted a census between 1990 and 2000. The currentness of a census, along with the availability of complementary data from surveys or registration systems, is one of many objective ways to judge the quality of demographic data. In some European countries registration systems offer complete information on population in the absence of a census. See *Primary data documentation* for the most recent census or survey year and for registration completeness.

Current population estimates for developing countries that lack recent census-based data, and pre- and postcensus estimates for countries with census data, are provided by national statistical offices, the United Nations Population Division, and other agencies. The standard estimation method requires fertility, mortality, and net migration data, which are often collected from sample surveys, some of which may be small or limited in coverage. The population estimates are the product of demographic modeling and so are susceptible to biases and errors because of shortcomings in the model as well as in the data. Population projections are made using the cohort component method.

The growth rate of the total population conceals the fact that different age groups may grow at very different rates. In many developing countries the population under 15 was earlier growing rapidly, but is now starting to shrink. Previously high fertility rates and declining mortality rates are now reflected in rapid growth of the working-age population.

The vital rates shown in the table are based on data derived from birth and death registration systems, censuses, and sample surveys conducted by national statistical offices, United Nations agencies, and other organizations. The estimates for 1999 for many countries are based on extrapolations of levels and trends measured in earlier years.

Vital registers are the preferred source of these data, but in many developing countries systems for registering births and deaths do not exist or are incomplete because of deficiencies in geographic coverage or coverage of events. Many developing countries carry out specialized household surveys that estimate vital rates by asking respondents about births and deaths in the recent past. Estimates derived in this way are subject to sampling errors as well as errors due to inaccurate recall by the respondents.

The United Nations Statistics Division monitors the completeness of vital registration systems. The share of countries with at least 90 percent complete vital registration increased from 45 percent in 1988 to 52 percent in 1999. Still, some of the most populous developing countries—China, India, Indonesia, Brazil, Pakistan, Bangladesh, Nigeria—do not have complete vital registration systems. Fewer than 30 percent of vital events worldwide are thought to be recorded.

International migration is the only other factor besides birth and death rates that directly determines a country's population growth. In the high-income countries about 40 percent of annual population growth in 1990–95 was due to migration, while in the developing countries migration reduced population growth by about 3 percent. Estimating international migration is difficult. At any time many people are located outside their home country as tourists, workers, or refugees or for other reasons. Standards relating to the duration and purpose of international moves that qualify as international migration vary, and accurate estimates require information on flows into and out of countries that is difficult to collect.

Definitions

• **Total population** of an economy includes all residents who are present regardless of legal status or citizenship—except for refugees not permanently settled in the country of asylum, who are generally considered part of the population of their country of origin. The indicators shown are midyear estimates for 1980 and 1999 and projections for 2015. • **Average annual population growth rate** is the exponential change for the period indicated. See *Statistical methods* for more information. • **Crude death rate** and **crude birth rate** are the number of deaths and the number of live births occurring during the year, per 1,000 population estimated at midyear. Subtracting the crude death rate from the crude birth rate provides the rate of natural increase, which is equal to the population growth rate in the absence of migration. • **Age dependency ratio** is the ratio of dependents—people younger than 15 and older than 64—to the working-age population—those ages 15–64.

Data sources

The World Bank's population estimates are produced by its Human Development Network and Development Data Group in consultation with its operational staff and country offices. Important inputs to the World Bank's demographic work come from the following sources: census reports and other statistical publications from national statistical offices; Demographic and Health Surveys conducted by national agencies, Macro International, and the U.S. Centers for Disease Control and Prevention; United Nations Statistics Division, *Population and Vital Statistics Report* (quarterly); United Nations Population Division, *World Population Prospects: The 1998 Revision;* Eurostat, *Demographic Statistics* (various years); Centro Latinoamericano de Demografía, *Boletín Demográfico* (various years); and U.S. Bureau of the Census, International Database.

2.2 | Labor force structure

	Population ages 15–64		Labor force						
	millions		Total millions			Average annual growth rate %		Female % of labor force	
	1980	1999	1980	1999	2010	1980–99	1999–2010	1980	1999
Albania	2	2	1	2	2	1.7	1.5	38.8	41.2
Algeria	9	18	5	10	15	3.8	3.5	21.4	27.0
Angola	4	6	3	6	8	2.6	3.0	47.0	46.3
Argentina	17	23	11	15	18	1.7	2.1	27.6	32.7
Armenia	2	3	1	2	2	1.4	1.3	47.9	48.5
Australia	10	13	7	10	11	1.9	0.8	36.8	43.5
Austria	5	6	3	4	4	0.6	0.0	40.5	40.3
Azerbaijan	4	5	3	4	4	1.4	1.9	47.5	44.4
Bangladesh	44	74	41	67	86	2.6	2.3	42.3	42.3
Belarus	6	7	5	5	5	0.2	0.1	49.9	48.9
Belgium	6	7	4	4	4	0.4	−0.2	33.9	40.8
Benin	2	3	2	3	4	2.7	2.8	47.0	48.3
Bolivia	3	5	2	3	4	2.6	2.5	33.3	37.7
Bosnia and Herzegovina	3	2	2	2	2	0.6	0.9	32.8	38.1
Botswana	0	1	0	1	1	3.0	1.0	50.1	45.4
Brazil	70	110	48	78	89	2.6	1.2	28.4	35.4
Bulgaria	6	6	5	4	4	−0.5	−0.8	45.3	48.2
Burkina Faso	3	6	4	5	7	1.9	1.8	47.6	46.5
Burundi	2	3	2	4	5	2.5	2.2	50.2	48.8
Cambodia	4	6	4	6	8	2.7	2.2	55.4	51.8
Cameroon	5	8	4	6	8	2.6	2.1	36.8	37.9
Canada	17	21	12	16	17	1.5	0.6	39.5	45.6
Central African Republic	1	2	..	..	..	..	..	..	..
Chad	2	3	2	4	5	2.5	2.9	43.4	44.6
Chile	7	10	4	6	8	2.5	1.9	26.3	33.2
China	586	844	539	751	819	1.7	0.8	43.2	45.2
Hong Kong, China	3	5	2	4	4	1.9	0.9	34.3	37.0
Colombia	16	26	9	18	23	3.4	2.2	26.2	38.4
Congo, Dem. Rep.	14	25	12	21	28	2.8	2.8	44.5	43.4
Congo, Rep.	1	1	1	1	2	2.7	2.8	42.4	43.5
Costa Rica	1	2	1	1	2	3.1	1.9	20.8	30.8
Côte d'Ivoire	4	8	3	6	8	3.4	1.7	32.2	33.3
Croatia	3	3	2	2	2	−0.2	−0.2	40.2	44.1
Cuba	6	8	4	6	6	2.1	0.7	31.4	39.2
Czech Republic	6	7	5	6	6	0.4	−0.4	47.1	47.3
Denmark	3	4	3	3	3	0.4	−0.5	44.0	46.4
Dominican Republic	3	5	2	4	5	2.9	2.2	24.7	30.4
Ecuador	4	8	3	5	6	3.3	2.7	20.1	27.7
Egypt, Arab Rep.	23	38	14	24	32	2.7	2.7	26.5	30.1
El Salvador	2	4	2	3	4	2.8	2.8	26.5	36.0
Eritrea	..	2	1	2	3	2.6	2.5	47.4	47.4
Estonia	1	1	1	1	1	−0.1	−0.2	50.6	49.0
Ethiopia	20	32	17	27	35	2.5	2.2	42.3	40.9
Finland	3	3	2	3	2	0.4	−0.5	46.5	48.0
France	34	39	24	26	27	0.6	0.3	40.1	44.9
Gabon	0	1	0	1	1	2.2	1.8	45.0	44.6
Gambia, The	0	1	0	1	1	3.4	2.4	44.8	45.0
Georgia	3	4	3	3	3	0.3	0.3	49.3	46.7
Germany	52	56	37	41	41	0.5	−0.1	40.1	42.2
Ghana	6	10	5	9	11	2.9	2.0	51.0	50.5
Greece	6	7	4	5	5	1.0	0.1	27.9	37.6
Guatemala	3	6	2	4	6	2.9	3.6	22.4	28.4
Guinea	2	4	2	3	4	2.1	2.0	47.1	47.2
Guinea-Bissau	0	1	0	1	1	1.8	1.9	39.9	40.5
Haiti	3	4	3	3	4	1.6	1.6	44.6	42.9
Honduras	2	3	1	2	3	3.6	3.5	25.2	31.4

Labor force structure | 2.2

	Population ages 15–64		Labor force						
	millions		Total millions			Average annual growth rate %		Female % of labor force	
	1980	1999	1980	1999	2010	1980–99	1999–2010	1980	1999
Hungary	7	7	5	5	5	–0.3	–0.5	43.3	44.7
India	394	609	300	441	541	2.0	1.9	33.7	32.2
Indonesia	83	133	59	99	124	2.8	2.0	35.2	40.6
Iran, Islamic Rep.	20	38	12	19	28	2.6	3.4	20.4	26.5
Iraq	7	13	4	6	9	3.0	2.8	17.3	19.4
Ireland	2	2	1	2	2	1.2	1.4	28.1	34.2
Israel	2	4	1	3	3	3.1	2.6	33.7	40.9
Italy	36	39	23	26	25	0.7	–0.3	32.9	38.3
Jamaica	1	2	1	1	2	1.8	1.4	46.3	46.2
Japan	79	87	57	68	66	0.9	–0.3	37.9	41.3
Jordan	1	3	1	1	2	5.2	3.5	14.7	23.9
Kazakhstan	9	10	7	7	8	0.2	0.5	47.6	46.9
Kenya	8	16	8	15	19	3.4	2.1	46.0	46.1
Korea, Dem. Rep.	10	16	8	12	13	2.5	0.5	44.8	43.3
Korea, Rep.	24	33	16	24	27	2.2	1.1	38.7	41.2
Kuwait	1	1	0	1	1	2.3	4.4	13.1	31.3
Kyrgyz Republic	2	3	2	2	3	1.6	2.0	47.5	47.2
Lao PDR	2	3	2	2	3	2.0	2.5	..	..
Latvia	2	2	1	1	1	–0.4	–0.4	50.8	50.4
Lebanon	2	3	1	1	2	2.9	2.6	22.6	29.3
Lesotho	1	1	1	1	1	2.3	1.3	37.9	36.9
Libya	2	3	1	2	2	2.6	2.4	18.6	22.6
Lithuania	2	2	2	2	2	0.3	0.2	49.7	48.0
Macedonia, FYR	1	1	1	1	1	0.8	0.6	36.1	41.5
Madagascar	5	8	4	7	10	2.6	2.9	45.2	44.7
Malawi	3	6	3	5	6	2.7	1.8	50.6	48.7
Malaysia	8	14	5	9	13	3.0	2.8	33.7	37.7
Mali	3	5	3	5	7	2.2	2.2	46.7	46.2
Mauritania	1	1	1	1	2	2.5	2.5	45.0	43.7
Mauritius	1	1	0	1	1	2.0	1.1	25.7	32.4
Mexico	35	60	22	39	50	3.1	2.2	26.9	32.9
Moldova	3	3	2	2	2	0.1	0.2	50.3	48.6
Mongolia	1	2	1	1	2	2.2	2.3	45.7	46.9
Morocco	10	18	7	11	15	2.5	2.5	33.5	34.7
Mozambique	6	9	7	9	11	1.5	1.9	49.0	48.4
Myanmar	19	30	17	24	28	1.7	1.6	43.7	43.4
Namibia	1	1	0	1	1	2.4	1.3	40.1	40.9
Nepal	8	13	7	11	14	2.3	2.5	38.8	40.5
Netherlands	9	11	6	7	8	1.4	0.2	31.5	40.4
New Zealand	2	2	1	2	2	1.9	0.7	34.3	44.8
Nicaragua	1	3	1	2	3	3.6	3.4	27.6	35.5
Niger	3	5	3	5	7	3.0	3.2	44.6	44.3
Nigeria	36	66	30	49	63	2.7	2.3	36.2	36.4
Norway	3	3	2	2	2	0.9	0.3	40.5	46.3
Oman	1	1	0	1	1	3.4	2.6	6.2	16.4
Pakistan	44	74	29	50	71	2.8	3.2	22.7	28.1
Panama	1	2	1	1	1	2.9	1.9	29.9	35.0
Papua New Guinea	2	3	2	2	3	2.2	2.1	41.7	42.1
Paraguay	2	3	1	2	3	2.9	2.9	26.7	29.8
Peru	9	15	5	9	13	2.9	2.7	23.9	31.0
Philippines	27	46	19	31	41	2.6	2.5	35.0	37.7
Poland	23	26	19	20	20	0.3	0.2	45.3	46.3
Portugal	6	7	5	5	5	0.5	0.0	38.7	43.9
Puerto Rico	2	3	1	1	2	1.9	1.2	31.8	36.9
Romania	14	15	11	11	11	–0.1	–0.1	45.8	44.5
Russian Federation	95	101	76	78	77	0.1	–0.1	49.4	49.0

	Population ages 15–64		Labor force						
	millions			Total millions		Average annual growth rate %		Female % of labor force	
	1980	1999	1980	1999	2010	1980–99	1999–2010	1980	1999
Rwanda	3	4	3	4	6	2.8	2.3	49.1	48.8
Saudi Arabia	5	12	3	7	10	4.6	3.3	7.6	15.5
Senegal	3	5	3	4	5	2.6	2.5	42.2	42.6
Sierra Leone	2	3	1	2	2	2.0	2.2	35.5	36.7
Singapore	2	2	1	2	2	2.9	1.3	34.6	39.1
Slovak Republic	3	4	2	3	3	0.9	0.2	45.3	47.8
Slovenia	1	1	1	1	1	0.3	–0.3	45.8	46.5
South Africa	16	26	10	17	18	2.5	0.9	35.1	37.7
Spain	23	27	14	17	17	1.1	0.0	28.3	37.0
Sri Lanka	9	13	5	8	10	2.2	1.7	26.9	36.4
Sudan	10	17	7	12	15	2.7	2.7	26.9	29.3
Sweden	5	6	4	5	5	0.7	–0.4	43.8	48.0
Switzerland	4	5	3	4	4	1.2	0.1	36.7	40.3
Syrian Arab Republic	4	9	2	5	7	3.6	3.8	23.5	26.7
Tajikistan	2	3	2	2	3	2.4	2.9	46.9	44.6
Tanzania	9	17	10	17	21	3.0	2.0	49.8	49.2
Thailand	26	42	24	36	41	2.1	1.0	47.4	46.3
Togo	1	2	1	2	2	2.7	2.2	39.3	40.0
Trinidad and Tobago	1	1	0	1	1	1.6	1.6	31.4	34.0
Tunisia	3	6	2	4	5	2.8	2.3	28.9	31.4
Turkey	25	43	19	31	37	2.6	1.8	35.5	37.3
Turkmenistan	2	3	1	2	3	2.9	2.2	47.0	45.8
Uganda	6	11	7	11	14	2.5	2.5	47.9	47.6
Ukraine	33	34	26	25	24	–0.2	–0.4	50.2	48.8
United Arab Emirates	1	2	1	1	2	4.8	2.0	5.1	14.5
United Kingdom	36	39	27	30	30	0.5	0.0	38.9	43.9
United States	151	179	110	143	157	1.4	0.9	41.0	45.8
Uruguay	2	2	1	2	2	1.4	1.0	30.8	41.5
Uzbekistan	9	14	6	10	13	2.4	2.4	48.0	46.8
Venezuela, RB	8	15	5	10	13	3.3	2.6	26.7	34.5
Vietnam	28	48	26	40	48	2.3	1.7	48.1	49.0
West Bank and Gaza	..	1	..	..	..	..	..	..	..
Yemen, Rep.	4	8	2	5	8	4.1	3.1	32.5	28.0
Yugoslavia, FR (Serb./Mont.)	6	7	5	5	5	0.6	0.3	38.7	42.8
Zambia	3	5	2	4	5	2.9	1.9	45.4	44.9
Zimbabwe	3	7	3	5	6	2.9	1.0	44.4	44.5
World	**2,595 s**	**3,761 s**	**2,036 s**	**2,895 s**	**3,373 s**	**1.9 w**	**1.4 w**	**39.1 w**	**40.6 w**
Low income	890	1,417	709	1,090	1,364	2.3	2.0	37.9	37.8
Middle income	1,200	1,749	969	1,370	1,556	1.8	1.2	40.2	42.0
Lower middle income	955	1,379	804	1,119	1,261	1.7	1.1	41.7	43.2
Upper middle income	245	369	165	251	295	2.2	1.5	32.7	36.5
Low & middle income	2,090	3,166	1,678	2,460	2,920	2.0	1.6	39.2	40.1
East Asia & Pacific	820	1,220	720	1,039	1,171	1.9	1.1	42.5	44.4
Europe & Central Asia	274	318	214	237	249	0.5	0.4	46.7	46.2
Latin America & Carib.	201	319	130	217	268	2.7	1.9	27.8	34.6
Middle East & N. Africa	92	172	54	97	135	3.0	3.0	23.8	27.3
South Asia	508	797	389	588	738	2.2	2.1	33.8	33.3
Sub-Saharan Africa	195	340	170	282	360	2.6	2.2	42.3	42.2
High income	505	595	358	435	453	1.0	0.4	38.4	43.1
Europe EMU	179	198	120	136	136	0.7	0.0	36.7	41.2

About the data

The labor force is the supply of labor available for the production of goods and services in an economy. It includes people who are currently employed and people who are unemployed but seeking work as well as first-time job-seekers. Not everyone who works is included, however. Unpaid workers, family workers, and students are among those usually omitted, and in some countries members of the military are not counted. The size of the labor force tends to vary during the year as seasonal workers enter and leave it.

Data on the labor force are compiled by the International Labour Organization (ILO) from censuses or labor force surveys. For international comparisons the most comprehensive source is labor force surveys. Despite the ILO's efforts to encourage the use of international standards, labor force data are not fully comparable because of differences among countries, and sometimes within countries, in their scope and coverage. In some countries data on the labor force refer to people above a specific age, while in others there is no specific age provision. The reference period of the census or survey is another important source of differences: in some countries data refer to people's status on the day of the census or survey or during a specific period before the inquiry date, while in others the data are recorded without reference to any period. In developing countries, where the household is often the basic unit of production and all members contribute to output, but some at low intensity or irregular intervals, the estimated labor force may be significantly smaller than the numbers actually working (ILO, *Yearbook of Labour Statistics 1997*).

The labor force estimates in the table were calculated by World Bank staff by applying economic activity rates from the ILO database to World Bank population estimates to create a series consistent with these population estimates. This procedure sometimes results in estimates of labor force size that differ slightly from those in the ILO's *Yearbook of Labour Statistics*. The population ages 15–64 is often used to provide a rough estimate of the potential labor force. But in many developing countries children under 15 work full or part time. And in some high-income countries many workers postpone retirement past age 65. As a result, labor force participation rates calculated in this way may systematically over- or underestimate actual rates.

In general, estimates of women in the labor force are lower than those of men and are not comparable internationally, reflecting the fact that for women, demographic, social, legal, and cultural trends and norms determine whether their activities are regarded as economic. In many countries large numbers of women work on farms or in other family enterprises without pay, while others work in or near their homes, mixing work and personal activities during the day. Countries differ in the criteria used to determine the extent to which such workers are to be counted as part of the labor force.

Table 2.2a

Change in female employment and labor force participation in selected countries in Europe and Central Asia, 1989–97
Percentage points

	Labor force participation rate	Employment rate
Estonia	–21	–27
Hungary	–27	–33
Latvia	–22	–34
Poland	–10	–10

Source: UNICEF 1999.

In some countries in Europe and Central Asia female employment has dropped more than female labor force participation. Since 1989 women have been laid off either in big waves, as in Central Europe, or gradually. Many women opted out of the labor force, but others remained active by looking for work, swelling the ranks of the unemployed. This rising unemployment among women has tempered the decline in their labor force participation.

Definitions

• **Population ages 15–64** is the number of people who could potentially be economically active. • **Total labor force** comprises people who meet the ILO definition of the economically active population: all people who supply labor for the production of goods and services during a specified period. It includes both the employed and the unemployed. While national practices vary in the treatment of such groups as the armed forces and seasonal or part-time workers, the labor force generally includes the armed forces, the unemployed, and first-time job-seekers, but excludes homemakers and other unpaid caregivers and workers in the informal sector. • **Average annual growth rate of the labor force** is calculated using the exponential endpoint method (see *Statistical methods* for more information). • **Females as a percentage of the labor force** show the extent to which women are active in the labor force.

Data sources

The population estimates are from the World Bank's population database. The economic activity rates are from the ILO database Estimates and Projections of the Economically Active Population, 1950–2010. The ILO publishes estimates of the economically active population in its *Yearbook of Labour Statistics*.

2.3 Employment by economic activity

	Agriculture				Industry				Services			
	Male % of male labor force		Female % of female labor force		Male % of male labor force		Female % of female labor force		Male % of male labor force		Female % of female labor force	
	1980	1996–98a	1980	1996–98a	1980	1996–98a	1980	1996–98a	1980	1996–98a	1980	1996–98a
Albania	54	..	62	..	28	..	17	..	18	..	21	..
Algeria	27	..	69	..	33	..	6	..	40	..	25	..
Angola	67	..	87	..	13	..	1	..	20	..	11	..
Argentina	17	2	3	0 b	40	33	18	12	44	65	79	88
Armenia	21	..	21	..	48	..	38	..	31	..	41	..
Australia	8	6	4	4	39	31	16	11	53	64	79	86
Austria	..	6	..	7	..	42	..	14	..	52	..	78
Azerbaijan	28	..	42	..	36	..	20	..	36	..	38	..
Bangladesh	67	54	81	78	5	11	14	8	29	34	5	11
Belarus	29	..	23	..	44	..	33	..	28	..	44	..
Belgium	..	..	..	..	..	..	..	..	..	..	..	..
Benin	66	..	69	..	10	..	4	..	24	..	27	..
Bolivia	52	58 c	28	2 c	21	40 c	19	16 c	27	58 c	53	82 c
Bosnia and Herzegovina	26	..	38	..	45	..	24	..	30	..	39	..
Botswana	6	..	3	..	41	..	8	..	53	..	89	..
Brazil	34	27	25	20	30	27	13	10	36	46	67	70
Bulgaria	..	..	..	..	..	..	..	..	..	..	..	..
Burkina Faso	92	..	93	..	3	..	2	..	5	..	5	..
Burundi	88	..	98	..	4	..	1	..	9	..	1	..
Cambodia	70	..	80	..	7	..	7	..	23	..	14	..
Cameroon	65	..	87	..	11	..	2	..	24	..	11	..
Canada	7	5	3	2	37	32	16	11	56	63	81	87
Central African Republic	79	..	90	..	5	..	1	..	15	..	9	..
Chad	82	..	95	..	6	..	0 b	..	12	..	4	..
Chile	22	19	3	5	27	31	16	14	51	49	81	82
China	..	..	..	..	..	..	..	..	..	..	..	..
Hong Kong, China	2	0 b	1	0 b	47	29	56	13	52	70	43	87
Colombia	2	1	1	1	39	31	26	21	59	68	74	78
Congo, Dem. Rep.	62	..	84	..	18	..	4	..	20	..	12	..
Congo, Rep.	42	..	81	..	20	..	2	..	38	..	17	..
Costa Rica	34	27	6	6	25	25	20	18	40	47	74	76
Côte d'Ivoire	60	..	75	..	10	..	5	..	30	..	20	..
Croatia	..	16	..	17	..	36	..	22	..	47	..	61
Cuba	30	..	10	..	32	..	22	..	39	..	68	..
Czech Republic	13	7	11	4	57	50	39	29	30	43	50	66
Denmark	11	5	4	2	41	37	16	15	48	58	80	83
Dominican Republic	40	27	11	2	26	27	16	21	34	46	73	77
Ecuador	44	11	22	2	21	26	15	14	34	63	63	84
Egypt, Arab Rep.	45	..	10	..	21	..	13	..	33	..	69	..
El Salvador	51	38	10	7	21	25	21	21	28	37	69	72
Eritrea	79	..	88	..	7	..	2	..	14	..	11	..
Estonia	19	12	12	6	50	42	36	24	31	46	52	69
Ethiopia	90	..	89	..	2	..	2	..	8	..	10	..
Finland	15	8	12	5	44	39	23	14	41	53	65	81
France	9	..	7	..	44	..	22	..	47	..	71	..
Gabon	59	..	74	..	18	..	6	..	24	..	21	..
Gambia, The	78	..	93	..	10	..	3	..	13	..	5	..
Georgia	31	..	34	..	33	..	21	..	..	..	45	..
Germany	..	3	..	3	..	45	..	19	..	52	..	79
Ghana	66	..	57	..	12	..	14	..	22	..	29	..
Greece	26	18	42	23	34	28	18	13	40	54	40	64
Guatemala	64	..	17	..	17	..	27	..	19	..	56	..
Guinea	86	..	97	..	2	..	1	..	12	..	3	..
Guinea-Bissau	81	..	98	..	3	..	0 b	..	17	..	3	..
Haiti	81	..	53	..	8	..	8	..	11	..	39	..
Honduras	63	49	40	8	17	21	9	27	20	30	51	66

Employment by economic activity | 2.3

	Agriculture				Industry				Services			
	Male % of male labor force		Female % of female labor force		Male % of male labor force		Female % of female labor force		Male % of male labor force		Female % of female labor force	
	1980	1996–98[a]	1980	1996–98[a]	1980	1996–98[a]	1980	1996–98[a]	1980	1996–98[a]	1980	1996–98[a]
Hungary	24	10	19	4	45	41	36	26	31	48	45	70
India	63	..	83	..	15	..	9	..	22	..	8	..
Indonesia	57	41	54	42	13	21	13	16	29	39	33	42
Iran, Islamic Rep.	36	..	50	..	28	..	17	..	35	..	33	..
Iraq	21	..	62	..	24	..	11	..	55	..	28	..
Ireland	..	13	..	3	..	38	..	17	..	48	..	80
Israel	8	3	4	1	39	36	16	13	52	60	79	85
Italy	13	7	16	7	42	38	28	22	45	55	56	72
Jamaica	47	29	23	10	20	25	8	9	33	46	69	82
Japan	9	5	13	6	40	39	28	23	51	56	58	71
Jordan	..	..	..	..	24	..	7	..	76	..	93	..
Kazakhstan	28	..	20	..	38	..	25	..	34	..	55	..
Kenya	23	..	25	..	24	..	9	..	53	..	65	..
Korea, Dem. Rep.	39	..	52	..	37	..	20	..	24	..	28	..
Korea, Rep.	31	11	39	14	32	34	24	19	37	55	37	67
Kuwait	2	..	0 [b]	..	36	..	3	..	62	..	97	..
Kyrgyz Republic	35	49	33	49	34	11	23	7	32	32	44	38
Lao PDR	77	..	82	..	7	..	4	..	16	..	13	..
Latvia	18	21	14	16	49	33	35	19	32	46	50	65
Lebanon	13	..	20	..	29	..	21	..	58	..	59	..
Lesotho	26	..	64	..	52	..	5	..	22	..	31	..
Libya	16	..	63	..	29	..	3	..	55	..	34	..
Lithuania	26	21	29	16	47	35	30	22	27	43	41	62
Macedonia, FYR	30	..	47	..	38	..	23	..	32	..	30	..
Madagascar	73	77 [d]	93	76 [d]	9	6 [d]	2	4 [d]	19	16 [d]	5	20 [d]
Malawi	78	..	96	..	10	..	1	..	12	..	3	..
Malaysia	34	21	44	15	26	34	20	28	40	46	36	57
Mali	86	..	92	..	2	..	1	..	12	..	7	..
Mauritania	65	..	79	..	11	..	2	..	24	..	19	..
Mauritius	29	..	30	..	19	..	40	..	47	..	31	..
Mexico	..	26	..	9	..	27	..	20	..	47	..	71
Moldova	49	..	38	..	32	..	21	..	19	..	41	..
Mongolia	43	..	36	..	21	..	21	..	36	..	43	..
Morocco	48	..	72	..	23	..	14	..	29	..	14	..
Mozambique	72	..	97	..	14	..	1	..	14	..	2	..
Myanmar	..	..	..	..	..	..	..	..	..	..	..	..
Namibia	52	..	42	..	22	..	10	..	27	..	47	..
Nepal	91	..	98	..	1	..	0 [b]	..	8	..	2	..
Netherlands	7	4	3	2	39	31	13	9	54	63	84	86
New Zealand	..	11	..	6	..	33	..	13	..	56	..	81
Nicaragua	..	..	..	..	..	..	..	..	..	..	..	..
Niger	7	..	6	..	69	..	29	..	25	..	66	..
Nigeria	52	..	57	..	10	..	5	..	38	..	38	..
Norway	10	6	6	3	40	35	13	10	50	59	80	87
Oman	52	..	24	..	21	..	33	..	27	..	43	..
Pakistan	..	41	..	66	..	20	..	11	..	39	..	23
Panama	37	28	6	2	21	22	12	11	39	50	81	87
Papua New Guinea	76	..	92	..	8	..	2	..	16	..	6	..
Paraguay	58	7	9	3	20	31	22	10	22	62	70	87
Peru	45	7	25	3	20	27	14	11	35	66	61	86
Philippines	60	47	37	27	16	18	15	12	25	35	48	61
Poland	..	19	..	19	..	41	..	21	..	39	..	60
Portugal	22	12	35	15	44	45	25	25	34	43	40	60
Puerto Rico	8	4	0 [b]	0 [b]	27	27	24	14	65	69	75	86
Romania	22	37	39	44	52	34	34	24	26	29	27	33
Russian Federation	19	..	13	..	50	..	37	..	31	..	50	..

	Agriculture				Industry				Services			
	Male % of male labor force		Female % of female labor force		Male % of male labor force		Female % of female labor force		Male % of male labor force		Female % of female labor force	
	1980	1996–98ᵃ	1980	1996–98ᵃ	1980	1996–98ᵃ	1980	1996–98ᵃ	1980	1996–98ᵃ	1980	1996–98ᵃ
Rwanda	88	..	98	..	5	..	1	..	7	..	1	..
Saudi Arabia	45	..	25	..	17	..	5	..	39	..	70	..
Senegal	74	..	90	..	9	..	2	..	17	..	8	..
Sierra Leone	63	..	82	..	20	..	4	..	17	..	14	..
Singapore	2	0 ᵇ	1	0 ᵇ	33	34	40	23	65	66	59	77
Slovak Republic	15	10	13	5	38	49	34	27	48	41	54	68
Slovenia	14	12	17	12	49	47	37	30	38	41	46	58
South Africa	18	..	16	..	45	..	16	..	37	..	68	..
Spain	20	9	18	6	42	40	21	14	39	51	60	81
Sri Lanka	44	38	51	49	19	23	18	22	30	37	28	27
Sudan	66	..	88	..	9	..	4	..	24	..	8	..
Sweden	8	4	3	1	45	38	16	12	47	58	81	87
Switzerland	8	5	5	4	47	35	23	14	46	59	72	83
Syrian Arab Republic	..	..	..	..	..	..	..	..	..	..	..	..
Tajikistan	36	..	54	..	29	..	16	..	35	..	30	..
Tanzania	80	..	92	..	7	..	2	..	13	..	7	..
Thailand	68	52	74	50	13	19	8	16	20	29	18	34
Togo	70	..	67	..	12	..	7	..	19	..	26	..
Trinidad and Tobago	11	11	9	3	44	37	21	13	45	52	70	83
Tunisia	33	..	53	..	30	..	32	..	37	..	16	..
Turkey	45	33	88	70	22	27	5	11	33	40	8	19
Turkmenistan	33	..	46	..	32	..	16	..	36	..	38	..
Uganda	84	..	91	..	6	..	2	..	10	..	8	..
Ukraine	26	..	24	..	46	..	33	..	28	..	44	..
United Arab Emirates	5	..	0 ᵇ	..	40	..	7	..	55	..	93	..
United Kingdom	4	2	1	1	48	38	23	13	49	60	76	86
United States	5	4	2	1	40	33	19	13	55	63	80	86
Uruguay	..	6	..	2	..	33	..	14	..	62	..	84
Uzbekistan	35	..	46	..	34	..	19	..	32	..	36	..
Venezuela, RB	20	16	2	2	31	29	18	13	49	55	79	85
Vietnam	71	70	75	71	16	12	10	9	13	18	15	20
West Bank and Gaza	..	..	..	..	..	..	..	..	..	..	..	..
Yemen, Rep.	60	..	98	..	19	..	1	..	21	..	1	..
Yugoslavia, FR (Serb./Mont.)	..	..	..	..	..	..	..	..	..	..	..	..
Zambia	69	..	85	..	13	..	3	..	19	..	13	..
Zimbabwe	29	..	50	..	31	..	8	..	40	..	42	..
World	.. w	.. w	47 w	.. w	.. w	.. w	.. w	.. w	.. w	.. w	.. w	.. w
Low income	62	..	74	..	15	..	10	..	23	..	16	..
Middle income	..	..	..	..	..	..	..	..	..	..	..	..
Lower middle income	..	..	..	..	..	..	..	..	..	..	..	..
Upper middle income	..	23	22	15	..	31	..	17	..	47	..	68
Low & middle income	..	..	..	..	..	..	..	..	..	..	..	..
East Asia & Pacific	..	..	..	..	..	..	..	..	..	..	..	..
Europe & Central Asia	26	..	27	..	44	..	31	..	31	..	42	..
Latin America & Carib.	..	23	17	13	..	27	..	14	..	50	..	73
Middle East & N. Africa	39	..	46	..	25	..	13	..	37	..	38	..
South Asia	64	..	83	..	14	..	10	..	23	..	8	..
Sub-Saharan Africa	62	..	74	..	14	..	5	..	24	..	22	..
High income	8	4	7	3	41	37	22	15	51	59	71	82
Europe EMU	..	6	..	5	..	41	..	18	..	53	..	77

a. Data are for the most recent year available. b. Less than 0.5. c. Break in series between 1980 and 1990. d. Data refer to 1999.

Employment by economic activity | 2.3

About the data

The International Labour Organization (ILO) classifies economic activity on the basis of the International Standard Industrial Classification (ISIC) of All Economic Activities. Because this classification is based on where work is performed (industry) rather than on what type of work is performed (occupation), all of an enterprise's employees are classified under the same industry, regardless of their trade or occupation. The categories should add up to 100 percent. Where they do not, the differences arise because of people who are not classifiable by economic activity.

Data on employment are drawn from labor force surveys, establishment censuses and surveys, administrative records of social insurance schemes, and official national estimates. The concept of employment generally refers to people above a certain age who worked, or who held a job, during a reference period. Employment data include both full-time and part-time workers. There are, however, many differences in how countries define and measure employment status, particularly for part-time workers, students, members of the armed forces, and household or contributing family workers. Where data are obtained from establishment surveys, they cover only employees; thus self-employed and contributing family workers are excluded. In such cases the employment share of the agricultural sector is underreported. Countries also take very different approaches to the treatment of unemployed people. In most countries unemployed people with previous job experience are classified according to their last job. But in some countries the unemployed and people seeking their first job are not classifiable by economic activity. Because of these differences, the size and distribution of employment by economic activity may not be fully comparable across countries (ILO, *Yearbook of Labour Statistics 1996*, p. 64).

The ILO's *Yearbook of Labour Statistics* reports data by major divisions of the ISIC revision 2 or ISIC revision 3. In this table the reported divisions or categories are aggregated into three broad groups: agriculture, industry, and services. An increasing number of countries report economic activity according to the ISIC. Where data are supplied according to national classifications, however, industry definitions and descriptions may differ. In addition, classification into broad groups may obscure fundamental differences in countries' industrial patterns.

The distribution of economic activity by gender reveals some interesting patterns. Agriculture accounts for the largest share of female employment in much of Africa and Asia. Services account for much of the increase in women's labor force participation in North Africa, Latin America and the Caribbean, and high-income economies. Worldwide, women are underrepresented in industry.

Segregating one sex in a narrow range of occupations significantly reduces economic efficiency by reducing labor market flexibility and thus the economy's ability to adapt to change. This segregation is particularly harmful for women, who have a much narrower range of labor market choices and lower levels of pay than men. But it is also detrimental to men when job losses are concentrated in industries dominated by men and job growth is centered in service occupations, where women often dominate, as has been the recent experience in many countries.

There are several explanations for the rising importance of service jobs for women. Many service jobs—such as nursing and social and clerical work—are considered "feminine" because of a perceived similarity to women's traditional roles. Women often do not receive the training needed to take advantage of changing employment opportunities. And the greater availability of part-time work in service industries may lure more women, although it is not clear whether this is a cause or an effect.

Definitions

• **Agriculture** includes hunting, forestry, and fishing, corresponding to division 1 (ISIC revision 2) or tabulation categories A and B (ISIC revision 3). • **Industry** includes mining and quarrying (including oil production), manufacturing, construction, electricity, gas, and water, corresponding to divisions 2–5 (ISIC revision 2) or tabulation categories C–F (ISIC revision 3). • **Services** include wholesale and retail trade and restaurants and hotels; transport, storage, and communications; financing, insurance, real estate, and business services; and community, social, and personal services—corresponding to divisions 6–9 (ISIC revision 2) or tabulation categories G–P (ISIC revision 3).

Data sources

The employment data are from the ILO database Key Indicators of the Labour Market (2000 issue).

Table 2.3a

Share of nonagricultural labor force in self-employment, 1970 and 1990
%

	Total nonagricultural labor force		Female nonagricultural labor force	
	1970	1990	1970	1990
North Africa	12	34	15	26
South America	29	41	28	36
South Asia	33	44	31	35
Eastern Europe	8	8	4	6
Northern and Western Europe	11	10	9	7

Source: United Nations 2000b.

Women's self-employment has increased where the number of self-employed workers has grown as a share of the nonagricultural labor force.

2.4 | Unemployment

	Unemployment						Long-term unemployment			Unemployment by level of educational attainment		
	Male % of male labor force		Female % of female labor force		Total % of total labor force		% of total unemployment			% of total unemployment		
							Male	Female	Total	Primary	Secondary	Tertiary
	1980–82	1996–98[a]	1980–82	1996–98[a]	1980–82	1996–98[a]	1996–98[a]	1996–98[a]	1996–98[a]	1996–98[a]	1996–98[a]	1996–98[a]
Albania	..	..	..	..	5.6	..	..	..	..	..	..	..
Algeria	..	26.9	..	24.0	..	28.7	..	..	..	..	..	..
Angola	..	..	..	..	..	..	..	..	..	..	..	..
Argentina	..	15.4	..	17.6	2.3	16.3	..	..	..	55.7	28.7	4.8
Armenia	..	4.9	..	15.0	..	9.3	..	..	..	..	..	..
Australia	5.1	8.2	7.9	7.7	6.1	8.0	33.1	27.5	30.8	53.3	32.1	11.8
Austria	1.6	4.0	2.3	4.6	1.9	4.2	28.9	28.4	28.7	37.0	59.1	3.9
Azerbaijan	..	0.9	..	1.4	..	1.1	..	..	..	2.1	50.7	47.2
Bangladesh	..	2.7	..	2.3	..	2.5	..	..	..	47.4	28.4	9.9
Belarus	..	..	..	..	..	2.3	..	..	..	8.0	16.8	75.2
Belgium	5.5	7.3	15.0	11.4	9.1	9.1	59.4	61.5	60.5	52.8	35.3	11.9
Benin	..	..	..	..	..	..	..	..	..	..	..	..
Bolivia	..	3.7	..	4.5	..	4.2	..	..	..	24.1	42.3	29.0
Bosnia and Herzegovina	..	..	..	..	..	..	..	..	..	..	..	..
Botswana	..	..	..	..	..	..	..	..	..	..	..	..
Brazil	2.8	6.4	2.8	10.0	2.8	7.8	..	..	..	..	..	..
Bulgaria	..	14.3	..	14.4	..	14.4	59.4	61.7	60.4	..	52.7	7.3
Burkina Faso	..	..	..	..	..	..	..	..	..	..	..	..
Burundi	..	..	..	..	..	..	..	..	..	..	..	..
Cambodia	..	..	..	..	..	..	..	..	..	..	..	..
Cameroon	..	..	..	..	..	..	..	..	..	..	..	..
Canada	6.9	8.5	8.4	8.1	7.5	8.3	14.5	10.2	12.5	33.5	31.2	35.3
Central African Republic	..	..	..	..	..	..	..	..	..	..	..	..
Chad	..	..	..	..	..	..	..	..	..	..	..	..
Chile	10.6	7.0	10.0	7.6	10.4	7.2	..	..	..	28.5	56.2	14.6
China	..	..	..	..	4.9	3.1	..	..	..	..	..	..
Hong Kong, China	3.9	5.1	3.4	4.0	3.8	4.7	..	..	..	..	..	..
Colombia	7.5	12.5	11.5	18.0	9.1	15.0	..	..	..	22.0	57.5	18.9
Congo, Dem. Rep.	..	..	..	..	..	..	..	..	..	..	..	..
Congo, Rep.	..	..	..	..	..	..	..	..	..	..	..	..
Costa Rica	5.3	4.4	7.8	8.0	5.9	5.6	..	..	..	70.7	16.2	9.8
Côte d'Ivoire	..	..	..	..	..	..	..	..	..	..	..	..
Croatia	3.4	11.9	8.2	12.1	5.3	11.4	..	..	..	21.7	69.3	8.3
Cuba	..	..	..	..	..	..	..	..	..	..	..	..
Czech Republic	..	3.8	..	5.8	..	4.7	31.3	29.9	30.5	24.8	71.4	3.7
Denmark	6.5	4.5	7.6	6.6	7.0	5.5	26.3	27.9	27.2	34.6	47.7	16.7
Dominican Republic	..	9.5	..	28.6	..	15.9	..	..	..	50.4	31.1	9.6
Ecuador	..	8.4	..	16.0	..	11.5	..	..	..	..	..	..
Egypt, Arab Rep.	3.9	..	19.2	..	5.2	..	..	..	..	..	..	..
El Salvador	..	9.5	..	5.3	12.9	8.0	..	..	..	53.9	18.8	8.3
Eritrea	..	..	..	..	..	..	..	..	..	..	..	..
Estonia	..	10.4	..	8.6	..	9.6	..	..	..	21.8	54.6	23.4
Ethiopia	..	..	..	..	..	..	..	..	..	..	..	..
Finland	4.7	10.7	4.7	11.9	4.7	11.3	33.9	28.9	31.4	38.9	51.6	9.9
France	4.3	10.2	9.5	13.8	6.4	11.8	39.1	43.3	41.2	..	..	..
Gabon	..	..	..	..	..	..	..	..	..	..	..	..
Gambia, The	..	..	..	..	..	..	..	..	..	..	..	..
Georgia	..	..	..	..	..	..	..	..	..	5.8	33.9	60.3
Germany	..	9.2	..	10.4	..	9.7	44.5	51.7	47.8	22.9	60.4	13.3
Ghana	..	..	..	..	..	..	..	..	..	..	..	..
Greece	3.3	6.6	5.7	15.9	2.4	10.3	45.8	62.2	55.7	35.2	40.5	23.2
Guatemala	..	..	..	..	..	..	..	..	..	..	..	..
Guinea	..	..	..	..	..	..	..	..	..	..	..	..
Guinea-Bissau	..	..	..	..	..	..	..	..	..	..	..	..
Haiti	..	..	..	..	..	..	..	..	..	..	..	..
Honduras	8.6	3.8	6.0	4.2	7.3	3.9	..	..	..	63.2	22.4	5.8

Unemployment | 2.4

	Unemployment						Long-term unemployment			Unemployment by level of educational attainment		
	Male % of male labor force		Female % of female labor force		Total % of total labor force		% of total unemployment			% of total unemployment		
							Male	Female	Total	Primary	Secondary	Tertiary
	1980–82	1996–98[a]	1980–82	1996–98[a]	1980–82	1996–98[a]	1996–98[a]	1996–98[a]	1996–98[a]	1996–98[a]	1996–98[a]	1996–98[a]
Hungary	..	8.5	..	7.0	..	7.8	52.6	49.2	51.3	39.0	57.2	3.7
India	..	..	..	..	..	..	..	..	..	..	..	..
Indonesia	..	3.3	..	5.1	..	5.5	..	..	..	37.4	49.0	8.5
Iran, Islamic Rep.	..	..	..	..	..	..	..	..	..	49.0	24.9	4.0
Iraq	..	..	..	..	..	..	..	..	..	..	..	..
Ireland	..	8.1	..	7.4	..	7.8	63.3	46.9	57.0	64.5	23.9	10.9
Israel	4.1	8.1	6.0	9.2	4.8	8.6	..	..	..	26.2	41.7	31.2
Italy	4.8	9.5	13.1	16.8	7.6	12.3	66.5	66.2	66.3	60.3	31.7	6.4
Jamaica	16.3	9.9	39.5	23.0	27.3	16.0	18.3	29.3	25.6	..	..	..
Japan	2.0	4.2	2.0	4.0	2.0	4.1	28.8	11.8	21.8	23.9	50.0	25.2
Jordan	..	..	..	..	..	..	..	..	..	50.2	14.8	32.4
Kazakhstan	..	..	..	..	..	13.7	..	..	..	..	..	..
Kenya	..	..	..	..	..	..	..	..	..	..	..	..
Korea, Dem. Rep.	..	..	..	..	..	..	..	..	..	..	..	..
Korea, Rep.	6.2	7.7	3.5	5.6	5.2	6.8	3.5	0.9	2.5	28.0	52.4	19.5
Kuwait	..	..	..	..	..	..	..	..	..	..	..	..
Kyrgyz Republic	..	..	..	..	..	..	..	..	..	34.9	55.5	9.6
Lao PDR	..	..	..	..	..	..	..	..	..	..	..	..
Latvia	..	13.5	..	14.1	..	13.8	63.1	63.0	63.0	..	72.2	7.9
Lebanon	..	..	..	..	..	..	..	..	..	..	..	..
Lesotho	..	..	..	..	..	..	..	..	..	..	..	..
Libya	..	..	..	..	..	..	..	..	..	16.7	55.9	27.3
Lithuania	..	14.5	..	12.4	..	13.5	..	..	..	..	..	..
Macedonia, FYR	15.6	35.0	32.8	44.5	22.0	38.8	..	..	..	..	..	..
Madagascar	..	..	..	..	..	..	..	..	..	..	..	..
Malawi	..	..	..	..	..	..	..	..	..	..	..	..
Malaysia	..	..	..	..	..	2.5	..	..	..	..	..	..
Mali	..	..	..	..	..	..	..	..	..	..	..	..
Mauritania	..	..	..	..	..	..	..	..	..	33.6	65.6	..
Mauritius	..	..	..	..	..	..	..	..	..	..	..	..
Mexico	..	2.0	..	2.8	..	2.3	2.1	2.6	2.3	18.4	52.7	18.5
Moldova	..	..	..	..	..	..	..	..	..	..	..	..
Mongolia	..	5.2	..	6.3	..	5.7	..	..	..	47.9	24.1	17.3
Morocco	..	15.8	..	23.0	..	17.8	..	..	..	..	..	..
Mozambique	..	..	..	..	..	..	..	..	..	..	..	..
Myanmar	..	..	..	..	..	..	..	..	..	..	..	..
Namibia	..	..	..	..	..	..	..	..	..	..	..	..
Nepal	..	..	..	..	..	..	..	..	..	..	..	..
Netherlands	4.4	3.5	5.2	5.5	4.6	4.4	49.9	48.5	49.1	30.3	32.9	14.2
New Zealand	..	7.6	..	7.4	..	7.5	22.2	16.1	19.5	18.7	3.9	35.0
Nicaragua	..	8.8	..	14.5	..	13.3	..	..	..	..	..	..
Niger	..	..	..	..	..	..	..	..	..	..	..	..
Nigeria	..	..	..	..	..	..	..	..	..	..	..	..
Norway	1.3	4.0	2.3	4.2	1.7	4.1	13.0	7.7	10.6	28.0	52.0	17.3
Oman	..	..	..	..	..	..	..	..	..	..	..	..
Pakistan	3.0	4.2	7.5	16.8	3.6	6.1	..	..	..	..	..	..
Panama	6.3	10.7	13.3	19.7	8.4	13.9	..	..	..	..	..	..
Papua New Guinea	..	..	..	..	..	..	..	..	..	..	..	..
Paraguay	3.8	7.8	4.8	8.6	4.1	8.2	..	..	..	15.2	55.9	27.9
Peru	..	6.5	..	9.3	..	7.7	..	..	..	..	..	..
Philippines	3.2	9.5	7.5	9.8	4.8	9.6	..	..	..	..	..	..
Poland	..	9.1	..	12.3	..	10.5	33.5	41.9	38.0	24.1	70.1	5.9
Portugal	4.1	3.9	13.0	6.2	7.8	5.0	53.4	57.7	55.6	73.9	14.9	5.8
Puerto Rico	19.5	14.4	12.3	11.8	17.1	13.3	..	..	..	..	..	..
Romania	..	6.5	..	6.1	..	6.3	44.1	50.1	47.0	21.3	72.1	5.9
Russian Federation	..	13.6	..	13.0	..	13.3	29.5	36.8	32.8	17.3	41.9	40.8

	Unemployment						Long-term unemployment			Unemployment by level of educational attainment		
	Male % of male labor force		Female % of female labor force		Total % of total labor force		% of total unemployment			% of total unemployment		
							Male	Female	Total	Primary	Secondary	Tertiary
	1980–82	1996–98[a]	1980–82	1996–98[a]	1980–82	1996–98[a]	1996–98[a]	1996–98[a]	1996–98[a]	1996–98[a]	1996–98[a]	1996–98[a]
Rwanda	..	..	..	..	..	..	..	..	..	..	..	..
Saudi Arabia	..	..	..	..	..	..	..	..	..	..	..	..
Senegal	..	..	..	..	..	..	..	..	..	..	..	..
Sierra Leone	..	..	..	..	..	..	..	..	..	..	..	..
Singapore	2.9	3.2	3.4	3.3	3.0	3.2	..	..	..	27.1	25.8	31.2
Slovak Republic	..	11.4	..	12.6	..	11.9	49.2	52.5	50.3	..	70.4	3.5
Slovenia	..	7.6	..	7.7	..	7.7	61.1	50.0	55.1	30.7	64.0	4.0
South Africa	..	..	..	..	..	5.1	..	..	..	..	..	..
Spain	10.8	13.8	12.8	26.6	11.4	18.8	49.9	60.4	55.5	54.1	18.6	20.6
Sri Lanka	..	7.1	..	16.2	..	10.6	..	..	..	51.8	..	48.0
Sudan	..	..	..	..	..	..	..	..	..	..	..	..
Sweden	1.7	6.9	2.3	6.0	2.0	6.5	31.8	26.9	29.6	30.4	53.3	14.9
Switzerland	0.2	3.2	0.3	4.1	0.2	3.6	25.5	32.8	28.5	..	..	..
Syrian Arab Republic	3.8	..	3.8	..	3.9	..	..	..	..	..	..	..
Tajikistan	..	2.4	..	2.9	..	2.7	..	..	..	10.6	83.2	6.3
Tanzania	..	..	..	..	..	..	..	..	..	..	..	..
Thailand	1.0	3.4	0.7	3.4	0.8	3.4	..	..	..	67.0	12.9	16.8
Togo	..	..	..	..	..	..	..	..	..	..	..	..
Trinidad and Tobago	8.0	11.3	14.0	18.9	10.0	14.2	24.0	39.9	31.7	40.5	57.5	1.3
Tunisia	..	..	..	..	..	..	..	..	..	..	..	..
Turkey	9.0	6.3	23.0	6.1	10.9	6.2	38.1	49.0	41.6	..	..	..
Turkmenistan	..	..	..	..	..	..	..	..	..	..	..	..
Uganda	..	..	..	..	..	..	..	..	..	..	..	..
Ukraine	..	11.9	..	10.8	..	11.3	..	..	..	5.4	28.7	65.9
United Arab Emirates	..	..	..	..	..	..	..	..	..	..	..	..
United Kingdom	8.3	6.8	4.8	5.3	6.8	6.1	44.9	27.8	38.6	26.2	30.5	36.3
United States	6.8	4.4	7.4	4.6	7.0	4.5	9.4	8.0	8.7	22.6	37.9	39.4
Uruguay	..	7.8	..	13.0	..	10.1	..	..	..	..	..	..
Uzbekistan	..	..	..	..	..	..	..	..	..	..	..	..
Venezuela, RB	..	9.8	..	14.2	5.9	11.4	..	..	..	60.0	22.6	13.1
Vietnam	..	..	..	..	..	..	..	..	..	..	..	..
West Bank and Gaza	..	..	..	..	..	..	..	..	..	..	..	..
Yemen, Rep.	..	..	..	..	..	..	..	..	..	..	..	..
Yugoslavia, FR (Serb./Mont.)	..	..	..	..	..	..	..	..	..	..	..	..
Zambia	..	..	..	..	..	..	..	..	..	..	..	..
Zimbabwe	..	..	..	..	..	..	..	..	..	41.1	52.7	0.1

a. Data are for the most recent year available.

Unemployment | 2.4

About the data

Unemployment and total employment in a country are the broadest indicators of economic activity as reflected by the labor market. The International Labour Organization (ILO) defines the unemployed as members of the economically active population who are without work but available for and seeking work, including people who have lost their jobs and those who have voluntarily left work. Some unemployment is unavoidable in all economies. At any time some workers are temporarily unemployed—between jobs as employers look for the right workers and workers search for better jobs. Such unemployment, often called frictional unemployment, results from the normal operation of labor markets.

Changes in unemployment over time may reflect changes in the demand for and supply of labor, but they may also reflect changes in reporting practices. Ironically, low unemployment rates can often disguise substantial poverty in a country, while high unemployment rates can occur in countries with a high level of economic development and low incidence of poverty. In countries without unemployment or welfare benefits, people eke out a living in the informal sector. In countries with well-developed safety nets, workers can afford to wait for suitable or desirable jobs. But high and sustained unemployment indicates serious inefficiencies in the allocation of resources.

The ILO definition of unemployment notwithstanding, reference periods, the criteria for those considered to be seeking work, and the treatment of people temporarily laid off and those seeking work for the first time vary across countries. In many developing countries it is especially difficult to measure employment and unemployment in agriculture. The timing of a survey, for example, can maximize the effects of seasonal unemployment in agriculture. And informal sector employment is difficult to quantify where informal activities are not registered and tracked.

Data on unemployment are drawn from labor force sample surveys and general household sample surveys, social insurance statistics, employment office statistics, and official estimates, which are usually based on information drawn from one or more of the above sources. Labor force surveys generally yield the most comprehensive data because they include groups—particularly people seeking work for the first time—not covered in other unemployment statistics. These surveys generally use a definition of unemployment that follows the international recommendations more closely than that used by other sources and therefore generate statistics that are more comparable internationally.

In contrast, the quality and completeness of data obtained from employment offices and social insurance programs vary widely. Where employment offices work closely with social insurance schemes, and registration with such offices is a prerequisite for receipt of unemployment benefits, the two sets of unemployment estimates tend to be comparable. Where registration is voluntary, and where employment offices function only in more populous areas, employment office statistics do not give a reliable indication of unemployment. Most commonly excluded from both these sources are discouraged workers who have given up their job search because they believe that no employment opportunities exist or do not register as unemployed after their benefits have been exhausted. Thus measured unemployment may be higher in economies that offer more or longer unemployment benefits.

Long-term unemployment is measured in terms of duration, that is, the length of time that an unemployed person has been without work and looking for a job. The underlying assumption is that shorter periods of joblessness are of less concern, especially when the unemployed are covered by unemployment benefits or similar forms of welfare support. The length of time a person has been unemployed is difficult to measure, because the ability to recall the length of that time diminishes as the period of joblessness extends. Women's long-term unemployment is likely to be lower in countries where women constitute a large share of the unpaid family workforce. Such women have more access than men to nonmarket work and are more likely to drop out of the labor force and not be counted as unemployed.

No data are given in the table for economies for which unemployment data are not consistently available or are deemed unreliable.

Definitions

• **Unemployment** refers to the share of the labor force without work but available for and seeking employment. Definitions of labor force and unemployment differ by country (see *About the data*). • **Long-term unemployment** refers to the number of people with continuous periods of unemployment extending for a year or longer, expressed as a percentage of the total unemployed. • **Unemployment by level of educational attainment** shows the unemployed by level of educational attainment, as a percentage of the total unemployed. The levels of educational attainment accord with the United Nations Educational, Cultural, and Scientific Organization's (UNESCO) International Standard Classification of Education.

Data sources

The unemployment data are from the ILO database Key Indicators of the Labour Market (2000 issue).

2.5 | Wages and productivity

	Average hours worked per week		Minimum wage $ per year		Agricultural wage $ per year		Labor cost per worker in manufacturing $ per year		Value added per worker in manufacturing $ per year	
	1980–84	1995–99[a]	1980–84	1995–99[a]	1980–84	1995–99[a]	1980–84	1995–99[a]	1980–84	1995–99[a]
Albania	..	..	..	..	..	..	..	..	..	..
Algeria	..	..	..	1,340	..	..	6,242	..	11,306	..
Angola	..	..	..	..	..	..	..	..	..	..
Argentina	41	40	..	2,400	..	..	6,768	7,338	33,694	37,480
Armenia	..	..	..	..	..	..	..	..	..	..
Australia	37	39	..	12,712	11,212	15,124	14,749	26,087	27,801	57,857
Austria	33	32	..	b	..	..	11,949	28,342	20,956	53,061
Azerbaijan	..	..	..	..	..	..	..	..	..	..
Bangladesh	..	52	..	492	192	360	556	671	1,820	1,711
Belarus	..	..	..	..	1,641	410	2,233	754	..	..
Belgium	..	38	7,661	15,882	6,399	..	12,805	24,132	25,579	58,678
Benin	..	..	..	..	..	..	..	..	..	..
Bolivia	..	46	..	529	..	..	4,432	2,343	21,519	26,282
Bosnia and Herzegovina	..	..	..	..	..	..	..	..	..	..
Botswana	45	..	894	961	650	1,223	3,250	2,884	7,791	..
Brazil	..	..	1,690	1,308	..	..	10,080	14,134	43,232	61,595
Bulgaria	..	..	..	573	..	1,372	2,485	1,179	..	..
Burkina Faso	..	..	695	585	..	..	3,282	..	15,886	..
Burundi	..	..	..	..	..	..	..	..	..	..
Cambodia	..	..	..	..	..	..	..	..	..	..
Cameroon	..	..	..	..	..	..	..	..	..	..
Canada	38	38	4,974	7,897	20,429	30,625	17,710	28,424	36,903	60,712
Central African Republic	..	..	..	..	..	..	..	..	..	..
Chad	..	..	..	..	..	..	..	..	..	..
Chile	43	45	663	1,781	..	..	6,234	5,822	32,805	32,977
China	..	..	..	..	349	325	472	729	3,061	2,885
Hong Kong, China	48	46	..	..	..	..	4,127	13,539	7,886	19,533
Colombia	..	..	..	1,128	..	..	2,988	2,507	15,096	17,061
Congo, Dem. Rep.	..	..	..	..	..	..	..	..	..	..
Congo, Rep.	..	..	..	..	..	..	..	..	..	..
Costa Rica	..	47	1,042	1,638	982	1,697	2,433	2,829	7,185	7,184
Côte d'Ivoire	..	..	1,246	871	..	..	5,132	9,995	16,158	..
Croatia	..	..	..	..	..	..	..	..	..	..
Cuba	..	..	..	..	..	..	..	..	..	..
Czech Republic	43	40	..	..	2,277	1,885	2,306	1,876	5,782	5,094
Denmark	..	37	9,170	19,933	..	..	16,169	29,235	27,919	49,273
Dominican Republic	44	44	..	1,439	..	..	2,191	1,806	8,603	..
Ecuador	..	..	1,637	492	..	..	5,065	3,738	12,197	9,747
Egypt, Arab Rep.	58	..	343	415	..	..	2,210	1,863	3,691	5,976
El Salvador	..	..	..	790	..	..	3,654	..	14,423	..
Eritrea	..	..	..	..	..	..	..	..	..	..
Estonia	..	..	..	..	..	..	..	..	..	..
Ethiopia	..	..	..	..	..	..	..	1,596	..	7,094
Finland	..	38	..	b	..	..	11,522	26,615	25,945	55,037
France	40	39	6,053	12,072	..	..	18,488	..	26,751	61,019
Gabon	..	..	..	..	..	..	..	..	..	..
Gambia, The	..	..	..	..	..	..	..	..	..	..
Georgia	..	..	..	..	..	..	..	..	..	..
Germany	41	40	..	b	..	..	15,708	33,226	34,945	79,616
Ghana	..	..	..	..	1,470	..	2,306	..	12,130	..
Greece	..	41	..	5,246	..	..	6,461	15,899	14,561	30,429
Guatemala	..	..	..	459	..	..	2,605	1,802	11,144	9,235
Guinea	40	..	..	..	..	..	..	..	..	..
Guinea-Bissau	48	..	..	..	..	..	..	..	..	..
Haiti	..	..	..	..	..	..	..	..	..	..
Honduras	..	44	..	..	1,623	..	2,949	2,658	7,458	7,427

Wages and productivity | 2.5

	Average hours worked per week		Minimum wage		Agricultural wage		Labor cost per worker in manufacturing		Value added per worker in manufacturing	
			$ per year		$ per year		$ per year		$ per year	
	1980–84	1995–99[a]	1980–84	1995–99[a]	1980–84	1995–99[a]	1980–84	1995–99[a]	1980–84	1995–99[a]
Hungary	35	33	..	1,132	1,186	1,766	1,410	2,777	4,307	6,106
India	46	..	..	408	205	245	1,035	1,192	2,108	3,118
Indonesia	..	..	..	241	..	..	898	1,008	3,807	5,139
Iran, Islamic Rep.	..	..	..	..	..	..	9,737	30,652	17,679	89,787
Iraq	..	..	..	..	..	..	4,624	13,288	13,599	34,316
Ireland	41	41	5,556	12,087	..	..	10,190	22,681	26,510	86,036
Israel	36	36	..	5,861	4,582	7,906	13,541	26,635	23,459	35,526
Italy	..	32	..	b	..	..	9,955	34,859	24,580	50,760
Jamaica	..	39	782	692	..	..	5,218	3,655	12,056	11,091
Japan	47	47	3,920	12,265	..	..	12,306	31,687	34,456	92,582
Jordan	..	50	b	b	..	..	4,643	2,082	16,337	11,906
Kazakhstan	..	..	..	..	..	..	..	..	..	..
Kenya	41	39	..	..	508	568	104	94	234	228
Korea, Dem. Rep.	..	..	..	..	..	..	..	..	..	..
Korea, Rep.	52	48	..	3,903	..	..	3,153	10,743	11,617	40,916
Kuwait	..	..	..	8,244	..	..	10,281	..	30,341	..
Kyrgyz Republic	..	..	..	65	1,695	168	2,287	687	..	..
Lao PDR	..	..	..	..	..	..	..	..	..	..
Latvia	..	..	..	..	..	..	..	366	..	..
Lebanon	..	..	..	..	..	..	..	..	..	..
Lesotho	..	45	..	..	..	..	1,442	..	6,047	..
Libya	..	..	..	..	..	..	8,648	..	21,119	..
Lithuania	..	..	..	..	..	..	..	..	..	..
Macedonia, FYR	..	..	..	..	..	..	..	..	..	..
Madagascar	..	40	..	..	..	..	1,575	..	3,542	..
Malawi	..	..	..	..	..	..	..	..	..	..
Malaysia	..	..	..	b	..	..	2,519	3,429	8,454	12,661
Mali	..	..	321	459	..	..	2,983	..	10,477	..
Mauritania	..	..	..	..	..	..	..	..	..	..
Mauritius	..	..	..	..	..	..	1,465	1,973	2,969	4,217
Mexico	43	45	1,343	768	1,031	908	3,772	7,607	17,448	25,931
Moldova	..	..	..	..	..	..	..	..	..	..
Mongolia	..	..	..	..	..	..	..	..	..	..
Morocco	..	..	..	1,672	..	..	2,583	3,391	6,328	9,089
Mozambique	..	..	..	..	..	..	..	..	..	..
Myanmar	..	..	..	..	..	..	..	..	..	..
Namibia	..	..	..	..	..	..	..	..	..	..
Nepal	..	..	..	..	..	..	371	..	1,523	..
Netherlands	40	40	9,074	15,170	..	..	18,891	34,326	27,491	56,801
New Zealand	39	39	3,309	9,091	..	..	10,605	23,767	16,835	32,723
Nicaragua	..	44	..	..	..	..	..	..	..	..
Niger	40	..	..	..	..	..	4,074	..	22,477	..
Nigeria	..	..	..	300	..	..	4,812	..	20,000	..
Norway	35	35	..	b	..	..	14,935	38,415	24,905	51,510
Oman	..	..	..	..	..	..	..	3,099	..	61,422
Pakistan	48	..	..	600	..	..	1,264	..	6,214	..
Panama	..	..	..	..	..	..	4,768	6,351	15,327	17,320
Papua New Guinea	44	..	..	..	..	..	4,825	..	13,563	..
Paraguay	36	39	..	..	1,606	1,210	2,509	3,241	..	14,873
Peru	48	..	..	..	..	944	2,988	..	15,962	..
Philippines	47	43	915	1,472	382	..	1,240	2,450	5,266	10,781
Poland	36	33	320	1,584	1,726	1,301	1,682	1,714	6,242	7,637
Portugal	39	40	1,606	4,086	..	..	3,115	7,577	7,161	17,273
Puerto Rico	..	..	..	..	..	..	..	..	..	..
Romania	..	40	..	..	1,669	1,864	1,739	1,190	..	3,482
Russian Federation	..	..	863	297	2,417	659	2,524	1,528	..	..

	Average hours worked per week		Minimum wage		Agricultural wage		Labor cost per worker in manufacturing		Value added per worker in manufacturing	
			$ per year		$ per year		$ per year		$ per year	
	1980–84	1995–99[a]	1980–84	1995–99[a]	1980–84	1995–99[a]	1980–84	1995–99[a]	1980–84	1995–99[a]
Rwanda	..	..	..	..	..	..	1,871	..	9,835	..
Saudi Arabia	..	..	..	..	..	..	9,814	..	..	..
Senegal	..	..	993	848	..	..	2,828	7,754	6,415	..
Sierra Leone	44	..	..	..	..	..	1,624	317	7,807	..
Singapore	46	47	..	..	..	4,856	5,576	21,317	16,442	40,674
Slovak Republic	..	..	..	..	..	..	..	..	..	..
Slovenia	..	..	..	..	..	..	..	9,632	..	12,536
South Africa	42	41	..	b	888	..	6,261	8,475	12,705	16,612
Spain	38	37	3,058	5,778	..	..	8,276	19,329	18,936	47,016
Sri Lanka	50	53	..	..	198	264	447	604	2,057	3,405
Sudan	..	..	..	..	..	..	..	..	..	..
Sweden	36	37	..	..	9,576	27,098	13,038	26,601	32,308	56,675
Switzerland	44	42	..	b	..	..	..	..	..	61,848
Syrian Arab Republic	..	..	..	..	..	..	2,844	4,338	9,607	9,918
Tajikistan	..	..	..	..	..	..	..	..	..	..
Tanzania	..	..	..	..	..	..	1,123	..	3,339	..
Thailand	48	..	..	1,083	..	..	2,305	2,705	11,072	19,946
Togo	..	..	..	..	..	..	..	..	..	..
Trinidad and Tobago	..	40	..	2,974	..	..	..	..	14,008	..
Tunisia	..	..	1,381	1,525	668	968	3,344	3,599	7,111	..
Turkey	..	48	594	1,254	1,015	2,896	3,582	7,958	13,994	32,961
Turkmenistan	..	..	..	..	..	..	..	..	..	..
Uganda	43	..	..	..	..	..	253	..	..	..
Ukraine	..	..	..	..	..	..	..	..	..	..
United Arab Emirates	..	..	..	..	..	..	6,968	..	20,344	..
United Kingdom	42	40	..	b	..	..	11,406	23,843	24,716	55,060
United States	40	41	6,006	8,056	..	..	19,103	28,907	47,276	81,353
Uruguay	48	42	1,262	1,027	1,289	..	4,128	3,738	13,722	16,028
Uzbekistan	..	..	..	..	..	..	..	..	..	..
Venezuela, RB	41	..	1,869	1,463	..	..	11,188	4,667	37,063	24,867
Vietnam	..	47	..	134	..	442	..	711	..	..
West Bank and Gaza	..	..	..	..	..	..	..	..	..	..
Yemen, Rep.	..	..	..	..	..	..	4,492	1,291	17,935	5,782
Yugoslavia, FR (Serb./Mont.)	..	..	..	..	..	..	..	..	..	..
Zambia	..	45	..	..	..	..	3,183	4,292	11,753	16,615
Zimbabwe	..	..	..	..	1,065	..	4,097	3,422	9,625	11,944

a. Figures in italics refer to 1990–94. b. Country has sectoral minimum wages but no minimum wage policy.

Wages and productivity | 2.5

Much of the available data on labor markets are collected through national reporting systems that depend on plant-level surveys. Even when these data are compiled and reported by international agencies such as the International Labour Organization or the United Nations Industrial Development Organization, differences in definitions, coverage, and units of account limit their comparability across countries. The indicators in this table are the result of a research project at the World Bank that has compiled results from more than 300 national and international sources in an effort to provide a set of uniform and representative labor market indicators. Nevertheless, many differences in reporting practices persist, some of which are described below.

Analyses of labor force participation, employment, and underemployment often rely on the number of hours of work per week. The indicator reported in the table is the time spent at the workplace working, preparing for work, or waiting for work to be supplied or for a machine to be fixed. It also includes the time spent at the workplace when no work is being performed but for which payment is made under a guaranteed work contract, or time spent on short periods of rest. Hours paid for but not spent at the place of work—such as paid annual and sick leave, paid holidays, paid meal breaks, and time spent in commuting between home and workplace—are not included. When this information is not available, the table reports the number of hours paid for, comprising the hours actually worked plus the hours paid for but not spent in the workplace. Data on hours worked are influenced by differences in methods of compilation and coverage as well as by national practices relating to the number of days worked and overtime, making comparisons across countries difficult.

Wages refer to remuneration in cash and in kind paid to employees at regular intervals. They exclude employers' contributions to social security and pension schemes as well as other benefits received by employees under these schemes. In some countries the national minimum wage represents a "floor," with higher minimum wages for particular occupations and skills set through collective bargaining. In those countries the agreements reached by employers associations and trade unions are extended by the government to all firms in the sector, or at least to large firms. Changes in the national minimum wage are generally associated with parallel changes in the minimum wages set through collective bargaining.

In many developing countries agricultural workers are hired on a casual or daily basis and lack any social security benefits. International comparisons of agricultural wages are subject to greater reservations than those of wages in other activities. The nature of the work carried out by different categories of agricultural workers and the length of the workday and workweek vary considerably from one country to another. Seasonal fluctuations in agricultural wages are more important in some countries than in others. And the methods followed in different countries for estimating the monetary value of payments in kind are not uniform.

Labor cost per worker in manufacturing is sometimes used as a measure of international competitiveness. The indicator reported in the table is the ratio of total compensation to the number of workers in the manufacturing sector. Compensation includes direct wages, salaries, and other remuneration paid directly by employers plus all contributions by employers to social security programs on behalf of their employees. But there are unavoidable differences in concepts and reference periods and in reporting practices. Remuneration for time not worked, bonuses and gratuities, and housing and family allowances should be considered part of the compensation costs, along with severance and termination pay. These indirect labor costs can vary substantially from country to country, depending on the labor laws and collective bargaining agreements in force.

International competitiveness also depends on productivity, which is often measured by value added per worker in manufacturing. The indicator reported in the table is the ratio of total value added in manufacturing to the number of employees engaged in that sector. Total value added is estimated as the difference between the value of industrial output and the value of materials and supplies for production (including fuel and purchased electricity) and cost of industrial services received.

Observations on labor costs and value added per worker are from plant-level surveys covering relatively large establishments, usually employing 10 or more workers and mostly in the formal sector. In high-income countries the coverage of these surveys tends to be quite good. In developing countries there is often a substantial bias toward very large establishments in the formal sector. As a result, the data may not be strictly comparable across countries. The data are converted into U.S. dollars using the average exchange rate for each year.

The data in the table are period averages and refer to workers of both sexes.

- **Average hours worked per week** refer to all workers (male and female) in nonagricultural activities or, if unavailable, in manufacturing. The data correspond to hours actually worked, to hours paid for, or to statutory hours of work in a normal workweek. • **Minimum wage** corresponds to the most general regime for nonagricultural activities. When rates vary across sectors, only that for manufacturing (or commerce, if the manufacturing wage is unavailable) is reported. • **Agricultural wage** is based on daily wages in agriculture. • **Labor cost per worker in manufacturing** is obtained by dividing the total payroll by the number of employees, or the number of people engaged, in manufacturing establishments.
- **Value added per worker in manufacturing** is obtained by dividing the value added of manufacturing establishments by the number of employees, or the number of people engaged, in those establishments.

The data in the table are drawn from Martin Rama and Raquel Artecona's "Database of Labor Market Indicators across Countries" (1999).

2.6 | Poverty

	National poverty line								International poverty line				
		Population below the poverty line				Population below the poverty line				Population below $1 a day	Poverty gap at $1 a day	Population below $2 a day	Poverty gap at $2 a day
	Survey year	Rural %	Urban %	National %	Survey year	Rural %	Urban %	National %	Survey year	%	%	%	%
Albania		..	..	..		..	..	..		..	..	..	..
Algeria	1988	16.6	7.3	12.2	1995	30.3	14.7	22.6	1995	<2	<0.5	15.1	3.6
Angola		..	..	..		..	..	..		..	..	..	..
Argentina	1991	..	..	25.5	1993	..	..	17.6		..	..	..	..
Armenia		..	..	..		..	..	..	1996	7.8	1.7	34.0	11.3
Australia		..	..	..		..	..	..		..	..	..	..
Austria		..	..	..		..	..	..		..	..	..	..
Azerbaijan	1995	..	..	68.1		..	..	..	1995	<2	<0.5	9.6	2.3
Bangladesh	1991–92	46.0	23.3	42.7	1995–96	39.8	14.3	35.6	1996	29.1	5.9	77.8	31.8
Belarus	1995	..	..	22.5		..	..	..	1998	<2	<0.5	<2	0.1
Belgium		..	..	..		..	..	..		..	..	..	..
Benin	1995	..	..	33.0		..	..	..		..	..	..	..
Bolivia	1993	..	29.3	..	1995	79.1	..	..	1997	29.4	15.2	51.4	27.8
Bosnia and Herzegovina		..	..	..		..	..	..		..	..	..	..
Botswana		..	..	..		..	..	..	1985–86	33.3	12.5	61.4	30.7
Brazil	1996	54.0	15.3	23.9	1998	51.4	13.7	22.0	1997	9.0	2.1	25.4	9.8
Bulgaria		..	..	..		..	..	..	1997	<2	<0.5	21.9	4.2
Burkina Faso		..	..	..		..	..	..	1994	61.2	25.5	85.8	50.9
Burundi	1990	..	..	36.2		..	..	..		..	..	..	..
Cambodia	1993–94	43.1	24.8	39.0	1997	40.1	21.1	36.1		..	..	..	..
Cameroon	1984	32.4	44.4	40.0		..	..	..		..	..	..	..
Canada		..	..	..		..	..	..		..	..	..	..
Central African Republic		..	..	..		..	..	..	1993	66.6	38.1	84.0	58.4
Chad	1995–96	67.0	63.0	64.0		..	..	..		..	..	..	..
Chile	1992	..	..	21.6	1994	..	..	20.5	1996	<2	<0.5	18.4	4.8
China	1996	7.9	<2	6.0	1998	4.6	<2	4.6	1998	18.5	4.2	53.7	21.0
Hong Kong, China		..	..	..		..	..	..		..	..	..	..
Colombia	1991	29.0	7.8	16.9	1992	31.2	8.0	17.7	1996	11.0	3.2	28.7	11.6
Congo, Dem. Rep.		..	..	..		..	..	..		..	..	..	..
Congo, Rep.		..	..	..		..	..	..		..	..	..	..
Costa Rica		..	..	..		..	..	..	1997	6.9	2.0	23.3	8.5
Côte d'Ivoire		..	..	..		..	..	..	1995	12.3	2.4	49.4	16.8
Croatia		..	..	..		..	..	..	1998	<2	<0.5	0.4	0.2
Cuba		..	..	..		..	..	..		..	..	..	..
Czech Republic		..	..	..		..	..	..	1996	<2	<0.5	<2	<0.5
Denmark		..	..	..		..	..	..		..	..	..	..
Dominican Republic	1989	27.4	23.3	24.5	1992	29.8	10.9	20.6	1996	3.2	0.7	16.0	5.0
Ecuador	1994	47.0	25.0	35.0		..	..	..	1995	20.2	5.8	52.3	21.2
Egypt, Arab Rep.	1995–96	23.3	22.5	22.9		..	..	..	1995	3.1	0.3	52.7	13.9
El Salvador	1992	55.7	43.1	48.3		..	..	..	1997	26.0	9.7	54.0	25.3
Eritrea		..	..	..		..	..	..		..	..	..	..
Estonia	1995	14.7	6.8	8.9		..	..	..	1998	<2	<0.5	5.2	0.8
Ethiopia		..	..	..		..	..	..	1995	31.3	8.0	76.4	32.9
Finland		..	..	..		..	..	..		..	..	..	..
France		..	..	..		..	..	..		..	..	..	..
Gabon		..	..	..		..	..	..		..	..	..	..
Gambia, The	1992	..	..	64.0		..	..	..	1992	53.7	23.3	84.0	47.5
Georgia	1997	9.9	12.1	11.1		..	..	..	1996	<2	<0.5	<2	<0.5
Germany		..	..	..		..	..	..		..	..	..	..
Ghana	1992	34.3	26.7	31.4		..	..	..	1998	38.8	3.4	74.6	16.1
Greece		..	..	..		..	..	..		..	..	..	..
Guatemala	1989	71.9	33.7	57.9		..	..	..	1998	10.0	2.2	33.8	11.8
Guinea	1994	..	..	40.0		..	..	..		..	..	..	..
Guinea-Bissau		..	..	..		..	..	..		..	..	..	..
Haiti	1987	..	..	65.0	1995	66.0	..	..		..	..	..	..
Honduras	1992	46.0	56.0	50.0	1993	51.0	57.0	53.0	1996	40.5	17.5	68.8	36.9

Poverty | 2.6

	National poverty line							International poverty line				
		Population below the poverty line				Population below the poverty line				Population below $1 a day	Poverty gap at $1 a day	Population below $2 a day	Poverty gap at $2 a day
	Survey year	Rural %	Urban %	National %	Survey year	Rural %	Urban %	National %	Survey year	%	%	%	%
Hungary	1989	..	..	1.6	1993	..	..	8.6	1998	<2	<0.5	7.3	1.7
India	1992	43.5	33.7	40.9	1994	36.7	30.5	35.0	1997	44.2	12.0	86.2	41.4
Indonesia	1996	..	..	15.7	1999	..	..	27.1	1999	7.7	1.0	55.3	16.5
Iran, Islamic Rep.		..	..	..		..	..	..		..	..	..	..
Iraq		..	..	..		..	..	..		..	..	..	..
Ireland		..	..	..		..	..	..		..	..	..	..
Israel		..	..	..		..	..	..		..	..	..	..
Italy		..	..	..		..	..	..		..	..	..	..
Jamaica	1992	..	..	34.2		..	..	..	1996	3.2	0.7	25.2	6.9
Japan		..	..	..		..	..	..		..	..	..	..
Jordan	1991	..	..	15.0	1997	..	..	11.7	1997	<2	<0.5	7.4	1.4
Kazakhstan	1996	39.0	30.0	34.6		..	..	..	1996	1.5	0.3	15.3	3.9
Kenya	1992	46.4	29.3	42.0		..	..	..	1994	26.5	9.0	62.3	27.5
Korea, Dem. Rep.		..	..	..		..	..	..		..	..	..	..
Korea, Rep.		..	..	..		..	..	..	1993	<2	<0.5	<2	<0.5
Kuwait		..	..	..		..	..	..		..	..	..	..
Kyrgyz Republic	1993	48.1	28.7	40.0	1997	64.5	28.5	51.0		..	..	..	..
Lao PDR	1993	53.0	24.0	46.1		..	..	..	1997	26.3	6.3	73.2	29.6
Latvia		..	..	..		..	..	..	1998	<2	<0.5	8.3	2.0
Lebanon		..	..	..		..	..	..		..	..	..	..
Lesotho	1993	53.9	27.8	49.2		..	..	..	1993	43.1	20.3	65.7	38.1
Libya		..	..	..		..	..	..		..	..	..	..
Lithuania		..	..	..		..	..	..	1996	<2	<0.5	7.8	2.0
Macedonia, FYR		..	..	..		..	..	..		..	..	..	..
Madagascar	1993–94	77.0	47.0	70.0		..	..	..	1997	63.4	26.9	89.0	53.2
Malawi	1990–91	..	..	54.0		..	..	..		..	..	..	..
Malaysia	1989	..	..	15.5		..	..	..		..	..	..	..
Mali		..	..	..		..	..	..	1994	72.8	37.4	90.6	60.5
Mauritania	1989–90	..	..	57.0		..	..	..	1995	28.6	9.1	68.7	29.6
Mauritius	1992	..	..	10.6		..	..	..		..	..	..	..
Mexico	1988	..	..	10.1		..	..	..	1996	12.2	3.5	34.8	13.2
Moldova	1997	26.7	..	23.3		..	..	..	1997	11.3	3.0	38.4	14.0
Mongolia	1995	33.1	38.5	36.3		..	..	..	1995	13.9	3.1	50.0	17.5
Morocco	1990–91	18.0	7.6	13.1	1998–99	27.2	12.0	19.0	1990–91	<2	<0.5	7.5	1.3
Mozambique		..	..	..		..	..	..	1996	37.9	12.0	78.4	36.8
Myanmar		..	..	..		..	..	..		..	..	..	..
Namibia		..	..	..		..	..	..	1993	34.9	14.0	55.8	30.4
Nepal	1995–96	44.0	23.0	42.0		..	..	..	1995	37.7	9.7	82.5	37.5
Netherlands		..	..	..		..	..	..		..	..	..	..
New Zealand		..	..	..		..	..	..		..	..	..	..
Nicaragua	1993	76.1	31.9	50.3		..	..	..		..	..	..	..
Niger	1989–93	66.0	52.0	63.0		..	..	..	1995	61.4	33.9	85.3	54.8
Nigeria	1985	49.5	31.7	43.0	1992–93	36.4	30.4	34.1	1997	70.2	34.9	90.8	59.0
Norway		..	..	..		..	..	..		..	..	..	..
Oman		..	..	..		..	..	..		..	..	..	..
Pakistan	1991	36.9	28.0	34.0		..	..	..	1996	31.0	6.2	84.7	35.0
Panama	1997	64.9	15.3	37.3		..	..	..	1997	10.3	3.2	25.1	10.2
Papua New Guinea		..	..	..		..	..	..		..	..	..	..
Paraguay	1991	28.5	19.7	21.8		..	..	..	1998	19.5	9.8	49.3	26.3
Peru	1994	67.0	46.1	53.5	1997	64.7	40.4	49.0	1996	15.5	5.4	41.4	17.1
Philippines	1994	53.1	28.0	40.6	1997	50.7	21.5	36.8		..	..	..	..
Poland	1993	..	..	23.8		..	..	..	1998	<2	<0.5	<2	<0.5
Portugal		..	..	..		..	..	..	1994	<2	<0.5	0.0	0.0
Puerto Rico		..	..	..		..	..	..		..	..	..	..
Romania	1994	27.9	20.4	21.5		..	..	..	1994	2.8	0.8	27.5	6.9
Russian Federation	1994	..	..	30.9		..	..	..	1998	7.1	1.4	25.1	8.7

	National poverty line								International poverty line				
		Population below the poverty line				Population below the poverty line				Population below $1 a day %	Poverty gap at $1 a day %	Population below $2 a day %	Poverty gap at $2 a day %
	Survey year	Rural %	Urban %	National %	Survey year	Rural %	Urban %	National %	Survey year				
Rwanda	1993	..	..	51.2		..	..	..	1983–85	35.7	7.7	84.6	36.7
Saudi Arabia		..	..	..		..	..	..		..	..	..	..
Senegal						..	..	..	1995	26.3	7.0	67.8	28.2
Sierra Leone	1989	76.0	53.0	68.0		..	..	..	1989	57.0	39.5	74.5	51.8
Singapore		..	..	..		..	..	..		..	..	..	..
Slovak Republic		..	..	..		..	..	..	1992	<2	<0.5	1.7	0.1
Slovenia		..	..	..		..	..	..	1998	<2	<0.5	<2	<0.5
South Africa		..	..	..		..	..	..	1993	11.5	1.8	35.8	13.4
Spain		..	..	..		..	..	..		..	..	..	..
Sri Lanka	1990–91	22.0	15.0	20.0	1995–96	27.0	15.0	25.0	1995	6.6	1.0	45.4	13.5
Sudan		..	..	..		..	..	..		..	..	..	..
Sweden		..	..	..		..	..	..		..	..	..	..
Switzerland		..	..	..		..	..	..		..	..	..	..
Syrian Arab Republic		..	..	..		..	..	..		..	..	..	..
Tajikistan		..	..	..		..	..	..		..	..	..	..
Tanzania	1991	..	..	51.1		..	..	..	1993	19.9	4.8	59.7	23.0
Thailand	1990	..	..	18.0	1992	15.5	10.2	13.1	1998	<2	<0.5	28.2	7.1
Togo	1987–89	..	..	32.3		..	..	..		..	..	..	..
Trinidad and Tobago	1992	20.0	24.0	21.0		..	..	..	1992	12.4	3.5	39.0	14.6
Tunisia	1985	29.2	12.0	19.9	1990	21.6	8.9	14.1	1995	<2	<0.5	10.0	2.3
Turkey		..	..	..		..	..	..	1994	2.4	0.5	18.0	5.0
Turkmenistan		..	..	..		..	..	..	1993	20.9	5.7	59.0	23.3
Uganda	1997–98	16.7	48.7	44.4	1999–2000	10.3	39.1	35.2		..	..	..	..
Ukraine	1995	..	..	31.7		..	..	..	1999	2.9	0.6	45.7	16.3
United Arab Emirates		..	..	..		..	..	..		..	..	..	..
United Kingdom		..	..	..		..	..	..		..	..	..	..
United States		..	..	..		..	..	..		..	..	..	..
Uruguay		..	..	..		..	..	..	1989	<2	<0.5	6.6	1.9
Uzbekistan		..	..	..		..	..	..	1993	3.3	0.5	26.5	7.3
Venezuela, RB	1989	..	..	31.3		..	..	..	1997	18.7	6.5	44.6	19.0
Vietnam	1993	57.2	25.9	50.9		..	..	..		..	..	..	..
West Bank and Gaza		..	..	..		..	..	..		..	..	..	..
Yemen, Rep.	1992	19.2	18.6	19.1		..	..	..	1998	15.7	4.5	45.2	15.0
Yugoslavia, FR (Serb./Mont.)		..	..	..		..	..	..		..	..	..	..
Zambia	1991	88.0	46.0	68.0	1993	..	..	86.0	1998	63.7	32.7	87.4	55.4
Zimbabwe	1990–91	31.0	10.0	25.5		..	..	..	1990–91	36.0	9.6	64.2	29.4

Poverty 2.6

International comparisons of poverty data entail both conceptual and practical problems. Different countries have different definitions of poverty, and consistent comparisons between countries can be difficult. Local poverty lines tend to have higher purchasing power in rich countries, where more generous standards are used than in poor countries. Is it reasonable to treat two people with the same standard of living—in terms of their command over commodities—differently because one happens to live in a better-off country? Can we hold the real value of the poverty line constant across countries, just as we do when making comparisons over time?

Poverty measures based on an international poverty line attempt to do this. The commonly used $1 a day standard, measured in 1985 international prices and adjusted to local currency using purchasing power parities (PPPs), was chosen for the World Bank's *World Development Report 1990: Poverty* because it is typical of the poverty lines in low-income countries. PPP exchange rates, such as those from the Penn World Tables or the World Bank, are used because they take into account the local prices of goods and services not traded internationally. But PPP rates were designed not for making international poverty comparisons but for comparing aggregates from national accounts. As a result, there is no certainty that an international poverty line measures the same degree of need or deprivation across countries.

Past editions of the *World Development Indicators* used PPPs from the Penn World Tables. Because the Penn World Tables updated to 1993 are not yet available, this year's edition (like last year's) uses 1993 consumption PPP estimates produced by the World Bank. The international poverty line, set at $1 a day in 1985 PPP terms, has been recalculated in 1993 PPP terms at about $1.08 a day. Any revisions in the PPP of a country to incorporate better price indexes can produce dramatically different poverty lines in local currency.

Problems also exist in comparing poverty measures within countries. For example, the cost of living is typically higher in urban than in rural areas. (Food staples, for example, tend to be more expensive in urban areas.) So the urban monetary poverty line should be higher than the rural poverty line. But it is not always clear that the difference between urban and rural poverty lines found in practice properly reflects the difference in the cost of living. In some countries the urban poverty line in common use has a higher real value—meaning that it allows the purchase of more commodities for consumption—than does the rural poverty line. Sometimes the difference has been so large as to imply that the incidence of poverty is greater in urban than in rural areas, even though the reverse is found when adjust-

ments are made only for differences in the cost of living. As with international comparisons, when the real value of the poverty line varies, it is not clear how meaningful such urban-rural comparisons are.

The problems of making poverty comparisons do not end there. More issues arise in measuring household living standards. The choice between income and consumption as a welfare indicator is one issue. Income is generally more difficult to measure accurately, and consumption accords better with the idea of the standard of living than does income, which can vary over time even if the standard of living does not. But consumption data are not always available, and when they are not there is little choice but to use income. There are still other problems. Household survey questionnaires can differ widely, for example, in the number of distinct categories of consumer goods they identify. Survey quality varies, and even similar surveys may not be strictly comparable.

Comparisons across countries at different levels of development also pose a potential problem, because of differences in the relative importance of consumption of nonmarket goods. The local market value of all consumption in kind (including consumption from own production, particularly important in underdeveloped rural economies) should be included in the measure of total consumption expenditure. Similarly, the imputed profit from production of nonmarket goods should be included in income. This is not always done, though such omissions were a far bigger problem in surveys before the 1980s. Most survey data now include valuations for consumption or income from own production. Nonetheless, valuation methods vary. For example, some surveys use the price in the nearest market, while others use the average farm gate selling price.

Whenever possible, consumption has been used as the welfare indicator for deciding who is poor. When only household income was available, average income has been adjusted to accord with either a survey-based estimate of mean consumption (when available) or an estimate based on consumption data from national accounts. This procedure adjusts only the mean, however; nothing can be done to correct for the difference in Lorenz (income distribution) curves between consumption and income.

Empirical Lorenz curves were weighted by household size, so they are based on percentiles of population, not households. In all cases the measures of poverty have been calculated from primary data sources (tabulations or household data) rather than existing estimates. Estimation from tabulations requires an interpolation method; the method chosen was Lorenz curves with flexible functional forms, which have proved reliable in past work.

• **Survey year** is the year in which the underlying data were collected. • **Rural poverty rate** is the percentage of the rural population living below the national rural poverty line. • **Urban poverty rate** is the percentage of the urban population living below the national urban poverty line. • **National poverty rate** is the percentage of the population living below the national poverty line. National estimates are based on population-weighted subgroup estimates from household surveys. • **Population below $1 a day** and **population below $2 a day** are the percentages of the population living on less than $1.08 a day and $2.15 a day at 1993 international prices (equivalent to $1 and $2 in 1985 prices, adjusted for purchasing power parity). Poverty rates are comparable across countries, but as a result of revisions in PPP exchange rates, they cannot be compared with poverty rates reported in previous editions for individual countries. • **Poverty gap** is the mean shortfall from the poverty line (counting the nonpoor as having zero shortfall), expressed as a percentage of the poverty line. This measure reflects the depth of poverty as well as its incidence.

The poverty measures are prepared by the World Bank's Development Research Group. The national poverty lines are based on the Bank's country poverty assessments. The international poverty lines are based on nationally representative primary household surveys conducted by national statistical offices or by private agencies under the supervision of government or international agencies and obtained from government statistical offices and World Bank country departments. The World Bank has prepared an annual review of poverty work in the Bank since 1993. The most recent is *Poverty Reduction and the World Bank: Progress in Fiscal 1999*.

2.7 | Social indicators of poverty

	Survey year	Infant mortality rate		Child immunization rate		Prevalence of child malnutrition		Low mother's body mass index		Total fertility rate	
		per 1,000 live births		% of children 12–23 months		% of children under five		% of women		births per woman	
		Poorest quintile	Richest quintile	Poorest quintile	Richest quintile	Poorest quintile	Richest quintile	Poorest quintile	Richest quintile	Poorest quintile	Richest quintile
Bangladesh	1996–97	96	57	47	67	60	28	64.4	32.6	3.8	2.2
Benin	1996	119	63	38	74	37	19	21.0	7.0	7.3	3.8
Bolivia	1998	107	26	22	31	17	3	0.5	2.2	7.4	2.1
Brazil	1996	83	29	57	74	12	3	8.8	5.4	4.8	1.7
Burkina Faso	1992–93	114	80	18	59	36	22	15.7	10.2	7.5	4.6
Cameroon	1991	104	51	27	64	25	6	..	..	6.2	4.8
Central African Republic	1994–95	132	54	18	64	37	20	16.3	11.2	5.1	4.9
Chad	1996–97	80	89	4	23	50	29	27.5	21.0	7.1	6.2
Colombia	1995	41	16	54	74	15	3	5.9	1.2	5.2	1.7
Côte d'Ivoire	1994	117	63	16	64	31[a]	13	11.0	5.7	6.4	3.7
Dominican Republic	1996	67	23	28	52	13	1	8.9	3.0	5.1	2.1
Egypt, Arab Rep.	1995–96	110	32	65	93	17	8	2.9	0.4	4.4	2.7
Ghana	1993	78	46	38	79	33	13	11.3	7.2	6.7	3.4
Guatemala	1995	57	35	41	43	35	7	4.2	2.0	8.0	2.4
Haiti	1994–95	94	74	19	44	39	10	24.9	9.3	7.0	2.3
India	1992–93	109	44	17	65	60	34	..	..	4.1	2.1
Indonesia	1997	78	23	43	72	..	..	..	..	3.3	2.0
Kazakhstan	1995	35	29	19	31	11	3	7.9	3.8	3.2	1.3
Kenya	1998	103	50	46	61	32	10	17.6	6.0	6.6	3.0
Kyrgyz Republic	1997	83	46	69	73	13	8	5.6	3.7	4.6	2.0
Madagascar	1997	119	58	22	66	45	32	24.3	15.1	8.1	3.4
Malawi	1992	141	106	73	89	34	17	14.1	6.0	7.2	6.1
Mali	1995–96	151	93	16	56	47	28	15.9	12.2	6.9	5.1
Morocco	1993	80	35	54	95	17	2	6.2	1.8	6.7	2.3
Mozambique	1997	188	95	20	85	37	14	17.2	4.2	5.2	4.4
Namibia	1992	64	57	54	63	36	13	19.3	5.3	6.9	3.6
Nepal	1996	96	64	32	71	53	28	25.7	21.4	6.2	2.9
Nicaragua	1997–98	51	26	61	73	18	4	4.0	4.1	6.6	1.9
Niger	1998	131	86	5	51	52	37	26.7	12.8	8.4	5.7
Nigeria	1990	102	69	14	58	40	22	..	..	6.6	4.7
Pakistan	1990–91	89	63	23	55	54	26	..	..	5.1	4.0
Paraguay	1990	43	16	20	53	6	1	..	..	7.9	2.7
Peru	1996	78	20	55	66	17	1	1.3	1.1	6.6	1.7
Philippines	1998	49	21	60	87	..	..	..	..	6.5	2.1
Senegal	1997	85	45	..	..	..	..	..	..	7.4	3.6
Tanzania	1996	87	65	57	83	40	18	12.2	7.1	7.8	3.9
Togo	1998	84	66	22	52	32	12	13.3	7.9	7.3	2.9
Turkey	1993	100	25	41	82	22	3	2.7	3.2	3.7	1.5
Uganda	1995	109	63	34	63	31	16	12.7	5.8	7.5	5.4
Uzbekistan	1996	50	47	83	77	25	12	11.4	5.7	4.4	2.1
Vietnam	1997	43	17	42	60	..	..	..	..	3.1	1.6
Zambia	1996	124	70	71	86	32	13	10.2	7.9	7.4	4.4
Zimbabwe	1994	52	42	72	86	19	9	5.7	1.2	6.2	2.8

a. The data contain large sampling errors because of the small number of cases.

Social indicators of poverty | 2.7

About the data

The data in the table describe the health status of individuals in different socioeconomic groups within countries. The data are from Demographic and Health Surveys conducted by Macro International with the support of the U.S. Agency for International Development. These large-scale household sample surveys, conducted periodically in about 50 developing countries, collect information on a large number of health, nutrition, and population measures as well as on respondents' social, demographic, and economic characteristics using a standard set of questionnaires.

In the table socioeconomic status is defined in terms of household assets, including ownership of consumer items, characteristics of the household's dwelling, and other characteristics related to wealth. Each household asset for which information was collected was assigned a weight generated through principal component analysis. The resulting scores were standardized and then used to create break points defining wealth quintiles, expressed as quintiles of individuals.

The choice of the asset index for defining socioeconomic status was based on pragmatic rather than conceptual considerations: Demographic and Health Surveys do not provide income or consumption data but do have detailed information on household ownership of consumer goods and access to a variety of goods and services. Like income or consumption, the asset index defines disparities in primarily economic terms. It therefore excludes other possibilities of disparities among groups, such as those based on gender, education, ethnic background, or other facets of social exclusion. To that extent the index provides only a partial view of the multidimensional concepts of poverty, inequality, and inequity.

The analysis has been carried out for 44 countries, with the results issued in country reports. The table shows the estimates for the poorest and richest quintiles only; the full set of estimates for more than 20 indicators is available in the country reports (see *Data sources*).

Definitions

• **Survey year** is the year in which the underlying data were collected. • **Infant mortality rate** is the number of infants dying before reaching one year of age, per 1,000 live births. The estimates are based on births in the 10 years preceding the survey and may therefore differ from the estimates in table 2.19. • **Child immunization rate** is the percentage of surviving children ages 12–23 months who received all the following vaccinations: one dose of measles and three doses of DPT (diphtheria, pertussis, and tetanus), BCG (Bacillus Camille Guerin), and OPV (oral polio vaccine). • **Prevalence of child malnutrition** is the percentage of children whose weight is more than two standard deviations below the median reference standard for their age as established by the U.S. National Center for Health Statistics, the U.S. Centers for Disease Control and Prevention, and the World Health Organization. The data are based on a sample of children who survived to age three, four, or five years, depending on the country. • **Low mother's body mass index** refers to the percentage of women whose body mass index (BMI) is less than 18.5, a cut-off point indicating acute malnutrition. The BMI is the weight in kilograms divided by the square of the height in meters. • **Total fertility rate** is the number of children that would be born to a woman if she were to live to the end of her childbearing years and bear children in accordance with current age-specific fertility rates. The estimates are based on births during the three years preceding the survey and may therefore differ from those in table 2.17.

Data sources

Data are from an analysis of Demographic and Health Surveys by the World Bank and Macro International. Country reports are available at www.worldbank.org/poverty/health/data/index.htm.

2.8 | Distribution of income or consumption

	Survey year	Gini index	Percentage share of income or consumption						
			Lowest 10%	Lowest 20%	Second 20%	Third 20%	Fourth 20%	Highest 20%	Highest 10%
Albania	..	..	..	..	..	..	..	..	..
Algeria	1995 [a,b]	35.3	2.8	7.0	11.6	16.1	22.7	42.6	26.8
Angola	..	..	..	..	..	..	..	..	..
Argentina	..	..	..	..	..	..	..	..	..
Armenia	1996 [a,b]	44.4	2.3	5.5	9.4	13.9	20.6	50.6	35.2
Australia	1994 [c,d]	35.2	2.0	5.9	12.0	17.2	23.6	41.3	25.4
Austria	1987 [c,d]	23.1	4.4	10.4	14.8	18.5	22.9	33.3	19.3
Azerbaijan	1995 [c,d]	36.0	2.8	6.9	11.5	16.1	22.3	43.3	27.8
Bangladesh	1995–96 [a,b]	33.6	3.9	8.7	12.0	15.7	20.8	42.8	28.6
Belarus	1998 [a,b]	21.7	5.1	11.4	15.2	18.2	21.9	33.3	20.0
Belgium	1992 [c,d]	25.0	3.7	9.5	14.6	18.4	23.0	34.5	20.2
Benin	..	..	..	..	..	..	..	..	..
Bolivia	1997 [c,d]	58.9	0.5	1.9	5.9	11.1	19.3	61.8	45.7
Bosnia and Herzegovina	..	..	..	..	..	..	..	..	..
Botswana	..	..	..	..	..	..	..	..	..
Brazil	1997 [c,d]	59.1	1.0	2.6	5.7	10.3	18.5	63.0	46.7
Bulgaria	1997 [c,d]	26.4	4.5	10.1	13.9	17.4	21.9	36.8	22.8
Burkina Faso	1994 [a,b]	48.2	2.2	5.5	8.7	12.0	18.7	55.0	39.5
Burundi	1992 [a,b]	33.3	3.4	7.9	12.1	16.3	22.1	41.6	26.6
Cambodia	1997 [a,b]	40.4	2.9	6.9	10.7	14.7	20.1	47.6	33.8
Cameroon	..	..	..	..	..	..	..	..	..
Canada	1994 [c,d]	31.5	2.8	7.5	12.9	17.2	23.0	39.3	23.8
Central African Republic	1993 [a,b]	61.3	0.7	2.0	4.9	9.6	18.5	65.0	47.7
Chad	..	..	..	..	..	..	..	..	..
Chile	1996 [c,d]	57.5	1.4	3.4	6.3	10.5	17.9	62.0	46.9
China	1998 [c,d]	40.3	2.4	5.9	10.2	15.1	22.2	46.6	30.4
Hong Kong, China	..	..	..	..	..	..	..	..	..
Colombia	1996 [c,d]	57.1	1.1	3.0	6.6	11.1	18.4	60.9	46.1
Congo, Dem. Rep.	..	..	..	..	..	..	..	..	..
Congo, Rep.	..	..	..	..	..	..	..	..	..
Costa Rica	1997 [c,d]	45.9	1.7	4.5	8.9	14.1	21.6	51.0	34.6
Côte d'Ivoire	1995 [a,b]	36.7	3.1	7.1	11.2	15.6	21.9	44.3	28.8
Croatia	1998 [c,d]	29.0	3.7	8.8	13.3	17.4	22.6	38.0	23.3
Cuba	..	..	..	..	..	..	..	..	..
Czech Republic	1996 [c,d]	25.4	4.3	10.3	14.5	17.7	21.7	35.9	22.4
Denmark	1992 [c,d]	24.7	3.6	9.6	14.9	18.3	22.7	34.5	20.5
Dominican Republic	1998 [c,d]	47.4	2.1	5.1	8.6	13.0	20.0	53.3	37.9
Ecuador	1995 [a,b]	43.7	2.2	5.4	9.4	14.2	21.3	49.7	33.8
Egypt, Arab Rep.	1995 [a,b]	28.9	4.4	9.8	13.2	16.6	21.4	39.0	25.0
El Salvador	1997 [c,d]	50.8	1.4	3.7	7.8	12.8	20.4	55.3	39.3
Eritrea	..	..	..	..	..	..	..	..	..
Estonia	1998 [c,d]	37.6	3.0	7.0	11.0	15.3	21.6	45.1	29.8
Ethiopia	1995 [a,b]	40.0	3.0	7.1	10.9	14.5	19.8	47.7	33.7
Finland	1991 [c,d]	25.6	4.2	10.0	14.2	17.6	22.3	35.8	21.6
France	1995 [c,d]	32.7	2.8	7.2	12.6	17.2	22.8	40.2	25.1
Gabon	..	..	..	..	..	..	..	..	..
Gambia, The	1992 [a,b]	47.8	1.5	4.4	9.0	13.5	20.4	52.8	37.6
Georgia	1996 [c,d]	37.1	2.3	6.1	11.4	16.3	22.7	43.6	27.9
Germany	1994 [c,d]	30.0	3.3	8.2	13.2	17.5	22.7	38.5	23.7
Ghana	1998 [a,b]	39.6	2.4	5.9	10.4	15.3	22.5	45.9	29.5
Greece	1993 [c,d]	32.7	3.0	7.5	12.4	16.9	22.8	40.3	25.3
Guatemala	1998 [c,d]	55.8	1.6	3.8	6.8	10.9	17.9	60.6	46.0
Guinea	1994 [a,b]	40.3	2.6	6.4	10.4	14.8	21.2	47.2	32.0
Guinea-Bissau	1991 [a,b]	56.2	0.5	2.1	6.5	12.0	20.6	58.9	42.4
Guyana	1993 [a,b]	40.2	2.4	6.3	10.7	15.0	21.2	46.9	32.0
Haiti	..	..	..	..	..	..	..	..	..
Honduras	1997 [c,d]	59.0	0.4	1.6	5.6	11.0	20.1	61.8	44.3

Distribution of income or consumption | 2.8

	Survey year	Gini index	Percentage share of income or consumption						
			Lowest 10%	Lowest 20%	Second 20%	Third 20%	Fourth 20%	Highest 20%	Highest 10%
Hungary	1998[a,b]	24.4	4.1	10.0	14.7	18.3	22.7	34.4	20.5
India	1997[a,b]	37.8	3.5	8.1	11.6	15.0	19.3	46.1	33.5
Indonesia	1999[a,b]	31.7	4.0	9.0	12.5	16.1	21.3	41.1	26.7
Iran, Islamic Rep.	..	..	..	..	..	..	..	..	..
Iraq		..	..	..	..	..	..	..	..
Ireland	1987[c,d]	35.9	2.5	6.7	11.6	16.4	22.4	42.9	27.4
Israel	1992[c,d]	35.5	2.8	6.9	11.4	16.3	22.9	42.5	26.9
Italy	1995[c,d]	27.3	3.5	8.7	14.0	18.1	22.9	36.3	21.8
Jamaica	1996[a,b]	36.4	2.9	7.0	11.5	15.8	21.8	43.9	28.9
Japan	1993[c,d]	24.9	4.8	10.6	14.2	17.6	22.0	35.7	21.7
Jordan	1997[a,b]	36.4	3.3	7.6	11.4	15.5	21.1	44.4	29.8
Kazakhstan	1996[a,b]	35.4	2.7	6.7	11.5	16.4	23.1	42.3	26.3
Kenya	1994[a,b]	44.5	1.8	5.0	9.7	14.2	20.9	50.2	34.9
Korea, Dem. Rep.		..	..	..	..	..	..	..	..
Korea, Rep.	1993[a,b]	31.6	2.9	7.5	12.9	17.4	22.9	39.3	24.3
Kuwait		..	..	..	..	..	..	..	..
Kyrgyz Republic	1997[c,d]	40.5	2.7	6.3	10.2	14.7	21.4	47.4	31.7
Lao PDR	1997[a,b]	37.0	3.2	7.6	11.4	15.3	20.8	45.0	30.6
Latvia	1998[c,d]	32.4	2.9	7.6	12.9	17.1	22.1	40.3	25.9
Lebanon		..	..	..	..	..	..	..	..
Lesotho	1986–87[a,b]	56.0	0.9	2.8	6.5	11.2	19.4	60.1	43.4
Libya		..	..	..	..	..	..	..	..
Lithuania	1996[a,b]	32.4	3.1	7.8	12.6	16.8	22.4	40.3	25.6
Luxembourg	1994[c,d]	26.9	4.0	9.4	13.8	17.7	22.6	36.5	22.0
Macedonia, FYR		..	..	..	..	..	..	..	..
Madagascar	1997[a,b]	46.0	2.2	5.4	9.2	13.4	19.9	52.0	37.3
Malawi		..	..	..	..	..	..	..	..
Malaysia	1997[c,d]	49.2	1.7	4.4	8.1	12.9	20.3	54.3	38.4
Mali	1994[a,b]	50.5	1.8	4.6	8.0	11.9	19.3	56.2	40.4
Mauritania	1995[a,b]	37.3	2.5	6.4	11.2	16.0	22.4	44.1	28.4
Mauritius		..	..	..	..	..	..	..	..
Mexico	1996[c,d]	51.9	1.6	4.0	7.6	12.2	19.6	56.7	41.1
Moldova	1997[c,d]	40.6	2.2	5.6	10.2	15.2	22.2	46.8	30.7
Mongolia	1995[a,b]	33.2	2.9	7.3	12.2	16.6	23.0	40.9	24.5
Morocco	1998–99[a,b]	39.5	2.6	6.5	10.6	14.8	21.3	46.6	30.9
Mozambique	1996–97[a,b]	39.6	2.5	6.5	10.8	15.1	21.1	46.5	31.7
Myanmar		..	..	..	..	..	..	..	..
Namibia		..	..	..	..	..	..	..	..
Nepal	1995–96[a,b]	36.7	3.2	7.6	11.5	15.1	21.0	44.8	29.8
Netherlands	1994[c,d]	32.6	2.8	7.3	12.7	17.2	22.8	40.1	25.1
New Zealand		..	..	..	..	..	..	..	..
Nicaragua	1998[a,b]	60.3	0.7	2.3	5.9	10.4	17.9	63.6	48.8
Niger	1995[a,b]	50.5	0.8	2.6	7.1	13.9	23.1	53.3	35.4
Nigeria	1996–97[a,b]	50.6	1.6	4.4	8.2	12.5	19.3	55.7	40.8
Norway	1995[c,d]	25.8	4.1	9.7	14.3	17.9	22.2	35.8	21.8
Oman		..	..	..	..	..	..	..	..
Pakistan	1996–97[a,b]	31.2	4.1	9.5	12.9	16.0	20.5	41.1	27.6
Panama	1997[a,b]	48.5	1.2	3.6	8.1	13.6	21.9	52.8	35.7
Papua New Guinea	1996[a,b]	50.9	1.7	4.5	7.9	11.9	19.2	56.5	40.5
Paraguay	1998[c,d]	57.7	0.5	1.9	6.0	11.4	20.1	60.7	43.8
Peru	1996[c,d]	46.2	1.6	4.4	9.1	14.1	21.3	51.2	35.4
Philippines	1997[a,b]	46.2	2.3	5.4	8.8	13.2	20.3	52.3	36.6
Poland	1998[a,b]	31.6	3.2	7.8	12.8	17.1	22.6	39.7	24.7
Portugal	1994–95[c,d]	35.6	3.1	7.3	11.6	15.9	21.8	43.4	28.4
Puerto Rico		..	..	..	..	..	..	..	..
Romania	1994[c,d]	28.2	3.7	8.9	13.6	17.6	22.6	37.3	22.7
Russian Federation	1998[a,b]	48.7	1.7	4.4	8.6	13.3	20.1	53.7	38.7

	Survey year	Gini index	Percentage share of income or consumption						
			Lowest 10%	Lowest 20%	Second 20%	Third 20%	Fourth 20%	Highest 20%	Highest 10%
Rwanda	1983–85[a,b]	28.9	4.2	9.7	13.2	16.5	21.6	39.1	24.2
Saudi Arabia		..	..	..	..	..	..	..	..
Senegal	1995[a,b]	41.3	2.6	6.4	10.3	14.5	20.6	48.2	33.5
Sierra Leone	1989[a,b]	62.9	0.5	1.1	2.0	9.8	23.7	63.4	43.6
Singapore		..	..	..	..	..	..	..	..
Slovak Republic	1992[c,d]	19.5	5.1	11.9	15.8	18.8	22.2	31.4	18.2
Slovenia	1998[c,d]	28.4	3.9	9.1	13.4	17.3	22.5	37.7	23.0
South Africa	1993–94[a,b]	59.3	1.1	2.9	5.5	9.2	17.7	64.8	45.9
Spain	1990[c,d]	32.5	2.8	7.5	12.6	17.0	22.6	40.3	25.2
Sri Lanka	1995[a,b]	34.4	3.5	8.0	11.8	15.8	21.5	42.8	28.0
St. Lucia	1995[c,d]	42.6	2.0	5.2	9.9	14.8	21.8	48.3	32.5
Sudan		..	..	..	..	..	..	..	..
Swaziland	1994[c,d]	60.9	1.0	2.7	5.8	10.0	17.1	64.4	50.2
Sweden	1992[c,d]	25.0	3.7	9.6	14.5	18.1	23.2	34.5	20.1
Switzerland	1992[c,d]	33.1	2.6	6.9	12.7	17.3	22.9	40.3	25.2
Syrian Arab Republic		..	..	..	..	..	..	..	..
Tajikistan		..	..	..	..	..	..	..	..
Tanzania	1993[a,b]	38.2	2.8	6.8	11.0	15.1	21.6	45.5	30.1
Thailand	1998[a,b]	41.4	2.8	6.4	9.8	14.2	21.2	48.4	32.4
Togo		..	..	..	..	..	..	..	..
Trinidad and Tobago	1992[c,d]	40.3	2.1	5.5	10.3	15.5	22.7	45.9	29.9
Tunisia	1995[a,b]	41.7	2.3	5.7	9.9	14.7	21.8	47.9	31.8
Turkey	1994[a,b]	41.5	2.3	5.8	10.2	14.8	21.6	47.7	32.3
Turkmenistan	1998[a,b]	40.8	2.6	6.1	10.2	14.7	21.5	47.5	31.7
Uganda	1996[a,b]	37.4	3.0	7.1	11.1	15.4	21.5	44.9	29.8
Ukraine	1999[a,b]	29.0	3.7	8.8	13.3	17.4	22.7	37.8	23.2
United Arab Emirates		..	..	..	..	..	..	..	..
United Kingdom	1991[c,d]	36.1	2.6	6.6	11.5	16.3	22.7	43.0	27.3
United States	1997[c,d]	40.8	1.8	5.2	10.5	15.6	22.4	46.4	30.5
Uruguay	1989[c,d]	42.3	2.1	5.4	10.0	14.8	21.5	48.3	32.7
Uzbekistan	1993[c,d]	33.3	3.1	7.4	12.0	16.7	23.0	40.9	25.2
Venezuela, RB	1997[a,b]	48.8	1.6	4.1	8.3	13.2	20.7	53.7	37.6
Vietnam	1998[a,b]	36.1	3.6	8.0	11.4	15.2	20.9	44.5	29.9
West Bank and Gaza		..	..	..	..	..	..	..	..
Yemen, Rep.	1998[a,b]	33.4	3.0	7.4	12.2	16.7	22.5	41.2	25.9
Yugoslavia, FR (Serb./Mont.)		..	..	..	..	..	..	..	..
Zambia	1998[a,b]	52.6	1.1	3.3	7.6	12.5	20.0	56.6	41.0
Zimbabwe	1990–91[a,b]	56.8	1.8	4.0	6.3	10.0	17.4	62.3	46.9

a. Refers to expenditure shares by percentiles of population. b. Ranked by per capita expenditure. c. Refers to income shares by percentiles of population. d. Ranked by per capita income.

Distribution of income or consumption | 2.8

Inequality in the distribution of income is reflected in the percentage shares of either income or consumption accruing to segments of the population ranked by income or consumption levels. The segments ranked lowest by personal income receive the smallest share of total income. The Gini index provides a convenient summary measure of the degree of inequality.

Data on personal or household income or consumption come from nationally representative household surveys. The data in the table refer to different years between 1985 and 1999. Footnotes to the survey year indicate whether the rankings are based on per capita income or consumption. Each distribution is based on percentiles of population—rather than of households—with households ranked by income or expenditure per person.

Where the original data from the household survey were available, they have been used to directly calculate the income (or consumption) shares by quintile. Otherwise, shares have been estimated from the best available grouped data.

The distribution indicators have been adjusted for household size, providing a more consistent measure of per capita income or consumption. No adjustment has been made for spatial differences in cost of living within countries, because the data needed for such calculations are generally unavailable. For further details on the estimation method for low- and middle-income economies see Ravallion and Chen (1996).

Because the underlying household surveys differ in method and in the type of data collected, the distribution indicators are not strictly comparable across countries. These problems are diminishing as survey methods improve and become more standardized, but achieving strict comparability is still impossible (see *About the data* for table 2.6).

Two sources of noncomparability should be noted. First, the surveys can differ in many respects, including whether they use income or consumption expenditure as the living standard indicator. The distribution of income is typically more unequal than the distribution of consumption. In addition, the definitions of income used usually differ among surveys. Consumption is usually a much better welfare indicator, particularly in developing countries. Second, households differ in size (number of members) and in the extent of income sharing among members. And individuals differ in age and consumption needs. Differences among countries in these respects may bias comparisons of distribution.

World Bank staff have made an effort to ensure that the data are as comparable as possible. Whenever possible, consumption has been used rather than income.

The income distribution and Gini indexes for high-income countries are calculated directly from the Luxembourg Income Study database, using an estimation method consistent with that applied for developing countries.

• **Survey year** is the year in which the underlying data were collected. • **Gini index** measures the extent to which the distribution of income (or, in some cases, consumption expenditure) among individuals or households within an economy deviates from a perfectly equal distribution. A Lorenz curve plots the cumulative percentages of total income received against the cumulative number of recipients, starting with the poorest individual or household. The Gini index measures the area between the Lorenz curve and a hypothetical line of absolute equality, expressed as a percentage of the maximum area under the line. Thus a Gini index of zero represents perfect equality, while an index of 100 implies perfect inequality. • **Percentage share of income or consumption** is the share that accrues to subgroups of population indicated by deciles or quintiles. Percentage shares by quintile may not sum to 100 because of rounding.

The data on distribution are compiled by the World Bank's Development Research Group using primary household survey data obtained from government statistical agencies and World Bank country departments. The data for high-income economies are from the Luxembourg Income Study database.

	Urban informal sector employment			Children 10–14 in the labor force		Pension contributors			Local health expenditure	
	% of urban employment			% of age group			% of labor force	% of working-age population		% of total health expenditure
	Male 1993–98[a]	Female 1993–98[a]	Total 1993–98[a]	1980	1999	Year			Period	
Albania	..	..	..	4	0	1995	32.0	31.0	..	..
Algeria	..	..	..	7	0	1997	31.0	23.0	1989–90	63.0
Angola	..	..	..	30	26	..	..	..	..	..
Argentina	..	..	46	8	3	1995	53.0	39.0	..	..
Armenia	..	..	..	0	0	1995	66.6	49.4	..	..
Australia	..	..	..	0	0	..	..	..	1986–88	36.2
Austria	..	..	..	0	0	1993	95.8	76.6	..	..
Azerbaijan	..	..	..	0	0	1996	52.0	46.0	..	..
Bangladesh	10	16	10	35	28	1993	3.5	2.6	1994–97	53.5
Belarus	..	..	..	0	0	1992	97.0	94.0	..	..
Belgium	..	..	..	0	0	1995	86.2	65.9	..	..
Benin	..	..	..	30	27	1996	4.8	..	1991–93	1.3
Bolivia	54	64	59	19	12	1999	14.8	13.3	1994–97	30.0
Bosnia and Herzegovina	..	..	..	1	..	..	..	..	..	..
Botswana	12	28	19	26	15	..	..	..	1994–97	20.3
Brazil	..	..	48	19	15	1996	36.0	31.0	..	..
Bulgaria	..	..	..	0	0	1994	64.0	63.0	1986–88	79.8
Burkina Faso	..	..	..	71	45	1993	3.1	3.0	1989–90	50.0
Burundi	..	..	..	50	49	1993	3.3	3.0	1989–90	74.0
Cambodia	..	..	..	27	24	..	..	..	1991–93	51.0
Cameroon	..	..	57	34	23	1993	13.7	11.5	1980–82	8.0
Canada	..	..	..	0	0	1992	91.9	80.2	..	..
Central African Republic	..	..	..	..	..	..	..	..	1989–90	21.0
Chad	..	..	..	42	37	1990	1.1	1.0	1989–90	24.0
Chile	32	27	30	0	0	1995	70.0	43.0	1983–85	55.9
China	..	..	..	30	9	1994	17.6	17.4	1983–85	40.0
Hong Kong, China	..	..	..	6	0	..	..	..	..	..
Colombia	54	53	53	12	6	1999	35.0	29.3	1989–90	14.4
Congo, Dem. Rep.	..	..	..	33	29	..	..	..	..	..
Congo, Rep.	..	..	..	27	26	1992	5.8	5.6	..	..
Costa Rica	..	..	40	10	4	1998	50.6	38.5	1989–90	1.7
Côte d'Ivoire	37	73	53	28	19	1997	9.3	9.1	1989–90	40.0
Croatia	6	7	6	0	0	1997	66.0	57.0	..	..
Cuba	..	..	..	0	0	..	..	..	1989–90	31.0
Czech Republic	..	..	..	0	0	1995	85.0	67.2	..	..
Denmark	..	..	..	0	0	1993	89.6	88.0	1986–88	27.5
Dominican Republic	..	..	..	25	14	1999	14.4	12.4	1994–97	16.0
Ecuador	39	42	40	9	5	1999	43.1	33.8	..	..
Egypt, Arab Rep.	..	..	..	18	10	1994	50.0	34.2	1994–97	64.0
El Salvador	..	..	..	17	14	1996	26.2	25.0	1994–97	88.0
Eritrea	..	..	..	44	39	..	..	..	..	..
Estonia	..	..	..	0	0	1995	76.0	67.0	..	..
Ethiopia	19	53	33	46	41	..	..	..	1989–90	60.0
Finland	..	..	..	0	0	1993	90.3	83.6	1983–85	25.8
France	..	..	..	0	0	1993	88.4	74.6	1986–88	51.4
Gabon	..	..	..	29	15	1991	7.3	7.0	1994–97	20.2
Gambia, The	66	83	72	44	34	..	..	..	1989–90	51.6
Georgia	..	..	..	0	0	1996	77.0	72.0	..	..
Germany	..	..	..	0	0	1995	94.2	82.3	..	..
Ghana	..	..	79	16	12	1993	7.2	9.0	1994–97	42.7
Greece	..	..	..	5	0	1996	88.0	73.0	1980–82	17.0
Guatemala	..	..	..	19	15	1999	22.8	19.3	1994–97	43.8
Guinea	..	..	..	41	32	1993	1.5	1.8	1994–97	66.0
Guinea-Bissau	..	..	..	43	37	..	..	..	1994–97	26.0
Haiti	..	..	..	33	23	..	..	..	1994–97	9.7
Honduras	..	..	49	14	7	1999	20.6	17.7	1989–90	40.6

Assessing vulnerability 2.9

	Urban informal sector employment			Children 10–14 in the labor force		Pension contributors			Local health expenditure	
	% of urban employment			% of age group				% of working-age population		% of total health expenditure
	Male 1993–98[a]	Female 1993–98[a]	Total 1993–98[a]	1980	1999	Year	% of labor force		Period	
Hungary	..	..	..	0	0	1996	77.0	65.0	1983–85	40.0
India	..	..	44	21	13	1992	10.6	7.9	1991–93	39.0
Indonesia	19	23	21	13	8	1995	8.0	7.0	1994–97	62.5
Iran, Islamic Rep.	3	90	18	14	3	1994	29.8	..	1994–97	37.0
Iraq	..	..	..	11	2	..	..	..	1983–85	40.0
Ireland	..	..	..	1	0	1992	79.3	64.7	..	..
Israel	..	..	..	0	0	1992	82.0	63.0	1986–88	49.0
Italy	..	..	..	2	0	1997	87.0	68.0	1983–85	31.0
Jamaica	26	21	24	0	0	1999	44.4	45.8	1994–97	20.0
Japan	..	..	..	0	..	1994	97.5	92.3	..	..
Jordan	..	..	..	4	0	1995	40.0	25.0	1994–97	9.0
Kazakhstan	..	..	17	0	0	1997	51.0	44.0	..	..
Kenya	..	..	58	45	40	1995	18.0	24.0	1991–93	21.0
Korea, Dem. Rep.	..	..	..	3	0	..	..	..	..	..
Korea, Rep.	..	..	..	0	0	1996	58.0	43.0	1994–97	100.0
Kuwait	..	..	..	0	0	..	..	..	1994–97	21.0
Kyrgyz Republic	..	..	12	0	0	1997	44.0	42.0	..	..
Lao PDR	..	..	..	31	26	..	..	..	1986–88	29.0
Latvia	..	..	9	0	0	1995	60.5	52.3	..	..
Lebanon	..	..	..	5	0	..	..	..	1980–82	4.0
Lesotho	..	..	..	28	21	..	..	..	1994–97	97.6
Libya	..	..	..	9	0	..	..	..	1994–97	37.0
Lithuania	12	5	9	0	0	..	..	..	..	..
Macedonia, FYR	..	..	..	1	0	1995	49.0	47.0	..	..
Madagascar	..	..	58	40	34	1993	5.4	4.8	1991–93	47.7
Malawi	..	..	..	45	32	..	..	..	1994–97	40.0
Malaysia	..	..	..	8	3	1993	48.7	37.8	1991–93	28.8
Mali	..	..	71	61	52	1990	2.5	2.0	1989–90	31.8
Mauritania	..	..	..	30	22	..	..	..	1994–97	32.5
Mauritius	..	..	..	5	2	..	..	..	1991–93	10.0
Mexico	28	26	27	9	5	1997	30.0	31.0	..	..
Moldova	..	..	..	3	0	..	..	..	..	..
Mongolia	..	..	..	4	1	..	..	..	1983–85	40.0
Morocco	..	..	..	21	2	1994	20.9	17.8	1991–93	23.1
Mozambique	..	..	..	39	33	..	..	..	1991–93	41.0
Myanmar	53	57	54	28	23	..	..	..	1986–88	35.0
Namibia	..	..	..	34	18	..	..	..	1991–93	35.0
Nepal	..	..	..	56	43	..	..	..	1994–97	21.4
Netherlands	..	..	..	0	0	1993	91.7	75.4	1989–90	8.3
New Zealand	..	..	..	0	0	..	..	..	1994–97	33.4
Nicaragua	..	..	..	19	12	1999	14.3	13.3	1994–97	37.3
Niger	..	..	..	48	44	1992	1.3	1.5	1986–88	56.0
Nigeria	..	..	..	29	24	1993	1.3	1.3	1988–92	14.8
Norway	..	..	..	0	0	1993	94.0	85.8	1983–85	10.4
Oman	..	..	..	6	0	..	..	..	1994–97	73.0
Pakistan	..	..	..	23	16	1993	3.5	2.1	1994–97	59.0
Panama	..	..	34	6	3	1998	51.6	40.7	1989–90	11.0
Papua New Guinea	..	..	..	28	18	..	..	..	1991–93	51.0
Paraguay	47	46	46	15	6	1997	31.0	29.0	1986–88	4.6
Peru	48	54	51	4	2	1997	20.0	16.0	1983–85	8.1
Philippines	16	19	17	14	6	1996	28.3	13.6	1989–90	72.3
Poland	14	11	13	0	0	1996	68.0	64.0	1986–88	15.2
Portugal	..	..	..	8	1	1996	84.3	80.0	1983–85	50.3
Puerto Rico	..	..	..	0	0	..	..	..	..	..
Romania	..	..	..	0	0	1994	55.0	48.0	..	..
Russian Federation	..	..	..	0	0	..	..	..	..	..

	Urban informal sector employment			Children 10–14 in the labor force		Pension contributors			Local health expenditure	
	% of urban employment			% of age group			% of labor force	% of working-age population		% of total health expenditure
	Male 1993–98[a]	Female 1993–98[a]	Total 1993–98[a]	1980	1999	Year			Period	
Rwanda	..	..	..	43	41	1993	9.3	13.3	1989–90	89.7
Saudi Arabia	..	..	..	5	0	..	..	..	1989–90	24.0
Senegal	..	..	..	43	28	1998	4.3	4.7	1989–90	41.0
Sierra Leone	..	..	..	19	14	..	..	..	1989–90	60.0
Singapore	..	..	..	2	0	1995	73.0	56.0	1991–93	13.7
Slovak Republic	25	11	19	0	0	1996	73.0	72.0	..	..
Slovenia	..	..	..	0	0	1995	86.0	68.7	..	..
South Africa	11	26	17	1	0	..	..	..	1994–97	1.5
Spain	..	..	..	0	0	1994	85.3	61.4	..	..
Sri Lanka	..	..	..	4	2	1992	28.8	20.8	1991–93	51.3
Sudan	..	..	..	33	28	1996	3.9	..	1994–97	50.0
Sweden	..	..	..	0	0	1994	91.1	88.9	1986–88	29.0
Switzerland	..	..	..	0	0	1994	98.1	96.8	..	..
Syrian Arab Republic	..	..	..	14	3	..	..	..	1986–88	47.0
Tajikistan	..	..	..	0	0	..	..	..	..	..
Tanzania	60	85	67	43	37	1996	2.0	2.0	1991–93	37.0
Thailand	75	79	6	25	13	1999	18.0	17.0	1991–93	35.6
Togo	..	..	..	36	27	1997	6.0	3.0	1991–93	49.0
Trinidad and Tobago	..	..	..	1	0	..	..	..	1994–97	34.5
Tunisia	..	..	..	6	0	1991	39.4	27.2	1991–93	30.0
Turkey	..	..	15	21	9	1990	34.6	..	1986–88	26.8
Turkmenistan	..	..	..	0	0	..	..	..	..	..
Uganda	68	81	84	49	44	1994	8.2	..	1994–97	60.0
Ukraine	5	5	5	0	0	1995	69.8	66.1	..	..
United Arab Emirates	..	..	..	0	0	..	..	..	..	..
United Kingdom	..	..	..	0	0	1994	89.7	84.5	1983–85	30.0
United States	..	..	..	0	0	1993	94.0	91.9	1986–88	39.0
Uruguay	33	27	30	4	1	1995	82.0	78.0	..	..
Uzbekistan	..	..	..	0	0	..	..	..	..	..
Venezuela, RB	44	40	42	4	0	1999	23.6	18.2	..	..
Vietnam	..	..	..	22	6	1998	8.4	10.0	1994–97	50.1
West Bank and Gaza	..	..	..	..	..	..	..	..	..	..
Yemen, Rep.	..	..	..	26	19	..	..	..	1994–97	17.0
Yugoslavia, FR (Serb./Mont.)	..	..	..	0	0	..	..	..	..	..
Zambia	..	..	81	19	16	1994	10.2	7.9	1994–97	12.0
Zimbabwe	..	..	..	37	28	..	..	..	1994–97	51.2

World	**20 w**	**12 w**
Low income	24	19
Middle income	21	7
Lower middle income	24	7
Upper middle income	9	6
Low & middle income	23	13
East Asia & Pacific	26	9
Europe & Central Asia	3	1
Latin America & Carib.	13	9
Middle East & N. Africa	14	5
South Asia	23	15
Sub-Saharan Africa	35	29
High income	0	0
Europe EMU	1	0

a. Data are for the most recent year available.

Assessing vulnerability | 2.9

As traditionally defined and measured, poverty is a static concept, and vulnerability a dynamic one. Vulnerability reflects a household's resilience in the face of shocks and the likelihood that a shock will lead to a decline in well-being. It is therefore primarily a function of a household's asset endowment and insurance mechanisms. Because poor people have fewer assets and less diversified sources of income than the better-off, fluctuations in income affect them more.

Poor households face many risks, and vulnerability is thus multidimensional. The indicators in the table focus on individual risks—informal sector employment, child labor, income insecurity in old age—and the extent to which publicly provided services may be capable of mitigating some of these risks. Poor people face labor market risks, often having to take up precarious, low-quality jobs in the informal sector and to increase their household's labor market participation through their children. Income security is a prime concern for the elderly. And affordable access to health care is a primary concern for all poor people, for whom illness and injury have both direct and opportunity costs.

For informal sector employment the most common sources of data are labor force and special informal sector surveys, based on a mixed household and enterprise survey approach or an economic or establishment census approach. Other sources include multipurpose household surveys, household income and expenditure surveys, surveys of household industries or economic activities, small and micro enterprise surveys, and official estimates. The international comparability of the data is affected by differences among countries in definitions and coverage and in the treatment of domestic workers and those who have a secondary job in the informal sector. The data in the table are based on national definitions of urban areas established by countries. For details on country definitions see the notes in the data source.

Reliable estimates of child labor are hard to obtain. In many countries child labor is officially presumed not to exist and so is not included in surveys or in official data. Underreporting also occurs because data exclude children engaged in agricultural or household activities with their families. Most child workers are in Asia. But the share of children working is highest in Africa, where, on average, one in three children ages 10–14 is engaged in some form of economic activity, mostly in agriculture (Fallon and Tzannatos 1998). Available statistics suggest that more boys than girls work. But the number of girls working is often underestimated because surveys exclude those working as unregistered domestic help or doing full-time household work to enable their parents to work outside the home.

Data on pension contributors come from national sources, the International Labour Organization, and International Monetary Fund country reports. Coverage by pension schemes may be broad or even universal where eligibility is determined by citizenship, residency, or income status. In contribution-related schemes, however, eligibility is usually restricted to individuals who have made contributions for a minimum number of years. Definitional issues—relating to the labor force, for example—may arise in comparing coverage by contribution-related schemes over time and across countries (for country-specific information see Palacios and Pallares-Miralles 2000). Coverage may be overstated in countries that do not attempt to count informal sector workers as part of the labor force.

Data on the share of national health expenditure devoted to local primary health care are reported to the World Health Organization by member states, primarily by ministries of health, finance, or regional development. Countries can achieve significant progress in health by providing universal access to affordable primary health care. The share of national health expenditure devoted to local health care represents the effort made to finance essential and accessible health care. The indicator does not take into account primary health care delivered by hospitals or central and regional activities to support and guide local health care. Nor does it indicate the quality or efficiency of health activities and services. Because each country defines local health care in the context of its own system, the data cannot be compared across countries.

• **Urban informal sector employment** is broadly characterized as employment in units in urban areas that produce goods or services on a small scale with the primary objective of generating employment and income for those concerned. These units typically operate at a low level of organization, with little or no division between labor and capital as factors of production. Labor relations are based on casual employment, kinship, or social relationships rather than contractual arrangements. • **Children 10–14 in the labor force** refer to the share of that age group active in the labor force. • **Pension contributors** refer to the share of the labor force or working-age population (here defined as ages 20–59) covered by a pension scheme. • **Local health expenditure** is the share of national health expenditure devoted to local primary health care. The data refer to first-level contact and include community health care, health center care, and dispensary care but not hospital care.

The data on urban informal sector employment are from the International Labour Organization (ILO) database Key Indicators of the Labour Market (2000 issue). The child labor force participation rates are from the ILO database Estimates and Projections of the Economically Active Population, 1950–2010. The data on pension contributors are drawn from Robert Palacios and Montserrat Pallares-Miralles's "International Patterns of Pension Provision" (2000). For updates and further notes and sources go to the World Bank's Web site on pensions (www.worldbank.org/pensions). The data on local health expenditure are from the World Health Organization's statistical information system.

	Public expenditure on pensions				Public expenditure on health			Public expenditure on education	
	Year	% of GDP	Year	Average pension % of per capita income	Year	% of GDP	Per capita PPP $	% of GNI 1994–97[a]	Per student % of GNI per capita 1994–97[a]
Albania	1995	5.1	..	..	1998	3.5	102	3.1	..
Algeria	1997	2.1	1991	75.0	1998	2.6	155	5.1	21.9
Angola	..	..	..	..	1991	3.9	66	..	..
Argentina	1994	6.2	..	..	1998	4.9	614	3.5	15.1
Armenia	1996	3.1	1996	18.7	1998	3.1	70	2.0	11.3
Australia	1995	4.6	1989	37.3	1998	5.9	1,372	5.4	19.8
Austria	1995	14.9	1993	69.3	1999	6.0	1,503	5.4	31.0
Azerbaijan	1996	2.5	1996	51.4	1997	1.2	27	3.0	13.6
Bangladesh	1992	0.0	..	..	1998	1.7	24	2.2	..
Belarus	1997	7.7	1995	31.2	1998	4.9	317	5.9	29.9
Belgium	1995	12.0	..	..	1998	7.9	1,933	3.1	14.5
Benin	1993	0.4	1993	189.7	1998	1.6	14	3.2	..
Bolivia	1995	2.5	..	..	1998	4.1	95	4.9	..
Bosnia and Herzegovina	..	..	..	..	..	..	..	..	..
Botswana	..	..	..	..	1998	2.5	165	8.6	29.0
Brazil	1996	4.9	..	..	1999	2.9	206	5.1	24.7
Bulgaria	1996	7.3	1995	39.3	1998	3.8	190	3.2	18.6
Burkina Faso	1992	0.3	1992	207.3	1999	1.4	14	1.5	..
Burundi	1991	0.2	1991	57.4	1998	0.6	4	4.0	..
Cambodia	..	..	..	..	1998	0.6	8	2.9	..
Cameroon	1993	0.4	..	..	1998	1.0	16	..	..
Canada	1995	5.4	1994	54.3	1999	6.3	1,663	6.9	30.2
Central African Republic	1990	0.3	..	..	1998	2.0	22	..	..
Chad	1997	0.1	..	..	1998	2.3	20	1.7	..
Chile	1993	5.8	1993	56.1	1998	2.7	238	3.6	14.6
China	1996	2.7	..	..	1997	2.0	62	2.3	13.8
Hong Kong, China					1996	2.1	494	2.9	..
Colombia	1994	1.1	1989	72.2	1998	5.2	305	4.1	18.4
Congo, Dem. Rep.	..	..	..	..	..	..	..	..	..
Congo, Rep.	1992	0.9	..	..	1998	2.0	16	6.1	..
Costa Rica	1996	3.8	1993	76.1	1998	5.2	394	5.4	..
Côte d'Ivoire	1997	0.3	..	..	1998	1.2	20	5.0	..
Croatia	1997	11.6	..	..	1997	8.1	555	5.3	26.3
Cuba	1992	12.6	..	..	1994	8.2	..	6.7	39.0
Czech Republic	1996	9.0	1996	37.0	1999	7.0	917	5.1	29.2
Denmark	1996	9.6	1994	46.7	1999	6.7	1,747	8.1	42.9
Dominican Republic	..	..	..	..	1998	1.9	196	2.3	8.4
Ecuador	1997	1.0	..	..	1998	1.7	53	3.5	14.1
Egypt, Arab Rep.	1994	2.5	1994	45.0	1997	1.8	57	4.8	..
El Salvador	1996	1.3	..	..	1998	2.6	107	2.5	9.7
Eritrea	..	..	..	..	1997	2.9	24	1.8	..
Estonia	1995	7.0	1995	56.7	1997	5.5	445	7.2	36.7
Ethiopia	1993	0.9	..	..	1998	1.7	10	4.0	63.1
Finland	1995	12.9	1994	57.4	1998	5.2	1,146	7.5	35.4
France	1995	13.3	..	..	1998	7.3	1,607	6.0	28.9
Gabon	..	..	..	..	1998	2.1	132	2.9	..
Gambia, The	..	..	..	..	1998	1.9	28	4.9	41.5
Georgia	1996	1.7	1996	12.6	1998	0.5	17	5.2	30.8
Germany	1995	12.0	1995	62.8	1999	7.9	1,872	4.8	27.5
Ghana	1993	0.1	..	..	1998	1.8	33	4.2	..
Greece	1993	11.9	1990	85.6	1998	4.7	686	3.1	17.5
Guatemala	1995	0.7	1995	27.6	1998	2.1	73	1.7	8.8
Guinea	..	..	..	..	1998	2.2	41	1.9	16.0
Guinea-Bissau	..	..	..	..	1994	1.1	585	..	..
Haiti	..	..	..	..	1998	1.4	21	..	..
Honduras	1994	0.6	..	..	1998	3.9	96	3.6	..

Enhancing security | 2.10

	Public expenditure on pensions				Public expenditure on health			Public expenditure on education	
	Year	% of GDP	Year	Average pension % of per capita income	Year	% of GDP	Per capita PPP $	% of GNI 1994–97[a]	Per student % of GNI per capita 1994–97[a]
Hungary	1996	9.7	1996	33.6	1998	5.2	562	4.6	25.8
India	..	..	..	..	1997	0.8	12	3.2	16.3
Indonesia	..	..	..	..	1999	0.7	21	1.4	6.0
Iran, Islamic Rep.	1994	1.5	..	..	1998	1.7	93	4.0	14.7
Iraq	..	..	..	..	1990	3.8	..	..	..
Ireland	1996	5.1	1993	77.9	1999	4.5	1,160	6.0	24.3
Israel	1996	5.9	1992	48.1	1998	6.0	1,083	7.6	27.2
Italy	1995	15.0	..	..	1999	5.6	1,245	4.9	30.0
Jamaica	1996	0.3	1989	25.9	1998	3.2	112	7.4	..
Japan	1995	6.6	1989	33.9	1998	5.9	1,444	3.6	19.9
Jordan	1995	4.2	1995	144.0	1998	5.3	139	6.8	25.7
Kazakhstan	1997	5.0	1996	18.8	1998	3.5	161	4.4	..
Kenya	1993	0.5	..	..	1998	2.4	24	6.5	..
Korea, Dem. Rep.	..	..	..	..	..	..	..	..	..
Korea, Rep.	1995	1.4	..	..	1998	2.3	330	3.7	..
Kuwait	1990	3.5	..	..	1997	2.9	697	5.0	21.2
Kyrgyz Republic	1997	6.4	1994	35.0	1998	2.9	70	5.3	31.7
Lao PDR	..	..	..	..	1998	1.2	17	2.1	12.1
Latvia	1995	10.2	1994	47.6	1999	4.3	269	6.3	35.9
Lebanon	..	..	..	..	1998	2.2	129	2.5	..
Lesotho	..	..	..	..	1995	3.4	52	8.4	37.1
Libya	..	..	..	..	..	..	..	..	..
Lithuania	1996	6.2	1995	21.3	1998	4.8	325	5.4	29.5
Macedonia, FYR	1998	8.7	1996	91.6	1998	5.5	244	5.1	..
Madagascar	1990	0.2	..	..	1998	1.1	9	1.9	..
Malawi	..	..	..	..	1998	2.8	16	5.4	17.4
Malaysia	1990	1.0	..	..	1998	1.4	109	4.9	20.7
Mali	1991	0.4	..	..	1998	5.1	15	2.2	25.2
Mauritania	1992	0.2	..	..	1998	1.4	21	5.1	35.8
Mauritius	1999	4.4	..	..	1998	1.8	162	4.6	22.9
Mexico	1996	0.4	..	..	1997	2.8	222	4.9	18.7
Moldova	1996	7.5	..	..	1998	6.4	134	10.6	53.5
Mongolia	..	..	..	..	1995	4.3	64	5.7	27.3
Morocco	1994	1.8	1994	118.0	1998	1.2	43	5.0	28.2
Mozambique	1996	0.0	..	..	1998	2.8	23	..	..
Myanmar	..	..	..	..	1998	0.2	..	1.2	6.5
Namibia	..	..	..	..	1998	4.1	218	9.1	27.1
Nepal	..	..	..	..	1998	1.3	15	3.2	14.5
Netherlands	1996	11.5	1989	48.5	1998	6.0	1,390	5.1	25.5
New Zealand	1995	6.5	..	..	1998	6.2	1,122	7.3	28.9
Nicaragua	1996	4.3	..	..	1998	8.3	180	3.9	15.2
Niger	1992	0.1	..	..	1998	1.2	9	2.3	..
Nigeria	1991	0.1	1991	40.5	1998	0.8	7	0.7	..
Norway	1995	8.9	1994	49.9	1998	7.4	2,043	7.4	38.6
Oman	..	..	..	..	1998	2.9	..	4.5	19.0
Pakistan	1993	0.9	..	..	1998	0.9	17	2.7	..
Panama	1996	4.3	..	..	1998	4.9	277	5.1	21.2
Papua New Guinea	..	..	..	..	1998	2.5	59	..	..
Paraguay	..	..	..	..	1998	1.7	75	4.0	15.9
Peru	1996	1.2	..	..	1998	2.4	108	2.9	..
Philippines	1993	1.0	..	..	1998	1.7	62	3.4	11.4
Poland	1995	14.4	1995	61.2	1999	4.5	374	7.5	32.6
Portugal	1995	9.9	1989	44.6	1998	5.2	793	5.8	24.1
Puerto Rico	..	..	..	..	..	..	..	..	..
Romania	1996	5.1	1994	34.1	1997	2.6	167	3.6	20.3
Russian Federation	1996	5.7	1995	18.3	1997	4.6	328	3.5	..

	Public expenditure on pensions				Public expenditure on health			Public expenditure on education	
	Year	% of GDP	Year	Average pension % of per capita income	Year	% of GDP	Per capita PPP $	% of GNI 1994–97[a]	Per student % of GNI per capita 1994–97[a]
Rwanda	..	..	..	..	1998	2.0	17	..	..
Saudi Arabia	..	..	..	..	1997	6.4	712	7.5	34.0
Senegal	1998	1.5	..	..	1998	2.6	36	3.7	..
Sierra Leone	..	..	..	..	1998	0.9	4	..	..
Singapore	1996	1.4	..	..	1998	1.2	279	3.0	..
Slovak Republic	1994	9.1	1994	44.5	1998	5.7	574	5.0	..
Slovenia	1996	13.6	1996	49.3	1998	6.6	991	5.7	13.7
South Africa	..	..	..	..	1998	3.3	290	7.9	23.0
Spain	1995	10.6	1995	54.1	1998	5.4	924	5.0	23.6
Sri Lanka	1996	2.4	..	..	1998	1.4	44	3.4	12.7
Sudan	..	..	..	..	1997	0.7	10	0.9	10.5
Sweden	1995	11.4	1994	78.0	1998	6.7	1,431	8.3	41.4
Switzerland	1995	12.6	1993	44.4	1998	7.6	2,011	5.4	32.5
Syrian Arab Republic	1991	0.5	..	..	1998	0.8	32	3.1	14.0
Tajikistan	1996	3.0	..	..	1998	5.2	54	2.2	..
Tanzania	..	..	..	..	1998	1.3	6	..	..
Thailand	..	..	..	..	1998	1.9	109	4.8	25.3
Togo	1997	0.6	1993	178.8	1998	1.3	18	4.5	..
Trinidad and Tobago	1996	0.6	..	..	1998	2.5	186	3.6	..
Tunisia	1991	2.6	1991	89.5	1998	2.2	125	7.7	27.5
Turkey	1995	3.7	1993	112.7	1997	2.9	193	2.2	16.4
Turkmenistan	1996	2.3	..	..	1998	4.1	115	..	..
Uganda	1997	0.8	..	..	1998	1.9	20	2.6	..
Ukraine	1996	8.6	1995	30.9	1998	3.6	120	7.3	..
United Arab Emirates	..	..	..	..	1998	0.8	145	1.8	..
United Kingdom	1995	10.2	..	..	1999	5.9	1,290	5.3	22.8
United States	1995	7.2	1989	33.0	1999	5.8	1,849	5.4	23.9
Uruguay	1996	15.0	1996	64.1	1998	1.9	169	3.3	15.5
Uzbekistan	1995	5.3	1995	45.8	1998	3.4	73	7.7	..
Venezuela, RB	1990	0.5	..	..	1998	2.6	153	5.2	..
Vietnam	1998	1.6	..	..	1998	0.8	13	3.0	12.8
West Bank and Gaza	..	..	..	..	1996	4.9	..	..	..
Yemen, Rep.	1994	0.1	..	..	1998	4.8	14	7.0	..
Yugoslavia, FR (Serb./Mont.)	..	..	..	..	..	..	..	..	..
Zambia	1993	0.1	..	..	1998	3.6	27	2.2	..
Zimbabwe	..	..	..	..	1997	2.9	81	..	..
World						2.6 w	317 w	4.6 m	23.3 m
Low income						1.2	20	3.2	16.3
Middle income						2.6	144	4.6	21.2
Lower middle income						2.4	101	4.1	18.5
Upper middle income						3.3	303	5.0	23.0
Low & middle income						2.0	87	4.0	20.5
East Asia & Pacific						1.7	64	3.0	12.8
Europe & Central Asia						4.0	262	5.1	29.3
Latin America & Carib.						3.3	226	3.9	15.4
Middle East & N. Africa						2.5	128	5.0	23.8
South Asia						0.9	15	3.2	14.5
Sub-Saharan Africa						1.8	40	4.0	26.2
High income						6.1	1,582	5.4	26.4
Europe EMU						6.6	1,473	5.3	26.5

a. Data are for the most recent year available.

Enhancing security | 2.10

About the data

Enhancing security for poor people means reducing their vulnerability to such risks as ill health, providing them the means to manage risk themselves, and strengthening market or public institutions for managing risk. The tools include microfinance programs, old age assistance and pensions, and public provision of basic health care and education.

Public interventions and institutions can provide services directly to poor people, although whether these work well for the poor is debated. State action is often ineffective, in part because governments can influence only a few of the many sources of well-being and in part because of difficulties in delivering goods and services. The effectiveness of public provision is further constrained by the fiscal resources at governments' disposal and the fact that state institutions may not be responsive to the needs of poor people.

Data on public pension spending are from national sources and include all government expenditures, including the administrative costs of pension programs. They cover noncontributory pensions or social assistance targeted to the elderly and disabled and spending by social insurance schemes for which contributions had previously been made. The pattern of spending in a country is correlated with its demographic structure—spending increases as the population ages.

The lack of consistent national health accounting systems in most developing countries makes cross-country comparisons of health spending difficult. Compiling estimates of public health expenditures is complicated in countries where state or provincial and local governments are involved in health care financing and delivery because the data on public spending often are not aggregated. The data in the table are the product of an effort to collect all available information on health expenditures from national and local government budgets, national accounts, household surveys, insurance publications, international donors, and existing tabulations.

The data on education spending in the table refer solely to public spending—government spending on public education plus subsidies for private education. The data generally exclude foreign aid for education. They may also exclude spending by religious schools, which play a significant role in many developing countries. Data for some countries and for some years refer to spending by the ministry of education only (excluding education expenditures by other ministries and departments, local authorities, and so on). The share of gross national income (GNI) devoted to education can be interpreted as reflecting a country's effort in education. It often bears a weak relationship to measures of output of the education system, as reflected in educational attainment. The pattern in this relationship suggests wide variations across countries in the efficiency with which the government's resources are translated into education outcomes.

Definitions

• **Public expenditure on pensions** includes all government expenditures on cash transfers to the elderly, the disabled, and survivors and the administrative costs of these programs. • **Average pension** is estimated by dividing total pension expenditure by the number of pensioners. • **Public expenditure on health** consists of recurrent and capital spending from government (central and local) budgets, external borrowings and grants (including donations from international agencies and nongovernmental organizations), and social (or compulsory) health insurance funds. • **Public expenditure on education** consists of public spending on public education plus subsidies to private education at the primary, secondary, and tertiary levels.

Data sources

The data on pension spending are drawn from Robert Palacios and Montserrat Pallares-Miralles's "International Patterns of Pension Provision" (2000). For updates and further notes and sources go to the World Bank's Web site on pensions (www.worldbank.org/pensions). The estimates of health expenditure come from the World Health Organization's *World Health Report 2000* and subsequent updates and from the Organisation for Economic Co-operation and Development for its member countries, supplemented by World Bank country and sector studies, including the Human Development Network's *Sector Strategy: Health, Nutrition, and Population* (World Bank 1997e). Data are also drawn from World Bank public expenditure reviews, the International Monetary Fund's Government Finance Statistics database, and other studies. The data on education expenditure were compiled using an electronic database of the United Nations Educational, Scientific, and Cultural Organization (UNESCO).

Table 2.10a

Access to prescribed medicines in the Russian Federation, 1994–96
% of people

	1994	1995	1996
Able to fill prescriptions	62	70	45
Unable to fill prescriptions	38	30	55
Because medicine was unavailable	76	59	45
Because medicine was too costly	23	32	48
For other reasons (such as lack of time)	2	9	8

Source: UNICEF 1999.

In many countries poverty and economic hardship have reduced people's access to health care services, often because of a decline in public health spending. Data from a large household survey in Russia show that lack of money is becoming the main obstacle to acquiring medicines.

2.11 | Education inputs

	Expenditure per student						Expenditure on teachers' compensation		Primary teachers with required academic qualifications	Primary pupil-teacher ratio
	Primary % of GNI per capita		Secondary % of GNI per capita		Tertiary % of GNI per capita		% of total current education expenditure		%	pupils per teacher
	1980	1997	1980	1997	1980	1997	1980	1997	1992–98[a]	1997
Albania	..	9.7	..	19.9	..	..	..	..	..	18
Algeria	8.9	..	23.9	..	..	..	63.6	74.3 [b]	93	27
Angola	..	..	..	..	..	..	..	..	..	29
Argentina	..	8.3	..	15.0	30.0	19.9	..	84.1	..	17
Armenia	..	..	..	..	..	25.8	..	..	89	19
Australia	..	14.9	44.5	16.8	51.1	29.7	..	54.1 [c]	..	18
Austria	15.7	21.7	..	24.7	37.4	35.3	53.1	61.7	..	12
Azerbaijan	..	20.6	..	..	..	15.3	..	..	100	20
Bangladesh	..	..	..	10.4	..	34.9	33.5	..	68	..
Belarus	..	46.8	..	30.1	..	18.6	..	..	100	19
Belgium	..	8.8	33.8	13.4	51.0	17.5	73.0	73.6 [d]	..	12
Benin	..	12.6	..	..	..	249.0	..	..	100	50
Bolivia	..	..	..	..	..	53.6	..	48.5	64	..
Bosnia and Herzegovina	..	..	..	..	..	..	..	..	84	..
Botswana	12.5	..	..	..	614.9	..	54.9	..	..	28
Brazil	9.6	11.1	11.0	..	58.7	..	..	..	83	24
Bulgaria	17.5	30.7	..	..	51.3	17.4	..	..	99	17
Burkina Faso	..	19.7	87.5	..	2,957.9	590.6	61.0	67.8	100	47
Burundi	..	..	222.2	..	1,479.9	..	74.3	..	..	42
Cambodia	..	..	..	..	..	..	..	..	91	44
Cameroon	..	..	..	..	401.2	..	65.4	..	90	49
Canada	..	..	..	..	38.7	39.8	52.2	57.3	..	16
Central African Republic	..	..	..	..	936.1	..	..	..	..	..
Chad	..	7.3	..	24.0	..	234.5	..	64.4	..	67
Chile	9.6	10.9	16.8	11.8	112.0	20.6	76.8	..	96	30
China	3.8	6.8	..	11.7	246.2	66.3	..	..	95	24
Hong Kong, China	7.7	..	8.2	..	4.2	..	72.9	..	..	..
Colombia	5.3	..	..	12.0	43.8	35.4	93.4	82.0 [e]	90	25
Congo, Dem. Rep.	..	..	..	..	747.9	..	..	..	..	45
Congo, Rep.	..	14.8	..	8.2	369.3	..	70.8	..	100	70
Costa Rica	..	..	25.7	23.2	75.8	..	50.2	..	86	29
Côte d'Ivoire	..	..	..	..	375.7	215.3	..	..	..	41
Croatia	..	..	..	..	..	..	..	..	94	19
Cuba	10.4	16.6	..	34.0	28.5	98.1	38.8	..	100	12
Czech Republic	..	16.4	..	21.5	..	34.9	..	44.4	..	18
Denmark	..	26.5	11.2	34.5	50.0	49.6	49.3	43.1	..	10
Dominican Republic	..	..	..	4.9	..	9.7	62.2	..	..	28
Ecuador	..	..	..	16.1	24.2	37.1	77.4	..	83	25
Egypt, Arab Rep.	..	..	..	..	57.7	..	..	..	100	23
El Salvador	..	7.1	..	5.5	141.7	7.7	..	..	..	33
Eritrea	..	9.2	..	9.9	..	..	..	..	..	44
Estonia	..	..	..	45.4	..	38.4	..	..	..	17
Ethiopia	18.6	31.6	..	71.9	1,118.6	869.0	68.4	..	..	43
Finland	20.4	22.8	..	27.5	37.3	45.6	50.5	47.7	..	18
France	12.0	15.8	20.2	26.8	29.3	28.0	68.1	..	..	19
Gabon	..	..	..	..	..	..	..	..	56	56
Gambia, The	18.7	15.9	..	29.2	..	268.5	..	..	100	30
Georgia	..	..	..	..	..	25.5	..	..	94	18
Germany	..	..	..	..	..	37.8	..	..	..	17
Ghana	..	..	..	..	..	26.7	60.0	..	..	33
Greece	7.0	..	9.5	15.1	30.1	22.3	84.8	..	..	14
Guatemala	4.8	6.2	10.4	5.2	..	31.1	..	62.8	..	35
Guinea	..	..	..	29.4	..	444.8	..	..	89	49
Guinea-Bissau	20.9	..	69.8	..	..	..	73.5	..	..	..
Haiti	6.5	..	..	..	130.0	..	66.9	..	86	35
Honduras	..	..	..	..	77.4	68.7	71.1	67.8	100	35

Education inputs | 2.11

	Expenditure per student						Expenditure on teachers' compensation		Primary teachers with required academic qualifications	Primary pupil-teacher ratio
	Primary % of GNI per capita		Secondary % of GNI per capita		Tertiary % of GNI per capita		% of total current education expenditure		%	pupils per teacher
	1980	1997	1980	1997	1980	1997	1980	1997	1992–98[a]	1997
Hungary	14.0	18.4	..	18.2	85.3	31.4	45.2	..	..	12
India	..	9.2	..	17.7	88.2	99.8	..	..	88	62
Indonesia	..	..	..	..	..	12.9	..	..	94	22
Iran, Islamic Rep.	16.2	..	..	11.0	67.5	7.6	..	47.4 [f]	38	30
Iraq	..	..	6.5	..	87.5	..	..	..	..	20
Ireland	11.6	13.7	24.3	21.9	60.1	36.3	67.6	73.6 [g]	100	22
Israel	16.2	16.3	..	21.8	74.1	37.1	51.2	..	..	14
Italy	..	22.3	..	28.5	..	21.2	..	67.3 [g]	..	11
Jamaica	13.9	13.7	22.0	..	202.8	..	65.6	64.1	100	31
Japan	14.8	18.9	16.6	19.0	21.0	13.9	49.8	..	..	19
Jordan	..	9.1	..	..	60.0	81.2	70.5	70.4	47	21
Kazakhstan	..	..	..	..	..	21.5	..	..	98	18
Kenya	15.4	..	..	..	928.1	..	..	..	..	31
Korea, Dem. Rep.	..	..	..	..	..	..	..	..	100	..
Korea, Rep.	11.7	17.9	9.3	12.9	16.1	6.0	69.2	..	100	31
Kuwait	13.2	21.5	..	5.5	37.4	87.9	46.5	..	100	14
Kyrgyz Republic	..	..	..	40.9	..	49.6	..	..	95	20
Lao PDR	..	5.0	..	12.9	..	63.8	..	67.1	87	30
Latvia	..	..	..	51.1	13.6	32.8	..	40.5	80	13
Lebanon	..	..	..	..	..	22.3	..	..	..	..
Lesotho	8.6	13.8	72.5	53.7	1,013.2	779.3	60.9	57.6	79	46
Libya	..	..	..	..	..	..	..	..	..	..
Lithuania	..	..	..	27.8	..	42.1	..	..	..	16
Macedonia, FYR	..	..	..	..	..	..	..	..	100	..
Madagascar	..	..	..	25.3	402.5	184.5	81.8	..	..	47
Malawi	7.6	8.9	..	27.1	1,833.7	1,593.3	43.4	..	..	59
Malaysia	..	9.8	..	18.4	148.9	57.3	57.5	58.6	..	19
Mali	31.8	15.2	..	29.2	3,631.8	379.3	51.0	..	..	80
Mauritania	30.4	12.0	..	60.4	..	205.9	..	..	..	50
Mauritius	..	10.3	..	15.9	344.1	62.8	31.4	..	100	24
Mexico	4.4	11.9	..	17.9	26.4	46.8	..	..	84	28
Moldova	..	..	..	..	..	63.2	..	..	..	23
Mongolia	..	..	..	..	..	37.8	..	..	97	31
Morocco	..	..	54.9	44.4	155.3	69.5	..	78.0	..	28
Mozambique	..	..	..	..	..	..	..	..	..	58
Myanmar	..	3.2	..	9.1	..	19.0	..	..	..	46
Namibia	..	20.9	..	36.1	..	101.7	..	..	25	..
Nepal	..	9.7	..	12.6	272.0	115.3	59.2	..	96	39
Netherlands	13.8	14.8	23.3	21.2	73.3	47.3	73.5	..	..	14
New Zealand	15.1	17.9	13.7	23.8	59.9	45.7	82.7	..	..	18
Nicaragua	9.5	14.5	10.7	7.6	42.3	..	66.7	..	63	36
Niger	..	29.3	..	74.4	1,538.5	..	..	..	..	41
Nigeria	4.7	..	..	..	529.6	..	..	..	90	34
Norway	..	30.6	14.9	17.8	38.2	46.7	..	..	..	7
Oman	..	13.3	..	22.0	..	26.7	..	..	..	26
Pakistan	..	..	..	..	..	..	..	..	99	40
Panama	..	..	11.5	11.5	29.8	40.2	65.3	51.2	100	..
Papua New Guinea	..	..	..	..	..	..	..	..	100	38
Paraguay	..	10.9	..	12.0	..	90.9	..	..	59	21
Peru	7.2	4.5	11.3	6.8	5.2	15.4	59.4	40.1	74	27
Philippines	..	9.8	4.3	9.4	13.8	14.4	..	..	100	35
Poland	8.1	17.6	14.8	17.1	47.3	27.2	..	..	..	15
Portugal	14.8	19.3	..	21.5	36.3	24.5	..	..	98	12
Puerto Rico	..	..	..	..	..	..	..	..	..	..
Romania	..	20.3	..	8.8	..	31.8	..	..	23	20
Russian Federation	..	..	..	..	..	..	..	..	..	20

	Expenditure per student						Expenditure on teachers' compensation		Primary teachers with required academic qualifications	Primary pupil-teacher ratio
	Primary % of GNI per capita		Secondary % of GNI per capita		Tertiary % of GNI per capita		% of total current education expenditure		%	pupils per teacher
	1980	1997	1980	1997	1980	1997	1980	1997	1992–98ª	1997
Rwanda	11.1	..	..	..	901.8	..	74.8	..	47	..
Saudi Arabia	..	..	..	..	109.2	72.3	..	..	100	13
Senegal	..	..	..	63.6	447.5	..	..	..	99	58
Sierra Leone	..	..	..	..	..	..	..	..	..	..
Singapore	..	..	..	..	40.6	28.0	47.5	..	..	25
Slovak Republic	..	22.3	..	..	..	30.8	..	37.9	79	20
Slovenia	..	20.1	..	7.5	..	37.5	..	62.2	..	14
South Africa	..	..	..	..	..	..	..	64.5 ᵇ	..	45
Spain	..	16.8	..	22.5	..	17.8	..	..	..	15
Sri Lanka	..	..	..	..	65.6	85.0	..	..	100	28
Sudan	..	5.0	..	4.1	527.9	..	..	..	..	29
Sweden	43.0	29.5	15.8	34.1	35.0	72.4	46.4	..	..	12
Switzerland	..	19.3	29.9	29.0	58.5	45.4	61.0	59.9	..	12
Syrian Arab Republic	..	..	15.1	16.6	74.7	..	57.8	..	..	23
Tajikistan	..	..	..	..	..	..	..	..	..	24
Tanzania	..	..	..	..	..	..	..	..	..	37
Thailand	8.8	12.2	..	10.9	60.1	26.2	80.3	56.8 ᶠ	84	..
Togo	8.3	9.9	..	26.7	891.5	495.2	68.3	74.2	..	46
Trinidad and Tobago	..	10.5	20.4	..	59.4	..	73.2	..	100	25
Tunisia	..	..	37.7	24.4	194.6	74.2	81.3	77.0	..	24
Turkey	..	9.0	..	9.2	95.0	51.1	..	..	100	24
Turkmenistan	..	..	..	..	..	..	..	..	..	..
Uganda	4.3	7.8	..	..	1,029.7	..	..	69.9	..	35
Ukraine	21.2	47.3	..	..	20.2	22.7	..	..	..	21
United Arab Emirates	..	..	..	..	..	..	..	30.2	..	16
United Kingdom	..	17.8	22.1	20.5	79.8	40.7	52.1	41.0 ʰ	..	19
United States	..	19.1	..	23.9	48.2	24.7	..	51.7	..	16
Uruguay	11.1	..	..	10.6	28.1	24.2	56.9	41.5	100	20
Uzbekistan	..	..	..	..	..	..	..	..	..	21
Venezuela, RB	5.7	2.1	23.1	4.8	71.1	..	68.8	21.4	..	21
Vietnam	..	7.7	..	8.0	..	77.9	..	66.0	77	33
West Bank and Gaza	..	..	..	..	..	..	..	..	..	..
Yemen, Rep.	..	..	..	..	..	..	..	..	74	30
Yugoslavia, FR (Serb./Mont.)	..	..	..	..	..	..	..	..	..	..
Zambia	10.6	5.0	..	..	605.5	356.2	52.6	..	71	39
Zimbabwe	19.4	19.4	..	34.8	324.5	340.3	75.2	91.1	100	39
World	.. m	.. m	.. m	.. m	60.1 m	.. m	64.5 m	62.0 m	89 m	34 w
Low income	..	..	..	..	..	..	66.7	67.5	88	50
Middle income	..	..	..	..	60.0	..	65.3	58.6	91	25
Lower middle income	..	..	..	..	67.4	..	65.6	64.1	91	24
Upper middle income	..	..	..	..	58.7	38.6	61.4	47.8	87	25
Low & middle income	..	..	..	..	..	..	65.5	64.4	89	36
East Asia & Pacific	..	8.9	..	..	..	41.7	69.2	62.3	94	25
Europe & Central Asia	..	..	..	..	..	..	45.2	40.5	..	20
Latin America & Carib.	..	..	..	9.9	51.3	..	66.7	57.0	84	25
Middle East & N. Africa	..	..	..	..	87.5	..	67.1	74.3	76	26
South Asia	..	..	..	..	88.2	85.0	46.4	..	87	59
Sub-Saharan Africa	..	..	..	..	915.3	..	65.4	67.8	..	40
High income	..	19.3	..	21.9	44.4	37.5	52.6	57.3	..	17
Europe EMU	..	16.8	..	24.7	37.4	35.8	67.9	67.4	..	16

a. Data are for the most recent year available. b. Not including tertiary education. c. Not including preprimary education. d. Flemish community only. e. Ministry of Education only. f. Not including expenditure on universities. g. Data refer to expenditure on public institutions only. h. Not including expenditure on independent private institutions.

Education inputs | 2.11

Data on education are compiled by the United Nations Educational, Scientific, and Cultural Organization (UNESCO) from official responses to surveys and from reports provided by education authorities in each country. Such data are used for monitoring, policymaking, and resource allocation. For a variety of reasons education statistics generally fail to provide a complete and accurate picture of a country's education system and should be interpreted with caution. Statistics often are out of date by two to three years. The information collected focuses more on inputs than on outcomes. And coverage, definitions, and data collection methods vary across countries and over time within countries. (For further discussion of the reliability of education data see Behrman and Rosenzweig 1994.)

The data on education spending in the table refer solely to public spending—government spending on public education plus subsidies for private education. The data generally exclude foreign aid for education. They may also exclude spending by religious schools, which play a significant role in many developing countries. Data for some countries and for some years refer to spending by the ministry of education only (excluding education expenditures by other ministries and departments, local authorities, and so on).

Many developing countries have sought to supplement public funds for education. Some countries have adopted tuition fees to recover part of the cost of providing education services or to encourage development of private schools. Charging fees raises difficult questions relating to equity, efficiency, access, and taxation, however, and some governments have used scholarships, vouchers, and other methods of public finance to counter this criticism. Data for a few countries include private spending, although national practices vary with respect to whether parents or schools pay for books, uniforms, and other supplies. For greater detail see the country- and indicator-specific notes in the source.

Well-trained and motivated teachers are a critical input to education, but they come at a cost: teachers' compensation (gross salaries and other benefits) typically accounts for two-thirds of education spending. Teachers are defined here as including both full- and part-time teaching staff and teachers assigned to nonteaching duties, but country reporting varies. Comparisons should thus be made with caution.

The share of teachers with required academic qualifications measures the quality of the teaching staff available in primary schools. It does not take account of competencies acquired by teachers through their professional experience or self-instruction, or of such factors as work experience, teaching methods and materials, or classroom conditions, all of which may affect the quality

of teaching. The qualifications are specified by the national authorities of each country and may not relate specifically to teaching. Since the indicator is based on minimum national qualifications, which may vary greatly, care should be taken in comparing across countries.

The comparability of pupil-teacher ratios across countries is affected by the definition of teachers, by whether teachers are assigned nonteaching duties, and by differences in class size by grade and in the number of hours taught. Moreover, the underlying enrollment levels are subject to a variety of reporting errors (for further discussion of enrollment data see *About the data* for table 2.12). While the pupil-teacher ratio is often used to compare the quality of schooling across countries, it is often only weakly related to the value added of schooling systems (Behrman and Rosenzweig 1994).

Table 2.11a

Reported deaths of teachers in Zambia and Zimbabwe, selected years, 1996–99

	1996	1998	1999
Zambia	680	1,331	..
Zimbabwe	950	1,250	1,403

Source: World Bank 2000a.

HIV/AIDS will affect the supply of education services in some countries through its effects on mortality rates. In Zambia the mortality rate among teachers in 1996–99 was more than 70 percent higher than the rate among the general adult population. In Zimbabwe it was similar to that for the general adult population.

• **Expenditure per student** is the public current spending on education divided by the total number of students by level, as a percentage of gross national income (GNI) per capita. • **Expenditure on teachers' compensation** is the public expenditure on teachers' gross salaries and other benefits as a percentage of the total public current spending on education. • **Primary teachers with required academic qualifications** refer to the percentage of primary school teachers with at least the minimum academic qualifications required by national public authorities for teaching in primary education. • **Primary pupil-teacher ratio** is the number of pupils enrolled in primary school divided by the number of primary school teachers (regardless of their teaching assignment).

International data on education are compiled by UNESCO's Institute for Statistics in cooperation with national commissions and national statistical services. Data on qualified teachers come from UNESCO's special data collection for the Education for All initiative. The remaining data in the table were compiled using an electronic database maintained by UNESCO.

	Gross enrollment ratio								Net enrollment ratio[a]			
	Preprimary % of relevant age group	Primary % of relevant age group		Secondary % of relevant age group		Tertiary % of relevant age group		Primary % of relevant age group		Secondary % of relevant age group	
	1997	1980	1997	1980	1997	1980	1997	1980	1997	1980	1997
Albania	40	113	107	67	38	5	12	..	102	..	..
Algeria	2	95	108	33	63	6	12	81	94	31	56
Angola	..	175	..	21	..	0 [b]	..	..	..	..	..
Argentina	56	106	111	56	73	22	36	..	104	..	..
Armenia	26	..	87	..	90	..	12	..	..	..	..
Australia	80	112	101	71	153 [c]	25	80	102	95	70	89
Austria	81	99	100	93	103	22	48	87	88	..	88
Azerbaijan	19	115	106	95	77	24	17	..	..	..	..
Bangladesh	..	61	..	18	..	3	..	..	..	..	..
Belarus	82	104	98	98	93	39	44	..	85	..	..
Belgium	118	104	103	91	146 [c]	26	56	97	98	..	88
Benin	3	67	78	16	18	1	3	..	63	..	..
Bolivia	..	87	..	37	..	15	..	79	..	16	..
Bosnia and Herzegovina	..	..	..	..	..	..	..	..	..	..	..
Botswana	..	91	108	19	65	1	6	76	81	14	45
Brazil	59	98	125	34	62	11	15	80	90	14	20
Bulgaria	63	98	99	85	77	16	41	96	92	73	74
Burkina Faso	2	18	40	3	..	0 [b]	1	15	31	..	..
Burundi	..	26	51	3	7	1	..	20	..	..	..
Cambodia	5	139	113	..	24	0 [b]	1	..	100	..	..
Cameroon	10	98	85	18	27	2	..	..	..	15	..
Canada	64	99	102	88	105	57	88	..	95	..	91
Central African Republic	..	71	..	14	..	1	..	56	..	..	..
Chad	1	..	58	..	10	..	1	..	46	..	6
Chile	98	109	101	53	75	12	32	..	89	..	58
China	29	113	123	46	70	2	6	..	102	..	..
Hong Kong, China	83	107	94	64	73	10	..	95	90	61	69
Colombia	33	112	113	39	67	9	17	..	85	..	46
Congo, Dem. Rep.	..	92	72	24	26	1	2	..	61	..	23
Congo, Rep.	..	141	114	74	53	5	..	96	..	..	..
Costa Rica	74	105	104	48	48	21	30	89	89	39	40
Côte d'Ivoire	3	75	71	19	25	3	6	..	55	..	..
Croatia	40	..	87	77	82	19	28	..	82	..	66
Cuba	88	106	106	81	81	17	12	95	101	..	..
Czech Republic	88	96	104	99	99	17	24	..	87	..	87
Denmark	83	96	102	105	121	28	48	96	99	88	88
Dominican Republic	33	118	94	42	54	..	23	..	..	..	22
Ecuador	59	118	127	53	50	35	..	..	97	..	..
Egypt, Arab Rep.	9	73	101	51	78	16	20	..	93	..	68
El Salvador	40	75	97	24	37	9	18	..	78	..	22
Eritrea	4	..	53	..	20	..	1	..	30	..	16
Estonia	68	103	94	127	104	25	42	..	87	..	83
Ethiopia	1	37	43	9	12	0 [b]	1	..	32	..	..
Finland	45	96	99	100	118	32	74	..	98	..	93
France	83	111	105	85	111	25	51	100	100	79	95
Gabon	..	174	162	34	56	..	8	..	..	..	..
Gambia, The	..	53	77	11	25	..	2	50	65	..	..
Georgia	30	93	88	109	77	30	42	..	87	..	74
Germany	89	..	104	..	104	..	47	..	86	..	88
Ghana	..	79	79	41	..	2	..	..	..	..	..
Greece	64	103	93	81	95	17	47	96	90	..	87
Guatemala	35	71	88	19	26	8	9	59	73	13	..
Guinea	5	36	54	17	14	5	1	..	42	..	..
Guinea-Bissau	..	68	62	6	..	..	..	47	..	3	..
Haiti	..	77	..	14	..	1	..	38	..	..	..
Honduras	14	98	111	30	..	8	10	78	..	..	..

Participation in education | 2.12

	Gross enrollment ratio						Net enrollment ratio[a]			
	Preprimary % of relevant age group	Primary % of relevant age group		Secondary % of relevant age group		Tertiary % of relevant age group		Primary % of relevant age group		Secondary % of relevant age group	
	1997	**1980**	**1997**	**1980**	**1997**	**1980**	**1997**	**1980**	**1997**	**1980**	**1997**
Hungary	109	96	103	70	98	14	24	95	97	..	86
India	5	83	100	30	49	5	7	..	..	..	..
Indonesia	19	107	113	29	56	4	11	88	95	..	42
Iran, Islamic Rep.	11	87	98	42	77	..	18	..	90	..	71
Iraq	7	113	85	57	42	9	..	99	76	47	..
Ireland	114	100	105	90	118	18	41	90	92	78	86
Israel	71	95	98	73	88	29	41	..	..	..	..
Italy	95	100	101	72	95	27	47	..	100	..	..
Jamaica	..	103	100	67	..	7	8	96	..	64	..
Japan	50	101	101	93	103	31	41	101	103	93	99
Jordan	19	82	71	59	57	13	18	73	68	53	41
Kazakhstan	29	85	98	93	87	34	33	..	..	..	..
Kenya	35	115	85	20	24	1	..	91	..	..	..
Korea, Dem. Rep.	..	..	..	..	..	..	..	..	..	..	..
Korea, Rep.	88	110	94	78	102	15	68	104	93	70	97
Kuwait	63	102	77	80	65	11	19	85	62	..	61
Kyrgyz Republic	7	116	104	110	79	16	12	..	95	..	..
Lao PDR	8	114	112	21	29	0 [b]	3	..	72	..	22
Latvia	47	102	96	99	84	24	33	..	90	..	79
Lebanon	75	111	111	59	81	30	27	..	76	..	..
Lesotho	..	104	108	18	31	1	2	67	70	13	18
Libya	..	125	..	76	..	8	..	..	..	62	..
Lithuania	40	79	98	114	86	35	31	..	..	..	81
Macedonia, FYR	26	100	99	61	63	28	20	..	95	..	56
Madagascar	5	130	92	..	16	3	2	..	61	..	..
Malawi	..	60	134	5	17	1	1	43	103	..	..
Malaysia	42	93	101	48	64	4	12	..	102	..	..
Mali	2	26	49	8	13	1	1	20	31	..	..
Mauritania	..	37	79	11	16	..	4	..	57	..	..
Mauritius	104	93	106	50	65	1	6	79	98	..	..
Mexico	73	120	114	49	64	14	16	..	101	..	51
Moldova	45	83	97	78	81	30	27	..	..	..	..
Mongolia	25	107	88	92	56	22	17	..	81	..	53
Morocco	68	83	86	26	39	6	11	62	74	20	..
Mozambique	..	99	60	5	7	0 [b]	1	36	40	..	6
Myanmar	..	91	121	22	30	5	5	..	..	..	..
Namibia	11	..	131	..	62	..	8	..	91	..	36
Nepal	..	86	113	22	42	3	5	..	..	..	..
Netherlands	100	100	108	93	132 [c]	29	47	93	100	81	91
New Zealand	76	111	101	83	113	27	63	..	100	81	90
Nicaragua	23	94	102	41	55	12	12	70	77	22	..
Niger	1	25	29	5	7	0 [b]	..	21	25	4	6
Nigeria	..	109	98	18	33	3	..	..	..	..	..
Norway	103	100	100	94	119	26	62	98	100	84	97
Oman	5	51	76	12	67	0 [b]	8	43	69	10	..
Pakistan	..	40	..	14	..	..	..	..	..	..	..
Panama	76	107	106	61	69	21	32	89	..	46	..
Papua New Guinea	1	59	80	12	14	2	3	..	..	..	..
Paraguay	61	106	111	27	47	9	10	89	91	..	38
Peru	40	114	123	59	73	17	26	86	91	..	55
Philippines	11	112	117	64	78	24	29	94	101	45	59
Poland	46	100	96	77	98	18	25	98	95	71	85
Portugal	61	123	128	37	111 [c]	11	39	99	..	..	78
Puerto Rico	..	..	..	..	..	42	..	..	..	..	..
Romania	53	104	104	94	78	12	23	..	95	..	73
Russian Federation	..	102	107	96	..	46	43	..	93	..	..

	Gross enrollment ratio							Net enrollment ratioᵃ			
	Preprimary % of relevant age group	Primary % of relevant age group		Secondary % of relevant age group		Tertiary % of relevant age group		Primary % of relevant age group		Secondary % of relevant age group	
	1997	1980	1997	1980	1997	1980	1997	1980	1997	1980	1997
Rwanda	..	63	..	3	..	0 ᵇ	..	59	..	..	..
Saudi Arabia	8	61	76	30	61	7	16	49	61	21	43
Senegal	2	46	71	11	16	3	3	37	60	..	..
Sierra Leone	..	52	..	14	..	1	..	..	..	..	..
Singapore	..	108	94	60	74	8	39	99	93	..	..
Slovak Republic	76	..	102	..	94	..	22	..	..	..	..
Slovenia	61	98	98	..	92	20	36	..	95	..	..
South Africa	35	90	133	..	95	..	19	..	..	..	58
Spain	74	109	107	87	120	23	51	102	105	74	..
Sri Lanka	..	103	109	55	75	3	5	..	..	..	..
Sudan	23	50	51	16	21	2	..	..	..	..	..
Sweden	73	97	107	88	140 ᶜ	31	50	..	102	..	99
Switzerland	95	84	97	94	100	18	33	79	90	78	84
Syrian Arab Republic	7	100	101	46	43	17	16	90	91	39	38
Tajikistan	10	..	95	..	78	24	20	..	..	..	..
Tanzania	..	93	67	3	6	0 ᵇ	1	68	48	..	..
Thailand	75	99	89	29	59	15	22	..	..	..	..
Togo	3	118	120	33	27	2	4	..	81	..	..
Trinidad and Tobago	..	99	99	69	74	4	8	90	88	..	..
Tunisia	11	102	118	27	64	5	14	82	98	23	..
Turkey	8	96	107	35	58	5	21	..	99	..	51
Turkmenistan	..	..	..	..	..	23	..	..	..	..	..
Uganda	..	50	74	5	12	1	2	..	..	..	..
Ukraine	..	102	..	94	..	42	42	..	..	..	..
United Arab Emirates	57	89	89	52	80	3	12	74	78	..	71
United Kingdom	30	103	116	84	129 ᶜ	19	52	97	99	79	92
United States	70	99	102	91	97	56	81	..	95	..	90
Uruguay	45	107	109	62	85	17	30	..	93	..	..
Uzbekistan	55	81	78	106	94	29	..	..	..	..	..
Venezuela, RB	44	93	91	21	40	21	..	82	84	14	22
Vietnam	40	109	114	42	57	2	7	95	..	..	..
West Bank and Gaza	..	..	..	..	..	..	..	..	..	..	..
Yemen, Rep.	..	..	70	..	34	..	4	..	..	..	..
Yugoslavia, FR (Serb./Mont.)	31	..	69	..	62	..	22	..	..	..	..
Zambia	..	90	89	16	27	2	3	77	75	..	16
Zimbabwe	..	85	112	8	50	1	7	..	..	..	..
World	**31 w**	**97 w**	**106 w**	**49 w**	**64 w**	**13 w**	**14 w**	**.. w**	**.. w**	**.. w**	**.. w**
Low income	..	83	97	29	46	6	8	..	..	..	..
Middle income	36	106	119	52	69	10	12	..	98	..	..
Lower middle income	31	107	120	52	70	9	9	..	99	..	..
Upper middle income	60	102	109	51	66	14	22	..	94	..	43
Low & middle income	25	96	107	42	59	8	10	..	..	..	..
East Asia & Pacific	33	111	119	44	69	4	8	..	100	..	..
Europe & Central Asia	..	99	100	86	..	31	32	..	92	..	..
Latin America & Carib.	57	105	113	42	60	14	17	..	91	..	33
Middle East & N. Africa	17	87	95	42	64	11	16	..	87	..	62
South Asia	5	77	100	27	49	5	7	..	..	..	..
Sub-Saharan Africa	..	81	78	15	27	1	..	..	..	..	..
High income	70	102	103	87	106	36	62	..	95	..	90
Europe EMU	87	106	104	81	108	25	49	..	94	..	91

a. Net enrollment ratios exceeding 100 percent indicate discrepancies between estimates of the school-age population and reported enrollment data. b. Less than 0.5. c. Includes training for the unemployed.

About the data

School enrollment data are reported to the United Nations Educational, Scientific, and Cultural Organization (UNESCO) by national education authorities. Enrollment ratios help to monitor two important issues for universal primary education, an international development goal that implies achieving a net primary enrollment ratio of 100 percent. Gross enrollment ratios help to assess whether an education system has sufficient capacity to meet the needs of universal primary education. And net enrollment ratios show the proportion of children of primary school age who are enrolled in school or who are out of school.

Enrollment ratios are a useful measure of participation in education, but they may also have significant limitations. Enrollment ratios are based on data collected during annual school surveys, which are typically conducted at the beginning of the school year. They do not reflect actual rates of attendance or dropouts during the school year. And school administrators may report exaggerated enrollments, especially if there is a financial incentive to do so. Often the number of teachers paid by the government is related to the number of pupils enrolled. Behrman and Rosenzweig (1994), comparing official school enrollment data for Malaysia in 1988 with gross school attendance rates from a household survey, found that the official statistics systematically overstated enrollment.

Overage or underage enrollments frequently occur, particularly when parents prefer, for cultural or economic reasons, to have children start school at other than the official age. Children's age at enrollment may be inaccurately estimated or misstated, especially in communities where registration of births is not strictly enforced. Parents who want to enroll their underage children in primary school may do so by overstating the age of the children. And in some education systems ages for children repeating a grade may be deliberately or inadvertently underreported.

As an international indicator, the gross primary enrollment ratio has been used to indicate broad levels of participation as well as school capacity. It has an inherent weakness: the length of primary education differs significantly across countries. A short duration tends to increase the ratio and a long duration to decrease it (in part because there are more dropouts among older children).

Other problems affecting cross-country comparisons of enrollment data stem from errors in estimates of school-age populations. Age-gender structures from censuses or vital registration systems, the primary sources of data on school-age populations, are commonly subject to underenumeration (especially of young children) aimed at circumventing laws or regulations; errors are also introduced when parents round up children's ages.

While census data are often adjusted for age bias, adjustments are rarely made for inadequate vital registration systems. Compounding these problems, pre- and postcensus estimates of school-age children are interpolations or projections based on models that may miss important demographic events (see the discussion of demographic data in *About the data* for table 2.1).

In using enrollment data, it is also important to consider repetition rates, which are quite high in some developing countries, leading to a substantial number of overage children enrolled in each grade and raising the gross enrollment ratio. A common error that may also distort enrollment ratios is the lack of distinction between new entrants and repeaters, which, other things equal, leads to underreporting of repeaters and overestimation of dropouts. Thus gross enrollment ratios provide an indication of the capacity of each level of the education system, but a high ratio does not necessarily indicate a successful education system. The net enrollment ratio excludes overage students in an attempt to capture more accurately the system's coverage and internal efficiency. It does not solve the problem completely, however, because some children fall outside the official school age because of late or early entry rather than because of grade repetition. The difference between gross and net enrollment ratios shows the incidence of overage and underage enrollments.

Table 2.12a

Average annual growth in the population of primary school age in selected Sub-Saharan African countries, with and without AIDS, 2000–10

%

	Without AIDS	With AIDS
Kenya	1.6	0.5
Uganda	3.4	3.0
Zambia	2.5	1.0
Zimbabwe	1.2	−0.8

Note: The projection without AIDS assumes that AIDS never existed; the projection with AIDS traces the historical development of AIDS and projects forward to 2010.
Source: World Bank 2000a.

HIV/AIDS affects the demand for education through its effects on the size of the school-age population—through high death rates among adults of reproductive age and through mother-child transmission during birth or through breastfeeding. The impact will be greatest where prevalence rates are high.

Definitions

• **Gross enrollment ratio** is the ratio of total enrollment, regardless of age, to the population of the age group that officially corresponds to the level of education shown. • **Net enrollment ratio** is the ratio of the number of children of official school age (as defined by the national education system) who are enrolled in school to the population of the corresponding official school age. Based on the International Standard Classification of Education, 1976 (ISCED76), • **Preprimary** education refers to the initial stage of organized instruction, designed primarily to introduce very young children to a school-type environment. • **Primary** education provides children with basic reading, writing, and mathematics skills along with an elementary understanding of such subjects as history, geography, natural science, social science, art, and music. • **Secondary** education completes the provision of basic education that began at the primary level, and aims at laying the foundations for lifelong learning and human development, by offering more subject- or skill-oriented instruction using more specialized teachers. • **Tertiary** education, whether or not leading to an advanced research qualification, normally requires, as a minimum condition of admission, the successful completion of education at the secondary level.

Data sources

The data are from an electronic database maintained by UNESCO.

2.13 | Education efficiency

	Net intake rate in grade 1		Percentage of cohort reaching grade 5				Repeaters				Coefficient of efficiency	
	% of school-age population		Male		Female		Primary % of total enrollment		Secondary % of total enrollment			ideal years to graduate as % of actual years
	Male 1994–99[a]	Female 1994–99[a]	1980	1997	1980	1997	1980	1997	1980	1997	Year	
Albania	..	..	..	81	..	83	..	5.3	..	15.9	1994	88.8 [b]
Algeria	82	79	90	93	85	95	11.7	10.5	8.5	19.6	1998	84.1
Angola	..	..	..	..	..	..	29.2	..	..	..	1997	51.6
Argentina	..	..	..	70	..	70	..	5.3	..	9.2	1997	89.0
Armenia	..	..	..	..	..	..	..	0.2	..	..	1995	99.8 [b]
Australia	..	..	..	..	..	..	..	..	..	..	..	..
Austria	..	..	..	..	..	..	..	..	..	..	..	..
Azerbaijan	81	82	..	..	..	..	..	0.4	..	1.0	1997	98.1
Bangladesh	65	65	18	..	26	..	17.8	..	..	..	1998	75.7
Belarus	..	..	..	..	..	..	0.3	0.9	..	0.7	1995	98.4 [b]
Belgium	..	..	75	..	77	..	19.4	..	..	..	..	..
Benin	50	36	59	64	62	57	19.6	25.1	..	..	1994	49.8 [b]
Bolivia	57	58	..	..	..	..	..	..	..	..	1998	54.9
Bosnia and Herzegovina	100	100	..	..	..	..	..	..	..	..	..	..
Botswana	62	63	80	87	84	93	2.9	3.3	..	2.8	1995	90.7 [b]
Brazil	..	..	..	..	..	..	20.2	18.4	7.3	10.8	1998	78.0
Bulgaria	98	98	..	..	..	..	1.7	3.4	0.1	2.0	1997	89.6
Burkina Faso	16	11	76	74	74	77	17.1	16.0	14.3	..	1997	67.7 [b]
Burundi	31	31	100	..	96	..	30.2	..	4.3	..	1997	69.3
Cambodia	63	62	..	51	..	46	..	26.3	..	6.8	1998	39.5
Cameroon	..	..	70	..	69	..	30.0	..	13.7	..	1998	63.7
Canada	..	..	..	..	..	..	..	..	..	..	..	..
Central African Republic	25	19	63	..	50	..	35.1	..	..	..	1990	20.0 [b]
Chad	..	..	..	62	..	53	..	32.0	..	18.4	1995	43.2 [b]
Chile	39 [c]	38 [c]	94	100	97	100	..	5.4	..	4.3	1995	91.7 [b]
China	115 [d]	117 [d]	..	93	..	94	..	1.6	..	0.2	1995	94.2 [b]
Hong Kong, China	..	..	98	..	99	..	3.6	1.1	6.5	..	..	..
Colombia	..	..	36	70	39	76	13.2	7.2	..	..	1997	71.3
Congo, Dem. Rep.	26	19	56	..	59	..	18.8	..	..	..	1992	58.3 [b]
Congo, Rep.	..	..	81	40	83	78	25.7	33.2	..	31.7	1994	34.2 [b]
Costa Rica	..	..	77	86	82	89	7.9	10.1	7.5	9.3	1997	83.6
Côte d'Ivoire	26	21	86	77	79	71	19.6	24.2	..	..	1996	59.0
Croatia	98	99	..	..	..	..	..	0.5	..	0.6	1997	100.0
Cuba	93	93	..	..	..	..	5.7	3.1	..	1.7	1997	94.8
Czech Republic	..	..	..	..	..	..	..	1.2	..	0.7	1994	98.2 [b]
Denmark	..	..	99	100	99	99	..	..	..	..	1994	100.0 [b]
Dominican Republic	..	..	..	..	..	..	18.0	..	..	..	..	..
Ecuador	85	84	..	84	..	86	9.7	3.5	..	..	1996	80.4
Egypt, Arab Rep.	88	85	92	..	88	..	7.9	6.5	..	..	1998	91.7
El Salvador	53	51	17	76	16	77	8.8	4.3	..	0.9	1995	63.5 [b]
Eritrea	22	19	..	73	..	67	..	20.5	..	15.0	1995	66.5 [b]
Estonia	..	..	..	96	..	97	..	2.8	..	3.4	1994	95.6 [b]
Ethiopia	..	..	50	51	51	50	12.2	7.8	..	18.0	1994	74.8 [b]
Finland	..	..	..	100	..	100	..	0.4	..	..	1994	99.6 [b]
France	..	..	..	..	..	..	..	..	9.3	8.1	1990	94.2 [b]
Gabon	52	57	57	58	56	61	34.8	34.9	..	22.0	1994	46.5 [b]
Gambia, The	38	37	74	78	71	83	12.4	12.7	2.1	..	1997	74.3
Georgia	95	92	..	..	..	..	..	0.4	..	0.5	1998	98.6 [b]
Germany	..	..	..	..	..	..	..	1.7	..	2.2	1994	98.0 [b]
Ghana	..	..	..	..	..	..	2.1	..	..	..	1990	87.5 [b]
Greece	..	..	99	..	98	..	1.1	..	3.9	..	1990	99.8 [b]
Guatemala	76	73	..	52	..	47	15.0	15.3	2.5	..	1998	51.1
Guinea	20	15	59	..	41	..	21.9	27.9	..	30.4	1997	53.1
Guinea-Bissau	..	..	25	..	17	..	28.9	..	14.5	..	..	..
Haiti	42	43	20	..	21	..	15.5	..	..	..	1997	47.0
Honduras	46	46	..	..	..	..	16.2	12.0	..	..	1997	61.4

Education efficiency | 2.13

	Net intake rate in grade 1		Percentage of cohort reaching grade 5				Repeaters				Coefficient of efficiency	
	% of school-age population						Primary % of total enrollment		Secondary % of total enrollment			ideal years to graduate as % of actual years
	Male	Female	Male		Female							
	1994–99[a]	1994–99[a]	1980	1997	1980	1997	1980	1997	1980	1997	Year	
Hungary	..	..	96	..	97	..	2.1	..	..	..	1991	93.7 [b]
India	74	61	..	..	..	..	..	3.7	..	..	1997	66.6
Indonesia	50	48	..	88	..	89	8.3	5.8	..	0.7	1997	88.3
Iran, Islamic Rep.	97	95	..	..	..	..	..	5.9	..	..	1997	92.1
Iraq	98	93	77	..	64	..	23.2	..	..	..	..	..
Ireland	99	98	..	..	..	..	..	1.7	..	2.2	1996	94.7
Israel	..	..	..	..	..	..	..	..	..	..	..	..
Italy	..	..	99	98	99	99	1.2	0.4	..	2.7	1994	99.6
Jamaica	75	75	91	..	91	..	3.9	..	2.1	..	1996	89.1
Japan	100	100	100	..	100	..	..	..	..	..	1998	99.9
Jordan	63	63	100	..	98	..	3.2	1.3	4.4	..	1997	97.3
Kazakhstan	99	98	..	..	..	..	..	0.6	..	0.8	1995	94.8
Kenya	..	..	60	..	62	..	12.9	..	..	..	..	..
Korea, Dem. Rep.	100	100	..	..	..	..	..	..	..	..	1998	100.0
Korea, Rep.	94	96	94	98	94	99	..	..	0.0 [e]	0.0 [e]	1997	97.2
Kuwait	63	62	..	..	..	..	6.2	3.4	7.0	5.4	1998	88.4
Kyrgyz Republic	97	96	..	..	..	..	..	0.4	..	0.6	1998	94.5
Lao PDR	55	53	..	57	..	54	..	23.4	..	5.5	1997	51.5
Latvia	80	84	..	..	..	..	..	2.5	..	1.3	1995	96.1 [b]
Lebanon	..	..	..	..	..	..	..	13.4	..	11.1	1998	40.0
Lesotho	23	26	50	55	68	71	20.7	20.1	..	..	1997	53.6
Libya	..	..	..	..	..	..	9.2	..	12.7	..	..	..
Lithuania	..	..	..	..	..	..	..	1.3	..	1.3	1995	98.1 [b]
Macedonia, FYR	77	75	..	95	..	95	..	0.5	..	..	1995	92.2 [b]
Madagascar	..	..	..	49	..	33	..	33.8	..	..	1996	25.9
Malawi	70	75	48	36	40	32	17.4	15.1	..	..	1994	43.9 [b]
Malaysia	96	97	97	..	97	..	..	..	..	..	1996	98.2
Mali	28	19	48	92	42	70	29.6	16.2	..	19.1	1995	66.4 [b]
Mauritania	30	30	..	61	..	68	14.0	15.8	..	12.7	1995	61.0 [b]
Mauritius	97	101 [d]	93	98	94	99	..	4.5	..	14.8	1998	98.3 [b]
Mexico	93	93	..	85	..	86	9.8	6.9	..	2.2	1998	93.8
Moldova	..	..	..	..	..	..	..	1.2	..	1.0	1995	97.3
Mongolia	83	78	..	..	..	..	1.1	0.7	..	0.2	1995	94.5 [b]
Morocco	55	51	79	76	78	74	29.5	12.3	14.9	19.1	1998	65.5
Mozambique	20	19	..	52	..	39	28.7	25.7	..	27.1	1998	38.1
Myanmar	..	..	..	..	..	..	..	..	..	..	1995	58.1 [b]
Namibia	56	59	..	..	..	..	..	11.7	..	9.3	1997	65.5 [b]
Nepal	59	48	..	..	..	..	..	..	..	..	1996	40.5
Netherlands	..	..	94	..	98	..	2.5	..	6.6	..	..	..
New Zealand	100	100	93	97	94	97	3.5	..	2.7	0.8	1995	93.0 [b]
Nicaragua	..	..	40	43	47	52	16.9	12.6	..	6.3	1994	52.8 [b]
Niger	23	15	74	72	72	73	14.3	13.0	6.6	20.4	1998	65.1
Nigeria	..	..	..	..	..	..	..	..	..	..	..	..
Norway	..	..	100	100	100	100	..	..	..	..	1994	100.0 [b]
Oman	75	75	96	96	87	96	12.4	9.2	..	12.9	1995	87.1 [b]
Pakistan	67 [f]	55 [f]	..	..	..	..	..	..	..	..	1997	68.3 [b]
Panama	77	77	74	..	79	..	12.7	..	10.3	..	..	..
Papua New Guinea	..	..	..	59	..	60	..	..	..	..	1996	67.5
Paraguay	69	73	58	77	58	80	13.6	9.1	..	3.0	1998	69.9
Peru	92	93	78	..	74	..	18.8	15.2	10.1	9.0	1998	80.3
Philippines	54	51	68	..	73	..	2.4	..	..	..	1990	76.1
Poland	..	..	..	..	..	..	2.2	1.3	0.4	..	1994	95.9 [b]
Portugal	100	100	..	..	..	..	19.5	..	..	..	1997	86.3
Puerto Rico	..	..	..	..	..	..	..	..	..	..	..	..
Romania	85	83	..	..	..	..	..	2.8	..	1.4	1996	92.7
Russian Federation	..	..	..	..	..	..	..	1.9	..	..	1993	97.0 [b]

2.13 | Education efficiency

	Net intake rate in grade 1		Percentage of cohort reaching grade 5				Repeaters				Coefficient of efficiency	
	% of school-age population		Male		Female		Primary % of total enrollment		Secondary % of total enrollment			ideal years to graduate as % of actual years
	Male 1994–99[a]	Female 1994–99[a]	1980	1997	1980	1997	1980	1997	1980	1997	Year	
Rwanda	..	..	69	..	74	..	5.7	..	..	..	1989	46.9
Saudi Arabia	87	75	82	87	86	92	15.7	7.6	14.8	9.2	1998	90.0 [b]
Senegal	..	..	89	89	82	85	15.6	13.3	..	14.7	1997	80.0 [b]
Sierra Leone	..	..	..	..	..	..	14.8	..	..	..	..	..
Singapore	..	..	100	..	99	..	6.6	..	..	..	..	..
Slovak Republic	..	..	..	..	..	..	..	2.1	..	..	1995	96.6 [b]
Slovenia	..	..	..	..	..	..	..	1.1	..	0.4	1995	98.9 [b]
South Africa	..	..	..	..	..	..	..	..	..	..	1990	75.1 [b]
Spain	..	..	95	..	94	..	6.4	..	8.8	..	1992	96.9 [b]
Sri Lanka	94	95	92	83	91	84	10.4	2.3	..	..	1997	90.4 [b]
Sudan	46	41	68	75	71	73	..	..	..	..	1998	68.4
Sweden	..	..	98	97	98	97	..	..	..	..	1994	99.3 [b]
Switzerland	98	99	75	..	74	..	2.0	1.6	2.9	..	1992	65.6 [b]
Syrian Arab Republic	91	88	93	93	88	94	8.1	7.3	13.9	..	1998	85.8 [b]
Tajikistan	..	..	..	..	..	..	..	0.5	..	0.4	..	..
Tanzania	13	15	89	78	90	84	1.2	2.1	..	..	1992	85.5 [b]
Thailand	37	37	..	..	..	..	8.3	..	..	..	1996	93.7
Togo	34	29	59	79	44	60	35.5	24.2	..	25.0	1997	44.9
Trinidad and Tobago	..	..	85	98	87	97	3.9	5.6	..	..	1997	93.0
Tunisia	87	85	89	90	84	92	20.6	16.1	7.4	17.2	1995	76.1 [b]
Turkey	..	..	..	..	..	..	..	4.9	..	..	1993	92.5 [b]
Turkmenistan	..	..	..	..	..	..	..	..	..	..	..	..
Uganda	93	90	82	..	73	..	10.3	..	..	..	..	..
Ukraine	..	..	..	..	..	..	0.3	..	..	..	1990	98.4
United Arab Emirates	96	98	100	83	100	84	9.0	4.2	..	7.8	1998	86.1 [b]
United Kingdom	..	..	..	..	..	..	..	..	..	..	..	..
United States	..	..	..	..	..	..	..	..	..	..	..	..
Uruguay	53	54	..	96	..	99	14.9	9.5	..	..	1995	88.0 [b]
Uzbekistan	..	..	..	..	..	..	..	0.2	..	..	1998	99.9
Venezuela, RB	62	62	..	86	..	92	10.7	10.3	6.6	4.7	1995	59.9 [b]
Vietnam	95	95	..	..	..	..	..	..	..	..	1997	79.6
West Bank and Gaza	..	..	..	..	..	..	..	..	..	..	..	..
Yemen, Rep.	83	51	..	..	..	..	..	..	..	..	1998	70.8
Yugoslavia, FR (Serb./Mont.)	..	..	..	..	..	..	..	1.0	..	..	1995	98.2 [b]
Zambia	40	45	88	..	82	..	1.9	2.8	..	..	1997	86.1
Zimbabwe	38	39	81	78	82	79	..	..	..	..	1998	86.1

World	.. w	.. w	.. w	.. w	.. w	.. w	.. w	.. w	.. w	.. w		
Low income	67	57	..	..	..	..	..	..	..	..		
Middle income	87	88	..	..	..	..	..	3.6	..	..		
Lower middle income	106 [d]	107 [d]	..	91	..	92	..	3.0	..	1.5		
Upper middle income	..	..	..	..	..	..	..	..	..	..		
Low & middle income	87		..	..	..	..	..	6.1	..	..		
East Asia & Pacific	102 [d]	103 [d]	..	92	..	93	..	2.5	..	0.3		
Europe & Central Asia	..	..	..	..	..	..	..	2.4	..	..		
Latin America & Carib.	..	..	..	..	..	..	15.3	12.9	..	..		
Middle East & N. Africa	86	83	88	..	84	..	12.1	..	..	..		
South Asia	73	60	..	..	..	..	..	3.7	..	..		
Sub-Saharan Africa	..	..	..	..	..	..	..	..	..	..		
High income	..	..	..	..	..	..	..	..	..	..		
Europe EMU	..	..	..	..	..	..	..	..	..	4.1		

a. Data are for the most recent year available. b. Primary school only. c. Not including special education. d. Ratios exceeding 100 percent indicate discrepancies between estimates of the school-age population and reported enrollment data. e. Less than 0.05. f. Does not include children in private institutions.

Education efficiency | **2.13**

About the data

Indicators of students' progress through school, estimated by the United Nations Educational, Scientific, and Cultural Organization (UNESCO), provide a measure of an education system's success in maintaining a flow of students from one grade to the next and thus in imparting a particular level of education.

Low net intake rates reflect the fact that many children do not enter school at the official age, even though school attendance is mandatory in most countries, at least through the primary level. In addition, students drop out of school for a variety of reasons, including discouragement over poor performance, the cost of schooling, and the opportunity cost of time spent in school. And students' progress to higher grades may be limited by the availability of teachers, classrooms, and educational materials.

The rate of progression—sometimes called the rate of persistence or survival—is estimated as the proportion of a single-year cohort of students that eventually reaches a particular grade of school. It measures the holding power and internal efficiency of an education system. Progression rates approaching 100 percent indicate a high level of retention and a low level of dropout.

Because tracking data for individual students generally are not available, aggregate student flows from one grade to the next are estimated using data on enrollment and repetition by grade for two consecutive years. This procedure, called the reconstructed cohort method, makes three simplifying assumptions: dropouts never return to school; promotion, repetition, and dropout rates remain constant over the entire period in which the cohort is enrolled in school; and the same rates apply to all pupils enrolled in a given grade, regardless of whether they previously repeated a grade (Fredricksen 1993). Given these assumptions, cross-country comparisons should be made with caution, because other flows—caused by new entrants, reentrants, grade skipping, migration, or school transfers during the school year—are not considered.

The percentage of the cohort reaching grade 5, rather than another grade, is shown because it is generally agreed that children who reach grade 5 should have acquired the basic literacy and numeracy skills that enable them to continue learning. This indicator provides no information on learning outcomes, however, and only indirectly reflects the quality of schooling. Assessing learning outcomes requires setting standards and measuring the attainment of those standards. National assessments are generally concerned with the performance not of individual students, but of all or part of the education system.

The repetition rate is often used to indicate the internal efficiency of the education system. Repeaters not only increase the cost of education for the family and for the school system, but also use up limited school resources. Countries have different policies on repetition and promotion of students; in some cases the number of repeaters is controlled because of limited capacity.

The coefficient of efficiency is a synthetic indicator of the internal efficiency of an education system, reflecting the combined impact on efficiency of repetition and dropout. The ideal value of the coefficient is 100 percent, corresponding to a situation in which all pupils complete the school cycle, neither repeating grades nor dropping out. A coefficient less than 100 percent indicates some level of resource waste.

Figure 2.13

The internal efficiency of education systems varies widely

Coefficient of efficiency (%), 1997–99

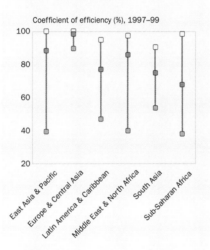

☐ Maximum
▧ Median
▨ Minimum

Source: UNESCO, Education for All Year 2000 Assessment database.

The coefficient of efficiency for an education system is 100 percent if children progress through school without repeating a grade or dropping out. This coefficient varies widely among countries and regions. Of the 66 countries for which data are available, 40 (61 percent) have coefficients exceeding 70 percent. Within countries and regions gender disparities in the internal efficiency of education systems are small, and in the majority of cases they favor girls.

Definitions

• **Net intake rate in grade 1** is the number of new entrants in the first grade of primary education who are of official primary school entrance age, expressed as a percentage of the population of the corresponding age.
• **Percentage of cohort reaching grade 5** is the share of children enrolled in the first grade of primary school who eventually reach grade 5. The estimate is based on the reconstructed cohort method (see *About the data*).
• **Repeaters** are the total number of students enrolled in the same grade as in the previous year, as a percentage of all students enrolled in that grade. • **Coefficient of efficiency** refers to the ideal number of pupil-years required to produce graduates from a given cohort (in the absence of repetition and dropout) as a percentage of the actual number of pupil-years spent to produce the same number of graduates.

Data sources

Data on net intake and coefficients of efficiency come from UNESCO's special data collection for the Education for All initiative. The remaining data in the table were compiled using an electronic database maintained by UNESCO.

2.14 | Education outcomes

	Adult illiteracy rate				Youth illiteracy rate				Expected years of schooling			
	Male % ages 15 and over		Female % ages 15 and over		Male % ages 15–24		Female % ages 15–24		Males		Females	
	1990	1999	1990	1999	1990	1999	1990	1999	1990	1997	1990	1997
Albania	14	9	32	23	3	1	7	3	..	..	..	..
Algeria	32	23	59	44	13	8	32	16	11	11	9	10
Angola	..	..	..	..	..	..	..	..	..	..	..	..
Argentina	4	3	4	3	2	2	2	1	..	..	..	..
Armenia	1	1	4	3	0 [a]	0 [a]	1	0 [a]	..	..	..	..
Australia	..	..	..	..	..	..	..	..	13	17	13	17
Austria	..	..	..	..	..	..	..	..	15	15	14	14
Azerbaijan	..	..	..	..	..	..	..	..	..	..	..	..
Bangladesh	54	48	77	71	45	40	68	61	6	..	4	..
Belarus	0 [a]	0 [a]	1	1	0 [a]	0 [a]	0 [a]	0 [a]	..	..	..	..
Belgium	..	..	..	..	..	..	..	..	14	17	14	17
Benin	59	45	84	76	37	23	74	63	..	..	..	..
Bolivia	13	8	30	21	4	2	11	7	..	..	..	..
Bosnia and Herzegovina	..	..	..	..	..	..	..	..	..	..	..	..
Botswana	34	26	30	21	21	16	13	8	10	11	11	11
Brazil	18	15	20	15	12	10	9	6	..	..	..	..
Bulgaria	2	1	4	2	1	0 [a]	1	1	12	12	12	12
Burkina Faso	75	67	92	87	64	55	86	78	3	..	2	..
Burundi	50	44	73	61	42	36	55	40	6	..	4	..
Cambodia	49	41	86	79	34	25	73	59	..	..	..	..
Cameroon	28	19	46	31	10	6	15	7	..	..	..	..
Canada	..	..	..	..	..	..	..	..	17	17	17	17
Central African Republic	53	41	79	67	34	25	61	43	..	..	..	..
Chad	63	50	81	68	42	28	62	42	..	..	..	..
Chile	6	4	6	5	2	2	2	1	..	13	..	13
China	14	9	33	25	3	1	8	4	..	..	..	..
Hong Kong, China	5	4	16	10	2	1	1	0 [a]	..	..	..	..
Colombia	11	9	12	9	6	4	4	3	..	..	..	..
Congo, Dem. Rep.	38	28	66	51	19	12	42	27	..	..	..	..
Congo, Rep.	23	13	42	27	5	2	10	4	..	..	..	..
Costa Rica	6	5	6	5	3	2	2	1	..	..	..	..
Côte d'Ivoire	56	46	77	63	40	31	59	42	..	..	..	..
Croatia	1	1	5	3	0 [a]	0 [a]	0 [a]	0 [a]	..	11	..	12
Cuba	5	3	5	4	1	0 [a]	1	0 [a]	12	..	13	..
Czech Republic	..	..	..	..	..	..	..	..	..	13	..	13
Denmark	..	..	..	..	..	..	..	..	14	15	14	15
Dominican Republic	20	17	21	17	13	10	12	9	..	..	..	..
Ecuador	10	7	15	11	4	3	5	4	..	..	..	..
Egypt, Arab Rep.	40	34	66	57	29	24	49	38	..	..	..	..
El Salvador	24	19	31	24	15	11	17	13	..	10	..	10
Eritrea	42	33	72	61	27	20	54	39	..	5	..	4
Estonia	..	..	..	..	..	..	..	..	12	12	12	13
Ethiopia	64	57	80	68	52	46	64	48	..	..	..	..
Finland	..	..	..	..	..	..	..	..	15	15	16	17
France	..	..	..	..	..	..	..	..	14	15	15	16
Gabon	..	..	..	..	..	..	..	..	..	..	..	..
Gambia, The	68	57	80	72	49	36	66	52	..	..	..	..
Georgia	..	..	..	..	..	..	..	..	..	11	..	11
Germany	..	..	..	..	..	..	..	..	15	16	15	16
Ghana	30	21	53	39	12	7	25	13	..	..	..	..
Greece	2	2	8	4	1	0 [a]	0 [a]	0 [a]	13	14	13	14
Guatemala	31	24	47	40	20	15	34	28	..	..	..	..
Guinea	..	..	..	..	..	..	..	..	..	..	..	..
Guinea-Bissau	54	42	89	82	30	19	79	68	..	..	..	..
Haiti	57	49	63	53	44	37	46	36	..	..	..	..
Honduras	31	26	32	26	23	19	20	16	..	..	..	..

Education outcomes | 2.14

	Adult illiteracy rate				Youth illiteracy rate				Expected years of schooling			
	Male % ages 15 and over		Female % ages 15 and over		Male % ages 15–24		Female % ages 15–24		Males		Females	
	1990	1999	1990	1999	1990	1999	1990	1999	1990	1997	1990	1997
Hungary	1	1	1	1	0 [a]	0 [a]	0 [a]	0 [a]	11	13	11	13
India	38	32	64	56	27	21	46	36	..	..	..	..
Indonesia	13	9	27	19	3	2	7	3	10	10	9	10
Iran, Islamic Rep.	27	17	45	31	8	4	18	9	..	12	..	11
Iraq	43	35	67	55	29	23	48	34	..	..	..	..
Ireland	..	..	..	..	..	..	..	..	12	14	13	14
Israel	3	2	9	6	1	0 [a]	2	0 [a]	..	..	..	..
Italy	2	1	3	2	0 [a]	0 [a]	0 [a]	0 [a]	..	..	..	..
Jamaica	22	18	14	10	13	10	5	3	11	..	11	..
Japan	..	..	..	..	..	..	..	..	..	..	..	..
Jordan	10	6	28	17	2	1	4	0 [a]	9	9	9	9
Kazakhstan	..	..	..	..	..	..	..	..	..	..	..	..
Kenya	19	12	39	25	7	4	13	6	..	..	..	..
Korea, Dem. Rep.	..	..	..	..	..	..	..	..	..	..	..	..
Korea, Rep.	2	1	7	4	0 [a]	0 [a]	0 [a]	0 [a]	14	15	13	14
Kuwait	20	16	27	21	12	9	13	7	7	9	7	9
Kyrgyz Republic	..	..	..	..	..	..	..	..	..	..	..	..
Lao PDR	47	37	80	68	28	18	62	44	9	9	9	7
Latvia	0 [a]	0 [a]	0 [a]	0 [a]	0 [a]	0 [a]	0 [a]	0 [a]	..	12	..	13
Lebanon	12	8	27	20	5	3	11	7	..	..	..	..
Lesotho	35	28	11	7	23	18	3	2	9	9	11	10
Libya	17	10	49	33	1	0 [a]	17	7	..	..	..	..
Lithuania	1	0 [a]	1	1	0 [a]	0 [a]	0 [a]	0 [a]	..	11	..	11
Macedonia, FYR	..	..	..	..	..	..	..	..	..	..	..	..
Madagascar	34	27	50	41	22	17	33	24	..	..	..	..
Malawi	31	26	64	55	24	20	49	40	..	..	..	..
Malaysia	13	9	25	17	5	3	6	3	..	..	..	..
Mali	67	53	81	67	46	29	63	42	3	..	1	..
Mauritania	53	48	74	69	44	39	65	60	..	..	..	..
Mauritius	15	12	25	19	9	7	9	6	..	..	..	..
Mexico	10	7	15	11	4	3	6	4	..	..	..	..
Moldova	1	1	4	2	0 [a]	0 [a]	0 [a]	0 [a]	..	..	..	..
Mongolia	35	27	59	48	21	16	39	27	..	7	..	9
Morocco	47	39	75	65	32	24	58	43	..	..	..	..
Mozambique	51	41	82	72	34	26	68	55	4	4	3	3
Myanmar	13	11	26	20	10	9	14	10	..	..	..	..
Namibia	23	18	28	20	14	10	11	7	..	..	..	..
Nepal	53	42	86	77	34	25	73	59	..	..	..	..
Netherlands	..	..	..	..	..	..	..	..	15	16	15	16
New Zealand	..	..	..	..	..	..	..	..	14	16	15	17
Nicaragua	36	33	34	30	32	29	28	24	..	..	..	..
Niger	82	77	95	92	75	68	91	87	..	..	..	..
Nigeria	41	29	62	46	19	11	34	18	..	..	..	..
Norway	..	..	..	..	..	..	..	..	14	15	15	16
Oman	33	21	62	40	5	1	25	5	..	..	..	..
Pakistan	50	41	79	70	36	24	67	52	..	..	..	..
Panama	10	8	12	9	4	3	5	4	..	..	..	..
Papua New Guinea	34	29	52	44	25	20	38	30	..	..	..	..
Paraguay	8	6	12	8	4	3	5	3	9	10	8	10
Peru	8	6	21	15	3	2	8	5	..	..	..	..
Philippines	7	5	8	5	3	2	3	1	..	..	..	..
Poland	0 [a]	0 [a]	1	0 [a]	0 [a]	0 [a]	0 [a]	0 [a]	12	13	12	13
Portugal	9	6	16	11	1	0 [a]	0 [a]	0 [a]	13	14	14	15
Puerto Rico	9	7	9	6	5	3	3	2	..	..	..	..
Romania	1	1	5	3	1	1	1	0 [a]	11	12	11	12
Russian Federation	0 [a]	0 [a]	1	1	0 [a]	0 [a]	0 [a]	0 [a]	..	..	..	..

	Adult illiteracy rate				Youth illiteracy rate				Expected years of schooling			
	Male % ages 15 and over		Female % ages 15 and over		Male % ages 15–24		Female % ages 15–24		Males		Females	
	1990	1999	1990	1999	1990	1999	1990	1999	1990	1997	1990	1997
Rwanda	37	27	56	41	22	15	33	20	..	..	..	..
Saudi Arabia	22	17	49	34	9	5	21	10	9	10	7	9
Senegal	62	54	81	73	50	41	70	59	..	..	..	..
Sierra Leone	..	..	..	..	..	..	..	..	..	..	..	..
Singapore	6	4	17	12	1	0 [a]	1	0 [a]	..	..	..	..
Slovak Republic	..	..	..	..	..	..	..	..	..	..	..	..
Slovenia	0 [a]	0 [a]	1	0 [a]	0 [a]	0 [a]	0 [a]	0 [a]	..	..	..	..
South Africa	18	14	20	16	11	9	12	9	13	14	13	14
Spain	2	2	5	3	0 [a]	0 [a]	0 [a]	0 [a]	..	..	..	..
Sri Lanka	7	6	15	11	4	3	6	4	..	..	..	..
Sudan	39	31	68	55	24	18	46	30	..	..	..	..
Sweden	..	..	..	..	..	..	..	..	13	15	13	16
Switzerland	..	..	..	..	..	..	..	..	14	15	13	14
Syrian Arab Republic	18	12	53	41	8	5	33	22	11	10	9	9
Tajikistan	1	1	3	1	0 [a]	0 [a]	0 [a]	0 [a]	..	..	..	..
Tanzania	23	16	49	34	10	7	22	12	..	..	..	..
Thailand	5	3	11	7	1	1	2	2	..	..	..	..
Togo	36	26	71	60	19	13	55	42	11	..	6	..
Trinidad and Tobago	6	5	11	8	3	2	4	3	11	..	11	..
Tunisia	28	20	54	41	7	3	25	12	11	..	10	..
Turkey	11	7	33	24	3	1	12	6	..	10	..	9
Turkmenistan	..	..	..	..	..	..	..	..	..	..	..	..
Uganda	31	23	57	45	20	15	39	29	..	..	..	..
Ukraine	0 [a]	0 [a]	1	1	0 [a]	0 [a]	0 [a]	0 [a]	..	..	..	..
United Arab Emirates	29	26	30	22	19	15	11	6	10	10	11	11
United Kingdom	..	..	..	..	..	..	..	..	14	16	14	17
United States	..	..	..	..	..	..	..	..	15	16	16	16
Uruguay	4	3	3	2	1	1	1	0 [a]	..	..	..	..
Uzbekistan	10	7	23	16	3	2	8	5	..	..	..	..
Venezuela, RB	10	7	12	8	5	3	3	2	..	..	..	..
Vietnam	6	5	13	9	5	3	5	3	..	..	..	..
West Bank and Gaza	..	..	..	..	..	..	..	..	..	..	..	..
Yemen, Rep.	45	33	87	76	26	18	75	56	..	..	..	..
Yugoslavia, FR (Serb./Mont.)	..	..	..	..	..	..	..	..	..	..	..	..
Zambia	22	15	41	30	14	10	24	15	..	8	..	7
Zimbabwe	13	8	25	16	3	2	9	5	..	..	..	..
World	.. w	.. w	.. w	.. w	.. w	0 .. w	.. w	.. w				
Low income	35	29	56	48	24	19	41	31				
Middle income	13	9	26	20	5	4	9	6				
Lower middle income	13	9	29	22	4	3	10	7				
Upper middle income	11	9	14	11	6	5	6	4				
Low & middle income	22	18	39	32	13	11	23	19				
East Asia & Pacific	13	8	29	22	3	2	8	4				
Europe & Central Asia	2	2	6	5	1	1	3	2				
Latin America & Carib.	14	11	17	13	8	6	8	6				
Middle East & N. Africa	33	25	59	47	18	13	37	24				
South Asia	41	34	66	58	29	23	50	41				
Sub-Saharan Africa	40	31	60	47	25	18	40	27				
High income	..	..	..	..	..	..	..	..				
Europe EMU	..	..	..	..	..	..	..	..				

a. Less than 0.5.

Many governments collect and publish statistics that indicate how their education systems are working and developing—statistics on enrollment and on such efficiency indicators as pupil-teacher ratios, repetition rates, and cohort progression through school. But until recently, despite an obvious interest in what education achieves, few systems in high-income or developing countries had systematically collected information on outcomes of education.

Basic student outcomes include achievements in reading and mathematics judged against established standards. In many countries national learning assessments are enabling ministries of education to monitor progress in these outcomes. Internationally, the United Nations Educational, Scientific, and Cultural Organization (UNESCO) has established literacy as an outcome indicator based on an internationally agreed definition. The rate of illiteracy is defined as the percentage of people who cannot, with understanding, read and write a short, simple statement about their everyday life. In practice, illiteracy is difficult to measure. To estimate illiteracy using such a definition requires census or survey measurements under controlled conditions. Many countries estimate the number of illiterate people from self-reported data, or by taking people with no schooling as illiterate.

Literacy statistics for most countries cover the population ages 15 and above, by five-year age groups, but some include younger ages or are confined to age ranges that tend to inflate literacy rates. As an alternative, UNESCO has proposed the narrower age range of 15–24, which better captures the ability of participants in the formal education system. The youth illiteracy rate reported in the table measures the accumulated outcomes of primary education over the previous 10 years or so by indicating the proportion of people who have passed through the primary education system (or never entered it) without acquiring basic literacy and numeracy skills. Reasons for this may include difficulties in attending school or dropping out before reaching grade 5 (see *About the data* for table 2.13) and thereby failing to achieve basic learning competencies.

The indicator expected years of schooling is an estimate of the total years of schooling that an average child at the age of school entry will receive, including years spent on repetition, given the current patterns of enrollment across cycles of education. It may also be interpreted as an indicator of the total education resources, measured in school years, that a child will acquire over his or her "lifetime" in school—or as an indicator of an education system's overall level of development.

Because the calculation of this indicator assumes that the probability of a child's being enrolled in school at any future age is equal to the current enrollment ratio for that age, it does not account for changes and trends in future enrollment ratios. The expected number of years and the expected number of grades completed are not necessarily consistent, because the first includes years spent in repetition. Comparability across countries and over time may be affected by differences in the length of the school year or changes in policies on automatic promotions and grade repetition.

Table 2.14a

Science achievement among eighth-grade students in selected economies, 1995 and 1999

Average science scale score

	1995	1999
Bulgaria	**545**	**518**
Canada	**514**	**533**
Czech Republic	555	539
Hong Kong, China	510	530
Hungary	**537**	**522**
Iran, Islamic Rep.	463	448
Latvia	**473**	**503**
Lithuania	**464**	**488**
Singapore	580	568
International average	518	521

Note: Figures in bold denote a statistically significant difference between 1995 and 1999.
Source: International Association for the Evaluation of Educational Achievement 2000.

Conducted in 41 countries in 1994–95, the Third International Mathematics and Science Study (TIMSS) was designed to provide a base from which policymakers, curriculum specialists, and researchers could assess the performance of their education systems. TIMSS 1999 (also known as TIMSS-Repeat), conducted in 38 countries, was designed to show trends in eighth-grade mathematics and science achievement.

In 16 of the 38 countries in 1999, boys had significantly higher average achievement in science than girls—on average, 29 percent of boys ranked in the top achievement quarter, compared with 21 percent of girls. A student's access to educational resources at home also mattered: students from homes with greater educational resources, including at least one parent who had finished university, had higher achievement in science than counterparts with fewer resources.

• **Adult illiteracy rate** is the percentage of people ages 15 and over who cannot, with understanding, read and write a short, simple statement about their everyday life. • **Youth illiteracy rate** is the illiteracy rate among people ages 15–24. • **Expected years of schooling** are the average number of years of formal schooling that children are expected to receive, including university education and years spent in repetition. They are the sum of the underlying age-specific enrollment ratios for primary, secondary, and tertiary education.

The data on illiteracy are based on UNESCO's 1999 literacy estimates and projections. The data on expected years of schooling are from UNESCO's *World Education Report 1998*.

2.15 Health expenditure, services, and use

	Health expenditure			Health expenditure per capita		Physicians		Hospital beds		Inpatient admission rate	Average length of stay	Outpatient visits per capita
	Public % of GDP 1990–98[a]	Private % of GDP 1990–98[a]	Total % of GDP 1990–98[a,b]	PPP $ 1990–98[a]	$ 1990–98[a]	per 1,000 people 1980	1990–98[a]	per 1,000 people 1980	1990–98[a]	% of population 1990–98[a]	days 1990–98[a]	1990–98[a]
Albania	3.5	0.5	4.0	116	36	..	1.4	..	3.2	..	13	2
Algeria	2.6	1.0	3.6	216	68	..	1.0	..	2.1	..	..	..
Angola	3.9	..	..	..	..	..	0.0 [c]	..	1.3	..	..	..
Argentina	4.9	5.4	10.3	1,291	852	..	2.7	..	3.3	..	..	..
Armenia	3.1	4.2	7.8	151	27	3.5	3.0	8.4	0.7	8	15	2
Australia	5.9	2.6	8.5	1,980	1,692	1.8	2.5	..	8.5	16	15	6
Austria	5.8	2.4	8.3	1,978	2,162	..	3.0	11.2	8.9	28	9	7
Azerbaijan	1.2	0.6	1.8	39	9	3.4	3.8	9.7	9.7	6	18	1
Bangladesh	1.7	1.9	3.6	51	12	0.1	0.2	0.2	0.3	..	..	..
Belarus	4.9	1.1	6.0	387	83	3.4	4.3	12.5	12.2	26	18	11
Belgium	7.9	1.0	8.9	2,172	2,184	2.5	3.4	..	7.2	20	11	8
Benin	1.6	1.6	3.2	29	12	0.1	0.1	1.5	0.2	..	..	..
Bolivia	4.1	2.4	6.4	150	69	..	1.3	..	1.7	..	..	..
Bosnia and Herzegovina	..	..	..	..	..	..	0.5	..	1.8	..	15	..
Botswana	2.5	1.6	4.1	267	127	0.1	0.2	2.4	1.6	..	..	..
Brazil	2.9	3.7	6.6	453	309	..	1.3	..	3.1	0 [d]	..	2
Bulgaria	3.8	0.8	4.7	230	69	2.5	3.5	11.1	10.6	18	14	6
Burkina Faso	1.2	2.7	3.9	36	10	0.0 [c]	0.0 [c]	..	1.4	2	3	0 [d]
Burundi	0.6	3.0	3.7	21	5	..	0.1	..	0.7	..	..	..
Cambodia	0.6	6.3	6.9	90	17	..	0.1	..	2.1	..	..	..
Cameroon	1.0	4.0	5.0	77	31	..	0.1	..	2.6	..	..	..
Canada	6.4	2.8	9.2	2,292	1,824	..	2.1	..	4.2	12	8	7
Central African Republic	2.0	1.0	3.0	33	9	0.0 [c]	0.1	1.6	0.9	..	..	..
Chad	2.3	0.6	2.9	25	7	..	0.0 [c]	..	0.7	..	..	..
Chile	2.7	3.1	5.9	511	289	..	1.1	3.4	2.7	..	..	..
China	2.0	2.6	4.5	143	33	0.9	2.0	2.0	2.9	4	13	..
Hong Kong, China	2.1	2.8	5.0	1,143	1,134	0.8	1.3	4.0	..	2	..	1
Colombia	5.2	4.2	9.3	553	227	..	1.1	1.6	1.5	..	..	..
Congo, Dem. Rep.	..	..	..	..	..	..	0.1	..	1.4	..	..	..
Congo, Rep.	2.0	3.8	5.8	46	41	..	0.3	..	3.4	..	..	..
Costa Rica	5.2	1.5	6.8	509	267	..	0.9	3.3	1.7	9	6	1
Côte d'Ivoire	1.2	2.6	3.8	62	29	..	0.1	..	0.8	..	..	..
Croatia	8.1	1.5	9.6	664	428	..	2.0	..	5.9	12	..	..
Cuba	8.2	0.9	9.1	..	83	..	5.3	..	5.1	..	..	..
Czech Republic	6.7	0.6	7.2	928	392	..	3.0	..	8.7	21	11	12
Denmark	6.8	1.5	8.3	2,102	2,732	..	2.9	..	4.6	20	7	6
Dominican Republic	1.9	3.0	4.8	246	93	..	2.2	..	1.5	..	..	..
Ecuador	1.7	2.0	3.6	115	59	..	1.7	1.9	1.6	..	..	..
Egypt, Arab Rep.	1.8	2.0	3.8	119	48	1.1	1.6	2.0	2.1	3	6	4
El Salvador	2.6	4.6	7.2	298	143	0.3	1.0	..	1.6	..	..	..
Eritrea	2.9	0.9	2.0	14	..	..	0.0 [c]	..	..	..	..	..
Estonia	5.5	1.4	6.9	553	219	4.2	3.1	12.4	7.4	18	9	6
Ethiopia	1.7	2.4	4.1	25	4	0.0 [c]	0.0 [c]	0.3	0.2	..	..	..
Finland	5.2	1.6	6.9	1,502	1,722	1.9	3.0	15.5	7.8	27	11	4
France	7.3	2.3	9.6	2,102	2,377	..	3.0	..	8.5	23	11	7
Gabon	2.1	1.0	3.1	198	121	..	0.2	..	3.2	..	..	..
Gambia, The	1.9	1.9	3.7	56	13	..	0.0 [c]	..	0.6	..	..	..
Georgia	0.5	1.7	2.2	73	14	4.8	3.8	10.7	4.8	5	11	1
Germany	7.9	2.7	10.6	2,424	2,769	2.2	3.5	..	9.3	22	12	7
Ghana	1.8	2.9	4.7	85	19	..	..	..	1.5	..	..	..
Greece	4.7	3.6	8.3	1,207	957	2.4	4.0	6.2	5.0	15	8	..
Guatemala	2.1	2.3	4.4	155	78	..	0.9	..	1.0	..	..	..
Guinea	2.2	1.4	3.6	68	19	..	0.2	..	0.6	..	..	..
Guinea-Bissau	1.1	..	..	..	..	0.1	0.2	1.8	1.5	..	..	..
Haiti	1.4	2.8	4.2	61	21	..	0.2	0.7	0.7	..	..	..
Honduras	3.9	4.7	8.6	210	74	..	0.8	1.3	1.1	..	..	..

Health expenditure, services, and use | 2.15

	Health expenditure			Health expenditure per capita		Physicians		Hospital beds		Inpatient admission rate	Average length of stay	Outpatient visits per capita
	Public % of GDP 1990–98[a]	Private % of GDP 1990–98[a]	Total % of GDP 1990–98[a,b]	PPP $ 1990–98[a]	$ 1990–98[a]	per 1,000 people 1980	1990–98[a]	per 1,000 people 1980	1990–98[a]	% of population 1990–98[a]	days 1990–98[a]	1990–98[a]
Hungary	5.2	2.0	6.4	660	290	2.5	3.5	9.1	8.3	24	9	15
India	0.8	4.2	5.4	94	20	0.4	0.4	0.8	0.8	..	..	..
Indonesia	0.7	0.8	1.6	44	8	..	0.2	..	0.7	..	..	..
Iran, Islamic Rep.	1.7	2.5	4.2	229	128	..	0.8	1.5	1.6	..	..	..
Iraq	3.8	1.8	5.6	..	..	0.6	0.6	1.9	1.5	..	..	..
Ireland	4.7	1.5	6.1	1,393	1,428	1.3	2.2	9.7	3.7	15	7	..
Israel	6.0	3.6	9.5	1,730	1,607	..	4.6	5.1	6.0	..	..	..
Italy	5.6	2.6	8.2	1,771	1,701	..	5.9	..	6.5	19	8	5
Jamaica	3.2	2.6	5.7	202	157	..	1.3	..	2.1	..	..	..
Japan	5.9	1.6	7.6	1,844	2,284	..	1.9	11.3	16.5	10	41	16
Jordan	5.3	3.8	9.1	296	123	0.8	1.7	1.3	1.8	11	4	3
Kazakhstan	3.5	2.4	5.9	273	86	3.2	3.5	13.2	8.5	15	16	0 [d]
Kenya	2.4	5.4	7.8	79	31	..	0.1	..	1.6	..	..	..
Korea, Dem. Rep.	..	..	..	..	..	..	..	..	..	..	..	..
Korea, Rep.	2.3	2.8	5.1	720	349	0.6	1.3	1.7	5.1	6	13	10
Kuwait	2.9	0.4	3.3	..	551	1.7	1.9	4.1	2.8	..	..	..
Kyrgyz Republic	2.9	1.6	4.5	109	15	2.9	3.1	12.0	9.5	21	15	1
Lao PDR	1.2	1.3	2.6	35	7	..	0.2	..	2.6	..	..	..
Latvia	4.2	2.6	6.7	410	167	4.1	3.4	13.7	10.3	21	14	4
Lebanon	2.2	7.6	9.8	430	361	..	2.3	..	2.7	17	4	..
Lesotho	3.4	2.2	..	..	..	..	0.1	..	..	..	..	..
Libya	..	..	..	..	..	1.3	1.3	..	4.3	..	..	..
Lithuania	4.8	1.5	6.3	429	183	3.9	3.9	12.1	9.6	24	12	7
Macedonia, FYR	5.5	1.0	6.5	288	113	..	2.3	..	5.2	10	15	3
Madagascar	1.1	1.0	2.1	16	5	..	0.3	..	0.9	..	..	..
Malawi	2.8	3.5	6.3	36	10	..	0.0 [c]	..	1.3	..	..	2
Malaysia	1.4	1.0	2.5	189	81	0.3	0.5	..	2.0	..	..	..
Mali	2.1	2.2	4.2	30	11	0.0 [c]	0.1	..	0.2	1	7	0 [d]
Mauritania	1.4	3.4	4.8	74	19	..	0.1	..	0.7	..	..	..
Mauritius	1.8	1.6	3.4	302	120	0.5	0.9	3.1	3.1	0 [d]	..	4
Mexico	2.8	1.9	4.7	371	202	..	1.6	..	1.1	6	4	2
Moldova	6.4	2.1	8.4	177	33	3.1	3.6	12.0	12.1	19	18	8
Mongolia	4.3	0.4	4.7	67	24	..	2.6	11.2	11.5	..	..	..
Morocco	1.2	3.2	4.4	134	49	..	0.4	..	1.0	3	7	..
Mozambique	2.8	0.7	3.5	28	8	0.0 [c]	..	1.1	0.9	..	..	..
Myanmar	0.2	1.6	1.8	..	102	..	0.3	0.9	0.6	..	..	..
Namibia	4.1	3.7	7.8	417	143	..	0.2	..	..	..	..	..
Nepal	1.3	4.2	5.4	66	11	0.0 [c]	0.0 [c]	0.2	0.2	..	..	..
Netherlands	6.0	2.5	8.6	1,974	2,140	..	2.6	12.5	11.3	11	34	6
New Zealand	6.2	1.8	8.1	1,454	1,128	1.6	2.3	..	6.2	14	7	..
Nicaragua	8.3	3.9	12.2	266	54	0.4	0.8	..	1.5	..	..	..
Niger	1.2	1.4	2.6	20	5	..	0.0 [c]	..	0.1	28	5	0 [d]
Nigeria	0.8	2.0	2.8	23	30	0.1	0.2	0.9	1.7	..	..	..
Norway	7.4	1.5	8.9	2,467	2,953	1.9	2.5	15.0	14.7	15	10	4
Oman	2.9	0.6	3.5	..	..	0.5	1.3	1.6	2.2	9	4	4
Pakistan	0.9	3.1	4.0	71	18	0.3	0.6	0.6	0.7	..	..	3
Panama	4.9	2.3	7.3	410	246	..	1.7	..	2.2	..	..	..
Papua New Guinea	2.5	0.7	3.2	75	27	0.1	0.1	5.5	4.0	..	..	..
Paraguay	1.7	3.6	5.2	233	86	..	1.1	..	1.3	..	..	..
Peru	2.4	3.7	6.1	278	141	0.7	0.9	..	1.5	1	6	2
Philippines	1.7	2.0	3.7	136	33	0.1	0.1	1.7	1.1	..	..	..
Poland	4.7	1.7	6.4	510	264	1.8	2.3	5.6	5.3	14	10	5
Portugal	5.2	3.0	7.5	1,137	803	..	3.1	..	4.0	12	9	3
Puerto Rico	..	6.5	..	..	..	..	1.8	..	3.3	..	..	..
Romania	2.6	1.5	4.1	265	63	1.5	1.8	8.8	7.6	18	10	5
Russian Federation	4.6	1.2	4.6	328	133	4.0	4.6	13.0	12.1	22	17	8

	Health expenditure			Health expenditure per capita		Physicians		Hospital beds		Inpatient admission rate	Average length of stay	Outpatient visits per capita
	Public % of GDP 1990–98[a]	Private % of GDP 1990–98[a]	Total % of GDP 1990–98[a,b]	PPP $ 1990–98[a]	$ 1990–98[a]	per 1,000 people 1980	per 1,000 people 1990–98[a]	per 1,000 people 1980	per 1,000 people 1990–98[a]	% of population 1990–98[a]	days 1990–98[a]	1990–98[a]
Rwanda	2.0	2.1	4.1	34	10	0.0[c]	0.0[c]	1.5	1.7	..	..	..
Saudi Arabia	6.4	1.6	8.0	890	611	..	1.7	..	2.3	11	4	1
Senegal	2.6	1.9	4.5	61	23	..	0.1	..	0.4	22	10	1
Sierra Leone	0.9	4.5	5.5	27	8	0.1	..	1.2	..	..	..	..
Singapore	1.2	2.1	3.2	777	841	0.9	1.4	4.2	3.6	12	..	..
Slovak Republic	5.7	1.5	7.2	728	285	..	3.0	..	7.5	20	11	4
Slovenia	6.6	0.9	7.6	1,126	746	..	2.1	7.0	5.7	16	11	..
South Africa	3.3	3.8	7.1	623	230	..	0.6	..	..	..	..	..
Spain	5.4	1.6	7.1	1,202	1,043	..	4.2	..	3.9	10	10	..
Sri Lanka	1.4	1.7	3.1	95	26	0.1	0.2	2.9	2.7	..	..	..
Sudan	0.7	2.7	3.4	48	126	0.1	0.1	0.9	1.1	..	..	..
Sweden	6.7	1.3	8.0	1,707	2,146	2.2	3.1	14.8	3.8	17	8	3
Switzerland	7.6	2.8	10.4	2,739	3,835	..	1.9	..	18.1	17	14	11
Syrian Arab Republic	0.8	1.6	2.4	90	116	0.4	1.3	1.1	1.4	..	..	..
Tajikistan	5.2	0.9	6.0	63	13	2.4	2.1	10.0	8.8	16	15	..
Tanzania	1.3	1.8	3.0	15	8	..	0.0[c]	1.4	0.9	..	..	..
Thailand	1.9	4.1	6.0	349	112	0.1	0.4	1.5	2.0	..	..	2
Togo	1.3	1.3	2.6	36	8	0.1	0.1	..	1.5	..	..	..
Trinidad and Tobago	2.5	1.8	4.3	323	204	0.7	0.8	..	5.1	..	..	..
Tunisia	2.2	2.9	5.1	287	108	0.3	0.7	2.1	1.7	8	..	..
Turkey	2.9	2.9	5.8	386	177	0.6	1.2	2.2	2.5	7	6	2
Turkmenistan	4.1	1.1	5.1	146	31	2.9	0.2	10.6	11.5	17	15	..
Uganda	1.9	4.1	5.9	65	18	..	0.0[c]	..	0.9	..	..	..
Ukraine	3.6	1.5	5.1	169	42	3.7	4.5	12.5	11.8	20	17	10
United Arab Emirates	0.8	7.4	8.2	1,495	1,428	1.1	1.8	2.8	2.6	11	5	..
United Kingdom	5.6	1.1	6.7	1,418	1,597	..	1.7	9.3	4.2	15	10	6
United States	5.8	7.2	13.0	3,950	4,108	1.8	2.7	5.9	3.7	13	7	6
Uruguay	1.9	7.2	9.1	823	621	..	3.7	..	4.4	..	..	..
Uzbekistan	3.4	0.6	4.1	87	..	2.9	3.3	11.5	8.3	19	14	..
Venezuela, RB	2.6	1.6	4.2	248	171	0.8	2.4	0.3	1.5	..	..	..
Vietnam	0.8	4.0	4.8	81	18	0.2	0.6	3.5	1.7	8	7	3
West Bank and Gaza	4.9	3.7	8.6	..	81	..	0.5	..	1.2	9	3	4
Yemen, Rep.	2.4	3.2	5.6	34	18	..	0.2	..	0.6	..	..	..
Yugoslavia, FR (Serb./Mont.)	..	..	..	..	..	..	2.0	..	5.3	8	12	2
Zambia	3.6	3.4	6.9	52	23	0.1	0.1	..	..	..	..	..
Zimbabwe	2.9	3.7	6.6	186	49	0.2	0.1	3.1	0.5	..	..	..
World	**2.6 w**	**3.0 w**	**5.5 w**	**561 w**	**489 w**	**1.0 w**	**1.5 w**	**3.4 w**	**3.3 w**	**9 w**	**13 w**	**6 w**
Low income	1.2	3.1	4.5	74	21	0.5	0.5	1.7	1.3	13	11	4
Middle income	2.5	2.6	5.0	267	117	1.2	1.8	3.4	3.4	6	12	4
Lower middle income	2.3	2.5	4.7	190	62	1.2	1.9	3.4	3.5	6	13	4
Upper middle income	3.4	2.9	6.2	549	318	..	1.6	..	3.3	6	8	4
Low & middle income	1.9	2.8	4.8	179	73	0.9	1.2	2.7	2.5	7	12	4
East Asia & Pacific	1.7	2.4	4.2	151	43	0.8	1.5	2.0	2.5	4	13	4
Europe & Central Asia	4.0	1.6	5.2	326	138	3.0	3.3	10.4	8.8	17	14	6
Latin America & Carib.	3.2	3.3	6.5	452	272	..	1.6	..	2.2	2	5	2
Middle East & N. Africa	2.3	2.3	4.6	228	126	..	1.0	..	1.8	5	6	3
South Asia	0.9	3.8	5.1	87	19	0.3	0.4	0.7	0.7	..	..	3
Sub-Saharan Africa	1.7	2.6	4.3	89	42	..	0.1	..	1.1	12	6	1
High income	6.0	3.7	9.7	2,587	2,702	..	2.8	..	7.2	15	14	8
Europe EMU	6.7	2.3	8.9	1,980	2,045	..	3.9	..	7.6	19	12	6

a. Data are for the most recent year available. b. Data may not sum to total because of rounding and because of differences in the year for which the most recent data are available. c. Less than 0.05. d. Less than 0.5.

Health expenditure, services, and use | 2.15

National health accounts track financial flows in the health sector, including both public and private expenditures. In contrast with high-income countries, few developing countries have health accounts that are methodologically consistent with national accounting approaches. The difficulties in creating national health accounts go beyond data collection. To establish a national health accounting system, a country needs to define the boundaries of the health care system and a taxonomy of health care delivery institutions. The accounting system should be comprehensive and standardized, providing not only accurate measurements of financial flows, but also information on the equity and efficiency of health financing to inform health policy.

The absence of consistent national health accounting systems in most developing countries makes cross-country comparisons of health spending difficult. Records of private out-of-pocket expenditures are often lacking. And compiling estimates of public health expenditures is complicated in countries where state or provincial and local governments are involved in health care financing and delivery because the data on public spending often are not aggregated. The data in the table are the product of an effort by the World Health Organization (WHO), the Organisation for Economic Co-operation and Development (OECD), and the World Bank to collect all available information on health expenditures from national and local government budgets, national accounts, household surveys, insurance publications, international donors, and existing tabulations.

Health service indicators (physicians and hospital beds per 1,000 people) and health care utilization indicators (inpatient admission rates, average length of stay, and outpatient visits) come from a variety of sources (see *Data sources*). Data are lacking for many countries, and for others comparability is limited by differences in definitions. In estimates of health personnel, for example, some countries incorrectly include retired physicians (because deletions are made only periodically) or those working outside the health sector. There is no universally accepted definition of hospital beds. Moreover, figures on physicians and hospital beds are indicators of availability, not of quality or use. They do not show how well trained the physicians are or how well equipped the hospitals or medical centers are. And physicians and hospital beds tend to be concentrated in urban areas, so these indicators give only a partial view of health services available to the entire population.

The average length of stay in hospitals is an indicator of the efficiency of resource use. Longer stays may reflect a waste of resources if patients are kept in hospitals beyond the time medically required, inflating demand for hospital beds and increasing hospital costs. Aside from differences in cases and financing methods, cross-country variations in average length of stay may result from differences in the role of hospitals. Many developing countries do not have separate extended care facilities, so hospitals become the source of both long-term and acute care. Other factors may also explain the variations. Data for some countries may not include all public and private hospitals. Admission rates may be overstated in some countries if outpatient surgeries are counted as hospital admissions. And in many countries outpatient visits, especially emergency visits, may result in double counting if a patient receives treatment in more than one department.

Table 2.15a

Responsiveness of national health systems to the population's expectations

	Score		Score
Malaysia	6.3	Senegal	5.0
Korea, Rep.	6.1	Brazil	4.8
Philippines	5.8	Bulgaria	4.4
Poland	5.7	Georgia	4.3
Indonesia	5.5	Burkina Faso	4.2
South Africa	5.4	Bangladesh	4.1
Ecuador	5.3	Nepal	3.8
Egypt, Arab Rep.	5.1	Uganda	3.7

Source: WHO, *World Health Report 2000.*

In *World Health Report 2000* the World Health Organization identified three overall goals for health systems: good health, fairness of financial contributions, and responsiveness to the population's expectations. To measure responsiveness, informants in 35 countries were asked to evaluate the health system's performance with respect to dignity, autonomy, confidentiality, prompt attention, quality of basic amenities, social support, and choice of provider. The results were combined in a composite score ranging from 0 to 10. The higher the score, the greater the health system's responsiveness.

• **Public health expenditure** consists of recurrent and capital spending from government (central and local) budgets, external borrowings and grants (including donations from international agencies and nongovernmental organizations), and social (or compulsory) health insurance funds. • **Private health expenditure** includes direct household (out-of-pocket) spending, private insurance, charitable donations, and direct service payments by private corporations. • **Total health expenditure** is the sum of public and private health expenditure. It covers the provision of health services (preventive and curative), family planning activities, nutrition activities, and emergency aid designated for health but does not include provision of water and sanitation. • **Physicians** are defined as graduates of any faculty or school of medicine who are working in the country in any medical field (practice, teaching, research). • **Hospital beds** include inpatient beds available in public, private, general, and specialized hospitals and rehabilitation centers. In most cases beds for both acute and chronic care are included. • **Inpatient admission rate** is the percentage of the population admitted to hospitals during a year. • **Average length of stay** is the average duration of inpatient hospital admissions. • **Outpatient visits per capita** are the number of visits to health care facilities per capita, including repeat visits.

The estimates of health expenditure come from the WHO's *World Health Report 2000* and subsequent updates and from the OECD for its member countries, supplemented by World Bank country and sector studies, including the Human Development Network's *Sector Strategy: Health, Nutrition, and Population* (World Bank 1997e). Data are also drawn from World Bank public expenditure reviews, the International Monetary Fund's Government Finance Statistics database, and other studies. The data on private expenditure are largely from household surveys and World Bank poverty assessments and sector studies. The data on physicians, hospital beds, and utilization of health services are from the WHO and OECD, supplemented by country data.

2.16 Disease prevention: coverage and quality

	Access to an improved water source		Access to improved sanitation facilities		Tetanus vaccinations	Child immunization rate		Access to essential drugs	Tuberculosis treatment success rate	DOTS detection rate
	% of population		% of population		% of pregnant women	% of children under 12 months		% of population	% of cases	% of cases
						Measles	DPT			
	1990	2000	1990	2000	1996–98[a]	1995–99[a]	1995–99[a]	1997	1990–97[a]	1995–97[a]
Albania	..	..	..	..	65	85	97	60	..	..
Algeria	..	94	..	73	52	78	83	95	86	97
Angola	..	38	..	44	24	57	22	20	..	70
Argentina	..	79	..	85	..	97	88	70	..	4
Armenia	..	..	..	..	..	91	91	40	77	49
Australia	100	100	100	100	..	89	88	100	..	..
Austria	100	100	100	100	..	90	90	100	..	..
Azerbaijan	..	..	..	..	..	87	93	86	7	..
Bangladesh	91	97	97	53	86	61	66	65	72	19
Belarus	..	100	..	..	..	98	97	70	..	..
Belgium	..	..	..	..	..	73	62	..	..	..
Benin	..	63	20	23	66	92	90	..	72	35
Bolivia	74	79	55	66	..	100	96	70	62	80
Bosnia and Herzegovina	..	..	..	..	..	83	90	..	..	..
Botswana	95	..	61	..	54	74	85	90	70	80
Brazil	82	87	72	77	..	90	83	40	..	..
Bulgaria	..	100	..	100	..	96	96	..	..	..
Burkina Faso	53	..	24	29	44	29	34	60	29	16
Burundi	65	..	89	..	9	47	63	20	45	25
Cambodia	..	30	..	18	31	63	64	30	94	50
Cameroon	52	62	87	92	38	46	48	..	..	..
Canada	100	100	100	100	..	96	97	100	..	..
Central African Republic	59	60	30	31	6	39	45	50	37	65
Chad	..	27	18	29	27	49	33	46	47	15
Chile	90	94	97	97	..	95	94	..	80	80
China	71	75	29	38	13	85	85	85	96	23
Hong Kong, China	..	..	..	..	..	82	88	..	..	..
Colombia	87	91	82	85	..	76	81	..	..	..
Congo, Dem. Rep.	..	45	..	20	10	15	25	..	80	46
Congo, Rep.	..	51	..	..	30	23	29	61	69	70
Costa Rica	..	98	..	96	..	92	93	100	..	..
Côte d'Ivoire	65	77	49	..	49	59	60	80	56	55
Croatia	..	95	..	100	..	92	93	100	..	..
Cuba	..	95	..	95	..	99	94	100	92	87
Czech Republic	..	..	..	..	..	97	98	..	66	53
Denmark	..	100	..	..	..	84	..	..	..	..
Dominican Republic	78	79	60	71	77	94	83	77	..	..
Ecuador	..	71	..	59	..	100	80	40	40	1
Egypt, Arab Rep.	94	95	87	94	61	96	95	..	81	10
El Salvador	..	74	..	83	..	75	94	80	..	45
Eritrea	..	46	..	13	34	55	56	57	..	3
Estonia	..	..	..	..	..	92	95	100	..	..
Ethiopia	22	24	13	15	30	53	64	..	72	24
Finland	100	100	100	100	..	98	100	98	..	..
France	..	..	..	..	..	84	98	..	..	..
Gabon	..	70	..	21	4	30	31	30	..	..
Gambia, The	..	62	..	37	96	87	90	90	80	75
Georgia	..	76	..	99	..	95	89	30	58	29
Germany	..	..	..	..	..	75	85	100	..	..
Ghana	56	64	60	63	18	73	72	..	51	33
Greece	..	..	..	..	..	88	88	100	..	..
Guatemala	78	92	77	85	..	93	86	50	81	52
Guinea	45	48	55	58	48	61	57	93	75	52
Guinea-Bissau	..	49	..	47	46	19	6	..	..	..
Haiti	46	46	25	28	..	85	59	30	..	2
Honduras	84	90	..	77	..	98	95	40	..	..

Disease prevention: coverage and quality | 2.16

	Access to an improved water source		Access to improved sanitation facilities		Tetanus vaccinations	Child immunization rate		Access to essential drugs	Tuberculosis treatment success rate	DOTS detection rate
	% of population		% of population		% of pregnant women	% of children under 12 months		% of population	% of cases	% of cases
	1990	2000	1990	2000	1996–98[a]	Measles 1995–99[a]	DPT 1995–99[a]	1997	1990–97[a]	1995–97[a]
Hungary	99	99	99	99	..	100	100	100	..	..
India	78	88	21	31	80	60	78	35	79	1
Indonesia	69	76	54	66	78	71	64	80	81	7
Iran, Islamic Rep.	86	95	81	81	75	99	100	85	87	7
Iraq	..	85	..	79	56	94	90	85	..	..
Ireland	..	..	..	..	..	..	76		..	..
Israel	..	..	..	..	..	94	93	..	..	100
Italy	..	..	..	..	..	55	70		82	9
Jamaica	..	71	..	84	..	82	81	95	72	81
Japan	..	..	..	..	..	94	70	100	..	..
Jordan	97	96	98	99	22	83	85	100	..	..
Kazakhstan	..	91	..	99	..	100	98	..	..	..
Kenya	40	49	84	86	64	79	79	35	77	55
Korea, Dem. Rep.	..	..	..	..	5	98	87	..	..	..
Korea, Rep.	..	92	..	63	..	85	74	..	71	56
Kuwait	..	..	..	..	8	96	94	..	..	..
Kyrgyz Republic	..	77	..	100	..	97	98	..	88	4
Lao PDR	..	90	..	46	32	71	56	..	55	32
Latvia	..	..	..	..	..	97	95	90	64	69
Lebanon	..	100	..	99	..	81	94	..	89	56
Lesotho	..	91	..	92	17	58	64	80	71	65
Libya	71	72	97	97	..	93	97	100	..	..
Lithuania	..	..	..	..	..	97	93	..	..	..
Macedonia, FYR	..	99	..	99	..	98	95	..	..	..
Madagascar	44	47	36	42	30	46	48	65	55	60
Malawi	49	57	73	77	81	79	81	..	68	50
Malaysia	..	..	..	..	71	88	89	70	69	70
Mali	55	65	70	69	19	57	52	60	65	17
Mauritania	37	37	30	33	63	56	19	100	96	40
Mauritius	100	100	100	99	78	79	85	..	..	..
Mexico	83	86	69	73	..	94	96	92	75	15
Moldova	..	100	..	..	..	99	97	25	..	..
Mongolia	..	60	..	30	..	86	90	60	78	30
Morocco	75	82	62	75	33	93	94	..	88	94
Mozambique	..	60	..	43	41	90	81	50	54	57
Myanmar	64	68	45	46	78	85	73	60	79	25
Namibia	72	77	33	41	70	65	72	80	54	74
Nepal	66	81	21	27	65	81	76	20	85	11
Netherlands	100	100	100	100	..	96	97	100	81	45
New Zealand	..	..	..	..	..	83	88	100	..	..
Nicaragua	70	79	76	84	23	97	83	46	79	90
Niger	53	59	15	20	41	25	21	..	57	21
Nigeria	49	57	60	63	29	26	21	10	32	10
Norway	100	100	..	..	..	93	95	100	80	90
Oman	37	39	84	92	96	90	99	90	87	83
Pakistan	84	88	34	61	58	81	80	65	70	2
Panama	..	87	..	94	..	90	92	80	..	..
Papua New Guinea	42	42	82	82	11	57	56	90	60	4
Paraguay	63	79	89	95	..	70	77	..	51	55
Peru	72	77	64	76	57	92	98	60	89	95
Philippines	87	87	74	83	46	87	87	95	82	3
Poland	..	..	..	..	..	97	98	..	..	..
Portugal	..	..	..	..	..	96	97	100	74	67
Puerto Rico	..	..	..	..	..	..	..	..	68	81
Romania	..	58	..	53	..	98	97	85	..	..
Russian Federation	..	99	..	..	..	98	97	..	62	1

	Access to an improved water source		Access to improved sanitation facilities		Tetanus vaccinations	Child immunization rate		Access to essential drugs	Tuberculosis treatment success rate	DOTS detection rate
	% of population		% of population		% of pregnant women	% of children under 12 months		% of population	% of cases	% of cases
						Measles	DPT			
	1990	2000	1990	2000	1996–98[a]	1995–99[a]	1995–99[a]	1997	1990–97[a]	1995–97[a]
Rwanda	..	41	..	8	43	78	85	60	61	45
Saudi Arabia	..	95	..	100	66	92	93	..	..	..
Senegal	72	78	57	70	34	48	52	..	41	62
Sierra Leone	..	28	..	28	42	68	56	..	74	37
Singapore	100	100	100	100	..	86	94	100	86	28
Slovak Republic	..	100	..	100	..	99	99	100	73	34
Slovenia	100	100	..	..	..	96	92	100	87	60
South Africa	..	86	..	86	26	82	76	80	69	6
Spain	..	..	..	..	..	93	94	100	..	..
Sri Lanka	66	83	82	83	78	95	99	95	80	71
Sudan	67	75	58	62	55	88	87	15	..	1
Sweden	100	100	100	100	..	96	99	..	..	..
Switzerland	100	100	100	100	79	81	..	100	..	..
Syrian Arab Republic	..	80	..	90	53	97	97	80	92	5
Tajikistan	..	..	..	..	..	90	96	..	..	..
Tanzania	50	54	88	90	27	78	82	..	76	55
Thailand	71	80	86	96	81	94	97	95	78	5
Togo	51	54	37	34	32	47	48	70	60	15
Trinidad and Tobago	..	86	..	88	..	88	90	..	..	..
Tunisia	80	..	76	..	50	93	100	51	..	..
Turkey	80	83	87	91	..	76	79	..	..	..
Turkmenistan	..	58	..	100	..	97	98	..	..	..
Uganda	44	50	84	75	38	53	51	70	33	65
Ukraine	..	..	..	..	87	99	99	..	..	..
United Arab Emirates	..	..	..	..	..	96	94	..	..	..
United Kingdom	100	100	100	100	..	91	93	..	..	..
United States	100	100	100	100	..	92	96	..	71	86
Uruguay	..	98	..	95	..	93	93	..	80	95
Uzbekistan	..	85	..	100	..	96	99	..	..	..
Venezuela, RB	..	84	..	74	..	79	79	90	80	75
Vietnam	48	56	73	73	83	93	93	85	90	77
West Bank and Gaza	..	..	..	..	31	93	94	..	..	..
Yemen, Rep.	66	69	39	45	26	74	72	50	76	30
Yugoslavia, FR (Serb./Mont.)	..	..	..	..	..	84	92	80	..	..
Zambia	52	64	63	78	37	72	92	..	..	..
Zimbabwe	77	85	64	68	58	79	81	70	..	..
World	**76 w**	**81 w**	**49 w**	**56 w**		**75 w**	**78 w**			
Low income	70	76	40	46		64	70			
Middle income	75	82	47	59		88	88			
Lower middle income	74	80	43	54		87	87			
Upper middle income	..	87	..	79		90	88			
Low & middle income	73	79	44	52		74	78			
East Asia & Pacific	70	75	38	48		83	82			
Europe & Central Asia	..	90	..	..		97	97			
Latin America & Carib.	81	85	72	78		90	87			
Middle East & N. Africa	85	89	78	83		91	92			
South Asia	79	87	31	36		63	75			
Sub-Saharan Africa	49	55	55	55		57	59			
High income	..	..	..	..		89	91			
Europe EMU	..	..	..	..		82	91			

a. Data are for the most recent year available.

Disease prevention: coverage and quality | 2.16

About the data

The indicators in the table are based on data provided to the World Health Organization (WHO) by member states as part of their efforts to monitor and evaluate progress in implementing national health strategies. Because reliable, observation-based statistical data for these indicators do not exist in some developing countries, the data are at times estimated.

People's health is influenced by the environment in which they live. Lack of clean water and basic sanitation is the main reason diseases transmitted by feces are so common in developing countries. Drinking water contaminated by feces deposited near homes and an inadequate water supply cause diseases accounting for 10 percent of the disease burden in developing countries (World Bank 1993c). The data on access to an improved water source measure the share of the population with ready access to water for domestic purposes. The data are based on surveys and estimates provided by governments to the WHO-UNICEF Joint Monitoring Programme. The coverage rates for water and sanitation are based on information from service users on the facilities their households actually use, rather than on information from service providers, who may include nonfunctioning systems. Access to drinking water from an improved source does not ensure that the water is adequate or safe, as these characteristics are not tested at the time of the surveys.

Neonatal tetanus is an important cause of infant mortality in some developing countries. It can be prevented through immunization of the mother during pregnancy. Recommended doses for full protection are generally two tetanus shots during the first pregnancy and one booster shot during each subsequent pregnancy, with five doses considered adequate for lifetime protection. Information on tetanus shots during pregnancy is collected through surveys in which pregnant respondents are asked to show antenatal cards on which tetanus shots have been recorded. Because not all women have antenatal cards, respondents are also asked about their receipt of these injections. Poor recall may result in a downward bias in estimates of the share of births protected. But in settings where receiving injections is common, respondents may erroneously report having received tetanus toxoid.

Governments in developing countries usually finance immunization against measles and diphtheria, pertussis (whooping cough), and tetanus (DPT) as part of the basic public health package, though they often rely on personnel with limited training to provide the vaccines. According to the World Bank's *World Development Report 1993: Investing in Health,* these diseases accounted for about 10 percent of the disease burden among children under five in 1990, compared with an expected 23 percent at 1970 levels of vaccination. In many develop-

ing countries, however, lack of precise information on the size of the cohort of children under one year of age makes immunization coverage difficult to measure.

Essential drugs are pharmaceutical products included by the WHO on a periodically updated list of safe and effective treatments for both communicable and noncommunicable diseases. They are cost-effective elements of a health system that can treat many common diseases and conditions, including, among many others, anemia, hypertension, tuberculosis, and malaria.

Data on the success rate of tuberculosis treatment are provided for countries that have implemented the recommended control strategy: directly observed treatment, short-course (DOTS). Countries that have not adopted DOTS or have only recently done so are omitted because of lack of data or poor comparability or reliability of reported results. The treatment success rate for tuberculosis provides a useful indicator of the quality of health services. A low rate or no success suggests that infectious patients may not be receiving adequate treatment. An essential complement to the tuberculosis treatment success rate is the DOTS detection rate, which indicates whether there is adequate coverage by the recommended case detection and treatment strategy. A country with a high treatment success rate may still face big challenges if its DOTS detection rate remains low.

Table 2.16a

Children receiving each dose of vaccine against diphtheria, pertussis, and tetanus in selected developing countries, various years
%

	DPT1	DPT2	DPT3
Bangladesh	84.9	79.2	69.3
Bolivia	81.6	66.8	48.6
Burkina Faso	78.3	64.5	41.0

Source: Demographic and Health Survey data.

A high "dropout" rate for DPT immunization (the proportion of children who receive the first dose but not the second, or the first and second but not the third) indicates a need to provide better information to caregivers, so that those who bring a child to a clinic for the first vaccination bring the child back for the rest.

Definitions

• **Access to an improved water source** refers to the percentage of the population with reasonable access to an adequate amount of water from an improved source, such as a household connection, public standpipe, borehole, protected well or spring, and rainwater collection. Unimproved sources include vendors, tanker trucks, and unprotected wells and springs. Reasonable access is defined as the availability of at least 20 liters a person a day from a source within one kilometer of the dwelling. • **Access to improved sanitation facilities** refers to the percentage of the population with at least adequate excreta disposal facilities (private or shared, but not public) that can effectively prevent human, animal, and insect contact with excreta. Improved facilities range from simple but protected pit latrines to flush toilets with a sewerage connection. To be effective, facilities must be correctly constructed and properly maintained. • **Tetanus vaccinations** refer to the percentage of pregnant women who receive two tetanus toxoid injections during their first pregnancy and one booster shot during each subsequent pregnancy. • **Child immunization rate** is the percentage of children under one year of age receiving vaccination coverage for four diseases—measles and diphtheria, pertussis (whooping cough), and tetanus (DPT). A child is considered adequately immunized against measles after receiving one dose of vaccine, and against DPT after receiving three doses. • **Access to essential drugs** refers to the percentage of the population for which a minimum of 20 of the most essential drugs are continuously available and affordable at public or private health facilities or drug outlets within one hour's walk. • **Tuberculosis treatment success rate** refers to the percentage of new, registered smear-positive (infectious) cases that were cured or in which a full-course treatment was completed. • **DOTS detection rate** is the percentage of estimated new infectious tuberculosis cases detected under the directly observed treatment, short-course (DOTS) case detection and treatment strategy.

Data sources

The table was produced using information provided to the WHO by countries, the WHO's EPI Information System, its Essential Drugs and Medicine Policy, and its *Global Tuberculosis Control Report 1999;* the United Nations Children's Fund's (UNICEF) *State of the World's Children 2001;* and the WHO and UNICEF's *Global Water Supply and Sanitation Assessment 2000 Report.*

2.17 | Reproductive health

	Total fertility rate		Adolescent fertility rate	Women at risk of unintended pregnancy	Contraceptive prevalence rate	Births attended by skilled health staff		Maternal mortality ratio	
	births per woman		births per 1,000 women ages 15–19	% of married women ages 15–49	% of women ages 15–49	% of total		per 100,000 live births Reported	Adjusted
	1980	1999	1999	1990–99[a]	1990–99[a]	1982	1996–99[a]	1990–99[a]	1995
Albania	3.6	2.4	12	..	..	..	..	..	31
Algeria	6.7	3.4	20	..	51	..	..	220	150
Angola	6.9	6.7	215	..	..	34	..	..	1,300
Argentina	3.3	2.5	63	..	..	..	..	38	85
Armenia	2.3	1.3	40	..	..	..	96	35	29
Australia	1.9	1.8	29	..	..	99	..	..	6
Austria	1.6	1.3	20	..	..	..	..	..	11
Azerbaijan	3.2	2.0	22	..	..	..	99	43	37
Bangladesh	6.1	3.2	140	15	54	2	14	440	600
Belarus	2.0	1.3	22	..	..	..	..	28	33
Belgium	1.7	1.6	11	..	..	..	..	..	8
Benin	7.0	5.6	109	21	16	..	60	500	880
Bolivia	5.5	4.0	80	26	49	..	59	390	550
Bosnia and Herzegovina	2.1	1.6	33	..	..	..	..	10	15
Botswana	6.1	4.1	74	..	..	..	..	330	480
Brazil	3.9	2.2	71	7	77	98	88	160	260
Bulgaria	2.0	1.1	45	..	..	..	..	15	23
Burkina Faso	7.5	6.6	139	26	12	12	27	..	1,400
Burundi	6.8	6.1	53	..	..	..	..	..	1,900
Cambodia	4.7	4.4	13	..	22	..	31	470	590
Cameroon	6.4	4.9	134	13	19	..	55	430	720
Canada	1.7	1.5	20	..	..	..	..	..	6
Central African Republic	5.8	4.7	128	16	15	..	..	1,100	1,200
Chad	6.9	6.3	185	9	4	24	11	830	1,500
Chile	2.8	2.2	44	..	..	92	100	20	33
China	2.5	1.9	15	..	85	..	..	55	60
Hong Kong, China	2.0	1.0	5	..	..	89	..	..	..
Colombia	3.9	2.7	84	8	72	..	..	80	120
Congo, Dem. Rep.	6.6	6.2	211	..	..	..	..	..	940
Congo, Rep.	6.3	5.9	138	..	..	..	..	..	1,100
Costa Rica	3.6	2.5	76	..	..	..	..	29	35
Côte d'Ivoire	7.4	4.9	128	43	15	..	47	600	1,200
Croatia	..	1.5	18	..	..	..	..	6	18
Cuba	2.0	1.6	67	..	..	..	..	27	24
Czech Republic	2.1	1.2	23	..	69	..	..	9	14
Denmark	1.5	1.8	9	..	..	..	..	10	15
Dominican Republic	4.2	2.8	12	13	64	..	96	230	110
Ecuador	5.0	3.1	73	..	66	62	..	160	210
Egypt, Arab Rep.	5.1	3.3	61	16	52	..	56	170	170
El Salvador	4.9	3.2	104	8	60	..	90	120	180
Eritrea	..	5.6	116	28	8	..	..	1,000	1,100
Estonia	2.0	1.2	25	..	..	..	..	50	80
Ethiopia	6.6	6.3	151	..	4	10	..	..	1,800
Finland	1.6	1.8	11	..	..	..	..	6	6
France	1.9	1.8	9	..	71	..	..	10	20
Gabon	4.5	5.1	162	..	..	..	..	600	620
Gambia, The	6.5	5.5	167	..	..	41	..	..	1,100
Georgia	2.3	1.3	31	21	41	..	..	70	22
Germany	1.4	1.4	14	..	..	..	..	8	12
Ghana	6.5	4.3	84	23	22	..	44	210	590
Greece	2.2	1.3	17	..	..	..	..	1	2
Guatemala	6.3	4.7	109	23	38	..	..	190	270
Guinea	6.1	5.3	162	24	6	..	35	670	1,200
Guinea-Bissau	5.8	5.5	183	..	..	..	..	910	910
Haiti	5.9	4.1	66	48	18	34	..	..	1,100
Honduras	6.5	4.0	108	11	50	..	55	110	220

Reproductive health | 2.17

	Total fertility rate		Adolescent fertility rate	Women at risk of unintended pregnancy	Contraceptive prevalence rate	Births attended by skilled health staff		Maternal mortality ratio	
	births per woman		births per 1,000 women ages 15–19	% of married women ages 15–49	% of women ages 15–49	% of total		per 100,000 live births Reported	Adjusted
	1980	1999	1999	1990–99[a]	1990–99[a]	1982	1996–99[a]	1990–99[a]	1995
Hungary	1.9	1.3	27	..	73	99	..	15	23
India	5.0	3.1	107	16	52	23	..	410	440
Indonesia	4.3	2.6	57	11	57	27	43	450	470
Iran, Islamic Rep.	6.7	2.7	48	..	73	..	..	37	130
Iraq	6.4	4.4	38	..	..	..	..	..	370
Ireland	3.2	1.9	14	..	60	..	..	6	9
Israel	3.2	2.9	20	..	..	..	..	5	8
Italy	1.6	1.2	8	..	..	..	..	7	11
Jamaica	3.7	2.5	92	15	65	86	95	120	120
Japan	1.8	1.4	3	..	..	..	..	8	12
Jordan	6.8	3.7	32	14	50	..	97	41	41
Kazakhstan	2.9	2.0	39	11	66	..	98	70	80
Kenya	7.8	4.5	106	24	39	..	44	590	1,300
Korea, Dem. Rep.	2.8	2.0	2	..	..	..	..	110	35
Korea, Rep.	2.6	1.6	4	..	..	70	..	20	20
Kuwait	5.3	2.7	32	..	..	..	98	5	25
Kyrgyz Republic	4.1	2.7	34	12	60	..	98	65	80
Lao PDR	6.7	5.4	41	..	25	..	..	650	650
Latvia	2.0	1.1	31	..	..	..	..	45	70
Lebanon	4.0	2.4	25	..	61	..	95	100	130
Lesotho	5.5	4.5	81	..	23	..	..	..	530
Libya	7.3	3.6	53	..	45	..	94	75	120
Lithuania	2.0	1.4	35	..	..	..	..	18	27
Macedonia, FYR	2.5	1.8	36	..	..	..	..	3	17
Madagascar	6.6	5.6	167	26	19	..	47	490	580
Malawi	7.6	6.3	151	36	22	..	..	620	580
Malaysia	4.2	3.0	24	..	..	88	..	39	39
Mali	7.1	6.4	174	26	7	14	24	580	630
Mauritania	6.3	5.3	129	..	..	..	58	550	870
Mauritius	2.7	2.0	40	..	75	..	..	50	45
Mexico	4.7	2.8	70	..	65	..	..	55	65
Moldova	2.4	1.7	53	..	74	..	..	42	65
Mongolia	5.3	2.7	50	10	60	..	..	150	65
Morocco	5.4	2.9	47	16	59	29	..	230	390
Mozambique	6.5	5.2	159	7	6	..	44	1,100	980
Myanmar	4.9	3.1	24	..	..	97	57	230	170
Namibia	5.9	4.7	101	22	29	..	..	230	370
Nepal	6.1	4.3	117	28	29	..	10	..	830
Netherlands	1.6	1.6	5	..	75	100	..	7	10
New Zealand	2.0	2.0	30	..	..	..	..	15	15
Nicaragua	6.3	3.6	130	15	60	..	65	150	250
Niger	7.4	7.3	211	17	8	26	18	590	920
Nigeria	6.9	5.2	114	22	6	..	..	700	1,100
Norway	1.7	1.8	12	..	..	..	..	6	9
Oman	9.9	4.5	61	..	..	..	..	19	120
Pakistan	7.0	4.8	100	32	24	..	..	..	200
Panama	3.7	2.5	77	..	..	80	..	70	100
Papua New Guinea	5.8	4.2	66	29	26	..	53	370	390
Paraguay	5.2	4.0	81	17	57	..	71	190	170
Peru	4.5	3.1	65	12	64	30	56	270	240
Philippines	4.8	3.5	43	26	47	..	56	170	240
Poland	2.3	1.4	21	..	..	..	..	8	12
Portugal	2.2	1.5	22	..	..	..	100	8	12
Puerto Rico	2.6	1.9	64	..	78	..	..	..	30
Romania	2.4	1.3	41	..	48	..	..	41	60
Russian Federation	1.9	1.3	42	..	34	..	99	50	75

	Total fertility rate		Adolescent fertility rate	Women at risk of unintended pregnancy	Contraceptive prevalence rate	Births attended by skilled health staff		Maternal mortality ratio	
	births per woman		births per 1,000 women ages 15–19	% of married women ages 15–49	% of women ages 15–49	% of total		per 100,000 live births	
								Reported	Adjusted
	1980	1999	1999	1990–99[a]	1990–99[a]	1982	1996–99[a]	1990–99[a]	1995
Rwanda	8.3	6.0	54	37	21	20	..	..	2,300
Saudi Arabia	7.3	5.5	107	..	21	..	91	..	23
Senegal	6.8	5.4	99	33	13	..	47	560	1,200
Sierra Leone	6.5	5.9	196	..	..	..	..	..	2,100
Singapore	1.7	1.5	9	..	..	100	100	6	9
Slovak Republic	2.3	1.4	62	..	..	..	..	9	14
Slovenia	2.1	1.2	65	..	..	..	..	11	17
South Africa	4.6	2.9	45	..	69	..	84	..	340
Spain	2.2	1.2	8	..	..	..	..	6	8
Sri Lanka	3.5	2.1	21	..	..	85	95	60	60
Sudan	6.5	4.5	53	25	10	23	..	500	1,500
Sweden	1.7	1.5	9	..	..	..	..	5	8
Switzerland	1.5	1.5	4	..	..	..	..	5	8
Syrian Arab Republic	7.4	3.7	41	..	45	43	..	110	200
Tajikistan	5.6	3.3	28	..	..	..	..	65	120
Tanzania	6.7	5.4	130	24	25	..	35	530	1,100
Thailand	3.5	1.9	76	..	72	40	..	44	44
Togo	6.8	5.1	84	..	24	..	51	480	980
Trinidad and Tobago	3.3	1.8	42	..	..	..	99	..	65
Tunisia	5.2	2.2	11	..	60	40	82	70	70
Turkey	4.3	2.4	56	11	64	70	81	130	55
Turkmenistan	4.9	2.8	16	..	..	..	..	65	65
Uganda	7.2	6.4	189	29	15	..	..	510	1,100
Ukraine	2.0	1.3	34	..	68	..	..	27	45
United Arab Emirates	5.4	3.3	71	..	..	..	..	3	30
United Kingdom	1.9	1.7	28	..	..	..	..	7	10
United States	1.8	2.1	50	..	64	..	99	8	12
Uruguay	2.7	2.3	66	..	..	..	..	26	50
Uzbekistan	4.8	2.7	42	14	56	..	98	21	60
Venezuela, RB	4.2	2.9	95	..	..	82	..	60	43
Vietnam	5.0	2.3	34	..	75	100	77	160	95
West Bank and Gaza	..	5.8	96	..	42	..	..	..	..
Yemen, Rep.	7.9	6.2	100	39	21	..	22	350	850
Yugoslavia, FR (Serb./Mont.)	2.3	1.7	32	..	..	..	93	10	15
Zambia	7.0	5.4	138	27	26	..	47	650	870
Zimbabwe	6.4	3.6	85	15	48	37	84	400	610
World	**3.7 w**	**2.7 w**	**68 w**		**50 w**				
Low income	5.3	3.7	104		23				
Middle income	3.2	2.2	39		53				
Lower middle income	3.0	2.1	33		53				
Upper middle income	3.7	2.4	58		65				
Low & middle income	4.1	2.9	73		49				
East Asia & Pacific	3.0	2.1	27		57				
Europe & Central Asia	2.5	1.6	39		64				
Latin America & Carib.	4.1	2.6	73		59				
Middle East & N. Africa	6.1	3.5	52		52				
South Asia	5.3	3.4	110		49				
Sub-Saharan Africa	6.6	5.3	130		21				
High income	1.8	1.7	25		75				
Europe EMU	1.8	1.4	11		75				

a. Data are for the most recent year available.

Reproductive health | 2.17

About the data

Reproductive health is a state of physical and mental well-being in relation to the reproductive system and its functions and processes. Means of achieving reproductive health include education and services during pregnancy and childbirth, provision of safe and effective contraception, and prevention and treatment of sexually transmitted diseases. Health conditions related to sex and reproduction have been estimated to account for 25 percent of the global disease burden in women (Murray and Lopez 1998). Reproductive health services will need to expand rapidly over the next two decades, when the number of women and men of reproductive age is projected to increase by more than 300 million.

Total and adolescent fertility rates are based on data on registered live births from vital registration systems or, in the absence of such systems, from censuses or sample surveys. As long as the surveys are fairly recent, the estimated rates are generally considered reliable measures of fertility in the recent past. In cases where no empirical information on age-specific fertility rates is available, a model is used to estimate the share of births to adolescents. For countries without vital registration systems, fertility rates for 1999 are generally based on extrapolations from trends observed in censuses or surveys from earlier years.

An increasing number of couples in the developing world want to limit or postpone childbearing but are not using effective contraceptive methods. These couples face the risk of unintended pregnancy, shown in the table as the percentage of married women of reproductive age who do not want to become pregnant but are not using contraception (Bulatao 1998). Information on this indicator is collected through surveys and excludes women not exposed to the risk of pregnancy because of postpartum anovulation, menopause, or infertility. Common reasons for not using contraception are lack of knowledge about contraceptive methods and concerns about their possible health side-effects.

Contraceptive prevalence reflects all methods—ineffective traditional methods as well as highly effective modern methods. Contraceptive prevalence rates are obtained mainly from Demographic and Health Surveys and contraceptive prevalence surveys (see *Primary data documentation* for the most recent survey year). Unmarried women are often excluded from such surveys, which may bias the estimates.

The share of births attended by skilled health staff is an indicator of a health system's ability to provide adequate care for pregnant women. Good antenatal and postnatal care improves maternal health and reduces maternal and infant mortality. But data may not reflect such improvements because health information systems are often weak, maternal deaths are underreported, and rates of maternal mortality are difficult to measure.

Maternal mortality ratios are generally of unknown reliability, as are many other cause-specific mortality indicators. Household surveys such as the Demographic and Health Surveys attempt to measure maternal mortality by asking respondents about survivorship of sisters. The main disadvantage of this method is that the estimates of maternal mortality that it produces pertain to 12 years or so before the survey, making them unsuitable for monitoring recent changes or observing the impact of interventions. In addition, measurement of maternal mortality is subject to many types of errors. Even in high-income countries with vital registration systems, misclassification of maternal deaths has been found to lead to serious underestimation. The maternal mortality ratios shown in the table as reported are estimates based on national surveys, vital registration, or surveillance or are derived from community and hospital records. Those shown as adjusted are based on a modeling exercise carried out by the World Health Organization (WHO) and United Nations Children's Fund (UNICEF). In this exercise maternal mortality was estimated with a regression model using information on fertility, birth attendants, and HIV prevalence. Neither set of ratios can be assumed to provide an accurate estimate of maternal mortality in any of the countries in the table.

Definitions

- **Total fertility rate** is the number of children that would be born to a woman if she were to live to the end of her childbearing years and bear children in accordance with current age-specific fertility rates. • **Adolescent fertility rate** is the number of births per 1,000 women ages 15–19. • **Women at risk of unintended pregnancy** are fertile, married women of reproductive age who do not want to become pregnant and are not using contraception. • **Contraceptive prevalence rate** is the percentage of women who are practicing, or whose sexual partners are practicing, any form of contraception. It is usually measured for married women ages 15–49 only. • **Births attended by skilled health staff** are the percentage of deliveries attended by personnel trained to give the necessary supervision, care, and advice to women during pregnancy, labor, and the postpartum period, to conduct deliveries on their own, and to care for newborns. • **Maternal mortality ratio** is the number of women who die during pregnancy and childbirth, per 100,000 live births.

Data sources

The data on reproductive health come from Demographic and Health Surveys, the WHO's *Coverage of Maternity Care* (1997) and other WHO sources, UNICEF, and national statistical offices.

Access to reproductive health care in selected developing countries, various years

	Contraceptive prevalence rate			Pregnant women receiving antenatal care			Births attended by skilled health staff		
	% of women ages 15–49			%			%		
	Poorest quintile	Richest quintile	Average	Poorest quintile	Richest quintile	Average	Poorest quintile	Richest quintile	Average
Bangladesh	39	49	42	14	59	26	2	30	8
Brazil	56	77	70	68	98	86	72	99	88
Cameroon	1	13	4	53	99	79	32	95	64
India	25	51	37	25	89	49	12	79	34
Indonesia	46	57	55	74	99	89	21	89	49
Kenya	13	49	32	88	96	92	23	78	44

Note: Households are grouped into quintiles by assets.

Source: World Bank and Macro International analysis based on Demographic and Health Survey data.

Women in the poorest households are less likely to use contraception and more likely to have unwanted or mistimed births than those in the richest households. And they are less likely to receive antenatal care and to have skilled health workers attending them in childbirth. The result is higher rates of maternal mortality among the poor.

2.18 | Health: risk factors and future challenges

	Years lived in poor health		Prevalence of anemia	Low-birthweight babies	Prevalence of child malnutrition		Consumption of iodized salt	Prevalence of smoking		Incidence of tuberculosis	Prevalence of HIV
	Males	Females			Weight for age	Height for age		Males	Females		
	% of lifespan	% of lifespan	% of pregnant women	% of births	% of children under 5	% of children under 5	% of households	% of adults	% of adults	per 100,000 people	% of adults
	1999	1999	1985–99[a]	1992–98[a]	1993–99[a]	1993–99[a]	1992–99[a]	1988–99[a]	1988–99[a]	1997	1999
Albania	13	13	..	8	8	15	..	44	6	28	0.01
Algeria	8	12	42	..	13	18	92	44	7	44	0.07
Angola	20	21	29	..	41	53	10	..	..	238	2.78
Argentina	10	11	26	7	2	5	90	47	34	56	0.69
Armenia	10	11	..	..	3	12	70	50	..	44	0.01
Australia	8	8	..	7	0	0	..	27	23	8	0.15
Austria	7	7	..	6	..	..	..	30	19	19	0.23
Azerbaijan	11	11	..	6	10	22	..	30	1	58	0.01
Bangladesh	13	14	53	50	56	55	55	4	10	246	0.02
Belarus	10	10	..	6	..	..	37	55	5	65	0.28
Belgium	8	8	..	6	..	..	..	31	26	16	0.15
Benin	18	20	41	9	29	25	79	37	..	220	2.45
Bolivia	14	13	54	9	8	27	91	43	18	253	0.10
Bosnia and Herzegovina	11	11	..	..	..	..	..	48	..	81	0.04
Botswana	18	18	..	..	17	29	27	21	..	503	35.80
Brazil	13	12	33	8	6	11	95	38	29	78	0.57
Bulgaria	9	9	..	7	..	..	..	49	24	43	0.01
Burkina Faso	20	22	24	..	33	33	23	..	..	155	6.44
Burundi	20	21	68	16	..	..	80	..	..	252	11.32
Cambodia	16	14	..	18	47	53	7	..	..	539	4.04
Cameroon	17	17	44	..	22	29	83	36	..	133	7.73
Canada	8	10	..	6	..	..	..	27	23	7	0.30
Central African Republic	18	19	67	..	23	28	87	..	..	237	13.84
Chad	18	20	37	..	39	40	55	24	..	205	2.69
Chile	10	11	13	5	1	2	100	26	18	29	0.19
China	10	11	52	..	9	16	91	63	4	113	0.07
Hong Kong, China	..	..	..	5	..	..	..	27	3	95	0.06
Colombia	11	12	24	17	8	15	92	24	21	55	0.31
Congo, Dem. Rep.	17	17	..	20	34	45	90	..	..	263	5.07
Congo, Rep.	19	22	..	..	..	..	..	..	..	277	6.43
Costa Rica	12	14	27	6	5	6	97	29	7	18	0.54
Côte d'Ivoire	11	10	34	..	24	24	90	42	2	290	10.76
Croatia	9	9	..	..	1	1	90	31	..	64	0.02
Cuba	8	10	47	8	..	..	45	48	26	18	0.03
Czech Republic	9	10	23	6	..	..	..	28	12	20	0.04
Denmark	8	8	..	..	..	..	..	32	30	11	0.17
Dominican Republic	13	14	..	14	6	11	13	24	17	114	2.80
Ecuador	11	12	17	17	..	..	99	47	18	165	0.29
Egypt, Arab Rep.	9	11	24	..	11	21	84	44	5	36	0.02
El Salvador	12	12	14	11	12	23	91	38	12	74	0.60
Eritrea	18	21	..	..	44	38	80	..	..	227	2.87
Estonia	10	10	..	..	..	..	..	48	22	52	0.04
Ethiopia	19	22	42	9	..	..	0	6	..	251	10.63
Finland	8	9	..	..	..	..	..	27	20	13	0.05
France	7	8	..	6	..	..	..	39	27	19	0.44
Gabon	15	15	..	..	..	..	..	..	..	174	4.16
Gambia, The	16	16	80	..	26	30	9	..	..	211	1.95
Georgia	9	10	..	..	3	12	..	60	15	67	0.01
Germany	9	8	..	..	..	..	..	43	30	15	0.10
Ghana	17	17	64	8	25	26	28	..	..	214	3.60
Greece	7	7	..	..	..	..	..	46	28	29	0.16
Guatemala	13	13	45	8	24	46	49	38	18	85	1.38
Guinea	20	21	..	13	..	..	37	60	44	171	1.54
Guinea-Bissau	18	20	74	..	..	..	..	..	..	181	2.50
Haiti	16	18	64	15	28	32	10	11	9	385	5.17
Honduras	12	12	14	9	25	39	80	36	11	96	1.92

Health: risk factors and future challenges | **2.18**

	Years lived in poor health		Prevalence of anemia	Low-birthweight babies	Prevalence of child malnutrition		Consumption of iodized salt	Prevalence of smoking		Incidence of tuber-culosis	Preva-lence of HIV
	Males	Females			Weight for age	Height for age		Males	Females		
	% of lifespan	% of lifespan	% of pregnant women	% of births	% of children under 5	% of children under 5	% of households	% of adults	% of adults	per 100,000 people	% of adults
	1999	1999	1985–99[a]	1992–98[a]	1993–99[a]	1993–99[a]	1992–99[a]	1988–99[a]	1988–99[a]	1997	1999
Hungary	9	10	..	8	..	..	..	44	27	47	0.05
India	11	13	88	34	45	43	70	45	7	187	0.70
Indonesia	12	12	64	15	34	42	64	69	3	285	0.05
Iran, Islamic Rep.	8	12	17	10	11	15	94	25	5	55	0.01
Iraq	10	12	18	24	..	..	10	40	5	160	0.01
Ireland	8	8	..	..	..	..	..	32	31	21	0.10
Israel	9	10	..	8	..	..	..	33	25	7	0.08
Italy	7	8	..	..	..	..	..	32	17	10	0.35
Jamaica	11	12	40	11	4	7	100	15	..	8	0.71
Japan	7	8	..	8	..	..	..	53	13	29	0.02
Jordan	8	12	50	2	5	8	95	44	5	11	0.02
Kazakhstan	12	12	27	9	8	16	53	33	..	104	0.04
Kenya	18	18	35	..	22	33	100	67	32	297	13.95
Korea, Dem. Rep.	11	12	71	..	32	15	5	42	..	178	0.01
Korea, Rep.	9	11	..	..	..	..	..	65	6	142	0.01
Kuwait	12	16	40	7	2	3	..	34	2	81	0.12
Kyrgyz Republic	13	14	..	6	11	25	27	60	16	99	0.01
Lao PDR	17	17	62	60	40	47	93	..	..	167	0.05
Latvia	10	10	..	4	..	..	..	53	18	82	0.11
Lebanon	8	11	49	19	3	12	92	53	..	26	0.09
Lesotho	17	18	7	..	16	44	73	39	1	407	23.57
Libya	8	12	..	..	5	15	90	..	..	19	0.05
Lithuania	9	13	..	4	..	..	..	41	9	80	0.02
Macedonia, FYR	11	11	..	8	6	7	..	51	18	47	0.00
Madagascar	19	23	..	15	40	48	73	..	..	205	0.14
Malawi	21	23	55	..	30	48	58	20	9	404	15.96
Malaysia	9	12	56	8	20	..	..	49	4	112	0.42
Mali	21	24	58	..	27	49	9	..	..	292	2.03
Mauritania	19	20	24	9	23	44	3	..	..	226	0.52
Mauritius	12	10	29	..	15	10	0	42	3	66	0.08
Mexico	12	12	41	9	8	18	97	..	..	41	0.29
Moldova	10	10	20	5	..	..	..	44	3	73	0.20
Mongolia	13	13	45	11	13	25	68	55	19	205	0.00
Morocco	10	11	45	4	..	..	..	..	..	122	0.03
Mozambique	19	20	58	..	26	36	62	..	..	255	13.22
Myanmar	12	12	58	..	28	42	65	71	52	171	1.99
Namibia	17	18	16	..	..	..	59	65	35	527	19.54
Nepal	14	14	65	23	47	48	55	20	15	211	0.29
Netherlands	7	8	..	..	..	..	..	37	30	10	0.19
New Zealand	9	10	..	6	..	..	..	26	24	5	0.06
Nicaragua	13	13	36	8	12	25	86	51	16	95	0.20
Niger	24	26	41	..	50	41	64	..	..	148	1.35
Nigeria	19	20	55	..	39	38	98	15	2	214	5.06
Norway	8	9	..	5	..	..	..	34	33	6	0.07
Oman	12	13	54	8	23	23	61	13	0	13	0.11
Pakistan	12	13	37	25	38	36	19	36	9	181	0.10
Panama	11	11	..	8	..	..	95	56	20	57	1.54
Papua New Guinea	15	14	16	16	..	..	..	76	80	250	0.22
Paraguay	13	12	44	9	..	..	83	24	7	73	0.11
Peru	12	12	53	6	8	26	93	42	16	265	0.35
Philippines	11	12	48	11	30	33	15	75	18	310	0.07
Poland	8	9	..	8	..	..	..	39	19	44	0.06
Portugal	8	9	..	7	..	..	..	30	7	55	0.74
Puerto Rico	..	..	..	14	..	..	..	22	10	10	..
Romania	10	10	31	10	..	..	..	43	15	121	0.02
Russian Federation	10	10	30	..	3	13	30	63	14	106	0.18

	Years lived in poor health		Prevalence of anemia	Low-birthweight babies	Prevalence of child malnutrition		Consumption of iodized salt	Prevalence of smoking		Incidence of tuberculosis	Prevalence of HIV
					Weight for age	Height for age					
	Males % of lifespan 1999	Females % of lifespan 1999	% of pregnant women 1985–99[a]	% of births 1992–98[a]	% of children under 5 1993–99[a]	% of children under 5 1993–99[a]	% of households 1992–99[a]	Males % of adults 1988–99[a]	Females % of adults 1988–99[a]	per 100,000 people 1997	% of adults 1999
Rwanda	20	23	..	..	27	42	95	7	4	276	11.21
Saudi Arabia	8	12	..	5	..	..	..	40	8	46	0.01
Senegal	19	19	26	..	22	23	9	32	5	223	1.77
Sierra Leone	22	27	31	..	..	..	75	19	..	315	2.99
Singapore	10	12	..	7	..	..	..	27	3	48	0.19
Slovak Republic	8	9	..	..	..	..	..	55	30	35	0.01
Slovenia	9	10	..	5	..	..	..	30	20	30	0.02
South Africa	18	18	37	..	9	23	62	..	..	394	19.94
Spain	7	8	..	..	..	..	..	42	25	61	0.58
Sri Lanka	10	10	39	18	33	20	47	..	..	48	0.07
Sudan	20	20	36	15	34	34	0	24	2	180	0.99
Sweden	8	9	..	..	..	..	..	17	22	5	0.08
Switzerland	8	9	..	5	..	..	..	38	27	11	0.46
Syrian Arab Republic	9	12	..	7	13	21	40	53	9	75	0.01
Tajikistan	15	15	50	..	..	..	20	..	..	87	0.01
Tanzania	19	21	59	..	31	43	74	50	12	307	8.10
Thailand	12	12	57	7	19	16	50	39	2	142	2.15
Togo	18	19	48	..	25	22	73	..	..	353	5.98
Trinidad and Tobago	9	10	53	14	..	..	..	..	..	11	1.05
Tunisia	7	11	38	16	9	23	98	61	4	40	0.04
Turkey	8	12	74	..	8	16	18	51	49	41	0.01
Turkmenistan	15	13	..	..	..	..	0	27	1	74	0.00
Uganda	21	23	30	..	26	38	69	52	17	312	8.30
Ukraine	9	9	..	8	..	..	4	49	21	61	0.96
United Arab Emirates	10	13	..	..	7	..	..	24	1	21	0.18
United Kingdom	7	8	..	6	..	..	..	29	28	18	0.11
United States	9	9	..	7	1	2	..	28	22	7	0.61
Uruguay	9	10	20	8	4	10	..	..	..	31	0.33
Uzbekistan	12	13	..	..	19	31	17	40	1	81	0.01
Venezuela, RB	11	12	29	12	8	15	90	42	39	42	0.49
Vietnam	12	13	..	11	37	39	89	73	4	189	0.24
West Bank and Gaza	..	..	..	6	..	..	..	40	3	26	..
Yemen, Rep.	13	14	..	26	46	52	39	60	29	111	0.01
Yugoslavia, FR (Serb./Mont.)	11	11	..	..	2	7	63	..	57	51	0.10
Zambia	21	21	34	10	24	42	90	35	10	576	19.95
Zimbabwe	18	19	..	11	16	21	80	34	1	543	25.06
World	11 w	12 w	55 w	.. w	.. w	.. w	67 w	47 w	12 w	136 w	1.05 w
Low income	13	15	69	..	..	..	61	43	9	212	2.01
Middle income	10	11	45	..	14	17	72	55	11	108	0.53
Lower middle income	10	11	47	..	9	18	69	58	8	112	0.18
Upper middle income	11	12	35	..	..	..	87	43	23	91	1.84
Low & middle income	12	13	55	..	..	..	67	50	10	157	1.19
East Asia & Pacific	11	11	54	..	12	23	74	63	6	150	0.22
Europe & Central Asia	10	10	39	..	..	..	25	51	21	75	0.18
Latin America & Carib.	12	12	34	10	9	16	89	37	25	81	0.58
Middle East & N. Africa	9	12	29	..	..	..	53	41	7	67	0.03
South Asia	11	13	79	34	47	43	66	40	8	193	0.56
Sub-Saharan Africa	19	20	45	..	..	..	60	..	..	267	8.38
High income	8	8	..	..	..	..	..	35	22	18	0.33
Europe EMU	8	8	..	..	..	..	..	38	25	22	0.31

a. Data are for the most recent year available.

About the data

The limited availability of data on health status is a major constraint in assessing the health situation in developing countries. Surveillance data are lacking for a number of major public health concerns. Estimates of prevalence and incidence are available for some diseases but are often unreliable and incomplete. National health authorities differ widely in their capacity and willingness to collect or report information. To compensate for the paucity of data and ensure reasonable reliability and international comparability, the World Health Organization (WHO) prepares estimates in accordance with epidemiological models and statistical standards.

An effort to summarize the overall health status of populations with one indicator was undertaken by the WHO as part of its *World Health Report 2000*. The WHO has developed a measure indicating the number of years lived in good health, known as disability-adjusted life expectancy (DALE). The measure, equal to life expectancy minus the number of years spent in poor health, is estimated from three sources: life table survival to each age, the assumed prevalence of each disability (diseases, disorders, or impairments), and weights assigned to the severity of the disability. Life table survivorship is available for some countries from vital registration and can be approximated with model life tables for others; prevalence of disability is much less available from measured sources. Time spent in poor health averages about 8 years, ranging between 6 and 10 years for most countries. The data are presented in the table as the share of the expected lifespan spent in poor health.

Adequate quantities of micronutrients (vitamins and minerals) are essential for healthy growth and development. Studies indicate that more people are deficient in iron (anemic) than any other micronutrient, and most are women of reproductive age. Anemia during pregnancy can harm both the mother and the fetus, causing loss of the baby, premature birth, or low birthweight. Estimates of the prevalence of anemia among pregnant women are generally drawn from clinical data, which suffer from two weaknesses: the sample is based on those who seek care and is therefore not random, and private clinics or hospitals may not be part of the reporting network.

Low birthweight, which is associated with maternal malnutrition, raises the risk of infant mortality and stunts growth in infancy and childhood. Estimates of low-birthweight infants are drawn mostly from hospital records. But many births in developing countries take place at home, and these births are seldom recorded. A hospital birth may indicate higher income and therefore better nutrition, or it could indicate a higher-risk birth, possibly skewing the data on birthweights downward. The data should therefore be treated with caution.

Estimates of child malnutrition, based on both weight for age (underweight) and height for age (stunting), are from national survey data. The proportion of children underweight is the most common indicator of malnutrition. Being underweight, even mildly, increases the risk of death and inhibits cognitive development in children. Moreover, it perpetuates the problem from one generation to the next, as malnourished women are more likely to have low-birthweight babies. Height for age reflects linear growth achieved pre- and postnatally, and a deficit indicates long-term, cumulative effects of inadequacies of health, diet, or care. It is often argued that stunting is a proxy for multifaceted deprivation.

Iodine deficiency is the single most important cause of preventable mental retardation, and it contributes significantly to the risk of stillbirth and miscarriage. Iodized salt is the best source of iodine, and a global campaign to iodize edible salt is significantly reducing the risks (UNICEF, *The State of the World's Children 1999*).

Smoking is the most common form of tobacco use in most countries, and the prevalence of smoking is therefore a good measure of the extent of the tobacco epidemic (Corrao and others 2000). While the prevalence of smoking has been declining in some high-income countries, it has been increasing in many low- and middle-income countries. Tobacco use causes heart and other vascular diseases, and cancers of the lung and other organs. Given the long delay between starting to smoke and the onset of disease, the health impact of smoking in developing countries will increase rapidly in the next few decades. Because the data present a one-time estimate, with no information on intensity of smoking or duration, they should be interpreted with caution. The data in the table are based on surveys and other studies compiled in *Tobacco Control Country Profiles* (Corrao and others 2000), issued for the 2000 World Conference on Tobacco or Health.

Tuberculosis is the main cause of death from a single infectious agent among adults in developing countries (WHO 1999). In high-income countries tuberculosis has reemerged largely as a result of cases among immigrants. The estimates of tuberculosis incidence in the table are based on a new approach in which reported cases are adjusted using the ratio of case notifications to the estimated share of cases detected by panels of 80 epidemiologists convened by the WHO.

Adult HIV prevalence rates reflect the rate of HIV infection in each country's population. The estimates of HIV prevalence are based on extrapolations from data collected through surveys and surveillance of small, non-representative groups.

Definitions

• **Years lived in poor health** show the difference between life expectancy at birth and disability-adjusted life expectancy (DALE), expressed as a percentage of life expectancy. • **Prevalence of anemia**, or iron deficiency, refers to the percentage of pregnant women with hemoglobin levels less than 11 grams per deciliter. • **Low-birthweight babies** are newborns weighing less than 2,500 grams, with the measurement taken within the first hours of life, before significant postnatal weight loss has occurred. • **Prevalence of child malnutrition** is the percentage of children under five whose weight for age and height for age are less than minus two standard deviations from the median for the international reference population ages 0–59 months. For children up to two years of age, height is measured by recumbent length. For older children, height is measured by stature while standing. The reference population, adopted by the WHO in 1983, is based on children from the United States, who are assumed to be well nourished. • **Consumption of iodized salt** refers to the percentage of households that use edible salt fortified with iodine. • **Prevalence of smoking** is the percentage of adult men and women who smoke cigarettes. The age range varies among countries, but in most is 18 and above or 15 and above. • **Incidence of tuberculosis** is the estimated number of new tuberculosis cases (pulmonary, smear positive, extra-pulmonary). • **Prevalence of HIV** refers to the percentage of people ages 15–49 who are infected with HIV.

Data sources

The data are drawn from a variety of sources, including the United Nations Administrative Committee on Coordination, Subcommittee on Nutrition's *Update on the Nutrition Situation;* the WHO's *World Health Report 2000* and *Global Tuberculosis Control Report 1999;* Corrao and others' *Tobacco Control Country Profiles* (2000); UNICEF's *State of the World's Children 2001;* the WHO and UNICEF's *Low Birth Weight: A Tabulation of Available Information* (1992); and UNAIDS and the WHO's *AIDS Epidemic Update* (2000).

2.19 | Mortality

	Life expectancy at birth		Infant mortality rate		Under-five mortality rate		Child mortality rate		Adult mortality rate		Survival to age 65	
	years		per 1,000 live births		per 1,000		Male per 1,000	Female per 1,000	Male per 1,000	Female per 1,000	Male % of cohort	Female % of cohort
	1980	1999	1980	1999	1980	1999	1988–99[a]	1988–99[a]	1999	1999	1999	1999
Albania	69	72	47	24	57	..	15	15	175	84	72	84
Algeria	59	71	98	34	139	39	..	..	153	117	73	79
Angola	41	47	154	127	261	208	..	..	427	375	36	41
Argentina	70	74	35	18	38	22	..	..	160	78	74	86
Armenia	73	74	26	14	..	18	..	..	159	77	74	87
Australia	74	79	11	5	13	5	..	..	108	55	83	91
Austria	73	78	14	4	17	5	..	..	121	59	81	91
Azerbaijan	68	71	30	16	..	21	..	..	205	98	68	83
Bangladesh	48	61	132	61	211	89	37	47	276	290	56	56
Belarus	71	68	16	11	..	14	..	..	335	115	55	81
Belgium	73	78	12	5	15	6	..	..	129	61	80	90
Benin	48	53	116	87	214	145	89	90	371	312	44	51
Bolivia	52	62	118	59	170	83	26	26	261	210	58	65
Bosnia and Herzegovina	70	73	31	13	..	18	..	..	166	90	74	85
Botswana	58	39	71	58	94	95	18	16	786	740	21	25
Brazil	63	67	70	32	81	40	8	9	256	139	59	77
Bulgaria	71	71	20	14	25	17	..	..	221	109	67	83
Burkina Faso	44	45	134	105	..	210	131	128	551	522	28	31
Burundi	47	42	122	105	193	176	101	114	582	546	26	30
Cambodia	39	54	201	100	330	143	..	..	364	315	44	50
Cameroon	50	51	103	77	173	154	69	75	477	419	43	48
Canada	75	79	10	5	13	6	..	..	106	53	83	92
Central African Republic	46	44	117	96	..	151	63	64	608	555	25	32
Chad	42	49	123	101	235	189	106	99	438	383	36	43
Chile	69	76	32	10	35	12	3	2	140	72	78	88
China	67	70	42	30	65	37	10	11	164	129	71	78
Hong Kong, China	74	80	11	3	..	5	..	..	106	54	84	91
Colombia	66	70	41	23	58	28	7	7	210	115	67	81
Congo, Dem. Rep.	49	46	112	85	210	161	..	..	515	482	37	42
Congo, Rep.	50	48	89	89	125	144	..	..	487	414	32	41
Costa Rica	73	77	19	12	29	14	..	..	116	68	81	89
Côte d'Ivoire	49	46	108	111	170	180	71	58	524	497	31	33
Croatia	70	73	21	8	23	9	..	..	194	76	69	86
Cuba	74	76	20	7	22	8	..	..	123	78	81	87
Czech Republic	70	75	16	5	19	5	..	..	173	81	74	87
Denmark	74	76	8	5	10	6	..	..	140	79	79	88
Dominican Republic	64	71	76	39	92	47	13	13	156	104	72	81
Ecuador	63	69	74	28	101	35	12	9	176	138	71	80
Egypt, Arab Rep.	56	67	120	47	175	61	22	28	193	168	67	72
El Salvador	57	70	84	30	120	36	17	20	207	125	67	79
Eritrea	44	50	..	60	..	105	89	78	484	431	33	40
Estonia	69	71	17	10	25	12	..	..	288	94	59	85
Ethiopia	42	42	155	104	213	180	..	..	567	523	26	30
Finland	73	77	8	4	9	5	..	..	136	59	79	91
France	74	79	10	5	13	5	..	..	124	50	81	92
Gabon	48	53	116	84	194	133	..	..	386	344	43	48
Gambia, The	40	53	159	75	216	110	83	79	411	349	42	49
Georgia	71	73	25	15	..	20	..	..	192	81	70	86
Germany	73	77	12	5	16	5	..	..	131	66	80	89
Ghana	53	58	94	57	157	109	53	51	316	272	53	58
Greece	74	78	18	6	23	7	..	..	114	61	82	90
Guatemala	57	65	84	40	121	52	15	18	288	186	57	70
Guinea	40	46	151	96	280	167	101	98	442	438	36	37
Guinea-Bissau	39	44	169	127	290	214	..	..	474	421	31	36
Haiti	51	53	123	70	200	118	59	58	438	344	40	50
Honduras	60	70	70	34	103	46	..	..	184	113	69	79

Mortality | 2.19

	Life expectancy at birth		Infant mortality rate		Under-five mortality rate		Child mortality rate		Adult mortality rate		Survival to age 65	
	years		per 1,000 live births		per 1,000		Male per 1,000 1988–99[a]	Female per 1,000 1988–99[a]	Male per 1,000	Female per 1,000	Male % of cohort	Female % of cohort
	1980	1999	1980	1999	1980	1999			1999	1999	1999	1999
Hungary	70	71	23	8	26	10	..	..	..	..	65	84
India	54	63	115	71	177	90	25	37	218	206	62	65
Indonesia	55	66	90	42	125	52	19	20	235	183	63	70
Iran, Islamic Rep.	60	71	87	26	126	33	..	..	156	139	73	78
Iraq	62	59	80	101	95	128	..	..	196	169	59	64
Ireland	73	76	11	6	14	7	..	..	124	71	79	88
Israel	73	78	16	6	19	8	..	..	110	67	83	89
Italy	74	78	15	5	17	6	..	..	116	54	81	91
Jamaica	71	75	33	20	39	24	..	..	137	84	78	86
Japan	76	81	8	4	11	4	..	..	97	45	85	93
Jordan	..	71	41	26	49	31	7	5	156	118	73	80
Kazakhstan	67	65	33	22	..	28	11	6	380	166	50	74
Kenya	55	48	75	76	115	118	36	38	591	546	35	39
Korea, Dem. Rep.	67	60	32	58	43	93	..	..	311	208	57	65
Korea, Rep.	67	73	26	8	27	9	..	..	198	93	70	85
Kuwait	71	77	27	11	35	13	..	..	122	64	80	89
Kyrgyz Republic	65	67	43	26	..	38	10	11	300	138	58	77
Lao PDR	45	54	127	93	200	143	..	..	376	317	44	51
Latvia	69	70	20	14	26	18	..	..	297	98	59	83
Lebanon	65	70	48	26	..	32	..	..	175	132	71	78
Lesotho	53	45	119	92	168	141	..	..	518	486	41	45
Libya	60	71	79	22	80	28	6	5	183	125	70	80
Lithuania	71	72	20	9	24	12	..	..	261	86	64	86
Macedonia, FYR	..	73	54	16	69	17	..	..	160	102	74	83
Madagascar	51	54	119	90	216	149	75	68	327	287	52	58
Malawi	44	39	169	132	265	227	126	114	548	541	27	28
Malaysia	67	72	30	8	42	10	4	4	183	111	72	82
Mali	42	43	184	120	..	223	136	138	470	406	35	41
Mauritania	47	54	120	88	175	142	..	..	346	294	46	52
Mauritius	66	71	32	19	40	23	..	..	207	113	69	84
Mexico	67	72	51	29	74	36	15	17	166	104	72	84
Moldova	66	67	35	17	..	22	..	..	310	173	57	74
Mongolia	58	67	82	58	..	73	27	22	199	168	66	73
Morocco	58	67	99	48	152	62	21	19	199	145	66	74
Mozambique	44	43	145	131	223	203	84	82	580	514	32	37
Myanmar	52	60	109	77	134	120	..	..	278	228	54	61
Namibia	53	50	90	63	114	108	30	34	524	475	37	40
Nepal	48	58	132	75	180	109	..	..	264	275	55	54
Netherlands	76	78	9	5	11	5	..	..	112	62	81	90
New Zealand	73	77	13	5	16	6	..	..	126	66	81	90
Nicaragua	59	69	84	34	143	43	12	11	200	137	67	77
Niger	42	46	135	116	317	252	184	202	468	374	33	42
Nigeria	46	47	99	83	196	151	118	102	444	390	39	45
Norway	76	78	8	4	11	4	..	..	111	58	83	91
Oman	60	73	41	17	95	24	..	..	139	103	77	83
Pakistan	55	63	127	90	161	126	22	37	186	153	64	69
Panama	70	74	32	20	36	25	..	..	140	83	76	85
Papua New Guinea	51	58	78	58	..	77	28	21	369	330	50	53
Paraguay	67	70	50	24	61	27	10	12	185	130	68	79
Peru	60	69	81	39	126	48	19	20	199	140	67	78
Philippines	61	69	52	31	81	41	21	19	193	146	68	76
Poland	70	73	26	9	..	10	..	..	227	88	70	86
Portugal	71	75	24	6	31	6	..	..	152	70	77	88
Puerto Rico	74	76	19	10	..	..	..	..	152	58	75	90
Romania	69	69	29	20	36	24	7	5	262	119	63	80
Russian Federation	67	66	22	16	..	20	3	2	382	138	51	79

2.19 Mortality

	Life expectancy at birth		Infant mortality rate		Under-five mortality rate		Child mortality rate		Adult mortality rate		Survival to age 65	
	years		per 1,000 live births		per 1,000		Male per 1,000 1988–99[a]	Female per 1,000 1988–99[a]	Male per 1,000	Female per 1,000	Male % of cohort	Female % of cohort
	1980	1999	1980	1999	1980	1999			1999	1999	1999	1999
Rwanda	46	40	128	123	..	203	87	73	604	566	24	27
Saudi Arabia	61	72	65	19	85	25	..	..	160	129	74	80
Senegal	45	52	117	67	..	124	76	74	459	389	38	46
Sierra Leone	35	37	190	168	336	283	..	..	544	483	23	28
Singapore	71	78	12	3	13	4	..	..	130	72	81	88
Slovak Republic	70	73	21	8	23	10	..	..	206	87	69	85
Slovenia	70	75	15	5	18	6	..	..	165	72	75	88
South Africa	57	48	67	62	91	76	..	..	601	533	47	57
Spain	76	78	12	5	16	6	..	..	127	55	81	91
Sri Lanka	68	73	34	15	48	19	10	9	150	96	76	84
Sudan	48	56	94	67	145	109	62	63	384	338	45	51
Sweden	76	79	7	4	9	4	..	..	101	57	84	91
Switzerland	76	80	9	5	11	5	..	..	104	49	84	92
Syrian Arab Republic	62	69	56	26	73	30	..	..	202	135	68	78
Tajikistan	66	69	58	20	..	34	..	..	232	140	64	77
Tanzania	50	45	108	95	176	152	61	58	542	500	30	34
Thailand	64	69	49	28	58	33	11	11	240	147	67	79
Togo	49	49	100	77	188	143	75	90	478	435	35	40
Trinidad and Tobago	68	73	35	16	40	20	4	3	180	132	74	83
Tunisia	62	73	69	24	100	30	19	19	159	133	73	79
Turkey	61	69	109	36	133	45	12	14	177	145	69	79
Turkmenistan	64	66	54	33	..	45	..	..	281	158	58	74
Uganda	48	42	116	88	180	162	82	72	597	590	25	25
Ukraine	69	67	17	14	..	17	..	..	346	134	53	79
United Arab Emirates	68	75	55	8	..	9	..	..	125	92	80	85
United Kingdom	74	77	12	6	14	6	..	..	119	66	81	89
United States	74	77	13	7	15	8	..	..	143	78	80	89
Uruguay	70	74	37	15	42	17	..	..	168	74	73	87
Uzbekistan	67	70	47	22	..	29	15	9	227	125	65	79
Venezuela, RB	68	73	36	20	42	23	..	..	155	88	74	85
Vietnam	63	69	57	37	105	42	..	..	205	144	66	76
West Bank and Gaza	..	72	..	23	..	26	10	7	164	106	73	82
Yemen, Rep.	49	56	141	79	198	97	33	36	307	283	48	51
Yugoslavia, FR (Serb./Mont.)	70	72	33	12	..	16	..	..	176	106	72	83
Zambia	50	38	90	114	149	187	96	93	607	597	25	25
Zimbabwe	55	40	80	70	108	118	26	26	569	526	31	35
World	**63 w**	**66 w**	**80 w**	**54 w**	**123 w**	**78 w**	**32 w**	**35 w**	**221 w**	**170 w**	**65 w**	**73 w**
Low income	53	59	112	77	177	116	45	51	288	258	55	60
Middle income	66	69	54	31	79	39	12	12	199	135	68	78
Lower middle income	66	69	55	32	84	40	12	13	191	133	69	78
Upper middle income	66	69	52	27	67	34	..	..	233	143	66	80
Low & middle income	60	64	86	59	135	85	32	35	239	190	62	69
East Asia & Pacific	65	69	55	35	82	44	12	13	184	141	69	76
Europe & Central Asia	68	69	41	21	..	26	..	..	289	127	60	80
Latin America & Carib.	65	70	61	30	80	38	13	14	207	122	67	80
Middle East & N. Africa	59	68	95	44	136	56	..	..	183	151	68	74
South Asia	54	63	119	74	180	99	26	38	223	212	61	65
Sub-Saharan Africa	48	47	114	92	189	161	92	86	499	453	36	41
High income	74	78	12	6	15	6	..	..	125	63	81	91
Europe EMU	74	78	13	5	16	5	..	..	125	58	80	91

a. Data are for the most recent year available.

Mortality | 2.19

Mortality rates for different age groups—infants, children, or adults—and overall indicators of mortality—life expectancy at birth or survival to a given age—are important indicators of health status in a country. Because data on the incidence and prevalence of diseases (morbidity data) frequently are unavailable, mortality rates are often used to identify vulnerable populations. And they are among the indicators most frequently used to compare levels of socioeconomic development across countries.

The main sources of mortality data are vital registration systems and direct or indirect estimates based on sample surveys or censuses. A "complete" vital registration system—one covering at least 90 percent of the population—is the best source of age-specific mortality data. But such systems are fairly uncommon in developing countries. Thus estimates must be obtained from sample surveys or derived by applying indirect estimation techniques to registration, census, or survey data. Survey data are subject to recall error, and surveys estimating infant deaths require large samples because households in which a birth or an infant death has occurred during a given year cannot ordinarily be preselected for sampling. Indirect estimates rely on estimated actuarial ("life") tables that may be inappropriate for the population concerned. Because life expectancy at birth is constructed using infant mortality data and model life tables, similar reliability issues arise for this indicator.

Life expectancy at birth and age-specific mortality rates for 1999 are generally estimates based on vital registration or the most recent census or survey available (see Primary data documentation). Extrapolations based on outdated surveys may not be reliable for monitoring changes in health status or for comparative analytical work.

Specific problems arise in calculating infant mortality rates in developing countries, where routine data collection in the health system often omits many infant deaths. In countries where civil registration of deaths is incomplete, especially in rural areas, many infants dying during the first weeks of life may not even have been registered as having been born. Rates based on civil registration in these countries, or on hospital data covering mainly urban areas, are therefore biased because they reflect the more privileged population. Infant and child mortality rates are higher for boys than for girls in countries in which parental gender preferences are absent. Child mortality captures the effect of gender discrimination better than does infant mortality, as malnutrition and medical interventions are more important in this age group. Where female child mortality is higher, as in some countries in South Asia, it is likely that girls have unequal access to resources.

Adult mortality rates have increased in many countries in Sub-Saharan Africa and Europe and Central Asia. In Sub-Saharan Africa the increase stems from AIDS-related mortality and affects both men and women. In Europe and Central Asia the causes are more diverse and affect men more. They include a high prevalence of smoking, a high-fat diet, excessive alcohol use, and stressful conditions related to the economic transition.

The percentage of a cohort surviving to age 65 reflects both child and adult mortality rates. Like life expectancy, it is a synthetic measure based on current age-specific mortality rates and used in the construction of life tables. It shows that even in countries where mortality is high, a certain share of the current birth cohort will live well beyond the life expectancy at birth, while in low-mortality countries close to 90 percent will reach at least age 65.

Figure 2.19

Progress in life expectancy has been uneven

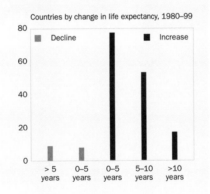

Countries by change in life expectancy, 1980–99

Note: The figure covers all countries for which data are available for 1980–99.
Source: World Bank data files.

Unweighted life expectancies for all countries represented in the figure increased by four years between 1980 and 1999, but the progress was unevenly distributed—in 17 countries life expectancy increased by more than 10 years, while in 17 others it declined. Among those with declining life expectancies, 9 are in Sub-Saharan Africa (with the largest declines) and 6 are in Europe and Central Asia (with declines of 0–2 years). The other two with declining life expectancies are Iraq and the Democratic Republic of Korea.

• **Life expectancy at birth** is the number of years a newborn infant would live if prevailing patterns of mortality at the time of its birth were to stay the same throughout its life. • **Infant mortality rate** is the number of infants dying before reaching the age of one year, per 1,000 live births in a given year. • **Under-five mortality rate** is the probability that a newborn baby will die before reaching age five, if subject to current age-specific mortality rates. • **Child mortality rate** is the probability of dying between the ages of one and five, if subject to current age-specific mortality rates. • **Adult mortality rate** is the probability of dying between the ages of 15 and 60—that is, the probability of a 15-year-old dying before reaching age 60, if subject to current age-specific mortality rates between ages 15 and 60. • **Survival to age 65** refers to the percentage of a cohort of newborn infants that would survive to age 65, if subject to current age-specific mortality rates.

The data are from the United Nations Statistics Division's *Population and Vital Statistics Report;* publications and other releases from country statistical offices; Demographic and Health Surveys from national sources and Macro International; and the United Nations Children's Fund's (UNICEF) *State of the World's Children 2000.*

ENVIRONMENT

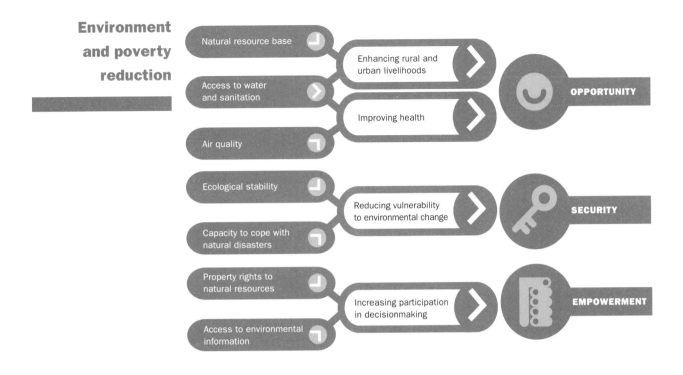

Environment and poverty reduction

Natural resource base

Access to water and sanitation

Air quality

Enhancing rural and urban livelihoods

Improving health

OPPORTUNITY

Ecological stability

Capacity to cope with natural disasters

Reducing vulnerability to environmental change

SECURITY

Property rights to natural resources

Access to environmental information

Increasing participation in decisionmaking

EMPOWERMENT

Environmental changes disproportionately affect poor people

Human development depends on the environment's providing a variety of goods and services—now and in the future. But the links between environmental conditions and human welfare are complex. Environmental changes can make poverty worse by compromising health, livelihoods, and protection from natural disasters. And economic growth can create new stresses on the environment as the demand for environmental resources rises and the damaging by-products of economic activity accumulate. But environmental resources are needed to promote economic growth and reduce poverty, and growth itself creates the means and the demand for an improved environment.

Long-term poverty reduction and sustainable economic growth are now being undermined by the continuing degradation of soils, the increasing scarcity of freshwater, the overexploitation of coastal ecosystems and fisheries, the loss of forest cover, long-term changes in the earth's climate, and the loss of biological diversity at the genetic, species, and ecosystem level. What is clear is that the roughly 2.8 billion poor and near-poor people in the world—those living on less than $2 a day—are disproportionately affected by these bad environmental conditions. They are particularly vulnerable to shocks from environmental change and natural catastrophes. Every year around 5 million people in developing countries die from waterborne diseases and polluted air. The livelihoods of around 1 billion rural people are at risk because of desertification and land degradation. And up to two-thirds of the world's people are likely to be affected by water scarcity.

Poverty, poor health, and environmental hazards

Diseases associated with environmental factors are highly concentrated among the poor.

• Sixty percent of all malaria deaths occur among the poorest fifth of the world's population.

• Half of all deaths from diarrhea also occur among the poorest fifth.

• Water-related diseases, such as cholera and diarrhea, claim an estimated 3 million lives in developing countries each year—the majority, children under five. In India alone,

nearly 1 million people die annually from waterborne diseases.

• Nearly 2 million women and children die annually in developing countries from exposure to indoor air pollution, including about 500,000 in India and 700,000 in China.

• More than 70 percent of freshwater sources are seriously contaminated or degraded, and withdrawals of groundwater exceed natural recharge rates by 160 billion cubic meters a year.

The environment affects the health of the poor . . .

The poor, with less access to improved water supplies, are more prone to water-related diseases

Environmental health risks account for a fifth of the total burden of disease in the developing world. Most of those suffering from such health risks—malaria, limited water supply, indoor air pollution, lack of sanitation—live in rural areas. But as a result of rapid urbanization and

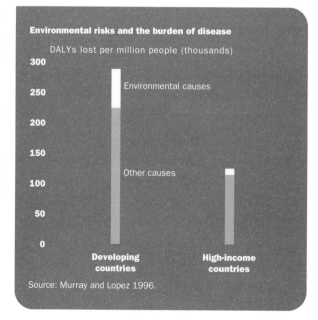

uncontrolled growth of urban slums, these health risks also affect the urban and semiurban poor.

It is difficult to quantify the health impacts from exposure to environmental hazards. But a standard measure of the burden of disease—disability-adjusted life years, or DALYs—has been established that quantifies illness and death from various causes. The concept combines life years lost as a result of premature death and years of healthy life lost as a result of illness or disability.

Poverty, natural resources, and livelihoods

The majority of poor people in most developing countries live in rural areas and depend on natural systems for their income. The very poor are often small farmers, landless laborers, or agricultural workers, dependent on such natural resources as water, soil, and fisheries for subsistence and income.

Most agricultural land in developing countries, however, has soil that is low quality or prone to degradation. About 1.2 billion hectares—almost 11 percent of the earth's

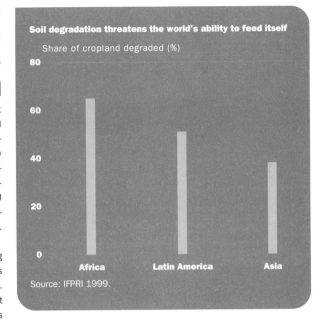

Soil degradation threatens the world's ability to feed itself

Share of cropland degraded (%)

Source: IFPRI 1999.

vegetated surface—have been degraded by human activity over the past 45 years, and an estimated 5–12 million hectares are lost annually to severe degradation in developing countries.

Soil degradation affects more than 900 million people in 100 countries. It appears to be most extensive in Africa, where it affects 65 percent of cropland area, compared with 51 percent in Latin America and 38 percent in Asia. Because grain production provides more than three-quarters of the world's food supply, soil degradation threatens the world's ability to meet its food requirements, expected to double in the next three to four decades.

. . . and their livelihoods

Land quality and food supply will be affected by a host of environmental changes

Global environmental changes will have many effects on agricultural productivity, which is already limited by climatic factors such as water availability and the length of the growing season. Changes in the climate system and atmospheric concentrations of carbon dioxide may have large regional effects on agricultural productivity. While changes in temperature and precipitation

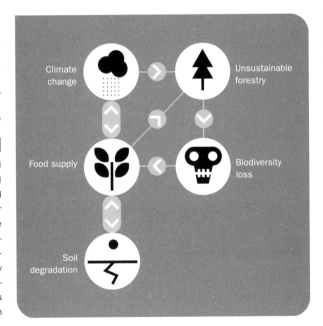

patterns may benefit agriculture in some areas, they may restrict it in others, increasing the risk of hunger and famine particularly in poor areas dependent on isolated agricultural systems, such as Sub-Saharan Africa, tropical areas of Latin America, some Pacific island nations, and South and East Asia.

Agriculture also faces a threat stemming from the need to feed a growing population: increasing need for cropland results in deforestation and loss of genetic diversity, with potentially harmful effects on agricultural sustainability.

Poverty and vulnerability

Poor countries, and poor people, are vulnerable to natural disasters, weather fluctuations, and changes in environmental conditions. Natural disasters such as floods, storms, droughts, and landslides affect poor people disproportionately. The poor tend to live in precarious housing, often located in environmentally vulnerable areas such as floodplains or steep slopes, putting them at greater risk from natural disasters and severe weather. More important, the poor have less capacity than the better-off to cope when disasters occur. They have less access to credit, and

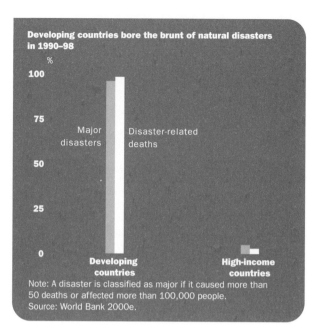

Developing countries bore the brunt of natural disasters in 1990–98

%

Major disasters

Disaster-related deaths

Developing countries

High-income countries

Note: A disaster is classified as major if it caused more than 50 deaths or affected more than 100,000 people.
Source: World Bank 2000e.

fewer assets to sell or consume in times of hardship. So the effects of natural disasters are often catastrophic for the poor.

People in low-income countries are four times as likely as people in high-income countries to die in a natural disaster. And the average costs of a natural disaster (as a share of GDP) are 20 percent higher in developing countries than in high-income economies.

Between 1990 and 1998, 94 percent of the world's 568 major natural disasters and more than 97 percent of all deaths related to natural disasters were in developing countries. The most densely populated areas of developing countries suffer the most.

The poor are more vulnerable to natural disasters, to weather fluctuations . . .

Natural disasters and the poor

Economic development is repeatedly interrupted by natural disasters—floods, droughts, landslides, windstorms, earthquakes, forest fires, and volcanic eruptions. Like economic crises, natural disasters can cause sharp increases in poverty and slow the pace of human development. And like economic crises, they hurt poor people in the short run and diminish their chances of escaping poverty in the long run.

In the past decade the incidence of natural disasters increased. Between 1988 and 1997 natural disasters claimed an estimated 50,000 lives a year and caused damage of more than $60 billion a year (World Bank 2000e). The full human and economic costs are even higher.

The heavily populated Asian region has suffered most, particularly in coastal areas. Between 1985 and 1999 Asia suffered 77 percent of all deaths, 90 percent of all homelessness, and 45 percent of all recorded economic losses due to disasters.

• In 1992 Hurricane Andrew hit the southeast coast of the United States and caused 32 deaths. In the same year a cyclone of similar intensity hit Bangladesh and caused 100,000 deaths.
• In October 1998 Hurricane Mitch—the deadliest Atlantic storm in 200 years—hit El Salvador, Guatemala, Honduras, and Nicaragua, dumping about 80 inches of rain. The hurricane caused some 10,000 deaths. The damage to the region is estimated at around $8.5 billion—more than the combined GDP of Honduras and Nicaragua.
• Over the past three decades disasters triggered by droughts, floods, and cyclones occurred five times as frequently, killed or affected 70 times as many peo-

ple, and caused twice as much damage worldwide as did earthquakes and volcanoes. The largest and best understood source of seasonal climate variation is El Niño and the Southern Oscillation (ENSO). The 1997–98 ENSO event was directly responsible for 22 disasters with total costs estimated at $25–36 billion. Whether the strong, frequent ENSO events of the 1990s signal the arrival of a new climate regime remains to be seen. But the capacity to predict natural disasters and mitigate their effects is typically limited in poor countries, exacerbating their impact.

Climate changes will affect human health

There is strong evidence that human-made pollution has contributed substantially to global warming and that the earth is likely to get hotter than previously predicted (IPCC 2001).

Long-term, human-induced changes in climate and associated rises in sea level are predicted to adversely affect human health, ecological systems, water resources, and socio-economic activities (including agriculture, forestry, fisheries, and human settlements).

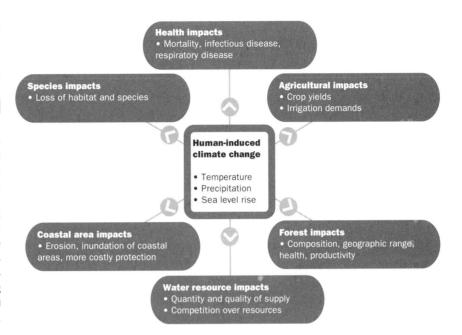

Health impacts
• Mortality, infectious disease, respiratory disease

Species impacts
• Loss of habitat and species

Agricultural impacts
• Crop yields
• Irrigation demands

Human-induced climate change
• Temperature
• Precipitation
• Sea level rise

Coastal area impacts
• Erosion, inundation of coastal areas, more costly protection

Forest impacts
• Composition, geographic range, health, productivity

Water resource impacts
• Quantity and quality of supply
• Competition over resources

. . . and to human-induced environmental changes

Economies cannot remain healthy with an unhealthy environment

Real, lasting poverty reduction is possible only if the environment is able to provide the services people depend on and if natural resources are used in a way that does not undermine long-term development. It is therefore imperative to integrate environmental concerns into poverty reduction and economic development strategies.

The critical links between poverty and the environment mean that a sound strategy to ensure environmental sustainability must be an important part of a poverty-focused development strategy. A development strategy should aim to:

• Improve people's health by reducing their exposure to environmental hazards such as indoor and urban air pollution, waterborne and vectorborne diseases, and toxic substances.

• Enhance the livelihoods of poor people who depend on land, water, forests, and biodiversity by helping them secure access to resources and creating circumstances in which they can manage those resources sustainably.

• Reduce people's vulnerability to environmental risks such as natural disasters, severe weather fluctuations, and the impacts of climate change by getting information to governments, the private sector, and poor communities and empowering them to adapt.

	Watershed	Watershed area	Countries within the watershed	Average population density	Use of watershed[a]							Water available
		sq. km		people per sq. km	Crop-land %	Forest %	Grassland %	Built-up area %	Irrigated area %	Arid area %	Wetlands %	cu. m per capita per year
Asia	Amu Darya	534,739	5	39	22.4	0.1	57.3	3.7	7.5	72.0	0.0	3,211
	Amur	1,929,955	3	34	18.4	53.8	8.8	2.6	0.8	15.1	4.4	4,917
	Brahmaputra	651,355	4	178	29.4	18.5	44.7	2.4	3.7	0.0	20.7	..
	Ganges	1,016,124	4	398	72.4	4.2	13.4	6.3	22.7	26.0	17.7	..
	Huang He	944,970	1	157	29.5	1.5	60.0	5.9	7.2	37.5	1.1	361
	Indus	1,081,718	4	163	30.0	0.4	46.4	4.6	24.1	62.6	4.2	830
	Kolyma	679,934	1	<1	0.0	0.7	45.3	0.3	0.0	0.0	1.0	722,456
	Lake Balkhash	512,015	2	11	23.2	4.0	61.1	1.5	1.9	91.6	4.7	439
	Lena	2,306,743	1	1	1.7	64.7	11.4	0.4	0.0	0.7	0.6	161,359
	Mekong	805,604	6	71	37.8	41.5	17.2	2.1	2.9	0.0	8.7	8,934
	Ob	2,972,493	4	10	36.9	33.9	16.0	3.0	0.5	42.5	11.2	14,937
	Syr Darya	782,617	4	28	22.2	2.4	67.4	3.2	5.4	88.5	2.0	1,171
	Tarim	1,152,448	2	7	2.3	0.0	35.3	0.3	0.6	61.4	16.3	754
	Tigris and Euphrates	765,742	4	57	25.4	1.2	47.7	6.2	9.1	90.9	2.9	2,189
	Yangtze	1,722,193	1	212	47.6	6.3	28.2	3.0	7.1	0.0	3.0	2,265
	Yenisey	2,554,388	2	3	12.8	39.7	32.4	1.3	0.0	10.9	2.7	79,083
Europe	Danube	795,656	13	102	66.9	18.2	3.2	10.7	5.2	2.6	1.4	2,519
	Dnieper	533,966	3	62	86.5	2.2	1.3	8.8	1.8	3.4	5.9	1,552
	Volga	1,410,951	2	42	60.2	22.5	7.3	8.2	0.4	19.6	1.1	4,260
Africa	Congo	3,730,881	9	15	7.2	44.0	45.4	0.2	0.0	0.0	9.0	22,752
	Lake Chad	2,497,738	8	12	3.1	0.2	45.2	0.2	0.0	82.8	8.2	7,922
	Niger	2,261,741	10	32	4.4	0.9	68.6	0.5	0.1	65.4	4.1	4,076
	Nile	3,254,853	10	44	10.7	2.0	53.0	1.0	1.4	67.4	6.1	2,207
	Okavango	721,258	4	2	5.5	1.7	91.1	0.2	0.0	75.8	4.1	..
	Orange	941,351	4	11	6.0	0.2	85.0	2.2	0.5	77.0	0.8	1,050
	Zambezi	1,332,412	8	18	19.9	4.0	72.0	0.7	0.1	8.8	7.6	..
North and Central America	Colorado	703,148	2	10	0.9	17.0	74.9	6.9	2.0	89.1	2.5	2,105
	Columbia	657,501	2	9	6.4	50.0	35.5	7.3	3.6	48.7	6.3	39,474
	Mackenzie	1,706,388	1	<1	2.6	66.0	14.7	1.9	0.0	0.0	48.9	408,243
	Mississippi	3,202,185	2	22	35.8	22.2	28.5	12.6	3.1	35.5	20.0	8,973
	Nelson-Saskatchewan	1,093,141	2	5	47.4	31.9	6.1	7.1	0.5	21.5	86.8	15,167
	Rio Grande	607,965	2	18	5.2	7.5	80.9	6.0	2.6	96.0	2.1	621
	St. Lawrence	1,049,636	2	43	16.4	43.5	0.1	14.5	0.2	0.0	47.2	9,095
	Yukon	847,620	2	<1	0.0	64.0	27.6	0.4	0.0	0.0	27.8	1,249,932
South America	Amazon	6,145,186	7	4	14.1	73.4	10.2	0.6	0.1	4.0	8.3	273,767
	Orinoco	953,675	2	17	7.6	50.5	37.8	2.6	0.2	8.5	15.3	90,482
	Paraná	2,582,704	4	27	43.3	18.1	33.0	4.2	0.5	9.9	10.9	8,025
	São Francisco	617,814	1	19	60.2	3.1	31.8	2.8	0.3	32.0	9.7	8,261
	Tocantins	764,213	1	5	61.5	9.9	26.2	1.3	0.0	0.0	19.1	103,383
Oceania	Murray-Darling	1,050,116	1	2	28.4	8.0	62.1	1.2	1.6	67.1	3.4	11,549

Note: Only watersheds exceeding 500,000 square kilometers are shown. For a more comprehensive list and underlying assumptions see the source.

a. Shares do not sum to 100 percent because some areas fall into more than one category or into a category not shown.

Source: World Resources Institute, UNEP, UNDP, and World Bank, *World Resources 2000–01: A Guide to the Global Environment.*

Table 3b   Global marine fisheries

	Average annual marine fish production								Year by which fully fished[c]	Discards
	Demersal[a] thousand metric tons				Pelagic[b] thousand metric tons					% of catch
	1965–67	1975–77	1985–87	1995–97	1965–67	1975–77	1985–87	1995–97		1988–92
Antarctic	5	83	73	11	0	0	1	0	1980	10
Arctic	0	0	0	0	..	..	..	..	..	..
Atlantic	7,816	9,009	8,718	7,445	9,181	12,305	9,864	10,528	1983	25
Indian	423	754	871	1,554	953	1,270	1,945	2,966	..	26
Mediterranean and the Black Sea	193	180	282	337	518	724	1,207	984	..	25
Pacific	3,953	8,481	10,910	10,078	14,638	13,044	24,783	28,356	1999	24
World	12,391	18,507	20,856	19,499	25,293	27,358	37,835	42,917	1999	25

Note: Production includes capture and aquaculture.
a. Demersal or bottom-dwelling fish species. b. Pelagic fish are those that live in ocean surface waters or open seas. c. Refers to the year in which the fishery yield reaches its peak.
Source: Food and Agriculture Organization data, as cited in World Resources Institute, UNEP, UNDP, and World Bank, *World Resources 2000–01: A Guide to the Global Environment.*

Table 3c   Atmospheric concentrations of greenhouse and ozone-depleting gases

	Carbon dioxide	Methane	Nitrous oxide	Carbon tetra-chloride	Methyl chloro-form	CFC-11[a]	CFC-12[a]	CFC-113[a]	Total gaseous chlorine[b]
	parts per million	parts per billion	parts per billion	parts per trillion	parts per trillion	parts per trillion	parts per trillion	parts per trillion	parts per trillion
1980	339	..	299	89	69	158	293	..	1,624
1981	340	..	299	90	75	166	305	..	1,692
1982	341	..	301	92	81	175	325	26	1,865
1983	343	..	302	93	85	182	341	28	1,939
1984	344	..	303	94	88	190	355	31	2,016
1985	346	..	304	96	92	200	376	36	2,121
1986	347	1,600	305	97	96	210	394	40	2,216
1987	349	1,610	305	99	98	221	413	48	2,322
1988	351	1,619	306	100	103	231	433	53	2,425
1989	353	1,641	306	100	107	240	452	59	2,524
1990	354	1,645	306	101	110	249	470	66	2,620
1991	355	1,657	307	101	113	254	484	71	2,685
1992	356	1,673	308	101	116	259	496	77	2,751
1993	357	1,671	308	101	112	260	503	80	2,764
1994	359	1,674	309	100	106	261	512	81	2,769
1995	361	1,681	309	99	97	261	518	82	2,753
1996	363	1,684	310	98	85	261	523	82	2,725
1997	364	1,690	311	97	73	260	528	83	2,693
1998	367	1,693	311	96	64	259	530	82	2,664

Note: All estimates are by volume.
a. CFC-11 (CCl_3F), CFC-12 (CCl_2F_2), and CFC-113 ($C_2Cl_3F_3$) are all chlorofluorocarbons. b. Total gaseous chlorine is calculated by multiplying the number of chlorine atoms in a unit of the chlorine-containing gases by the concentration of that gas. Chlorine acts as a catalyst in the destruction of ozone.
Source: Carbon Dioxide Information Analysis Center data, as cited in World Resources Institute, UNEP, UNDP, and World Bank, *World Resources 2000–01: A Guide to the Global Environment.*

3.1 Rural environment and land use

	Rural population			Rural population density	Land area	Land use					
	% of total	% of total	average annual % growth	people per sq. km of arable land	thousand sq. km	Arable land % of land area		Permanent cropland % of land area		Other % of land area	
	1980	1999	1980–99	1998	1998	1980	1998	1980	1998	1980	1998
Albania	66	59	0.6	345	27	21.4	21.1	4.3	4.5	74.4	74.5
Algeria	57	40	0.7	159	2,382	2.9	3.2	0.3	0.2	96.8	96.6
Angola	79	66	2.1	268	1,247	2.3	2.4	0.4	0.4	97.3	97.2
Argentina	17	10	–1.2	15	2,737	9.1	9.1	0.8	0.8	90.1	90.1
Armenia	34	30	0.4	234	28	..	17.6	..	2.3	..	80.1
Australia	14	15	1.7	5	7,682	5.7	7.0	0.0	0.0	94.2	93.0
Austria	35	35	0.4	205	83	18.6	16.9	1.2	1.0	80.2	82.1
Azerbaijan	47	43	0.9	205	87	..	19.3	..	3.0	..	77.7
Bangladesh	86	76	1.4	1,204	130	68.3	61.4	2.0	2.6	29.6	36.0
Belarus	44	29	–1.9	49	207	..	29.8	..	0.6	..	69.6
Belgium	5	3	–2.5	35	33 ª	23.2 ª	24.8 ª	0.4 ª	0.6 ª	76.4 ª	74.6ª
Benin	73	58	1.8	207	111	13.6	15.4	0.8	1.4	85.7	83.3
Bolivia	55	38	0.3	156	1,084	1.7	1.8	0.2	0.2	98.1	98.0
Bosnia and Herzegovina	65	57	–0.9	436	51	..	9.8	..	2.9	..	87.3
Botswana	85	50	0.2	231	567	0.7	0.6	0.0	0.0	99.3	99.4
Brazil	34	19	–1.3	62	8,457	4.6	6.3	1.2	1.4	94.2	92.3
Bulgaria	39	31	–1.6	60	111	34.6	38.8	3.2	2.0	62.2	59.2
Burkina Faso	92	82	1.8	260	274	10.0	12.4	0.1	0.2	89.8	87.4
Burundi	96	91	2.3	779	26	35.8	30.0	10.1	12.9	54.0	57.2
Cambodia	88	84	2.7	263	177	11.3	21.0	0.4	0.6	88.3	78.4
Cameroon	69	52	1.3	127	465	12.7	12.8	2.2	2.6	85.1	84.6
Canada	24	23	0.8	15	9,221	4.9	4.9	0.0	0.0	95.0	95.0
Central African Republic	65	59	1.8	108	623	3.0	3.1	0.1	0.1	96.9	96.8
Chad	81	77	2.4	159	1,259	2.5	2.8	0.0	0.0	97.5	97.2
Chile	19	15	0.2	111	749	5.1	2.6	0.3	0.4	94.6	96.9
China[b]	80	68	0.4	689	9,327	10.4	13.3	0.4	1.2	89.3	85.5
Hong Kong, China	9	0	..	0	1	7.0	5.1	1.0	1.0	92.0	93.9
Colombia	36	27	0.4	529	1,039	3.6	2.0	1.4	2.0	95.0	96.0
Congo, Dem. Rep.	71	70	3.1	506	2,267	2.9	3.0	0.4	0.5	96.6	96.5
Congo, Rep.	59	38	0.6	630	342	0.4	0.5	0.1	0.1	99.5	99.4
Costa Rica	57	52	1.9	824	51	5.5	4.4	4.4	5.5	90.1	90.1
Côte d'Ivoire	65	56	2.6	290	318	6.1	9.3	7.2	13.8	86.6	76.9
Croatia	50	43	–1.0	133	56	..	26.1	..	2.3	..	71.6
Cuba	32	25	–0.6	77	110	23.9	33.1	6.4	7.6	69.7	59.3
Czech Republic	25	25	0.0	84	77	..	40.1	..	3.0	..	56.9
Denmark	16	15	–0.3	33	42	62.3	55.7	0.3	0.2	37.4	44.0
Dominican Republic	50	36	0.3	280	48	22.1	22.1	7.2	9.9	70.6	68.0
Ecuador	53	38	0.6	302	277	5.6	5.7	3.3	5.2	91.1	89.2
Egypt, Arab Rep.	56	55	2.1	1,197	995	2.3	2.8	0.2	0.5	97.5	96.7
El Salvador	58	54	1.1	582	21	26.9	27.0	11.7	12.1	61.4	60.9
Eritrea	87	82	2.4	638	101	..	4.9	..	0.0	..	95.0
Estonia	30	31	0.0	40	42	..	26.5	..	0.4	..	73.1
Ethiopia	90	83	2.3	513	1,000	..	9.9	..	0.6	..	89.4
Finland	40	33	–0.6	81	305	7.8	7.1	0.0	0.0	92.2	92.9
France	27	25	0.0	79	550	31.8	33.4	2.5	2.1	65.7	64.5
Gabon	50	20	–2.0	76	258	1.1	1.3	0.6	0.7	98.2	98.1
Gambia, The	80	68	2.7	430	10	15.5	19.5	0.4	0.5	84.1	80.0
Georgia	48	40	–0.7	279	70	..	11.3	..	4.1	..	84.6
Germany	17	13	–1.4	89	349	34.4	34.0	1.4	0.7	64.1	65.3
Ghana	69	62	2.4	319	228	8.4	15.8	7.5	7.5	84.2	76.7
Greece	42	40	0.2	149	129	22.5	22.1	7.9	8.5	69.6	69.4
Guatemala	63	61	2.4	482	108	11.7	12.5	4.4	5.0	83.9	82.4
Guinea	81	68	1.6	550	246	2.9	3.6	1.8	2.4	95.4	94.0
Guinea-Bissau	83	77	1.7	298	28	9.1	10.7	1.1	1.8	89.9	87.6
Haiti	76	65	1.1	895	28	19.8	20.3	12.5	12.7	67.7	67.0
Honduras	65	48	1.4	179	112	13.9	15.1	1.8	3.1	84.3	81.7

Rural environment and land use | 3.1

	Rural population			Rural population density	Land area	Land use					
	% of total		average annual % growth	people per sq. km of arable land	thousand sq. km	Arable land % of land area		Permanent cropland % of land area		Other % of land area	
	1980	**1999**	**1980–99**	**1998**	**1998**	**1980**	**1998**	**1980**	**1998**	**1980**	**1998**
Hungary	43	36	–1.2	76	92	54.4	52.2	3.3	2.4	42.2	45.4
India	77	72	1.6	438	2,973	54.8	54.3	1.8	2.7	43.4	43.0
Indonesia	78	60	0.4	695	1,812	9.9	9.9	4.4	7.2	85.6	82.9
Iran, Islamic Rep.	50	37	1.1	145	1,622	8.0	10.4	0.5	1.2	91.5	88.4
Iraq	35	24	1.0	104	437	12.0	11.9	0.4	0.8	87.6	87.3
Ireland	45	41	0.1	114	69	16.1	19.7	0.0	0.0	83.9	80.3
Israel	11	9	1.1	153	21	15.8	17.0	4.3	4.2	80.0	78.8
Italy	33	33	0.1	231	294	32.2	28.2	10.0	9.4	57.7	62.5
Jamaica	53	44	0.1	664	11	12.5	16.1	9.7	9.2	77.8	74.7
Japan	24	21	–0.2	599	377	12.9	12.0	1.6	1.0	85.5	87.0
Jordan	40	26	1.9	485	89	3.4	2.9	0.4	1.5	96.2	95.6
Kazakhstan	46	44	–0.3	22	2,671	..	11.2	..	0.1	..	88.7
Kenya	84	68	1.9	494	569	6.7	7.0	0.8	0.9	92.5	92.1
Korea, Dem. Rep.	43	40	1.1	548	120	13.4	14.1	2.4	2.5	84.2	83.4
Korea, Rep.	43	19	–3.3	532	99	20.9	17.3	1.4	2.0	77.8	80.7
Kuwait	10	3	–5.4	821	18	0.1	0.3	0.0	0.1	99.9	99.6
Kyrgyz Republic	62	66	1.9	235	192	..	7.0	..	0.4	..	92.6
Lao PDR	87	77	1.8	483	231	2.9	3.5	0.1	0.2	97.0	96.3
Latvia	32	31	–0.4	41	62	..	29.7	..	0.5	..	69.8
Lebanon	26	11	–2.9	262	10	20.5	17.6	8.9	12.5	70.6	69.9
Lesotho	87	73	1.4	466	30	9.6	10.7	..	..	..	..
Libya	31	13	–1.5	39	1,760	1.0	1.0	0.2	0.2	98.8	98.8
Lithuania	39	32	–0.6	40	65	..	45.4	..	0.9	..	53.6
Macedonia, FYR	47	38	–0.6	133	25	..	23.1	..	1.9	..	75.0
Madagascar	82	71	2.1	408	582	4.3	4.4	0.9	0.9	94.8	94.7
Malawi	91	76	2.0	437	94	16.1	19.9	0.9	1.3	83.0	78.7
Malaysia	58	43	1.1	537	329	3.0	5.5	11.6	17.6	85.4	76.9
Mali	82	71	1.7	160	1,220	1.6	3.8	0.0	0.0	98.3	96.2
Mauritania	73	44	0.0	233	1,025	0.2	0.5	0.0	0.0	99.8	99.5
Mauritius	58	59	1.1	684	2	49.3	49.3	3.4	3.0	47.3	47.8
Mexico	34	26	0.5	98	1,909	12.1	13.2	0.8	1.1	87.1	85.7
Moldova	60	54	–0.2	129	33	..	54.5	..	11.7	..	33.8
Mongolia	48	37	0.5	67	1,567	0.8	0.8	0.0	0.0	99.2	99.2
Morocco	59	45	0.5	140	446	16.9	20.2	1.1	2.1	82.0	77.6
Mozambique	87	61	0.0	339	784	3.7	4.0	0.3	0.3	96.0	95.7
Myanmar	76	73	1.3	340	658	14.6	14.5	0.7	0.9	84.8	84.6
Namibia	77	70	2.1	143	823	0.8	1.0	0.0	0.0	99.2	99.0
Nepal	94	88	2.2	700	143	16.0	20.3	0.2	0.5	83.8	79.2
Netherlands	12	11	0.1	186	34	23.3	26.7	0.9	1.0	75.8	72.3
New Zealand	17	14	0.2	35	268	9.3	5.8	3.7	6.4	86.9	87.8
Nicaragua	47	44	2.1	87	121	9.5	20.2	1.5	2.4	89.1	77.4
Niger	87	80	2.8	163	1,267	2.8	3.9	0.0	0.0	97.2	96.1
Nigeria	73	57	1.6	248	911	30.6	31.0	2.8	2.8	66.6	66.3
Norway	30	25	–0.4	123	307	2.7	3.0	..	..	..	..
Oman	69	18	–3.1	2,785	212	0.1	0.1	0.1	0.2	99.8	99.7
Pakistan	72	64	1.9	394	771	25.9	27.8	0.4	0.8	73.7	71.4
Panama	50	44	1.3	244	74	5.8	6.7	1.6	2.1	92.5	91.2
Papua New Guinea	87	83	2.0	6,379	453	0.0	0.1	1.1	1.3	98.9	98.5
Paraguay	58	45	1.5	108	397	4.1	5.5	0.3	0.2	95.6	94.2
Peru	35	28	0.7	189	1,280	2.5	2.9	0.3	0.4	97.2	96.7
Philippines	63	42	0.2	573	298	17.5	18.4	14.8	15.1	67.7	66.5
Poland	42	35	–0.5	97	304	48.0	46.0	1.1	1.2	50.9	52.8
Portugal	71	37	–3.3	206	92	26.5	20.5	7.8	7.7	65.7	71.8
Puerto Rico	33	25	–0.4	2,990	9	8.3	3.7	7.3	5.1	84.3	91.2
Romania	51	44	–0.7	106	230	42.7	40.5	2.9	2.2	54.4	57.3
Russian Federation	30	23	1.2	27	16,889	..	7.5	..	0.1	..	92.4

	Rural population			Rural population density	Land area	Land use					
	% of total		average annual % growth	people per sq. km of arable land	thousand sq. km	Arable land % of land area		Permanent cropland % of land area		Other % of land area	
	1980	1999	1980–99	1998	1998	1980	1998	1980	1998	1980	1998
Rwanda	95	94	2.4	929	25	30.8	33.2	10.3	10.1	58.9	56.6
Saudi Arabia	34	15	–0.3	82	2,150	0.9	1.7	0.0	0.1	99.1	98.2
Senegal	64	53	1.7	219	193	12.2	11.6	0.0	0.2	87.8	88.2
Sierra Leone	76	64	1.3	649	72	6.3	6.8	0.7	0.8	93.0	92.5
Singapore	0	0	..	0	1	3.3	1.6	9.8	0.0	86.9	98.4
Slovak Republic	48	43	–0.2	157	48	..	30.6	..	2.8	..	66.6
Slovenia	52	50	0.0	427	20	..	11.5	..	2.7	..	85.8
South Africa	52	50	2.0	140	1,221	10.2	12.1	0.7	0.8	89.1	87.1
Spain	27	23	–0.7	63	499	31.1	28.6	9.9	9.6	59.0	61.8
Sri Lanka	78	77	1.2	1,664	65	13.2	13.4	15.9	15.8	70.9	70.8
Sudan	80	65	1.2	112	2,376	5.2	7.0	0.0	0.1	94.8	92.9
Sweden	17	17	0.3	53	412	7.2	6.8	..	..	..	..
Switzerland	43	32	–0.9	553	40	9.9	10.5	0.5	0.6	89.6	88.9
Syrian Arab Republic	53	46	2.3	151	184	28.5	25.6	2.5	4.2	69.1	70.2
Tajikistan	66	73	2.9	583	141	..	5.4	..	0.9	..	93.7
Tanzania	85	68	1.8	595	884	3.5	4.2	1.0	1.0	95.5	94.7
Thailand	83	79	1.1	281	511	32.3	32.9	3.5	7.0	64.2	60.1
Togo	77	67	2.2	137	54	35.9	40.4	1.6	1.8	62.6	57.7
Trinidad and Tobago	37	26	–0.8	460	5	13.6	14.6	9.0	9.2	77.4	76.2
Tunisia	49	35	0.4	116	155	20.5	18.7	9.7	12.9	69.7	68.5
Turkey	56	26	–2.1	70	770	32.9	31.8	4.1	3.3	63.0	65.0
Turkmenistan	53	55	2.9	160	470	..	3.5	..	0.1	..	96.4
Uganda	91	86	2.4	357	200	20.4	25.3	8.0	8.8	71.6	65.9
Ukraine	38	32	–0.9	49	579	..	56.7	..	1.7	..	41.6
United Arab Emirates	29	15	1.7	1,017	84	0.2	0.5	0.1	0.5	99.7	99.0
United Kingdom	11	11	0.0	100	242	28.7	25.9	0.3	0.2	71.1	73.9
United States	26	23	0.4	36	9,159	20.6	19.3	0.2	0.2	79.2	80.5
Uruguay	15	9	–2.0	24	175	8.0	7.2	0.3	0.3	91.7	92.5
Uzbekistan	59	63	2.5	335	414	..	10.8	..	0.9	..	88.3
Venezuela, RB	21	13	–0.1	117	882	3.2	3.0	0.9	1.0	95.9	96.0
Vietnam	81	80	1.9	1,080	325	18.2	17.5	1.9	4.8	79.8	77.7
West Bank and Gaza	..	..	..	..	..	..	..	..	..	..	..
Yemen, Rep.	81	76	3.3	838	528	2.6	2.8	0.2	0.2	97.2	96.9
Yugoslavia, FR (Serb./Mont.)	54	48	–0.2	..	..	28.0	..	2.9	..	69.1	..
Zambia	60	60	2.9	111	743	6.9	7.1	0.0	0.0	93.1	92.9
Zimbabwe	78	65	1.9	240	387	6.5	8.3	0.3	0.3	93.3	91.3
World	**60 w**	**54 w**	**0.9 w**	**520 w**	**130,079 s**	**10.2 w**	**10.6 w**	**0.9 w**	**1.0 w**	**88.9 w**	**88.4 w**
Low income	76	69	1.6	507	33,008	11.8	13.0	1.0	1.4	87.1	85.5
Middle income	62	50	0.3	584	66,146	7.9	8.9	1.0	1.0	91.0	90.1
Lower middle income	69	57	0.4	631	43,918	9.6	9.7	1.1	0.9	89.3	89.3
Upper middle income	36	25	–0.4	193	22,228	6.1	7.2	1.0	1.2	92.9	91.6
Low & middle income	68	59	1.0	542	99,154	9.5	10.3	1.0	1.2	89.4	88.5
East Asia & Pacific	78	66	0.5	691	15,969	10.1	12.0	1.5	2.6	88.4	85.4
Europe & Central Asia	41	33	–0.6	125	23,742	37.1	11.7	3.1	0.4	59.8	87.9
Latin America & Carib.	35	25	0.0	252	20,062	5.8	6.7	1.1	1.3	93.1	92.1
Middle East & N. Africa	52	42	1.4	534	10,995	4.5	5.2	0.4	0.7	94.9	93.9
South Asia	78	72	1.6	537	4,781	42.5	42.4	1.5	2.1	56.1	55.5
Sub-Saharan Africa	77	66	2.0	369	23,605	5.5	6.5	0.7	0.8	93.7	92.5
High income	25	23	–0.1	175	30,925	12.0	11.8	0.5	0.5	87.5	87.7
Europe EMU	26	22	–0.6	141	2,401	26.5	25.6	4.5	4.1	68.9	70.3

a. Includes Luxembourg. b. Includes Taiwan, China.

About the data

Indicators of rural development are sparse, as few indicators are disaggregated between rural and urban areas (for some that are, see tables 2.6, 3.5, and 3.10). This table shows indicators of rural population and land use. Rural population is approximated as the midyear nonurban population.

The data in the table show that land use patterns are changing. They also indicate major differences in resource endowments and uses among countries. True comparability of the data is limited, however, by variations in definitions, statistical methods, and the quality of data collection. Countries use different definitions of rural population and land use, for example. The Food and Agriculture Organization (FAO), the primary compiler of these data, occasionally adjusts its definitions of land use categories and sometimes revises earlier data. (In 1985, for example, the FAO began to exclude from cropland land used for shifting cultivation but currently lying fallow.) And following FAO practice, this year's edition of the *World Development Indicators*, like the previous two, breaks down the category *cropland*, used in earlier editions, into *arable land* and *permanent cropland*. Because the data reflect changes in data reporting procedures as well as actual changes in land use, apparent trends should be interpreted with caution.

Satellite images show land use that differs from that given by ground-based measures in both area under cultivation and type of land use. Furthermore, land use data in countries such as India are based on reporting systems that were geared to the collection of tax revenue. Because taxes on land are no longer a major source of government revenue, the quality and coverage of land use data (except for cropland) have declined. Data on forest area, aggregated in the category *other*, may be particularly unreliable because of differences in definitions and irregular surveys (see *About the data* for table 3.4).

Box 3.1

Monitoring rural development

Successful development requires understanding where and how people live and work, gathering data and formulating strategies, and closely monitoring policy outcomes at both national and subnational levels. And monitoring requires meaningful data broken down along rural and urban lines.

Because the characteristics of urban and rural development differ, disaggregated data are also needed to inform the design and implementation of policies. But there are immense problems in the availability, quality, and reliability of rural data in most developing countries. Many countries have poor capacity for collecting and analyzing data, and the data that do exist are of poor quality and often not comparable.

Rural development is the outcome of all productive activities in rural areas—agricultural and nonagricultural—that improve the livelihood and well-being of rural people. Thus a comprehensive strategy is required that links rural development and rural well-being. The World Bank is preparing an update of its rural development strategy, which has as its overarching goal reducing rural poverty.

Progress in rural poverty reduction—as a proxy for rural well-being—will be monitored using a set of indicators consistent with the international development goals (see the *World view* section). The indicators capture different aspects of poverty, reflecting the need to tackle poverty not only by increasing incomes but also by enhancing equity and improving access to basic services. For some of these indicators data are not being collected or are not being disaggregated. So the challenge is to ensure systematic collection and disaggregation.

The indicators for monitoring progress in rural poverty reduction include the following:
- Rural headcount index (the percentage of the rural population in extreme poverty, living on less than $1 a day).
- Agricultural GDP.
- Rural gender development index (the percentage of rural women and girls with access to health services and the ability to read and write).
- Rural malnutrition rate (the percentage of rural children who are malnourished).
- Rural illiteracy rate (the percentage of rural people ages 15–50 who cannot read or write).
- Rural under-five mortality rate.
- Percentage of the rural population with access to potable water.
- Percentage of the rural population with access to sanitation.
- Rural roads usable for vehicles year-round.
- Percentage of rural households with access to credit services from financial institutions.

Definitions

- **Rural population** is calculated as the difference between the total population and the urban population (see *Definitions* for tables 2.1 and 3.10). • **Rural population density** is the rural population divided by the arable land area. • **Land area** is a country's total area, excluding area under inland water bodies, national claims to continental shelf, and exclusive economic zones. In most cases the definition of inland water bodies includes major rivers and lakes. (See table 1.1 for the total surface area of countries.) • **Land use** is broken into three categories. • **Arable land** includes land defined by the FAO as land under temporary crops (double-cropped areas are counted once), temporary meadows for mowing or for pasture, land under market or kitchen gardens, and land temporarily fallow. Land abandoned as a result of shifting cultivation is excluded. • **Permanent cropland** is land cultivated with crops that occupy the land for long periods and need not be replanted after each harvest, such as cocoa, coffee, and rubber. This category includes land under flowering shrubs, fruit trees, nut trees, and vines, but excludes land under trees grown for wood or timber. • **Other land** includes forest and woodland as well as logged-over areas to be forested in the near future. Also included are uncultivated land, grassland not used for pasture, wetlands, wastelands, and built-up areas—residential, recreational, and industrial lands and areas covered by roads and other fabricated infrastructure.

Data sources

The data on urban population shares used to estimate rural population come from the United Nations Population Division's *World Urbanization Prospects: The 1999 Revision*. The total population figures are World Bank estimates. The data on land area and land use are from the FAO's electronic files and are published in its *Production Yearbook*. The FAO gathers these data from national agencies through annual questionnaires and by analyzing the results of national agricultural censuses.

3.2 Agricultural inputs

	Arable land		Irrigated land		Land under cereal production		Fertilizer consumption		Agricultural machinery			
	hectares per capita		% of cropland		thousand hectares		hundreds of grams per hectare of arable land		Tractors per 1,000 agricultural workers		Tractors per 100 hectares of arable land	
	1979–81	1996–98	1979–81	1996–98	1979–81	1998–2000	1979–81	1996–98	1979–81	1996–98	1979–81	1996–98
Albania	0.22	0.17	53.0	48.5	367	214	1,556	212	15	10	173	141
Algeria	0.37	0.26	3.4	6.9	2,968	2,478	277	101	27	39	68	121
Angola	0.41	0.26	2.2	2.1	705	888	49	15	4	3	35	34
Argentina	0.89	0.70	5.7	5.7	11,067	10,261	46	330	132	190	73	112
Armenia	..	0.13	..	51.2	..	183	..	54	..	70	..	354
Australia	2.97	2.80	3.5	4.6	15,986	16,197	269	406	751	704	75	61
Austria	0.20	0.17	0.2	0.3	1,062	817	2,615	1,836	945	1,617	2,084	2,527
Azerbaijan	..	0.21	..	75.1	..	577	..	128	..	34	..	195
Bangladesh	0.10	0.06	17.1	44.8	10,823	11,227	459	1,460	0	0	5	7
Belarus	..	0.61	..	1.8	..	2,295	..	1,371	..	121	..	158
Belgiumª	0.08	0.08	1.7	4.2	426	333	5,323	3,834	917	1,186	1,416	1,326
Benin	0.43	0.29	0.3	0.6	525	836	11	212	0	0	1	1
Bolivia	0.35	0.24	6.6	6.2	559	776	23	53	4	4	21	31
Bosnia and Herzegovina	..	0.14	..	0.3	..	370	..	326	..	270	..	580
Botswana	0.44	0.22	0.5	0.3	153	87	32	110	9	20	54	175
Brazil	0.32	0.33	3.3	4.1	20,612	16,908	915	1,020	31	58	139	151
Bulgaria	0.43	0.51	28.3	17.9	2,110	1,938	2,334	417	66	68	161	58
Burkina Faso	0.39	0.32	0.4	0.7	2,026	2,999	26	115	0	0	0	6
Burundi	0.22	0.12	4.5	6.7	203	202	11	25	0	0	1	2
Cambodia	0.29	0.33	5.8	7.1	1,241	2,010	45	26	0	0	6	3
Cameroon	0.68	0.43	0.2	0.5	1,021	1,045	56	63	0	0	1	1
Canada	1.86	1.52	1.3	1.6	19,561	17,444	416	591	824	1,678	144	156
Central African Republic	0.81	0.56	..	..	194	156	5	2	0	0	0	0
Chad	0.70	0.49	0.4	0.6	907	1,897	6	35	0	0	1	0
Chile	0.34	0.14	31.1	78.4	820	574	338	2,225	43	52	90	256
China	0.10	0.10	45.1	38.3	94,647	90,212	1,494	2,860	2	1	76	56
Hong Kong, China	0.00	0.00	37.5	31.0	0	0	..	..	0	0	10	7
Colombia	0.13	0.05	7.7	20.7	1,361	1,041	812	2,826	8	6	77	105
Congo, Dem. Rep.	0.25	0.14	0.1	0.1	1,115	2,118	12	3	0	0	3	4
Congo, Rep.	0.08	0.06	0.6	0.5	19	3	27	255	2	1	49	41
Costa Rica	0.12	0.06	12.1	24.9	136	91	2,650	7,972	22	22	210	311
Côte d'Ivoire	0.24	0.21	1.0	1.0	1,008	1,621	261	333	1	1	16	13
Croatia	..	0.30	..	0.2	..	608	..	1,606	..	13	..	21
Cuba	0.27	0.33	22.9	19.4	224	209	2,024	580	78	96	259	214
Czech Republic	..	0.30	..	0.7	..	1,614	..	1,048	..	167	..	275
Denmark	0.52	0.44	14.5	20.2	1,818	1,509	2,453	1,827	973	1,133	708	597
Dominican Republic	0.19	0.13	11.7	17.1	149	149	572	937	3	4	20	23
Ecuador	0.20	0.13	24.8	28.8	419	854	471	955	6	7	40	57
Egypt, Arab Rep.	0.06	0.05	100.0	99.8	2,007	2,631	2,864	3,858	4	11	158	318
El Salvador	0.12	0.10	4.3	4.4	422	457	1,376	1,619	5	4	61	61
Eritrea	..	0.11	..	5.3	..	412	..	140	..	0	..	11
Estonia	..	0.77	..	0.4	..	357	..	247	..	519	..	451
Ethiopia	..	0.17	..	1.8	..	6,852	..	159	..	0	..	3
Finland	0.50	0.42	2.5	3.0	1,190	1,160	2,022	1,442	721	1,196	892	907
France	0.32	0.31	4.6	9.7	9,804	9,141	3,260	2,708	737	1,256	836	698
Gabon	0.42	0.28	2.4	3.0	6	18	20	8	5	7	43	46
Gambia, The	0.26	0.16	0.6	1.0	54	122	136	59	0	0	3	2
Georgia	..	0.14	..	43.8	..	372	..	442	..	31	..	212
Germany	0.15	0.14	3.7	4.0	7,692	6,904	4,249	2,423	624	960	1,340	950
Ghana	0.18	0.19	0.2	0.2	902	1,317	104	54	1	1	19	11
Greece	0.30	0.27	24.2	35.2	1,600	1,286	1,927	1,811	120	289	485	843
Guatemala	0.19	0.13	5.0	6.6	716	686	726	1,604	3	2	32	32
Guinea	0.16	0.13	7.9	6.4	708	743	16	35	0	0	2	6
Guinea-Bissau	0.32	0.26	6.0	4.9	142	134	24	13	0	0	1	1
Haiti	0.10	0.07	7.9	8.2	416	448	62	165	0	0	3	2
Honduras	0.44	0.28	4.1	3.7	421	500	163	720	5	7	21	30

Agricultural inputs | 3.2

	Arable land		Irrigated land		Land under cereal production		Fertilizer consumption		Agricultural machinery			
	hectares per capita		% of cropland		thousand hectares		hundreds of grams per hectare of arable land		Tractors per 1,000 agricultural workers		Tractors per 100 hectares of arable land	
	1979–81	1996–98	1979–81	1996–98	1979–81	1998–2000	1979–81	1996–98	1979–81	1996–98	1979–81	1996–98
Hungary	0.47	0.47	3.6	4.2	2,878	2,598	2,906	929	59	162	111	191
India	0.24	0.17	22.8	33.6	104,349	101,190	345	976	2	6	24	91
Indonesia	0.12	0.09	16.2	15.5	11,825	15,298	645	1,434	0	1	5	39
Iran, Islamic Rep.	0.36	0.28	35.5	39.8	8,062	7,954	430	691	17	38	57	136
Iraq	0.40	0.24	32.1	63.6	2,159	2,927	172	702	23	74	44	95
Ireland	0.33	0.37	..	..	425	283	5,373	5,135	606	993	1,289	1,239
Israel	0.08	0.06	49.3	45.5	129	57	2,384	3,423	294	323	809	699
Italy	0.17	0.14	19.3	24.5	5,082	4,128	2,295	2,169	370	950	1,117	1,774
Jamaica	0.06	0.07	10.1	9.1	4	2	1,231	1,353	9	11	208	177
Japan	0.04	0.04	56.0	54.6	2,724	2,054	4,131	3,278	209	681	2,723	4,830
Jordan	0.14	0.06	11.0	19.4	158	76	404	890	48	30	153	188
Kazakhstan	..	1.98	..	7.3	..	11,466	..	29	..	70	..	34
Kenya	0.23	0.14	0.9	1.5	1,692	1,851	160	357	1	1	17	36
Korea, Dem. Rep.	0.09	0.07	58.9	73.0	1,625	1,330	4,688	826	13	19	275	441
Korea, Rep.	0.05	0.04	59.6	60.5	1,689	1,173	3,920	5,358	1	50	14	779
Kuwait	0.00	0.00	83.3	81.0	0	1	4,500	2,944	3	11	220	129
Kyrgyz Republic	..	0.29	..	75.4	..	645	..	293	..	36	..	142
Lao PDR	0.21	0.17	15.4	18.9	751	708	40	91	0	0	8	11
Latvia	..	0.72	..	1.1	..	434	..	191	..	330	..	326
Lebanon	0.07	0.04	28.3	37.6	34	39	1,663	3,249	28	111	141	300
Lesotho	0.22	0.16	..	..	203	178	150	182	6	6	47	62
Libya	0.58	0.35	10.7	22.2	538	319	357	320	101	296	134	187
Lithuania	..	0.79	..	0.3	..	1,046	..	448	..	294	..	267
Macedonia, FYR	..	0.30	..	8.5	..	222	..	747	..	398	..	902
Madagascar	0.28	0.18	21.5	35.0	1,309	1,373	31	45	1	1	11	14
Malawi	0.25	0.18	1.1	1.4	1,155	1,547	203	294	0	0	8	8
Malaysia	0.07	0.08	6.7	4.8	729	702	4,273	6,940	4	23	77	238
Mali	0.31	0.46	4.5	3.0	1,346	2,422	61	92	0	1	5	6
Mauritania	0.14	0.20	22.8	9.8	125	235	57	60	1	1	13	7
Mauritius	0.10	0.09	15.0	17.6	0	0	2,547	3,480	4	6	33	37
Mexico	0.34	0.27	20.3	23.8	9,356	11,061	570	658	16	20	54	68
Moldova	..	0.41	..	14.1	..	865	..	668	..	82	..	257
Mongolia	0.71	0.57	3.0	6.4	559	261	83	33	32	21	82	53
Morocco	0.39	0.33	15.0	12.7	4,414	5,166	268	357	7	10	34	47
Mozambique	0.24	0.19	2.1	3.2	1,077	1,816	107	21	1	1	20	18
Myanmar	0.28	0.22	10.4	15.5	5,133	6,302	111	182	1	0	9	9
Namibia	0.64	0.50	0.6	0.9	195	298	..	..	10	11	39	39
Nepal	0.16	0.13	22.5	38.2	2,251	3,283	98	383	0	0	10	16
Netherlands	0.06	0.06	58.5	60.4	225	203	8,620	5,547	561	603	2,238	1,789
New Zealand	0.80	0.41	5.2	8.7	193	130	1,965	4,218	619	437	367	488
Nicaragua	0.39	0.53	6.0	3.2	266	378	392	198	6	7	19	11
Niger	0.62	0.51	0.7	1.3	3,872	7,532	10	7	0	0	0	0
Nigeria	0.39	0.24	0.7	0.8	6,048	18,440	59	59	1	2	3	10
Norway	0.20	0.21	..	..	311	334	3,146	2,203	824	1,306	1,603	1,584
Oman	0.01	0.01	92.7	98.4	2	3	840	3,779	1	1	76	94
Pakistan	0.24	0.17	72.7	81.2	10,693	12,489	525	1,178	5	12	50	150
Panama	0.22	0.18	5.0	4.9	166	188	692	753	27	20	122	100
Papua New Guinea	0.01	0.01	..	..	2	2	3,827	2,283	1	1	699	193
Paraguay	0.52	0.43	3.4	2.9	307	554	44	233	14	24	45	75
Peru	0.19	0.15	32.3	28.9	732	1,091	381	498	5	3	37	25
Philippines	0.11	0.08	12.8	15.6	6,790	6,299	636	1,283	1	1	20	21
Poland	0.41	0.36	0.7	0.7	7,875	8,577	2,393	1,148	112	285	425	936
Portugal	0.25	0.19	20.1	24.0	1,099	590	1,113	1,288	72	219	351	802
Puerto Rico	0.02	0.01	27.2	51.3	1	0	..	..	..	..	..	..
Romania	0.44	0.41	21.9	30.6	6,340	5,496	1,448	392	39	88	150	176
Russian Federation	..	0.86	..	3.8	..	46,809	..	112	..	101	..	72

	Arable land		Irrigated land		Land under cereal production		Fertilizer consumption		Agricultural machinery			
	hectares per capita		% of cropland		thousand hectares		hundreds of grams per hectare of arable land		Tractors per 1,000 agricultural workers		Tractors per 100 hectares of arable land	
	1979–81	1996–98	1979–81	1996–98	1979–81	1998–2000	1979–81	1996–98	1979–81	1996–98	1979–81	1996–98
Rwanda	0.15	0.10	0.4	0.4	239	219	3	4	0	0	1	1
Saudi Arabia	0.20	0.19	28.9	42.3	388	588	228	865	2	12	10	26
Senegal	0.42	0.25	2.6	3.1	1,216	1,218	104	108	0	0	2	2
Sierra Leone	0.14	0.10	4.1	5.4	434	279	58	62	0	0	6	2
Singapore	0.00	0.00	..	..	..	..	22,333	25,183	3	20	220	650
Slovak Republic	..	0.27	..	11.2	..	898	..	751	..	92	..	175
Slovenia	..	0.12	..	0.7	..	95	..	3,086	..	3,604	..	4,311
South Africa	0.45	0.36	8.4	8.5	6,760	4,742	874	534	94	60	140	68
Spain	0.42	0.36	14.8	19.0	7,391	6,652	1,012	1,474	200	576	335	583
Sri Lanka	0.06	0.05	28.3	32.1	864	890	1,800	2,517	4	2	141	81
Sudan	0.66	0.60	14.4	11.5	4,447	7,973	51	42	2	2	8	6
Sweden	0.36	0.32	..	..	1,505	1,221	1,654	1,068	715	989	623	590
Switzerland	0.06	0.06	6.2	5.6	172	188	4,623	4,529	494	635	2,428	2,675
Syrian Arab Republic	0.60	0.32	9.6	21.3	2,642	3,075	250	737	29	66	54	188
Tajikistan	..	0.13	..	80.9	..	405	..	841	..	37	..	395
Tanzania	0.16	0.12	3.1	3.3	2,834	3,201	110	86	1	1	35	20
Thailand	0.35	0.28	16.4	23.1	10,625	11,425	177	925	1	10	11	123
Togo	0.75	0.52	0.3	0.3	416	765	14	76	0	0	0	0
Trinidad and Tobago	0.06	0.06	1.7	2.5	4	4	1,064	1,406	50	53	337	358
Tunisia	0.51	0.31	4.9	7.7	1,416	1,273	212	365	30	38	79	121
Turkey	0.57	0.42	9.6	14.8	13,499	13,655	529	751	38	60	169	330
Turkmenistan	..	0.35	..	106.2	..	612	..	982	..	81	..	307
Uganda	0.32	0.25	0.1	0.1	752	1,382	1	2	0	1	6	9
Ukraine	..	0.65	..	7.3	..	12,040	..	165	..	87	..	109
United Arab Emirates	0.01	0.02	237.7	88.9	0	1	2,250	7,775	6	4	106	69
United Kingdom	0.12	0.11	2.0	1.7	3,930	3,290	3,185	3,588	726	898	742	800
United States	0.83	0.65	10.8	12.0	72,639	59,953	1,092	1,135	1,230	1,515	253	271
Uruguay	0.48	0.39	5.4	13.5	614	578	564	1,102	171	173	236	262
Uzbekistan	..	0.19	..	88.3	..	1,455	..	1,623	..	59	..	380
Venezuela, RB	0.19	0.12	10.0	15.4	814	668	711	1,058	50	59	133	185
Vietnam	0.11	0.07	25.6	42.0	5,962	8,228	302	2,933	1	4	38	206
West Bank and Gaza	..	..	..	..	..	..	..	..	..	..	..	..
Yemen, Rep.	0.16	0.09	19.9	31.0	865	694	93	111	3	2	33	40
Yugoslavia, FR (Serb./Mont.)	0.73	..	1.9	..	4,310	..	1,261	..	140	..	616	..
Zambia	0.89	0.56	0.4	0.9	595	693	145	94	3	2	9	11
Zimbabwe	0.36	0.28	3.1	3.5	1,633	1,784	610	537	7	6	66	69

World	0.25 w	0.24 w	17.7 w	19.5 w	588,514 s	679,938 s	870 w	988 w	19 w	20 w	172 w	187 w
Low income	0.22	0.18	19.9	26.1	199,694	258,543	290	632	2	5	20	69
Middle income	0.18	0.23	23.4	19.7	233,799	287,355	985	1,081	8	11	103	126
Lower middle income	0.14	0.21	31.1	22.7	169,290	225,384	1,004	1,135	5	7	83	96
Upper middle income	0.32	0.29	10.3	11.8	64,509	61,971	952	936	39	81	137	205
Low & middle income	0.20	0.21	21.7	22.4	433,493	545,899	645	891	5	8	62	102
East Asia & Pacific	0.12	0.11	36.9	37.1	141,593	143,963	1,155	2,332	2	2	55	72
Europe & Central Asia	0.16	0.59	10.6	10.4	37,380	115,149	1,445	339	..	100	223	167
Latin America & Carib.	0.32	0.27	11.8	13.7	49,759	47,697	586	811	25	35	95	118
Middle East & N. Africa	0.29	0.20	25.8	36.2	25,655	27,225	422	699	12	24	61	121
South Asia	0.23	0.16	28.7	40.8	132,128	131,768	360	975	2	5	25	89
Sub-Saharan Africa	0.32	0.25	4.0	4.2	46,978	80,097	158	135	3	2	23	17
High income	0.46	0.41	9.8	11.2	155,021	134,039	1,314	1,264	519	927	387	430
Europe EMU	0.23	0.21	12.5	16.3	34,399	30,223	2,739	2,295	452	855	896	953

a. Includes Luxembourg.

Agricultural inputs | 3.2

Agricultural activities provide developing countries with food and revenue, but they also can degrade natural resources. Poor farming practices can cause soil erosion and loss of fertility. Efforts to increase productivity through the use of chemical fertilizers, pesticides, and intensive irrigation have environmental costs and health impacts. Excessive use of chemical fertilizers can alter the chemistry of soil. Pesticide poisoning is common in developing countries. And salinization of irrigated land diminishes soil fertility. Thus inappropriate use of inputs for agricultural production has far-reaching effects.

This table provides indicators of major inputs to agricultural production: land, fertilizers, and agricultural machinery. There is no single correct mix of inputs: appropriate levels and application rates vary by country and over time, depending on the type of crops, the climate and soils, and the production process used.

The data shown here and in table 3.3 are collected by the Food and Agriculture Organization (FAO) through annual questionnaires. The FAO tries to impose standard definitions and reporting methods, but exact consistency across countries and over time is not possible. Data on agricultural employment in particular should be used with caution. In many countries much agricultural employment is informal and unrecorded, including substantial work performed by women and children.

Fertilizer consumption measures the quantity of plant nutrients in the form of nitrogen, potassium, and phosphorous compounds available for direct application. Consumption is calculated as production plus imports minus exports. Traditional nutrients—animal and plant manures—are not included. Because some chemical compounds used for fertilizers have other industrial applications, the consumption data may overstate the quantity available for crops.

To smooth annual fluctuations in agricultural activity, the indicators in the table have been averaged over three years.

Figure 3.2a

Farm workers in developing countries must rely on their own labor

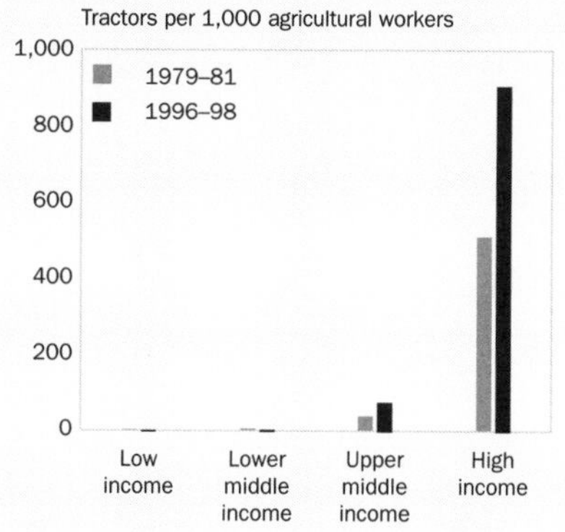

Tractors per 1,000 agricultural workers

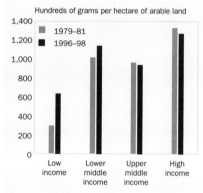

Source: Table 3.2.

Figure 3.2b

Fertilizer consumption is growing rapidly in low-income countries

Hundreds of grams per hectare of arable land

Source: Table 3.2.

• **Arable land** includes land defined by the FAO as land under temporary crops (double-cropped areas are counted once), temporary meadows for mowing or for pasture, land under market or kitchen gardens, and land temporarily fallow. Land abandoned as a result of shifting cultivation is excluded. • **Irrigated land** refers to areas purposely provided with water, including land irrigated by controlled flooding. Cropland refers to arable land and land used for permanent crops (see table 3.1). • **Land under cereal production** refers to harvested areas, although some countries report only sown or cultivated area. • **Fertilizer consumption** measures the quantity of plant nutrients used per unit of arable land. Fertilizer products cover nitrogenous, potash, and phosphate fertilizers (including ground rock phosphate). The time reference for fertilizer consumption is the crop year (July through June). • **Agricultural machinery** refers to wheel and crawler tractors (excluding garden tractors) in use in agriculture at the end of the calendar year specified or during the first quarter of the following year.

The data are from electronic files that the FAO makes available to the World Bank. Data on arable land, irrigated land, and land under cereal production are published in the FAO's *Production Yearbook*.

3.3 Agricultural output and productivity

	Crop production index		Food production index		Livestock production index		Cereal yield		Agricultural productivity	
									Agriculture value added per worker	
							kilograms per hectare		1995 $	
	1989–91 = 100		1989–91 = 100		1989–91 = 100					
	1979–81	**1998–2000**	**1979–81**	**1998–2000**	**1979–81**	**1998–2000**	**1979–81**	**1998–2000**	**1979–81**	**1997–99**
Albania	..	..	..	..	..	..	2,500	2,664	1,184	1,934
Algeria	77.5	125.8	67.6	131.1	55.0	125.3	656	761	1,416	1,876
Angola	101.9	148.1	90.0	144.0	83.8	135.6	526	646	..	126
Argentina	83.5	159.5	91.7	137.9	100.8	105.6	2,186	3,446	7,155	9,983
Armenia	..	97.4	..	78.0	..	64.8	..	1,536	..	5,180
Australia	79.8	163.9	92.0	137.7	85.6	111.4	1,321	2,031	19,914	*31,432*
Austria	92.8	102.4	92.2	106.0	94.5	107.8	4,131	5,654	15,422	28,410
Azerbaijan	..	45.8	..	63.1	..	74.1	..	2,000	..	837
Bangladesh	80.0	110.4	79.2	114.5	81.3	136.2	1,938	2,786	217	292
Belarus	..	86.3	..	61.7	..	60.9	..	1,949	..	3,744
Belgiumª	84.9	138.6	88.5	114.7	88.8	114.3	4,861	7,676	23,704	48,529
Benin	53.8	175.6	63.1	152.4	69.0	123.6	698	1,056	311	558
Bolivia	71.6	151.0	70.9	137.3	75.5	127.1	1,183	1,531	..	1,054
Bosnia and Herzegovina	..	..	..	..	..	..	..	3,490	..	8,471
Botswana	86.3	70.6	87.2	97.6	87.5	101.3	203	196	630	681
Brazil	75.3	121.6	69.5	136.8	67.9	149.9	1,496	2,660	2,048	4,300
Bulgaria	107.7	67.8	105.5	72.2	96.3	64.0	3,853	2,833	2,754	6,007
Burkina Faso	59.3	146.1	62.7	136.0	59.9	136.2	575	887	134	162
Burundi	79.9	89.7	79.9	90.3	82.3	81.6	1,081	1,283	177	140
Cambodia	55.2	136.0	48.9	139.2	27.3	149.1	1,025	1,867	..	*406*
Cameroon	87.3	124.6	80.5	126.2	61.1	119.6	849	1,395	834	1,072
Canada	77.6	128.9	79.6	128.8	88.4	132.8	2,173	3,034	14,161	*34,922*
Central African Republic	102.8	128.2	79.7	132.8	48.9	129.2	529	1,057	377	460
Chad	67.1	177.9	90.8	155.2	120.4	114.0	587	642	155	220
Chile	70.7	126.3	71.5	133.1	75.8	143.7	2,124	4,540	3,174	4,997
China	67.1	141.5	60.9	168.5	45.4	209.9	3,027	4,879	161	316
Hong Kong, China	133.6	59.3	99.8	49.5	194.3	44.6	1,712	..	..	..
Colombia	84.1	100.0	75.5	118.5	72.6	125.8	2,452	3,056	3,034	3,454
Congo, Dem. Rep.	73.0	89.0	72.2	92.1	77.7	103.7	807	781	270	*283*
Congo, Rep.	84.6	114.5	82.3	118.0	80.7	129.0	838	687	385	498
Costa Rica	70.6	133.6	73.0	132.3	77.2	120.9	2,498	3,350	3,130	4,973
Côte d'Ivoire	73.8	132.4	70.8	131.2	74.7	127.6	867	1,138	1,074	1,104
Croatia	..	87.2	..	72.8	..	54.4	..	4,445	..	7,123
Cuba	84.3	54.5	90.1	58.2	96.0	63.6	2,458	1,876	..	..
Czech Republic	..	90.3	..	83.6	..	76.6	..	4,174	..	5,091
Denmark	65.2	95.6	83.2	107.9	95.0	118.5	4,040	6,120	19,350	52,809
Dominican Republic	96.5	89.2	85.2	100.7	68.8	121.1	3,024	3,804	1,842	2,710
Ecuador	78.2	122.4	77.4	134.1	73.0	150.1	1,633	2,057	1,206	1,789
Egypt, Arab Rep.	75.5	141.4	68.2	149.5	66.5	156.0	4,053	7,041	721	1,222
El Salvador	120.4	108.5	90.8	121.2	88.8	127.9	1,702	1,966	1,925	1,690
Eritrea	..	174.1	..	133.9	..	105.7	..	794	..	..
Estonia	..	66.8	..	45.2	..	39.7	..	1,516	..	3,646
Ethiopia	..	121.6	..	119.9	..	116.2	..	1,141	..	144
Finland	76.3	86.4	93.5	89.7	107.1	93.3	2,511	2,763	18,547	36,384
France	87.4	112.1	93.8	107.6	97.8	105.0	4,700	7,272	19,318	*50,171*
Gabon	76.3	118.2	79.0	113.9	86.5	118.3	1,718	1,728	1,814	1,889
Gambia, The	79.5	114.1	82.8	115.9	94.4	117.3	1,284	1,101	325	222
Georgia	..	61.1	..	78.3	..	83.5	..	1,485	..	1,952
Germany	89.8	114.0	91.3	94.4	98.7	86.0	4,166	6,480	..	28,924
Ghana	67.0	174.1	68.7	163.5	79.7	101.8	807	1,306	670	554
Greece	86.8	107.0	91.2	100.1	99.9	97.8	3,090	3,512	8,600	*12,711*
Guatemala	89.6	120.9	69.7	124.5	76.0	129.0	1,578	1,726	2,143	2,099
Guinea	89.7	142.4	96.3	143.1	116.4	139.6	958	1,313	..	284
Guinea-Bissau	64.8	119.4	68.3	119.9	78.4	120.5	711	1,421	221	306
Haiti	103.4	86.6	101.3	95.4	100.2	128.8	1,009	914	578	392
Honduras	90.4	116.5	88.2	112.4	80.8	130.1	1,170	1,167	696	1,008

Agricultural output and productivity | 3.3

	Crop production index		Food production index		Livestock production index		Cereal yield		Agricultural productivity	
	1989–91 = 100		1989–91 = 100		1989–91 = 100		kilograms per hectare		Agriculture value added per worker 1995 $	
	1979–81	1998–2000	1979–81	1998–2000	1979–81	1998–2000	1979–81	1998–2000	1979–81	1997–99
Hungary	93.3	79.3	90.7	74.7	94.1	69.2	4,519	4,503	3,390	*4,860*
India	70.9	122.1	68.1	124.6	62.2	133.5	1,324	2,293	272	395
Indonesia	66.2	117.4	62.8	119.1	47.2	125.4	2,837	3,915	609	742
Iran, Islamic Rep.	56.7	151.3	61.1	151.4	68.2	146.1	1,108	1,959	2,197	3,679
Iraq	74.7	82.9	78.0	77.7	81.4	65.0	832	558	..	..
Ireland	93.9	110.0	83.3	110.3	83.3	110.6	4,733	6,883	..	..
Israel	98.2	105.6	85.6	110.7	78.4	116.0	1,840	1,373	..	..
Italy	106.1	105.6	101.4	104.8	93.0	104.2	3,548	5,040	10,016	23,906
Jamaica	98.6	122.9	86.0	120.8	73.9	119.6	1,667	1,197	893	1,229
Japan	107.9	88.3	94.0	92.4	85.1	94.2	5,252	5,971	15,698	*30,620*
Jordan	54.6	116.7	57.5	139.6	51.5	198.8	521	772	1,158	1,434
Kazakhstan	..	65.6	..	58.9	..	45.3	..	975	..	*1,414*
Kenya	74.5	107.1	67.6	104.5	60.2	104.1	1,364	1,411	262	226
Korea, Dem. Rep.	..	..	..	..	..	..	3,694	3,037	..	..
Korea, Rep.	87.8	106.4	77.6	112.3	52.6	150.3	4,986	6,400	3,800	12,252
Kuwait	37.1	163.8	91.4	185.7	106.6	180.5	3,124	2,556	..	..
Kyrgyz Republic	..	129.1	..	114.5	..	78.7	..	2,521	..	3,430
Lao PDR	73.5	139.5	70.3	144.3	56.0	161.6	1,402	2,925	..	*558*
Latvia	..	70.2	..	45.8	..	35.9	..	1,982	..	2,523
Lebanon	52.0	137.6	59.2	142.7	100.5	161.6	1,307	2,428	..	*28,243*
Lesotho	95.1	115.9	89.1	99.9	87.7	88.5	977	974	723	544
Libya	76.3	132.9	78.7	161.6	68.4	174.4	430	761	..	..
Lithuania	..	74.4	..	66.6	..	58.1	..	2,179	..	3,192
Macedonia, FYR	..	108.7	..	95.8	..	85.1	..	3,093	..	2,141
Madagascar	83.1	104.2	84.5	108.5	89.5	105.7	1,664	1,891	197	184
Malawi	85.7	149.1	93.2	153.2	78.2	112.4	1,161	1,514	109	138
Malaysia	74.7	111.2	55.4	134.1	41.4	149.1	2,828	2,826	3,939	6,578
Mali	54.5	145.1	76.7	127.2	94.5	123.3	804	1,163	241	279
Mauritania	62.1	152.3	86.5	107.3	89.4	100.9	384	928	299	469
Mauritius	93.3	93.3	89.7	103.1	64.0	135.3	2,536	5,193	3,087	5,330
Mexico	86.5	121.6	83.8	128.4	83.5	134.6	2,164	2,640	1,482	1,742
Moldova	..	54.7	..	45.2	..	36.3	..	2,485	..	1,277
Mongolia	44.6	35.4	88.1	87.8	93.2	92.3	573	735	932	1,193
Morocco	54.8	95.3	55.9	100.1	59.8	106.7	811	780	1,146	1,651
Mozambique	109.6	143.2	100.1	131.0	85.8	103.2	603	919	..	136
Myanmar	89.0	152.8	88.2	148.6	89.1	143.0	2,521	3,104	..	..
Namibia	80.2	111.6	107.2	97.3	115.6	95.5	377	296	862	1,248
Nepal	62.7	120.8	65.9	121.4	77.3	123.5	1,615	2,008	162	189
Netherlands	79.8	107.5	86.5	100.3	88.3	99.9	5,696	7,343	23,907	51,594
New Zealand	74.4	134.1	90.7	125.4	95.5	116.4	4,089	6,343	18,066	27,083
Nicaragua	124.1	134.2	117.8	131.2	139.7	116.7	1,475	1,694	1,620	1,919
Niger	90.1	151.3	97.9	140.1	109.7	120.4	440	377	222	205
Nigeria	51.4	155.4	57.2	152.2	84.3	125.6	1,265	1,208	414	641
Norway	91.2	87.8	92.1	96.3	95.2	100.1	3,634	4,002	17,013	32,848
Oman	60.4	113.8	62.5	114.9	61.6	104.0	982	2,173	..	..
Pakistan	65.6	124.5	66.4	143.3	59.5	150.4	1,608	2,255	394	626
Panama	97.1	97.2	85.6	107.2	71.3	121.9	1,524	1,971	2,122	2,580
Papua New Guinea	86.5	109.7	86.2	113.1	85.0	136.6	2,087	4,170	649	808
Paraguay	58.7	110.4	61.0	132.8	62.1	129.4	1,535	2,159	2,641	3,512
Peru	82.2	162.9	77.3	161.7	78.0	150.5	1,946	2,911	1,194	1,569
Philippines	88.4	112.9	86.1	128.4	73.3	173.5	1,611	2,420	1,347	1,342
Poland	84.6	85.8	87.9	88.7	98.0	87.2	2,345	2,885	..	1,554
Portugal	85.0	85.0	72.2	94.1	71.8	117.1	1,102	2,762	..	*7,621*
Puerto Rico	131.2	62.9	99.7	81.7	90.3	87.5	8,925	4,000	..	..
Romania	114.1	90.3	113.0	93.4	110.0	87.9	2,854	2,543	..	3,228
Russian Federation	..	66.8	..	60.8	..	51.6	..	1,181	..	2,282

	Crop production index		Food production index		Livestock production index		Cereal yield		Agricultural productivity	
							kilograms per hectare		Agriculture value added per worker 1995 $	
	1989–91 = 100		1989–91 = 100		1989–91 = 100					
	1979–81	1998–2000	1979–81	1998–2000	1979–81	1998–2000	1979–81	1998–2000	1979–81	1997–99
Rwanda	84.3	88.1	85.3	91.5	81.0	108.8	1,134	930	371	234
Saudi Arabia	27.2	94.3	26.7	88.5	32.8	147.7	820	4,147	2,167	10,930
Senegal	77.2	102.9	74.0	114.2	65.1	138.0	690	721	336	307
Sierra Leone	80.3	81.0	84.5	85.3	84.1	109.0	1,249	1,116	368	379
Singapore	595.0	48.2	154.3	41.5	173.7	39.5	..	..	13,937	42,903
Slovak Republic	..	..	..	..	..	..	..	4,225	..	3,491
Slovenia	..	93.5	..	104.6	..	108.7	..	5,379	..	30,136
South Africa	95.0	105.3	92.6	103.3	89.7	96.5	2,105	2,313	2,899	4,070
Spain	83.0	107.6	82.1	110.7	84.2	122.5	1,986	3,221	..	21,687
Sri Lanka	99.3	113.9	98.3	115.7	93.2	133.5	2,462	3,190	638	734
Sudan	131.1	162.0	105.4	155.9	89.3	146.3	645	519	..	..
Sweden	92.1	93.8	100.1	101.1	103.8	104.1	3,595	4,570	18,128	34,285
Switzerland	95.5	99.4	95.8	97.1	98.8	94.1	4,883	6,269	..	..
Syrian Arab Republic	100.4	158.4	94.2	150.5	72.2	133.0	1,156	1,331	..	..
Tajikistan	..	62.5	..	58.1	..	36.9	..	1,268	..	..
Tanzania	81.8	100.3	76.7	105.6	69.3	118.2	1,063	1,295	..	188
Thailand	78.9	112.9	80.0	113.1	64.9	127.0	1,911	2,442	634	939
Togo	70.4	148.1	77.0	140.7	51.9	128.8	729	931	345	543
Trinidad and Tobago	119.9	101.4	101.9	105.0	84.3	100.9	3,167	2,936	..	2,463
Tunisia	68.5	116.6	66.6	127.0	60.7	147.2	828	1,168	1,743	3,047
Turkey	76.6	114.7	75.8	113.0	80.4	109.1	1,869	2,287	1,860	1,858
Turkmenistan	..	80.3	..	132.1	..	134.4	..	2,514	..	856
Uganda	67.5	118.8	70.4	116.5	84.8	119.8	1,555	1,371	..	350
Ukraine	..	58.2	..	49.1	..	45.8	..	2,032	..	1,383
United Arab Emirates	38.9	275.3	48.8	250.6	45.3	170.8	2,224	1,455	..	..
United Kingdom	80.1	102.8	92.0	98.7	98.1	97.9	4,792	6,975	21,177	34,730
United States	98.6	121.9	94.5	122.9	89.0	120.0	4,151	5,794	..	..
Uruguay	86.8	149.7	87.0	137.8	85.9	122.2	1,644	3,456	6,240	8,679
Uzbekistan	..	88.4	..	116.5	..	115.8	..	2,585	..	1,621
Venezuela, RB	76.3	105.7	80.2	117.3	84.9	118.8	1,904	3,089	3,935	5,125
Vietnam	66.7	158.5	63.8	152.2	52.9	163.7	2,049	3,955	..	236
West Bank and Gaza	..	..	..	..	..	..	..	..	..	..
Yemen, Rep.	82.3	129.1	75.0	130.9	68.9	138.6	1,038	1,005	..	355
Yugoslavia, FR (Serb./Mont.)	96.3	..	94.3	..	94.2	..	3,601	..	..	..
Zambia	65.7	88.5	74.0	99.7	86.2	113.2	1,676	1,391	143	218
Zimbabwe	77.9	121.2	81.3	108.2	83.3	113.7	1,359	1,184	308	369
World	**79.1 w**	**123.6 w**	**78.8 w**	**127.9 w**	**79.6 w**	**129.4 w**	**1,608 w**	**2,067 w**	**.. w**	**.. w**
Low income	71.6	124.4	70.7	126.5	68.4	131.2	1,083	1,301	..	346
Middle income	74.5	128.2	72.0	141.4	69.3	153.7	1,844	2,390	..	..
Lower middle income	72.1	132.3	68.2	150.5	59.8	176.0	1,778	2,046	..	..
Upper middle income	80.7	117.3	79.5	122.7	82.3	122.8	1,936	2,800	..	..
Low & middle income	73.5	126.8	71.5	136.3	69.1	148.0	1,426	1,811	..	..
East Asia & Pacific	69.0	135.4	63.8	156.4	48.0	197.6	2,116	2,982	..	..
Europe & Central Asia	..	..	..	..	..	..	2,854	2,407	..	2,220
Latin America & Carib.	80.3	124.3	78.3	131.2	79.8	131.9	1,842	2,488	..	..
Middle East & N. Africa	66.1	131.3	64.8	134.0	64.1	136.8	965	1,411	..	..
South Asia	71.9	121.3	69.6	125.7	64.0	136.2	1,510	2,274	265	..
Sub-Saharan Africa	75.4	128.5	78.3	124.7	84.1	114.2	895	1,120	418	380
High income	93.5	115.7	92.1	112.9	91.1	109.9	3,253	4,002	..	..
Europe EMU	91.0	108.6	91.4	103.4	93.8	101.0	4,109	6,067	..	..

a. Includes Luxembourg.

Agricultural output and productivity | 3.3

About the data

The agricultural production indexes in the table are prepared by the Food and Agriculture Organization (FAO). The FAO obtains data from official and semiofficial reports of crop yields, area under production, and livestock numbers. If data are not available, the FAO makes estimates. The indexes are calculated using the Laspeyres formula: production quantities of each commodity are weighted by average international commodity prices in the base period and summed for each year. Because the FAO's indexes are based on the concept of agriculture as a single enterprise, estimates of the amounts retained for seed and feed are subtracted from the production data to avoid double counting. The resulting aggregate represents production available for any use except as seed and feed. The FAO's indexes may differ from other sources because of differences in coverage, weights, concepts, time periods, calculation methods, and use of international prices.

To ease cross-country comparisons, the FAO uses international commodity prices to value production. These prices, expressed in international dollars (equivalent in purchasing power to the U.S. dollar), are derived using a Geary-Khamis formula applied to agricultural outputs (see Inter-Secretariat Working Group on National Accounts 1993, sections 16.93–96). This method assigns a single price to each commodity so that, for example, one metric ton of wheat has the same price regardless of where it was produced. The use of international prices eliminates fluctuations in the value of output due to transitory movements of nominal exchange rates unrelated to the purchasing power of the domestic currency.

Data on cereal yield may be affected by a variety of reporting and timing differences. The FAO allocates production data to the calendar year in which the bulk of the harvest took place. But most of a crop harvested near the end of a year will be used in the following year. Cereal crops harvested for hay or harvested green for food, feed, or silage and those used for grazing are generally excluded. But millet and sorghum, which are grown as feed for livestock and poultry in Europe and North America, are used as food in Africa, Asia, and countries of the former Soviet Union. So some cereal crops are excluded from the data for some countries and included elsewhere, depending on their use.

Agricultural productivity is measured by value added per unit of input. (For further discussion of the calculation of value added in national accounts see *About the data* for tables 4.1 and 4.2.) Agricultural value added includes that from forestry and fishing. Thus interpre-

tations of land productivity should be made with caution. To smooth annual fluctuations in agricultural activity, the indicators in the table have been averaged over three years.

Figure 3.3

The land area under cereal production has grown in low-income countries . . .

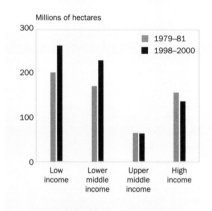

Millions of hectares

Legend: 1979–81, 1998–2000

. . . but cereal yields, though growing worldwide, still lag

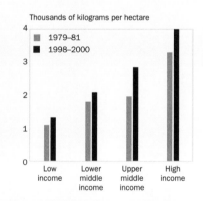

Thousands of kilograms per hectare

Legend: 1979–81, 1998–2000

Source: Tables 3.2 and 3.3.

Definitions

• **Crop production index** shows agricultural production for each period relative to the base period 1989–91. It includes all crops except fodder crops. The regional and income group aggregates for the FAO's production indexes are calculated from the underlying values in international dollars, normalized to the base period 1989–91. The data in this table are three-year averages. However, missing observations have not been estimated or imputed. • **Food production index** covers food crops that are considered edible and that contain nutrients. Coffee and tea are excluded because, although edible, they have no nutritive value. • **Livestock production index** includes meat and milk from all sources, dairy products such as cheese, and eggs, honey, raw silk, wool, and hides and skins. • **Cereal yield,** measured in kilograms per hectare of harvested land, includes wheat, rice, maize, barley, oats, rye, millet, sorghum, buckwheat, and mixed grains. Production data on cereals refer to crops harvested for dry grain only. Cereal crops harvested for hay or harvested green for food, feed, or silage and those used for grazing are excluded. • **Agricultural productivity** refers to the ratio of agricultural value added, measured in constant 1995 U.S. dollars, to the number of workers in agriculture.

Data sources

The agricultural production indexes are prepared by the FAO and published annually in its *Production Yearbook*. The FAO makes these data and the data on cereal yield and agricultural employment available to the World Bank in electronic files that may contain more recent information than the published versions. For sources of agricultural value added see table 4.2.

3.4 | Deforestation and biodiversity

	Forest area		Average annual deforestation		Mammals		Birds		Higher plants[a]		Nationally protected areas	
	thousand sq. km 2000	% of total land area 2000	sq. km 1990–2000	Decline in forest area % 1990–2000	Species 1996[b]	Threatened species 2000[b]	Species 1996[b]	Threatened species 2000[b]	Species 1997[b]	Threatened species 1997[b]	thousand sq. km 1999[b]	% of total land area 1999[b]
Albania	10	36.2	78	0.8	68	3	230	3	3,031	79	0.8	2.9
Algeria	21	0.9	−266	−1.3	92	13	192	6	3,164	141	58.9	2.5
Angola	698	56.0	1,242	0.2	276	18	765	15	5,185	30	81.8	6.6
Argentina	346	12.7	2,851	0.8	320	32	897	39	9,372	247	46.6	1.7
Armenia	4	12.4	−42	−1.3	..	7	..	4	..	31	2.1	7.4
Australia	1,581	20.6	0	0.0	252	63	649	35	15,638	2,245	563.9	7.3
Austria	39	47.0	−77	−0.2	83	9	213	3	3,100	23	23.4	28.3
Azerbaijan	11	12.6	−130	−1.3	..	13	..	8	..	28	4.8	5.5
Bangladesh	13	10.2	−165	−1.3	109	21	295	23	5,000	24	1.0	0.8
Belarus	94	45.3	−2,562	−3.2	..	5	221	3	..	1	8.6	4.1
Belgium	7	22.2	13	0.2	58	11	180	2	1,550	2	0.8	..
Benin	27	24.0	699	2.3	188	7	307	2	2,201	4	7.8	7.1
Bolivia	531	48.9	1,611	0.3	316	23	..	27	17,367	227	156.0	14.4
Bosnia and Herzegovina	23	44.6	0	0.0	..	10	..	3	..	64	0.2	0.4
Botswana	124	21.9	1,184	0.9	164	5	386	7	2,151	7	105.0	18.5
Brazil	5,325	63.0	22,264	0.4	394	79	1,492	113	56,215	1,358	355.5	4.2
Bulgaria	37	33.4	−204	−0.6	81	15	240	10	3,572	106	4.9	4.4
Burkina Faso	71	25.9	152	0.2	147	7	335	2	..	..	28.6	10.5
Burundi	1	3.7	147	9.0	107	5	451	7	..	..	1.4	5.5
Cambodia	93	52.9	561	0.6	123	21	307	19	..	5	28.6	16.2
Cameroon	239	51.3	2,218	0.9	297	37	690	15	8,260	89	21.0	4.5
Canada	2,446	26.5	0	0.0	193	14	426	8	3,270	278	921.0	10.0
Central African Republic	229	36.8	300	0.1	209	12	537	3	3,602	1	51.1	8.2
Chad	127	10.1	817	0.6	134	17	370	5	1,600	12	114.9	9.1
Chile	155	20.7	203	0.1	91	21	296	21	5,284	329	141.3	18.9
China	1,635	17.5	−18,063	−1.2	394	76	1,100	73	32,200	312	598.1	6.4
Hong Kong, China	..	..	..	..	24	1	76	11	1,984	9	0.4	40.4
Colombia	496	47.8	1,905	0.4	359	36	1,695	77	51,220	712	93.6	9.0
Congo, Dem. Rep.	1,352	59.6	5,324	0.4	415	40	929	28	11,007	78	101.9	4.5
Congo, Rep.	221	64.6	175	0.1	200	12	449	4	6,000	3	15.4	4.5
Costa Rica	20	38.5	158	0.8	205	14	600	13	12,119	527	7.0	13.7
Côte d'Ivoire	71	22.4	2,649	3.1	230	17	535	12	3,660	94	19.9	6.3
Croatia	18	31.9	−20	−0.1	..	9	224	4	..	6	3.7	6.6
Cuba	23	21.4	−277	−1.3	31	11	137	18	6,522	888	19.1	17.4
Czech Republic	26	34.1	−5	0.0	..	8	199	2	..	81	12.2	15.8
Denmark	5	10.7	−10	−0.2	43	5	196	1	1,450	2	13.7	32.3
Dominican Republic	14	28.4	0	0.0	20	5	136	15	5,657	136	12.2	25.2
Ecuador	106	38.1	1,372	1.2	302	31	1,388	62	19,362	824	119.3	43.1
Egypt, Arab Rep.	1	0.1	−20	−3.4	98	12	153	7	2,076	82	7.9	0.8
El Salvador	1	5.8	72	4.6	135	2	251	0	2,911	42	0.1	0.5
Eritrea	16	15.7	54	0.3	112	12	319	7	..	..	5.0	5.0
Estonia	21	48.7	−125	−0.6	65	5	213	3	..	2	5.1	12.1
Ethiopia	46	4.6	403	0.8	255	34	626	16	6,603	163	55.2	5.5
Finland	219	72.0	−80	0.0	60	6	248	3	1,102	6	18.2	6.0
France	153	27.9	−616	−0.4	93	18	269	5	4,630	195	58.8	10.7
Gabon	218	84.7	101	0.0	190	15	466	6	6,651	91	7.2	2.8
Gambia, The	5	48.1	−45	−1.0	108	3	280	2	974	1	0.2	2.0
Georgia	30	42.9	0	0.0	..	14	..	3	..	29	1.9	2.7
Germany	107	30.7	0	0.0	76	12	239	5	2,682	14	94.2	27.0
Ghana	63	27.8	1,200	1.7	222	13	529	8	3,725	103	11.0	4.8
Greece	36	27.9	−300	−0.9	95	14	251	7	4,992	571	3.1	2.4
Guatemala	29	26.3	537	1.7	250	6	458	6	8,681	355	18.2	16.8
Guinea	69	28.2	347	0.5	190	11	409	10	3,000	39	1.6	0.7
Guinea-Bissau	22	77.8	216	0.9	108	2	243	0	..	..	0.0	0.0
Haiti	1	3.2	70	5.7	3	4	75	14	5,242	100	0.1	0.4
Honduras	54	48.1	590	1.0	173	9	422	5	5,680	96	11.1	9.9

Deforestation and biodiversity | 3.4

	Forest area		Average annual deforestation		Mammals		Birds		Higher plants[a]		Nationally protected areas	
	thousand sq. km **2000**	% of total land area **2000**	sq. km **1990–2000**	Decline in forest area % **1990–2000**	Species **1996**[b]	Threatened species **2000**[b]	Species **1996**[b]	Threatened species **2000**[b]	Species **1997**[b]	Threatened species **1997**[b]	thousand sq. km **1999**[b]	% of total land area **1999**[b]
Hungary	18	19.9	−72	−0.4	72	9	205	8	2,214	30	6.3	6.8
India	641	21.6	−381	−0.1	316	86	923	70	16,000	1,236	142.9	4.8
Indonesia	1,050	58.0	13,124	1.2	436	140	1,519	113	29,375	264	192.3	10.6
Iran, Islamic Rep.	73	4.5	0	0.0	140	23	323	13	8,000	2	83.0	5.1
Iraq	8	1.8	0	0.0	81	10	172	11	..	..	0.0	0.0
Ireland	7	9.6	−170	−3.0	25	5	142	1	950	1	0.6	0.9
Israel	1	6.4	−50	−4.9	92	14	180	12	2,317	32	3.1	15.0
Italy	100	34.0	−295	−0.3	90	14	234	5	5,599	311	21.5	7.3
Jamaica	3	30.0	54	1.5	24	5	113	12	3,308	744	0.0	0.0
Japan	241	64.0	−34	0.0	132	37	250	34	5,565	707	25.5	6.8
Jordan	1	1.0	0	0.0	71	8	141	8	2,100	9	3.0	3.4
Kazakhstan	121	4.5	−2,390	−2.2	..	18	..	15	..	71	73.4	2.7
Kenya	171	30.0	931	0.5	359	51	844	24	6,506	240	35.0	6.1
Korea, Dem. Rep.	82	68.2	0	0.0	..	13	115	19	2,898	4	3.1	2.6
Korea, Rep.	63	63.3	49	0.1	49	13	112	25	2,898	66	6.8	6.9
Kuwait	0	0.3	−2	−5.2	21	1	20	7	..	..	0.3	1.7
Kyrgyz Republic	10	5.2	−228	−2.6	..	7	..	4	..	34	6.9	3.6
Lao PDR	126	54.4	527	0.4	172	27	487	19	..	2	0.0	0.0
Latvia	29	47.1	−127	−0.4	83	5	217	3	..	..	7.8	12.6
Lebanon	0	3.5	1	0.3	54	6	154	7	3,000	5	0.0	0.0
Lesotho	0	0.5	0	0.0	33	3	58	7	1,591	21	0.1	0.3
Libya	4	0.2	−47	−1.4	76	9	91	1	1,825	57	1.7	0.1
Lithuania	20	30.8	−48	−0.2	68	5	202	4	..	1	6.5	10.0
Macedonia, FYR	9	35.6	0	0.0	..	11	..	3	..	..	1.8	7.1
Madagascar	117	20.2	1,174	0.9	148	50	256	27	9,505	306	18.2	3.1
Malawi	26	27.6	707	2.4	195	8	521	11	3,765	61	10.6	11.3
Malaysia	193	58.7	2,377	1.2	286	47	501	37	15,500	490	14.8	4.5
Mali	132	10.8	993	0.7	137	13	397	4	1,741	15	45.3	3.7
Mauritania	3	0.3	98	2.7	61	10	273	2	1,100	3	17.5	1.7
Mauritius	0	7.9	1	0.6	4	4	27	9	750	294	0.1	4.9
Mexico	552	28.9	6,306	1.1	450	69	769	39	26,071	1,593	71.0	3.7
Moldova	3	9.9	−7	−0.2	68	3	177	5	..	5	0.4	1.2
Mongolia	106	6.8	600	0.5	134	12	390	16	2,272	0	161.3	10.3
Morocco	30	6.8	12	0.0	105	16	210	9	3,675	186	3.2	0.7
Mozambique	306	39.0	637	0.2	179	15	498	16	5,692	89	47.8	6.1
Myanmar	344	52.3	5,169	1.4	251	36	867	35	7,000	32	1.7	0.3
Namibia	80	9.8	734	0.9	154	14	469	11	3,174	75	106.2	12.9
Nepal	39	27.3	783	1.8	167	27	611	26	6,973	20	11.1	7.8
Netherlands	4	11.1	−10	−0.3	55	11	191	4	1,221	1	2.4	7.1
New Zealand	79	29.7	−390	−0.5	10	8	150	62	2,382	211	63.3	23.6
Nicaragua	33	27.0	1,172	3.0	200	6	482	5	7,590	98	9.0	7.4
Niger	13	1.0	617	3.7	131	11	299	3	1,170	..	96.9	7.6
Nigeria	135	14.8	3,984	2.6	274	25	681	9	4,715	37	30.2	3.3
Norway	89	28.9	−310	−0.4	54	10	243	2	1,715	12	93.7	30.5
Oman	0	0.0	0	0.0	56	9	107	10	1,204	30	34.3	16.1
Pakistan	25	3.2	304	1.1	151	18	375	17	4,950	14	37.2	4.8
Panama	29	38.6	519	1.6	218	20	732	16	9,915	1,302	14.2	19.1
Papua New Guinea	306	67.6	1,129	0.4	214	58	644	32	11,544	92	0.1	0.0
Paraguay	234	58.8	1,230	0.5	305	9	556	26	7,851	129	14.0	3.5
Peru	652	50.9	2,688	0.4	344	47	1,538	73	18,245	906	34.6	2.7
Philippines	58	19.4	887	1.4	153	50	395	67	8,931	360	14.5	4.9
Poland	93	30.6	−110	−0.1	84	15	227	4	2,450	27	29.1	9.6
Portugal	37	40.1	−570	−1.7	63	17	207	7	5,050	269	5.9	6.4
Puerto Rico	2	25.8	5	0.2	16	2	105	8	2,493	223	0.1	1.1
Romania	64	28.0	−147	−0.2	84	17	247	8	3,400	99	10.7	4.6
Russian Federation	8,514	50.4	−1,353	0.0	269	42	628	38	..	214	516.7	3.1

3.4 Deforestation and biodiversity

	Forest area		Average annual deforestation		Mammals		Birds		Higher plants[a]		Nationally protected areas	
	thousand sq. km 2000	% of total land area 2000	sq. km 1990–2000	Decline in forest area % 1990–2000	Species 1996[b]	Threatened species 2000[b]	Species 1996[b]	Threatened species 2000[b]	Species 1997[b]	Threatened species 1997[b]	thousand sq. km 1999[b]	% of total land area 1999[b]
Rwanda	3	12.4	150	3.9	151	8	513	9	..	..	3.6	14.6
Saudi Arabia	15	0.7	0	0.0	77	7	155	15	2,028	7	49.6	2.3
Senegal	62	32.2	450	0.7	155	11	384	4	2,086	31	21.8	11.3
Sierra Leone	11	14.7	361	2.9	147	11	466	10	2,090	29	0.8	1.1
Singapore	0	3.3	0	0.0	45	3	118	7	2,168	29	0.0	0.0
Slovak Republic	20	42.5	−69	−0.3	..	9	209	4	..	65	10.5	21.8
Slovenia	11	55.0	−22	−0.2	69	9	207	1	..	13	1.1	5.5
South Africa	89	7.3	80	0.1	247	41	596	28	23,420	2,215	65.8	5.4
Spain	144	28.8	−860	−0.6	82	24	278	7	5,050	985	42.2	8.4
Sri Lanka	19	30.0	348	1.6	88	20	250	14	3,314	455	8.6	13.3
Sudan	616	25.9	9,589	1.4	267	24	680	6	3,137	10	86.4	3.6
Sweden	271	65.9	−6	0.0	60	8	249	2	1,750	13	36.2	8.8
Switzerland	12	30.3	−43	−0.4	75	6	193	2	3,030	30	7.1	18.0
Syrian Arab Republic	5	2.5	0	0.0	63	4	204	8	3,000	8	0.0	0.0
Tajikistan	4	2.8	−20	−0.5	..	9	..	7	..	50	5.9	4.2
Tanzania	388	43.9	913	0.2	316	43	822	33	10,008	436	138.2	15.6
Thailand	148	28.9	1,124	0.7	265	34	616	37	11,625	385	70.7	13.8
Togo	5	9.4	209	3.4	196	9	391	0	2,201	4	4.3	7.9
Trinidad and Tobago	3	50.5	22	0.8	100	1	260	1	2,259	21	0.2	3.9
Tunisia	5	3.3	−11	−0.2	78	11	173	5	2,196	24	0.4	0.3
Turkey	102	13.3	−220	−0.2	116	17	302	11	8,650	1,876	10.7	1.4
Turkmenistan	38	8.0	0	0.0	..	13	..	6	..	17	19.8	4.2
Uganda	42	21.0	913	2.0	338	19	830	13	5,406	15	19.1	9.6
Ukraine	96	16.5	−310	−0.3	..	17	263	8	..	52	9.0	1.6
United Arab Emirates	3	3.8	−78	−2.8	25	3	67	8	..	..	0.0	0.0
United Kingdom	26	10.7	−200	−0.8	50	12	230	2	1,623	18	50.6	20.9
United States	2,260	24.7	−3,880	−0.2	428	37	650	55	19,473	4,669	1,226.7	13.4
Uruguay	13	7.4	−501	−5.0	81	6	237	11	2,278	15	0.5	0.3
Uzbekistan	20	4.8	−46	−0.2	..	11	..	9	..	41	8.2	2.0
Venezuela, RB	495	56.1	2,175	0.4	305	25	1,181	24	21,073	426	319.8	36.3
Vietnam	98	30.2	−516	−0.5	213	37	535	35	10,500	341	9.9	3.0
West Bank and Gaza	..	..	..	..	..	1	..	1	..	..	..	..
Yemen, Rep.	4	0.9	92	1.8	66	4	143	12	..	..	0.0	0.0
Yugoslavia, FR (Serb./Mont.)	29	..	14	0.0	..	11	..	5	5,351	155	3.3	..
Zambia	312	42.0	8,509	2.4	229	12	605	11	4,747	12	63.6	8.6
Zimbabwe	190	49.2	3,199	1.5	270	12	532	10	4,440	100	30.7	7.9
World	**38,609 s**	**29.7 w**	**90,399 s**	**0.2 w**							**8,546.8 s**	**6.6 w**
Low income	8,840	26.8	71,466	0.8							1,852.4	5.6
Middle income	21,791	32.9	26,930	0.1							3,400.2	5.1
Lower middle income	13,966	31.7	−10,268	−0.1							2,098.8	4.8
Upper middle income	7,825	35.2	37,198	0.5							1,301.4	5.9
Low & middle income	30,630	30.9	98,396	0.3							5,252.6	5.3
East Asia & Pacific	4,341	27.2	7,048	0.2							1,102.2	6.9
Europe & Central Asia	9,464	39.7	−8,143	−0.1							771.3	3.2
Latin America & Carib.	9,440	47.1	45,878	0.5							1,456.3	7.3
Middle East & N. Africa	168	1.5	−239	−0.1							242.1	2.2
South Asia	782	16.3	889	0.1							213.0	4.5
Sub-Saharan Africa	6,436	27.3	52,963	0.8							1,467.7	6.2
High income	7,979	26.1	−7,997	−0.1							3,294.2	10.8
Europe EMU	898	37.4	−2,675	−0.3							270.2	11.4

a. Flowering plants only. b. Data may refer to earlier years. They are the most recent reported by the World Conservation Monitoring Centre in 2000.

Deforestation and biodiversity | 3.4

About the data

The estimates of forest area are from the Food and Agriculture Organization's (FAO) *State of the World's Forests 2001,* which provides information on forest cover as of 2000 and a revised estimate of forest cover in 1990. The current survey is the latest global forest assessment and the first to use a uniform global definition of forest. According to this assessment, the global rate of net deforestation has slowed to 9 million hectares a year, a rate 20 percent lower than that previously reported.

No breakdown of forest cover between natural forest and plantation is shown in the table because of space limitations. (This breakdown is provided by the FAO only for developing countries.) For this reason the deforestation data in the table may underestimate the rate at which natural forest is disappearing in some countries.

Deforestation is a major cause of loss of biodiversity, and habitat conservation is vital for stemming this loss. Conservation efforts traditionally have focused on protected areas, which have grown substantially in recent decades. Measures of species richness are one of the most straightforward ways to indicate the importance of an area for biodiversity. The number of small plants and animals is usually estimated by sampling of plots. It is also important to know which aspects are under the most immediate threat. This, however, requires a large amount of data and time-consuming analysis. For this reason global analyses of the status of threatened species have been carried out for few groups of organisms. Only for birds has the status of all species been assessed. An estimated 45 percent of mammal species remain to be assessed. For plants the World Conservation Union's (IUCN) *1997 IUCN Red List of Threatened Plants* provides the first-ever comprehensive listing of threatened species on a global scale, the result of more than 20 years' work by botanists from around the world. Nearly 34,000 plant species, 12.5 percent of the total, are threatened with extinction.

The table shows information on protected areas, numbers of certain species, and numbers of those species under threat. The World Conservation Monitoring Centre (WCMC) compiles these data from a variety of sources. Because of differences in definitions and reporting practices, cross-country comparability is limited. Compounding these problems, available data cover different periods.

Nationally protected areas are areas of at least 1,000 hectares that fall into one of five management categories defined by the WCMC:

- Scientific reserves and strict nature reserves with limited public access.

- National parks of national or international significance (not materially affected by human activity).
- Natural monuments and natural landscapes with unique aspects.
- Managed nature reserves and wildlife sanctuaries.
- Protected landscapes and seascapes (which may include cultural landscapes).

Designating land as a protected area does not necessarily mean that protection is in force, however. For small countries that may only have protected areas smaller than 1,000 hectares, this size limit in the definition will result in an underestimate of the extent and number of protected areas.

Threatened species are defined according to the IUCN's classification categories: endangered (in danger of extinction and unlikely to survive if causal factors continue operating), vulnerable (likely to move into the endangered category in the near future if causal factors continue operating), rare (not endangered or vulnerable, but at risk), indeterminate (known to be endangered, vulnerable, or rare but not enough information is available to say which), out of danger (formerly included in one of the above categories but now considered relatively secure because appropriate conservation measures are in effect), and insufficiently known (suspected but not definitely known to belong to one of the above categories).

Figures on species are not necessarily comparable across countries because taxonomic concepts and coverage vary. And while the number of birds and mammals is fairly well known, it is difficult to make an accurate count of plants. Although the data in the table should be interpreted with caution, especially for numbers of threatened species (where our knowledge is very incomplete), they do identify countries that are major sources of global biodiversity and show national commitments to habitat protection.

Definitions

- **Forest area** is land under natural or planted stands of trees, whether productive or not. • **Average annual deforestation** refers to the permanent conversion of natural forest area to other uses, including shifting cultivation, permanent agriculture, ranching, settlements, and infrastructure development. Deforested areas do not include areas logged but intended for regeneration or areas degraded by fuelwood gathering, acid precipitation, or forest fires. Negative numbers indicate an increase in forest area. • **Mammals** exclude whales and porpoises. • **Birds** are listed for countries included within their breeding or wintering ranges. • **Higher plants** refer to native vascular plant species. • **Threatened species** are the number of species classified by the IUCN as endangered, vulnerable, rare, indeterminate, out of danger, or insufficiently known. • **Nationally protected areas** are totally or partially protected areas of at least 1,000 hectares that are designated as national parks, natural monuments, nature reserves or wildlife sanctuaries, protected landscapes and seascapes, or scientific reserves with limited public access. The data do not include sites protected under local or provincial law. Total land area is used to calculate the percentage of total area protected (see table 3.1).

Data sources

The forestry data are from the FAO's *State of the World's Forests 2001.* The data on species are from the WCMC's *Biodiversity Data Sourcebook* (1994) and the IUCN's *2000 IUCN Red List of Threatened Animals* and *1997 IUCN Red List of Threatened Plants.* The data on protected areas are from the WCMC's Protected Areas Data Unit.

3.5 Freshwater

	Freshwater resources			Annual freshwater withdrawals					Access to an improved water source			
	Internal flows billion cu. m 1999	Flows from other countries billion cu. m 1999	Total resources per capita cu. m[a] 1999	billion cu. m[b]	% of total renewable resources[a,b]	% for agriculture[c]	% for industry[c]	% for domestic[c]	Urban % of population 1990	Urban % of population 2000	Rural % of population 1990	Rural % of population 2000
Albania	27	15.7	12,621	1.4	3.3	71	0	29	..	..	..	..
Algeria	14	0.4	477	4.5	31.5	60[d]	15[d]	25[d]	..	98	..	88
Angola	184	..	14,890	0.5	0.3	76[d]	10[d]	14[d]	..	34	..	40
Argentina	360	..	9,841	28.6	7.9	75	9	16	..	85	..	30
Armenia	9	1.5	2,783	2.9	27.6	66	4	30	..	..	..	..
Australia	352	0.0	18,559	15.1	4.3	70	6	12	100	100	100	100
Austria	55	29.0	10,381	2.2	2.7	9	60	31	100	100	100	100
Azerbaijan	8	22.2	3,796	16.5	54.6	70	25	5	..	..	..	..
Bangladesh	105	1,105.6	9,482	14.6	1.2	86	2	12	98	99	89	97
Belarus	37	20.8	5,781	2.7	4.7	35	43	22	..	100	..	100
Belgium	..	..	..	..	..	..	..	..	..	..	..	..
Benin	10	15.5	4,220	0.2	0.6	67[d]	10[d]	23[d]	..	74	..	55
Bolivia	316	..	38,830	1.4	0.4	48	20	32	92	93	52	55
Bosnia and Herzegovina	36	2.0	9,662	..	..	..	..	..	..	..	..	..
Botswana	3	11.8	9,256	0.1	0.7	48[d]	20[d]	32[d]	100	100	91	..
Brazil	5,418	..	32,256	54.9	1.0	61	18	21	93	95	50	54
Bulgaria	18	..	2,193	..	..	..	..	..	..	100	..	100
Burkina Faso	18	..	1,592	0.4	2.2	81[d]	0[d]	19[d]	74	84	50	..
Burundi	4	..	539	0.1	2.8	64[d]	0[d]	36[d]	94	96	63	..
Cambodia	121	355.6	40,505	0.5	0.1	94	1	5	..	53	..	25
Cameroon	268	0.0	18,243	0.4	0.1	35[d]	19[d]	46[d]	76	82	36	42
Canada	2,740	52.0	91,567	45.1	1.6	9	80	11	100	100	99	99
Central African Republic	141	..	39,833	0.1	0.0	73[d]	6[d]	21[d]	80	80	46	46
Chad	15	28.0	5,744	0.2	0.4	82[d]	2[d]	16[d]	..	31	..	26
Chile	928	..	61,793	21.4	2.3	84	11	5	98	99	48	66
China	2,812	17.2	2,257	525.5	18.6	77	18	5	99	94	60	66
Hong Kong, China	..	..	..	..	..	..	..	..	..	..	..	..
Colombia	2,133	..	51,349	8.9	0.4	37	4	59	95	98	68	73
Congo, Dem. Rep.	935	84.0	20,472	0.4	0.0	23[d]	16[d]	61[d]	..	89	..	26
Congo, Rep.	222	610.0	291,000	0.0	0.0	11[d]	27[d]	62[d]	..	71	..	17
Costa Rica	112	..	31,318	5.8	5.1	80	7	13	..	98	..	98
Côte d'Ivoire	77	1.0	4,998	0.7	0.9	67[d]	11[d]	22[d]	89	90	49	65
Croatia	38	33.7	15,995	0.1	0.1	..	50	50	..	..	..	..
Cuba	38	..	3,400	5.2	13.7	51	0	49	..	99	..	82
Czech Republic	15	1.0	1,557	2.5	15.8	1	57	39	..	..	..	..
Denmark	6	..	1,127	0.9	14.8	16	9	53	..	100	..	100
Dominican Republic	21	..	2,499	8.3	39.7	89	1	11	83	83	70	70
Ecuador	442	..	35,611	17.0	3.8	82	6	12	..	81	..	51
Egypt, Arab Rep.	2	56.0	930	55.1	94.5	86[d]	8[d]	6[d]	97	96	91	94
El Salvador	18	..	2,876	0.7	4.1	46	20	34	..	88	47	61
Eritrea	3	6.0	2,205	..	..	..	..	..	..	63	..	42
Estonia	13	0.1	8,874	0.2	1.3	5	39	56	..	..	..	..
Ethiopia	110	0.0	1,752	2.2	2.0	86[d]	3[d]	11[d]	77	77	13	13
Finland	107	3.0	21,293	2.4	2.2	0	82	17	100	100	100	100
France	180	11.0	3,258	40.6	21.3	12	73	15	..	..	..	..
Gabon	164	0.0	135,716	0.1	0.0	6[d]	22[d]	72[d]	..	73	..	55
Gambia, The	3	5.0	6,395	0.0	0.4	91[d]	2[d]	7[d]	..	80	..	53
Georgia	58	5.2	11,610	3.5	5.5	59	20	21	..	..	..	..
Germany	107	71.0	2,168	46.3	26.0	0	86	14	..	..	..	..
Ghana	30	22.9	2,832	0.3	0.6	52[d]	13[d]	35[d]	83	87	43	49
Greece	54	15.0	6,548	7.0	10.2	81	3	16	..	..	..	..
Guatemala	134	..	12,121	1.2	0.9	74	17	9	88	97	72	88
Guinea	226	0.0	31,170	0.7	0.3	87[d]	3[d]	10[d]	72	72	36	36
Guinea-Bissau	16	11.0	22,791	0.0	0.1	36[d]	4[d]	60[d]	..	29	..	55
Haiti	12	..	1,551	1.0	8.1	94	1	5	55	49	42	45
Honduras	96	..	15,211	1.5	1.6	91	5	4	90	97	79	82

	Freshwater resources			Annual freshwater withdrawals					Access to an improved water source			
	Internal flows billion cu. m 1999	Flows from other countries billion cu. m 1999	Total resources per capita cu. m[a] 1999	billion cu. m[b]	% of total renewable resources[a,b]	% for agriculture[c]	% for industry[c]	% for domestic[c]	Urban % of population 1990	2000	Rural % of population 1990	2000
Hungary	6	114.0	11,919	6.3	5.2	5	70	14	100	100	98	98
India	1,261	647.2	1,913	500.0	26.2	92	3	5	92	92	73	86
Indonesia	2,838	..	13,709	74.3	2.6	93	1	6	90	91	60	65
Iran, Islamic Rep.	129	..	2,040	70.0	54.5	92	2	6	95	99	75	89
Iraq	35	..	1,544	42.8	121.6	92	5	3	..	96	..	48
Ireland	49	3.0	13,859	1.2	2.3	15	21	40	..	..	..	..
Israel	1	0.3	180	1.7	155.5	64[d]	7[d]	29[d]	..	..	..	..
Italy	161	6.8	2,906	57.5	34.4	45	37	18	..	..	..	..
Jamaica	9	..	3,618	0.9	9.6	77	7	15	..	81	..	59
Japan	430	0.0	3,397	91.4	21.3	64	17	19	..	..	..	..
Jordan	1	..	148	1.0	140.0	75	3	22	99	100	92	84
Kazakhstan	75	34.2	7,342	33.7	30.7	81	17	2	..	98	..	82
Kenya	20	10.0	1,027	2.0	6.8	76[d]	4[d]	20[d]	89	87	25	31
Korea, Dem. Rep.	67	10.1	3,293	14.2	18.4	73	16	11	..	100	..	100
Korea, Rep.	65	4.9	1,490	23.7	33.9	63	11	26	..	97	..	71
Kuwait	0	0.0	0	0.5	..	60	2	37	..	..	..	..
Kyrgyz Republic	47	..	9,559	10.1	21.7	94	3	3	..	98	..	66
Lao PDR	190	91.2	55,251	1.0	0.4	82	10	8	..	59	..	100
Latvia	17	18.7	14,561	0.3	0.8	13	32	55	..	..	..	..
Lebanon	5	0.0	1,124	1.3	26.9	68	4	28	..	100	..	100
Lesotho	5	0.0	2,470	0.1	1.0	56[d]	22[d]	22[d]	..	98	..	88
Libya	1	0.0	148	3.9	486.3	87[d]	4[d]	9[d]	72	72	68	68
Lithuania	16	9.3	6,732	0.3	1.0	3	16	81	..	..	..	..
Macedonia, FYR	6	1.0	3,464	..	..	..	..	..	..	..	..	..
Madagascar	337	0.0	22,391	19.7	5.8	99[d]	0[d]	1[d]	85	85	31	31
Malawi	18	1.1	1,724	0.9	5.1	86[d]	3[d]	10[d]	90	95	43	44
Malaysia	580	..	25,539	12.7	2.2	76	13	11	..	..	..	94
Mali	60	40.0	9,449	1.4	1.4	97[d]	1[d]	2[d]	65	74	52	61
Mauritania	0	11.0	4,387	16.3	143.0	92	2	6	34	34	40	40
Mauritius	2	0.0	1,873	0.4	16.4	77[d]	7[d]	16[d]	100	100	100	100
Mexico	409	49.0	4,742	77.8	17.0	78	5	17	92	94	61	63
Moldova	1	10.7	2,733	3.0	25.3	26	65	9	..	100	..	100
Mongolia	35	..	14,632	0.4	1.2	53	27	20	..	77	..	30
Morocco	30	0.0	1,062	11.1	36.8	92[d]	3[d]	5[d]	94	100	58	58
Mozambique	100	116.0	12,486	0.6	0.3	89	2[d]	9[d]	..	86	..	43
Myanmar	881	128.2	22,404	4.0	0.4	90	3	7	88	88	56	60
Namibia	6	39.3	26,744	0.3	0.5	68[d]	3[d]	29[d]	98	100	63	67
Nepal	198	12.0	8,989	29.0	13.8	99	0	1	96	85	63	80
Netherlands	11	80.0	5,758	7.8	8.6	0	68	16	100	100	100	100
New Zealand	327	..	85,811	2.0	0.6	55	13	9	100	100	..	..
Nicaragua	190	..	38,668	1.3	0.7	84	2	14	93	95	44	59
Niger	4	29.0	3,097	0.5	1.5	82[d]	2[d]	16[d]	65	70	51	56
Nigeria	221	59.0	2,260	4.0	1.4	54[d]	15[d]	31[d]	78	81	33	39
Norway	382	11.0	88,117	2.0	0.5	3	68	27	100	100	100	100
Oman	1	..	426	1.2	120.0	94	2	5	41	41	30	30
Pakistan	85	170.3	1,892	155.6	61.0	97	2	2	96	96	79	84
Panama	147	..	52,437	1.6	1.1	70	2	28	..	88	..	86
Papua New Guinea	801	..	170,258	0.1	0.0	49	22	29	88	88	32	32
Paraguay	94	..	17,541	0.4	0.5	78	7	15	80	95	47	58
Peru	1,746	..	69,203	19.0	1.1	86	7	7	84	87	47	51
Philippines	479	..	6,450	55.4	11.6	88	4	8	94	92	81	80
Poland	55	8.0	1,630	12.1	19.2	3	67	20	..	..	..	..
Portugal	37	35.0	7,208	7.3	10.1	53	40	8	..	..	..	..
Puerto Rico	..	..	..	..	..	..	..	..	..	..	..	..
Romania	37	..	1,648	..	..	..	..	..	..	91	..	16
Russian Federation	4,314	184.5	30,767	77.1	1.7	20	62	19	..	100	..	96

3.5 | Freshwater

	Freshwater resources			Annual freshwater withdrawals					Access to an improved water source			
	Internal flows billion cu. m 1999	Flows from other countries billion cu. m 1999	Total resources per capita cu. m[a] 1999	billion cu. m[b]	% of total renewable resources[a,b]	% for agriculture[c]	% for industry[c]	% for domestic[c]	Urban % of population 1990	2000	Rural % of population 1990	2000
Rwanda	6	..	758	0.8	12.2	94 [d]	1 [d]	5 [d]	..	60	..	40
Saudi Arabia	2	0.0	119	17.0	708.3	90	1	9	..	100	..	64
Senegal	26	13.0	4,243	1.5	3.8	92 [d]	3 [d]	5 [d]	90	92	60	65
Sierra Leone	160	0.0	32,328	0.4	0.2	89 [d]	4 [d]	7 [d]	..	23		31
Singapore	..	..	..	..	..	..	..	..	100	100	..	..
Slovak Republic	13	70.0	15,382	1.4	1.7	8	50	39	..	100	..	100
Slovenia	19	0.0	9,318	0.5	2.7	..	50	50	100	100	100	100
South Africa	45	5.2	1,187	13.3	26.6	72 [d]	11 [d]	17 [d]	..	92	..	80
Spain	112	0.3	2,844	35.5	31.7	68	18	13	..	..	..	..
Sri Lanka	50	..	2,634	9.8	19.5	96	2	2	90	91	59	80
Sudan	35	119.0	5,312	17.8	11.6	94 [d]	1 [d]	5 [d]	86	86	60	69
Sweden	178	0.0	20,096	2.7	1.5	4	30	35	100	100	100	100
Switzerland	40	13.0	7,427	2.6	4.9	0	58	42	100	100	100	100
Syrian Arab Republic	7	37.7	2,845	14.4	32.2	94	2	4	..	94	..	64
Tajikistan	66	13.3	12,763	11.9	14.9	92	4	4	..	..	..	..
Tanzania	80	9.0	2,703	1.2	1.3	89 [d]	2 [d]	9 [d]	80	80	42	42
Thailand	210	199.9	6,804	33.1	8.1	91	4	5	83	89	68	77
Togo	12	0.5	2,628	0.1	0.8	25 [d]	13 [d]	62 [d]	82	85	38	38
Trinidad and Tobago	..	..	..	..	..	..	..	..	..	..	..	..
Tunisia	4	0.6	434	2.8	69.0	86 [d]	2 [d]	13 [d]	94	..	61	..
Turkey	196	7.6	3,162	35.5	17.4	73 [d]	11 [d]	16 [d]	82	82	76	84
Turkmenistan	1	44.1	9,520	23.8	52.3	98	1	1	..	..	..	..
Uganda	39	27.0	3,073	0.2	0.3	60	8	32	80	72	40	46
Ukraine	53	86.5	2,795	26.0	18.6	30	52	18	..	..	..	..
United Arab Emirates	0	0.0	71	2.1	1,055.0	67	9	24	..	..	..	..
United Kingdom	145	2.0	2,471	9.3	6.4	2	8	65	100	100	100	100
United States	2,460	18.0	8,906	447.7	18.1	27 [d]	65 [d]	8 [d]	100	100	100	100
Uruguay	59	..	17,809	4.2	7.1	91	3	6	..	98	..	93
Uzbekistan	16	..	668	58.0	356.1	94	2	4	..	96	..	78
Venezuela, RB	846	..	35,686	4.1	0.5	46	10	44	..	88	..	58
Vietnam	367	524.7	11,497	54.3	6.1	86	10	4	81	81	40	50
West Bank and Gaza	..	..	..	..	..	..	..	..	..	..	..	..
Yemen, Rep.	4	..	241	2.9	71.5	92	1	7	85	85	60	64
Yugoslavia, FR (Serb./Mont.)	44	144.0	17,709	..	..	..	..	..	..	..	..	..
Zambia	80	35.8	11,739	1.7	1.5	77 [d]	7 [d]	16 [d]	88	88	28	48
Zimbabwe	14	5.9	1,680	1.2	6.1	79 [d]	7 [d]	14 [d]	99	100	68	77
World	**42,809 s**	**5,979.4 s**	**8,241 w**	..	..	**70 w**	**22 w**	**8 w**	**94 w**	**93 w**	**64 w**	**71 w**
Low income	10,450	4,539.8	6,205	..	..	87	8	5	89	89	64	71
Middle income	24,227	1,088.5	9,537	..	..	74	13	12	95	93	62	68
Lower middle income	15,052	790.8	7,585	..	..	75	15	10	96	93	62	68
Upper middle income	9,175	297.7	16,744	..	..	73	10	17	..	94	..	68
Low & middle income	34,677	5,628.3	7,949	..	..	82	10	7	93	92	63	70
East Asia & Pacific	9,445	1,331.8	..	..	..	80	14	6	96	93	60	67
Europe & Central Asia	5,221	848.1	12,797	..	..	63	26	11	..	..	..	..
Latin America & Carib.	13,987	49.0	27,919	..	..	74	9	18	92	93	56	62
Middle East & N. Africa	235	94.7	1,145	..	..	89	4	6	93	96	76	80
South Asia	1,849	1,945.1	2,854	..	..	93	2	4	93	92	75	85
Sub-Saharan Africa	3,941	1,359.6	8,257	..	..	87	4	9	81	82	37	41
High income	8,132	351.1	..	..	..	30	59	11	..	..	..	..
Europe EMU	819	239.8	3,769	..	..	21	63	16	..	..	..	..

a. River flows from other countries are included when available, but river outflows are not, because of data unreliability. b. Data refer to any year from 1980 to 1999. c. Unless otherwise noted, sectoral withdrawal shares are estimated for 1987. d. Data refer to a year other than 1987 (see *Primary data documentation*).

Freshwater | 3.5

The data on freshwater resources are based on estimates of runoff into rivers and recharge of groundwater. These estimates are based on different sources and refer to different years, so cross-country comparisons should be made with caution. Because the data are collected intermittently, they may hide significant variations in total renewable water resources from one year to the next. The data also fail to distinguish between seasonal and geographic variations in water availability within countries. Data for small countries and countries in arid and semiarid zones are less reliable than those for larger countries and countries with greater rainfall. Finally, caution is also needed in comparing data on annual freshwater withdrawals, which are subject to variations in collection and estimation methods.

This year's table shows both internal freshwater resources and the river flows arising outside countries. Because the data on total freshwater resources include river flows entering a country while river flows out of the country are not deducted (because of data unreliability), they overestimate the availability of water from international river ways. This can be important in water-short countries, notably in the Middle East.

The data on access to an improved water source measure the share of the population with reasonable and ready access to an adequate amount of safe water for domestic purposes. An improved source can be any form of collection or piping used to make water regularly available. While information on access to an improved water source is widely used, it is extremely subjective, and such terms as *safe, improved, adequate,* and *reasonable* may have very different meanings in different countries despite official World Health Organization definitions (see *Definitions*). Even in high-income countries treated water may not always be safe to drink. While access to safe water is equated with connection to a public supply system, this does not take account of variations in the quality and cost (broadly defined) of the service once connected. Thus cross-country comparisons must be made cautiously. Changes over time within countries may result from changes in definitions or measurements.

- **Freshwater resources** refer to total renewable resources, broken down between internal flows of rivers and groundwater from rainfall in the country, and river flows from other countries. Freshwater resources per capita are calculated using the World Bank's population estimates (see table 2.1). • **Annual freshwater withdrawals** refer to total water withdrawal, not counting evaporation losses from storage basins. Withdrawals also include water from desalination plants in countries where they are a significant source. Withdrawal data are for single years between 1980 and 1999 unless otherwise indicated. Withdrawals can exceed 100 percent of total renewable resources where extraction from nonrenewable aquifers or desalination plants is considerable or where there is significant water reuse. Withdrawals for agriculture and industry are total withdrawals for irrigation and livestock production and for direct industrial use (including withdrawals for cooling thermoelectric plants). Withdrawals for domestic uses include drinking water, municipal use or supply, and use for public services, commercial establishments, and homes. For most countries sectoral withdrawal data are estimated for 1987. • **Access to an improved water source** refers to the percentage of the population with reasonable access to an adequate amount of water from an improved source, such as a household connection, public standpipe, borehole, protected well or spring, or rainwater collection. Unimproved sources include vendors, tanker trucks, and unprotected wells and springs. Reasonable access is defined as the availability of at least 20 liters a person a day from a source within one kilometer of the dwelling.

Figure 3.5a

Most people without access to an improved water source live in Asia

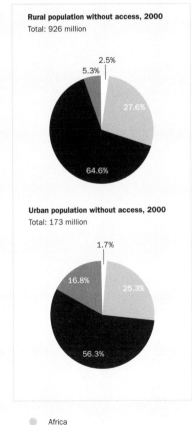

Rural population without access, 2000
Total: 926 million

2.5%
5.3%
27.6%
64.6%

Urban population without access, 2000
Total: 173 million

1.7%
16.8%
25.3%
56.3%

○ Africa
● Asia
● Latin America & Caribbean
  Europe

Source: World Health Organization data.

Figure 3.5b

Agriculture accounts for most freshwater withdrawals in developing countries—industry in high-income countries

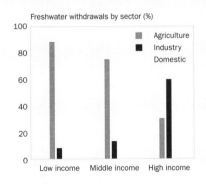

Freshwater withdrawals by sector (%)

■ Agriculture
■ Industry
■ Domestic

Note: Data are for the most recent year available (see table 3.5).
Source: Table 3.5.

The data on freshwater resources and withdrawals are compiled by the World Resources Institute from various sources and published in *World Resources 1998–99* and *World Resources 2000–01* (produced in collaboration with the United Nations Environment Programme, United Nations Development Programme, and World Bank). The data on access to an improved water source come from the World Health Organization.

3.6 Water pollution

	Emissions of organic water pollutants				Industry shares of emissions of organic water pollutants							
	kilograms per day		kilograms per day per worker		Primary metals %	Paper and pulp %	Chemicals %	Food and beverages %	Stone, ceramics, and glass %	Textiles %	Wood %	Other %
	1980	1998[a]	1980	1998[a]	1998[a]	1998[a]	1998[a]	1998[a]	1998[a]	1998[a]	1998[a]	1998[a]
Albania	..	5,844	..	0.24	22.9	1.5	6.2	62.0	0.4	4.7	0.7	1.5
Algeria	60,290	102,969	0.19	0.25	44.6	..	3.8	40.8	0.4	8.0	2.5	..
Angola	..	1,472	..	0.20	7.6	3.0	9.1	65.9	0.3	5.5	4.4	4.1
Argentina	244,711	186,844	0.18	0.21	6.3	12.6	8.1	59.4	0.2	7.4	1.5	4.6
Armenia		12,858	..	0.23	..	..	..	..	..	..	..	..
Australia	204,333	173,269	0.18	0.19	12.4	22.8	6.7	43.5	0.2	5.3	2.8	6.3
Austria	108,416	78,040	0.16	0.14	13.1	19.5	9.1	36.1	0.3	6.7	4.3	10.9
Azerbaijan	..	45,025	..	0.17	11.6	2.5	12.0	49.0	0.2	18.1	1.0	5.6
Bangladesh	66,713	186,852	0.16	0.16	2.8	6.8	3.5	34.2	0.1	50.9	0.6	1.1
Belarus	..	..	..	..	..	..	..	..	..	..	..	..
Belgium	136,452	113,460	0.16	0.16	14.4	17.7	11.6	36.8	0.2	8.8	2.0	8.4
Benin	1,646	..	0.28		..	..	..	..	..	..	..	..
Bolivia	9,343	10,251	0.22	0.23	4.7	13.8	6.5	61.8	0.3	9.0	2.6	1.2
Bosnia and Herzegovina	..	8,903	..	0.18	20.5	13.1	6.6	33.3	0.2	17.6	5.8	2.8
Botswana	1,307	4,386	0.24	0.18	0.0	11.5	2.8	67.5	0.0	12.5	2.1	3.7
Brazil	866,790	690,876	0.16	0.19	19.0	12.6	9.3	41.6	0.2	10.9	1.6	4.8
Bulgaria	152,125	103,132	0.13	0.16	3.7	8.5	10.5	49.6	0.1	17.8	2.3	7.5
Burkina Faso	2,385	..	0.29		..	..	..	..	..	..	..	..
Burundi	769	1,644	0.22	0.24	0.0	8.3	4.7	67.8	0.1	16.7	1.6	0.8
Cambodia	..	12,078	..	0.16	0.0	3.4	3.3	59.2	0.6	24.7	5.8	3.1
Cameroon	14,569	12,367	0.29	0.23	3.1	5.9	21.9	61.2	0.0	3.2	4.4	0.3
Canada	330,241	298,209	0.18	0.17	9.6	29.6	9.2	34.2	0.1	5.6	3.8	8.0
Central African Republic	861	670	0.26	0.17	..	..	..	..	..	..	..	..
Chad	..	..	..		..	..	..	..	..	..	..	..
Chile	44,371	74,810	0.21	0.23	8.4	11.6	8.9	59.7	0.1	5.9	2.7	2.7
China	3,377,105	8,491,856	0.14	0.14	18.8	12.5	13.2	30.1	0.6	15.3	1.1	8.4
Hong Kong, China	102,002	35,961	0.11	0.15	1.4	36.0	4.2	20.6	0.1	29.9	0.2	7.6
Colombia	96,055	111,545	0.19	0.20	3.1	15.6	10.8	51.5	0.2	14.7	1.0	3.1
Congo, Dem. Rep.	..	..	..	..	..	..	..	..	..	..	..	..
Congo, Rep.	1,039	..	0.21		..	..	..	..	..	..	..	..
Costa Rica	..	32,658	..	0.21	1.1	10.5	6.6	61.7	0.1	15.8	1.5	2.6
Côte d'Ivoire	15,414	12,401	0.23	0.24	..	5.5	7.1	71.9	0.0	8.6	5.9	1.0
Croatia	..	48,447	..	0.17	7.2	14.4	8.6	45.2	0.2	14.6	3.8	6.0
Cuba	120,703	..	0.24		..	..	..	..	..	..	..	..
Czech Republic	..	165,993	..	0.14	16.3	6.4	7.9	43.8	0.3	10.5	3.8	11.0
Denmark	65,465	92,733	0.17	0.18	2.1	29.0	7.7	46.3	0.2	3.6	3.0	8.2
Dominican Republic	54,935	..	0.38		..	..	..	..	..	..	..	..
Ecuador	25,297	31,974	0.23	0.26	2.3	11.8	6.9	68.9	0.1	6.6	1.7	1.6
Egypt, Arab Rep.	169,146	225,843	0.19	0.18	13.0	6.5	11.3	46.3	0.3	18.0	0.6	4.1
El Salvador	9,390	21,833	0.24	0.19	1.9	8.9	7.0	47.0	0.1	33.1	0.5	1.5
Eritrea	..	..	..	..	..	..	..	..	..	..	..	..
Estonia	..	..	..	..	..	..	..	..	..	..	..	..
Ethiopia	16,754	..	0.22		..	..	..	..	..	..	..	..
Finland	92,275	63,662	0.17	0.19	9.0	41.6	5.6	30.2	0.2	3.0	3.9	6.5
France	729,776	585,382	0.14	0.15	11.6	21.2	10.8	37.7	0.2	6.0	1.8	10.8
Gabon	2,661	1,886	0.15	0.26	0.0	6.0	4.9	79.7	0.1	1.2	6.9	1.2
Gambia, The	549	832	0.30	0.34	0.0	15.3	1.9	77.8	0.1	2.6	1.9	0.4
Georgia	..	..	..	..	..	..	..	..	..	..	..	..
Germany	..	811,316	..	0.12	12.7	16.8	15.5	30.6	0.3	4.8	2.2	17.2
Ghana	15,868	14,449	0.20	0.17	9.8	16.9	10.5	39.5	0.2	9.1	12.4	1.7
Greece	65,304	58,134	0.17	0.20	6.0	12.4	8.8	53.6	0.3	14.1	1.4	3.5
Guatemala	20,856	18,728	0.25	0.28	5.1	7.1	6.1	72.7	0.1	7.1	0.8	1.0
Guinea	..	..	..	..	..	..	..	..	..	..	..	..
Guinea-Bissau	..	..	..	..	..	..	..	..	..	..	..	..
Haiti	4,734	..	0.19		..	..	..	..	..	..	..	..
Honduras	13,067	34,036	0.23	0.20	1.1	7.8	3.9	55.5	0.1	26.8	4.0	0.8

	Emissions of organic water pollutants				Industry shares of emissions of organic water pollutants							
	kilograms per day		kilograms per day per worker		Primary metals %	Paper and pulp %	Chemicals %	Food and beverages %	Stone, ceramics, and glass %	Textiles %	Wood %	Other %
	1980	1998ª	1980	1998ª	1998ª	1998ª	1998ª	1998ª	1998ª	1998ª	1998ª	1998ª
Hungary	201,888	140,894	0.15	0.17	10.0	10.1	8.1	49.9	0.2	12.6	1.9	7.3
India	1,422,564	1,760,353	0.21	0.19	15.0	8.4	8.9	49.1	0.2	12.9	0.3	5.3
Indonesia	214,010	347,083	0.22	0.16	5.3	19.0	10.5	32.7	0.1	23.7	3.0	5.8
Iran, Islamic Rep.	72,334	101,900	0.15	0.17	20.6	8.0	8.0	39.7	0.5	17.3	0.7	5.4
Iraq	32,986	19,617	0.19	0.16	8.8	14.1	15.1	39.4	0.7	16.7	0.3	4.8
Ireland	43,544	34,176	0.19	0.16	1.8	17.3	11.3	51.3	0.2	6.2	2.0	10.0
Israel	39,113	54,149	0.15	0.16	3.7	19.7	9.4	43.9	0.2	12.1	1.8	9.3
Italy	442,712	359,578	0.13	0.13	12.1	16.0	11.8	28.7	0.3	16.1	2.5	12.6
Jamaica	11,123	17,507	0.25	0.29	6.9	7.2	3.8	70.8	0.1	9.8	1.3	..
Japan	1,456,016	1,391,281	0.14	0.14	8.4	22.0	8.8	39.5	0.2	6.4	1.8	12.9
Jordan	4,146	16,142	0.17	0.18	3.9	16.2	14.5	51.4	0.5	7.2	3.3	3.0
Kazakhstan	..	..	..	..	..	..	..	..	..	..	..	..
Kenya	26,834	49,125	0.19	0.24	4.1	12.2	5.9	68.4	0.1	8.8	1.9	..
Korea, Dem. Rep.	..	..	..	..	..	..	..	..	..	..	..	..
Korea, Rep.	281,900	317,903	0.14	0.12	11.9	17.6	11.6	26.4	0.3	15.8	1.5	14.9
Kuwait	6,921	8,303	0.16	0.15	3.2	4.6	13.9	51.3	0.5	16.6	3.7	6.2
Kyrgyz Republic	..	20,700	..	0.16	13.7	0.2	0.9	54.8	0.4	21.0	1.0	8.0
Lao PDR	..	..	..	..	..	..	..	..	..	..	..	..
Latvia	..	27,357	..	0.18	2.8	11.8	4.5	58.2	0.1	11.0	5.9	5.7
Lebanon	14,586	..	0.20	..	..	..	..	..	..	..	..	..
Lesotho	993	3,123	0.24	0.16	1.2	4.0	0.7	39.7	0.1	51.3	0.6	2.3
Libya	3,532	..	0.21	..	..	..	..	..	..	..	..	..
Lithuania	..	38,570	..	0.17	1.6	10.8	5.1	55.5	0.2	17.1	4.4	5.2
Macedonia, FYR	..	23,490	..	0.18	11.7	9.6	6.2	45.0	0.1	20.9	1.7	4.9
Madagascar	9,131	..	0.23	..	..	..	..	..	..	..	..	..
Malawi	12,224	9,055	0.32	0.26	0.0	12.6	5.1	67.7	0.1	11.6	1.7	1.1
Malaysia	77,215	166,577	0.15	0.12	6.7	14.5	16.7	31.8	0.3	8.2	6.7	15.2
Mali	..	..	..	..	..	..	..	..	..	..	..	..
Mauritania	..	..	..	..	..	..	..	..	..	..	..	..
Mauritius	9,224	16,524	0.21	0.16	1.1	5.7	2.3	38.0	0.1	50.8	0.8	1.2
Mexico	130,993	158,505	0.22	0.17	8.2	8.6	14.0	54.8	0.2	5.4	0.4	8.3
Moldova	..	34,234	..	0.29	0.2	4.0	1.4	81.7	0.2	10.8	1.3	0.5
Mongolia	9,254	7,939	0.19	0.18	1.8	4.3	0.9	64.2	0.3	24.6	4.9	..
Morocco	26,598	86,320	0.15	0.18	0.7	7.7	7.1	53.5	0.3	27.3	0.9	2.5
Mozambique	..	495	..	0.16	3.1	..	4.1	..	0.1	..	..	1.2
Myanmar	..	4,479	..	0.09	11.4	6.8	29.6	18.5	1.5	3.9	27.1	1.2
Namibia	..	7,350	..	0.35	0.0	5.0	1.6	90.4	0.1	1.2	0.9	0.8
Nepal	18,692	26,550	0.25	0.14	1.5	8.1	3.9	43.3	1.2	39.3	1.7	1.0
Netherlands	165,416	122,843	0.18	0.18	7.7	26.3	12.2	41.8	0.2	2.5	1.2	8.1
New Zealand	59,012	50,706	0.21	0.22	4.6	19.6	4.9	58.6	0.1	4.9	3.1	4.2
Nicaragua	9,647	..	0.28	..	..	..	..	..	..	..	..	..
Niger	372	..	0.19	..	..	..	..	..	..	..	..	..
Nigeria	72,082	53,646	0.17	0.18	0.9	31.2	6.5	37.4	0.2	10.6	10.4	2.9
Norway	67,897	52,745	0.19	0.21	6.3	35.4	3.4	43.6	0.1	1.8	3.4	6.2
Oman	..	4,602	..	0.17	5.5	14.8	6.5	50.7	0.8	14.5	3.7	3.5
Pakistan	75,125	114,726	0.17	0.18	14.1	5.8	7.3	39.5	0.2	30.1	0.3	2.7
Panama	8,121	11,754	0.26	0.31	1.6	11.1	5.5	75.3	0.2	5.6	0.5	0.3
Papua New Guinea	4,365	..	0.22	..	..	..	..	..	..	..	..	..
Paraguay	..	3,250	..	0.28	2.3	9.9	6.0	73.6	0.3	6.7	0.3	0.9
Peru	50,367	51,828	0.18	0.21	9.6	12.0	8.4	53.0	0.2	12.3	1.6	2.9
Philippines	182,052	178,239	0.19	0.18	5.2	9.8	7.3	54.5	0.2	16.4	2.0	4.6
Poland	580,869	386,376	0.14	0.16	14.9	4.7	6.6	49.7	0.4	13.0	1.9	8.7
Portugal	105,441	137,314	0.15	0.14	3.5	14.2	5.1	38.9	0.4	26.6	4.8	6.4
Puerto Rico	24,034	17,494	0.16	0.16	0.9	10.9	17.7	40.2	0.2	19.6	1.3	9.1
Romania	343,145	333,168	0.12	0.12	17.1	6.7	9.0	34.3	0.3	18.5	4.8	9.4
Russian Federation	..	1,531,501	..	0.16	18.0	6.9	9.5	46.6	0.4	7.5	2.5	8.6

3.6 Water pollution

	Emissions of organic water pollutants				Industry shares of emissions of organic water pollutants							
	kilograms per day		kilograms per day per worker		Primary metals %	Paper and pulp %	Chemicals %	Food and beverages %	Stone, ceramics, and glass %	Textiles %	Wood %	Other %
	1980	1998[a]	1980	1998[a]	1998[a]	1998[a]	1998[a]	1998[a]	1998[a]	1998[a]	1998[a]	1998[a]
Rwanda	..	..	..	..	..	..	..	..	..	..	..	..
Saudi Arabia	18,181	24,436	0.12	0.14	4.4	15.9	21.1	45.1	1.0	3.8	2.0	6.8
Senegal	9,865	10,488	0.31	0.31	0.0	6.3	8.8	78.8	0.0	4.6	0.1	1.3
Sierra Leone	1,612	4,170	0.24	0.32	..	9.6	3.0	82.3	0.1	2.0	2.2	0.8
Singapore	28,558	33,661	0.10	0.10	2.0	27.9	15.4	19.8	0.2	3.8	1.6	29.3
Slovak Republic	..	61,108	..	0.15	18.3	13.3	9.8	36.8	0.3	9.3	2.8	9.4
Slovenia	..	38,187	..	0.17	28.9	17.4	8.5	24.4	0.2	12.7	2.1	5.9
South Africa	237,599	241,922	0.17	0.17	11.6	16.4	9.1	42.4	0.2	10.5	3.3	6.4
Spain	376,253	348,262	0.16	0.16	6.4	18.7	8.7	45.0	0.3	9.4	3.7	7.8
Sri Lanka	30,086	83,850	0.18	0.17	1.0	6.5	6.0	47.5	0.2	36.6	1.0	1.2
Sudan	..	..	..	..	..	..	..	..	..	..	..	..
Sweden	130,439	91,248	0.15	0.16	10.9	37.7	7.7	27.3	0.1	1.5	3.2	11.6
Switzerland	..	123,752	..	0.17	24.9	23.6	10.4	25.0	0.2	3.2	4.2	8.7
Syrian Arab Republic	36,262	21,421	0.19	0.22	2.9	1.5	8.4	68.3	0.3	17.2	0.3	1.1
Tajikistan	..	..	..	..	..	..	..	..	..	..	..	..
Tanzania	21,084	32,508	0.21	0.26	4.7	10.8	5.0	65.2	0.1	11.8	1.4	1.2
Thailand	213,271	355,819	0.22	0.16	6.1	5.3	5.3	42.2	0.2	35.4	1.5	3.9
Togo	963	..	0.27	..	..	..	..	..	..	..	..	..
Trinidad and Tobago	7,835	11,787	0.18	0.28	4.4	10.9	6.7	72.6	0.1	2.9	1.3	1.2
Tunisia	20,294	45,613	0.16	0.16	5.5	7.2	6.4	41.4	0.3	34.2	1.5	3.4
Turkey	160,173	186,269	0.20	0.16	10.7	7.0	7.6	42.9	0.3	25.5	1.0	5.1
Turkmenistan	..	..	..	..	..	..	..	..	..	..	..	..
Uganda	..	..	..	..	..	..	..	..	..	..	..	..
Ukraine	..	518,996	..	0.17	22.2	3.3	6.9	51.4	0.4	5.9	1.8	8.2
United Arab Emirates	4,524	..	0.15	..	..	..	..	..	..	..	..	..
United Kingdom	964,510	611,743	0.15	0.15	7.6	28.0	11.8	32.7	0.2	6.6	2.4	10.7
United States	2,742,993	2,577,002	0.14	0.15	8.5	32.4	10.2	28.1	0.2	6.3	2.8	11.6
Uruguay	34,270	27,722	0.21	0.25	1.4	11.3	5.9	67.6	0.2	11.1	0.7	1.9
Uzbekistan	..	..	..	..	..	..	..	..	..	..	..	..
Venezuela, RB	84,797	92,026	0.20	0.21	14.1	11.5	9.9	51.8	0.2	7.3	1.7	3.4
Vietnam	..	..	..	..	..	..	..	..	..	..	..	..
West Bank and Gaza	..	..	..	..	..	..	..	..	..	..	..	..
Yemen, Rep.	..	7,823	..	0.25	0.0	9.1	12.9	71.1	0.3	4.9	1.0	0.9
Yugoslavia, FR (Serb./Mont.)	..	119,790	..	0.16	10.1	12.4	7.7	44.0	0.3	15.5	2.1	8.0
Zambia	13,605	11,433	0.23	0.22	3.4	10.8	7.3	63.6	0.2	9.3	2.9	2.4
Zimbabwe	32,681	32,956	0.20	0.20	13.6	11.3	5.6	48.0	0.2	15.2	2.9	3.1

Note: Industry shares may not sum to 100 percent because data may be from different years.
a. Data refer to any year from 1993 to 1998.

Water pollution | 3.6

About the data

Emissions of organic pollutants from industrial activities are a major cause of degradation of water quality. Water quality and pollution levels are generally measured in terms of concentration, or load—the rate of occurrence of a substance in an aqueous solution. Polluting substances include organic matter, metals, minerals, sediment, bacteria, and toxic chemicals. This table focuses on organic water pollution resulting from industrial activities. Because water pollution tends to be sensitive to local conditions, the national-level data in the table may not reflect the quality of water in specific locations.

The data in the table come from an international study of industrial emissions that may be the first to include data from developing countries (Hettige, Mani, and Wheeler 1998). Unlike estimates from earlier studies based on engineering or economic models, these estimates are based on actual measurements of plant-level water pollution. The focus is on organic water pollution measured in terms of biochemical oxygen demand (BOD) because the data for this indicator are the most plentiful and the most reliable for cross-country comparisons of emissions. BOD measures the strength of an organic waste in terms of the amount of oxygen consumed in breaking it down. A sewage overload in natural waters exhausts the water's dissolved oxygen content. Wastewater treatment, by contrast, reduces BOD.

Data on water pollution are more readily available than other emissions data because most industrial pollution control programs start by regulating emissions of organic water pollutants. Such data are fairly reliable because sampling techniques for measuring water pollution are more widely understood and much less expensive than those for air pollution.

In their study Hettige, Mani, and Wheeler (1998) used plant- and sector-level information on emissions and employment from 13 national environmental protection agencies and sector-level information on output and employment from the United Nations Industrial Development Organization (UNIDO). Their econometric analysis found that the ratio of BOD to employment in each industrial sector is about the same across countries. This finding allowed the authors to estimate BOD loads across countries and over time. The estimated BOD intensities per unit of employment were multiplied by sectoral employment numbers from UNIDO's industry database for 1980–98. The sectoral emissions estimates were then totaled to get daily emissions of organic water pollutants in kilograms per day for each country and year. The data in the table were derived by updating these estimates through 1998.

Figure 3.6

Emissions of organic water pollutants have been rising in developing countries . . .

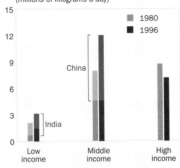

Emissions of organic water pollutants (millions of kilograms a day)

. . . with China and India making the biggest contributions

Contributions to global emissions of organic water pollutants, 1996

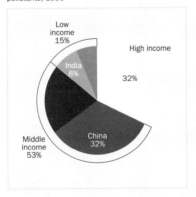

Source: Table 3.6 and World Bank database.

Definitions

• **Emissions of organic water pollutants** are measured in terms of biochemical oxygen demand, which refers to the amount of oxygen that bacteria in water will consume in breaking down waste. This is a standard water treatment test for the presence of organic pollutants. Emissions per worker are total emissions divided by the number of industrial workers. • **Industry shares of emissions of organic water pollutants** refer to emissions from manufacturing activities as defined by two-digit divisions of the International Standard Industrial Classification (ISIC) revision 2: primary metals (ISIC division 37), paper and pulp (34), chemicals (35), food and beverages (31), stone, ceramics, and glass (36), textiles (32), wood (33), and other (38 and 39).

Data sources

Indicators for 1980–93 were drawn from a 1998 study by Hemamala Hettige, Muthukumara Mani, and David Wheeler, "Industrial Pollution in Economic Development: Kuznets Revisited" (available on the Web at www.worldbank.org/nipr). These indicators were then updated through 1998 by the World Bank's Development Research Group using the same methodology as the initial study. Sectoral employment numbers are from UNIDO's industry database.

3.7 Energy production and use

	Commercial energy production		Commercial energy use			Commercial energy use per capita			Net energy imports[a]	
	thousand metric tons of oil equivalent		thousand metric tons of oil equivalent		average annual % growth	kg of oil equivalent		average annual % growth	% of commercial energy use	
	1980	1998	1980	1998	1980–98	1980	1998	1980–98	1980	1998
Albania	3,428	864	3,049	947	–7.0	1,142	284	–8	–12	9
Algeria	66,741	132,332	12,089	26,506	3.7	648	898	1.0	–452	–399
Angola	11,301	43,035	4,437	7,147	2.9	632	595	–0.2	–155	–502
Argentina	38,813	80,657	41,868	62,349	2.3	1,490	1,726	0.9	7	–29
Armenia	1,263	547	1,070	1,939	..	346	511	..	..	72
Australia	86,096	212,012	70,372	105,009	2.4	4,790	5,600	1.0	–22	–102
Austria	7,655	8,999	23,450	28,815	1.5	3,105	3,567	1.1	67	69
Azerbaijan	14,821	16,178	15,001	12,372	..	2,433	1,564	..	..	–31
Bangladesh	9,234	16,725	10,930	19,965	3.6	126	159	1.4	16	16
Belarus	2,566	3,395	2,385	26,470	..	247	2,614	..	..	87
Belgium	7,986	12,810	46,100	58,349	1.8	4,682	5,719	1.5	83	78
Benin	1,212	1,947	1,363	2,240	2.6	394	377	–0.5	11	13
Bolivia	4,241	5,837	2,287	4,621	3.0	427	581	0.7	–85	–26
Bosnia and Herzegovina	..	684	..	1,950	..	..	517	..	..	65
Botswana	..	..	..	..	..	..	..	..	..	..
Brazil	62,083	126,065	111,262	174,964	2.6	914	1,055	0.9	44	28
Bulgaria	7,737	10,116	28,673	19,963	–2.5	3,235	2,418	–2.1	73	49
Burkina Faso	..	..	..	..	..	..	..	..	..	..
Burundi	..	..	..	..	..	..	..	..	..	..
Cambodia	..	..	..	..	..	..	..	..	..	..
Cameroon	6,707	12,965	3,676	6,183	2.6	425	432	–0.3	–82	–110
Canada	207,417	365,674	193,000	234,325	1.6	7,848	7,747	0.4	–7	–56
Central African Republic	..	..	..	..	..	..	..	..	..	..
Chad	..	..	..	..	..	..	..	..	..	..
Chile	5,801	7,905	9,662	23,630	5.8	867	1,594	4.1	40	67
China	608,625	1,020,270	593,118	1,031,410	3.8	604	830	2.4	–3	1
Hong Kong, China	39	48	5,439	16,593	5.9	1,079	2,497	4.5	99	100
Colombia	18,040	74,422	19,349	30,713	2.8	680	753	0.8	7	–142
Congo, Dem. Rep.	8,697	13,546	8,706	13,711	3.0	322	284	–0.3	0	1
Congo, Rep.	3,970	14,160	845	1,206	2.0	506	433	–0.8	–370	–1,074
Costa Rica	767	1,102	1,527	2,781	4.0	669	789	1.5	50	60
Côte d'Ivoire	..	..	..	..	..	..	..	..	..	..
Croatia	..	3,956	..	8,136	..	..	1,808	..	..	51
Cuba	4,227	4,448	14,910	11,858	–2.0	1,536	1,066	–2.8	72	62
Czech Republic	41,000	30,555	47,252	41,034	–1.1	4,618	3,986	–1.2	13	26
Denmark	896	20,177	19,734	20,804	0.9	3,852	3,925	0.7	95	3
Dominican Republic	1,332	1,433	3,464	5,583	2.4	608	676	0.3	62	74
Ecuador	11,755	22,514	5,191	8,973	2.6	652	737	0.3	–126	–151
Egypt, Arab Rep.	34,168	57,464	15,970	41,798	4.7	391	679	2.4	–114	–37
El Salvador	1,913	1,987	2,537	3,860	2.0	553	640	0.5	25	49
Eritrea	..	..	..	..	..	..	..	..	..	..
Estonia	6,951	2,920	6,275	4,835	..	4,240	3,335	..	..	40
Ethiopia	10,588	16,379	11,157	17,429	2.4	296	284	–0.4	5	6
Finland	6,912	13,591	25,413	33,459	1.7	5,317	6,493	1.3	73	59
France	46,829	125,528	190,111	255,674	2.0	3,528	4,378	1.6	75	51
Gabon	9,441	18,892	1,493	1,668	–0.4	2,160	1,413	–3.4	–532	–1,033
Gambia, The	..	..	..	..	..	..	..	..	..	..
Georgia	1,504	729	4,474	2,526	..	882	464	..	..	71
Germany	185,628	131,412	360,441	344,506	–0.1	4,603	4,199	–0.5	48	62
Ghana	3,305	5,705	4,027	7,270	3.7	375	396	0.6	18	22
Greece	3,696	9,892	15,960	26,976	3.1	1,655	2,565	2.6	77	63
Guatemala	2,503	4,739	3,754	6,258	3.1	550	579	0.5	33	24
Guinea	..	..	..	..	..	..	..	..	..	..
Guinea-Bissau	..	..	..	..	..	..	..	..	..	..
Haiti	1,877	1,626	2,099	2,072	–0.1	392	271	–2.0	11	22
Honduras	1,315	1,897	1,892	3,333	3.1	530	542	0.0	31	43

Energy production and use | 3.7

	Commercial energy production		Commercial energy use			Commercial energy use per capita			Net energy imports[a]	
	thousand metric tons of oil equivalent		thousand metric tons of oil equivalent		average annual % growth	kg of oil equivalent		average annual % growth	% of commercial energy use	
	1980	1998	1980	1998	1980–98	1980	1998	1980–98	1980	1998
Hungary	14,957	11,849	28,961	25,255	–1.0	2,705	2,497	–0.7	48	53
India	222,418	413,055	242,592	475,788	3.9	353	486	1.9	8	13
Indonesia	128,403	211,522	59,561	123,074	4.7	402	604	2.9	–116	–72
Iran, Islamic Rep.	84,001	232,481	38,918	102,148	6.1	995	1,649	3.5	–116	–128
Iraq	136,643	110,824	12,030	29,972	4.8	925	1,342	1.7	–1,036	–270
Ireland	1,894	2,465	8,485	13,251	2.5	2,495	3,570	2.2	78	81
Israel	153	619	8,563	18,873	5.2	2,208	3,165	2.6	98	97
Italy	19,644	29,049	138,629	167,933	1.4	2,456	2,916	1.3	86	83
Jamaica	224	655	2,378	4,058	3.8	1,115	1,575	2.8	91	84
Japan	43,247	109,965	346,492	510,106	2.7	2,967	4,035	2.3	88	78
Jordan	1	295	1,714	4,887	5.1	786	1,063	0.6	100	94
Kazakhstan	76,799	64,086	76,799	39,037	..	5,163	2,590	..	..	–64
Kenya	7,891	11,609	9,791	14,527	2.2	589	505	–0.9	19	20
Korea, Dem. Rep.	..	..	..	..	..	..	..	..	..	..
Korea, Rep.	9,644	27,738	41,238	163,375	9.5	1,082	3,519	8.3	77	83
Kuwait	91,636	114,225	12,248	14,598	–0.2	8,908	7,823	–0.7	–648	–682
Kyrgyz Republic	2,190	1,227	1,717	2,921	..	473	609	..	..	58
Lao PDR	..	..	..	..	..	..	..	..	..	..
Latvia	261	1,774	566	4,275	..	222	1,746	..	..	59
Lebanon	178	200	2,480	5,288	4.6	826	1,256	2.6	93	96
Lesotho	..	..	..	..	..	..	..	..	..	..
Libya	96,550	76,524	7,193	12,420	3.8	2,364	2,343	0.7	–1,242	–516
Lithuania	..	4,510	..	9,347	..	..	2,524	..	..	52
Macedonia, FYR	..	..	..	..	..	..	..	..	..	..
Madagascar	..	..	..	..	..	..	..	..	..	..
Malawi	..	..	..	..	..	..	..	..	..	..
Malaysia	18,202	74,912	12,215	43,623	7.9	888	1,967	5.1	–49	–72
Mali	..	..	..	..	..	..	..	..	..	..
Mauritania	..	..	..	..	..	..	..	..	..	..
Mauritius	..	..	..	..	..	..	..	..	..	..
Mexico	149,359	228,187	98,898	147,834	2.1	1,464	1,552	0.2	–51	–54
Moldova	35	63	..	4,053	..	..	943	..	..	98
Mongolia	..	..	..	..	..	..	..	..	..	..
Morocco	877	753	4,778	9,344	4.2	247	336	2.1	82	92
Mozambique	7,413	6,945	8,074	6,863	–1.0	668	405	–2.6	8	–1
Myanmar	9,513	12,405	9,430	13,631	2.0	279	307	0.5	–1	9
Namibia	..	..	..	..	..	..	..	..	..	..
Nepal	4,630	6,886	4,805	7,831	2.7	331	343	0.2	4	12
Netherlands	71,830	62,495	65,000	74,408	1.5	4,594	4,740	0.9	–11	16
New Zealand	5,488	13,837	9,251	17,159	3.8	2,972	4,525	2.7	41	19
Nicaragua	910	1,458	1,566	2,651	2.8	536	553	0.0	42	45
Niger	..	..	..	..	..	..	..	..	..	..
Nigeria	148,479	184,847	52,846	86,489	2.6	743	716	–0.3	–181	–114
Norway	55,716	206,667	18,792	25,423	1.7	4,593	5,736	1.3	–196	–713
Oman	15,090	52,202	996	7,285	11.5	905	3,165	7.0	–1,415	–617
Pakistan	20,997	42,351	25,472	57,854	4.8	308	440	2.2	18	27
Panama	529	642	1,865	2,383	1.7	957	862	–0.2	72	73
Papua New Guinea	..	..	..	..	..	..	..	..	..	..
Paraguay	1,605	6,868	2,089	4,277	4.5	671	819	1.5	23	–61
Peru	14,655	11,964	11,700	14,400	1.1	675	581	–0.9	–25	17
Philippines	10,670	17,818	21,212	38,313	3.7	439	526	1.4	50	53
Poland	121,848	86,703	123,465	96,440	–1.3	3,470	2,494	–1.8	1	10
Portugal	1,481	2,315	10,291	21,849	4.4	1,054	2,192	4.4	86	89
Puerto Rico	..	..	..	..	..	..	..	..	..	..
Romania	52,587	28,241	65,110	39,611	–3.0	2,933	1,760	–3.0	19	29
Russian Federation	748,647	928,987	763,707	581,774	..	5,494	3,963	..	..	–60

	Commercial energy production		Commercial energy use			Commercial energy use per capita			Net energy imports[a]	
	thousand metric tons of oil equivalent		thousand metric tons of oil equivalent		average annual % growth	kg of oil equivalent		average annual % growth	% of commercial energy use	
	1980	1998	1980	1998	1980–98	1980	1998	1980–98	1980	1998
Rwanda	..	..	..	..	..	..	..	..	..	..
Saudi Arabia	533,071	505,121	35,357	103,230	5.2	3,773	5,244	0.9	−1,408	−389
Senegal	1,046	1,653	1,921	2,822	2.2	347	312	−0.5	46	41
Sierra Leone	..	..	..	..	..	..	..	..	..	..
Singapore	..	24	6,062	24,299	9.8	2,656	7,681	7.8	..	100
Slovak Republic	3,416	4,833	20,810	16,906	−1.4	4,175	3,136	−1.8	84	71
Slovenia	1,623	2,891	4,313	6,649	..	2,269	3,354	..	..	57
South Africa	73,169	144,405	65,417	110,986	2.2	2,372	2,681	−0.1	−12	−30
Spain	15,644	31,920	68,583	112,782	3.1	1,834	2,865	2.8	77	72
Sri Lanka	3,209	4,319	4,536	7,300	2.3	308	389	0.9	29	41
Sudan	7,089	13,527	8,406	14,899	2.8	450	526	0.5	16	9
Sweden	16,133	34,155	40,984	52,472	1.3	4,932	5,928	0.8	61	35
Switzerland	7,030	11,163	20,861	26,605	1.5	3,301	3,742	0.8	66	58
Syrian Arab Republic	9,502	35,411	5,348	17,346	5.6	614	1,133	2.3	−78	−104
Tajikistan	1,986	1,268	1,650	3,255	..	416	532	..	..	61
Tanzania	9,502	13,931	10,280	14,660	2.0	553	456	−1.1	8	5
Thailand	11,182	39,347	22,808	68,971	7.9	488	1,153	6.4	51	43
Togo	..	..	..	..	..	..	..	..	..	..
Trinidad and Tobago	13,141	14,651	3,873	8,950	3.9	3,580	6,964	3.0	−239	−64
Tunisia	6,966	7,113	3,907	7,582	3.7	612	812	1.5	−78	6
Turkey	17,190	28,649	31,314	72,512	4.9	704	1,144	2.9	45	60
Turkmenistan	8,034	17,411	7,948	11,122	..	2,778	2,357	..	..	−57
Uganda	..	..	..	..	..	..	..	..	..	..
Ukraine	109,708	80,415	97,893	142,939	..	1,956	2,842	..	..	44
United Arab Emirates	89,716	144,935	6,112	27,336	8.7	5,860	10,035	3.1	−1,368	−430
United Kingdom	197,864	274,230	201,299	232,879	1.1	3,574	3,930	0.8	2	−18
United States	1,553,260	1,695,430	1,811,650	2,181,800	1.4	7,973	7,937	0.4	14	22
Uruguay	763	1,262	2,641	2,992	1.3	906	910	0.6	71	58
Uzbekistan	4,615	50,334	4,821	46,278	..	302	1,930	..	..	−9
Venezuela, RB	139,392	230,563	34,962	56,543	2.5	2,317	2,433	0.0	−299	−308
Vietnam	18,364	42,668	19,573	33,695	3.0	364	440	0.9	6	−27
West Bank and Gaza	..	..	..	..	..	..	..	..	..	..
Yemen, Rep.	60	19,565	1,424	3,333	4.5	167	201	0.5	96	−487
Yugoslavia, FR (Serb./Mont.)	..	..	..	..	..	..	..	..	..	..
Zambia	4,198	5,657	4,551	6,088	1.3	793	630	−1.6	8	7
Zimbabwe	5,793	8,235	6,570	10,065	2.8	937	861	−0.2	12	18
World	6,882,644 s	9,611,004 s	6,902,381 t	9,345,307 t	2.9 w	1,627 w	1,659 w	0.9 w	.. w	.. w
Low income	797,751	1,290,575	648,676	1,178,897	5.1	442	550	2.5	−23	−10
Middle income	3,302,896	4,605,397	2,481,018	3,409,502	4.4	1,246	1,311	2.3	−33	−35
Lower middle income	1,944,378	2,867,598	1,779,108	2,282,178	5.4	1,126	1,116	3.2	−9	−26
Upper middle income	1,358,518	1,737,798	701,910	1,127,324	2.7	1,713	2,025	1.0	−94	−54
Low & middle income	4,100,647	5,895,972	3,129,694	4,588,399	4.6	905	967	2.3	−31	−29
East Asia & Pacific	814,603	1,446,679	779,155	1,516,091	4.5	571	857	3.0	−5	5
Europe & Central Asia	1,241,543	1,380,292	1,332,941	1,215,898	7.8	3,349	2,637	..	7	−14
Latin America & Carib.	475,245	830,882	379,775	585,082	2.4	1,070	1,183	0.5	−24	−42
Middle East & N. Africa	988,969	1,237,344	145,929	378,338	5.1	838	1,344	2.3	−577	−228
South Asia	260,487	483,335	288,334	568,738	4.0	325	445	1.9	10	−15
Sub-Saharan Africa	319,801	517,440	203,560	324,252	2.3	727	700	−0.5	−57	−60
High income	2,781,997	3,715,032	3,772,688	4,756,908	1.7	4,796	5,366	1.0	27	22
Europe EMU	365,532	420,629	940,146	1,114,343	1.2	3,408	3,834	0.9	61	62

a. A negative value indicates that a country is a net exporter.

Energy production and use | 3.7

In developing countries growth in commercial energy use is closely related to growth in the modern sectors—industry, motorized transport, and urban areas—but commercial energy use also reflects climatic, geographic, and economic factors (such as the relative price of energy). Commercial energy use has been growing rapidly in low- and middle-income countries, but high-income countries still use more than five times as much on a per capita basis. Because commercial energy is widely traded, it is necessary to distinguish between its production and its use. Net energy imports show the extent to which an economy's use exceeds its domestic production. High-income countries are net energy importers; middle-income countries have been their main suppliers.

Energy data are compiled by the International Energy Agency (IEA) and the United Nations Statistics Division (UNSD). IEA data for non-OECD countries are based on national energy data adjusted to conform with annual questionnaires completed by OECD member governments. UNSD data are primarily from responses to questionnaires sent to national governments, supplemented by official national statistical publications and by data from intergovernmental organizations. When official data are not available, the UNSD prepares estimates based on the professional and commercial literature. This variety of sources affects the cross-country comparability of data.

Commercial energy use refers to the use of domestic primary energy before transformation to other end-use fuels (such as electricity and refined petroleum products). It includes energy from combustible renewables and waste, which comprises solid biomass and animal products, gas and liquid from biomass, industrial waste, and municipal waste. Biomass is defined as any plant matter used directly as fuel or converted into fuel, heat, or electricity. (The data series published in *World Development Indicators 1998* and earlier editions did not include energy from combustible renewables and waste.) All forms of commercial energy—primary energy and primary electricity—are converted into oil equivalents. To convert nuclear electricity into oil equivalents, a notional thermal efficiency of 33 percent is assumed; for hydroelectric power 100 percent efficiency is assumed.

Figure 3.7a

High-income countries, with 15 percent of the world's population, use half its commercial energy . . .

Commercial energy use, 1998

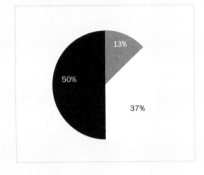

● Low income
　Middle income
● High income

. . . and 10 times as much per capita as low-income countries

Energy use per capita (thousands of kg of oil equiv.)

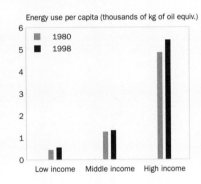

Source: Table 3.7.

• **Commercial energy production** refers to commercial forms of primary energy—petroleum (crude oil, natural gas liquids, and oil from nonconventional sources), natural gas, and solid fuels (coal, lignite, and other derived fuels)—and primary electricity, all converted into oil equivalents (see *About the data*).
• **Commercial energy use** refers to apparent consumption, which is equal to indigenous production plus imports and stock changes, minus exports and fuels supplied to ships and aircraft engaged in international transport (see *About the data*). • **Net energy imports** are calculated as energy use less production, both measured in oil equivalents. A negative value indicates that the country is a net exporter.

The data on commercial energy production and use are primarily from IEA electronic files and from the United Nations Statistics Division's *Energy Statistics Yearbook*. The IEA's data are published in its annual publications, *Energy Statistics and Balances of Non-OECD Countries*, *Energy Statistics of OECD Countries*, and *Energy Balances of OECD Countries*.

Figure 3.7b

High-income countries depend on imports for roughly a quarter of their energy

Net energy imports (% of commercial energy use)

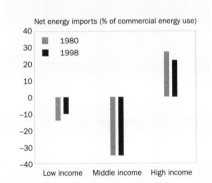

Source: Table 3.7.

3.8 | Energy efficiency and emissions

	GDP per unit of energy use		Traditional fuel use		Carbon dioxide emissions					
	PPP $ per kg oil equivalent		% of total energy use		Total million metric tons		Per capita metric tons		kg per PPP $ of GDP	
	1980	1998	1980	1997	1980	1997	1980	1997	1980	1997
Albania	..	10.3	13.1	7.3	5.3	1.7	2.0	0.5	..	0.2
Algeria	5.0	5.4	1.9	1.5	68.2	98.7	3.7	3.4	1.1	0.7
Angola		3.8	64.9	69.7	5.4	5.3	0.8	0.5	..	0.2
Argentina	4.7	7.3	5.9	4.0	111.0	140.6	4.0	3.9	0.6	0.3
Armenia	..	4.3	..	0.0	..	2.9	..	0.8	..	0.4
Australia	2.1	4.1	3.8	4.4	205.5	319.6	14.0	17.2	1.4	0.8
Austria	3.5	6.7	1.2	4.7	55.1	62.6	7.3	7.8	0.7	0.3
Azerbaijan	..	1.5	..	0.0	..	32.0	..	4.1	..	1.9
Bangladesh	4.5	8.9	81.3	46.0	7.8	24.6	0.1	0.2	0.2	0.1
Belarus	..	2.5	..	0.8	..	62.3	..	6.1	..	1.0
Belgium	2.4	4.3	0.2	1.6	131.0	106.5	13.3	10.5	1.2	0.4
Benin	1.3	2.4	85.4	89.2	0.7	1.0	0.2	0.2	0.4	0.2
Bolivia	3.4	4.0	19.3	14.0	4.7	11.3	0.9	1.4	0.6	0.6
Bosnia and Herzegovina	..	..	..	10.1	..	4.5	..	1.2	..	..
Botswana	..	..	35.7	..	1.0	3.4	1.1	2.2	0.6	0.3
Brazil	4.4	6.5	35.5	28.7	197.0	307.2	1.6	1.9	0.4	0.3
Bulgaria	0.9	2.0	0.5	1.3	77.9	50.3	8.8	6.1	3.0	1.2
Burkina Faso	..	..	91.3	87.1	0.4	1.0	0.1	0.1	0.1	0.1
Burundi			97.0	94.2	0.1	0.2	0.0	0.0	0.1	0.1
Cambodia	..	..	100.0	89.3	0.3	0.5	0.0	0.0	..	0.0
Cameroon	2.8	3.5	51.7	69.2	4.1	2.7	0.5	0.2	0.4	0.1
Canada	1.5	3.2	0.4	4.7	426.1	496.6	17.3	16.6	1.5	0.7
Central African Republic	..	..	88.9	87.5	0.1	0.2	0.0	0.1	0.1	0.1
Chad	..	..	95.9	97.6	0.2	0.1	0.0	0.0	0.1	0.0
Chile	3.1	5.4	12.3	11.3	28.3	60.1	2.5	4.1	0.9	0.5
China	0.8	4.0	8.4	5.7	1,516.6	3,593.5	1.5	2.9	3.3	0.9
Hong Kong, China	6.4	8.5	0.9	0.7	17.1	23.8	3.4	3.7	0.5	0.2
Colombia	4.1	7.9	15.9	17.7	42.0	71.9	1.5	1.8	0.5	0.3
Congo, Dem. Rep.	3.5	2.8	73.9	91.7	3.7	2.3	0.1	0.1	0.1	0.1
Congo, Rep.	0.8	1.8	77.8	53.0	0.4	0.3	0.2	0.1	0.6	0.1
Costa Rica	5.7	9.5	26.3	54.2	2.7	5.4	1.2	1.6	0.3	0.2
Côte d'Ivoire	..	..	52.8	91.5	5.3	13.3	0.6	0.9	0.5	0.6
Croatia	..	3.9	..	3.2	..	20.1	..	4.4	..	0.6
Cuba	..	..	27.9	30.2	32.4	26.0	3.3	2.3	..	..
Czech Republic	..	3.2	0.6	1.6	..	125.2	..	12.2	..	0.9
Denmark	..	6.4	0.4	5.9	63.9	57.7	12.5	10.9	..	0.4
Dominican Republic	3.7	7.5	27.5	14.3	6.9	14.0	1.2	1.7	0.5	0.4
Ecuador	3.0	4.3	26.7	17.5	14.1	21.7	1.8	1.8	0.9	0.6
Egypt, Arab Rep.	3.5	4.7	4.7	3.2	46.7	118.3	1.1	2.0	0.8	0.6
El Salvador	4.3	6.5	52.9	34.5	2.4	5.9	0.5	1.0	0.2	0.2
Eritrea	..	..	..	96.0	..	..	..	..	..	..
Estonia	..	2.5	..	13.8	..	19.1	..	13.1	..	1.6
Ethiopia	*1.4*	2.1	89.6	95.9	1.9	3.8	0.0	0.1	*0.1*	0.1
Finland	1.8	3.4	4.3	6.5	55.8	56.6	11.7	11.0	1.2	0.5
France	2.9	5.0	1.3	5.7	497.2	349.8	9.2	6.0	0.9	0.3
Gabon	1.9	4.5	30.8	32.9	5.0	3.4	7.2	3.0	1.7	0.5
Gambia, The	..	..	72.7	78.6	0.2	0.2	0.2	0.2	0.3	0.1
Georgia	..	7.1	..	1.0	..	4.5	..	0.8	..	0.3
Germany	..	5.5	0.3	1.3	..	851.5	..	10.4	..	0.5
Ghana	2.9	4.6	43.7	78.1	2.6	4.8	0.2	0.3	0.2	0.1
Greece	4.2	5.7	3.0	4.5	58.1	87.2	6.0	8.3	0.9	0.6
Guatemala	4.1	6.1	54.6	62.0	4.8	8.3	0.7	0.8	0.3	0.2
Guinea	..	..	71.4	74.2	0.9	1.1	0.2	0.2	..	0.1
Guinea-Bissau	..	..	80.0	57.1	0.1	0.2	0.2	0.2	0.5	0.2
Haiti	3.7	5.3	80.7	74.7	0.9	1.4	0.2	0.2	0.1	0.1
Honduras	2.9	4.5	55.3	54.8	2.3	4.6	0.6	0.8	0.4	0.3

Energy efficiency and emissions | 3.8

	GDP per unit of energy use		Traditional fuel use		Carbon dioxide emissions					
	PPP $ per kg oil equivalent		% of total energy use		Total million metric tons		Per capita metric tons		kg per PPP $ of GDP	
	1980	1998	1980	1997	1980	1997	1980	1997	1980	1997
Hungary	2.0	4.3	2.0	1.6	84.8	59.6	7.9	5.9	1.4	0.6
India	1.9	4.3	31.5	20.7	356.1	1,065.4	0.5	1.1	0.8	0.5
Indonesia	2.2	4.6	51.5	29.3	97.5	251.5	0.7	1.3	0.8	0.4
Iran, Islamic Rep.	2.9	3.3	0.4	0.7	120.0	296.9	3.1	4.9	1.1	0.9
Iraq	..	..	0.3	0.1	46.7	92.3	3.6	4.2	..	..
Ireland	2.3	6.4	0.0	0.2	26.1	37.3	7.7	10.2	1.3	0.5
Israel	3.6	5.7	0.0	0.0	22.1	60.4	5.7	10.4	0.7	0.6
Italy	3.9	7.4	0.8	1.0	392.7	424.7	7.0	7.4	0.7	0.3
Jamaica	1.9	2.2	5.0	6.0	8.5	11.0	4.0	4.3	1.9	1.2
Japan	3.3	6.0	0.1	1.6	964.2	1,204.2	8.3	9.6	0.8	0.4
Jordan	3.3	3.6	0.0	0.0	5.2	15.7	2.4	3.5	0.9	0.9
Kazakhstan	..	1.8	..	0.2	..	123.0	..	8.0	..	1.7
Kenya	1.1	2.0	76.8	80.3	6.8	7.2	0.4	0.3	0.7	0.2
Korea, Dem. Rep.	..	..	3.1	1.4	128.9	260.5	7.3	11.4	..	..
Korea, Rep.	2.8	4.0	4.0	2.4	132.9	457.4	3.5	9.9	1.2	0.6
Kuwait	1.3	2.1	0.0	0.0	25.4	51.0	18.5	28.2	1.6	1.8
Kyrgyz Republic	..	4.0	..	0.0	..	6.8	..	1.4	..	0.6
Lao PDR	..	..	72.3	88.7	0.2	0.4	0.1	0.1	..	0.1
Latvia	19.6	3.4	..	26.2	..	8.3	..	3.3	..	0.6
Lebanon	..	3.7	2.4	2.5	6.9	17.7	2.3	4.3	..	1.0
Lesotho	..	..	..	..	..	..	..	..	..	..
Libya	..	..	2.3	0.9	28.5	43.5	9.4	8.4	..	..
Lithuania	..	2.7	..	6.3	..	15.1	..	4.1	..	0.6
Macedonia, FYR	..	..	..	6.1	..	10.9	..	5.5	..	1.2
Madagascar	..	..	78.4	84.3	1.6	1.2	0.2	0.1	0.3	0.1
Malawi	..	..	90.6	88.6	0.8	0.8	0.1	0.1	0.3	0.1
Malaysia	2.7	3.9	15.7	5.5	29.1	137.2	2.1	6.3	0.9	0.7
Mali	..	..	86.7	88.9	0.4	0.5	0.1	0.0	0.1	0.1
Mauritania	..	..	0.0	0.0	0.6	3.0	0.4	1.2	0.4	0.8
Mauritius	..	..	59.1	36.1	0.6	1.7	0.6	1.5	0.3	0.2
Mexico	3.1	5.2	5.0	4.5	259.6	379.7	3.8	4.0	0.8	0.5
Moldova	..	2.2	..	0.5	..	10.4	..	2.4	..	1.1
Mongolia	..	..	14.4	4.3	6.9	7.8	4.1	3.3	3.6	2.1
Morocco	6.8	10.2	5.2	4.0	17.7	35.9	0.9	1.3	0.5	0.4
Mozambique	0.6	2.0	43.7	91.4	3.3	1.2	0.3	0.1	0.7	0.1
Myanmar	..	..	69.3	60.5	5.0	8.8	0.1	0.2	..	..
Namibia	..	..	..	..	..	..	..	..	..	..
Nepal	1.5	3.5	94.2	89.6	0.6	2.2	0.0	0.1	0.1	0.1
Netherlands	2.2	4.9	0.0	1.1	154.5	163.6	10.9	10.5	1.1	0.5
New Zealand	..	4.0	0.2	0.8	17.9	31.6	5.8	8.4	..	0.5
Nicaragua	3.6	4.0	49.2	42.2	2.1	3.2	0.7	0.7	0.4	0.3
Niger	..	..	79.5	80.6	0.6	1.1	0.1	0.1	0.2	0.2
Nigeria	0.8	1.2	66.8	67.8	69.1	83.7	1.0	0.7	1.6	0.9
Norway	2.4	4.8	0.4	1.1	91.5	68.5	22.4	15.6	2.0	0.6
Oman	..	..	0.0	..	5.9	18.4	5.3	8.2	..	..
Pakistan	2.1	4.0	24.4	29.5	33.3	98.2	0.4	0.8	0.6	0.4
Panama	3.2	6.5	26.6	14.4	3.7	8.0	1.9	2.9	0.6	0.5
Papua New Guinea	..	..	65.4	62.5	1.8	2.5	0.6	0.5	0.5	0.2
Paraguay	4.2	5.4	62.0	49.6	1.6	4.1	0.5	0.8	0.2	0.2
Peru	4.6	7.8	15.2	24.6	24.7	30.1	1.4	1.2	0.5	0.3
Philippines	5.6	7.0	37.0	26.9	38.8	81.7	0.8	1.1	0.3	0.3
Poland	..	3.2	0.4	0.8	465.4	357.0	13.1	9.2	..	1.2
Portugal	5.6	7.0	1.2	0.9	29.9	53.8	3.1	5.4	0.5	0.4
Puerto Rico	..	..	0.0	..	14.7	17.1	4.6	4.5	..	..
Romania	1.6	3.5	1.3	5.7	199.6	111.3	9.0	4.9	1.9	0.8
Russian Federation	..	1.7	..	0.8	..	1,444.5	..	9.8	..	1.4

	GDP per unit of energy use		Traditional fuel use		Carbon dioxide emissions					
	PPP $ per kg oil equivalent		% of total energy use		Total million metric tons		Per capita metric tons		kg per PPP $ of GDP	
	1980	1998	1980	1997	1980	1997	1980	1997	1980	1997
Rwanda	..	..	89.8	88.3	0.3	0.5	0.1	0.1	0.1	0.1
Saudi Arabia	3.0	2.1	0.0	0.0	132.2	273.7	14.1	14.3	1.2	1.3
Senegal	2.3	4.4	50.8	56.2	3.0	3.5	0.5	0.4	0.7	0.3
Sierra Leone	..	..	90.0	86.1	0.6	0.5	0.2	0.1	0.3	0.2
Singapore	2.3	3.1	0.4	0.0	31.1	81.9	12.9	21.9	2.2	1.1
Slovak Republic	..	3.2	..	0.5	..	38.1	..	7.1	..	0.7
Slovenia	..	4.4	..	1.5	..	15.5	..	7.8	..	0.5
South Africa	2.7	3.3	4.9	43.4	214.9	321.5	7.8	7.9	1.2	0.9
Spain	3.8	5.9	0.4	1.3	214.0	257.7	5.7	6.6	0.8	0.4
Sri Lanka	3.5	8.0	53.5	46.5	3.7	8.1	0.3	0.4	0.2	0.1
Sudan	..	..	86.9	75.1	3.4	3.8	0.2	0.1	..	..
Sweden	2.1	3.6	7.7	17.9	72.6	48.6	8.7	5.5	0.8	0.3
Switzerland	4.4	7.0	0.9	6.0	43.0	42.6	6.8	6.0	0.5	0.2
Syrian Arab Republic	2.9	3.3	0.0	0.0	20.3	49.9	2.3	3.3	1.3	1.0
Tajikistan	..	..	..	..	..	5.6	..	0.9	..	..
Tanzania	..	1.1	92.0	91.4	2.0	2.9	0.1	0.1	..	0.2
Thailand	3.0	5.1	40.3	24.6	42.7	226.8	0.9	3.8	0.6	0.6
Togo	..	..	35.7	71.9	0.8	1.0	0.3	0.2	0.2	0.2
Trinidad and Tobago	1.3	1.1	1.4	0.8	16.8	22.3	15.5	17.4	3.4	2.4
Tunisia	4.0	6.9	16.1	12.5	10.3	18.8	1.6	2.0	0.7	0.4
Turkey	3.6	5.8	20.5	3.1	82.8	216.0	1.9	3.5	0.7	0.5
Turkmenistan	..	1.2	..	..	..	31.0	..	6.7	..	2.5
Uganda	..	..	93.6	89.7	0.6	1.2	0.1	0.1	0.1	0.1
Ukraine	..	1.2	..	0.5	..	370.5	..	7.3	..	2.1
United Arab Emirates	4.4	1.8	0.0	..	37.1	82.5	35.6	32.0	1.4	1.6
United Kingdom	..	5.4	0.0	3.3	591.2	527.1	10.5	8.9	..	0.4
United States	1.6	3.8	1.3	3.8	4,609.4	5,467.1	20.3	20.1	1.6	0.7
Uruguay	5.0	9.9	11.1	21.0	6.2	5.7	2.1	1.8	0.5	0.2
Uzbekistan	..	1.1	..	0.0	..	104.8	..	4.4	..	2.1
Venezuela, RB	1.7	2.4	0.9	0.7	92.0	191.2	6.1	8.4	1.5	1.4
Vietnam	..	4.0	49.1	37.8	17.1	45.5	0.3	0.6	..	0.4
West Bank and Gaza	..	..	..	..	..	..	..	..	..	..
Yemen, Rep.	..	3.7	0.0	1.4	..	16.7	..	1.0	..	1.3
Yugoslavia, FR (Serb./Mont.)	..	..	..	1.5	..	50.2	..	4.7	..	..
Zambia	0.9	1.2	37.4	72.7	3.6	2.6	0.6	0.3	0.9	0.4
Zimbabwe	1.5	3.3	27.6	25.2	9.9	18.8	1.4	1.6	1.0	0.6
World	**2.1 w**	**4.2 w**	**7.4 w**	**8.2 w**	**14,014.6 s**	**23,868.2 s**	**3.5 w**	**4.1 w**	**1.2 w**	**0.6 w**
Low income	..	3.4	46.4	29.8	794.9	2,527.5	0.5	1.1	0.7	0.6
Middle income	2.2	3.9	10.4	7.3	4,304.9	10,006.0	2.4	3.8	1.3	0.8
Lower middle income	1.6	3.6	10.7	5.7	2,457.8	6,957.9	1.7	3.4	1.7	0.9
Upper middle income	3.3	4.3	8.6	10.6	1,847.0	3,048.1	4.6	5.5	1.0	0.6
Low & middle income	..	3.7	18.5	12.9	5,099.8	12,533.6	1.5	2.5	1.2	0.7
East Asia & Pacific	..	..	15.1	9.7	2,019.6	5,075.6	1.4	2.8	2.0	0.8
Europe & Central Asia	..	2.3	3.2	1.3	915.8	3,285.6	..	6.9	2.1	1.2
Latin America & Carib.	3.7	5.7	18.4	16.0	884.7	1,356.4	2.5	2.8	0.6	0.4
Middle East & N. Africa	3.4	3.5	1.6	1.1	517.8	1,113.6	3.1	4.0	1.1	0.9
South Asia	2.0	4.5	34.2	23.8	403.4	1,200.5	0.4	0.9	0.7	0.5
Sub-Saharan Africa	..	..	47.2	63.5	358.4	501.8	1.0	0.8	0.9	0.6
High income	2.2	4.6	1.0	3.4	8,914.8	11,334.6	12.6	12.8	1.2	0.5
Europe EMU	3.1	5.6	0.7	2.5	1,569.7	2,378.6	7.9	8.2	0.9	0.4

Energy efficiency and emissions | 3.8

About the data

The ratio of GDP to energy use provides a measure of energy efficiency. To produce comparable and consistent estimates of real GDP across countries relative to physical inputs to GDP—that is, units of energy use—GDP is converted to international dollars using purchasing power parity (PPP) rates. Differences in this ratio over time and across countries reflect in part structural changes in the economy, changes in the energy efficiency of particular sectors, and differences in fuel mixes.

The data on traditional fuel are from the United Nations Statistics Division's *Energy Statistics Yearbook*. This series differs from those published in *World Development Indicators 1999* and previous editions, which came from other sources.

Carbon dioxide (CO_2) emissions, largely a by-product of energy production and use (see table 3.7), account for the largest share of greenhouse gases, which are associated with global warming. Anthropogenic CO_2 emissions result primarily from fossil fuel combustion and cement manufacturing. In combustion, different fossil fuels release different amounts of CO_2 for the same level of energy use. Burning oil releases about 50 percent more CO_2 than burning natural gas, and burning coal releases about twice as much. Cement manufacturing releases about half a metric ton of CO_2 for each ton of cement produced.

The Carbon Dioxide Information Analysis Center (CDIAC), sponsored by the U.S. Department of Energy, calculates annual anthropogenic emissions of CO_2. These calculations are derived from data on fossil fuel consumption, based on the World Energy Data Set maintained by the United Nations Statistics Division, and from data on world cement manufacturing, based on the Cement Manufacturing Data Set maintained by the U.S. Bureau of Mines. Emissions of CO_2 are often calculated and reported in terms of their content of elemental carbon. For this table these values were converted to the actual mass of CO_2 by multiplying the carbon mass by 3.664 (the ratio of the mass of carbon to that of CO_2).

Although the estimates of global CO_2 emissions are probably within 10 percent of actual emissions (as calculated from global average fuel chemistry and use), country estimates may have larger error bounds. Trends estimated from a consistent time series tend to be more accurate than individual values. Each year the CDIAC recalculates the entire time series from 1950 to the present, incorporating its most recent findings and the latest corrections to its database. Estimates do not include fuels supplied to ships and aircraft engaged in international transport because of

the difficulty of apportioning these fuels among the countries benefiting from that transport.

Figure 3.8

Carbon dioxide emissions have been rising globally . . .

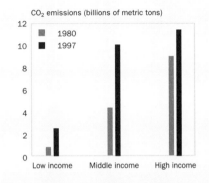

CO₂ emissions (billions of metric tons)

. . . even on a per capita basis . . .

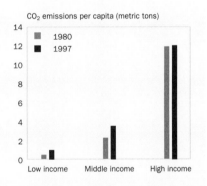

CO₂ emissions per capita (metric tons)

. . . but production is becoming cleaner

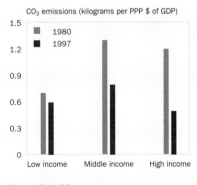

CO₂ emissions (kilograms per PPP $ of GDP)

Source: Table 3.8.

Definitions

• **GDP per unit of energy use** is the PPP GDP per kilogram of oil equivalent of commercial energy use. PPP GDP is gross domestic product converted to international dollars using purchasing power parity rates. An international dollar has the same purchasing power over GDP as a U.S. dollar has in the United States. • **Traditional fuel use** includes estimates of the consumption of fuelwood, charcoal, bagasse, and animal and vegetable wastes. Total energy use comprises commercial energy use (see table 3.7) and traditional fuel use. • **Carbon dioxide emissions** are those stemming from the burning of fossil fuels and the manufacture of cement. They include carbon dioxide produced during consumption of solid, liquid, and gas fuels and gas flaring.

Data sources

The underlying data on commercial energy production and use are from electronic files of the International Energy Agency. The data on traditional fuel use are from the United Nations Statistics Division's *Energy Statistics Yearbook*. The data on CO_2 emissions are from the Carbon Dioxide Information Analysis Center, Environmental Sciences Division, Oak Ridge National Laboratory, in the U.S. state of Tennessee.

3.9 | Sources of electricity

	Electricity production (billion kwh)		Hydropower %		Coal %		Oil %		Gas %		Nuclear power %	
	1980	1998	1980	1998	1980	1998	1980	1998	1980	1998	1980	1998
Albania	3.7	5.1	79.4	95.9	..	..	20.6	4.1	..	..	..	..
Algeria	7.1	23.6	3.6	3.2	..	..	12.2	3.5	84.1	93.3	..	..
Angola	0.7	1.1	88.1	90.0	..	..	11.9	10.0	..	..	..	..
Argentina	39.7	74.2	38.1	35.9	2.1	1.9	31.6	5.3	22.0	46.6	5.9	10.0
Armenia	13.0	6.2	12.0	24.8	..	..	54.8	0.3	..	49.5	33.2	25.7
Australia	95.2	194.3	13.6	8.1	73.3	80.0	5.4	1.1	7.3	9.0	..	..
Austria	41.6	55.9	69.1	66.5	7.0	9.1	14.0	5.6	9.2	15.8	..	..
Azerbaijan	15.0	18.0	7.3	10.8	..	..	92.7	69.5	..	19.7	..	..
Bangladesh	2.4	12.9	24.8	6.7	..	..	26.6	8.7	48.6	84.6	..	..
Belarus	34.1	25.3	0.1	0.1	..	..	99.9	14.8	..	85.1	..	..
Belgium	53.1	82.1	0.5	0.5	29.4	20.6	34.7	3.1	11.2	18.3	23.6	56.2
Benin	0.0	0.1	..	3.2	..	..	100.0	96.8	..	..	..	..
Bolivia	1.6	3.7	68.2	41.2	..	..	10.3	2.2	18.4	54.8	..	..
Bosnia and Herzegovina	..	2.5	..	61.2	..	33.7	..	5.1	..	..	..	..
Botswana	..	..	..	..	..	..	..	..	..	..	..	..
Brazil	139.4	321.6	92.5	90.6	2.4	2.2	3.8	3.9	..	0.3	..	1.0
Bulgaria	34.8	41.5	10.7	7.5	49.2	44.8	22.5	2.2	..	4.9	17.7	40.7
Burkina Faso	..	..	..	..	..	..	..	..	..	..	..	..
Burundi	..	..	..	..	..	..	..	..	..	..	..	..
Cambodia	..	..	..	..	..	..	..	..	..	..	..	..
Cameroon	1.5	3.3	93.9	98.9	..	..	6.1	1.1	..	..	..	..
Canada	373.3	561.7	67.3	59.1	16.0	19.1	3.7	3.3	2.5	4.6	10.2	12.7
Central African Republic	..	..	..	..	..	..	..	..	..	..	..	..
Chad	..	..	..	..	..	..	..	..	..	..	..	..
Chile	11.8	35.5	67.0	47.0	16.1	32.9	14.7	5.1	1.3	11.8	..	..
China	300.6	1,166.2	19.4	17.8	59.9	75.9	20.5	4.5	0.2	0.6	..	1.2
Hong Kong, China	12.6	31.4	..	..	22.6	65.1	100.0	1.1	..	33.8	..	..
Colombia	20.4	45.9	70.0	67.0	7.7	8.8	1.8	0.7	19.3	22.0	..	..
Congo, Dem. Rep.	4.4	5.7	95.5	97.8	..	..	4.5	2.2	..	..	..	..
Congo, Rep.	0.2	0.3	62.2	99.4	..	..	35.9	0.3	1.9	0.3	..	..
Costa Rica	2.2	5.8	95.2	81.6	..	..	4.3	6.8	..	..	..	..
Côte d'Ivoire	..	..	..	..	..	..	..	..	..	..	..	..
Croatia	..	10.9	..	50.2	..	4.9	..	32.5	..	12.3	..	..
Cuba	9.9	14.1	1.0	0.7	..	..	89.7	93.7	..	0.2	..	..
Czech Republic	52.7	64.6	4.6	2.2	84.8	71.6	9.6	1.0	1.1	3.2	..	20.4
Denmark	26.8	41.1	0.1	0.1	81.8	57.6	18.0	12.1	..	19.9	..	..
Dominican Republic	3.3	7.6	18.8	18.8	..	4.5	78.8	76.3	..	..	..	..
Ecuador	3.4	9.9	25.9	66.0	..	..	74.1	34.0	..	..	..	..
Egypt, Arab Rep.	18.9	63.0	51.8	19.4	..	..	27.7	30.2	20.5	50.4	..	..
El Salvador	1.5	3.8	63.7	40.8	..	..	2.7	46.8	..	..	..	..
Eritrea	..	..	..	..	..	..	..	..	..	..	..	..
Estonia	18.9	8.5	..	0.0	..	93.5	100.0	2.9	..	3.4	..	..
Ethiopia	0.7	1.6	70.2	97.0	..	..	27.6	3.0	..	..	..	..
Finland	40.7	70.2	25.1	21.4	42.6	19.3	10.8	1.6	4.2	12.6	17.2	31.1
France	256.9	506.9	26.9	12.2	27.2	7.4	18.9	2.3	2.7	1.0	23.8	76.5
Gabon	0.5	1.0	49.1	71.3	..	..	50.9	18.4	..	10.3	..	..
Gambia, The	..	..	..	..	..	..	..	..	..	..	..	..
Georgia	14.7	8.1	..	79.0	..	..	..	2.9	..	18.1	..	..
Germany	466.3	552.4	4.1	3.1	62.9	54.2	5.7	1.2	14.2	9.8	11.9	29.3
Ghana	5.3	7.3	99.2	99.6	..	..	0.8	0.4	..	..	..	..
Greece	22.7	46.2	15.0	8.0	44.8	70.3	40.1	17.5	..	3.7	..	..
Guatemala	1.8	4.5	12.9	77.1	..	..	70.5	22.9	..	..	..	..
Guinea	..	..	..	..	..	..	..	..	..	..	..	..
Guinea-Bissau	..	..	..	..	..	..	..	..	..	..	..	..
Haiti	0.3	0.7	70.1	46.2	..	..	26.1	53.8	..	..	..	..
Honduras	0.9	3.5	86.3	55.4	..	..	13.7	44.6	..	..	..	..

Sources of electricity | 3.9

	Electricity production		Sources of electricity[a]									
	billion kwh		Hydropower %		Coal %		Oil %		Gas %		Nuclear power %	
	1980	1998	1980	1998	1980	1998	1980	1998	1980	1998	1980	1998
Hungary	23.9	37.2	0.5	0.4	50.4	26.0	13.9	16.0	35.2	20.0	0.0	37.5
India	119.3	494.0	39.0	16.8	51.2	75.4	6.4	0.7	0.8	4.7	2.5	2.3
Indonesia	8.4	77.9	16.0	12.4	..	28.8	84.0	21.1	..	34.4	..	..
Iran, Islamic Rep.	22.4	103.4	25.1	6.8	..	..	50.1	13.7	24.8	79.5	..	..
Iraq	11.4	30.3	6.1	1.9	..	..	93.9	98.1	..	..	..	..
Ireland	10.6	20.9	7.9	4.4	16.4	40.4	60.4	23.2	15.2	30.8	..	..
Israel	12.4	38.0	0.0	0.1	18.1	69.8	100.0	30.1	..	0.1	..	..
Italy	183.5	253.6	24.7	16.3	9.9	11.0	57.0	42.3	5.0	27.9	1.2	..
Jamaica	1.5	6.5	8.3	1.4	..	..	87.9	92.9	..	..	..	..
Japan	572.5	1,036.2	15.4	8.9	9.6	19.1	46.2	16.4	14.2	21.1	14.4	32.1
Jordan	1.1	6.7	..	0.2	..	..	100.0	89.6	..	10.2	..	..
Kazakhstan	61.5	49.1	9.3	12.5	..	72.0	90.7	7.3	..	8.2	..	..
Kenya	1.5	4.8	71.1	68.2	..	..	28.9	24.2	..	..	..	..
Korea, Dem. Rep.	..	..	..	..	..	..	..	..	..	..	..	..
Korea, Rep.	37.2	235.3	5.3	1.8	6.7	42.8	78.7	6.1	..	11.2	9.3	38.1
Kuwait	9.0	30.0	..	..	..	..	20.1	70.1	79.9	29.9	..	..
Kyrgyz Republic	9.2	11.6	..	85.6	..	..	..	..	..	..	..	..
Lao PDR	..	..	..	..	..	..	..	..	..	..	..	..
Latvia	4.7	5.7	..	75.2	..	1.7	..	5.3	..	17.7	..	..
Lebanon	2.8	8.4	30.9	9.4	..	..	69.1	90.6	..	..	..	..
Lesotho	..	..	..	..	..	..	..	..	..	..	..	..
Libya	4.8	19.5	..	..	..	..	100.0	100.0	..	..	..	..
Lithuania	11.7	17.2	4.0	2.4	..	..	96.0	16.9	..	1.7	..	79.0
Macedonia, FYR	..	..	..	..	..	..	..	..	..	..	..	..
Madagascar	..	..	..	..	..	..	..	..	..	..	..	..
Malawi	..	..	..	..	..	..	..	..	..	..	..	..
Malaysia	10.0	60.7	13.9	8.0	..	3.2	84.9	19.0	1.2	69.8	..	..
Mali	..	..	..	..	..	..	..	..	..	..	..	..
Mauritania	..	..	..	..	..	..	..	..	..	..	..	..
Mauritius	..	..	..	..	..	..	..	..	..	..	..	..
Mexico	67.0	182.3	25.2	13.5	0.0	9.8	57.9	55.4	15.5	13.1	..	5.1
Moldova	15.4	4.6	2.6	1.8	..	10.9	97.4	4.8	..	82.4	..	..
Mongolia	..	..	..	..	..	..	..	..	..	..	..	..
Morocco	5.2	14.1	28.9	12.4	19.5	55.3	51.6	32.3	..	..	..	..
Mozambique	0.5	6.9	65.2	99.4	17.5	..	17.3	0.5	..	0.0	..	..
Myanmar	1.5	4.6	53.5	31.8	2.0	..	31.3	7.1	13.2	61.0	..	..
Namibia	..	..	..	..	..	..	..	..	..	..	..	..
Nepal	0.2	1.3	94.4	90.5	..	..	5.6	9.5	..	..	..	..
Netherlands	64.8	91.2	..	0.1	13.7	29.9	38.4	3.9	39.8	57.0	6.5	4.2
New Zealand	22.6	37.6	83.6	64.9	1.9	3.9	0.2	0.0	7.5	23.2	..	..
Nicaragua	1.1	2.3	51.3	13.1	..	..	45.3	73.8	..	..	..	..
Niger	..	..	..	..	..	..	..	..	..	..	..	..
Nigeria	7.1	15.7	39.0	35.6	0.4	..	45.1	24.6	15.5	39.8	..	..
Norway	83.8	116.1	99.8	99.4	0.0	0.2	0.1	0.0	..	0.2	..	..
Oman	0.8	8.2	..	..	..	..	21.5	16.6	78.5	83.4	..	..
Pakistan	15.0	62.2	58.2	35.5	0.2	0.7	1.1	38.2	40.5	25.0	0.0	0.6
Panama	2.0	4.4	49.4	48.7	..	..	48.4	50.7	..	..	..	..
Papua New Guinea	..	..	..	..	..	..	..	..	..	..	..	..
Paraguay	0.8	50.9	80.0	99.9	..	..	11.1	0.0	..	..	..	..
Peru	10.0	18.6	69.8	74.3	..	..	27.4	21.0	1.7	4.0	..	..
Philippines	18.0	41.2	19.6	12.3	1.0	22.9	67.9	43.1	..	0.1	..	..
Poland	120.9	140.8	1.9	1.6	94.7	96.3	2.9	1.3	0.1	0.2	..	..
Portugal	15.2	38.9	52.7	33.4	2.3	31.0	42.9	27.5	..	5.2	..	..
Puerto Rico	..	..	..	..	..	..	..	..	..	..	..	..
Romania	67.5	53.5	18.7	35.3	31.4	28.0	9.6	7.7	40.2	19.0	..	9.9
Russian Federation	804.9	826.2	16.1	19.2	..	19.4	77.2	6.1	..	42.7	6.7	12.6

	Electricity production billion kwh		Sources of electricity[a] Hydropower %		Coal %		Oil %		Gas %		Nuclear power %	
	1980	1998	1980	1998	1980	1998	1980	1998	1980	1998	1980	1998
Rwanda	..	..	..	..	..	..	..	..	..	..	..	..
Saudi Arabia	20.5	116.5	..	..	..	..	58.5	62.8	41.5	37.2	..	..
Senegal	0.6	1.3	..	..	..	..	100.0	94.9	..	5.1	..	..
Sierra Leone	..	..	..	..	..	..	..	..	..	..	..	..
Singapore	7.0	28.6	..	..	..	..	100.0	82.1	..	17.0	..	..
Slovak Republic	20.0	25.2	11.3	17.1	37.9	23.5	17.9	4.9	10.2	9.4	22.7	45.2
Slovenia	8.0	13.7	..	25.1	..	35.5	..	2.6	..	0.1	..	36.7
South Africa	99.0	202.8	1.0	0.7	99.0	92.6	0.0	..	..	..	..	6.7
Spain	109.2	193.5	27.1	17.6	30.0	32.6	35.2	9.0	2.7	8.4	4.7	30.5
Sri Lanka	1.7	5.7	88.7	68.9	..	..	11.3	31.1	..	..	..	..
Sudan	0.8	2.0	70.0	53.0	..	..	30.0	47.0	..	..	..	..
Sweden	96.3	158.2	61.1	47.0	0.2	2.0	10.4	2.1	..	0.3	27.5	46.5
Switzerland	48.2	61.7	68.1	54.2	0.1	..	1.0	0.6	0.6	1.4	29.8	41.9
Syrian Arab Republic	4.0	18.3	64.7	41.1	..	..	31.9	26.4	3.4	32.4	..	..
Tajikistan	13.6	14.4	93.4	98.1	..	..	6.6	..	..	1.9	..	..
Tanzania	0.8	2.2	86.4	96.5	..	..	13.6	3.5	..	..	..	..
Thailand	14.4	90.1	8.8	5.7	9.8	18.3	81.4	20.7	9.9	51.7	..	..
Togo	..	..	..	..	..	..	..	..	..	..	..	..
Trinidad and Tobago	2.0	5.2	..	..	..	..	2.3	..	96.5	99.7	..	..
Tunisia	2.9	9.1	0.8	0.8	..	..	64.5	14.5	34.7	84.7	..	..
Turkey	23.3	111.0	48.8	38.0	25.6	32.1	25.1	7.1	..	22.4	..	..
Turkmenistan	6.7	9.4	0.1	0.1	..	..	99.9	..	..	99.9	..	..
Uganda	..	..	..	..	..	..	..	..	..	..	..	..
Ukraine	236.0	172.8	5.7	9.2	..	26.5	88.3	4.0	..	16.8	6.0	43.5
United Arab Emirates	6.3	33.4	..	..	..	..	3.7	7.9	96.3	92.1	..	..
United Kingdom	284.1	356.6	1.4	1.5	73.2	34.5	11.7	1.6	0.7	32.5	13.0	28.1
United States	2,427.3	3,803.7	11.5	7.7	51.2	52.7	10.8	3.9	15.3	14.7	11.0	18.8
Uruguay	4.6	9.6	76.3	95.7	..	..	23.5	4.0	..	..	..	..
Uzbekistan	33.9	45.9	..	12.5	..	4.1	..	11.9	..	71.5	..	..
Venezuela, RB	35.8	80.9	40.7	71.6	..	..	32.4	3.3	26.9	25.1	..	..
Vietnam	3.6	21.7	41.8	51.2	39.9	16.1	18.3	16.9	0.4	15.9	..	..
West Bank and Gaza	..	..	..	..	..	..	..	..	..	..	..	..
Yemen, Rep.	0.5	2.5	..	..	..	..	100.0	100.0	..	..	..	..
Yugoslavia, FR (Serb./Mont.)	..	..	..	..	..	..	..	..	..	..	..	..
Zambia	9.5	7.9	98.8	99.5	0.7	0.5	0.5	0.0	..	..	..	..
Zimbabwe	4.5	6.6	88.3	28.5	11.7	71.5	..	..	..	..	..	..
World	8,176.6 s	14,223.4 s	20.4 w	17.8 w	33.2 w	38.4 w	28.3 w	8.9 w	8.8 w	16.2 w	8.7 w	17.2 w
Low income	547.8	1,037.4	25.0	22.5	11.5	43.5	57.8	8.4	1.7	16.6	3.9	8.6
Middle income	2,227.4	4,544.3	21.7	23.4	23.1	37.9	47.1	12.1	4.6	18.9	3.2	6.9
Lower middle income	1,511.1	2,883.7	18.7	21.1	15.1	41.2	58.6	9.8	3.2	22.0	4.0	5.3
Upper middle income	716.3	1,660.7	28.0	27.3	39.8	32.2	22.9	16.1	7.4	13.7	1.4	9.7
Low & middle income	2,775.2	5,581.8	22.4	23.2	20.8	39.0	49.2	11.4	4.0	18.5	3.3	7.2
East Asia & Pacific	393.8	1,697.6	17.8	14.7	47.1	61.2	33.4	7.9	0.3	9.2	0.9	6.1
Europe & Central Asia	1,640.1	1,715.3	13.5	18.0	13.6	30.3	65.4	6.6	2.3	30.0	5.1	14.9
Latin America & Carib.	360.7	891.3	60.3	61.2	2.1	4.7	25.7	18.4	9.8	11.3	0.6	2.2
Middle East & N. Africa	104.6	431.2	20.4	7.1	1.0	1.8	52.5	43.2	26.2	47.9	..	..
South Asia	138.5	575.9	41.6	19.3	44.1	64.7	6.3	5.3	5.9	8.7	2.2	2.1
Sub-Saharan Africa	137.6	270.4	23.3	18.3	71.7	71.2	4.2	2.9	0.8	2.4	..	5.0
High income	5,401.4	8,641.6	19.4	14.2	39.5	38.0	17.6	7.2	11.3	14.8	11.5	23.6
Europe EMU	1,242.9	1,866.0	17.0	11.9	37.2	27.4	22.9	9.0	10.0	12.8	11.9	36.5

a. Shares may not sum to 100 percent because other sources of generated electricity (such as geothermal, solar, and wind) are not shown.

Use of energy in general, and access to electricity in particular, are important in improving people's standard of living. But electricity generation also can damage the environment. Whether such damage occurs depends largely on how electricity is generated. For example, burning coal releases twice as much carbon dioxide—a major contributor to global warming—as does burning an equivalent amount of natural gas (see *About the data* for table 3.8). Nuclear energy does not generate carbon dioxide emissions, but it produces other dangerous waste products. The table provides information on electricity production by source. Shares may not sum to 100 percent because some sources of generated electricity (such as geothermal, solar, and wind) are not shown.

The International Energy Agency (IEA) compiles data on energy inputs used to generate electricity. IEA data for non-OECD countries are based on national energy data adjusted to conform with annual questionnaires completed by OECD member governments. In addition, estimates are sometimes made to complete major aggregates from which key data are missing, and adjustments are made to compensate for differences in definitions. The IEA makes these estimates in consultation with national statistical offices, oil companies, electricity utilities, and national energy experts.

The IEA occasionally revises its time series to reflect political changes. Since 1990, for example, it has constructed energy statistics for countries of the former Soviet Union. In addition, energy statistics for other countries have undergone continuous changes in coverage or methodology as more detailed energy accounts have become available in recent years. Breaks in series are therefore unavoidable.

Figure 3.9

The gap in electricity production is widening

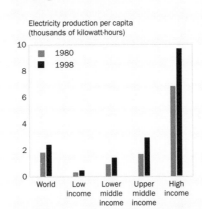

Electricity production per capita (thousands of kilowatt-hours)

■ 1980
■ 1998

Source: Tables 2.1 and 3.9.

Electricity production per person in high-income countries was 20 times as much as in low-income countries in 1980—and 22 times as much in 1998.

• **Electricity production** is measured at the terminals of all alternator sets in a station. In addition to hydropower, coal, oil, gas, and nuclear power generation, it covers generation by geothermal, solar, wind, and tide and wave energy, as well as that from combustible renewables and waste. Production includes the output of electricity plants designed to produce electricity only as well as that of combined heat and power plants. • **Sources of electricity** refer to the inputs used to generate electricity: hydropower, coal, oil, gas, and nuclear power. Hydropower refers to electricity produced by hydroelectric power plants, oil refers to crude oil and petroleum products, gas refers to natural gas but not natural gas liquids, and nuclear power refers to electricity produced by nuclear power plants.

The data on electricity production are from the IEA's electronic files and its annual publications, *Energy Statistics and Balances of Non-OECD Countries*, *Energy Statistics of OECD Countries*, and *Energy Balances of OECD Countries*.

3.10 Urbanization

	Urban population				Population in urban agglomerations of more than one million			Population in largest city		Access to improved sanitation facilities			
	millions		% of total population		% of total population			% of urban population		Urban % of population		Rural % of population	
	1980	1999	1980	1999	1980	2000	2015	1980	2000	1990	2000	1990	2000
Albania	0.9	1.4	34	41	..	..	..	..	..	..	..	..	..
Algeria	8.1	17.8	44	60	8	6	7	17	10	..	90	..	47
Angola	1.5	4.1	21	34	13	21	26	63	62	..	70	..	30
Argentina	23.3	32.8	83	90	42	41	40	43	38	..	89	..	48
Armenia	2.0	2.7	66	70	34	34	35	51	48	..	..	..	..
Australia	12.6	16.1	86	85	61	56	55	26	23	100	100	100	100
Austria	4.9	5.2	65	65	27	26	26	42	40	100	100	100	100
Azerbaijan	3.3	4.5	53	57	26	24	25	48	42	..	..	..	..
Bangladesh	12.5	30.6	14	24	6	13	18	26	39	78	82	27	44
Belarus	5.4	7.1	57	71	14	18	20	24	25	..	..	..	..
Belgium	9.4	9.9	95	97	12	11	11	13	11	..	..	..	..
Benin	0.9	2.5	27	42	..	..	..	..	..	46	46	6	6
Bolivia	2.4	5.0	46	62	14	18	20	30	27	77	82	28	38
Bosnia and Herzegovina	1.5	1.7	36	43	..	..	..	..	..	..	..	..	..
Botswana	0.1	0.8	15	50	..	..	..	..	..	84	..	44	..
Brazil	80.5	135.6	66	81	32	34	34	16	13	84	85	37	40
Bulgaria	5.4	5.7	61	69	12	15	16	20	21	..	100	..	100
Burkina Faso	0.6	2.0	9	18	..	..	..	44	54	88	88	14	16
Burundi	0.2	0.6	4	9	..	..	..	..	..	67	79	90	..
Cambodia	0.8	1.8	12	16	..	..	..	44	51	..	58	..	10
Cameroon	2.7	7.1	31	48	11	21	27	19	23	99	99	79	85
Canada	18.6	23.5	76	77	32	37	38	16	20	100	100	99	99
Central African Republic	0.8	1.4	35	41	..	..	..	..	..	43	43	23	23
Chad	0.8	1.8	19	23	..	..	..	40	57	70	81	4	13
Chile	9.0	12.8	81	85	33	36	37	41	43	98	98	93	93
China	192.3	396.4	20	32	13	14	17	6	3	57	68	18	24
Hong Kong, China	4.6	6.7	92	100	91	100	100	100	100	..	..	..	..
Colombia	18.2	30.5	64	73	26	32	35	20	20	95	97	53	51
Congo, Dem. Rep.	7.8	14.9	29	30	8	10	12	28	33	..	53	..	6
Congo, Rep.	0.7	1.8	41	62	27	42	46	65	67	..	14	..	..
Costa Rica	1.0	1.7	43	48	..	..	..	61	52	..	98	..	96
Côte d'Ivoire	2.8	7.1	35	46	15	21	26	44	47	78	..	30	..
Croatia	2.3	2.6	50	57	..	..	..	28	41	72	..	28	..
Cuba	6.6	8.4	68	75	20	20	20	29	27	..	96	..	91
Czech Republic	7.6	7.7	75	75	12	12	12	15	16	..	..	..	..
Denmark	4.3	4.5	84	85	27	26	26	32	30	..	..	..	..
Dominican Republic	2.9	5.4	51	64	34	60	65	50	65	66	75	52	64
Ecuador	3.7	8.0	47	64	23	32	37	29	29	..	70	..	37
Egypt, Arab Rep.	17.9	28.2	44	45	23	23	24	38	37	96	98	80	91
El Salvador	1.9	2.8	42	46	16	22	25	39	48	..	88	..	78
Eritrea	0.3	0.7	14	18	..	..	..	..	..	..	66	..	1
Estonia	1.0	1.0	70	69	..	..	..	..	..	..	93	..	..
Ethiopia	4.0	10.8	11	17	3	4	6	30	23	58	58	6	6
Finland	2.9	3.4	60	67	13	23	25	22	33	100	100	100	100
France	39.5	44.2	73	75	21	21	20	23	22	..	..	..	..
Gabon	0.3	1.0	50	80	..	..	..	..	..	..	25	..	4
Gambia, The	0.1	0.4	20	32	..	..	..	..	..	..	41	..	35
Georgia	2.6	3.3	52	60	22	24	26	42	40	..	..	..	..
Germany	64.7	71.7	83	87	39	41	43	10	9	..	..	..	..
Ghana	3.4	7.1	31	38	9	10	14	30	27	59	62	61	64
Greece	5.6	6.3	58	60	31	30	30	54	49	..	..	..	..
Guatemala	2.6	4.4	37	40	11	28	32	29	70	94	98	66	76
Guinea	0.9	2.3	19	32	12	25	32	65	75	94	94	41	41
Guinea-Bissau	0.1	0.3	17	23	..	..	..	..	..	..	88	..	34
Haiti	1.3	2.7	24	35	13	22	29	55	62	48	50	15	16
Honduras	1.2	3.3	35	52	..	..	..	33	28	85	94	..	57

Urbanization | 3.10

	Urban population				Population in urban agglomerations of more than one million			Population in largest city		Access to improved sanitation facilities			
	millions		% of total population		% of total population			% of urban population		Urban % of population		Rural % of population	
	1980	1999	1980	1999	1980	2000	2015	1980	2000	1990	2000	1990	2000
Hungary	6.1	6.4	57	64	19	18	19	34	28	100	100	98	98
India	158.8	280.1	23	28	8	10	12	5	6	58	73	8	14
Indonesia	32.9	82.5	22	40	8	10	12	18	13	76	87	44	52
Iran, Islamic Rep.	19.7	39.7	50	63	21	23	23	26	18	86	86	74	74
Iraq	8.5	17.4	66	76	29	31	34	39	27	..	93	..	31
Ireland	1.9	2.2	55	59	..	..	..	48	44	..	..	..	..
Israel	3.4	5.6	89	91	37	35	33	41	38	..	..	..	..
Italy	37.6	38.6	67	67	24	19	21	14	11	..	..	..	..
Jamaica	1.0	1.4	47	56	..	..	..	..	..	..	98	..	65
Japan	89.0	99.6	76	79	34	38	39	25	26	..	..	..	..
Jordan	1.3	3.5	60	74	29	29	32	49	39	100	100	95	98
Kazakhstan	8.0	8.4	54	56	6	8	8	12	15	..	100	..	98
Kenya	2.7	9.5	16	32	5	8	10	32	23	94	96	81	81
Korea, Dem. Rep.	10.1	14.0	57	60	10	14	15	18	22	..	99	..	100
Korea, Rep.	21.7	38.0	57	81	40	47	45	38	26	..	76	..	4
Kuwait	1.2	1.9	90	97	60	60	53	67	61	*100*	..	100	..
Kyrgyz Republic	1.4	1.6	38	34	..	..	..	..	..	83	100	..	100
Lao PDR	0.4	1.2	13	23	..	..	..	..	..	..	84	..	34
Latvia	1.7	1.7	68	69	..	..	..	49	46	93	..	..	..
Lebanon	2.2	3.8	74	89	40	47	48	55	53	..	100	..	87
Lesotho	0.2	0.6	13	27	..	..	..	..	..	..	93	..	92
Libya	2.1	4.7	69	87	26	33	32	38	38	97	97	96	96
Lithuania	2.1	2.5	61	68	..	..	..	..	..	..	..	..	..
Macedonia, FYR	1.0	1.2	54	62	..	..	..	..	..	..	..	..	..
Madagascar	1.6	4.4	18	29	6	10	13	33	33	70	70	25	30
Malawi	0.6	2.5	9	24	..	..	..	..	..	96	96	70	70
Malaysia	5.8	12.9	42	57	7	6	6	16	10	..	..	..	98
Mali	1.2	3.1	19	29	..	..	..	40	35	95	93	62	58
Mauritania	0.4	1.5	27	56	..	..	..	..	..	44	44	19	19
Mauritius	0.4	0.5	42	41	..	..	..	..	..	100	100	100	99
Mexico	44.8	71.7	66	74	28	28	25	31	25	85	87	28	32
Moldova	1.6	2.0	40	46	..	..	..	..	..	..	100	..	..
Mongolia	0.9	1.5	52	63	..	..	..	..	..	..	46	..	2
Morocco	8.0	15.6	41	55	15	18	20	26	22	95	100	31	42
Mozambique	1.6	6.7	13	39	6	17	21	47	43	..	69	..	26
Myanmar	8.1	12.3	24	27	7	9	11	27	33	65	65	38	39
Namibia	0.2	0.5	23	30	..	..	..	..	..	84	96	14	17
Nepal	0.9	2.7	7	12	..	..	..	..	..	68	75	16	20
Netherlands	12.5	14.1	88	89	14	14	14	8	8	100	100	100	100
New Zealand	2.6	3.3	83	86	..	..	..	30	34	*100*	..	..	..
Nicaragua	1.5	2.7	50	56	..	..	..	36	34	97	96	53	68
Niger	0.7	2.1	13	20	..	..	..	..	..	71	79	4	5
Nigeria	19.1	53.4	27	43	8	12	15	23	24	77	85	51	45
Norway	2.9	3.4	71	75	..	..	..	22	29	100	..	..	..
Oman	0.3	1.9	32	82	..	..	..	..	..	98	98	61	61
Pakistan	23.2	49.1	28	36	15	21	25	22	23	78	94	13	42
Panama	1.0	1.6	50	56	..	..	..	62	71	*100*	99	68	87
Papua New Guinea	0.4	0.8	13	17	..	..	..	..	..	92	92	80	80
Paraguay	1.3	3.0	42	55	22	23	26	52	41	92	95	87	95
Peru	11.2	18.3	65	72	25	29	30	39	40	81	90	26	40
Philippines	18.1	42.8	38	58	14	16	17	33	25	85	92	64	71
Poland	20.7	25.2	58	65	18	18	18	16	14	..	..	..	..
Portugal	2.9	6.3	29	63	19	57	68	46	59	..	..	..	..
Puerto Rico	2.1	2.9	67	75	34	35	36	51	47	..	..	..	..
Romania	10.9	12.6	49	56	9	9	10	18	16	..	86	..	10
Russian Federation	97.0	113.1	70	77	18	19	21	8	8	..	..	..	..

3.10 Urbanization

	Urban population				Population in urban agglomerations of more than one million			Population in largest city		Access to improved sanitation facilities			
	millions		% of total population			% of total population		% of urban population		Urban % of population		Rural % of population	
	1980	1999	1980	1999	1980	2000	2015	1980	2000	1990	2000	1990	2000
Rwanda	0.2	0.5	5	6	..	..	..	..	..	..	12	..	8
Saudi Arabia	6.2	17.2	66	85	19	25	24	16	19	*100*	100	..	100
Senegal	2.0	4.3	36	47	17	22	26	48	46	86	94	38	48
Sierra Leone	0.8	1.8	24	36	..	..	..	..	..	..	23	..	31
Singapore	2.3	3.9	100	100	100	100	100	100	100	100	100	..	..
Slovak Republic	2.6	3.1	52	57	..	..	..	..	..	..	100	..	100
Slovenia	0.9	1.0	48	50	..	..	..	..	..	100	..	..	..
South Africa	13.3	21.1	48	50	27	32	36	12	14	..	99	..	73
Spain	27.2	30.5	73	77	20	17	18	16	13				
Sri Lanka	3.2	4.4	22	23	..	..	..	..	..	93	91	79	83
Sudan	3.7	10.2	20	35	6	9	11	31	25	87	87	48	48
Sweden	6.9	7.4	83	83	17	18	19	20	21	100	100	100	100
Switzerland	3.6	4.8	57	68	..	..	..	20	20	100	100	100	100
Syrian Arab Republic	4.1	8.5	47	54	28	28	31	34	27	..	98	..	81
Tajikistan	1.4	1.7	34	28	..	..	..	..	..	..	..	..	..
Tanzania	2.7	10.4	15	32	5	12	18	30	25	97	98	86	86
Thailand	7.9	12.8	17	21	10	12	15	59	56	97	97	83	96
Togo	0.6	1.5	23	33	..	..	..	..	..	71	69	24	17
Trinidad and Tobago	0.7	1.0	63	74	..	..	..	..	..	*100*	..	97	..
Tunisia	3.3	6.1	52	65	18	20	21	35	30	97	..	48	..
Turkey	19.5	47.7	44	74	19	27	30	23	19	98	98	70	70
Turkmenistan	1.3	2.1	47	45	..	..	..	..	..	..	..	..	..
Uganda	1.1	3.0	9	14	..	..	..	42	39	96	96	82	72
Ukraine	30.9	33.9	62	68	14	15	17	7	8	..	..	..	..
United Arab Emirates	0.7	2.4	72	85	..	..	..	31	37	*100*	..	77	..
United Kingdom	50.0	53.2	89	89	25	23	23	15	14	100	100	100	100
United States	167.6	214.2	74	77	38	38	37	9	8	100	100	100	100
Uruguay	2.5	3.0	85	91	42	37	35	49	41	..	96	..	89
Uzbekistan	6.5	9.1	41	37	11	9	8	28	23	..	100	..	100
Venezuela, RB	12.0	20.5	79	87	28	29	30	21	15	..	75	..	69
Vietnam	10.3	15.2	19	20	14	13	14	34	30	86	86	70	70
West Bank and Gaza	..	..	..	..	..	..	..	..	..	..	..	..	..
Yemen, Rep.	1.6	4.2	19	24	..	..	..	15	30	80	87	27	31
Yugoslavia, FR (Serb./Mont.)	4.5	5.5	46	52	11	14	15	24	27	..	..	..	..
Zambia	2.3	3.9	40	40	9	16	22	23	37	86	99	48	64
Zimbabwe	1.6	4.1	22	35	9	14	20	39	41	98	99	51	51

World	**1,748.5 s**	**2,776.6 s**	**39 w**	**46 w**	**.. w**	**.. w**	**.. w**	**18 w**	**17 w**	**78 w**	**84 w**	**29 w**	**36 w**
Low income	389.7	760.0	24	31	..	..	..	16	18	68	79	25	31
Middle income	775.8	1,328.1	38	50	..	..	..	19	16	75	82	29	38
Lower middle income	506.2	897.0	31	43	..	..	..	16	14	70	80	29	36
Upper middle income	269.6	431.2	64	75	..	..	..	25	21	..	87	..	54
Low & middle income	1,165.5	2,088.1	32	41	..	..	..	18	17	72	81	27	34
East Asia & Pacific	310.2	633.0	22	34	..	..	..	15	10	63	74	28	35
Europe & Central Asia	249.3	315.5	59	67	..	..	..	15	15	..	..	..	..
Latin America & Carib.	233.5	380.7	65	75	29	32	32	27	25	85	87	39	49
Middle East & N. Africa	83.8	169.0	48	58	21	22	23	30	25	92	94	63	67
South Asia	201.2	372.7	22	28	8	12	14	9	12	63	76	12	21
Sub-Saharan Africa	87.5	217.3	23	34	..	..	..	28	29	80	81	47	41
High income	583.0	688.5	75	77	..	..	..	17	17	..	..	..	..
Europe EMU	204.5	227.8	74	78	26	27	28	16	15	..	..	..	..

About the data

The population of a city or metropolitan area depends on the boundaries chosen. For example, in 1990 Beijing, China, contained 2.3 million people in 87 square kilometers of "inner city" and 5.4 million in 158 square kilometers of "core city." The population of "inner city and inner suburban districts" was 6.3 million, and that of "inner city, inner and outer suburban districts, and inner and outer counties" was 10.8 million. (For most countries, the last definition is used.)

Estimates of the world's urban population would change significantly if China, India, and a few other populous nations were to change their definition of urban centers. According to China's State Statistical Bureau, by the end of 1996 urban residents accounted for about 43 percent of China's population, while in 1994 only 20 percent of the population was considered urban. Besides the continuous migration of people from rural to urban areas, one of the main reasons for this shift

was the rapid growth in the hundreds of towns reclassified as cities in recent years. Because the estimates in the table are based on national definitions of what constitutes a city or metropolitan area, cross-country comparisons should be made with caution.

To estimate urban populations, the United Nations' ratios of urban to total population were applied to the World Bank's estimates of total population (see table 2.1).

The urban population with access to improved sanitation facilities is defined as those with access to at least adequate excreta disposal facilities that can effectively prevent human, animal, and insect contact with excreta. The rural population with access is included to allow comparison of rural and urban access. This definition and the definition of urban areas vary, however, so comparisons between countries can be misleading (see *Definitions* for table 2.16).

Definitions

• **Urban population** is the midyear population of areas defined as urban in each country and reported to the United Nations (see *About the data*). • **Population in urban agglomerations of more than one million** is the percentage of a country's population living in metropolitan areas that in 1990 had a population of more than one million. • **Population in largest city** is the percentage of a country's urban population living in that country's largest metropolitan area. • **Access to improved sanitation facilities** refers to the percentage of the urban or rural population with access to at least adequate excreta disposal facilities (private or shared, but not public) that can effectively prevent human, animal, and insect contact with excreta. Improved facilities range from simple but protected pit latrines to flush toilets with a sewerage connection. To be effective, facilities must be correctly constructed and properly maintained.

Data sources

The data on urban population and the population in urban agglomerations and in the largest city come from the United Nations Population Division's *World Urbanization Prospects: The 1999 Revision.* The total population figures are World Bank estimates. The data on access to sanitation in urban and rural areas are from the World Health Organization.

Figure 3.10

Most without access to improved sanitation facilities live in Asia
Population without access to improved sanitation facilities

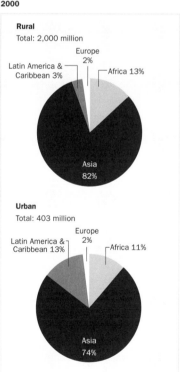

Source: World Health Organization data.

Many more people gained access to improved sanitation facilities between 1990 and 2000—the number with access rose from 2.9 billion to 3.7 billion. But 2.4 billion still lack access, most of them in Asia. To reach the target of universal coverage by 2025, additional sanitation services will need to be provided for more than 4 billion people.

3.11 | Urban environment

	City	Urban population	Average house-hold income	House price to income ratio	Work trips by public trans-portation	Travel time to work	Households with access to services				Wastewater treated
							Potable water %	Sewerage connection %	Electricity %	Telephone %	%
		thousands 1998	$ 1998[a]	1998[a]	% 1998[a]	minutes 1998[a]	1998[a]	1998[a]	1998[a]	1998[a]	1998[a]
Algeria	Algiers	2,562	..	..	..	75	..	..	..	..	80
Argentina	Buenos Aires	2,996	13,026	5.1	59	42	100	98	100	70	0
	Córdoba	132	6,448	6.8	44	32	99	40	99	80	49
	Rosario	1,248	7,500	5.7	..	22	98	67	93	76	1
Armenia	Yerevan	1,250	1,205	4.0	84	30	98	98	100	88	36
Bangladesh	Chittagong	2,301	1,875	8.1	27	45	44	..	95	0	0
	Dhaka	10,000	1,920	16.7	9	45	60	22	90	7	0
	Sylhet	242	1,584	6.0	10	50	29	0	93	40	0
	Tangail	152	1,500	13.9		30	12	0	90	12	0
Barbados	Bridgetown	..	17,179	4.4	..	..	98	5	99	78	7
Belize	Belize City	55	660		..	..	..	..	..	..	..
Bolivia	Santa Cruz de la Sierra	1,065 [b]	2,987	29.3	..	29	53	33	98	59	53
Bosnia and Herzegovina	Sarajevo	522 [b]	..	..	100	12	95	90	100	..	..
Brazil	Belém	1,638 [b]	..	..	..	..	..	..	..	..	..
	Icapui	..	1,357	4.5	..	30	88	..	90	33	..
	Maranguape	..			30	20	73	..	..	..	..
	Porto Alegre	3	..	..	..	..	99	87	100	..	..
	Recife	3,088	4,115	12.5	46	35	89	41	100	29	33
	Rio de Janeiro	10,192	..	..	..	..	88	80	10	..	..
	Santo André	1,658	1,516	23.4	43	40	98	95	100	79	..
Bulgaria	Bourgas	..	1,815	5.1	61	32	100	93	100	..	93
	Sofia	1,200	1,518	13.2	79	32	95	91	100	89	94
	Troyan	24	2,492	3.7	44	22	99	82	100	45	..
	Veliko Tarnovo	..	2,243	5.4	46	30	98	98	100	96	50
Burkina Faso	Bobo-Dioulasso	..	..	..	..	..	24	..	29	6	..
	Koudougou	..	..	..	..	..	30	..	26	7	0
	Ouagadougou	1,130 [b]	..	..	2	..	30	..	47	11	19
Burundi	Bujumbura	373	1,342	..	48	25	26	62	57	19	21
Cambodia	Phnom Penh	1,000	3,584	8.9	0	45	45	75	76	40	0
Cameroon	Douala	1,148	850	13.4	..	40	34	1	95	9	5
	Yaoundé	968	..	..	42	45	34	1	95	9	24
Canada	Hull	254	..	..	16	..	100	100	100	100	100
Central African Rep.	Bangui	..	167	..	66	60	31	..	18	11	0
Chad	N'Djamena	998 [b]	..	..	35	..	42	0	13	6	21
Chile	Gran Concepción	..	..	..	57	35	100	91	95	69	6
	Santiago de Chile	5,737	..	..	60	38	100	99	99	73	3
	Tome	..	..	..	..	..	92	52	98	58	57
	Valparaiso	851	..	..	55	..	98	92	97	63	100
	Viña del Mar	851	..	..	..	..	97	97	98	65	93
Colombia	Armenia	..	3,429	5.0	42	60	90	50	99	97	0
	Marinilla	170	1,848	8.5	18	15	98	93	100	65	..
	Medellín	2,901	..	..	38	35	100	99	100	87	..
Congo, Dem. Rep.	Kinshasa	5,398	..	..	72	57	72	0	66	1	..
Congo, Rep.	Brazzaville	989	..	..	55	20	56	0	52	18	..
Côte d'Ivoire	Abidjan	3,201	3,440	14.5	..	45	26	15	41	5	45
Croatia	Zagreb	1,164	7,674	7.8	56	31	98	97	100	94	..
Cuba	Baracoa	..	..	..	..	..	83	3	93	32	..
	Camagüey	..	..	..	2	60	72	47	97	..	..
	Cienfuegos	..	1,848	4.0	..	80	100	73	100	9	2
	Havana	..	1,800	8.5	58	83	100	85	100	14	..
	Pinar del Río	..	..	..	..	80	97	48	100	..	..
	Santa Clara	..	..	..	7	48	95	42	100	43	..
Czech Republic	Brno	..	..	..	50	25	100	96	100	69	100
	Prague	1,193	6,880	..	55	22	99	100	100	100	..
Dominican Republic	Santiago de los Caballeros	691	..	..	..	30	75	80	..	71	80
Ecuador	Ambato	286	..	..	..	..	90	81	91	87	0

Urban environment | 3.11

	City	Urban population	Average house-hold income	House price to income ratio	Work trips by public trans-portation	Travel time to work	Households with access to services				Wastewater treated
							Potable water %	Sewerage connection %	Electricity %	Telephone %	
		thousands 1998	$ 1998[a]	1998[a]	% 1998[a]	minutes 1998[a]	1998[a]	1998[a]	1998[a]	1998[a]	% 1998[a]
	Cuenca	..	2,160	4.6	..	25	97	92	97	48	82
	Guayaquil	2,317	5,391	3.4	89	45	70	42	..	44	9
	Manta	126	..	..	..	30	70	52	98	40	..
	Puyo	40	3,972	2.1	..	15	80	30	90	60	..
	Quito	1,531	3,760	2.4	..	33	85	70	96	55	..
	Tena	..	960	6.3	..	5	80	60	..	..	0
El Salvador	San Salvador	1,863	6,080	3.5	..	..	82	80	98	70	..
Estonia	Riik	..	..	..	..	..	92	90	98	55	..
	Tallin	397 [b]	5,825	6.4	..	35	98	98	100	86	100
Gabon	Libreville	523 [b]	375	..	80	30	55	0	95	45	44
Gambia, The	Banjul	50	900	11.4	55	22	23	12	24	..	..
Georgia	Tbilisi	1,310 [b]	915	9.4	..	..	..	98	100	58	0
Ghana	Accra	1,500	825	14.0	54	21	..	..	..	..	0
	Kumasi	780	795	13.7	51	21	65	..	95	51	..
Guatemala	Quezaltenango	333	4,645	4.3	..	15	60	55	80	40	..
Guinea	Conakry	1,824 [b]	..	..	26	45	30	32	54	6	..
Indonesia	Jakarta	9,489	1,366	14.6	..	..	50	65	99	..	16
	Semarang	1,076	756	..	..	..	34	..	85	..	0
	Surabaya	2,373	1,167	3.4	18	35	41	56	89	71	0
Iraq	Baghdad	4,797 [b]	..	..	..	..	..	..	..	..	..
Italy	Aversa	..	..	..	..	..	..	..	..	..	90
Jamaica	Kingston	655 [b]	..	..	..	..	97	..	88	..	20
	Montego Bay	..	..	..	..	..	78	..	86	..	15
Jordan	Amman	1,621	5,801	6.1	21	25	98	81	99	62	54
Kenya	Kisumu	134	1,365	8.5	43	24	38	31	49	..	65
	Mombasa	..	..	..	47	20	..	..	..	..	50
	Nairobi	2,310 [b]	..	..	71	57	89	..	..	..	52
Korea, Rep.	Hanam	124	19,933	3.7	..	..	81	68	100	100	81
	Pusan	3,843	19,933	4.0	39	42	98	69	100	100	69
	Seoul	10,389	23,500	5.7	71	60	100	99	100	..	99
Kuwait	Kuwait City	1,165 [b]	38,700	6.5	21	10	100	98	100	98	..
Kyrgyz Republic	Bishkek	619	..	..	95	35	30	23	100	20	15
Lao PDR	Vientiane	562	1,077	23.2	2	27	87	..	100	87	20
Latvia	Riga	775 [b]	3,206	15.6	..	..	95	93	100	70	..
Lebanon	Sin El Fil	..	7,200	8.3	50	10	80	30	98	80	..
Liberia	Monrovia	651	1,250	28.0	80	60	..	..	..	..	0
Libya	Tripoli	1,773	48,073	0.8	18	20	97	90	99	6	40
Lithuania	Vilnius	578	3,754	20.0	52	37	89	89	100	77	54
Madagascar	Antananarivo	1,507 [b]	..	..	..	..	..	..	..	..	..
Malawi	Lilongwe	765 [b]	..	..	27	5	65	12	50	10	0
Malaysia	Penang	..	3,193	7.2	55	40	99	..	100	98	20
Mauritania	Nouakchott	881 [b]	1,600	5.4	45	50	..	..	..	..	..
Mexico	Ciudad Juarez	1,018	..	..	24	23	89	77	96	45	..
Moldova	Chisinau	..	..	..	80	23	100	95	100	83	71
Mongolia	Ulaanbaatar	627	991	7.8	80	30	60	60	100	90	96
Morocco	Casablanca	3,292	..	..	..	30	83	93	91	..	..
	Rabat	646	..	..	40	20	93	97	52	..	..
Myanmar	Yangon	3,692	..	8.3	69	45	78	81	85	17	0
Nicaragua	Leon	..	..	..	..	15	78	..	84	21	..
Niger	Niamey	731 [b]	..	..	..	30	33	0	51	4	..
Nigeria	Ibadan	1,731 [b]	..	..	46	45	26	12	41	..	..
	Lagos	13,427 [b]	..	..	48	60	..	..	41	..	..
Oman	Muscat	887	..	..	..	20	80	90	89	53	..
Panama	Colón	132	2,820	14.2	..	15	..	..	..	..	0
Paraguay	Asunción	1,262 [b]	13,279	10.7	..	25	46	8	86	17	0
Peru	Cajamarca	..	5,160	3.9	..	20	86	69	81	38	62

	City	Urban population	Average house-hold income	House price to income ratio	Work trips by public trans-portation	Travel time to work	Households with access to services				Wastewater treated
							Potable water %	Sewerage connection %	Electricity %	Telephone %	
		thousands 1998	$ 1998[a]	1998[a]	% 1998[a]	minutes 1998[a]	1998[a]	1998[a]	1998[a]	1998[a]	% 1998[a]
	Huanuco	747	500	30.0	..	20	57	28	80	32	..
	Huaras	54	1,200	6.7	..	15	..	..	71	..	..
	Iquitos	347	3,600	5.6	25	10	73	60	82	62	..
	Lima	7,431	3,179	10.4	82	..	75	71	99	..	4
	Tacna	..	2,000	4.0	..	25	65	58	74	16	64
	Tumbes	..	..	..	..	20	60	35	80	25	..
Philippines	Cebu	2,189	490	13.3	..	35	41	92	80	25	..
Poland	Bydgoszcz	..	4,730	4.3	35	18	95	87	100	85	28
	Gdansk	893 [b]	5,200	4.4	56	20	99	94	100	56	100
	Katowice	3,487 [b]	7,818	1.7	29	36	99	94	100	75	67
	Poznan	..	6,918	5.8	51	25	95	96	100	86	78
Qatar	Doha	391 [b]	..	..	..	..	..	..	..	..	..
Russian Federation	Astrakhan	..	1,614	5.0	66	35	81	79	100	51	92
	Belgorod	..	2,072	4.0	..	25	90	89	100	51	96
	Kostroma	..	1,505	6.9	68	20	88	84	100	46	96
	Moscow	9,321 [b]	5,251	5.1	85	62	100	100	100	100	98
	Nizhny Novgorod	1,458 [b]	1,811	6.9	79	35	98	98	100	64	98
	Novomoscowsk	..	1,693	4.2	61	25	99	93	100	62	97
	Omsk	1,216 [b]	2,346	3.9	86	43	87	87	100	41	89
	Pushkin	..	2,680	9.6	60	15	99	99	100	89	100
	Surgut	..	5,602	4.5	81	57	98	98	100	50	93
	Veliky Novgorod	..	2,420	3.4	75	30	97	97	100	51	95
Rwanda	Kigali	358	2,336	11.4	32	45	36	20	57	6	20
Samoa	Apia	34	3,000	10.0	..	..	60	0	98	96	0
Singapore	Singapore	3,164	27,047	3.1	53	30	100	100	100	100	100
Slovenia	Ljubljana	273	10,320	7.8	20	30	100	100	100	97	98
Spain	Madrid	4,577	..	..	16	32	..	..	..	..	100
	Pamplona	..	..	..	..	..	100	..	100	..	79
Sweden	Amal	13	16,720	2.9	..	..	100	100	100	..	100
	Stockholm	736	21,430	6.0	48	28	100	100	100	..	100
	Umea	104	18,200	5.3	..	16	100	100	100	..	100
Switzerland	Basel	170	42,000	12.3	..	..	100	100	100	99	100
Syrian Arab Republic	Damascus	2,335	1,061	10.3	33	40	98	71	95	10	3
Thailand	Bangkok	5,647	8,521	8.8	28	60	99	100	100	60	..
	Chiang Mai	499	3,384	6.8	5	30	95	60	100	75	70
Togo	Lomé	663	..	..	40	30	..	70	51	18	..
Trinidad and Tobago	Port of Spain	..	..	..	44	..	..	..	..	..	..
Tunisia	Tunis	2,023	6,286	5.0	..	..	75	47	95	27	83
Turkey	Ankara	2,837	4,677	4.5	..	32	97	98	100	..	80
Uganda	Entebbe	65	960	10.4	65	20	48	13	42	0	30
	Jinja	92	650	15.4	49	12	65	43	55	5	30
Uruguay	Montevideo	1,670	14,748	5.6	60	45	98	79	100	75	34
West Bank and Gaza	Gaza	367	4,620	5.4	..	..	85	38	99	38	..
Yemen, Rep.	Aden	1,200	..	..	78	20	..	..	96	..	30
	Sana'a	1,200	..	..	78	20	30	9	96	..	30
Yugoslavia, FR (Serb./Mont.)	Belgrade	1,182	2,922	13.5	72	40	95	86	100	86	20
Zimbabwe	Bulawayo	900	..	..	75	15	100	100	98	..	80
	Chegutu	..	1,445	3.4	20	22	100	68	9	3	69
	Gweru	..	..	..	..	15	100	100	90	61	95
	Harare	1,634	..	..	32	45	100	100	88	42	..
	Mutare	149	..	..	70	20	88	88	74	4	100

a. Data are preliminary. b. Data refer to 2000 and are from the United Nations Population Division's *World Urbanization Prospects: The 1999 Revision.*

Urban environment | 3.11

Despite the importance of cities and urban agglomerations as home to almost half the world's people, data on many aspects of urban life are sparse. Compiling comparable data has been difficult, and the available indicators have been scattered among international agencies with different mandates. Even within cities it is difficult to assemble an integrated data set. Urban areas are often spread across many jurisdictions, with no single agency responsible for collecting and reporting data for the entire area. Adding to the difficulties of data collection are gaps and overlaps in the data collection and reporting responsibilities of different administrative units. Creating a comprehensive, comparable international data set is further complicated by differences in the definition of an urban area and by uneven data quality.

The United Nations Global Plan of Action calls for monitoring the changing role of the world's cities and human settlements. The international agency with the mandate to assemble information on urban areas is the United Nations Centre for Human Settlements (UNCHS, or Habitat). Its Urban Indicators Programme is intended to provide data for monitoring and evalu-

ating the performance of urban areas and for developing government policies and strategies. These data are collected through questionnaires completed by city officials in more than a hundred countries. The table shows selected indicators for more than 160 cities from the UNCHS data set. A few more indicators are included on the *World Development Indicators* CD-ROM. These data are still preliminary and are undergoing further validation.

The UNCHS selection of cities does not reflect population weights or the economic importance of cities and is therefore biased toward smaller cities. Moreover, it is based on demand for participation in the Urban Indicators Programme. As a result, the database excludes a large number of major cities. The table reflects this bias, as well as the criterion of data availability for the indicators shown in the table.

The data should be used with care. Because different data collection methods and definitions may have been used, comparisons can be misleading. In addition, the definitions used here for urban population and access to potable water are more stringent than those used for tables 3.5 and 3.10 (see *Definitions*).

• **Urban population** refers to the population of the urban agglomeration, a contiguous inhabited territory without regard to administrative boundaries. • **Average household income** is the average of the household income in all five quintiles, based on survey data. It is the total income of all household members from all sources, including wages, pensions or benefits, business earnings, rents, and the value of any business or subsistence products consumed (for example, foodstuffs). • **House price to income ratio** is the average house price divided by the average household income. • **Work trips by public transportation** are the percentage of trips to work made by bus or minibus, tram, or train. Buses or minibuses refer to road vehicles other than cars taking passengers on a fare-paying basis. Other means of transport commonly used in developing countries, such as taxi, ferry, rickshaw, or animal, are not included. • **Travel time to work** is the average time in minutes, for all modes, for a one-way trip to work. Train and bus times include average walking and waiting times, and car times include parking and walking to the workplace. • **Households with access to services** are the percentage of households in formal settlements with access to potable water and connections to sewerage, electricity, and telephone. Households with access to potable water are those having access to safe or potable drinking water within 200 meters of the dwelling. Potable water is water that is free from contamination and safe to drink without further treatment. • **Wastewater treated** is the percentage of all wastewater undergoing some form of treatment.

The data in the table are from the Global Urban Indicators database of the UNCHS.

Table 3.11a

Population of the world's 10 largest metropolitan areas in 1000, 1900, 2000, and 2015
Millions

| | 1000 | | 1900 | | 2000 | | 2015 |
City	Population	City	Population	City	Population	City	Population
Cordova	0.45	London	6.5	Tokyo	26.4	Tokyo	26.4
Kaifeng	0.40	New York	4.2	Mexico City	18.1	Mumbai	26.1
Constantinople	0.30	Paris	3.3	Mumbai	18.1	Lagos	23.2
Angkor	0.20	Berlin	2.7	São Paulo	17.8	Dhaka	21.1
Kyoto	0.18	Chicago	1.7	New York	16.6	São Paulo	20.4
Cairo	0.14	Vienna	1.7	Lagos	13.4	Karachi	19.2
Baghdad	0.13	Tokyo	1.5	Los Angeles	13.1	Mexico City	19.2
Nishapur	0.13	St. Petersburg	1.4	Calcutta	12.9	New York	17.4
Hasa	0.11	Manchester	1.4	Shanghai	12.9	Jakarta	17.3
Anhilvada	0.10	Philadelphia	1.4	Buenos Aires	12.6	Calcutta	17.3

Source: O'Meara 1999; and United Nations Population Division 2000.

3.12 | Traffic and congestion

	Motor vehicles				Passenger cars		Two-wheelers		Road traffic		Fuel prices	
	per 1,000 people		per kilometer of road		per 1,000 people		per 1,000 people		million vehicle kilometers		Super $ per liter	Diesel $ per liter
	1990	1999	1990	1999	1990	1999	1990	1999	1990	1999	2000	2000
Albania	11	44	*3*	8	2	29	3	1	..	..	0.57	0.30
Algeria	..	53	..	14	..	25	..	..	..	..	0.27	0.15
Angola	19	20	..	3	15	18	..	..	..	..	0.30	0.15
Argentina	181	*181*	27	30	134	*140*	1	*1*	43,119	*27,458*	1.07	0.52
Armenia	..	..	..	..	..	..	..	..	..	..	0.55	0.31
Australia	530	*601*	11	*12*	450	*485*	18	*16*	138,501	..	0.57	0.57
Austria	421	536	30	22	387	496	71	77	..	..	0.82	0.74
Azerbaijan	52	49	7	14	36	38	5	1	..	..	*0.46*	*0.22*
Bangladesh	1	*1*	0	*1*	0	*0*	1	*1*	..	..	0.46	0.29
Belarus	60	136	13	21	59	135	..	53	10,026	5,462	*0.34*	*0.13*
Belgium	423	497	30	*34*	385	448	14	25	..	*158,759*	0.96	0.78
Benin	3	8	2	7	2	7	34	44	..	..	0.48	0.39
Bolivia	41	*48*	6	*7*	25	*29*	9	*9*	1,139	*1,730*	0.80	0.50
Bosnia and Herzegovina	114	*30*	24	*5*	101	*27*	..	..	..	..	0.68	0.57
Botswana	18	70	3	11	10	30	..	1	..	..	0.42	0.39
Brazil	88	*79*	8	*6*	..	..	..	..	..	..	0.92	0.34
Bulgaria	163	266	39	58	146	233	55	63	..	..	0.70	0.58
Burkina Faso	4	6	3	5	2	4	9	10	..	..	0.68	0.46
Burundi	..	..	..	..	..	..	..	..	..	..	1.01	0.71
Cambodia	1	6	0	2	0	5	9	41	314	1,407	0.61	0.44
Cameroon	10	*12*	3	..	6	7	..	..	..	..	0.56	0.47
Canada	605	*581*	20	*19*	468	*459*	12	*11*	..	..	0.58	0.47
Central African Republic	1	1	0	*0*	1	0	0	0	1,494	*1,250*	*0.81*	*0.65*
Chad	2	4	0	*1*	1	2	0	1	..	..	0.68	0.60
Chile	81	135	13	26	52	88	2	2	..	..	0.64	0.47
China	5	8	4	7	1	3	3	8	..	..	0.40	0.45
Hong Kong, China	66	78	253	*276*	42	58	4	5	8,192	10,781	1.46	0.80
Colombia	..	51	..	19	..	43	8	12	50,945	*41,587*	0.49	0.35
Congo, Dem. Rep.	..	..	..	..	..	..	..	..	..	..	1.00	0.93
Congo, Rep.	18	*20*	3	*4*	12	*14*	..	..	..	..	0.53	0.30
Costa Rica	88	138	7	14	56	91	14	23	..	507,796	0.65	0.44
Côte d'Ivoire	24	*32*	6	*9*	15	*20*	..	..	..	..	0.76	0.51
Croatia	..	..	..	..	..	..	..	..	..	..	0.76	0.60
Cuba	37	*32*	*16*	*6*	*18*	*16*	19	16	..	..	*0.50*	*0.18*
Czech Republic	246	363	46	29	228	335	113	78	..	..	0.77	0.68
Denmark	368	411	27	31	320	352	9	12	36,304	44,845	1.01	0.90
Dominican Republic	74	*47*	48	*30*	21	28	..	..	..	..	0.71	0.39
Ecuador	35	*46*	8	*13*	31	*41*	2	2	10,306	*16,335*	0.31	0.18
Egypt, Arab Rep.	29	*30*	33	28	21	23	6	7	..	..	0.26	0.10
El Salvador	33	61	14	*36*	17	30	0	5	2,002	4,244	0.67	0.40
Eritrea	1	2	1	*1*	1	2	..	..	..	..	0.56	0.33
Estonia	211	378	22	11	154	318	66	1	..	*5,982*	0.60	0.55
Ethiopia	1	*1*	2	3	1	*1*	0	*0*	..	4	0.46	0.27
Finland	441	462	29	31	386	403	12	35	39,750	46,010	1.06	0.84
France	494	564	32	37	405	469	*55*	..	422,000	516,300	0.99	0.82
Gabon	31	*36*	4	5	19	22	..	..	..	..	0.53	0.37
Gambia, The	13	*15*	5	7	7	8	..	..	..	..	0.64	0.47
Georgia	107	58	27	16	89	45	5	1	4,620	..	*0.46*	*0.25*
Germany	405	*529*	53	*66*	386	*508*	18	36	446,000	583,100	0.91	0.78
Ghana	..	8	..	4	..	5	..	..	..	..	0.20	0.19
Greece	248	*348*	22	28	171	254	120	203	..	..	0.72	0.71
Guatemala	..	57	..	45	..	52	..	12	..	4,547	0.53	0.42
Guinea	4	5	1	1	2	2	..	..	..	..	0.85	0.69
Guinea-Bissau	7	11	2	3	4	6	..	..	..	..	..	..
Haiti	..	7	..	13	..	4	..	..	..	..	0.64	0.35
Honduras	22	61	9	28	..	52	..	14	3,288	..	0.62	0.46

	Motor vehicles				Passenger cars		Two-wheelers		Road traffic		Fuel prices	
	per 1,000 people		per kilometer of road		per 1,000 people		per 1,000 people		million vehicle kilometers		Super $ per liter	Diesel $ per liter
	1990	1999	1990	1999	1990	1999	1990	1999	1990	1999	2000	2000
Hungary	212	272	21	15	188	238	16	14	22,898	..	0.81	0.79
India	4	8	2	3	2	5	15	27	..	..	0.60	0.39
Indonesia	16	25	10	14	7	14	34	62	..	..	0.17	0.06
Iran, Islamic Rep.	34	41	14	15	25	30	36	43	..	..	0.05	0.02
Iraq	14	51	6	23	1	36	..	..	..	..	0.03	0.01
Ireland	270	305	10	12	227	272	6	6	24,205	28,390	0.72	0.72
Israel	210	270	74	102	174	220	8	12	18,212	34,344	1.14	0.64
Italy	529	591	99	52	476	539	45	66	344,726	..	0.97	0.83
Jamaica	..	50	..	7	..	41	..	..	..	..	0.62	0.49
Japan	469	560	52	61	283	395	146	115	628,581	746,054	1.06	0.76
Jordan	60	68	26	44	..	49	0	0	1,098	2,154	0.45	0.15
Kazakhstan	76	86	8	12	50	66	..	10	18,248	3,215	0.36	0.29
Kenya	12	13	5	6	10	10	1	1	5,170	6,200	0.71	0.60
Korea, Dem. Rep.	..	..	..	..	..	..	..	..	..	..	0.73	0.41
Korea, Rep.	79	238	60	120	48	167	32	59	30,464	67,266	0.92	0.66
Kuwait	..	408	..	156	..	317	..	..	..	..	0.21	0.18
Kyrgyz Republic	44	39	10	9	44	39	..	1	5,220	..	0.44	0.33
Lao PDR	9	4	3	1	6	3	18	49	..	..	0.41	0.32
Latvia	135	258	6	9	106	216	76	8	3,932	..	0.67	0.58
Lebanon	321	336	183	205	300	313	13	15	..	..	0.53	0.31
Lesotho	10	19	4	8	3	6	..	..	..	..	0.50	0.47
Libya	..	230	..	48	..	159	..	0	..	..	0.25	0.16
Lithuania	159	322	12	16	132	294	52	5	..	..	0.66	0.55
Macedonia, FYR	132	153	30	35	121	139	1	1	3,102	4,247	0.76	0.56
Madagascar	6	8	2	2	4	4	..	..	41,500	..	0.76	0.45
Malawi	4	6	4	4	2	3	..	..	..	..	0.69	0.68
Malaysia	124	200	26	69	101	170	167	224	..	..	0.28	0.16
Mali	3	5	2	3	2	3	..	..	..	..	0.70	0.43
Mauritania	9	12	3	4	6	8	..	..	..	..	0.67	0.40
Mauritius	59	98	35	60	44	73	54	96	..	..	..	..
Mexico	119	151	41	44	82	102	3	3	55,095	..	0.61	0.45
Moldova	53	70	17	24	48	54	45	25	..	538	0.45	0.40
Mongolia	21	30	1	1	6	17	22	11	340	38	0.38	0.38
Morocco	37	52	15	26	28	41	1	1	..	..	0.82	0.53
Mozambique	4	1	2	0	3	0	..	..	1,889	..	0.56	0.54
Myanmar	..	2	..	2	..	1	..	..	..	..	..	..
Namibia	72	85	1	2	40	47	1	1	1,896	2,317	0.47	0.44
Nepal	..	..	..	..	..	..	..	..	..	..	0.63	0.37
Netherlands	405	427	58	58	368	383	44	25	90,150	109,955	1.03	0.78
New Zealand	524	540	19	22	436	481	24	12	..	..	0.48	0.34
Nicaragua	19	10	5	8	10	3	3	2	108	523	0.62	0.54
Niger	6	6	4	5	5	4	..	..	178	240	0.68	0.48
Nigeria	30	24	21	14	12	8	5	4	..	..	0.27	0.27
Norway	458	505	22	25	380	407	48	54	..	30,152	1.19	1.15
Oman	130	142	9	9	83	97	3	2	..	..	0.31	0.29
Pakistan	6	8	4	4	4	5	8	15	18,933	218,779	0.53	0.27
Panama	75	113	18	27	60	83	2	3	..	..	0.53	0.41
Papua New Guinea	..	26	..	6	..	7	..	..	..	..	0.53	0.34
Paraguay	..	24	..	4	..	14	..	..	..	..	0.72	0.34
Peru	..	43	..	13	..	27	..	..	..	..	0.80	0.54
Philippines	10	31	4	11	7	10	6	14	6,189	9,548	0.37	0.28
Poland	168	286	18	29	138	240	36	37	59,608	174,000	0.76	0.65
Portugal	222	348	34	..	162	310	5	77	28,623	93,020	0.77	0.54
Puerto Rico	..	282	..	74	..	232	..	..	..	..	0.34	0.32
Romania	72	154	11	17	56	133	13	14	23,907	36,884	0.46	0.35
Russian Federation	87	153	14	39	65	120	..	..	..	63,450	0.33	0.29

	Motor vehicles				Passenger cars		Two-wheelers		Road traffic		Fuel prices	
	per 1,000 people		per kilometer of road		per 1,000 people		per 1,000 people		million vehicle kilometers		Super $ per liter	Diesel $ per liter
	1990	1999	1990	1999	1990	1999	1990	1999	1990	1999	2000	2000
Rwanda	2	4	1	2	1	2	..	..	..	..	0.89	0.84
Saudi Arabia	165	157	19	20	98	93	0	0	..	..	0.24	0.10
Senegal	11	14	6	8	8	10	0	0	..	..	0.73	0.52
Sierra Leone	10	6	4	2	7	4	2	2	996	529	0.00	0.00
Singapore	146	164	142	170	101	119	45	41	..	..	0.84	0.38
Slovak Republic	194	260	57	33	163	229	61	8	..	10,387	0.69	0.68
Slovenia	306	455	42	45	289	418	8	5	5,620	9,042	0.63	0.66
South Africa	139	143	26	..	97	94	8	4	..	..	0.50	0.50
Spain	360	472	43	53	309	389	79	34	100,981	201,896	0.73	0.65
Sri Lanka	20	34	4	56	6	15	23	40	3,468	15,630	0.66	0.27
Sudan	9	12	21	28	8	10	..	..	..	..	0.28	0.24
Sweden	464	478	29	20	426	437	11	29	61,040	66,806	0.94	0.80
Switzerland	491	527	46	53	449	486	114	104	48,660	54,112	0.78	0.84
Syrian Arab Republic	26	30	10	11	10	9	..	..	..	..	0.44	0.13
Tajikistan	3	1	1	1	0	0	..	..	..	..	0.45	0.55
Tanzania	5	5	2	2	1	1	..	..	..	..	0.75	0.73
Thailand	46	106	36	97	14	28	86	174	45,769	99,900	0.39	0.35
Togo	23	27	11	15	15	19	8	14	..	..	0.48	0.40
Trinidad and Tobago	..	115	..	18	..	96	..	..	..	..	0.39	0.20
Tunisia	48	64	19	25	23	30	..	..	..	..	0.49	0.29
Turkey	50	85	8	14	34	63	10	15	27,041	49,846	0.88	0.66
Turkmenistan	..	..	..	..	..	..	..	..	..	..	0.02	0.02
Uganda	2	6	..	..	1	2	0	3	..	..	0.86	0.75
Ukraine	63	93	19	27	63	104	..	49	59,500	61,200	0.37	0.30
United Arab Emirates	121	103	52	52	97	82	..	..	..	..	0.25	0.26
United Kingdom	400	418	64	67	341	373	14	12	399,000	404,500	1.17	1.22
United States	758	760	30	32	573	478	17	14	2,527,441	2,536,555	0.47	0.48
Uruguay	138	174	45	63	122	158	74	110	..	..	1.19	0.53
Uzbekistan	..	..	..	..	..	..	..	..	..	..	0.43	0.28
Venezuela, RB	..	88	..	21	..	68	..	..	..	563	0.12	0.08
Vietnam	..	..	..	..	..	..	45	45	..	..	0.38	0.27
West Bank and Gaza	..	..	..	..	..	..	..	..	..	..	0.01	0.00
Yemen, Rep.	34	34	8	8	14	15	..	..	8,681	11,476	0.21	0.06
Yugoslavia, FR (Serb./Mont.)	137	190	31	42	133	176	3	4	..	..	0.56	0.56
Zambia	14	26	3	4	8	17	..	..	..	..	1.00	1.00
Zimbabwe	..	32	..	19	..	29	..	32	..	..	0.85	0.72
World	**118 w**	**122 w**			**91 w**	**90 w**					**0.54 m**	**0.35 m**
Low income	9	10	..	..	4	5	..	..	..	..	0.49	0.31
Middle income	41	59	..	..	24	37	..	..	..	..	0.48	0.32
Lower middle income	17	35	..	..	11	22	..	..	..	..	0.49	0.30
Upper middle income	123	147	..	..	99	119	..	..	..	..	0.47	0.35
Low & middle income	26	38	..	..	16	24	..	..	..	..	0.49	0.31
East Asia & Pacific	11	22	..	..	4	7	..	..	..	..	0.31	0.25
Europe & Central Asia	97	154	..	..	82	128	..	..	..	..	0.51	0.35
Latin America & Carib.	100	91	..	..	..	68	..	..	..	..	0.47	0.30
Middle East & N. Africa	48	59	..	..	32	40	..	..	..	..	0.31	0.17
South Asia	4	7	..	..	2	4	..	..	..	..	0.57	0.25
Sub-Saharan Africa	21	23	..	..	14	14	..	..	..	..	0.59	0.43
High income	527	582	..	..	396	415	..	..	..	..	0.86	0.69
Europe EMU	435	513	..	..	387	409	..	..	..	..	1.04	0.79

Traffic and congestion | 3.12

About the data

Traffic congestion in urban areas constrains economic productivity, damages people's health, and degrades the quality of their lives. The particulate air pollution emitted by motor vehicles—the dust and soot in exhaust—is proving to be far more damaging to human health than was once believed. (For information on suspended particulates and other air pollutants see table 3.13.)

In recent years ownership of passenger cars has increased, and the expansion of economic activity has led to the transport by road of more goods and services over greater distances (see table 5.8). These developments have increased demand for roads and vehicles, adding to urban congestion, air pollution, health hazards, traffic accidents, and injuries.

Congestion, the most visible cost of expanding vehicle ownership, is reflected in the indicators in the table. Other relevant indicators—such as average vehicle speed in major cities or the cost of traffic congestion, which takes a heavy toll on economic productivity—are not included here because data are incomplete or difficult to compare.

The data in the table—except for those on fuel prices—are compiled by the International Road Federation (IRF) through questionnaires sent to national organizations. The IRF uses a hierarchy of sources to gather as much information as possible. The primary sources are national road associations. Where such an association is lacking or does not respond, other agencies are contacted, including road directorates, ministries of transport or public works, and central statistical offices. As a result, the compiled data are of uneven quality. The coverage of each indicator may differ across countries because of differences in definitions. Comparability also is limited when time-series data are reported. Moreover, the data do not capture the quality or age of vehicles or the condition or width of roads. Thus comparisons over time and between countries should be made with caution.

The data on fuel prices are compiled by the German Agency for Technical Cooperation (GTZ) from its global network of regional offices and representatives, as well as other sources, including the Allgemeiner Deutscher Automobil Club (for Europe) and a project of the Latin American Energy Organization (OLADE, for Latin America). Local prices have been converted to U.S. dollars using the exchange rate on the survey date as listed in the international monetary table of the *Financial Times*. For countries with multiple exchange rates, the market, parallel, or black market rate was used rather than the official exchange rate.

Definitions

- **Motor vehicles** include cars, buses, and freight vehicles but not two-wheelers. Population figures refer to the midyear population in the year for which data are available. Roads refer to motorways, highways, main or national roads, and secondary or regional roads. A motorway is a road specially designed and built for motor traffic that separates the traffic flowing in opposite directions. • **Passenger cars** refer to road motor vehicles, other than two-wheelers, intended for the carriage of passengers and designed to seat no more than nine people (including the driver). • **Two-wheelers** refer to mopeds and motorcycles. • **Road traffic** is the number of vehicles multiplied by the average distances they travel. • **Fuel prices** refer to the pump prices of the most widely sold grade of gasoline and of diesel fuel. Prices have been converted from the local currency to U.S. dollars (see *About the data*).

Data sources

The data on vehicles and traffic are from the IRF's electronic files and its annual *World Road Statistics*. The data on fuel prices are from the GTZ's *Fuel Prices and Taxation* (1999) and the electronic update for 2000.

Table 3.12a

The top 10 vehicle-owning economies in 1999

	Vehicles per 1,000 people
United States	760
Australia	601
Italy	591
Canada	581
France	564
Japan	560
New Zealand	540
Austria	536
Germany	529
Switzerland	527
World	122
Low income	10
Middle income	59
High income	582
Europe EMU	513

Source: Table 3.12.

Table 3.12b

The 10 economies with the highest fuel prices in 2000—and the 10 with the lowest

$ per liter of super grade gasoline

	Fuel price		Fuel price		Fuel price
Hong Kong, China	1.46	Turkmenistan	0.02	World (median)	0.54
Norway	1.19	Iraq	0.03	Low income	0.49
Uruguay	1.19	Iran, Islamic Rep.	0.05	Middle income	0.48
United Kingdom	1.17	Venezuela, RB	0.12	High income	0.86
Israel	1.14	Indonesia	0.17	Europe EMU	1.04
Argentina	1.07	Ghana	0.20		
Finland	1.06	Kuwait	0.21		
Japan	1.06	Yemen, Rep.	0.21		
Netherlands	1.03	Saudi Arabia	0.24		
Burundi	1.01	United Arab Emirates	0.25		

Source: Table 3.12.

3.13 Air pollution

	City	City population	Total suspended particulates	Sulfur dioxide	Nitrogen dioxide
		thousands **2000**	micrograms per cubic meter **1995[a]**	micrograms per cubic meter **1998[b]**	micrograms per cubic meter **1998[b]**
Argentina	Córdoba	1,423	97	..	97
Australia	Melbourne	3,187	35	0	30
	Perth	1,313	45	5	19
	Sydney	3,664	54	28	81
Austria	Vienna	2,070	47	14	42
Belgium	Brussels	1,122	78	20	48
Brazil	Rio de Janeiro	10,582	139	129	..
	São Paulo	17,755	86	43	83
Bulgaria	Sofia	1,192	195	39	122
Canada	Montreal	3,448	34	10	42
	Toronto	4,651	36	17	43
	Vancouver	2,033	29	14	37
Chile	Santiago	5,538	..	29	81
China	Anshan	1,453	305	115	88
	Beijing	10,839	377	90	122
	Changchun	3,093	381	21	64
	Chengdu	3,294	366	77	74
	Chongqing	5,312	320	340	70
	Dalian	2,628	185	61	100
	Guangzhou	3,893	295	57	136
	Guiyang	2,533	330	424	53
	Harbin	2,928	359	23	30
	Jinan	2,568	472	132	45
	Kunming	1,701	253	19	33
	Lanzhou	1,730	732	102	104
	Liupanshui	2,023	408	102	..
	Nanchang	1,722	279	69	29
	Pinxiang	1,502	276	75	..
	Quingdao	2,316	..	190	64
	Shanghai	12,887	246	53	73
	Shenyang	4,828	374	99	73
	Taiyuan	2,415	568	211	55
	Tianjin	9,156	306	82	50
	Urumqi	1,643	515	60	70
	Wuhan	5,169	211	40	43
	Zhengzhou	2,070	474	63	95
	Zibo	2,675	453	198	43
Colombia	Bogotá	6,288	120	..	..
Croatia	Zagreb	1,060	71	31	..
Cuba	Havana	2,256	..	1	5
Czech Republic	Prague	1,226	59	14	33
Denmark	Copenhagen	1,388	61	7	54
Ecuador	Guayaquil	2,293	127	15	..
	Quito	1,754	175	22	..
Egypt, Arab Rep.	Cairo	10,552	..	69	..
Finland	Helsinki	1,167	40	4	35
France	Paris	9,624	14	14	57
Germany	Berlin	3,324	50	18	26
	Frankfurt	3,687	36	11	45
	Munich	2,294	45	8	53
Ghana	Accra	1,976	137	..	..
Greece	Athens	3,116	178	34	64
Hungary	Budapest	1,825	63	39	51
Iceland	Reykjavik	168	24	5	42
India	Ahmedabad	4,160	299	30	21
	Bangalore	5,561	123	..	..

About the data

In many towns and cities exposure to air pollution is the main environmental threat to human health. Winter smog—made up of soot, dust, and sulfur dioxide—has long been associated with temporary spikes in the number of deaths. Long-term exposure to high levels of soot and small particles in the air also contributes to a wide range of chronic respiratory diseases and exacerbates heart disease and other conditions. Particulate pollution, on its own or in combination with sulfur dioxide, leads to an enormous burden of ill health, causing at least 500,000 premature deaths and 4–5 million new cases of chronic bronchitis each year (World Bank 1992).

Emissions of sulfur dioxide and nitrogen oxides lead to the deposition of acid rain and other acidic compounds over long distances—often more than 1,000 kilometers from their source. Acid deposition changes the chemical balance of soils and can lead to the leaching of trace minerals and nutrients critical to trees and plants. The links between forest damage and acid deposition are complex. Direct exposure to high levels of sulfur dioxide or acid deposition can cause defoliation and dieback.

Where coal is the primary fuel for power plants, steel mills, industrial boilers, and domestic heating, the result is usually high levels of urban air pollution—especially particulates and sometimes sulfur dioxide—and, if the sulfur content of the coal is high, widespread acid deposition. Where coal is not an important primary fuel or is used by plants with effective dust control, the worst emissions of air pollutants stem from the combustion of petroleum products.

The data on air pollution are based on reports from urban monitoring sites. Annual means (measured in micrograms per cubic meter) are average concentrations observed at these sites. Coverage is not comprehensive because not all cities have monitoring systems. For example, data are reported for just 5 cities in Africa but for more than 87 cities in China. Pollutant concentrations are sensitive to local conditions, and even in the same city different monitoring sites may register different concentrations. Thus these data should be considered only a general indication of air quality in each city, and cross-country comparisons should be made with caution. World Health Organization (WHO) annual mean guidelines for air quality standards are 90 micrograms per cubic meter for total suspended particulates, and 50 for sulfur dioxide and nitrogen dioxide.

Air pollution | 3.13

	City	City population	Total suspended particulates	Sulfur dioxide	Nitrogen dioxide
		thousands **2000**	micrograms per cubic meter **1995ᵃ**	micrograms per cubic meter **1998ᵇ**	micrograms per cubic meter **1998ᵇ**
	Calcutta	12,918	375	49	34
	Chennai	6,002	130	15	17
	Delhi	11,695	415	24	41
	Hyderabad	6,842	152	12	17
	Kanpur	2,450	459	15	14
	Lucknow	2,568	463	26	25
	Mumbai	18,066	240	33	39
	Nagpur	2,062	185	6	13
	Pune	3,489	208	..	..
Indonesia	Jakarta	11,018	271	..	..
Iran, Islamic Rep.	Tehran	7,225	248	209	..
Ireland	Dublin	985	..	20	..
Italy	Milan	4,251	77	31	248
	Rome	2,688	73	..	..
	Torino	1,294	151	..	..
Japan	Osaka	11,013	43	19	63
	Tokyo	26,444	49	18	68
	Yokohama	3,178	..	100	13
Kenya	Nairobi	2,310	69	..	..
Korea, Rep.	Pusan	3,830	94	60	51
	Seoul	9,888	84	44	60
	Taegu	2,675	72	81	62
Malaysia	Kuala Lumpur	1,378	85	24	..
Mexico	Mexico City	18,131	279	74	130
Netherlands	Amsterdam	1,144	40	10	58
New Zealand	Auckland	1,102	26	3	20
Norway	Oslo	970	15	8	43
Philippines	Manila	10,870	200	33	..
Poland	Lodz	1,055	..	21	43
	Warsaw	2,269	..	16	32
Portugal	Lisbon	3,826	61	8	52
Romania	Bucharest	2,054	82	10	71
Russian Federation	Moscow	9,321	100	109	..
	Omsk	1,216	100	20	34
Singapore	Singapore	3,567	..	20	30
Slovak Republic	Bratislava	460	62	21	27
South Africa	Cape Town	2,993	..	21	72
	Durban	1,335	..	31	..
	Johannesburg	2,335	..	19	31
Spain	Barcelona	2,819	117	11	43
	Madrid	4,072	42	24	66
Sweden	Stockholm	1,583	9	3	20
Switzerland	Zurich	983	31	11	39
Thailand	Bangkok	7,281	223	11	23
Turkey	Ankara	3,203	57	55	46
	Istanbul	9,451	..	120	..
Ukraine	Kiev	2,670	100	14	51
United Kingdom	Birmingham	2,272	..	9	45
	London	7,640	..	25	77
	Manchester	2,252	..	26	49
United States	Chicago	6,951	..	14	57
	Los Angeles	13,140	..	9	74
	New York	16,640	..	26	79
Venezuela, RB	Caracas	3,151	53	33	57

a. Data are for the most recent year available in 1990–95. Most are for 1995. b. Data are for the most recent year available in 1990–98. Most are for 1995.

Definitions

• **City population** is the number of residents of the city as defined by national authorities and reported to the United Nations. • **Total suspended particulates** refer to smoke, soot, dust, and liquid droplets from combustion that are in the air. Particulate levels indicate the quality of the air people are breathing and the state of a country's technology and pollution controls. • **Sulfur dioxide** (SO_2) is an air pollutant produced when fossil fuels containing sulfur are burned. It contributes to acid rain and can damage human health, particularly that of the young and the elderly. • **Nitrogen dioxide** (NO_2) is a poisonous, pungent gas formed when nitric oxide combines with hydrocarbons and sunlight, producing a photochemical reaction. These conditions occur in both natural and anthropogenic activities. NO_2 is emitted by bacteria, nitrogenous fertilizers, aerobic decomposition of organic matter in oceans and soils, combustion of fuels and biomass, and motor vehicles and industrial activities.

Data sources

The data in the table are from the WHO's Healthy Cities Air Management Information System and the World Resources Institute, which relies on various national sources as well as, among others, the United Nations Environment Programme and WHO's *Urban Air Pollution in Megacities of the World* (1992), the Organisation for Economic Co-operation and Development's *OECD Environmental Data: Compendium 1999*, the U.S. Environmental Protection Agency's *National Air Quality and Emissions Trends Report 1995* and AIRS Executive International database, the *China Environmental Yearbook 1997*, and the United Nations Centre for Human Settlements' (UNCHS) Urban Indicators database.

3.14 | Government commitment

	Environmental strategy or action plan	Country environmental profile	Biodiversity assessment, strategy, or action plan	Participation in treaties[a]				
				Climate change	Ozone layer	CFC control	Law of the Sea[b]	Biological diversity
Albania	1993	..	..	1995	2000	2000	..	1994
Algeria	..	..	..	1994	1993	1993	1996	1995
Angola	..	..	..	2000 c	2000	2000	1994	1998
Argentina	1992	..	..	1994	1990	1990	1996	1995
Armenia	..	..	..	1994	2000	2000	..	1993
Australia	1992	..	1994	1994	1987	1989	1995	1993
Austria	..	..	..	1994	1987	1989	1995	1994
Azerbaijan	..	..	..	1995	1996	1996	..	2000 c
Bangladesh	1991	1989	1990	1994	1990	1990	..	1994
Belarus	..	..	..	2000 c	1986	1989	..	1993
Belgium	..	..	..	1996	1989	1989	..	1997
Benin	1993	..	..	1994	1993	1993	..	1994
Bolivia	1994	1986	1988	1995	1995	1995	1995	1995
Bosnia and Herzegovina	..	..	..	..	1992	1992	1994	..
Botswana	1990	1986	1991	1994	1992	1992	1994	1996
Brazil	..	..	1988	1994	1990	1990	1994	1994
Bulgaria	..	..	1994	1995	1991	1991	1996	1996
Burkina Faso	1993	1994	..	1994	1989	1989	..	1993
Burundi	1994	1981	1989	1997	1997	1997	..	1997
Cambodia	1999	..	..	1996	..	..	..	1995
Cameroon	..	1989	1989	1995	1989	1989	1994	1995
Canada	1990	..	1994	1994	1986	1988	..	1993
Central African Republic	..	..	..	1995	1993	1993	..	1995
Chad	1990	1982	..	1994	1989	1994	..	1994
Chile	..	1987	1993	1995	1990	1990	..	1994
China	1994	..	1994	1994	1989	1991	1996	1993
Hong Kong, China	..	..	..	..	..	..	..	..
Colombia	..	1990	1988	1995	1990	1994	..	1995
Congo, Dem. Rep.	..	1986	1990	1995	1995	1995	1994	1995
Congo, Rep.	..	..	1990	1997	1995	1995	..	1996
Costa Rica	1990	1987	1992	1994	1991	1991	1994	1994
Côte d'Ivoire	1994	..	1991	1995	1993	1993	1994	1995
Croatia	..	..	..	1996	1992	1992	1994	1997
Cuba	..	..	..	1994	1992	1992	1994	1994
Czech Republic	1994	..	..	1994	1993	1993	1996	1994
Denmark	1994	..	..	1994	1988	1989	..	1994
Dominican Republic	..	1984	1995	1999	1993	1993	..	1996
Ecuador	1993	1987	1995	1994	1990	1990	..	1993
Egypt, Arab Rep.	1992	1992	1988	1995	1988	1988	1994	1994
El Salvador	1994	1985	1988	1996	1993	1993	..	1994
Eritrea	1995	..	..	1995	..	..	..	1996
Estonia	1998	..	..	1994	1997	1997	..	1994
Ethiopia	1994	..	1991	1994	1995	1995	..	1994
Finland	1995	..	..	1994	1986	1989	1996	1994
France	1990	..	..	1994	1988	1989	1996	1994
Gabon	..	..	1990	1998	1994	1994	..	1997
Gambia, The	1992	1981	1989	1994	1990	1990	1994	1994
Georgia	1998	..	..	1994	1996	1996	1996	1994
Germany	..	..	..	1994	1988	1989	1994	1994
Ghana	1992	1985	1988	1995	1989	1989	1994	1994
Greece	..	..	..	1994	1989	1989	1995	1994
Guatemala	1994	1984	1988	1996	1987	1990	..	1995
Guinea	1994	1983	1988	1994	1992	1992	1994	1993
Guinea-Bissau	1993	..	1991	1996	..	..	1994	1996
Haiti	..	1985	..	1996	2000	2000	1996	1996
Honduras	1993	1989	..	1996	1994	1994	1994	1995

Table 3.14a

Status of national environmental action plans

Completed

Albania	Guinea	Niger
Armenia	Guinea-Bissau	Nigeria
Azerbaijan	Guyana	Pakistan
Bangladesh	Haiti	Papua New Guinea
Belarus	Honduras	Philippines
Benin	Hungary	Poland
Bhutan	India	Romania
Bolivia	Indonesia	Russian Federation
Botswana	Iran, Islamic Rep.	Rwanda
Bulgaria	Kazakhstan	São Tomé and Principe
Burkina Faso	Kenya	Senegal
Burundi	Kiribati	Seychelles
Cambodia	Kyrgyz Rep.	Sierra Leone
Cameroon	Lao PDR	Slovak Rep.
Cape Verde	Latvia	Slovenia
China	Lebanon	Solomon Islands
Comoros	Lesotho	South Africa
Congo, Dem. Rep.	Lithuania	Sri Lanka
Congo, Rep.	Macedonia, FYR	St. Kitts and Nevis
Costa Rica	Madagascar	Swaziland
Côte d'Ivoire	Malawi	Syrian Arab Rep.
Czech Republic	Maldives	Tanzania
Egypt, Arab Rep.	Mali	Togo
El Salvador	Mauritania	Tonga
Equatorial Guinea	Mauritius	Tunisia
Eritrea	Mexico	Turkey
Estonia	Moldova	Uganda
Ethiopia	Mongolia	Ukraine
Gabon	Montserrat	Uzbekistan
Gambia, The	Mozambique	Vanuatu
Georgia	Namibia	Vietnam
Ghana	Nepal	Yemen, Rep.
Grenada	Nicaragua	Zambia

Being prepared

Central African Rep.	Ecuador	Tajikistan
Croatia	Korea, Rep.	Turkmenistan
Dominican Rep.	Malaysia	Zimbabwe
	Paraguay	

Note: Status is as of December 2000.

Source: World Bank regional data; and World Resources Institute, International Institute for Environment and Development, and IUCN, *1996 World Directory of Country Environmental Studies*.

	Environmental strategy or action plan	Country environmental profile	Biodiversity assessment, strategy, or action plan	Participation in treaties[a]				
				Climate change	Ozone layer	CFC control	Law of the Sea[b]	Biological diversity
Hungary	1995	..	..	1994	1988	1989	..	1994
India	1993	1989	1994	1994	1991	1992	1995	1994
Indonesia	1992	1994	1993	1994	1992	1992	1994	1994
Iran, Islamic Rep.	..	..	..	1996	1991	1991	..	1996
Iraq	..	..	..	..	..	..	1994	..
Ireland	..	..	..	1994	1988	1989	..	1996
Israel	..	..	..	1994	1992	1992	..	1995
Italy	..	..	..	1994	1988	1989	1995	1994
Jamaica	1994	1987	..	1995	1993	1993	1994	1995
Japan	..	..	..	1994	1988	1988	1996	1993
Jordan	1991	1979	..	1994	1989	1989	1995	1994
Kazakhstan	..	..	..	1995	1998	1998	..	1994
Kenya	1994	1989	1992	1994	1989	1989	1994	1994
Korea, Dem. Rep.	..	..	..	1995	1995	1995	..	1995
Korea, Rep.	..	..	..	1994	1992	1992	1996	1995
Kuwait	..	..	..	1995	1993	1993	1994	..
Kyrgyz Republic	1995	..	..	2000 c	2000	2000	..	1996
Lao PDR	1995	..	..	1995	1998	1998	..	1996
Latvia	..	..	..	1995	1995	1995	..	1996
Lebanon	..	..	..	1995	1993	1993	1995	1995
Lesotho	1989	1982	..	1995	1994	1994	..	1995
Libya	..	..	..	1999 c	1990	1990	..	..
Lithuania	..	..	..	1995	1995	1995	..	1996
Macedonia, FYR	1997	..	..	1998	1994	1994	1994	1997 c
Madagascar	1988	..	1991	1996	1997	1997	..	1996
Malawi	1994	1982	..	1994	1991	1991	.	1994
Malaysia	1991	1979	1988	1994	1989	1989	1997	1994
Mali	..	1991	1989	1995	1995	1995	1994	1995
Mauritania	1988	1984	..	1994	1994	1994	1996	1996
Mauritius	1990	..	..	1994	1992	1992	1994	1993
Mexico	..	..	1988	1994	1987	1988	1994	1993
Moldova	..	..	..	1995	1997	1997	..	1996
Mongolia	1995	..	..	1994	1996	1996	..	1993
Morocco	..	1980	1988	1996	1996	1996	..	1995
Mozambique	1994	..	..	1995	1994	1994	..	1995
Myanmar	..	1982	1989	1995	1994	1994	1996	1995
Namibia	1992	..	..	1995	1993	1993	1994	1997
Nepal	1993	1983	..	1994	1994	1994	..	1994
Netherlands	1994	..	..	1994	1988	1989	1996	1994
New Zealand	1994	..	..	1994	1987	1988	1996	1993
Nicaragua	1994	1981	..	1996	1993	1993	..	1996
Niger	..	1985	1991	1995	1993	1993	..	1995
Nigeria	1990	..	1992	1994	1989	1989	1994	1994
Norway	..	..	1994	1994	1986	1988	1996	1993
Oman	..	1981	..	1995	1999	1999	1994	1995
Pakistan	1994	1994	1991	1994	1993	1993	..	1994
Panama	1990	1980	..	1995	1989	1989	1996	1995
Papua New Guinea	1992	1994	1993	1994	1993	1993	..	1993
Paraguay	..	1985	..	1994	1993	1993	1994	1994
Peru	..	1988	1988	1994	1989	1993	..	1993
Philippines	1989	1992	1989	1994	1991	1991	1994	1994
Poland	1993	..	1991	1994	1990	1990	..	1996
Portugal	1995	..	..	1994	1989	1989	..	1994
Puerto Rico	..	..	..	..	..	..	..	..
Romania	..	..	..	1994	1993	1993	1997	1994
Russian Federation	..	..	1994	1995	1986	1989	..	1995

Table 3.14b

States that have signed the Convention on Climate Change

Antigua and Barbuda[a]	Greece	Palau[a]
Argentina	Guatemala[a]	Panama[a]
Australia	Guinea[a]	Papua New Guinea
Austria	Honduras	Paraguay[a]
Azerbaijan[a]	Indonesia	Peru
Bahamas, The[a]	Ireland	Philippines
Barbados[a]	Israel	Poland
Belgium	Italy	Portugal
Bolivia[a]	Jamaica[a]	Romania
Brazil	Japan	Russian Federation
Bulgaria	Kazakhstan	Samoa
Canada	Kiribati[a]	Seychelles
Chile	Korea, Rep.	Slovak Republic
China	Latvia	Slovenia
Cook Islands	Liechtenstein	Solomon Islands
Costa Rica	Lithuania	Spain
Croatia	Luxembourg	St. Lucia
Cuba	Malaysia	St. Vincent and the
Cyprus[a]	Maldives[a]	Grenadines
Czech Republic	Mali	Sweden
Denmark	Malta	Switzerland
Ecuador	Marshall	Thailand
Egypt, Arab Rep.	Islands	Trinidad and Tobago[a]
El Salvador[a]	Mexico	Turkmenistan[a]
Equatorial	Micronesia[a]	Tuvalu[a]
Guinea[a]	Monaco	Ukraine
Estonia	Mongolia[a]	United Kingdom
Fiji[a]	Netherlands	United States
Finland	New Zealand	Uruguay
France	Nicaragua[a]	Uzbekistan[a]
Georgia[a]	Niger	Vietnam
Germany	Niue[a]	Zambia
	Norway	

Note: Status is as of November 2000.

a. Ratification or accession signed.

Source: Secretariat of the United Nations Framework Convention on Climate Change.

3.14 | Government commitment

	Environmental strategy or action plan	Country environmental profile	Biodiversity assessment, strategy, or action plan	Participation in treaties[a]				
				Climate change	Ozone layer	CFC control	Law of the Sea[b]	Biological diversity
Rwanda	1991	1987	..	1998	..	..	..	1996
Saudi Arabia	..	..	..	1995	1993	1993	..	..
Senegal	1984	1990	1991	1995	1993	1993	1994	1995
Sierra Leone	1994	..	..	1995	..	..	1995	1995
Singapore	1993	1988	1995	1997	1989	1989	1994	1996
Slovak Republic	..	..	..	1994	1993	1993	1996	1994
Slovenia	..	..	..	1996	1992	1992	1994	1996
South Africa	1993	..	..	1997	1990	1990	1994	1996
Spain	..	..	..	1994	1988	1989	..	1994
Sri Lanka	1994	1983	1991	1994	1990	1990	1994	1994
Sudan	..	1989	..	1994	1993	1993	1994	1996
Sweden	..	..	..	1994	1987	1988	1996	1994
Switzerland	..	..	..	1994	1988	1989	..	1995
Syrian Arab Republic	..	1981	..	1996	1990	1990	..	1996
Tajikistan	..	..	..	1998	1996	1998	..	1997
Tanzania	1994	1989	1988	1996	1993	1993	1994	1996
Thailand	..	1992	..	1995	1989	1989	..	..
Togo	1991	..	..	1995	1991	1991	1994	1996
Trinidad and Tobago	..	..	..	1994	1989	1989	1994	1996
Tunisia	1994	1980	1988	1994	1989	1989	1994	1993
Turkey	1998	1982	..	..	1991	1991	..	1997
Turkmenistan	..	..	..	1995	1994	1994	..	1996
Uganda	1994	1982	1988	1994	1988	1988	1994	1993
Ukraine	..	..	..	1997	1986	1988	..	1995
United Arab Emirates	..	..	..	1996	1990	1990	..	2000 c
United Kingdom	1995	..	1994	1994	1987	1989	..	1994
United States	1995	..	1995	1994	1986	1988	..	1993
Uruguay	..	..	..	1994	1989	1991	1994	1994
Uzbekistan	1999	..	..	1994	1993	1993	..	1995
Venezuela, RB	..	..	..	1995	1988	1989	..	1994
Vietnam	1996	..	1993	1995	1994	1994	1994	1995
West Bank and Gaza	..	..	..	..	..	..	..	..
Yemen, Rep.	..	1990	1992	1996	1996	1996	1994	1996
Yugoslavia, FR (Serb./Mont.)	..	..	..	1997	1990	1991	..	..
Zambia	1994	1988	..	1994	1990	1990	1994	1993
Zimbabwe	1987	1982	..	1994	1993	1993	1994	1995

a. The years shown refer to the year the treaty entered into force in the country. b. Convention became effective 16 November 1994. c. Ratification of the treaty.

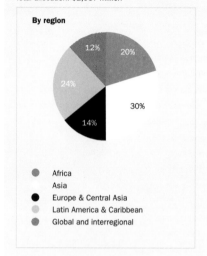

Figure 3.14

A global focus on biodiversity and climate change

Allocation of funds by the Global Environment Facility, February 1995–June 2000
Total allocation: $2,937 million

By region
- Africa
- Asia
- Europe & Central Asia
- Latin America & Caribbean
- Global and interregional

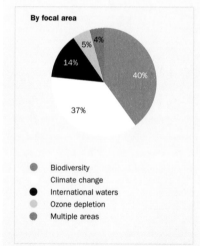

By focal area
- Biodiversity
- Climate change
- International waters
- Ozone depletion
- Multiple areas

Source: Global Environment Facility data.

Government commitment | 3.14

National environmental strategies and participation in international treaties on environmental issues provide some evidence of government commitment to sound environmental management. But the signing of these treaties does not always imply ratification. Nor does it guarantee that governments will comply with treaty obligations.

In many countries efforts to halt environmental degradation have failed, primarily because governments have neglected to make this issue a priority, a reflection of competing claims on scarce resources. To address this problem, many countries are preparing national environmental strategies—some focusing narrowly on environmental issues, others integrating environmental, economic, and social concerns. Among such initiatives are conservation strategies and environmental action plans. Some countries have also prepared country environmental profiles and biological diversity strategies and profiles.

National conservation strategies—promoted by the World Conservation Union (IUCN)—provide a comprehensive, cross-sectoral analysis of conservation and resource management issues to help integrate environmental concerns with the development process. Such strategies discuss current and future needs, institutional capabilities, prevailing technical conditions, and the status of natural resources in a country.

National environmental action plans (NEAPs), supported by the World Bank and other development agencies, describe a country's main environmental concerns, identify the principal causes of environmental problems, and formulate policies and actions to deal with them (table 3.14a). The NEAP is a continuing process in which governments develop comprehensive environmental policies, recommend specific actions, and outline the investment strategies, legislation, and institutional arrangements required to implement them.

Country environmental profiles identify how national economic and other activities can stay within the constraints imposed by the need to conserve natural resources. Some profiles consider issues of equity, justice, and fairness. Biodiversity profiles—prepared by the World Conservation Monitoring Centre and the IUCN—provide basic background on species diversity, protected areas, major ecosystems and habitat types, and legislative and administrative support. In an effort to establish a scientific baseline for measuring progress in biodiversity conservation, the United Nations Environment Programme (UNEP) coordinates global biodiversity assessments.

To address global issues, many governments have also signed international treaties and agreements launched in the wake of the 1972 United Nations Conference on Human Environment in Stockholm and the 1992 United Nations Conference on Environment and Development (the Earth Summit) in Rio de Janeiro:

- The Framework Convention on Climate Change aims to stabilize atmospheric concentrations of greenhouse gases at levels that will prevent human activities from interfering dangerously with the global climate.

- The Vienna Convention for the Protection of the Ozone Layer aims to protect human health and the environment by promoting research on the effects of changes in the ozone layer and on alternative substances (such as substitutes for chlorofluorocarbons) and technologies, monitoring the ozone layer, and taking measures to control the activities that produce adverse effects.

- The Montreal Protocol for CFC Control requires that countries help protect the earth from excessive ultraviolet radiation by cutting chlorofluorocarbon consumption by 20 percent over their 1986 level by 1994 and by 50 percent over their 1986 level by 1999, with allowances for increases in consumption by developing countries.

- The United Nations Convention on the Law of the Sea, which became effective in November 1994, establishes a comprehensive legal regime for seas and oceans, establishes rules for environmental standards and enforcement provisions, and develops international rules and national legislation to prevent and control marine pollution.

- The Convention on Biological Diversity promotes conservation of biodiversity among nations through scientific and technological cooperation, access to financial and genetic resources, and transfer of ecologically sound technologies.

To help developing countries comply with their obligations under these agreements, the Global Environment Facility (GEF) was created to focus on global improvement in biodiversity, climate change, international waters, and ozone layer depletion. The UNEP, United Nations Development Programme (UNDP), and World Bank manage the GEF according to the policies of its governing body of country representatives. The World Bank is responsible for the GEF Trust Fund and is chair of the GEF.

- **Environmental strategies and action plans** provide a comprehensive, cross-sectoral analysis of conservation and resource management issues to help integrate environmental concerns with the development process. They include national conservation strategies, national environmental action plans, national environmental management strategies, and national sustainable development strategies. The year shown for a country refers to the year in which a strategy or action plan was adopted. • **Country environmental profiles** identify how national economic and other activities can stay within the constraints imposed by the need to conserve natural resources. The year shown for a country refers to the year in which a profile was completed. • **Biodiversity assessments, strategies, and action plans** include biodiversity profiles (see *About the data*). • **Participation in treaties** covers five international treaties (see *About the data*). • **Climate change** refers to the Framework Convention on Climate Change (signed in New York in 1992). • **Ozone layer** refers to the Vienna Convention for the Protection of the Ozone Layer (signed in 1985). • **CFC control** refers to the Montreal Protocol for CFC Control (formally, the Protocol on Substances That Deplete the Ozone Layer, signed in 1987). • **Law of the Sea** refers to the United Nations Convention on the Law of the Sea (signed in Montego Bay, Jamaica, in 1982). • **Biological diversity** refers to the Convention on Biological Diversity (signed at the Earth Summit in Rio de Janeiro in 1992). The year shown for a country refers to the year in which a treaty entered into force in that country.

The data are from the Secretariat of the United Nations Framework Convention on Climate Change; the Ozone Secretariat of the UNEP; the World Resources Institute; the UNEP; the U.S. National Aeronautics and Space Administration's Socioeconomic Data and Applications Center (SEDAC), Center for International Earth Science Information Network (CIESIN); the World Resources Institute, International Institute for Environment and Development, and IUCN's *1996 World Directory of Country Environmental Studies;* and the World Bank's *1998 Catalog: Operational Documents as of July 31, 1998.*

3.15 Toward a measure of genuine savings

	Gross domestic savings	Consumption of fixed capital	Net domestic savings	Education expenditure	Energy depletion	Mineral depletion	Net forest depletion	Carbon dioxide damage	Genuine domestic savings
	% of GDP 1999	% of GDP 1999	% of GDP 1999	% of GDP 1999	% of GDP 1999	% of GDP 1999	% of GDP 1999	% of GDP 1999	% of GDP 1999
Albania	−1.7	8.5	−10.1	2.9	0.7	0.0	0.0	0.3	−8.3
Algeria	31.7	8.9	22.8	4.3	19.8	0.1	0.0	1.2	6.0
Angola	32.5	7.9	24.6	1.5	17.5	0.0	0.0	0.4	8.2
Argentina	17.2	10.9	6.2	3.1	0.8	0.1	0.0	0.3	8.2
Armenia	−9.3	7.4	−16.7	1.9	0.0	0.0	0.0	1.1	−15.9
Australia	22.1	14.0	8.1	5.3	0.5	1.2	0.0	0.5	11.2
Austria	23.9	12.4	11.4	4.9	0.0	0.0	0.0	0.2	16.1
Azerbaijan	22.8	11.1	11.7	2.6	33.3	0.0	0.0	5.4	−24.4
Bangladesh	16.7	6.3	10.4	1.8	0.2	0.0	2.0	0.3	9.6
Belarus	21.1	9.6	11.5	5.5	0.0	0.0	0.0	1.5	15.5
Belgium	25.2	9.9	15.3	3.1	0.0	0.0	0.0	0.3	18.1
Benin	6.4	7.2	−0.8	2.7	0.0	0.0	0.3	0.3	1.3
Bolivia	9.2	8.4	0.8	5.4	1.3	0.7	0.0	0.8	3.4
Bosnia and Herzegovina	0.1	8.5	−8.4	..	0.0	0.0	..	0.9	..
Botswana	14.2	14.9	−0.7	7.4	0.0	0.2	0.0	0.3	6.2
Brazil	19.3	10.2	9.1	4.7	0.8	0.6	0.0	0.2	12.2
Bulgaria	11.3	8.9	2.4	3.2	0.9	0.3	0.0	2.3	2.0
Burkina Faso	9.8	6.5	3.2	1.4	0.0	0.0	3.9	0.2	0.4
Burundi	−0.4	5.6	−6.0	3.0	0.0	0.2	4.0	0.2	−7.3
Cambodia	5.5	6.7	−1.2	1.8	0.0	0.0	..	0.1	..
Cameroon	18.9	7.8	11.2	2.2	5.2	0.0	0.0	0.2	7.9
Canada	23.1	12.2	11.0	6.3	2.6	0.1	0.0	0.5	14.0
Central African Republic	7.2	6.8	0.4	1.6	0.0	0.0	0.0	0.1	1.8
Chad	−3.0	6.4	−9.4	2.0	0.0	0.0	0.0	0.0	−7.4
Chile	23.0	10.3	12.7	3.3	0.0	4.2	0.0	0.4	11.3
China	40.1	8.3	31.8	2.0	1.4	0.3	0.3	2.5	29.4
Hong Kong, China	30.6	12.3	18.3	2.8	0.0	0.0	0.0	0.1	21.0
Colombia	11.3	9.5	1.8	3.0	5.2	0.1	0.0	0.4	−0.9
Congo, Dem. Rep.	9.0	5.7	3.3	..	0.0	0.0	..	..	..
Congo, Rep.	29.7	8.0	21.7	4.9	25.6	0.0	0.0	0.3	0.7
Costa Rica	23.8	10.2	13.6	4.5	0.0	0.0	0.8	0.2	17.1
Côte d'Ivoire	23.1	8.0	15.1	4.2	0.0	0.0	0.1	0.7	18.5
Croatia	15.8	10.4	5.4	..	0.5	0.0	0.0	0.6	..
Cuba	..	..	..	..	0.0	0.0	0.0	..	..
Czech Republic	26.9	10.5	16.4	4.5	0.1	0.0	0.0	1.3	19.5
Denmark	23.8	14.2	9.6	8.1	0.1	0.0	0.0	0.2	17.4
Dominican Republic	16.7	5.9	10.8	2.0	0.0	0.5	0.0	0.5	11.8
Ecuador	24.2	8.9	15.3	3.0	10.3	0.0	0.0	0.7	7.2
Egypt, Arab Rep.	14.4	8.8	5.6	4.5	2.2	0.1	0.1	0.8	7.0
El Salvador	4.2	9.4	−5.3	2.1	0.0	0.0	1.3	0.3	−4.6
Eritrea	−21.3	6.1	−27.4	1.6	0.0	0.0	0.0	..	−25.8
Estonia	18.8	10.0	8.9	6.2	0.0	0.0	0.0	2.2	12.9
Ethiopia	2.7	5.5	−2.8	2.7	0.0	0.0	10.8	0.4	−11.3
Finland	27.7	15.4	12.3	6.9	0.0	0.0	0.0	0.3	18.9
France	21.5	12.4	9.2	5.6	0.0	0.0	0.0	0.1	14.6
Gabon	34.8	10.0	24.9	1.9	15.2	0.0	0.0	0.4	11.1
Gambia, The	1.7	6.9	−5.2	3.5	0.0	0.0	2.8	0.3	−4.8
Georgia	−2.2	7.8	−10.0	0.0	0.0	0.0	0.0	0.9	−10.9
Germany	23.3	12.4	10.8	4.3	0.0	0.0	0.0	0.2	14.9
Ghana	6.2	7.2	−1.0	4.3	0.0	1.0	3.1	0.4	−1.1
Greece	15.8	6.4	9.4	2.3	0.0	0.0	0.0	0.4	11.2
Guatemala	9.1	9.0	0.1	1.5	0.7	0.0	1.7	0.3	−1.0
Guinea	15.4	7.5	7.9	0.0	0.0	2.6	1.3	0.2	3.8
Guinea-Bissau	−2.2	6.2	−8.4	2.5	0.0	0.0	0.0	0.5	−6.4
Haiti	−4.2	1.8	−6.0	1.6	0.0	0.0	4.6	0.2	−9.1
Honduras	19.1	5.4	13.7	3.4	0.0	0.2	0.0	0.5	16.5

Toward a measure of genuine savings | 3.15

	Gross domestic savings	Consumption of fixed capital	Net domestic savings	Education expenditure	Energy depletion	Mineral depletion	Net forest depletion	Carbon dioxide damage	Genuine domestic savings
	% of GDP 1999	% of GDP 1999	% of GDP 1999	% of GDP 1999	% of GDP 1999	% of GDP 1999	% of GDP 1999	% of GDP 1999	% of GDP 1999
Hungary	26.3	10.4	15.9	4.4	0.2	0.0	0.0	0.8	19.4
India	20.0	9.4	10.6	3.3	1.3	0.3	1.7	1.5	9.0
Indonesia	31.6	7.9	23.8	0.6	6.1	1.0	0.6	0.9	15.8
Iran, Islamic Rep.	22.9	9.3	13.6	3.2	20.6	0.2	0.0	1.5	–5.5
Iraq	..	..	..	..	0.0	0.0	0.0	..	..
Ireland	37.1	9.3	27.8	4.8	0.0	0.1	0.0	0.3	32.2
Israel	11.4	13.9	–2.5	6.1	0.0	0.0	0.0	0.2	3.2
Italy	22.3	12.1	10.2	4.6	0.0	0.0	0.0	0.8	14.5
Jamaica	16.6	9.7	6.9	6.5	0.0	1.7	0.0	0.2	10.9
Japan	27.7	16.0	11.8	4.7	0.0	0.0	0.0	0.2	16.3
Jordan	2.6	9.0	–6.5	5.5	0.0	0.9	0.0	1.1	–3.0
Kazakhstan	22.8	8.6	14.2	4.4	21.6	0.0	0.0	5.2	–8.2
Kenya	6.8	7.1	–0.2	6.0	0.0	0.0	4.8	0.4	0.6
Korea, Dem. Rep.	..	..	..	..	0.0	0.0	0.0	..	..
Korea, Rep.	33.6	11.1	22.6	0.0	0.0	0.0	0.0	0.6	21.9
Kuwait	22.3	11.8	10.5	1.7	41.7	0.0	0.0	..	–29.5
Kyrgyz Republic	3.2	6.7	–3.5	5.1	0.0	0.0	0.0	3.2	–1.5
Lao PDR	13.4	6.8	6.6	1.8	0.0	0.1	..	0.2	..
Latvia	15.4	9.5	5.9	6.2	0.0	0.0	0.0	0.9	11.2
Lebanon	–12.8	10.3	–23.1	1.4	0.0	0.0	0.0	..	..
Lesotho	–34.6	7.3	–41.9	8.1	0.0	0.0	4.1	..	..
Libya	..	..	..	..	0.0	0.0	0.0	..	..
Lithuania	12.6	9.7	2.9	5.2	0.0	0.0	0.0	0.8	7.2
Macedonia, FYR	7.1	9.0	–1.9	..	0.0	0.0	0.0	1.9	..
Madagascar	5.0	6.6	–1.6	1.8	0.0	0.0	0.0	0.2	0.0
Malawi	–0.6	6.1	–6.7	3.8	0.0	0.0	5.7	0.3	–9.0
Malaysia	47.3	9.9	37.4	4.0	5.7	0.0	0.6	0.9	34.1
Mali	10.1	6.6	3.6	2.2	0.0	0.0	0.0	0.1	5.6
Mauritania	7.2	7.1	0.1	3.6	0.0	18.0	0.0	1.9	–16.3
Mauritius	22.7	10.0	12.7	3.2	0.0	0.0	0.0	0.2	15.6
Mexico	21.9	10.5	11.4	4.4	4.0	0.1	0.0	0.5	11.3
Moldova	7.3	6.7	0.5	8.7	0.0	0.0	0.0	4.6	4.6
Mongolia	20.8	7.0	13.8	..	0.0	7.3	0.0	5.2	..
Morocco	20.1	8.6	11.5	4.6	0.0	0.6	0.0	0.6	14.8
Mozambique	6.7	6.5	0.2	3.5	0.0	0.0	2.1	0.2	1.3
Myanmar	10.2	2.5	7.7	..	0.0	0.0	..	..	..
Namibia	9.3	14.3	–5.1	8.6	0.0	0.3	0.0	..	3.3
Nepal	13.3	4.4	8.9	2.1	0.0	0.0	9.7	0.2	1.1
Netherlands	26.7	12.4	14.3	5.1	0.0	0.0	0.0	0.3	19.2
New Zealand	19.7	9.5	10.2	6.3	0.6	0.1	0.0	0.3	15.4
Nicaragua	–12.0	7.7	–19.7	2.4	0.0	0.1	0.0	0.9	–18.3
Niger	3.8	6.3	–2.5	3.0	0.0	0.0	4.2	0.3	–4.0
Nigeria	18.4	6.8	11.7	0.7	28.5	0.0	0.8	1.5	–18.3
Norway	30.3	16.2	14.2	6.8	1.5	0.0	0.0	0.3	19.2
Oman	..	10.8	..	3.5	31.9	0.0	0.0	..	..
Pakistan	10.1	7.4	2.7	2.4	1.9	0.0	1.6	1.0	0.6
Panama	24.0	6.9	17.1	4.4	0.0	0.0	0.0	0.4	21.2
Papua New Guinea	20.9	8.3	12.6	..	10.0	8.6	0.0	0.4	..
Paraguay	9.3	8.8	0.5	3.5	0.0	0.0	0.0	0.3	3.7
Peru	19.7	9.3	10.4	2.5	0.5	0.8	0.0	0.3	11.3
Philippines	19.6	8.4	11.3	3.0	0.0	0.1	1.3	0.6	12.3
Poland	20.0	10.1	9.9	5.0	0.3	0.2	0.1	1.5	12.9
Portugal	16.2	5.1	11.1	5.5	0.0	0.0	0.1	0.3	16.3
Puerto Rico	..	6.8	..	..	0.0	0.0	0.0	..	..
Romania	15.7	8.9	6.8	3.3	2.3	0.1	0.0	1.7	6.1
Russian Federation	33.0	9.6	23.4	3.7	12.8	0.0	0.0	2.0	12.2

	Gross domestic savings	Consumption of fixed capital	Net domestic savings	Education expenditure	Energy depletion	Mineral depletion	Net forest depletion	Carbon dioxide damage	Genuine domestic savings
	% of GDP 1999	% of GDP 1999	% of GDP 1999	% of GDP 1999	% of GDP 1999	% of GDP 1999	% of GDP 1999	% of GDP 1999	% of GDP 1999
Rwanda	−1.3	6.5	−7.9	3.3	0.0	0.0	3.0	0.2	−7.8
Saudi Arabia	31.3	10.7	20.6	6.4	39.1	0.0	0.0	1.1	−13.3
Senegal	12.6	7.5	5.1	3.4	0.0	0.2	0.0	0.5	7.8
Sierra Leone	−6.0	6.2	−12.2	1.0	0.0	0.1	2.9	0.3	−14.5
Singapore	51.7	12.5	39.3	2.4	0.0	0.0	0.0	0.5	41.2
Slovak Republic	26.5	10.0	16.5	4.2	0.0	0.0	0.0	1.2	19.6
Slovenia	23.9	17.6	6.3	5.2	0.0	0.0	0.0	0.5	11.1
South Africa	18.2	11.7	6.5	6.8	0.0	0.9	0.6	1.4	10.4
Spain	23.4	11.8	11.7	4.5	0.0	0.0	0.0	0.3	15.9
Sri Lanka	19.8	5.1	14.7	2.6	0.0	0.0	1.5	0.3	15.5
Sudan	..	7.0	..	..	0.0	0.1	0.0	0.2	..
Sweden	22.4	12.5	9.9	7.3	0.0	0.1	0.0	0.1	17.1
Switzerland	25.0	12.8	12.2	5.1	0.0	0.0	0.0	0.1	17.2
Syrian Arab Republic	18.2	3.5	14.7	2.0	19.0	0.1	0.0	1.6	−3.9
Tajikistan	13.5	7.2	6.3	2.0	0.2	0.0	0.0	..	..
Tanzania	2.2	6.7	−4.5	3.4	0.0	0.1	0.2	0.2	−1.5
Thailand	33.4	9.2	24.1	3.3	0.1	0.0	0.7	0.9	25.8
Togo	3.6	7.2	−3.6	4.3	0.0	0.7	0.3	0.4	−0.7
Trinidad and Tobago	26.6	10.5	16.1	3.2	12.6	0.0	0.0	2.0	4.8
Tunisia	24.4	9.4	15.1	6.3	1.9	0.6	0.3	0.6	18.1
Turkey	19.6	6.5	13.1	3.2	0.2	0.0	0.0	0.6	15.4
Turkmenistan	26.0	8.1	17.9	..	44.0	0.0	0.0	6.1	..
Uganda	4.9	6.8	−2.0	2.2	0.0	0.0	2.2	0.1	−2.1
Ukraine	20.9	8.0	12.8	5.9	6.1	0.0	0.0	5.2	7.5
United Arab Emirates	..	11.9	..	1.8	23.9	0.0	0.0	0.9	..
United Kingdom	15.9	12.4	3.5	4.7	0.3	0.0	0.0	0.2	7.7
United States	18.4	12.8	5.6	4.7	0.7	0.0	0.0	0.4	9.2
Uruguay	13.6	10.7	3.0	3.0	0.0	0.0	0.6	0.2	5.2
Uzbekistan	15.8	7.9	7.8	7.7	16.6	0.0	0.0	3.6	−4.6
Venezuela, RB	22.2	6.8	15.4	4.9	17.4	0.3	0.0	1.0	1.7
Vietnam	23.2	7.1	16.1	2.8	4.0	0.1	2.5	0.9	11.4
West Bank and Gaza	−18.7	8.9	−27.6	..	0.0	0.0	0.0	..	..
Yemen, Rep.	11.8	7.2	4.6	5.1	27.6	0.0	0.0	1.5	−19.4
Yugoslavia, FR (Serb./Mont.)	..	..	..	..	0.0	0.0	0.0	..	..
Zambia	−1.1	6.9	−8.1	1.9	0.0	2.1	0.0	0.5	−8.7
Zimbabwe	11.0	7.4	3.6	7.0	0.1	2.8	0.6	2.0	5.1
World	**24.7 w**	**12.3 w**	**12.3 w**	**4.5 w**	**1.3 w**	**0.1 w**	**0.1 w**	**0.5 w**	**15.0 w**
Low income	20.3	8.3	12.0	2.9	3.8	0.3	1.5	1.4	7.8
Middle income	26.1	9.6	16.6	3.5	4.2	0.3	0.1	1.1	14.3
Lower middle income	29.8	8.6	21.2	2.9	4.5	0.2	0.2	1.7	17.5
Upper middle income	23.4	10.4	13.0	3.9	3.9	0.3	0.1	0.6	12.0
Low & middle income	25.2	9.4	15.9	3.4	4.1	0.3	0.4	1.2	13.3
East Asia & Pacific	36.1	9.0	27.1	1.7	1.3	0.2	0.4	1.7	25.2
Europe & Central Asia	24.6	9.1	15.6	4.1	6.0	0.0	0.0	1.7	11.9
Latin America & Carib.	19.2	10.0	9.1	4.1	2.8	0.4	0.0	0.4	9.6
Middle East & N. Africa	24.2	9.3	15.0	4.7	19.7	0.1	0.0	1.1	−1.3
South Asia	18.3	8.8	9.5	3.1	1.0	0.2	1.8	1.3	8.3
Sub-Saharan Africa	15.3	9.3	6.0	4.7	4.2	0.6	1.1	0.9	3.9
High income	22.7	13.1	9.6	4.8	0.5	0.0	0.0	0.3	13.5
Europe EMU	23.2	12.1	11.1	4.8	..	0.0	0.0	0.2	..

Toward a measure of genuine savings | 3.15

About the data

Genuine domestic savings are derived from standard national accounting measures of gross domestic savings by making four types of adjustments. First, estimates of capital consumption of produced assets are deducted to obtain net domestic savings. Then current expenditures on education are added to net domestic savings as an approximate value of investments in human capital (in standard national accounting these expenditures are treated as consumption). Next, estimates of the depletion of a variety of natural resources are deducted to reflect the decline in asset values associated with their extraction and harvest. Finally, a deduction is made for damage from carbon dioxide emissions.

There are important gaps in the accounting of natural resource depletion and costs of pollution. Key estimates missing on the resource side include the value of fossil water extracted from aquifers, depletion and degradation of soils, and net depletion of fish stocks. The most important pollutants affecting human health and economic assets are also excluded, because no internationally comparable data are widely available on damage from particulate emissions, ground-level ozone, or acid rain.

Estimates of resource depletion are based on the calculation of unit resource rents. An economic rent represents an excess return to a given factor of production—that is, in this case the returns from resource depletion are higher than the normal rate of return on capital. Because natural resources are fixed in extent (at least for a given state of technology), resource rents will persist over time; in contrast, for produced goods and services competitive forces will expand supply until economic profits are driven to zero. For each type of resource and each country, unit resource rents are derived by taking the difference between world prices and the average unit extraction or harvest costs (including a "normal" return on capital). Unit rents are then multiplied by the physical quantity extracted or harvested in order to arrive at a depletion figure. This figure is one of a range of depletion estimates that are possible, depending on the assumptions made about future quantities, prices, and costs, and there is reason to believe that it is at the high end of the range. Some of the largest depletion estimates in the table should therefore be viewed with caution.

A positive depletion figure for forest resources implies that the harvest rate exceeds the rate of natural growth, and a negative figure that growth exceeds harvest. In principle, there should be an addition to savings in countries where growth exceeds harvest, but there is good reason to believe that most of this net growth is in forested areas that cannot be exploited economi-

cally at present. The average world prices used to estimate unit rents on timber are probably too high for countries with low-grade timber resources, so some of the net forest depletion estimates, especially those for Sub-Saharan Africa, should be viewed with caution. In addition, because the depletion estimates reflect only timber values, they ignore all the external benefits associated with standing forests.

Pollution damage is calculated as the marginal social cost associated with a unit of pollution multiplied by the increase in the stock of pollutant in the receiving medium. For carbon dioxide the unit damage figure represents the present value of damage to economic assets and decline in human welfare over the time the unit of pollution remains in the atmosphere.

Figure 3.15

Genuine domestic savings—a proxy for economic sustainability

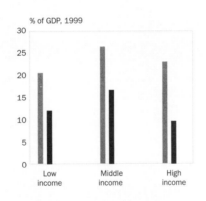

% of GDP, 1999

- Gross domestic savings
- Net domestic savings
  Genuine domestic savings

Source: Table 3.15.

When expenditure on education, depletion of natural resources, and damage from carbon dioxide are accounted for in estimating domestic savings, the results are lower than traditional estimates, particularly for low- and middle-income countries, where expenditure on education is relatively low. This measure of genuine domestic savings can serve as a proxy for the sustainability of economic activities.

Definitions

- **Gross domestic savings** are calculated as the difference between GDP and public and private consumption. • **Consumption of fixed capital** represents the replacement value of capital used up in the process of production. • **Net domestic savings** are equal to gross domestic savings less the value of consumption of fixed capital. • **Education expenditure** refers to the current operating expenditures in education, including wages and salaries and excluding capital investments in buildings and equipment. • **Energy depletion** is equal to the product of unit resource rents and the physical quantities of energy extracted. It covers crude oil, natural gas, and coal. • **Mineral depletion** is equal to the product of unit resource rents and the physical quantities of minerals extracted. It refers to bauxite, copper, iron, lead, nickel, phosphate, tin, gold, and silver. • **Net forest depletion** is calculated as the product of unit resource rents and the excess of roundwood harvest over natural growth. • **Carbon dioxide damage** is estimated to be $20 per ton of carbon (the unit damage) times the number of tons of carbon emitted. • **Genuine domestic savings** are equal to net domestic savings plus education expenditure and minus energy depletion, mineral depletion, net forest depletion, and carbon dioxide damage.

Data sources

Gross domestic savings are derived from the World Bank's national accounts data files, described in the *Economy* section. Consumption of fixed capital is from the United Nations Statistics Division's *National Accounts Statistics: Main Aggregates and Detailed Tables, 1997,* extrapolated to 1999. The education expenditure data are from the United Nations Statistics Division's *Statistical Yearbook 1997,* extrapolated to 1999. The wide range of data sources and estimation methods used to arrive at resource depletion estimates are described in a World Bank working paper, "Estimating National Wealth" (Kunte and others 1998). The unit damage figure for carbon dioxide emissions is from Fankhauser (1995).

ECONOMY

End of a long decline?

Year-to-year growth in GDP per capita shows high volatility, especially for low-income economies. The world average, which includes the high-income economies, shows that until recently the trend had been steadily downward. Beginning in the early 1990s growth in developing economies accelerated, except for an interruption caused by the financial crisis of 1997–98.

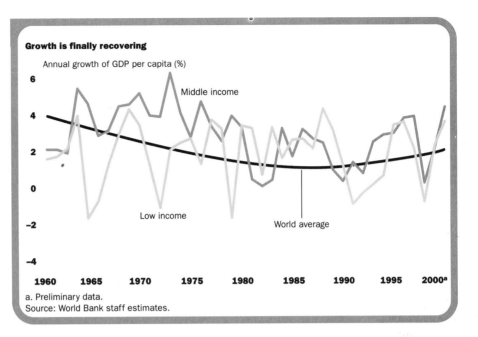

Growth is finally recovering

Annual growth of GDP per capita (%)

Middle income

Low income

World average

a. Preliminary data.
Source: World Bank staff estimates.

New opportunities for growth

Economic growth does not follow a smooth path, but for most of the second half of the 20th century growth was slowing—in both high-income and developing economies. Why the slowdown? Growth opportunities from postwar reconstruction ran out. The costs of transition from colonial to independent and, in many cases, socialized economies mounted. Old technologies yielded fewer productivity gains. The energy crises of the 1970s interrupted growth in oil consuming countries. The growth slowdown contributed to increasing debt in developing economies, which, combined with poor macroeconomic management, left many with poorer prospects and fewer opportunities for investment.

Now, at the beginning of the 21st century, there are signs of a resumption of faster growth. Will it continue? Favorable signs include market reforms, a wave of productivity growth based on information and communications technology, and a new recognition in many countries of the need to create an environment that encourages investment and growth. Continuing growth in high-income economies can help to stimulate growth in developing countries. But even well-managed economies may suffer setbacks. Changes in a country's terms of trade or in the demand for its exports can have a profound and unexpected effect on growth. So can natural disasters and man-made environmental change. Overcoming adversity requires good policies and good governance, creating an atmosphere for growth to continue.

Accelerating growth

Increasing growth can have a profound effect on the welfare of people in the space of one generation. Even a small improvement can make a difference. Increasing annual growth from 1.75 percent to 3.5 percent reduces the time needed to double output from 40 years to 20. China, growing 6 percent a year over the past 40 years, has increased its GDP per capita more than sevenfold.

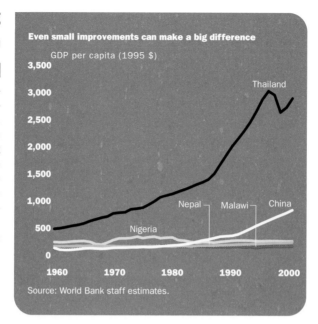

Even small improvements can make a big difference

GDP per capita (1995 $)

Thailand
Nepal Malawi China
Nigeria

1960 1970 1980 1990 2000

Source: World Bank staff estimates.

Per capita output in Malawi and Nepal was similar to China's in the 1960s. But over the past 40 years these two economies have grown only 1 percent a year, adding only 50 percent to their per capita output. At their current rate of growth it will take another 30 years or more to double their 1960 GDP per capita.

Both Thailand and Nigeria started the period ahead of China. But while Thailand grew at an average annual rate of 5 percent, Nigeria grew at only 0.2 percent. China's GDP per capita is now more than three times Nigeria's.

The importance of growth

Income and poverty

Measures of per capita income are more than a way of keeping score. Poverty rates fall as income rises. But poverty rates differ dramatically among countries with the same level of gross national income (GNI) per capita.

Growth in the average does not guarantee growth for all. Whether growth helps to reduce poverty depends on how the growth is distributed. A continuing challenge in development is how to ensure that poor people are not left behind.

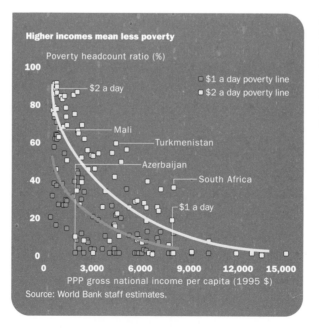

Higher incomes mean less poverty

Poverty headcount ratio (%)

□ $1 a day poverty line
■ $2 a day poverty line

$2 a day
Mali
Turkmenistan
Azerbaijan
South Africa
$1 a day

PPP gross national income per capita (1995 $)

Source: World Bank staff estimates.

Countries with higher average incomes—as measured by GNI per capita—have lower poverty rates. But the dispersion of poverty rates at the same average income level tells us that greater equity in the distribution of the benefits of growth is necessary—and possible.

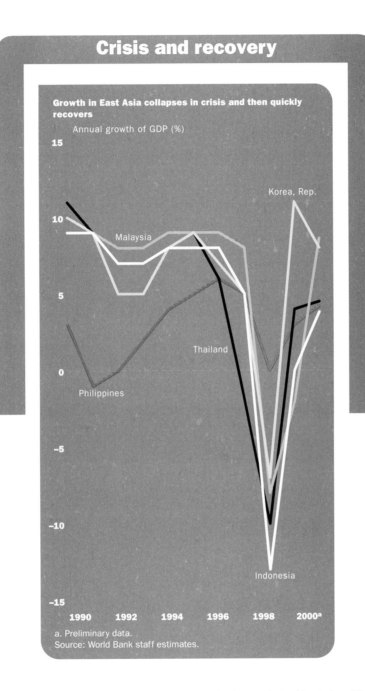

Crisis and recovery

Growth in East Asia collapses in crisis and then quickly recovers

Annual growth of GDP (%)

Korea, Rep.

Malaysia

Thailand

Philippines

Indonesia

1990 1992 1994 1996 1998 2000ᵃ

a. Preliminary data.
Source: World Bank staff estimates.

For many countries spurts of growth have been followed by steep declines. Bad luck or bad policies? Perhaps both. Trade shocks, natural disasters, and "contagious" changes in investors' expectations can all knock a country off its growth path. In East Asia years of spectacular growth may have bred complacency toward shaky financial institutions, speculative bubbles in real estate, and, in some places, endemic corruption. When Thailand overplayed its hand in the currency markets, a rapid loss of confidence spread to neighboring countries and then around the world.

The financial crisis that began in Asia in 1997 was particularly alarming because it struck hardest at some of the largest and fastest growing developing countries, threatening to reverse years of economic and social progress. Brazil, Russia, and Mexico were among the countries outside Asia affected by the spreading crisis. As currencies came under pressure and investors withdrew, growth rates fell, but recovery began in the following year. Mexico, which had a severe financial crisis in 1995, was the least affected. The hardest hit countries in Asia were beginning to show signs of recovery by 1999. Preliminary estimates for 2000 show that all have returned to positive growth.

Lower inflation

Inflation rates and the variability of inflation rates have declined everywhere, though interrupted by an outburst of high inflation in the transition economies of Europe and Central Asia.

Over the past decade many countries that had experienced years of high and volatile inflation adopted policies leading to greater macroeconomic stability.

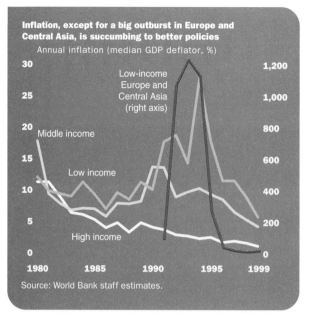

Inflation, except for a big outburst in Europe and Central Asia, is succumbing to better policies

Annual inflation (median GDP deflator, %)

Source: World Bank staff estimates.

Of the 12 countries with the highest average inflation in the 1980s, 10 (including Argentina and Brazil) reduced inflation to less than 20 percent a year by the end of the 1990s. The exceptions were two war-torn countries: the Democratic Republic of the Congo and Sierra Leone. The same 10 countries have experienced a resurgence of economic growth.

But inflation remains a threat. Five countries had average price increases of more than 100 percent a year between 1995 and 1999: Angola, Belarus, Bulgaria, the Democratic Republic of the Congo, and Turkmenistan.

Better policies

Fiscal stability

One thing countries can do to stabilize their economies is to bring public spending in line with resources. Small deficits are sustainable in a growing economy, but large deficits increase the likelihood of inflation, make markets less stable, and tend to crowd out private economic activity.

The figure shows the 12 developing countries with the largest budget deficits in 1990. Most had succeeded in shrinking their deficits by 1998.

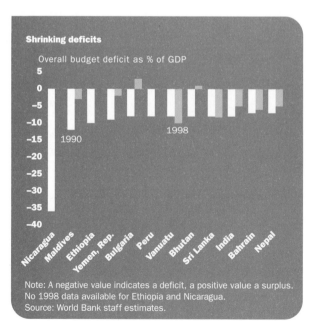

Shrinking deficits

Overall budget deficit as % of GDP

Note: A negative value indicates a deficit, a positive value a surplus. No 1998 data available for Ethiopia and Nicaragua.
Source: World Bank staff estimates.

In 1990, 22 low- and middle-income economies (31 percent of those reporting data) had deficits exceeding 5 percent of GDP. In 1998 only 16 (25 percent) did. High-income countries have also become more deficit-conscious. In 1990 Belgium, Greece, and Italy were all running large deficits. Under pressure from the European Monetary Union, they brought their deficits below 5 percent of GDP by 1998. Still, fiscal discipline is hard to maintain. In 1998 the Republic of Congo, Lebanon, and Mongolia all ran deficits above 10 percent of GDP.

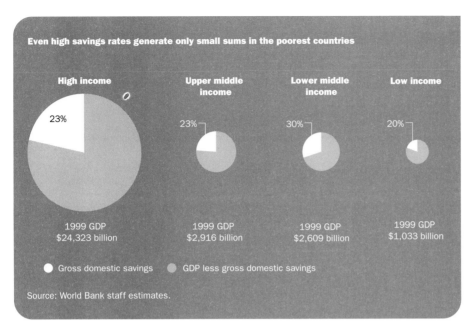

Even high savings rates generate only small sums in the poorest countries

High income	Upper middle income	Lower middle income	Low income
23%	23%	30%	20%
1999 GDP $24,323 billion	1999 GDP $2,916 billion	1999 GDP $2,609 billion	1999 GDP $1,033 billion

● Gross domestic savings　　● GDP less gross domestic savings

Source: World Bank staff estimates.

A large piece of a small pie

Many developing countries have maintained high savings rates. Throughout the 1990s the median savings rate in China was 42 percent of GDP, in Gabon almost 40 percent, and in Bhutan 38 percent. On average, the lower-middle-income economies were the best savers, putting aside almost 30 percent of their combined GDP. But even high savings rates can raise only small sums in desperately poor economies. The high-income economies are by far the largest suppliers of savings.

Savings and investment

The savings-investment gap

Savings provide the funds for investment. When an economy cannot save enough from its own income to finance investment, it must seek other sources of funds: foreign investors, bank loans, private transfers, and development assistance.

Through much of the 1990s the high-income countries had a large surplus of savings over investment—funds available for use in developing countries. Meanwhile, the middle-income economies began to grow—and by the end of the decade they too were generating a surplus of savings.

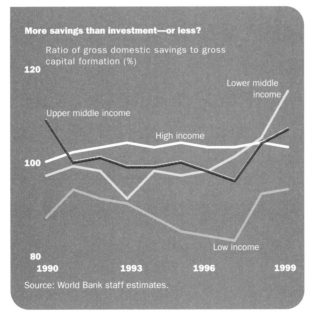

More savings than investment—or less?

Ratio of gross domestic savings to gross capital formation (%)

Source: World Bank staff estimates.

For the poorest countries the challenge was to make themselves attractive to foreign investors or to appeal to donors for a share of the shrinking pool of development assistance.

The data in the figure reflect actual investment (including changes in inventories) and understate the need for new investment in poor countries. Many projects with great potential benefits are never begun because investors are reluctant to provide funds in an investment climate that appears uncertain because of bad policies, poor governance, or civil disorder. Such risks lead to underinvestment.

	Gross domestic product		Exports of goods and services		Imports of goods and services		GDP deflator		Current account balance		Gross international reserves	
	annual % growth		annual % growth		annual % growth		% growth		% of GDP		$ millions	months of import coverage
	1999	2000	1999	2000	1999	2000	1999	2000	1999	2000	2000	2000
Algeria	3.3	3.5	6.2	6.3	1.8	3.5	10.9	20.0	..	16.2	..	..
Argentina	−3.2	0.0	−1.2	14.4	−10.9	−1.0	−2.0	−0.9	−4.3	−3.5	28,969	7.5
Armenia	3.3	4.5	17.5	18.7	0.1	3.9	0.1	−0.7	−16.6	−16.1	344	3.8
Azerbaijan	7.4	8.1	67.1	7.9	−17.3	6.2	−5.1	12.8	−27.6	−2.8	..	..
Bangladesh	4.9	5.5	6.0	9.7	5.0	4.3	4.6	3.8	−0.9	0.0	1,621	2.1
Bolivia	0.6	2.5	−9.7	9.8	−15.7	9.0	2.8	5.4	−6.7	−6.3	1,114	5.5
Brazil	0.8	4.0	12.0	20.1	−14.8	34.9	5.9	8.7	−3.3	−4.5	45,680	6.1
Bulgaria	2.4	4.5	−5.2	6.6	5.1	4.4	3.1	9.9	−5.5	−4.2	3,395	5.0
Cameroon	4.4	4.2	12.9	−5.4	3.1	10.6	−1.2	1.2	−4.3	−1.4	11	0.0
Chile	−1.1	5.5	6.9	12.0	−14.3	15.0	3.6	3.8	−0.1	−2.4	15,497	7.4
China	7.1	8.0	8.3	24.2	19.2	31.2	−2.3	1.0	1.6	1.1	166,216	7.4
Colombia	−4.3	3.0	4.7	7.3	−13.5	17.0	12.5	10.0	−0.1	−2.3	9,458	6.2
Congo, Rep.	−3.0	3.8	19.0	20.0	14.5	37.5	22.4	35.1	..	−6.6	..	..
Costa Rica	8.0	1.4	21.0	−0.4	3.1	1.3	12.6	12.0	−4.3	−5.2	1,143	1.6
Côte d'Ivoire	2.8	2.2	2.4	0.6	3.9	6.8	1.3	1.9	0.3	−5.0	..	..
Croatia	−0.3	3.5	−0.2	7.8	−3.9	−0.7	4.0	6.1	−7.5	−4.4	3,278	3.7
Dominican Republic	8.3	8.4	7.8	15.6	6.5	15.8	6.4	8.0	−2.5	−6.0	407	0.5
Ecuador	−7.3	2.0	−0.4	−4.8	−39.0	8.2	62.0	95.0	5.0	5.7	2,532	4.8
Egypt, Arab Rep.	6.0	6.5	9.2	14.1	12.8	9.4	1.7	4.0	−1.9	−1.0	..	..
El Salvador	3.4	2.5	14.9	17.2	5.8	12.5	0.5	3.5	−1.9	−2.1	1,904	4.2
Estonia	−1.1	5.0	−2.4	3.4	−6.4	10.4	3.9	6.5	−5.6	−6.5	851	2.2
Ghana	4.4	3.7	12.8	5.7	12.0	−15.9	14.0	19.5	−9.9	−8.5	..	..
Guatemala	3.6	3.0	3.6	12.8	0.7	6.1	5.2	6.0	−5.6	−4.7	1,659	3.5
Honduras	−1.9	5.0	−9.4	14.6	6.7	8.0	11.1	9.3	−3.9	−10.5	1,246	4.2
India	6.5	6.7	1.7	7.0	−3.6	7.7	3.3	7.0	−0.8	−1.5	37,811	5.5
Indonesia	0.3	3.8	−31.6	10.0	−40.7	9.1	12.8	6.4	4.1	4.8	41,350	7.6
Iran, Islamic Rep.	2.5	5.2	5.6	13.0	−7.1	16.0	23.7	18.0	..	−1.6	9,088	..
Jamaica	−0.4	1.5	0.2	4.4	−2.4	4.3	7.9	11.6	−3.7	−4.8	1,082	2.7
Jordan	3.1	3.2	1.1	3.5	0.3	2.3	−1.6	4.6	4.8	2.8	3,449	7.2
Kazakhstan	1.7	8.0	18.7	16.6	−1.5	11.9	8.2	13.7	−1.1	6.2	2,707	3.5
Kenya	1.3	−0.2	−5.2	−1.2	−6.5	10.8	6.8	7.4	0.1	−9.0	734	2.0
Latvia	0.1	5.0	−5.8	14.0	−6.2	10.5	2.0	3.0	−10.3	−7.1	..	..
Lithuania	−4.2	2.3	−18.3	16.4	−14.0	−1.8	3.3	2.3	−11.2	−6.2	..	..
Macedonia, FYR	2.7	6.0	2.5	17.4	−2.8	20.3	−0.3	3.2	−3.2	−6.5	677	3.5

continues on page 192

Table 4b	Key macroeconomic indicators											
	Nominal exchange rate			Real effective exchange rate		Money and quasi money		Gross domestic credit		Real interest rate		Short-term debt[a]
	local currency units per $	% change		1995 = 100		annual % growth		annual % growth		%	%	% of exports
	2000	1999	2000	1999	2000	1999	2000	1999	2000	1999	2000	1999
Algeria	79.1	14.8	14.3	110.2	106.9	13.7	..	25.2	..	0.1	..	1.4
Argentina	1.0	0.0	0.0	..	..	4.1	0.0	3.2	−1.5	13.3	10.0	93.0
Armenia	552.2	0.3	5.4	115.3	101.6	14.0	35.6	3.7	4.2	38.8	..	10.0
Azerbaijan	4,558.0	12.5	4.5	..	..	21.5	..	−9.3	..	..	..	2.2
Bangladesh	54.0	5.2	5.9	..	..	15.5	18.9	15.4	14.7	9.1	10.0	3.3
Bolivia	6.4	6.1	6.7	118.4	119.5	5.7	−3.2	5.5	0.1	31.7	28.6	91.1
Brazil	2.0	48.0	9.3	..	..	7.4	−1.1	2.3	4.4	..	..	48.5
Bulgaria	2.1	16.2	8.0	118.7	117.2	11.9	32.5	0.2	6.5	9.4	6.8	6.2
Cameroon	705.0	16.1	8.0	109.7	90.8	13.3	15.7	8.9	−2.8	23.5	19.5	56.6
Chile	572.7	11.9	8.0	105.4	104.3	14.8	6.0	8.8	11.8	8.8	11.5	26.8
China	8.3	0.0	0.0	106.9	110.2	14.7	14.0	12.1	10.2	8.4	..	7.7
Colombia	2,187.0	24.3	16.7	102.7	98.9	13.8	16.9	8.1	10.9	15.9	..	25.7
Congo, Rep.	705.0	16.1	8.0	..	..	19.9	17.2	9.2	−10.5	−0.3	31.8	59.5
Costa Rica	318.0	9.9	6.7	103.5	109.6	21.7	17.3	1.1	13.1	11.6	..	9.2
Côte d'Ivoire	705.0	16.1	8.0	103.5	94.8	−1.7	..	−1.2	..	..	..	22.7
Croatia	8.2	22.4	6.6	97.1	101.3	−1.8	27.4	−1.5	2.0	10.5	..	7.8
Dominican Republic	16.7	1.6	4.0	105.6	114.1	23.7	−28.3	28.1	..	17.5	24.2	10.8
Ecuador	25,000.0	196.6	23.5	80.3	89.3	99.2	..	144.7	..	1.3	..	19.5
Egypt, Arab Rep.	3.7	0.5	8.4	..	..	5.7	11.4	12.5	9.9	11.0	8.9	22.3
El Salvador	8.8	0.0	0.0	..	..	9.1	..	11.7	..	14.9	..	22.8
Estonia	16.8	16.0	8.1	..	..	24.7	25.9	10.2	28.1	4.6	3.5	30.5
Ghana	7,142.9	48.3	107.1	125.7	77.9	16.2	22.6	54.5	46.0	..	..	27.0
Guatemala	7.7	14.2	−1.1	..	..	12.5	51.6	17.8	30.3	13.6	..	34.0
Honduras	15.1	5.0	4.4	..	..	24.7	33.8	6.1	34.3	17.1	..	16.8
India	46.8	2.4	7.5	..	..	17.2	10.7	17.9	10.3	9.0	5.1	6.0
Indonesia	9,595.0	−11.7	35.4	..	..	12.5	..	21.4	..	13.2	..	34.1
Iran, Islamic Rep.	2,262.9	0.1	29.1	245.4	328.0	21.5	20.5	23.4	14.4	..	..	17.8
Jamaica	44.6	11.4	11.3	..	..	12.2	9.4	18.1	7.1	17.7	15.2	18.1
Jordan	0.7	0.0	0.0	..	..	15.5	6.0	3.0	0.8	..	..	16.0
Kazakhstan	144.5	64.9	4.6	..	..	84.4	78.1	40.6	63.6	..	..	6.7
Kenya	78.0	17.8	7.0	..	..	6.0	5.7	9.6	3.4	14.6	15.2	30.8
Latvia	0.6	2.5	5.1	..	..	8.3	30.7	14.8	43.7	12.0	..	36.1
Lithuania	4.0	0.0	0.0	..	..	7.7	16.4	19.1	−1.2	9.5	..	12.6
Macedonia, FYR	70.2	16.4	22.5	73.8	74.4	32.0	..	12.1	..	20.8	..	4.4

continues on page 193

	Gross domestic product		Exports of goods and services		Imports of goods and services		GDP deflator		Current account balance		Gross international reserves	
	annual % growth		annual % growth		annual % growth		% growth		% of GDP		$ millions	months of import coverage
	1999	2000	1999	2000	1999	2000	1999	2000	1999	2000	2000	2000
Malawi	4.0	3.2	−18.6	−4.8	8.1	−8.6	42.2	28.8	..	−15.3	..	..
Malaysia	5.8	8.6	13.4	19.8	10.8	26.5	−0.2	2.5	15.9	3.0	34,975	3.8
Mauritius	3.4	7.7	1.4	7.2	4.6	3.4	5.8	6.2	−1.2	−0.7	867	3.3
Mexico	3.5	7.2	13.9	16.2	12.8	20.4	16.2	9.5	−2.9	−3.2	33,595	1.9
Moldova	−4.4	0.0	−24.5	13.9	−39.5	13.8	39.9	31.3	−2.0	−8.5	206	2.4
Morocco	−0.7	0.9	6.7	−1.3	5.3	−3.8	0.9	2.5	−0.5	−2.5	..	..
Nicaragua	7.0	5.5	6.7	23.3	19.8	2.5	11.2	10.9	−25.9	−28.6	567	3.1
Nigeria	1.0	3.4	−12.2	2.6	18.6	13.0	12.9	9.5	1.4	0.6	..	..
Pakistan	4.0	6.3	−2.4	14.2	−6.2	1.0	4.7	2.7	−3.8	−1.7	2,805	2.4
Panama	3.0	2.3	6.9	5.8	5.3	−0.8	−0.7	1.5	−13.9	−10.4	938	1.8
Papua New Guinea	3.2	5.4	3.7	2.4	4.2	10.0	12.2	5.6	3.3	−4.4	361	1.6
Paraguay	−0.8	1.0	−16.9	−1.3	−26.7	−4.5	3.9	11.0	−3.0	−1.5	797	2.7
Peru	1.4	4.0	5.4	9.5	−17.1	10.0	3.8	4.4	−3.5	−2.8	8,976	9.1
Philippines	3.2	4.2	3.6	6.6	−2.8	0.2	8.2	4.2	10.3	4.7	16,302	3.9
Poland	4.1	4.5	−2.6	6.2	1.0	−1.3	6.8	9.0	−8.0	−7.1	31,142	7.0
Romania	−3.2	2.2	8.9	..	−4.8	..	46.4	..	−3.8	..	..	..
Russian Federation	3.2	6.5	−0.6	−0.8	−27.1	16.3	63.4	34.5	5.2	19.2	25,960	4.5
Slovak Republic	1.9	1.8	3.6	18.5	−6.1	10.1	6.6	6.9	−5.9	−2.8	5,136	3.9
South Africa	1.2	2.2	0.0	4.7	−7.0	8.6	6.9	6.0	−0.4	−0.7	11,319	3.7
Sri Lanka	4.3	6.0	4.0	8.9	7.0	12.0	4.6	3.6	−3.1	−5.3	1,041	1.5
Sudan	5.2	7.9	..	..	..	..	16.0	12.0	−16.0	..	240	0.8
Swaziland	2.0	3.0	3.3	5.9	−0.1	0.9	9.5	8.0	1.4	−3.3	397	2.6
Syrian Arab Republic	5.3	5.5	4.7	3.9	5.3	5.5	5.7	5.2	1.0	−1.7	..	..
Thailand	4.2	4.5	8.9	19.9	20.2	19.6	−2.6	1.0	10.0	9.0	36,343	5.4
Trinidad and Tobago	6.8	5.6	−5.9	9.3	−7.5	9.4	5.2	3.4	..	7.4	1,365	3.9
Tunisia	6.2	5.0	5.3	5.7	7.9	6.6	3.5	3.2	−2.1	−3.2	..	..
Turkey	−5.1	7.2	−7.0	10.9	−3.7	30.6	56.2	56.3	−0.7	−5.1	37,205	6.6
Uganda	7.4	5.1	33.0	−10.8	2.8	−0.1	4.4	3.5	−11.6	−12.9	824	5.1
Ukraine	−0.4	5.5	−7.9	15.7	−19.1	13.0	24.4	25.0	4.3	5.1	1,551	1.0
Uruguay	−3.2	0.5	−8.6	6.0	−4.3	5.5	4.8	5.0	−2.9	−2.9	2,745	6.6
Uzbekistan	4.4	3.0	−1.9	−6.6	−12.2	−13.4	38.5	44.5	−0.1	1.2	1,100	4.4
Venezuela, RB	−7.2	2.5	−11.1	1.8	−21.0	27.0	27.6	13.4	3.6	10.8	21,263	10.4
Zambia	2.4	4.0	4.9	5.6	1.6	9.4	21.7	25.9	..	−16.3	..	..
Zimbabwe	0.1	−5.5	5.3	−16.1	2.3	−20.5	48.1	59.9	..	−2.0	..	..

Note: Data for 2000 are the latest preliminary estimates and may differ from those in earlier World Bank publications.
Source: World Bank staff estimates.

Table 4b	**Key macroeconomic indicators**											
	Nominal exchange rate			**Real effective exchange rate**		**Money and quasi money**		**Gross domestic credit**		**Real interest rate**		**Short-term debt[a]**
	local currency units per $ 2000	% change 1999	% change 2000	1995 = 100 1999	2000	annual % growth 1999	2000	annual % growth 1999	2000	% 1999	% 2000	% of exports 1999
Malawi	80.1	5.8	72.4	111.7	114.4	26.5	46.4	75.3	36.0	8.0	40.5	11.2
Malaysia	3.8	0.0	0.0	83.3	88.0	16.9	11.9	0.3	8.5	7.5	5.2	7.7
Mauritius	27.9	2.8	9.5	..	..	15.2	11.4	9.0	6.2	15.0	14.5	21.3
Mexico	9.6	–3.6	0.6	..	..	11.8	–1.5	0.9	2.3	8.4	12.6	15.1
Moldova	12.4	39.3	6.8	100.1	114.8	42.9	41.7	18.1	14.4	–3.1	..	4.6
Morocco	10.6	9.0	5.3	106.1	112.2	10.2	..	2.6	..	12.5	..	1.4
Nicaragua	13.1	10.0	6.0	104.4	116.8	18.8	..	15.9	..	9.8	..	79.4
Nigeria	..	347.5	..	78.9	92.9	31.7	..	179.6	..	6.5	..	43.4
Pakistan	58.0	12.9	12.1	92.4	93.7	4.3	11.3	4.5	12.6	..	..	19.0
Panama	1.0	0.0	0.0	..	..	8.5	..	19.0	..	10.8	..	8.8
Papua New Guinea	..	28.6	..	84.2	113.7	9.2	..	1.8	..	5.9	..	11.4
Paraguay	3,526.9	17.2	5.9	97.7	101.9	11.7	9.4	1.9	20.2	25.4	..	21.4
Peru	3.5	11.1	0.5	..	..	14.5	–0.6	19.7	1.8	25.9	1.0	70.6
Philippines	50.0	3.2	24.0	96.4	83.2	16.3	16.9	2.3	12.0	3.3	7.0	12.2
Poland	4.1	18.4	–0.1	112.3	123.3	19.4	14.4	20.5	7.9	9.5	15.4	14.5
Romania	25,926.0	66.7	42.0	97.8	108.7	44.9	37.4	20.1	3.3	..	..	9.4
Russian Federation	28.2	30.8	4.3	80.9	102.4	56.7	60.2	34.1	12.0	–14.5	3.9	18.4
Slovak Republic	49.8	14.5	17.0	100.0	107.6	11.6	..	1.4	..	13.6	..	12.7
South Africa	7.6	5.0	23.0	84.6	79.4	10.9	7.5	9.0	14.8	10.4	8.0	39.7
Sri Lanka	82.6	5.7	14.4	..	..	12.4	..	18.5	..	2.3	..	13.9
Sudan	257.4	8.4	–0.1	..	..	23.5	..	22.8	..	..	..	689.2
Swaziland	7.6	5.0	23.0	..	..	15.6	–9.5	6.6	44.6	5.0	8.6	4.5
Syrian Arab Republic	11.2	0.0	0.0	..	..	13.4	..	–4.5	..	..	..	107.1
Thailand	42.3	2.1	12.8	..	..	5.4	3.6	–4.2	–8.3	11.9	4.7	31.4
Trinidad and Tobago	6.3	–4.5	0.0	110.2	119.7	4.2	..	1.6	..	11.3	..	24.1
Tunisia	..	13.8	..	101.5	103.5	18.9	16.5	13.6	19.1	..	..	15.9
Turkey	683,746.0	72.2	32.8	..	..	98.3	54.2	82.5	61.6	..	..	44.6
Uganda	1,766.7	10.5	17.3	90.1	92.1	13.6	13.8	15.3	108.8	16.5	16.8	18.2
Ukraine	5.4	52.2	4.2	126.6	118.9	41.3	39.2	30.5	18.5	24.5	20.3	1.8
Uruguay	12.5	7.4	7.7	112.5	116.1	13.1	8.0	12.8	7.6	46.2	39.3	42.4
Uzbekistan	..	..	..	..	..	..	..	..	..	..	..	22.6
Venezuela, RB	699.8	14.8	7.9	152.3	169.2	20.9	36.6	7.2	7.0	3.5	..	9.4
Zambia	4,002.6	14.5	59.1	111.9	106.3	27.7	48.5	13.3	36.8	15.4	25.9	11.8
Zimbabwe	55.1	2.1	44.4	..	..	35.9	..	6.8	..	4.9	..	29.1

Note: Data for 2000 are preliminary and may not cover the entire year.
a. More recent data on short-term debt are available on a Web site maintained by the Bank for International Settlements, the International Monetary Fund, the Organisation for Economic Co-operation and Development, and the World Bank: www.oecd.org/dac/debt.
Source: International Monetary Fund, *International Financial Statistics;* and World Bank, Debtor Reporting System.

4.1 | Growth of output

	Gross domestic product		Agriculture		Industry		Manufacturing		Services	
	average annual % growth		average annual % growth		average annual % growth		average annual % growth		average annual % growth	
	1980–90	1990–99	1980–90	1990–99	1980–90	1990–99	1980–90	1990–99	1980–90	1990–99
Albania	1.5	3.2	1.9	6.4	2.1	–2.1	..	–8.4	–0.4	5.3
Algeria	2.7	1.6	4.6	3.0	2.3	1.0	3.3	–5.7	3.6	2.7
Angola	3.4	0.4	0.5	–3.0	6.4	2.7	–11.1	–1.4	1.3	–2.4
Argentina	–0.7	4.9	0.7	3.8	–1.3	4.6	–0.8	3.5	0.0	5.0
Armenia	..	–3.2	..	0.2	..	–7.0	..	–8.0	..	–4.2
Australia	3.5	4.1	3.4	3.0	2.9	3.0	1.9	2.0	3.7	4.4
Austria	2.2	1.9	1.1	0.4	1.9	1.9	2.7	1.5	2.5	2.0
Azerbaijan	..	–9.6	..	2.3	..	3.9	..	..	..	3.1
Bangladesh	4.3	4.7	2.7	2.5	4.9	7.4	3.0	7.5	4.4	4.4
Belarus	..	–3.0	..	–5.4	..	–3.8	..	–2.6	..	–1.3
Belgium	2.0	1.7	2.0	1.2	2.2	1.5	2.7	1.4	1.8	1.8
Benin	2.5	4.7	5.1	5.3	3.4	3.8	5.1	5.6	0.7	4.4
Bolivia	–0.2	4.2	..	3.2	..	4.5	..	4.5	..	4.5
Bosnia and Herzegovina	..	35.2	..	14.8	..	35.2	..	31.3	..	46.6
Botswana	10.3	4.3	3.3	0.3	10.2	2.8	8.7	3.9	11.7	6.3
Brazil	2.7	3.0	2.8	3.3	2.0	2.7	1.6	2.1	3.3	3.0
Bulgaria	3.4	–2.7	–2.1	0.3	5.2	–4.7	..	..	4.5	–1.8
Burkina Faso	3.6	3.8	3.1	3.3	3.8	4.0	2.0	4.2	4.6	3.6
Burundi	4.4	–2.9	3.1	–2.0	4.5	–6.7	5.7	–8.0	5.6	–2.5
Cambodia	..	4.8	..	2.1	..	9.6	..	8.2	..	6.9
Cameroon	3.4	1.3	2.2	5.3	5.9	–2.0	5.0	0.1	2.1	0.1
Canada	3.3	2.7	2.0	1.0	2.8	2.6	3.3	3.8	3.4	2.5
Central African Republic	1.4	1.8	1.6	3.7	1.4	0.4	5.0	–0.4	1.0	–0.7
Chad	6.1	2.1	2.3	4.8	8.1	1.0	..	..	6.7	1.2
Chile	4.2	7.2	5.9	1.1	3.5	6.2	3.4	5.0	2.9	7.6
China	10.1	10.7	5.9	4.3	11.1	14.4	10.4	13.9	13.5	9.2
Hong Kong, China	6.9	3.9	..	..	..	..	..	..	..	..
Colombia	3.6	3.3	2.9	–2.0	5.0	1.4	3.5	–2.9	3.1	5.2
Congo, Dem. Rep.	1.6	–5.1	2.5	2.9	0.9	–11.7	1.6	–13.4	1.3	–15.2
Congo, Rep.	3.3	–0.5	3.4	1.7	5.2	–0.2	6.8	–1.9	2.1	–1.4
Costa Rica	3.0	5.1	3.1	4.3	2.8	6.0	3.0	6.4	3.1	4.7
Côte d'Ivoire	0.7	3.7	0.3	3.1	4.4	6.4	3.0	5.1	–0.3	2.8
Croatia	..	0.2	..	–2.5	..	–3.6	..	–4.9	..	1.9
Cuba	..	..	..	..	..	..	..	..	..	..
Czech Republic	..	0.8	..	2.7	..	–0.1	..	..	..	1.0
Denmark	2.0	2.4	2.6	3.2	2.0	1.6	1.6	1.2	1.9	2.6
Dominican Republic	3.1	5.8	0.4	3.6	3.6	6.9	2.9	4.8	3.5	5.7
Ecuador	2.0	2.2	4.4	2.3	1.2	3.0	0.0	2.4	1.7	1.6
Egypt, Arab Rep.	5.4	4.4	2.7	3.1	5.2	3.9	..	6.0	6.6	4.8
El Salvador	0.2	5.0	–1.1	1.1	0.1	5.4	–0.2	5.3	0.7	5.8
Eritrea	..	5.0	..	..	..	..	..	..	..	..
Estonia	2.2	–1.3	..	–3.7	..	–4.2	..	..	..	1.1
Ethiopia	1.1	4.6	0.2	2.3	0.4	5.7	–0.9	5.8	3.1	6.8
Finland	3.3	2.4	–0.4	1.1	3.3	4.2	3.4	5.8	3.6	1.8
France	2.4	1.5	1.3	1.9	1.4	0.9	..	..	3.0	1.5
Gabon	0.9	3.2	1.2	–1.9	1.5	2.6	1.8	0.6	0.1	4.5
Gambia, The	3.6	2.8	0.9	2.0	4.7	0.7	7.8	0.8	2.7	4.1
Georgia	..	..	..	..	..	..	..	..	..	..
Germany[a]	2.2	1.3	1.7	2.2	1.2	0.3	..	–0.3	2.9	1.9
Ghana	3.0	4.3	1.0	3.4	3.3	2.4	3.9	–4.5	5.7	5.8
Greece	1.8	2.2	–0.1	0.4	1.3	1.1	0.5	0.5	2.7	2.4
Guatemala	0.8	4.2	1.2	2.8	–0.2	4.4	0.0	2.8	0.9	4.7
Guinea	..	4.2	..	4.5	..	4.5	..	3.7	..	3.3
Guinea-Bissau	4.0	0.3	4.7	3.4	2.2	2.3	..	4.1	3.5	–7.3
Haiti	0.0	–1.3	–0.1	–3.8	–1.7	0.7	–1.7	–7.3	0.9	–0.2
Honduras	2.7	3.3	2.7	2.3	3.3	3.6	3.7	3.9	2.5	3.8

Growth of output | 4.1

	Gross domestic product		Agriculture		Industry		Manufacturing		Services	
	average annual % growth		average annual % growth		average annual % growth		average annual % growth		average annual % growth	
	1980–90	1990–99	1980–90	1990–99	1980–90	1990–99	1980–90	1990–99	1980–90	1990–99
Hungary	1.3	1.0	1.7	–3.1	0.2	2.3	..	7.2	2.1	0.8
India	5.8	6.0	3.1	3.4	6.9	6.7	7.4	7.5	7.0	7.8
Indonesia	6.1	4.7	3.4	2.3	6.9	6.5	12.6	7.6	7.0	4.0
Iran, Islamic Rep.	1.7	3.6	4.5	4.2	3.3	2.8	4.5	4.8	–1.0	4.7
Iraq	–6.8	..	..	..	..	..	..	..	..	..
Ireland	3.2	6.9	..	..	..	..	..	..	..	..
Israel	3.5	5.2	..	..	..	..	..	..	..	..
Italy	2.4	1.4	0.5	1.7	1.7	1.1	2.3	1.5	2.9	1.5
Jamaica	2.0	0.3	0.6	2.2	2.4	–0.5	2.7	–1.9	1.8	0.6
Japan	4.0	1.3	1.3	–1.6	4.2	0.8	4.8	1.2	4.0	2.2
Jordan	2.5	5.3	6.8	–1.9	1.7	5.6	0.5	5.9	2.2	5.1
Kazakhstan	..	–5.9	..	–13.4	..	–10.1	..	..	..	2.2
Kenya	4.2	2.2	3.3	1.4	3.9	1.9	4.9	2.4	4.9	3.5
Korea, Dem. Rep.	..	..	..	..	..	..	..	..	..	..
Korea, Rep.	9.4	5.7	2.8	2.1	12.0	6.2	13.0	7.1	8.9	5.8
Kuwait	1.3	..	14.7	..	1.0	..	2.3	..	2.1	..
Kyrgyz Republic	..	–5.4	..	0.8	..	–12.3	..	–3.7	..	–7.4
Lao PDR	3.7	6.6	3.5	4.6	6.1	11.8	8.9	12.6	3.3	6.5
Latvia	3.7	–4.8	2.8	–8.2	4.6	–10.5	4.6	–9.8	3.4	1.9
Lebanon	..	7.7	..	..	..	..	..	..	..	..
Lesotho	4.6	4.4	2.8	2.0	5.5	6.3	8.5	7.9	4.0	5.2
Libya	–5.7	..	..	..	..	..	..	..	..	..
Lithuania	..	–4.0	..	–1.1	..	–8.1	..	–10.3	..	–0.1
Macedonia, FYR	..	–0.8	..	–0.4	..	–4.7	..	..	..	0.1
Madagascar	1.1	1.7	2.5	1.5	0.9	1.9	2.1	0.6	0.3	1.9
Malawi	2.5	3.6	2.0	7.5	2.9	1.2	3.6	–2.7	3.3	3.5
Malaysia	5.3	7.3	3.4	0.2	6.8	8.8	9.3	9.7	4.9	8.0
Mali	0.8	3.6	3.3	2.8	4.3	6.5	6.8	3.2	1.9	2.7
Mauritania	1.8	4.2	1.7	5.0	4.9	2.6	–2.1	–0.9	0.4	4.7
Mauritius	6.2	5.1	2.9	–0.7	10.3	5.4	11.1	5.4	5.5	6.3
Mexico	1.1	2.7	0.8	1.6	1.1	3.5	1.5	4.0	1.4	2.5
Moldova	..	–11.0	..	–16.0	..	–15.9	..	..	..	1.8
Mongolia	5.4	0.7	1.4	3.1	6.6	–1.0	..	..	5.9	1.3
Morocco	4.2	2.3	6.7	–0.4	3.0	3.1	4.1	2.6	4.2	2.6
Mozambique	–0.1	6.2	6.6	5.5	–4.5	12.6	..	17.6	9.1	2.7
Myanmar	0.6	6.3	0.5	4.9	0.5	10.1	–0.2	6.7	0.8	6.6
Namibia	1.3	3.4	1.9	3.8	–0.6	2.5	3.7	4.3	2.3	3.4
Nepal	4.6	4.9	4.0	2.4	8.7	7.2	9.3	9.5	3.9	6.3
Netherlands	2.3	2.7	3.4	2.1	1.6	1.7	2.3	2.2	2.6	3.1
New Zealand	1.7	3.1	3.8	2.7	1.1	2.4	..	..	2.9	3.7
Nicaragua	–1.9	3.2	–2.2	5.3	–2.3	3.9	–3.2	1.6	–1.5	1.5
Niger	–0.1	2.4	1.7	3.6	–1.7	1.8	–2.7	2.3	–0.7	1.6
Nigeria	1.6	2.4	3.3	2.9	–1.1	1.7	0.7	2.0	3.7	2.8
Norway	2.8	3.8	0.1	2.6	4.0	4.2	..	..	2.3	3.6
Oman	8.4	5.9	7.9	..	10.3	..	20.6	..	5.9	..
Pakistan	6.3	3.8	4.3	4.4	7.3	4.0	7.7	3.7	6.8	4.4
Panama	0.5	4.2	2.5	2.1	–1.3	5.9	0.4	3.5	0.7	4.1
Papua New Guinea	1.9	4.7	1.8	4.4	1.9	6.7	0.1	6.3	2.0	3.3
Paraguay	2.5	2.4	3.6	2.8	0.3	3.0	4.0	0.5	3.1	1.9
Peru	–0.1	5.0	3.0	5.6	0.1	6.2	–0.2	4.2	–0.4	4.2
Philippines	1.0	3.2	1.0	1.4	–0.9	3.2	0.2	2.9	2.8	4.0
Poland	..	4.5	..	–0.1	..	3.8	..	..	..	4.1
Portugal	3.1	2.5	..	0.4	..	2.7	..	..	..	2.4
Puerto Rico	4.0	3.1	1.8	..	3.6	..	1.5	..	4.6	..
Romania	0.5	–0.8	..	0.1	..	–0.8	..	..	..	–0.8
Russian Federation	..	–6.1	..	–7.9	..	–9.6	..	..	..	–2.2

	Gross domestic product		Agriculture		Industry		Manufacturing		Services	
	average annual % growth		average annual % growth		average annual % growth		average annual % growth		average annual % growth	
	1980–90	1990–99	1980–90	1990–99	1980–90	1990–99	1980–90	1990–99	1980–90	1990–99
Rwanda	2.2	–1.5	0.5	–3.9	2.5	2.0	2.6	6.1	5.5	–1.2
Saudi Arabia	0.0	1.6	13.4	0.7	–2.3	1.5	7.5	2.7	1.3	2.2
Senegal	3.1	3.3	2.8	1.4	4.3	4.4	4.6	3.7	2.8	3.6
Sierra Leone	1.2	–4.7	3.1	1.0	1.7	–4.6	..	5.0	–2.7	–10.8
Singapore	6.7	8.0	–6.2	0.4	5.3	7.9	6.6	6.7	7.6	8.0
Slovak Republic	2.0	1.8	1.6	0.7	2.0	–3.5	..	..	0.8	6.8
Slovenia	..	2.4	..	0.2	..	2.5	..	3.8	..	3.8
South Africa	1.0	1.9	2.9	1.0	0.7	0.9	1.1	1.1	2.4	2.4
Spain	3.0	2.2	..	..	..	..	..	..	..	..
Sri Lanka	4.0	5.3	2.2	1.8	4.6	7.0	6.3	8.2	4.7	6.1
Sudan	0.4	8.2	–0.6	14.3	2.5	5.6	3.4	2.8	1.7	3.3
Sweden	2.3	1.6	1.5	0.0	2.8	3.0	..	..	2.6	1.2
Switzerland	2.0	0.6	..	..	..	..	..	..	..	..
Syrian Arab Republic	1.5	5.7	–0.6	..	6.6	..	..	..	0.1	..
Tajikistan	..	..	..	..	..	..	..	..	..	..
Tanzania[b]	..	2.8	..	3.2	..	2.5	..	2.3	..	2.4
Thailand	7.6	4.7	3.9	2.5	9.8	5.7	9.5	6.7	7.3	4.4
Togo	1.7	2.4	5.6	4.5	1.1	2.5	1.7	2.4	–0.3	0.6
Trinidad and Tobago	–0.8	2.7	3.9	1.8	–5.5	2.8	–10.1	4.0	–2.4	2.6
Tunisia	3.3	4.6	2.8	2.1	3.1	4.5	3.7	5.4	3.5	5.3
Turkey	5.4	3.8	1.3	1.4	7.8	4.3	7.9	5.1	4.4	3.9
Turkmenistan	..	–6.8	..	–8.2	..	–5.5	..	..	..	–6.9
Uganda	2.9	7.2	2.1	3.7	5.0	12.7	3.7	14.2	2.8	8.1
Ukraine	..	–10.7	..	–6.3	..	–13.5	..	–13.4	..	–3.1
United Arab Emirates	–2.1	2.9	9.6	..	–4.2	..	3.1	..	3.6	..
United Kingdom	3.2	2.5	2.1	–0.2	3.1	1.3	..	..	3.1	3.1
United States	3.6	3.3	..	..	..	..	..	..	..	..
Uruguay	0.5	3.8	0.1	3.7	–0.2	1.3	0.4	0.1	1.0	5.0
Uzbekistan	..	–1.2	..	–0.4	..	–4.0	..	..	..	–0.4
Venezuela, RB	1.1	1.7	3.1	1.5	1.7	3.3	4.4	1.3	0.5	0.2
Vietnam	4.6	8.1	4.3	4.9	..	12.5	..	..	..	8.1
West Bank and Gaza	..	3.7	..	–3.8	..	1.2	..	4.0	..	3.8
Yemen, Rep.	..	3.2	..	5.0	..	7.9	..	4.5	..	–1.6
Yugoslavia, FR (Serb./Mont.)	..	..	..	..	..	..	..	..	..	..
Zambia	1.0	0.2	3.6	9.4	1.0	–3.9	4.0	0.7	0.5	0.3
Zimbabwe	3.6	2.8	3.1	4.6	3.2	0.7	2.8	0.7	3.1	3.5
World	**3.4 w**	**2.5 w**	**2.7 w**	**1.7 w**	**.. w**	**2.3 w**	**.. w**	**.. w**	**.. w**	**.. w**
Low income	4.7	3.2	3.0	2.5	5.4	2.8	7.7	2.7	5.6	4.7
Middle income	3.3	3.5	3.6	2.0	3.7	4.3	4.6	6.3	3.6	3.7
Lower middle income	4.2	3.4	4.0	2.0	6.2	5.1	6.7	9.0	5.3	3.6
Upper middle income	2.6	3.6	2.9	2.0	2.4	3.6	3.4	4.1	2.9	3.8
Low & middle income	3.5	3.5	3.4	2.2	3.9	4.0	4.9	5.8	3.9	3.8
East Asia & Pacific	8.0	7.5	4.4	3.3	9.5	9.8	10.4	10.2	8.8	6.5
Europe & Central Asia	..	–2.3	..	–2.9	..	–3.2	..	..	..	0.8
Latin America & Carib.	1.7	3.4	2.3	2.3	1.4	3.4	1.3	2.6	1.8	3.5
Middle East & N. Africa	2.0	3.0	5.6	2.6	0.4	2.1	..	2.8	2.2	3.4
South Asia	5.6	5.6	3.2	3.4	6.8	6.5	7.1	7.0	6.5	6.9
Sub-Saharan Africa	1.7	2.2	2.3	2.7	1.2	1.5	1.7	1.6	2.4	2.4
High income	3.4	2.3	..	..	..	1.4	..	..	..	..
Europe EMU	..	1.8	..	2.2	..	1.2	..	0.8	..	1.9

a. Data prior to 1990 refer to the Federal Republic of Germany before unification. b. Data cover mainland Tanzania only.

Growth of output | 4.1

An economy's growth is measured by the change in the volume of its output or in the real incomes of persons resident in the economy. The 1993 United Nations System of National Accounts (1993 SNA) offers three plausible indicators from which to calculate growth: the volume of gross domestic product, real gross domestic income, and real gross national income. The volume of GDP is the sum of value added, measured at constant prices, by households, government, and the enterprises operating in the economy. This year's edition of the *World Development Indicators* continues to follow the practice of past editions, measuring the growth of the economy by the change in GDP measured at constant prices.

Each industry's contribution to the growth in the economy's output is measured by the growth in value added by the industry. In principle, value added in constant prices can be estimated by measuring the quantity of goods and services produced in a period, valuing them at an agreed set of base year prices, and subtracting the cost of inputs, also in constant prices. This double deflation method, recommended by the 1993 SNA and its predecessors, requires detailed information on the structure of prices of inputs and outputs.

In many industries, however, value added is extrapolated from the base year using single volume indexes of outputs or, more rarely, inputs. Particularly in the service industries, including most of government, value added in constant prices is often imputed from labor inputs, such as real wages or the number of employees. In the absence of well-defined measures of output, measuring the growth of services remains difficult.

Moreover, technical progress can lead to improvements in production and in the quality of goods and services that if not properly accounted for can distort measures of value added and thus of growth. When inputs are used to estimate output, as is the case for nonmarket services, unmeasured technical progress leads to underestimates of the volume of output. Similarly, unmeasured changes in the quality of goods and services produced lead to underestimates of the value of output and value added. The result can be underestimates of growth and productivity change, and overestimates of inflation.

Informal economic activities pose a particular measurement problem, especially in developing countries, where much economic activity may go unrecorded. Obtaining a complete picture of the economy requires estimating household outputs produced for local sale and home use, barter exchanges, and illicit or deliberately unreported activity. How consistent and complete such estimates will be depends on the skill and methods of the compiling statisticians and the resources available to them.

Rebasing national accounts

When countries rebase their national accounts, they update the weights assigned to various components to reflect better the current pattern of production (or consumption). The new base year should represent normal operation of the economy—that is, it should be a year without major shocks or distortions—but the choice of base year is often arbitrary. Some developing countries have not rebased their national accounts for many years. Using an old base year can be misleading because implicit price and volume weights become progressively less relevant and useful.

To obtain comparable series of constant price data, the World Bank rescales GDP and value added by industrial origin to a common reference year, currently 1995. This process gives rise to a discrepancy between the rescaled GDP and the sum of the rescaled components. Because allocating the discrepancy would give rise to distortions in the growth rates, the discrepancy is left unallocated. As a result, the weighted average of the growth rates of the components generally will not equal the GDP growth rate.

Growth rates of GDP and its components are calculated using constant price data in the local currency. Regional and income group growth rates are calculated after converting local currencies to constant price U.S. dollars using an exchange rate in the common reference year. The growth rates in the table are annual average compound growth rates. Methods of computing growth rates and the alternative conversion factor are described in *Statistical methods*.

Changes in the System of National Accounts

For the first time, this year's edition of the *World Development Indicators* uses terminology in line with the 1993 SNA. Most countries continue to compile their national accounts according to the System of National Accounts version 3, referred to as the 1968 SNA, but more and more are adopting the 1993 SNA. Countries that use the 1993 SNA are identified in *Primary data documentation*. A few low-income countries still use concepts from older SNA guidelines, including valuations such as factor cost, in describing major economic aggregates.

• **Gross domestic product** (GDP) at purchaser prices is the sum of the gross value added by all resident producers in the economy plus any product taxes and minus any subsidies not included in the value of the products. It is calculated without making deductions for depreciation of fabricated assets or for depletion and degradation of natural resources. Value added is the net output of an industry after adding up all outputs and subtracting intermediate inputs. The industrial origin of value added is determined by the International Standard Industrial Classification (ISIC) revision 3. • **Agriculture** corresponds to ISIC divisions 1–5 and includes forestry and fishing. • **Industry** comprises mining, manufacturing (also reported as a separate subgroup), construction, electricity, water, and gas (ISIC divisions 10–45). • **Manufacturing** refers to industries belonging to divisions 15–37. • **Services** correspond to ISIC divisions 50–99.

The national accounts data for most developing countries are collected from national statistical organizations and central banks by visiting and resident World Bank missions. The data for high-income economies come from Organisation for Economic Co-operation and Development (OECD) data files; for information on the OECD's national accounts series see its *Main Economic Indicators* (monthly). The World Bank rescales constant price data to a common reference year. The complete national accounts time series is available on the *World Development Indicators 2001* CD-ROM. The United Nations Statistics Division publishes detailed national accounts for United Nations member countries in *National Accounts Statistics: Main Aggregates and Detailed Tables* and publishes updates in the *Monthly Bulletin of Statistics*.

4.2 | Structure of output

	Gross domestic product $ millions		Agriculture value added % of GDP		Industry value added % of GDP		Manufacturing value added % of GDP		Services value added % of GDP	
	1990	1999	1990	1999	1990	1999	1990	1999	1990	1999
Albania	2,102	3,676	36	53	48	26	42	12	16	21
Algeria	61,902	47,872	14	11	45	51	12	10	41	38
Angola	10,260	8,545	18	7	41	77	5	4	41	16
Argentina	141,352	283,166	8	5	36	28	27	18	56	67
Armenia	4,124	1,845	17	29	52	33	33	23	31	39
Australia	310,041	404,033	3	3	26	25	13	13	70	72
Austria	162,288	208,173	5	2	31	29	21	19	64	69
Azerbaijan	9,837	4,004	..	23	..	35	..	5	..	41
Bangladesh	30,129	45,961	29	25	21	24	13	15	50	50
Belarus	34,911	26,815	24	13	47	42	39	35	29	45
Belgium	197,787	248,404	2	1	28	25	21	18	69	73
Benin	1,845	2,369	36	38	13	14	8	8	51	48
Bolivia	4,868	8,323	26	18	20	18	17	15	54	64
Bosnia and Herzegovina	..	4,387	..	15	..	27	..	21	..	58
Botswana	3,766	5,996	5	4	56	45	5	5	39	51
Brazil	464,989	751,505	8	9	39	31	25	23	53	61
Bulgaria	20,726	12,403	18	15	51	23	..	15	31	62
Burkina Faso	2,765	2,580	32	31	22	28	16	22	45	40
Burundi	1,132	714	56	52	19	17	13	9	25	30
Cambodia	1,115	3,117	56	51	11	15	5	6	33	35
Cameroon	11,152	9,187	25	44	29	19	15	10	46	38
Canada	572,673	634,898	2	..	29	..	16	..	69	..
Central African Republic	1,488	1,053	48	55	20	20	11	9	33	25
Chad	1,739	1,530	29	36	18	15	14	12	53	49
Chile	30,323	67,469	9	8	41	34	20	16	50	57
China	354,644	989,465	27	18	42	49	33	38	31	33
Hong Kong, China	74,784	158,943	0	0	25	15	18	6	74	85
Colombia	40,274	86,605	17	13	38	26	21	14	45	61
Congo, Dem. Rep.	9,348	5,584	30	58	28	17	11	..	42	25
Congo, Rep.	2,799	2,217	13	10	41	49	8	7	46	41
Costa Rica	7,188	15,148	18	11	29	37	22	30	53	53
Côte d'Ivoire	10,796	11,206	32	26	23	26	21	21	44	48
Croatia	18,156	20,426	10	9	34	32	28	20	56	59
Cuba	..	..	..	..	..	..	..	..	..	..
Czech Republic	34,880	53,111	8	4	49	43	..	..	43	53
Denmark	133,361	174,280	4	2	23	21	16	14	73	76
Dominican Republic	7,074	17,398	13	11	31	34	18	17	55	54
Ecuador	10,686	18,991	13	12	38	37	19	21	49	50
Egypt, Arab Rep.	43,130	89,148	19	17	29	32	18	20	52	51
El Salvador	4,807	12,467	17	10	26	29	22	23	57	60
Eritrea	437	645	29	17	19	29	13	15	52	54
Estonia	6,760	5,233	17	6	50	26	42	15	34	69
Ethiopia	6,842	6,439	49	52	13	11	8	7	38	37
Finland	136,794	129,661	6	3	29	28	20	21	65	68
France	1,215,893	1,432,323	3	3	27	23	..	..	70	74
Gabon	5,952	4,352	7	8	43	41	6	5	50	51
Gambia, The	317	393	29	31	13	13	7	6	58	56
Georgia	..	2,737	..	36	..	13	..	8	..	51
Germany	1,770,368	2,111,940	1	1	33	28	26	21	64	71
Ghana	5,886	7,774	45	36	17	25	10	9	38	39
Greece	84,925	125,088	10	7	26	20	15	11	65	72
Guatemala	7,650	18,215	26	23	20	20	15	13	54	57
Guinea	2,818	3,482	24	24	33	37	5	4	43	39
Guinea-Bissau	244	218	61	62	19	12	8	10	21	26
Haiti	2,981	4,302	33	29	22	22	16	7	45	48
Honduras	3,049	5,387	22	16	26	32	16	20	51	52

Structure of output 4.2

	Gross domestic product		Agriculture value added		Industry value added		Manufacturing value added		Services value added	
	$ millions		% of GDP		% of GDP		% of GDP		% of GDP	
	1990	1999	1990	1999	1990	1999	1990	1999	1990	1999
Hungary	33,056	48,436	15	6	39	34	23	25	46	61
India	316,211	447,292	31	28	28	26	17	16	41	46
Indonesia	114,427	142,511	19	19	39	43	21	25	41	37
Iran, Islamic Rep.	120,404	110,791	24	21	29	31	12	17	48	48
Iraq	48,657		..	..	..	..	..	..	..	..
Ireland	47,301	93,410	8	5	32	34	..	..	60	62
Israel	52,490	100,840	..	..	..	..	..	..	..	..
Italy	1,102,435	1,170,971	3	3	31	26	22	19	66	71
Jamaica	4,239	6,889	6	7	43	32	20	14	50	61
Japan	2,970,043	4,346,922	3	2	41	36	28	24	56	62
Jordan	4,020	8,073	8	2	28	26	15	16	64	72
Kazakhstan	40,304	15,842	27	11	45	32	9	..	29	57
Kenya	8,533	10,638	29	23	19	16	12	11	52	61
Korea, Dem. Rep.	..	..	..	..	..	..	..	..	..	..
Korea, Rep.	252,622	406,940	9	5	43	44	29	32	48	51
Kuwait	18,428	29,572	1	..	52	..	12	..	47	..
Kyrgyz Republic	..	1,251	34	38	36	27	28	12	30	36
Lao PDR	865	1,432	61	53	15	22	10	17	24	25
Latvia	12,490	6,260	22	4	46	28	34	15	32	68
Lebanon	2,838	17,229	..	12	..	27	..	17	..	61
Lesotho	622	874	23	18	34	38	..	..	43	44
Libya	..	..	..	..	..	..	..	..	..	..
Lithuania	13,254	10,634	27	9	31	32	21	18	42	59
Macedonia, FYR	..	3,452	8	12	47	35	..	..	45	53
Madagascar	3,081	3,721	32	30	14	14	12	11	53	56
Malawi	1,803	1,810	45	38	29	18	19	14	26	45
Malaysia	44,024	79,039	15	11	42	46	24	32	43	43
Mali	2,421	2,670	46	47	16	17	9	4	39	37
Mauritania	1,020	958	30	25	29	29	10	10	42	46
Mauritius	2,642	4,244	12	6	32	33	24	25	56	61
Mexico	262,710	483,737	8	5	28	28	21	21	64	67
Moldova	10,583	1,160	43	25	33	22	..	15	24	53
Mongolia	..	916	15	32	41	30	..	..	44	39
Morocco	25,821	34,998	18	15	32	33	18	17	50	53
Mozambique	2,512	3,979	37	33	18	25	10	13	44	42
Myanmar	..	..	57	60	11	9	8	7	32	31
Namibia	2,340	3,075	12	13	38	33	14	15	50	55
Nepal	3,628	4,995	52	42	16	21	6	9	32	37
Netherlands	295,961	393,692	4	3	28	24	18	16	68	74
New Zealand	43,103	54,651	7	..	26	..	18	..	67	..
Nicaragua	1,009	2,268	31	32	21	23	17	14	48	46
Niger	2,481	2,018	35	41	16	17	7	6	49	42
Nigeria	28,472	35,045	33	39	41	33	6	5	26	28
Norway	115,453	152,943	3	2	31	31	..	..	66	67
Oman	10,535	14,962	3	..	58	..	4	..	39	..
Pakistan	40,010	58,154	26	27	25	23	17	16	49	49
Panama	5,313	9,557	9	7	15	17	9	8	76	76
Papua New Guinea	3,221	3,586	29	30	30	46	9	8	41	24
Paraguay	5,265	7,741	28	29	25	26	17	14	47	45
Peru	26,294	51,933	7	7	38	38	27	24	55	55
Philippines	44,331	76,559	22	18	34	30	25	21	44	52
Poland	61,197	155,166	8	3	48	31	..	18	44	65
Portugal	70,936	113,716	8	4	29	27	..	..	64	69
Puerto Rico	30,604	47,624	1	..	42	..	40	..	57	..
Romania	38,299	34,027	20	16	50	31	..	22	30	53
Russian Federation	579,068	401,442	17	7	48	38	..	..	35	56

	Gross domestic product		Agriculture value added		Industry value added		Manufacturing value added		Services value added	
	$ millions		% of GDP		% of GDP		% of GDP		% of GDP	
	1990	1999	1990	1999	1990	1999	1990	1999	1990	1999
Rwanda	2,584	1,956	33	46	25	20	19	12	42	34
Saudi Arabia	104,670	139,383	6	7	50	48	8	10	43	45
Senegal	5,698	4,752	20	18	19	26	13	17	61	56
Sierra Leone	897	669	47	43	20	27	4	4	33	31
Singapore	36,638	84,945	0	0	35	36	27	26	65	64
Slovak Republic	15,485	19,712	7	4	59	32	..	22	33	64
Slovenia	12,673	20,011	6	4	46	38	35	28	49	58
South Africa	111,997	131,127	5	4	40	32	24	19	55	64
Spain	513,665	595,927	..	4	..	28	..	..	..	69
Sri Lanka	8,032	15,958	26	21	26	27	15	16	48	52
Sudan	13,167	9,718	..	40	..	18	..	9	..	42
Sweden	237,928	238,682	3	..	28	..	..	..	69	..
Switzerland	228,415	258,550	..	..	..	..	..	..	..	..
Syrian Arab Republic	12,309	19,380	29	..	24	..	..	..	48	..
Tajikistan	..	1,870	..	19	..	25	..	21	..	57
Tanzania[a]	4,259	8,760	46	45	18	15	9	7	36	40
Thailand	85,345	124,369	12	10	37	40	27	32	50	50
Togo	1,628	1,405	34	41	23	21	10	9	44	38
Trinidad and Tobago	5,068	6,869	3	2	47	40	9	8	52	58
Tunisia	12,291	20,944	16	13	30	28	17	18	54	59
Turkey	150,721	185,691	18	16	30	24	20	15	52	60
Turkmenistan	..	3,204	32	27	30	45	..	34	38	28
Uganda	4,304	6,411	57	44	11	18	6	9	32	38
Ukraine	91,327	38,653	26	13	45	38	36	33	30	49
United Arab Emirates	34,132	47,234	2	..	64	..	8	..	35	..
United Kingdom	987,641	1,441,787	2	1	31	25	..	..	67	74
United States	5,750,800	9,152,098	..	..	..	..	..	..	..	..
Uruguay	9,287	20,805	9	6	35	27	28	17	56	67
Uzbekistan	23,673	17,705	33	33	33	24	..	11	34	43
Venezuela, RB	48,593	102,222	5	5	50	36	20	14	44	59
Vietnam	6,472	28,682	37	25	23	34	19	18	40	40
West Bank and Gaza	..	4,222	..	9	..	29	..	16	..	62
Yemen, Rep.	4,660	6,825	25	17	28	40	10	11	47	42
Yugoslavia, FR (Serb./Mont.)	..	..	..	..	..	..	..	..	..	..
Zambia	3,288	3,150	21	25	49	24	14	12	30	51
Zimbabwe	8,784	5,608	16	20	33	25	23	17	50	55
World	**21,728,147 t**	**30,876,254 t**	**6** w	**5** w	**34** w	**31** w	**..** w	**..** w	**59** w	**63** w
Low income	878,364	1,033,244	29	26	31	30	18	19	40	44
Middle income	3,520,734	5,518,746	13	10	39	36	25	25	47	54
Lower middle income	1,808,310	2,608,902	21	14	39	39	26	28	40	46
Upper middle income	1,728,727	2,915,898	8	6	40	33	24	23	52	60
Low & middle income	4,393,226	6,551,527	16	12	38	35	24	24	46	53
East Asia & Pacific	927,038	1,894,945	20	14	40	45	29	33	40	41
Europe & Central Asia	1,244,658	1,097,780	17	10	44	33	..	..	40	56
Latin America & Carib.	1,136,103	2,052,720	9	8	36	30	24	21	55	62
Middle East & N. Africa	402,940	613,765	15	14	38	38	12	14	47	48
South Asia	404,001	581,186	31	27	27	26	17	16	43	47
Sub-Saharan Africa	297,444	324,097	18	15	34	29	17	16	48	56
High income	17,320,028	24,323,287	..	..	..	..	..	..	..	..
Europe EMU	5,656,919	6,535,484	3	2	30	27	..	..	67	71

a. Data cover mainland Tanzania only.

About the data

A country's gross domestic product (GDP) represents the sum of value added by all producers in that country. Value added is the value of the gross output of producers less the value of intermediate goods and services consumed in production, excluding the consumption of fixed capital in the production process. Since 1968 the System of National Accounts has called for estimates of value added to be valued at either basic prices (excluding net taxes on products) or producer prices (including net taxes on products paid by the producers, but excluding sales or value added taxes). Both valuations exclude transport charges that are invoiced separately by the producers. Some countries, however, report data at purchaser prices—the prices at which final sales are made (including transport charges)—which may affect estimates of the distribution of output. Total GDP as shown in the table and elsewhere in this book is measured at purchaser prices. Value added by industry is normally measured at basic prices. When value added is measured at producer prices, this is noted in *Primary data documentation*.

While GDP estimates based on the production approach are generally more reliable than estimates compiled from the income or expenditure side, different countries use different definitions, methods, and reporting standards. World Bank staff review the quality of national accounts data and sometimes make adjustments to increase consistency with international guidelines. Nevertheless, significant discrepancies remain between international standards and actual practice. Many statistical offices, especially those in developing countries, face severe limitations in the resources, time, training, and budgets required to produce reliable and comprehensive series of national accounts statistics.

Data problems in measuring output

Among the difficulties faced by compilers of national accounts is the extent of unreported economic activity in the informal or secondary economy. In developing countries a large share of agricultural output is either not exchanged (because it is consumed within the household) or not exchanged for money.

Agricultural production often must be estimated indirectly, using a combination of methods involving estimates of inputs, yields, and area under cultivation. This approach sometimes leads to crude approximations that can differ from the true values over time and across crops for reasons other than climatic conditions or farming techniques. Similarly, agricultural inputs that cannot easily be allocated to specific outputs are frequently "netted

out" using equally crude and ad hoc approximations. For further discussion of the measurement of agricultural production see *About the data* for table 3.3.

Industrial output ideally should be measured through regular censuses and surveys of firms. But in most developing countries such surveys are infrequent, so survey results must be extrapolated using an appropriate indicator. The choice of sampling unit, which may be the enterprise (where responses may be based on financial records) or the establishment (where production units may be recorded separately), also affects the quality of the data. Moreover, much industrial production is organized in unincorporated or owner-operated ventures that are not captured by surveys aimed at the formal sector. Even in large industries, where regular surveys are more likely, evasion of excise and other taxes lowers the estimates of value added. Such problems become more acute as countries move from state control of industry to private enterprise, because new firms enter business and growing numbers of established firms fail to report. In accordance with the System of National Accounts, output should include all such unreported activity as well as the value of illegal activities and other unrecorded, informal, or small-scale operations. Data on these activities need to be collected using techniques other than conventional surveys of firms.

In industries dominated by large organizations and enterprises, such as public utilities, data on output, employment, and wages are usually readily available and reasonably reliable. But in the service industry the many self-employed workers and one-person businesses are sometimes difficult to locate, and they have little incentive to respond to surveys, let alone report their full earnings. Compounding these problems are the many forms of economic activity that go unrecorded, including the work that women and children do for little or no pay. For further discussion of the problems of using national accounts data see Srinivasan (1994) and Heston (1994).

Dollar conversion

To produce national accounts aggregates that are internationally comparable, the value of output must be converted to a common currency. The World Bank conventionally uses the U.S. dollar and applies the average official exchange rate reported by the International Monetary Fund for the year shown. An alternative conversion factor is applied if the official exchange rate is judged to diverge by an exceptionally large margin from the rate effectively applied to transactions in foreign currencies and traded products.

Definitions

• **Gross domestic product** (GDP) at purchaser prices is the sum of the gross value added by all resident producers in the economy plus any product taxes and minus any subsidies not included in the value of the products. It is calculated without making deductions for depreciation of fabricated assets or for depletion and degradation of natural resources. Value added is the net output of an industry after adding up all outputs and subtracting intermediate inputs. The industrial origin of value added is determined by the International Standard Industrial Classification (ISIC) revision 3. • **Value added** is the net output of an industry after adding up all outputs and subtracting intermediate inputs. The industrial origin of value added is determined by the International Standard Industrial Classification (ISIC) revision 3. • **Agriculture** corresponds to ISIC divisions 1–5 and includes forestry and fishing. • **Industry** comprises mining, manufacturing (also reported as a separate subgroup), construction, electricity, water, and gas (ISIC divisions 10–45). • **Manufacturing** refers to industries belonging to divisions 15–37. • **Services** correspond to ISIC divisions 50–99.

Data sources

The national accounts indicators for most developing countries are collected from national statistical organizations and central banks by visiting and resident World Bank missions. The data for high-income economies come from Organisation for Economic Co-operation and Development (OECD) data files; see the OECD's *Main Economic Indicators* (monthly). The United Nations Statistics Division publishes detailed national accounts for United Nations member countries in *National Accounts Statistics: Main Aggregates and Detailed Tables* and publishes updates in the *Monthly Bulletin of Statistics*.

4.3 Structure of manufacturing

	Value added in manufacturing		Food, beverages, and tobacco		Textiles and clothing		Machinery and transport equipment		Chemicals		Other manufacturing[a]	
	$ millions		% of total		% of total		% of total		% of total		% of total	
	1990	1998	1990	1998	1990	1998	1990	1998	1990	1998	1990	1998
Albania	878	365	24	..	33	..	..	..	..	..	44	..
Algeria	6,151	4,597	13	34	17	8	..	..	..	..	70	58
Angola	513	407	..	..	..	..	..	..	..	..	..	..
Argentina	37,868	53,322	20	..	10	..	13	..	12	..	46	..
Armenia	1,243	377	..	..	..	..	..	..	..	..	..	..
Australia	39,600	46,658	18	..	6	..	20	..	7	..	48	..
Austria	34,289	39,783	15	15	7	4	28	31	7	8	43	42
Azerbaijan	..	404	..	..	..	..	..	..	..	..	..	..
Bangladesh	3,839	6,887	24	..	38	..	7	..	17	..	15	..
Belarus	13,325	7,820	..	..	..	..	..	..	..	..	..	..
Belgium	41,171	45,826	17	19	7	6	..	..	13	16	62	59
Benin	145	190	..	..	..	..	..	..	..	..	..	..
Bolivia	826	1,259	28	34	5	5	1	1	3	5	63	55
Bosnia and Herzegovina	..	884	12	..	15	..	18	..	7	..	49	..
Botswana	184	236	51	..	12	..	..	..	..	..	36	..
Brazil	90,052	151,198	..	..	..	..	..	..	..	..	..	..
Bulgaria	..	2,082	22	20	9	10	19	5	5	..	45	65
Burkina Faso	423	522	..	..	..	..	..	..	..	..	..	..
Burundi	134	62	..	..	..	..	..	..	..	..	..	..
Cambodia	58	178	..	..	..	..	..	..	..	..	..	..
Cameroon	1,581	905	44	33	8	13	1	1	5	6	41	47
Canada	88,928	101,004	15	14	6	5	26	32	10	9	44	40
Central African Republic	154	91	57	..	6	..	2	..	6	..	28	..
Chad	239	170	..	..	..	..	..	..	..	..	..	..
Chile	5,613	11,773	25	29	7	4	5	5	10	11	52	50
China	117,151	355,272	15	17	15	12	24	30	13	15	34	26
Hong Kong, China	12,626	9,435	8	12	36	27	21	27	2	2	33	33
Colombia	8,034	13,612	..	..	..	..	..	..	..	..	..	..
Congo, Dem. Rep.	1,029	..	..	..	..	..	..	..	..	..	..	..
Congo, Rep.	234	147	..	..	..	..	..	..	..	..	..	..
Costa Rica	1,393	2,976	47	46	8	6	7	9	9	13	30	26
Côte d'Ivoire	2,257	2,281	..	42	..	10	..	3	..	12	..	33
Croatia	4,770	3,476	22	..	15	..	20	..	8	..	36	..
Cuba	..	..	..	..	..	..	..	..	..	..	..	..
Czech Republic	..	..	..	..	..	..	..	..	..	..	..	..
Denmark	20,757	25,319	22	20	4	3	24	25	12	15	39	37
Dominican Republic	1,270	2,741	..	..	..	..	..	..	..	..	..	..
Ecuador	2,068	4,315	..	..	..	..	..	..	..	..	..	..
Egypt, Arab Rep.	7,296	14,403	..	..	..	..	..	..	..	..	..	..
El Salvador	1,044	2,566	36	35	14	29	4	5	24	7	23	24
Eritrea	49	87	..	..	..	..	..	..	..	..	..	..
Estonia	2,679	765	..	..	..	..	..	..	..	..	..	..
Ethiopia	497	426	62	52	21	18	1	2	2	4	14	23
Finland	27,533	28,637	13	..	4	..	24	..	8	..	52	..
France	..	..	13	14	6	5	31	30	9	9	41	42
Gabon	332	227	..	..	..	..	..	..	..	..	..	..
Gambia, The	18	21	..	..	..	..	..	..	..	..	..	..
Georgia	..	244	..	..	..	..	..	..	..	..	..	..
Germany	456,313	461,905	..	..	..	..	..	..	..	..	..	..
Ghana	575	672	..	36	..	5	..	2	..	10	..	48
Greece	12,523	12,810	22	26	20	12	12	14	10	13	36	35
Guatemala	1,151	2,619	..	..	..	..	..	..	..	..	..	..
Guinea	126	151	..	..	..	..	..	..	..	..	..	..
Guinea-Bissau	19	19	..	..	..	..	..	..	..	..	..	..
Haiti	469	275	51	46	9	19	..	..	..	..	40	34
Honduras	443	836	..	..	..	..	..	..	..	..	..	..

Structure of manufacturing | 4.3

	Value added in manufacturing		Food, beverages, and tobacco		Textiles and clothing		Machinery and transport equipment		Chemicals		Other manufacturing[a]	
	$ millions		% of total		% of total		% of total		% of total		% of total	
	1990	1998	1990	1998	1990	1998	1990	1998	1990	1998	1990	1998
Hungary	6,613	9,958	14	19	9	8	26	25	12	7	39	41
India	48,793	59,654	12	10	15	10	25	25	14	21	34	33
Indonesia	23,643	23,774	27	16	15	18	12	20	9	9	37	36
Iran, Islamic Rep.	14,503	19,684	..	..	..	..	..	..	..	..	..	..
Iraq	..	..	20	..	16	..	4	..	11	..	49	..
Ireland	..	..	27	20	4	2	29	34	16	25	24	18
Israel	..	..	14	12	9	9	32	32	9	5	37	42
Italy	247,930	235,087	8	..	13	..	34	..	7	..	37	..
Jamaica	827	965	41	48	5	7	..	..	..	..	54	46
Japan	837,191	895,426	9	10	5	4	40	39	10	10	37	36
Jordan	520	1,025	28	28	7	6	4	5	15	17	47	45
Kazakhstan	2,136	..	..	..	..	..	..	..	..	..	..	..
Kenya	862	1,026	38	48	10	7	10	10	9	8	33	27
Korea, Dem. Rep.	..	..	..	..	..	..	..	..	..	..	..	..
Korea, Rep.	72,837	97,866	11	9	12	9	32	41	9	10	36	32
Kuwait	2,142	2,913	4	5	3	4	2	3	3	4	88	84
Kyrgyz Republic	..	131	..	..	..	..	..	..	..	..	..	..
Lao PDR	85	209	..	..	..	..	..	..	..	..	..	..
Latvia	4,150	939	..	39	..	12	..	15	..	6	..	29
Lebanon	..	2,615	..	..	..	..	..	..	..	..	..	..
Lesotho	..	..	..	..	..	..	..	..	..	..	..	..
Libya	..	..	..	..	..	..	..	..	..	..	..	..
Lithuania	2,730	1,784	..	..	..	..	..	..	..	..	..	..
Macedonia, FYR	..	..	20	32	26	18	14	15	9	11	31	24
Madagascar	337	365	..	..	..	..	..	..	..	..	..	..
Malawi	313	217	38	..	10	..	1	..	18	..	33	..
Malaysia	10,665	20,774	13	9	6	4	31	42	11	8	39	36
Mali	200	101	..	..	..	..	..	..	..	..	..	..
Mauritania	94	84	..	..	..	..	..	..	..	..	..	..
Mauritius	524	857	30	29	46	47	2	2	4	4	17	17
Mexico	49,992	80,990	22	21	5	3	24	30	18	18	32	29
Moldova	..	241	..	..	..	..	..	..	..	..	..	..
Mongolia	..	..	33	23	37	63	1	0	1	1	27	12
Morocco	4,753	6,088	22	35	17	18	8	8	12	16	41	23
Mozambique	235	374	..	..	..	..	..	..	..	..	..	..
Myanmar	..	..	..	..	..	..	..	..	..	..	..	..
Namibia	292	437	..	..	..	..	..	..	..	..	..	..
Nepal	209	435	37	35	31	34	1	3	5	6	25	23
Netherlands	52,805	60,905	21	25	3	2	25	25	16	14	35	34
New Zealand	7,665	10,881	28	31	8	..	13	14	7	13	44	43
Nicaragua	170	322	..	..	..	..	..	..	..	..	..	..
Niger	163	128	..	..	..	..	..	..	..	..	..	..
Nigeria	1,562	1,665	15	..	46	..	13	..	4	..	22	..
Norway	..	..	18	..	2	..	25	..	9	..	46	..
Oman	396	..	..	18	..	10	..	6	..	6	..	61
Pakistan	6,184	9,137	24	..	27	..	9	..	15	..	25	..
Panama	502	754	51	54	8	7	2	..	8	7	31	32
Papua New Guinea	289	351	..	..	..	..	..	..	..	..	..	..
Paraguay	883	1,332	55	..	16	..	..	..	..	..	29	..
Peru	7,090	13,178	23	..	11	..	8	..	9	..	49	..
Philippines	11,008	14,254	39	33	11	9	13	15	12	13	26	29
Poland	..	29,811	21	29	9	7	26	23	7	7	37	35
Portugal	..	..	15	14	21	17	13	15	6	5	45	49
Puerto Rico	12,126	..	16	13	5	4	18	15	44	54	17	13
Romania	..	10,494	19	..	18	..	14	..	4	..	45	..
Russian Federation	..	..	..	20	..	3	..	22	..	8	..	47

	Value added in manufacturing		Food, beverages, and tobacco		Textiles and clothing		Machinery and transport equipment		Chemicals		Other manufacturing[a]	
	$ millions		% of total		% of total		% of total		% of total		% of total	
	1990	1998	1990	1998	1990	1998	1990	1998	1990	1998	1990	1998
Rwanda	473	245	..	..	..	..	..	..	..	..	..	..
Saudi Arabia	7,962	12,550	..	..	..	..	..	..	..	..	..	..
Senegal	747	739	60	44	3	5	5	3	9	26	23	21
Sierra Leone	31	26	..	..	..	..	..	..	..	..	..	..
Singapore	9,968	19,092	4	3	3	1	53	60	10	10	29	25
Slovak Republic	..	4,732	..	..	..	..	..	..	..	..	..	..
Slovenia	4,008	4,711	..	12	..	11	..	15	..	12	..	50
South Africa	24,040	23,255	14	17	8	7	18	19	9	10	50	48
Spain	..	..	..	16	..	7	..	25	..	10	..	43
Sri Lanka	1,077	2,285	..	..	..	..	..	..	..	..	..	..
Sudan	..	868	..	..	..	..	..	..	..	..	..	..
Sweden	..	..	10	8	2	1	32	39	9	10	47	42
Switzerland	..	..	10	9	4	3	34	27	..	..	53	60
Syrian Arab Republic	..	..	..	..	..	..	..	..	..	..	..	..
Tajikistan	..	321	..	..	..	..	..	..	..	..	..	..
Tanzania[b]	361	573	51	..	3	..	6	..	11	..	28	..
Thailand	23,217	34,360	24	..	30	..	19	..	2	..	26	..
Togo	162	133	..	..	..	..	..	..	..	..	..	..
Trinidad and Tobago	438	508	30	24	3	1	3	2	19	38	44	34
Tunisia	2,075	3,644	19	21	20	26	5	6	4	8	52	39
Turkey	26,896	27,957	16	12	15	18	16	18	10	10	43	42
Turkmenistan	..	774	..	..	..	..	..	..	..	..	..	..
Uganda	230	547	..	..	..	..	..	..	..	..	..	..
Ukraine	31,489	10,880	..	..	..	..	..	..	..	..	..	..
United Arab Emirates	2,643	..	..	..	..	..	..	..	..	..	..	..
United Kingdom	..	..	13	12	5	5	32	32	11	11	38	40
United States	..	..	12	11	5	4	31	39	12	11	40	35
Uruguay	2,597	4,181	..	..	..	..	..	..	..	..	..	..
Uzbekistan	..	1,719	..	..	..	..	..	..	..	..	..	..
Venezuela, RB	9,809	13,657	17	28	5	5	5	10	9	12	64	45
Vietnam	1,219	..	..	..	..	..	..	..	..	..	..	..
West Bank and Gaza	..	595	..	..	..	..	..	..	..	..	..	..
Yemen, Rep.	449	773	..	..	..	..	..	..	..	..	..	..
Yugoslavia, FR (Serb./Mont.)	..	..	21	29	14	9	22	16	7	11	35	34
Zambia	408	372	44	..	11	..	7	..	9	..	29	..
Zimbabwe	1,799	1,004	28	34	19	15	9	7	6	5	38	38

World	.. w	.. w										
Low income	149,047	141,155
Middle income	695,467	1,253,770
Lower middle income	333,543	709,366
Upper middle income	406,057	552,503
Low & middle income	850,114	1,392,307
East Asia & Pacific	263,090	557,376
Europe & Central Asia	..	..
Latin America & Carib.	256,552	384,079
Middle East & N. Africa	..	77,513
South Asia	61,070	79,682
Sub-Saharan Africa	42,341	40,633
High income	..	..
Europe EMU	1,213,523	1,239,919

a. Includes unallocated data. b. Data cover mainland Tanzania only.

Structure of manufacturing | 4.3

The data on the distribution of manufacturing value added by industry are provided by the United Nations Industrial Development Organization (UNIDO). UNIDO obtains data on manufacturing value added from a variety of national and international sources, including the United Nations Statistics Division, the World Bank, the Organisation for Economic Co-operation and Development, and the International Monetary Fund. To improve comparability over time and across countries, UNIDO supplements these data with information from industrial censuses, statistics supplied by national and international organizations, unpublished data that it collects in the field, and estimates by the UNIDO Secretariat. Nevertheless, coverage may be less than complete, particularly for the informal sector. To the extent that direct information on inputs and outputs is not available, estimates may be used that may result in errors in industry totals. Moreover, countries use different reference periods (calendar or fiscal year) and valuation methods (basic, producer, or purchaser prices) to estimate value added. (See also *About the data* for table 4.2.)

The data on manufacturing value added in U.S. dollars are from the World Bank's national accounts files. These figures may differ from those used by UNIDO to calculate the shares of value added by industry. Thus estimates of value added in a particular industry group calculated by applying the shares to total value added will not match those from UNIDO sources.

The classification of manufacturing industries in the table accords with the United Nations International Standard Industrial Classification (ISIC) revision 2. First published in 1948, the ISIC has its roots in the work of the League of Nations Committee of Statistical Experts. The committee's efforts, interrupted by the second world war, were taken up by the United Nations Statistical Commission, which at its first session appointed a committee on industrial classification. The ISIC has been revised at approximately 20-year intervals. The last revision, ISIC revision 3, was completed in 1989. Revision 2 is still widely used for compiling cross-country data, however, and concordances matching ISIC categories to national systems of classification and to related systems such as the Standard International Trade Classification (SITC) are readily available.

In establishing a classification system, compilers must define both the types of activities to be described and the organizational units whose activities are to be reported. There are many possibilities, and the choices made affect how the resulting statistics can be interpreted and how useful they are in analyzing economic behavior. The ISIC emphasizes commonalities in the production process and is explicitly not intended to mea-

sure outputs (for which there is a newly developed Central Product Classification). Nevertheless, the ISIC views an activity as defined by "a process resulting in a homogeneous set of products" (United Nations 1990 [ISIC, series M, no. 4, rev. 3], p. 9). Firms typically use a multitude of processes to produce a final product. For example, an automobile manufacturer engages in forging, welding, and painting as well as advertising, accounting, and many other service activities. In some cases the processes may be carried out by different technical units within the larger enterprise, but collecting data at such a detailed level is not practical. Nor would it be useful to record production data at the very highest level of a large, multiplant, multiproduct firm. The ISIC has therefore adopted as the definition of an establishment "an enterprise or part of an enterprise which independently engages in one, or predominantly one, kind of economic activity at or from one location . . . for which data are available . . ." (United Nations 1990, p. 25). By design, this definition matches the reporting unit required for the production accounts of the United Nations System of National Accounts.

Figure 4.3

Manufacturing growth resumed in East Asia in 1999

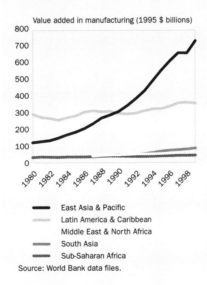

Value added in manufacturing (1995 $ billions)

East Asia & Pacific
Latin America & Caribbean
Middle East & North Africa
South Asia
Sub-Saharan Africa

Source: World Bank data files.

Latin America saw a slowdown in manufacturing, however, while South Asia continued to make modest gains.

• **Value added in manufacturing** is the sum of gross output less the value of intermediate inputs used in production for industries classified in ISIC major division 3. • **Food, beverages, and tobacco** comprise ISIC division 31. • **Textiles and clothing** comprise ISIC division 32. • **Machinery and transport equipment** comprise ISIC groups 382–84. • **Chemicals** comprise ISIC groups 351 and 352. • **Other manufacturing** includes wood and related products (ISIC division 33), paper and related products (ISIC division 34), petroleum and related products (ISIC groups 353–56), basic metals and mineral products (ISIC divisions 36 and 37), fabricated metal products and professional goods (ISIC groups 381 and 385), and other industries (ISIC group 390). When data for textiles and clothing, machinery and transport equipment, or chemicals are shown in the table as not available, they are included in other manufacturing.

The data on value added in manufacturing in U.S. dollars are from the World Bank's national accounts files. The data used to calculate shares of value added by industry are provided to the World Bank in electronic files by UNIDO. The most recent published source is UNIDO's *International Yearbook of Industrial Statistics 2000*. The ISIC system is described in the United Nations' *International Standard Industrial Classification of All Economic Activities, Third Revision* (1990). The discussion of the ISIC draws on Jacob Ryten's paper "Fifty Years of ISIC: Historical Origins and Future Perspectives" (1998).

4.4 Growth of merchandise trade

	Export volume		Import volume		Export value		Import value		Net barter terms of trade	
	average annual % growth		average annual % growth		average annual % growth		average annual % growth		1995 = 100	
	1980–90	1990–98	1980–90	1990–98	1980–90	1990–98	1980–90	1990–98	1990	1998
Albania[a]	..	..	..	..	..	13.8	..	20.6	..	..
Algeria	4.1	1.2	–4.8	1.5	–4.0	–0.7	–2.9	1.0	126	117
Angola	6.0	4.5	–0.3	–1.0	6.4	1.2	1.0	–0.3	145	70
Argentina	5.0	10.3	–6.9	23.0	2.1	12.8	–6.5	24.6	94	104
Armenia[a]	..	..	..	..	..	–15.1	..	–3.4	..	..
Australia[a]	6.3	7.7	6.0	9.0	6.6	6.0	6.4	7.3	117	100
Austria[a]	6.6	..	5.7	..	10.2	6.3	8.7	4.8	..	..
Azerbaijan[a]	..	..	..	..	..	–11.3	..	3.2	..	..
Bangladesh	7.6	13.3	1.8	10.0	7.6	12.3	3.7	11.1	115	103
Belarus[a]	..	..	..	..	..	21.7	..	23.7	..	..
Belgium[a,b]	4.5	6.2	4.0	5.4	7.8	6.3	6.4	4.5	100	99
Benin	4.9	24.8	–10.3	10.6	10.9	28.2	–4.9	13.0	100	95
Bolivia	3.1	2.7	–1.2	10.9	–1.9	5.1	–0.3	12.7	115	110
Bosnia and Herzegovina	..	..	..	..	..	..	..	..	..	..
Botswana	13.2	5.5	11.3	1.0	17.9	4.4	9.4	2.0	110	91
Brazil	6.3	4.4	0.7	22.7	5.1	7.4	–1.9	17.5	60	105
Bulgaria[a]	..	..	..	..	–12.3	4.3	–14.0	5.9	100	..
Burkina Faso	–0.3	14.3	3.2	3.0	7.9	15.7	3.8	4.0	91	91
Burundi	3.5	3.2	1.0	–0.2	2.5	–2.3	2.2	–7.4	75	74
Cambodia	..	..	..	..	..	..	..	..	..	..
Cameroon	8.3	–1.3	4.8	3.8	2.4	0.2	0.1	2.8	88	102
Canada[a]	6.4	9.0	7.4	8.7	6.8	8.2	7.9	7.4	100	96
Central African Republic	–0.2	30.0	4.9	7.2	3.3	13.2	8.6	2.7	124	64
Chad	8.7	5.0	10.9	–2.7	9.4	5.2	12.7	–0.3	116	103
Chile	7.2	9.9	3.4	13.9	8.1	9.7	2.7	13.7	83	75
China[†]	13.9	10.7	15.8	9.4	12.9	15.8	13.6	13.5	101	109
Hong Kong, China	10.8	10.0	9.3	11.3	16.8	10.5	15.0	11.8	101	103
Colombia	7.9	5.2	–2.1	16.0	7.7	7.8	0.0	16.4	95	105
Congo, Dem. Rep.	3.8	–9.3	4.4	–12.7	3.3	–6.8	3.6	–6.5	108	83
Congo, Rep.	5.5	9.7	–2.1	10.6	0.4	6.9	–0.5	10.8	121	82
Costa Rica	3.7	13.7	5.2	16.5	4.6	18.7	4.5	16.0	72	104
Côte d'Ivoire	2.6	4.7	–2.0	5.7	1.7	6.3	–1.4	5.7	82	97
Croatia[a]	..	..	..	..	..	2.2	..	10.7	..	..
Cuba	–1.1	–10.5	–0.3	–0.7	–0.9	–9.7	1.7	–1.2	96	102
Czech Republic[a]	..	..	..	..	..	12.2	..	14.7	..	..
Denmark[a]	4.1	5.3	3.1	6.4	8.4	4.7	6.3	5.4	100	99
Dominican Republic	–0.9	1.8	0.8	13.0	–2.1	4.0	3.3	14.4	94	105
Ecuador	7.5	8.7	–1.8	10.9	–0.4	8.6	–1.3	14.3	141	100
Egypt, Arab Rep.	1.8	1.9	–2.7	6.7	–3.7	2.9	1.4	8.8	86	84
El Salvador	–4.6	2.8	4.6	10.5	–4.6	12.4	2.4	12.7	69	105
Eritrea	..	..	..	..	..	..	..	..	..	..
Estonia[a]	..	..	..	..	..	30.5	..	38.4	..	..
Ethiopia	0.0	8.8	3.5	1.2	–0.8	15.2	4.3	9.8	90	89
Finland[a]	2.3	9.3	4.4	4.3	7.4	8.7	6.9	4.7	102	97
France[a]	3.6	5.8	3.7	4.9	7.5	4.4	6.5	2.7	94	100
Gabon	2.5	5.5	–3.5	2.7	–3.9	4.4	1.1	4.0	135	97
Gambia, The	–1.8	–14.0	–5.3	–2.0	1.2	–12.5	2.8	0.0	102	101
Georgia	..	..	..	..	..	..	..	..	..	..
Germany[a,c]	4.5	5.6	4.9	3.8	9.2	4.4	7.1	3.7	102	99
Ghana	–14.7	12.9	–17.5	8.2	0.3	11.8	2.8	8.0	100	98
Greece[a]	5.0	8.9	6.4	8.9	5.8	4.3	6.6	1.7	92	90
Guatemala	–0.5	7.0	2.9	10.8	–2.2	11.6	0.6	12.4	75	96
Guinea	..	7.4	..	–0.7	3.5	5.5	9.9	1.1	143	89
Guinea-Bissau	–2.3	15.7	–1.8	–0.5	3.9	13.3	3.6	2.3	143	87
Haiti	–0.3	0.9	–4.6	13.3	–1.1	0.2	–2.9	12.9	116	105
Honduras	4.0	3.5	1.6	12.2	1.6	10.1	0.6	14.1	81	118
† Data for Taiwan, China	16.6	3.1	17.6	5.3	14.8	7.5	12.3	9.3	101	108

Growth of merchandise trade | 4.4

	Export volume		Import volume		Export value		Import value		Net barter terms of trade	
	average annual % growth		average annual % growth		average annual % growth		average annual % growth		1995 = 100	
	1980–90	1990–98	1980–90	1990–98	1980–90	1990–98	1980–90	1990–98	1990	1998
Hungary[a]	3.4	6.0	1.3	8.7	1.4	10.5	0.1	12.4	100	..
India	–3.4	2.7	–2.8	5.4	7.3	10.4	4.3	10.3	79	104
Indonesia	8.7	9.0	1.9	6.8	–0.3	9.4	2.7	6.2	102	95
Iran, Islamic Rep.	10.5	–1.7	–2.9	–10.5	1.2	–2.3	–0.4	–8.0	170	110
Iraq	1.7	10.7	–12.0	–1.2	–4.5	7.2	–9.9	–0.2	121	76
Ireland[a]	9.3	14.2	4.8	10.6	12.8	14.0	7.0	11.1	107	102
Israel[a]	6.9	9.6	5.8	9.3	8.3	11.2	5.9	9.5	97	107
Italy[a]	4.3	6.3	5.3	3.8	8.7	6.1	6.9	2.8	98	109
Jamaica	1.1	5.1	3.1	10.2	0.7	3.6	2.9	9.8	105	100
Japan[a]	5.1	1.9	6.6	5.3	8.9	4.5	5.1	5.0	73	97
Jordan	1.8	4.0	–4.4	4.9	6.1	8.2	–1.9	6.2	85	108
Kazakhstan[a]	..	..	..	..	..	14.5	..	3.6	..	..
Kenya	1.7	6.0	2.0	11.4	–1.0	10.0	1.5	8.8	68	106
Korea, Dem. Rep.	..	..	..	..	..	..	..	..	..	..
Korea, Rep.	11.5	9.1	10.9	4.9	15.0	10.9	11.9	7.9	98	83
Kuwait	–2.0	19.2	–6.3	9.4	–7.6	19.7	–4.1	8.9	95	77
Kyrgyz Republic[a]	..	..	..	..	..	12.1	..	15.7	..	..
Lao PDR[a]	..	..	..	..	11.0	22.5	6.6	19.8	..	..
Latvia[a]	..	6.2	..	..	..	15.1	..	29.5	..	..
Lebanon	–3.6	3.9	–7.4	13.4	–3.6	6.6	–5.4	14.7	105	117
Lesotho	4.1	10.9	0.4	1.4	3.3	16.6	1.8	3.5	84	101
Libya	0.4	–5.3	–6.2	–2.0	–7.0	–5.2	–4.0	0.9	145	101
Lithuania[a]	..	..	..	..	..	15.9	..	23.9	..	..
Macedonia, FYR[a]	..	..	..	..	..	2.0	..	4.9	..	..
Madagascar	–3.1	–8.5	–4.7	–2.2	–1.0	–2.5	–2.8	0.3	99	125
Malawi	2.4	4.8	–0.2	1.5	2.0	3.5	3.2	0.7	116	109
Malaysia	14.6	15.5	6.0	13.2	8.6	14.1	7.7	12.3	102	99
Mali	4.4	10.9	3.1	6.8	6.0	7.3	2.9	5.7	122	99
Mauritania	3.8	6.6	–0.6	8.9	7.9	3.6	0.5	6.5	96	99
Mauritius	10.5	4.5	11.1	3.2	14.4	5.1	12.8	4.8	108	101
Mexico	17.6	15.3	3.0	11.3	8.2	16.1	8.6	13.5	113	99
Moldova[a]	..	..	..	..	..	9.1	..	11.8	..	..
Mongolia	3.1	..	0.9	..	5.0	–2.3	5.0	–2.2	..	..
Morocco	4.5	10.2	4.1	6.4	6.1	8.8	3.6	5.7	101	103
Mozambique	–9.6	10.8	–2.8	–4.2	–9.6	6.1	0.1	–1.4	161	104
Myanmar	–4.1	8.2	–6.5	22.9	–7.9	14.2	–4.5	27.5	94	99
Namibia[a]	..	..	..	..	..	4.6	..	0.9	..	..
Nepal[a]	..	..	..	..	8.1	8.2	6.9	11.5	..	..
Netherlands[a]	4.5	6.9	4.5	6.3	4.6	6.6	4.4	6.1	98	99
New Zealand[a]	3.5	5.0	4.3	6.0	6.2	5.4	5.4	7.0	103	98
Nicaragua	–4.8	13.1	–3.5	10.3	–5.8	14.1	–3.1	11.2	119	102
Niger	–5.2	2.8	–5.2	5.4	–5.4	–0.3	–3.5	0.7	90	100
Nigeria	–4.4	5.5	–21.4	7.1	–8.5	2.1	–15.6	7.6	162	87
Norway[a]	4.1	7.6	3.4	7.8	5.3	4.5	6.2	5.3	112	98
Oman	6.7	4.6	–1.7	5.0	2.9	3.4	0.7	8.3	158	97
Pakistan	–0.3	–3.7	–5.3	–3.0	8.1	6.6	3.0	4.5	91	114
Panama	–0.5	6.2	–6.7	9.4	–0.5	10.5	–3.6	9.2	69	109
Papua New Guinea	1.3	–5.3	..	..	4.9	6.8	1.3	1.2	..	..
Paraguay	12.8	0.0	10.4	16.3	11.6	4.6	4.2	16.2	87	113
Peru	2.7	8.6	–2.0	13.5	–1.5	10.5	1.3	14.5	93	90
Philippines	–7.5	18.6	–7.8	17.1	3.9	20.3	2.9	17.6	90	103
Poland[a]	4.8	8.4	1.5	20.4	1.4	10.7	–3.2	22.1	91	99
Portugal[a]	11.9	..	15.1	..	15.1	6.4	10.3	5.1	100	..
Puerto Rico	..	..	..	..	..	..	..	..	..	..
Romania[a]	..	..	..	..	–4.0	9.1	–3.8	7.6	..	..
Russian Federation[a]	..	..	..	..	..	12.8	..	12.9	..	..

	Export volume		Import volume		Export value		Import value		Net barter terms of trade	
	average annual % growth		average annual % growth		average annual % growth		average annual % growth		1995 = 100	
	1980–90	1990–98	1980–90	1990–98	1980–90	1990–98	1980–90	1990–98	1990	1998
Rwanda	4.7	–11.2	2.3	0.7	1.1	–4.9	3.3	–1.1	50	134
Saudi Arabia	–6.3	4.3	–8.4	–2.0	–13.4	1.9	–6.1	0.8	168	88
Senegal	1.2	5.5	0.4	3.0	3.5	5.0	1.4	3.2	116	108
Sierra Leone	0.0	–32.4	–5.8	–8.2	–2.4	–30.3	–8.7	–4.6	120	99
Singapore	13.5	16.2	9.9	12.9	9.9	12.8	8.0	11.1	111	100
Slovak Republic[a]	..	..	..	..	..	12.2	..	15.9	..	..
Slovenia[a]	..	..	..	..	..	10.9	..	11.8	..	..
South Africa[a,d]	3.3	7.4	–0.8	7.9	0.7	3.2	–1.3	8.3	98	100
Spain[a]	3.0	12.1	8.4	6.0	10.8	10.0	10.6	5.6	96	103
Sri Lanka	–4.0	1.5	–6.2	3.9	5.5	12.9	2.7	11.1	83	122
Sudan	–3.0	14.3	–7.7	14.7	–2.5	9.6	–6.4	13.8	123	93
Sweden[a]	4.4	1.2	5.0	1.7	8.0	7.0	6.7	4.7	97	99
Switzerland[a]	3.7	..	4.3	..	9.5	3.2	8.8	1.9	..	..
Syrian Arab Republic	6.6	0.4	–11.7	5.9	2.4	–0.8	–8.5	7.0	131	90
Tajikistan	..	..	..	..	..	..	..	..	..	..
Tanzania	–2.5	6.6	–3.6	–2.0	–4.1	10.2	–2.1	2.2	105	91
Thailand	11.2	5.1	8.8	–1.0	14.0	12.5	12.7	7.1	103	93
Togo	–1.1	9.6	0.6	6.1	1.2	9.1	2.0	6.3	127	125
Trinidad and Tobago	–11.5	2.7	–20.2	10.2	–9.9	4.8	–12.0	11.7	117	105
Tunisia	5.0	4.8	1.7	3.4	3.5	7.2	2.7	6.0	103	101
Turkey	..	10.3	..	11.1	14.0	10.5	9.3	12.1	104	102
Turkmenistan	..	..	..	..	..	..	..	..	..	..
Uganda	–5.9	19.5	–9.6	31.7	–4.5	21.6	0.6	30.5	74	78
Ukraine[a]	..	..	..	..	..	10.9	..	14.9	..	..
United Arab Emirates	8.5	9.9	–1.3	9.4	–1.2	7.5	0.7	12.7	174	98
United Kingdom[a]	4.5	6.4	6.7	5.5	5.9	6.4	8.5	5.6	101	103
United States[a]	3.6	6.8	7.2	8.5	5.7	7.9	8.2	9.0	98	104
Uruguay	4.4	7.4	1.2	13.7	4.5	7.8	–1.2	13.9	100	103
Uzbekistan	..	..	..	..	..	..	..	..	..	..
Venezuela, RB	3.1	6.1	–3.8	3.8	–4.6	3.7	–3.2	5.5	142	83
Vietnam	..	..	..	..	..	..	..	..	..	..
West Bank and Gaza	..	..	..	..	..	..	..	..	..	..
Yemen, Rep.[a]	..	..	..	..	..	23.7	..	–2.4	..	..
Yugoslavia, FR (Serb./Mont.)	..	..	..	..	..	..	..	..	..	..
Zambia	–0.2	3.2	–1.5	–2.4	1.4	0.3	–3.5	–1.9	131	84
Zimbabwe	3.6	8.4	4.3	10.0	2.5	5.2	0.3	6.2	100	105

a. Data are from the International Monetary Fund's International Financial Statistics database. b. Includes Luxembourg. c. Data prior to 1990 refer to the Federal Republic of Germany before unification. d. Data refer to the South African Customs Union (Botswana, Lesotho, Namibia, South Africa, and Swaziland).

Data on international trade in goods are recorded in each country's balance of payments and by customs services. While the balance of payments focuses on the financial transactions that accompany trade, customs data record the direction of trade and the physical quantities and value of goods entering or leaving the customs area. Customs data may differ from those recorded in the balance of payments because of differences in valuation and the time of recording. The 1993 System of National Accounts and the fifth edition of the International Monetary Fund's (IMF) *Balance of Payments Manual* (1993) have attempted to reconcile the definitions and reporting standards for international trade statistics, but differences in sources, timing, and national practices limit comparability. Real growth rates derived from trade volume indexes and terms of trade based on unit price indexes may therefore differ from those derived from national accounts aggregates.

Trade in goods, or merchandise trade, includes all goods that add to or subtract from an economy's material resources. Currency in circulation, titles of ownership, and securities are excluded, but monetary gold is included. Trade data are collected on the basis of a country's customs area, which in most cases is the same as its geographic area. Goods provided as part of foreign aid are included, but goods destined for extraterritorial agencies (such as embassies) are not.

Collecting and tabulating trade statistics is difficult. Some developing countries lack the capacity to report timely data. As a result, it is necessary to estimate their trade from the data reported by their partners. (For further discussion of the use of partner country reports see *About the data* for table 6.2.) In some cases economic or political concerns may lead national authorities to suppress or misrepresent data on certain trade flows, such as oil, military equipment, or the exports of a dominant producer. In other cases reported trade data may be distorted by deliberate under- or overinvoicing to effect capital transfers or avoid taxes. And in some regions smuggling and black market trading result in unreported trade flows.

By international agreement customs data are reported to the United Nations Statistics Division, which maintains the Commodity Trade (COMTRADE) database. The United Nations Conference on Trade and Development (UNCTAD) compiles a variety of international trade statistics, including price and volume indexes, based on the COMTRADE data. The IMF and the World Trade Organization also compile data on trade prices and volumes. The growth rates and terms of trade for low- and middle-income economies shown in this table were calculated from index numbers compiled by UNCTAD. Volume measures for high-income economies were derived by deflating the value of trade using deflators from the IMF's *International Financial Statistics*. In some cases price and volume indexes from different sources may vary significantly as a result of differences in estimation procedures. All indexes are rescaled to a 1995 base year. Terms of trade were computed from the same indicators.

The terms of trade measure the relative prices of a country's exports and imports. There are a number of ways to calculate terms of trade. The most common is the net barter, or commodity, terms of trade, constructed as the ratio of the export price index to the import price index. When the net barter terms of trade increase, a country's exports are becoming more valuable or its imports cheaper.

• **Growth rates of export and import volumes** are average annual growth rates calculated for low- and middle-income economies from UNCTAD's quantum index series and for high-income economies from export and import data deflated by the IMF's trade price deflators. • **Growth rates of export and import values** are average annual growth rates calculated from UNCTAD's value indexes or from current values of merchandise exports and imports. • **Net barter terms of trade** are calculated as the ratio of the export price index to the corresponding import price index measured relative to the base year 1995.

The main source of trade data for developing countries is UNCTAD's annual *Handbook of International Trade and Development Statistics*. The IMF's *International Financial Statistics* includes data on the export and import values and deflators for high-income and selected developing economies.

4.5 | Structure of merchandise exports

	Merchandise exports		Food		Agricultural raw materials		Fuels		Ores and metals		Manufactures	
	$ millions		% of total		% of total		% of total		% of total		% of total	
	1990	1999	1990	1999	1990	1999	1990	1999	1990	1999	1990	1999
Albania	230	270	..	10	..	9	..	1	..	13	..	68
Algeria	12,930	11,900	0	0	0	0	96	96	0	0	3	3
Angola	3,910	3,950	0	..	0	..	93	..	6	..	0	..
Argentina	12,353	23,333	56	50	4	2	8	12	2	4	29	32
Armenia	..	235	..	8	..	4	..	9	..	13	..	63
Australia	39,752	56,080	20	24	10	6	18	19	16	17	16	29
Austria	41,265	63,467	3	5	4	3	1	1	3	3	88	83
Azerbaijan	..	930	..	8	..	9	..	69	..	1	..	13
Bangladesh	1,556	5,180	14	7	7	2	1	0	..	0	77	91
Belarus	..	5,920	..	7	..	3	..	9	..	1	..	75
Belgium[a]	117,703	178,811	9	10	2	1	3	2	4	3	77	78
Benin	288	389	..	15	..	80	..	1	..	0	..	3
Bolivia	926	1,045	19	26	8	4	25	6	44	23	5	41
Bosnia and Herzegovina	..	650	..	..	..	..	..	..	..	..	..	..
Botswana	1,784	2,500	..	..	..	..	..	..	..	..	..	..
Brazil	31,414	48,011	28	29	3	5	2	1	14	10	52	54
Bulgaria	5,030	4,060	..	14	..	0	..	8	..	11	..	61
Burkina Faso	152	254	..	..	..	..	..	..	..	..	..	..
Burundi	75	55	..	..	..	..	..	..	..	..	..	..
Cambodia	86	320	..	..	..	..	..	..	..	..	..	..
Cameroon	2,002	1,430	20	..	14	..	50	..	7	..	9	..
Canada	127,629	238,446	9	7	9	7	10	9	9	4	59	67
Central African Republic	120	195	..	..	..	..	..	..	..	..	..	..
Chad	188	305	..	..	..	..	..	..	..	..	..	..
Chile	8,372	15,616	24	29	9	9	1	0	55	43	11	17
China[†]	62,091	195,150	13	6	3	1	8	2	2	2	72	88
Hong Kong, China[b]	82,390	174,408	3	2	0	0	0	0	1	2	95	95
Colombia	6,766	11,576	33	24	4	5	37	40	0	1	25	31
Congo, Dem. Rep.	999	600	..	..	..	..	..	..	..	..	..	..
Congo, Rep.	981	1,650	..	..	..	..	..	..	..	..	..	..
Costa Rica	1,448	6,577	58	29	5	2	1	0	1	1	27	68
Côte d'Ivoire	3,072	4,077	..	..	..	..	..	..	..	..	..	..
Croatia	..	4,268	..	10	..	5	..	8	..	2	..	76
Cuba	5,100	1,630	..	..	..	..	..	..	..	..	..	..
Czech Republic	12,170	26,855	..	4	..	3	..	3	..	2	..	88
Denmark	36,870	49,043	27	21	3	3	3	4	1	1	60	66
Dominican Republic	2,170	5,200	21	..	0	..	0	..	0	..	78	..
Ecuador	2,714	4,451	44	53	1	5	52	33	0	0	2	9
Egypt, Arab Rep.	2,585	3,559	10	9	10	8	29	37	9	4	42	37
El Salvador	582	1,164	57	42	1	1	2	5	3	3	38	50
Eritrea	..	..	..	..	..	..	..	..	..	..	..	..
Estonia	..	2,940	..	11	..	11	..	4	..	5	..	69
Ethiopia	..	420	..	..	..	..	..	..	..	..	..	..
Finland	26,571	41,677	2	2	10	7	1	2	4	3	83	85
France	216,588	300,362	16	12	2	1	2	2	3	2	77	81
Gabon	2,204	2,600	..	..	..	..	..	..	..	..	..	..
Gambia, The	40	27	..	90	..	4	..	0	..	0	..	5
Georgia	..	240	..	..	..	..	..	..	..	..	..	..
Germany	421,100	541,514	5	4	1	1	1	1	3	2	89	84
Ghana	897	1,820	51	55	15	11	9	5	17	8	8	20
Greece	8,105	11,130	30	28	3	4	7	10	7	7	54	50
Guatemala	1,163	2,398	67	58	6	4	2	4	0	1	24	34
Guinea	671	880	..	8	..	1	..	0	..	71	..	20
Guinea-Bissau	19	49	..	..	..	..	..	..	..	..	..	..
Haiti	160	196	14	15	1	0	0	..	0	0	85	84
Honduras	831	1,249	82	62	4	5	1	0	4	1	9	32
† Data for Taiwan, China	67,142	121,637	4	1	2	1	1	1	1	1	93	95

Structure of merchandise exports | 4.5

	Merchandise exports		Food		Agricultural raw materials		Fuels		Ores and metals		Manufactures	
	$ millions		% of total		% of total		% of total		% of total		% of total	
	1990	1999	1990	1999	1990	1999	1990	1999	1990	1999	1990	1999
Hungary	10,000	25,015	23	9	3	1	3	2	6	2	63	85
India	17,975	36,560	16	17	4	2	3	0	5	2	71	76
Indonesia	25,675	48,665	11	12	5	4	44	23	4	5	35	54
Iran, Islamic Rep.	16,870	16,200	..	..	..	..	..	..	..	..	..	..
Iraq	12,380	9,700	..	..	..	..	..	..	..	..	..	..
Ireland	23,743	70,387	22	9	2	1	1	0	1	0	70	85
Israel	12,080	25,794	8	3	3	2	1	1	2	1	87	93
Italy	170,304	230,613	6	7	1	1	2	1	1	1	88	89
Jamaica	1,135	1,131	19	24	0	0	1	0	10	6	69	70
Japan	287,581	419,363	1	1	1	0	0	0	1	1	96	94
Jordan	1,064	1,782	11	17	0	0	0	0	38	27	51	56
Kazakhstan	..	5,590	..	8	..	2	..	42	..	22	..	25
Kenya	1,031	1,850	49	58	6	8	13	8	3	3	29	23
Korea, Dem. Rep.	1,857	560	..	..	..	..	..	..	..	..	..	..
Korea, Rep.	65,016	144,745	3	2	1	1	1	4	1	1	94	91
Kuwait	7,042	12,181	1	0	0	0	93	79	0	0	6	20
Kyrgyz Republic	..	455	..	16	..	6	..	12	..	6	..	20
Lao PDR	78	311	..	..	..	..	..	..	..	..	..	..
Latvia	..	1,725	..	6	..	30	..	3	..	4	..	57
Lebanon	494	677	..	..	..	..	..	..	..	..	..	..
Lesotho	59	200	..	..	..	..	..	..	..	..	..	..
Libya	13,877	9,090	0	0	0	0	94	95	0	0	5	5
Lithuania	..	3,005	..	11	..	3	..	15	..	2	..	67
Macedonia, FYR	..	1,200	..	16	..	2	..	1	..	9	..	72
Madagascar	319	243	73	36	4	6	1	2	8	4	14	50
Malawi	417	390	93	..	2	..	0	..	0	..	5	..
Malaysia	29,416	84,455	12	8	14	3	18	7	2	1	54	80
Mali	359	536	36	5	62	94	..	0	0	0	2	1
Mauritania	469	455	..	..	..	..	..	..	..	..	..	..
Mauritius	1,194	1,546	32	24	1	0	1	0	0	0	66	75
Mexico	40,711	136,703	12	5	2	1	38	7	6	1	43	85
Moldova	..	470	..	68	..	2	..	0	..	3	..	27
Mongolia	660	336	..	..	..	..	..	..	..	..	..	..
Morocco	4,265	7,350	26	31	3	3	4	2	15	15	52	49
Mozambique	126	250	..	..	..	..	..	..	..	..	..	..
Myanmar	325	1,125	51	..	36	..	0	..	2	..	10	..
Namibia	1,085	1,650	..	..	..	..	..	..	..	..	..	..
Nepal	210	590	13	6	3	0	..	..	0	0	83	90
Netherlands	131,775	200,357	20	17	4	4	10	7	3	2	59	70
New Zealand	9,488	12,452	47	46	18	13	4	2	6	5	23	33
Nicaragua	330	544	77	88	14	2	0	1	1	0	8	9
Niger	282	276	..	29	..	1	..	0	..	67	..	2
Nigeria	13,670	11,300	1	0	1	0	97	99	0	0	1	1
Norway	34,047	44,884	7	9	2	1	48	50	10	7	33	27
Oman	5,508	7,231	1	4	0	0	92	77	1	1	5	17
Pakistan	5,589	8,884	9	13	10	1	1	1	0	0	79	84
Panama	340	822	75	72	1	1	0	9	1	2	21	17
Papua New Guinea	1,144	1,877	22	48	9	8	0	0	58	35	10	9
Paraguay	959	741	52	70	38	14	0	0	0	0	10	15
Peru	3,230	6,114	21	30	3	3	10	5	47	40	18	21
Philippines	8,068	36,650	19	5	2	1	2	1	8	1	38	41
Poland	14,320	27,405	13	9	3	2	11	5	9	5	59	77
Portugal	16,417	23,901	7	7	6	3	3	2	3	1	80	87
Puerto Rico	..	..	..	..	..	..	..	..	..	..	..	..
Romania	4,960	8,505	1	6	3	5	18	5	4	5	73	78
Russian Federation	..	74,300	..	1	..	4	..	41	..	11	..	25

	Merchandise exports		Food		Agricultural raw materials		Fuels		Ores and metals		Manufactures	
	$ millions		% of total		% of total		% of total		% of total		% of total	
	1990	1999	1990	1999	1990	1999	1990	1999	1990	1999	1990	1999
Rwanda	110	70	..	..	..	..	..	..	..	..	..	..
Saudi Arabia	44,417	50,500	1	1	0	0	92	85	0	1	7	13
Senegal	761	1,010	53	13	3	3	12	17	9	10	23	57
Sierra Leone	138	7	..	..	..	..	..	..	..	..	..	..
Singapore[b]	52,752	114,689	5	3	3	1	18	8	2	1	72	86
Slovak Republic	..	10,245	..	3	..	2	..	5	..	3	..	82
Slovenia	..	8,604	..	4	..	2	..	1	..	4	..	90
South Africa[c]	23,549	26,707	8	10	4	3	7	10	11	21	22	55
Spain	55,642	110,100	15	15	2	1	5	2	2	2	75	78
Sri Lanka	1,983	4,599	34	21	6	2	1	0	2	0	54	75
Sudan	374	755	61	67	38	27	..	0	0	0	1	3
Sweden	57,540	84,878	2	3	7	4	3	2	3	2	83	83
Switzerland	63,784	80,365	3	3	1	1	0	0	3	4	94	92
Syrian Arab Republic	4,212	3,464	14	15	5	5	45	68	1	1	36	7
Tajikistan	..	690	..	..	..	..	..	..	..	..	..	..
Tanzania	415	575	..	70	..	13	..	0	..	1	..	16
Thailand	23,070	58,392	29	17	5	3	1	2	1	1	63	74
Togo	268	220	23	21	21	34	0	0	45	27	9	18
Trinidad and Tobago	2,080	2,240	5	8	0	0	67	54	1	0	27	37
Tunisia	3,526	5,910	11	11	1	1	17	7	2	1	69	80
Turkey	12,959	26,028	22	15	3	1	2	1	4	2	68	78
Turkmenistan	..	1,600	..	..	..	..	..	..	..	..	..	..
Uganda	147	517	..	78	..	18	..	0	..	1	..	3
Ukraine	..	11,580	..	..	..	..	..	..	..	..	..	..
United Arab Emirates	20,730	29,500	8	..	1	..	5	..	39	..	46	..
United Kingdom	185,172	268,998	7	6	1	0	8	6	3	2	79	83
United States	393,592	695,215	11	8	4	2	3	2	3	2	74	83
Uruguay	1,693	2,232	40	51	21	9	0	1	0	0	39	38
Uzbekistan	..	2,000	..	..	..	..	..	..	..	..	..	..
Venezuela, RB	17,497	19,852	2	3	0	0	80	81	7	4	10	12
Vietnam	2,404	11,523	..	..	..	..	..	..	..	..	..	..
West Bank and Gaza	..	..	..	..	..	..	..	..	..	..	..	..
Yemen, Rep.	692	2,300	75	5	10	1	8	93	7	0	1	1
Yugoslavia, FR (Serb./Mont.)	..	1,515	..	..	..	..	..	..	..	..	..	..
Zambia	1,309	680	..	..	..	..	..	..	..	..	..	..
Zimbabwe	1,726	2,110	44	51	7	10	1	2	16	11	31	27
World	3,345,931 t	5,442,226 t	10 w	8 w	3 w	2 w	8 w	5 w	4 w	3 w	74 w	79 w
Low income	108,352	174,118	15	15	4	3	28	20	4	3	48	52
Middle income	623,420	1,219,963	15	10	4	2	19	12	5	4	54	68
Lower middle income	267,420	537,227	19	9	4	2	11	15	4	5	59	62
Upper middle income	353,899	682,729	12	10	4	2	24	10	6	4	51	73
Low & middle income	731,493	1,394,132	15	10	4	2	20	14	5	4	54	66
East Asia & Pacific	220,817	585,640	12	7	5	2	10	5	2	2	68	81
Europe & Central Asia[d]	..	247,686	..	6	..	3	..	20	..	7	..	56
Latin America & Carib.	143,367	294,456	26	24	4	3	24	17	12	6	34	51
Middle East & N. Africa	127,739	135,767	3	4	1	1	78	73	3	1	17	21
South Asia	27,675	56,164	16	16	5	2	2	0	4	2	71	79
Sub-Saharan Africa	66,687	74,350	13	15	3	4	28	29	7	14	20	39
High income	2,610,188	4,048,257	8	7	3	2	5	3	3	2	79	82
Europe EMU	1,104,079	1,583,193	10	9	2	1	3	2	2	2	81	82

Note: Components may not sum to 100 percent because of unclassified trade.

a. Includes Luxembourg. b. Includes re-exports. c. Data on total merchandise exports for 1990 refer to the South African Customs Union (Botswana, Lesotho, Namibia, South Africa, and Swaziland); those for 1999 refer to South Africa only. Data on export commodity shares refer to the South African Customs Union. d. Data for 1999 include the intratrade of the Baltic states and the Commonwealth of Independent States.

Structure of merchandise exports | 4.5

About the data

Data on merchandise trade come from customs reports of goods entering an economy or from reports of the financial transactions related to merchandise trade recorded in the balance of payments. Because of differences in timing and definitions, estimates of trade flows from customs reports are likely to differ from those based on the balance of payments. Moreover, several international agencies process trade data, each making estimates to correct for unreported or misreported data, and this leads to other differences in the available data.

The most detailed source of data on international trade in goods is the Commodity Trade (COMTRADE) database maintained by the United Nations Statistics Division. The International Monetary Fund (IMF) also collects customs-based data on exports and imports of goods. The value of exports is recorded as the cost of the goods delivered to the frontier of the exporting country for shipment—the f.o.b. (free on board) value. Many countries report trade data in U.S. dollars. When countries report in local currency, the United Nations Statistics Division applies the average official exchange rate for the period shown.

Countries may report trade according to the general or special system of trade (see *Primary data documentation*). Under the general system exports comprise outward-moving goods that are (a) goods wholly or partly produced in the country; (b) foreign goods, neither transformed nor declared for domestic consumption in the country, that move outward from customs storage; and (c) goods previously included as imports for domestic consumption but subsequently exported without transformation. Under the special system exports comprise categories a and c. In some compilations categories b and c are classified as re-exports. Because of differences in reporting practices, data on exports may not be fully comparable across economies.

The data on total exports of goods (merchandise) in this table come from the World Trade Organization (WTO). The WTO uses two main sources, national statistical offices and the IMF's *International Financial Statistics*. It supplements these with the COMTRADE database and publications or databases of regional organizations, specialized agencies, and economic groups (such as the Commonwealth of Independent States, the Economic Commission for Latin America and the Caribbean, Eurostat, the Food and Agriculture Organization, the Organisation for Economic Co-operation and Development, and the Organization of Petroleum Exporting Countries). It also consults private sources, such as country reports of the Economist Intelligence Unit and press clippings. In recent years country Web sites and direct contacts through email have helped to improve the collection of up-to-date statistics for many countries, reducing the

proportion of estimated figures. The WTO database now covers most of the major traders in Africa, Asia, and Latin America, which together with the high-income countries account for nearly 90 percent of total world trade. There has also been a remarkable improvement in the availability of recent, reliable, and standardized figures for countries in Europe and Central Asia.

The shares of exports by major commodity group were estimated by World Bank staff from the COMTRADE database. The values of total exports reported here have not been fully reconciled with the estimates of exports of goods and services from the national accounts (shown in table 4.9) or those from the balance of payments (table 4.15).

The classification of commodity groups is based on the Standard International Trade Classification (SITC) revision 1. Most countries now report using later revisions of the SITC or the Harmonized System. Concordance tables are used to convert data reported in one system of nomenclature to another. The conversion process may introduce some errors of classification, but conversions from later to early systems are generally reliable. Shares may not sum to 100 percent because of unclassified trade.

Definitions

• **Merchandise exports** show the f.o.b. value of goods provided to the rest of the world valued in U.S. dollars. • **Food** comprises the commodities in SITC sections 0 (food and live animals), 1 (beverages and tobacco), and 4 (animal and vegetable oils and fats) and SITC division 22 (oil seeds, oil nuts, and oil kernels). • **Agricultural raw materials** comprise SITC section 2 (crude materials except fuels) excluding divisions 22, 27 (crude fertilizers and minerals excluding coal, petroleum, and precious stones), and 28 (metalliferous ores and scrap). • **Fuels** comprise SITC section 3 (mineral fuels). • **Ores and metals** comprise the commodities in SITC divisions 27, 28, and 68 (nonferrous metals). • **Manufactures** comprise the commodities in SITC sections 5 (chemicals), 6 (basic manufactures), 7 (machinery and transport equipment), and 8 (miscellaneous manufactured goods), excluding division 68.

Data sources

The WTO publishes data on world trade in its *Annual Report*. Estimates of total exports of goods are also published in the IMF's *International Financial Statistics* and *Direction of Trade Statistics* and in the United Nations Statistics Division's *Monthly Bulletin of Statistics*. The United Nations Conference on Trade and Development (UNCTAD) publishes data on the structure of exports and imports in its *Handbook of International Trade and Development Statistics*. Tariff line records of exports and imports are compiled in the United Nations Statistics Division's COMTRADE database.

Figure 4.5

Manufactured exports are increasingly important for developing countries

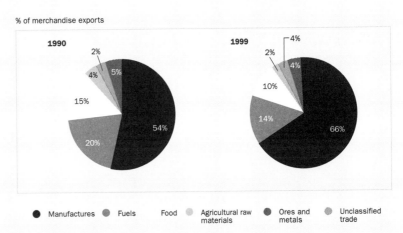

% of merchandise exports

● Manufactures ● Fuels ● Food ● Agricultural raw materials ● Ores and metals ● Unclassified trade

Source: World Trade Organization data.

In the past decade developing countries' manufactured exports increased from 54 percent to 66 percent of their merchandise exports.

	Merchandise imports		Food		Agricultural raw materials		Fuels		Ores and metals		Manufactures	
	$ millions		% of total		% of total		% of total		% of total		% of total	
	1990	1999	1990	1999	1990	1999	1990	1999	1990	1999	1990	1999
Albania	380	1,230	..	27	..	1	..	4	..	1	..	67
Algeria	9,712	9,800	24	27	5	3	1	2	2	1	68	67
Angola	1,578	2,330	..	..	..	..	..	..	..	..	..	..
Argentina	4,076	25,466	4	5	4	2	8	3	6	2	78	89
Armenia	..	800	..	26	..	1	..	22	..	1	..	50
Australia	42,032	69,113	5	5	2	1	5	6	1	1	80	86
Austria	49,146	68,763	5	6	3	3	6	4	4	3	81	84
Azerbaijan	..	1,035	..	16	..	2	..	6	..	1	..	75
Bangladesh	3,432	7,740	19	15	5	5	16	7	..	..	56	69
Belarus	..	6,665	..	12	..	3	..	23	..	4	..	56
Belgium[a]	119,702	166,798	10	10	2	2	8	5	6	4	68	78
Benin	265	643	..	25	..	5	..	21	..	1	..	49
Bolivia	687	1,755	12	8	2	1	1	5	1	1	85	86
Bosnia and Herzegovina	..	2,500	..	..	..	..	..	..	..	..	..	..
Botswana	1,946	2,300	..	..	..	..	..	..	..	..	..	..
Brazil	22,524	51,747	9	8	3	2	27	11	5	3	56	76
Bulgaria	5,100	5,475	8	9	3	3	36	30	4	6	49	50
Burkina Faso	536	670	..	..	..	..	..	..	..	..	..	..
Burundi	231	118	..	..	..	..	..	..	..	..	..	..
Cambodia	164	680	..	..	..	..	..	..	..	..	..	..
Cameroon	1,400	1,500	19	..	0	..	2	..	1	..	78	..
Canada	123,244	220,183	6	5	2	1	6	3	3	2	81	86
Central African Republic	154	300	..	..	..	..	..	..	..	..	..	..
Chad	285	320	..	..	..	..	..	..	..	..	..	..
Chile	7,678	15,137	4	7	2	1	16	9	1	1	75	81
China[†]	53,345	165,788	9	4	6	4	2	5	3	5	80	80
Hong Kong, China	84,725	180,716	8	5	2	1	2	2	2	2	85	90
Colombia	5,590	10,659	7	13	4	2	6	3	3	2	77	79
Congo, Dem. Rep.	887	370	..	..	..	..	..	..	..	..	..	..
Congo, Rep.	621	520	..	..	..	..	..	..	..	..	..	..
Costa Rica	1,990	6,320	8	7	2	1	10	5	2	2	66	69
Côte d'Ivoire	2,097	3,270	..	..	..	..	..	..	..	..	..	..
Croatia	..	7,774	..	8	..	1	..	11	..	2	..	73
Cuba	4,600	3,990	..	..	..	..	..	..	..	..	..	..
Czech Republic	12,880	28,825	..	6	..	2	..	7	..	3	..	82
Denmark	33,333	44,319	12	12	3	3	7	3	2	2	73	77
Dominican Republic	3,006	8,200	..	..	..	..	..	..	..	..	..	..
Ecuador	1,861	3,017	9	11	3	3	2	7	3	1	84	76
Egypt, Arab Rep.	9,216	16,022	32	23	7	4	3	6	2	3	56	59
El Salvador	1,263	3,130	14	19	3	2	16	11	4	1	63	66
Eritrea	..	..	..	..	..	..	..	..	..	..	..	..
Estonia	..	4,115	..	13	..	3	..	7	..	3	..	74
Ethiopia	1,081	1,740	..	..	..	..	..	..	..	..	..	..
Finland	27,001	31,507	5	6	2	3	12	9	4	5	76	75
France	234,436	290,098	10	9	3	2	10	7	4	3	74	80
Gabon	918	1,200	..	..	..	..	..	..	..	..	..	..
Gambia, The	199	192	..	44	..	2	..	6	..	1	..	46
Georgia	..	600	..	..	..	..	..	..	..	..	..	..
Germany	355,686	472,456	10	8	3	2	8	6	4	3	72	70
Ghana	1,205	3,505	11	12	1	3	17	18	0	1	70	66
Greece	19,777	30,215	15	13	3	2	8	6	3	2	70	76
Guatemala	1,649	4,382	10	13	2	1	17	10	2	1	69	75
Guinea	723	1,100	..	24	..	1	..	10	..	1	..	64
Guinea-Bissau	68	95	..	..	..	..	..	..	..	..	..	..
Haiti	332	1,025	..	..	..	..	..	..	..	..	..	..
Honduras	935	2,728	10	17	1	1	16	9	1	1	71	72
† Data for Taiwan, China	54,831	110,698	7	5	5	2	11	8	6	5	69	79

Structure of merchandise imports | 4.6

	Merchandise imports		Food		Agricultural raw materials		Fuels		Ores and metals		Manufactures	
	$ millions		% of total		% of total		% of total		% of total		% of total	
	1990	1999	1990	1999	1990	1999	1990	1999	1990	1999	1990	1999
Hungary	10,340	28,010	8	3	4	1	14	6	4	2	70	86
India	23,642	44,589	3	9	4	3	27	20	8	6	51	54
Indonesia	21,837	24,004	5	11	5	7	9	10	4	3	77	69
Iran, Islamic Rep.	15,716	13,200	..	..	..	..	..	..	..	..	..	..
Iraq	7,660	6,800	..	..	..	..	..	..	..	..	..	..
Ireland	20,669	46,396	11	7	2	1	6	3	2	1	76	81
Israel	16,793	33,160	8	6	2	1	9	7	3	2	77	83
Italy	181,968	216,938	12	10	6	4	11	7	5	4	64	73
Jamaica	1,859	2,587	15	17	1	2	20	13	1	1	61	65
Japan	235,368	311,262	15	15	7	4	25	16	9	6	44	58
Jordan	2,600	3,728	26	23	2	2	18	9	1	2	51	63
Kazakhstan	..	3,685	..	10	..	1	..	9	..	3	..	77
Kenya	2,125	2,850	9	12	3	3	20	16	2	1	66	68
Korea, Dem. Rep.	2,930	880	..	..	..	..	..	..	..	..	..	..
Korea, Rep.	69,844	119,750	6	6	8	4	16	19	7	7	63	64
Kuwait	3,972	7,617	17	17	1	1	1	1	2	2	79	79
Kyrgyz Republic	..	600	..	14	..	1	..	20	..	2	..	64
Lao PDR	201	525	..	..	..	..	..	..	..	..	..	..
Latvia	..	2,945	..	12	..	2	..	11	..	2	..	73
Lebanon	2,529	6,207	..	..	..	..	..	..	..	..	..	..
Lesotho	672	910	..	..	..	..	..	..	..	..	..	..
Libya	5,336	5,850	23	23	2	1	0	0	1	1	74	75
Lithuania	..	4,835	..	12	..	6	..	14	..	2	..	65
Macedonia, FYR	..	1,913	..	16	..	3	..	9	..	3	..	47
Madagascar	571	514	11	14	1	1	17	24	1	0	69	60
Malawi	581	660	9	20	1	1	11	10	1	1	78	68
Malaysia	29,258	64,966	7	6	1	1	5	3	4	3	82	85
Mali	619	751	26	19	1	1	19	21	1	1	53	58
Mauritania	387	330	..	..	..	..	..	..	..	..	..	..
Mauritius	1,618	2,138	12	14	3	2	8	7	1	1	76	76
Mexico	43,548	148,741	15	5	4	1	4	2	3	2	75	86
Moldova	..	570	..	6	..	2	..	37	..	1	..	53
Mongolia	924	426	..	..	..	..	..	..	..	..	..	..
Morocco	6,800	10,610	10	17	6	6	17	17	6	4	61	58
Mozambique	878	1,040	..	..	..	..	..	..	..	..	..	..
Myanmar	270	2,300	13	..	1	..	5	..	0	..	81	..
Namibia	1,163	1,700	..	..	..	..	..	..	..	..	..	..
Nepal	686	1,390	15	12	7	7	9	24	2	3	67	42
Netherlands	126,098	187,632	13	11	2	2	10	7	3	3	71	77
New Zealand	9,501	14,301	7	8	1	1	8	6	3	2	81	83
Nicaragua	638	1,846	19	19	1	1	19	8	1	1	59	71
Niger	388	396	..	39	..	4	..	15	..	2	..	41
Nigeria	5,627	10,370	6	27	1	1	0	2	2	3	67	67
Norway	27,231	34,041	6	7	2	2	4	3	6	5	82	82
Oman	2,681	4,674	19	23	1	1	4	1	1	2	69	70
Pakistan	7,546	10,782	17	17	4	6	21	21	4	2	54	53
Panama	1,539	3,516	12	12	1	0	16	12	1	1	70	75
Papua New Guinea	1,193	1,188	18	19	0	1	7	3	1	1	73	76
Paraguay	1,352	2,275	8	17	0	0	14	12	1	1	77	70
Peru	3,470	8,060	24	15	2	2	12	10	1	1	61	73
Philippines	13,041	32,546	10	8	2	2	15	8	3	3	53	60
Poland	11,570	45,910	8	7	3	2	22	7	4	3	63	80
Portugal	25,263	38,589	12	12	4	3	11	7	3	2	71	76
Puerto Rico	..	..	..	..	..	..	..	..	..	..	..	..
Romania	7,600	10,390	12	7	4	1	38	10	6	3	39	77
Russian Federation	42,970	41,100	..	19	..	1	..	2	..	2	..	42

4.6 | Structure of merchandise imports

	Merchandise imports ($ millions)		Food (% of total)		Agricultural raw materials (% of total)		Fuels (% of total)		Ores and metals (% of total)		Manufactures (% of total)	
	1990	1999	1990	1999	1990	1999	1990	1999	1990	1999	1990	1999
Rwanda	288	270	..	..	..	..	..	..	..	..	..	..
Saudi Arabia	24,069	28,032	15	16	1	1	0	0	3	3	81	78
Senegal	1,219	1,530	29	29	2	2	16	10	2	2	51	57
Sierra Leone	149	81	..	..	..	..	..	..	..	..	..	..
Singapore	60,899	111,060	6	4	2	0	16	9	2	2	73	83
Slovak Republic	6,670	11,245	..	6	..	2	..	9	..	3	..	74
Slovenia	..	9,952	..	6	..	3	..	6	..	4	..	80
South Africa[b]	18,399	26,696	5	5	2	1	1	10	2	2	77	73
Spain	87,715	144,750	11	10	3	2	12	7	4	3	71	77
Sri Lanka	2,685	5,893	19	15	2	1	13	6	2	1	65	77
Sudan	618	1,412	13	15	1	1	20	10	0	1	66	72
Sweden	54,264	68,455	6	7	2	2	9	6	3	3	79	77
Switzerland	69,681	79,921	6	6	2	1	5	3	3	4	84	85
Syrian Arab Republic	2,400	3,832	31	22	2	3	3	3	1	1	62	59
Tajikistan	..	665	..	..	..	..	..	..	..	..	..	..
Tanzania	1,027	1,781	..	16	..	2	..	8	..	1	..	72
Thailand	33,379	50,305	5	5	5	3	9	10	4	3	75	78
Togo	581	620	22	17	1	1	8	40	1	2	67	40
Trinidad and Tobago	1,262	2,470	19	11	1	1	11	21	6	1	62	64
Tunisia	5,542	8,476	11	8	4	3	9	7	4	2	72	80
Turkey	22,302	40,404	8	5	4	3	21	11	5	4	61	73
Turkmenistan	..	1,080	..	9	..	1	..	7	..	2	..	81
Uganda	213	1,341	..	14	..	2	..	12	..	2	..	69
Ukraine	..	11,845	..	..	..	..	..	..	..	..	..	..
United Arab Emirates	11,199	28,870	14	..	1	..	3	..	4	..	77	..
United Kingdom	222,977	320,251	10	9	3	2	6	3	4	3	75	83
United States	516,987	1,059,126	6	5	2	2	13	8	3	2	73	80
Uruguay	1,343	3,357	7	11	4	3	18	11	2	1	69	74
Uzbekistan	..	2,250	..	..	..	..	..	..	..	..	..	..
Venezuela, RB	7,335	14,789	11	13	4	2	3	2	4	2	77	81
Vietnam	2,752	11,600	..	..	..	..	..	..	..	..	..	..
West Bank and Gaza	..	..	..	..	..	..	..	..	..	..	..	..
Yemen, Rep.	1,571	2,380	27	35	1	2	40	6	1	1	31	55
Yugoslavia, FR (Serb./Mont.)	..	3,215	..	..	..	..	..	..	..	..	..	..
Zambia	1,220	720	..	..	..	..	..	..	..	..	..	..
Zimbabwe	1,847	2,830	4	9	3	2	16	12	2	3	73	75
World	3,419,027 t	5,617,403 t	9 w	8 w	3 w	2 w	11 w	7 w	4 w	3 w	71 w	76 w
Low income	113,973	176,435	7	14	3	4	17	13	..	..	64	63
Middle income	573,092	1,178,715	9	8	4	3	11	8	4	4	70	73
Lower middle income	280,285	516,326	10	9	5	3	9	7	3	3	70	70
Upper middle income	294,734	662,343	8	6	4	3	12	10	4	4	70	76
Low & middle income	685,604	1,355,237	9	8	4	3	11	9	4	3	70	73
East Asia & Pacific	230,492	477,062	7	5	5	3	9	10	4	5	73	74
Europe & Central Asia[c]	162,413	269,386	..	10	..	2	..	7	..	3	..	67
Latin America & Carib.	121,185	329,087	11	9	3	2	13	7	3	2	69	80
Middle East & N. Africa	101,723	126,197	19	..	3	..	6	..	3	..	69	..
South Asia	39,173	71,402	9	12	4	4	23	18	6	5	54	56
Sub-Saharan Africa	56,115	81,936	..	11	..	2	..	10	..	2	..	71
High income	2,722,931	4,262,008	9	8	3	2	11	7	4	3	72	77
Europe EMU	1,114,349	1,504,975	10	9	3	2	9	6	4	3	72	75

Note: Components may not sum to 100 percent because of unclassified trade.

a. Includes Luxembourg. b. Data on total merchandise imports for 1990 refer to the South African Customs Union (Botswana, Lesotho, Namibia, South Africa, and Swaziland); those for 1999 refer to South Africa only. Data on import commodity shares refer to the South African Customs Union. c. Data for 1999 include the intratrade of the Baltic states and the Commonwealth of Independent States.

Structure of merchandise imports | 4.6

Data on imports of goods are derived from the same sources as data on exports. In principle, world exports and imports should be identical. Similarly, exports from an economy should equal the sum of imports by the rest of the world from that economy. But differences in timing and definitions result in discrepancies in reported values at all levels. For further discussion of indicators of merchandise trade see *About the data* for tables 4.4 and 4.5.

The value of imports is generally recorded as the cost of the goods when purchased by the importer plus the cost of transport and insurance to the frontier of the importing country—the c.i.f. (cost, insurance, and freight) value. A few countries, including Australia, Canada, and the United States, collect import data on an f.o.b. (free on board) basis and adjust them for freight and insurance costs. Many countries collect and report trade data in U.S. dollars. When countries report in local currency, the United Nations Statistics Division applies the average official exchange rate for the period shown.

Countries may report trade according to the general or special system of trade (see *Primary data documentation*). Under the general system imports include goods imported for domestic consumption and imports into bonded warehouses and free trade zones. Under the special system imports comprise goods imported for domestic consumption (including transformation and repair) and withdrawals for domestic consumption from bonded warehouses and free trade zones. Goods transported through a country en route to another are excluded.

The data on total imports of goods (merchandise) in this table come from the World Trade Organization (WTO). The WTO uses two main sources, national statistical offices and the International Monetary Fund's (IMF) *International Financial Statistics*. It supplements these with the Commodity Trade (COMTRADE) database maintained by the United Nations Statistics Division and publications or databases of regional organizations, specialized agencies, and economic groups (such as the Commonwealth of Independent States, the Economic Commission for Latin America and the Caribbean, Eurostat, the Food and Agriculture Organization, the Organisation for Economic Co-operation and Development, and the Organization of Petroleum Exporting Countries). It also consults private sources, such as country reports of the Economist Intelligence Unit and press clippings. In recent years country Web sites and direct contacts through email have helped to improve the collection of up-to-date statistics for many countries, reducing the proportion of estimated figures. The WTO database now covers most of the major traders in

Africa, Asia, and Latin America, which together with the high-income countries account for nearly 90 percent of total world trade. There has also been a remarkable improvement in the availability of recent, reliable, and standardized figures for countries in Europe and Central Asia.

The shares of imports by major commodity group were estimated by World Bank staff from the COMTRADE database. The values of total imports reported here have not been fully reconciled with the estimates of imports of goods and services from the national accounts (shown in table 4.9) or those from the balance of payments (table 4.15).

The classification of commodity groups is based on the Standard International Trade Classification (SITC) revision 1. Most countries now report using later revisions of the SITC or the Harmonized System. Concordance tables are used to convert data reported in one system of nomenclature to another. The conversion process may introduce some errors of classification, but conversions from later to early systems are generally reliable. Shares may not sum to 100 percent because of unclassified trade.

- **Merchandise imports** show the c.i.f. value of goods purchased from the rest of the world valued in U.S. dollars. • **Food** comprises the commodities in SITC sections 0 (food and live animals), 1 (beverages and tobacco), and 4 (animal and vegetable oils and fats) and SITC division 22 (oil seeds, oil nuts, and oil kernels). • **Agricultural raw materials** comprise SITC section 2 (crude materials except fuels) excluding divisions 22, 27 (crude fertilizers and minerals excluding coal, petroleum, and precious stones), and 28 (metalliferous ores and scrap). • **Fuels** comprise SITC section 3 (mineral fuels). • **Ores and metals** comprise the commodities in SITC divisions 27, 28, and 68 (nonferrous metals). • **Manufactures** comprise the commodities in SITC sections 5 (chemicals), 6 (basic manufactures), 7 (machinery and transport equipment), and 8 (miscellaneous manufactured goods), excluding division 68.

The WTO publishes data on world trade in its *Annual Report*. Estimates of total imports of goods are also published in the IMF's *International Financial Statistics* and *Direction of Trade Statistics* and in the United Nations Statistics Division's *Monthly Bulletin of Statistics*. The United Nations Conference on Trade and Development (UNCTAD) publishes data on the structure of exports and imports in its *Handbook of International Trade and Development Statistics*. Tariff line records of exports and imports are compiled in the United Nations Statistics Division's COMTRADE database.

4.7 | Structure of service exports

	Commercial service exports		Transport		Travel		Other	
	$ millions		% of total		% of total		% of total	
	1990	**1999**	**1990**	**1999**	**1990**	**1999**	**1990**	**1999**
Albania	32	253	18.8	6.3	12.5	83.4	68.8	10.7
Algeria	479	..	41.5	..	13.4	..	44.9	..
Angola	65	116	49.2	36.2	20.0	0.9	30.8	62.9
Argentina	2,264	4,288	51.1	22.9	39.9	67.6	9.1	9.5
Armenia	..	132	..	40.2	..	25.8	..	34.1
Australia	9,833	17,201	35.5	25.7	43.2	47.5	21.4	26.8
Austria	22,755	30,510	6.4	14.1	59.0	36.2	34.6	49.6
Azerbaijan	..	320	..	40.3	..	39.1	..	20.6
Bangladesh	296	266	12.8	35.3	6.4	18.8	80.4	45.9
Belarus	..	870	..	56.9	..	2.5	..	40.6
Belgium[a]	24,690	39,036	29.7	25.3	15.1	18.6	55.2	56.1
Benin	109	102	33.9	24.5	50.5	54.9	16.5	20.6
Bolivia	133	243	36.1	21.8	43.6	35.8	20.3	42.8
Bosnia and Herzegovina	..	..	..	..	..	..	..	..
Botswana	183	241	20.2	18.3	63.9	72.6	15.3	9.1
Brazil	3,706	6,763	36.4	24.9	37.3	19.9	26.3	55.2
Bulgaria	837	1,756	27.5	29.6	38.2	53.0	34.3	17.4
Burkina Faso	34	..	..	..	..	..	29.4	..
Burundi	7	3	42.9	66.7	42.9	33.3	14.3	33.3
Cambodia	50	118	..	39.8	100.0	45.8	0.0	14.4
Cameroon	369	..	42.5	..	14.4	..	43.1	..
Canada	18,350	34,161	23.0	18.5	34.7	29.8	42.3	51.7
Central African Republic	17	..	52.9	..	17.6	..	35.3	..
Chad	23	..	..	..	34.8	..	47.8	..
Chile	1,786	3,701	40.0	41.3	29.7	28.4	30.2	30.3
China[†]	5,748	23,695	47.1	10.2	30.2	59.5	22.7	30.3
Hong Kong, China	18,128	34,853	38.6	32.7	29.4	19.6	32.1	47.7
Colombia	1,548	1,840	31.3	28.3	26.2	50.4	42.5	21.3
Congo, Dem. Rep.	127	..	..	..	5.5	..	74.8	..
Congo, Rep.	65	45	53.8	57.8	12.3	6.7	33.8	37.8
Costa Rica	583	1,315	16.3	15.1	48.9	68.6	34.8	16.4
Côte d'Ivoire	425	454	62.4	26.4	12.0	23.1	25.4	50.4
Croatia	..	3,708	..	12.5	..	67.5	..	20.1
Cuba	743	2,350	..	..	74.0	77.3	26.1	22.7
Czech Republic	..	6,807	..	22.7	..	44.6	..	32.7
Denmark	12,731	15,823	32.5	43.4	26.2	22.6	41.3	33.9
Dominican Republic	1,086	2,829	5.6	2.5	66.9	89.2	27.5	8.3
Ecuador	508	763	47.6	37.6	37.0	45.0	15.4	17.4
Egypt, Arab Rep.	4,813	9,276	50.1	28.7	22.9	42.1	27.1	29.3
El Salvador	301	328	26.2	13.7	25.2	27.7	48.5	58.5
Eritrea	..	..	..	..	..	..	..	..
Estonia	200	1,486	75.0	47.0	13.5	36.9	11.5	16.0
Ethiopia	261	348	80.5	51.7	1.9	10.6	17.2	37.4
Finland	4,562	6,574	38.4	30.2	25.8	23.1	35.7	46.7
France	66,274	82,577	24.5	24.2	30.6	38.0	44.9	37.8
Gabon	214	..	33.2	..	1.4	..	65.0	..
Gambia, The	55	94	9.1	9.6	87.3	77.7	3.6	12.8
Georgia	..	278	..	24.8	..	67.6	..	7.6
Germany	51,605	79,305	28.6	25.2	27.9	21.1	43.5	53.7
Ghana	79	162	49.4	56.2	5.1	11.7	45.6	32.7
Greece	6,514	10,068	4.9	1.9	39.7	40.9	55.4	57.2
Guatemala	313	653	7.3	12.7	37.7	54.5	55.0	32.8
Guinea	91	36	14.3	52.8	33.0	5.6	53.8	41.7
Guinea-Bissau	4	6	..	..	0.0	50.0	75.0	33.3
Haiti	43	178	18.6	2.2	79.1	63.5	2.3	34.3
Honduras	121	416	34.7	13.7	24.0	43.5	40.5	42.8
† Data for Taiwan, China	6,937	14,518	33.5	24.8	25.1	24.5	41.4	50.7

Structure of service exports 4.7

	Commercial service exports ($ millions)		Transport (% of total)		Travel (% of total)		Other (% of total)	
	1990	1999	1990	1999	1990	1999	1990	1999
Hungary	2,677	5,608	1.6	10.2	36.8	60.6	61.6	29.3
India	4,609	13,940	20.8	13.5	33.8	21.6	45.4	64.9
Indonesia	2,488	4,624	2.8	..	86.5	98.2	10.7	1.8
Iran, Islamic Rep.	343	*902*	10.5	*46.5*	8.2	*1.3*	81.3	*52.2*
Iraq	..	..	..	..	..	..	..	..
Ireland	3,286	13,985	31.0	9.8	44.4	18.3	24.5	71.9
Israel	4,546	10,309	30.8	21.9	30.7	28.8	38.5	49.4
Italy	48,579	61,177	21.0	*16.0*	33.9	*44.7*	45.2	*39.3*
Jamaica	976	1,795	18.0	14.8	76.9	69.8	4.9	15.4
Japan	41,384	60,313	42.9	38.0	8.7	5.7	48.4	56.3
Jordan	1,430	1,770	25.9	16.8	35.7	49.5	38.3	33.7
Kazakhstan	..	933	..	45.1	..	38.9	..	16.0
Kenya	774	632	32.0	45.9	60.1	47.6	7.8	6.6
Korea, Dem. Rep.	..	..	..	..	..	..	..	..
Korea, Rep.	9,155	24,822	34.7	45.2	34.5	22.8	30.7	32.0
Kuwait	1,054	1,416	87.5	76.8	12.5	17.2	0.0	6.0
Kyrgyz Republic	..	*58*	..	*32.8*	..	*13.8*	..	*53.4*
Lao PDR	11	*116*	72.7	*16.4*	27.3	*81.9*	0.0	*1.7*
Latvia	*290*	1,026	*94.8*	69.5	*2.4*	11.5	*2.4*	19.0
Lebanon	..	..	..	..	..	..	..	..
Lesotho	34	*46*	14.7	*2.2*	50.0	*52.2*	35.3	*45.7*
Libya	83	*37*	84.3	*67.6*	7.2	*24.3*	8.4	*8.1*
Lithuania	..	1,083	..	36.8	..	50.8	..	12.4
Macedonia, FYR	..	*130*	..	*46.9*	..	*11.5*	..	*41.5*
Madagascar	129	*264*	31.8	*23.1*	31.0	*34.8*	36.4	*42.4*
Malawi	37	..	45.9		43.2		10.8	
Malaysia	3,769	11,986	31.8	23.5	44.7	29.3	23.6	47.2
Mali	71	*62*	31.0	*37.1*	53.5	*41.9*	14.1	*21.0*
Mauritania	14	*24*	35.7	*4.2*	64.3	*83.3*	0.0	*16.7*
Mauritius	478	1,065	32.8	23.2	51.0	51.0	15.9	25.8
Mexico	7,222	11,829	12.4	11.6	76.5	64.1	11.1	24.3
Moldova	..	105	..	43.8	..	36.2	..	20.0
Mongolia	48	73	41.7	39.7	10.4	49.3	47.9	12.3
Morocco	1,871	2,818	9.6	16.9	68.4	69.5	22.0	13.6
Mozambique	103	*286*	..	..	..	..	38.8	*79.7*
Myanmar	93	*529*	10.8	*6.2*	21.5	*32.1*	68.8	*61.6*
Namibia	106	*315*	..	..	81.1	*91.4*	18.9	*8.6*
Nepal	166	454	3.6	12.3	65.7	38.1	30.7	49.6
Netherlands	29,621	54,232	43.7	37.6	14.0	12.9	42.3	49.5
New Zealand	2,415	4,246	43.4	30.8	42.7	51.7	13.9	17.5
Nicaragua	34	255	20.6	11.0	35.3	49.0	47.1	40.0
Niger	22	..	..	..	59.1	..	36.4	..
Nigeria	965	980	3.8	12.0	2.6	5.5	93.6	82.4
Norway	12,452	15,406	68.7	62.0	12.6	14.9	18.7	23.1
Oman	68	*18*	14.7	..	85.3	..	0.0	*0.0*
Pakistan	1,218	*1,446*	59.3	*57.1*	12.1	*7.1*	28.7	*35.8*
Panama	907	1,589	64.8	56.5	19.0	24.4	16.2	19.2
Papua New Guinea	198	248	11.1	4.4	12.1	2.4	76.8	93.1
Paraguay	404	412	18.3	12.6	21.0	19.7	60.4	67.7
Peru	715	1,608	43.4	16.0	30.3	62.4	26.3	21.6
Philippines	2,897	4,778	8.5	12.0	16.1	53.4	75.4	34.5
Poland	3,200	9,961	57.3	21.4	11.2	43.4	31.5	35.2
Portugal	5,054	8,280	15.7	17.5	70.4	61.7	14.0	20.8
Puerto Rico	..	..	..	..	..	..	..	..
Romania	610	1,342	50.5	39.9	17.4	18.8	32.1	41.3
Russian Federation	..	9,087	..	33.3	..	41.1	..	25.6

	Commercial service exports $ millions		Transport % of total		Travel % of total		Other % of total	
	1990	1999	1990	1999	1990	1999	1990	1999
Rwanda	31	31	58.1	29.0	32.3	61.3	9.7	9.7
Saudi Arabia	3,031	5,156	..	..	..	..	..	..
Senegal	356	329	19.1	16.1	42.7	49.2	38.2	34.7
Sierra Leone	45	..	8.9	..	75.6	..	13.3	..
Singapore	12,719	23,612	17.5	19.4	36.6	22.0	45.9	58.6
Slovak Republic	..	2,058		32.3		24.8		42.9
Slovenia	..	1,947	..	26.6		51.6		21.8
South Africa	3,442	4,780	33.7	22.4	53.3	52.8	13.0	24.7
Spain	27,649	53,001	17.2	14.2	67.2	61.1	15.6	24.7
Sri Lanka	425	888	39.8	45.0	30.1	25.9	30.1	29.1
Sudan	134	82	14.2	3.7	15.7	2.4	70.1	92.7
Sweden	13,453	18,408	35.8	24.8	21.7	22.8	42.6	52.4
Switzerland	18,232	26,319	11.8	11.8	40.6	29.1	47.5	59.2
Syrian Arab Republic	740	1,551	29.7	14.2	43.2	76.7	27.0	9.0
Tajikistan	..	..	..	..	..	..	..	..
Tanzania	131	636	19.8	11.3	36.6	74.8	43.5	13.8
Thailand	6,292	14,142	21.1	17.9	68.7	50.3	10.2	31.9
Togo	114	65	27.2	16.9	50.9	16.9	21.9	66.2
Trinidad and Tobago	322	574	50.6	35.2	29.5	35.0	19.9	29.8
Tunisia	1,575	2,763	23.0	21.7	64.8	66.1	12.2	12.2
Turkey	7,882	16,031	11.7	11.9	40.9	32.5	47.4	55.6
Turkmenistan	..	269	..	..	..	12.3	..	37.9
Uganda	21	165	..	13.3	..	81.8	100.0	4.8
Ukraine	..	3,771	..	81.5	..	8.7	..	9.8
United Arab Emirates	..	..	..	..	..	..	..	..
United Kingdom	53,172	101,517	25.5	18.5	29.3	22.8	45.2	58.7
United States	132,184	253,358	28.2	19.1	37.6	34.4	34.1	46.5
Uruguay	460	1,269	37.0	18.1	51.7	51.5	11.3	30.4
Uzbekistan	..	..	..	..	..	..	..	..
Venezuela, RB	1,121	1,297	40.9	21.4	44.2	74.1	14.9	4.5
Vietnam	182	2,609	..	..	..	..	..	..
West Bank and Gaza	..	..	..	..	..	..	..	..
Yemen, Rep.	82	166	26.8	21.7	48.8	38.6	24.4	39.8
Yugoslavia, FR (Serb./Mont.)	..	..	..	..	..	..	..	..
Zambia	94	..	69.1	..	13.8	..	18.1	..
Zimbabwe	253	646	44.3	23.1	25.3	49.1	30.4	27.9
World	**766,353 s**	**1,271,417 s**	**28.0 w**	**23.4 w**	**34.7 w**	**32.2 w**	**37.3 w**	**44.4 w**
Low income	14,336	26,674	25.2	22.4	38.0	35.0	36.8	42.7
Middle income	96,783	210,882	29.0	23.1	43.3	44.8	27.7	32.1
Lower middle income	50,219	101,958	26.9	18.4	41.9	49.6	31.2	31.9
Upper middle income	46,564	108,924	31.5	27.7	44.8	40.1	23.6	32.3
Low & middle income	111,119	237,556	28.5	23.0	42.6	43.7	28.9	33.3
East Asia & Pacific	31,386	84,486	28.7	23.3	44.3	44.5	27.0	32.3
Europe & Central Asia	21,612	65,147	25.0	25.8	35.8	40.0	39.3	34.2
Latin America & Carib.	25,942	40,581	27.7	20.7	52.0	51.5	20.3	27.8
Middle East & N. Africa	15,595	23,831	33.8	24.8	40.3	51.7	26.0	23.6
South Asia	6,815	14,660	27.9	13.8	30.1	22.1	42.0	64.1
Sub-Saharan Africa	9,769	8,851	32.1	22.6	38.6	45.7	29.3	31.7
High income	655,234	1,033,861	27.9	23.4	33.3	29.5	38.8	47.1
Europe EMU	284,075	389,641	24.6	23.4	36.0	32.8	39.3	43.8

Note: Shares may not sum to 100 percent because of rounding.
a. Includes Luxembourg.

Balance of payments statistics, the main source of information on international trade in services, have many weaknesses. Some large economies—such as the former Soviet Union—did not report data on trade in services until recently. Disaggregation of important components may be limited, and it varies significantly across countries. There are inconsistencies in the methods used to report items. And the recording of major flows as net items is common (for example, insurance transactions are often recorded as premiums less claims). These factors contribute to a downward bias in the value of the service trade reported in the balance of payments.

Efforts are being made to improve the coverage, quality, and consistency of these data. Eurostat and the Organisation for Economic Co-operation and Development, for example, are working together to improve the collection of statistics on trade in services in member countries. In addition, the International Monetary Fund (IMF) has implemented the new classification of trade in services introduced in the fifth edition of its *Balance of Payments Manual* (1993).

Still, difficulties in capturing all the dimensions of international trade in services mean that the record is likely to remain incomplete. Cross-border intrafirm service transactions, which are usually not captured in the balance of payments, are increasing rapidly as foreign direct investment expands and electronic networks become pervasive. One example of such transactions is transnational corporations' use of mainframe computers around the clock for data processing, exploiting time zone differences between their home country and the host countries of their affiliates. Another important dimension of service trade not captured by conventional balance of payments statistics is establishment trade—sales in the host country by foreign affiliates. By contrast, cross-border intrafirm transactions in merchandise may be reported as exports or imports in the balance of payments.

The data on exports of services in this table and on imports of services in table 4.8, unlike those in editions before 2000, include only commercial services and exclude the category "government services not included elsewhere." The data are compiled by the World Trade Organization (WTO) from balance of payments statistics provided by the IMF and from national statistics. Estimates of missing data provided by the WTO are used to compute regional and income group aggregates but are not shown in the tables. Data on total trade in goods and services from the IMF's Balance of Payments database are shown in table 4.15.

Figure 4.7

Travel services are the most important service export from developing countries

% of commercial service exports, 1999

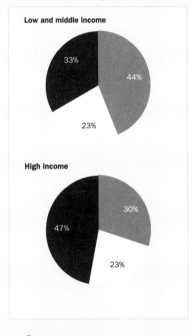

Low and middle income
33%
44%
23%

High income
30%
47%
23%

● Travel services
  Transport services
● Other commercial services
Source: World Trade Organization data.

• **Commercial service exports** are total service exports minus exports of government services not included elsewhere. International transactions in services are defined by the IMF's *Balance of Payments Manual* (1993) as the economic output of intangible commodities that may be produced, transferred, and consumed at the same time. Definitions may vary among reporting economies. • **Transport** covers all transport services (sea, air, land, internal waterway, space, and pipeline) performed by residents of one economy for those of another and involving the carriage of passengers, movement of goods (freight), rental of carriers with crew, and related support and auxiliary services. Excluded are freight insurance, which is included in insurance services; goods procured in ports by nonresident carriers and repairs of transport equipment, which are included in goods; repairs of railway facilities, harbors, and airfield facilities, which are included in construction services; and rental of carriers without crew, which is included in other services. • **Travel** covers goods and services acquired from an economy by travelers in that economy for their own use during visits of less than one year for business or personal purposes. Travel services include the goods and services consumed by travelers, such as meals, lodging, and transport (within the economy visited). • **Other commercial services** include such activities as insurance and financial services, international telecommunications, and postal and courier services; computer data; news-related service transactions between residents and nonresidents; construction services; royalties and license fees; miscellaneous business, professional, and technical services; and personal, cultural, and recreational services.

The data on exports of commercial services come from the WTO. Selected data appear in the WTO's *Annual Report.* The IMF publishes balance of payments data in its *International Financial Statistics* and *Balance of Payments Statistics Yearbook.*

4.8 | Structure of service imports

	Commercial service imports		Transport		Travel		Other	
	$ millions		% of total		% of total		% of total	
	1990	**1999**	**1990**	**1999**	**1990**	**1999**	**1990**	**1999**
Albania	29	152	27.6	59.2	*1.1*	7.9	72.4	32.9
Algeria	1,155	..	58.1	..	12.9	..	29.0	..
Angola	1,288	1,789	38.3	17.9	3.0	4.2	58.8	77.9
Argentina	2,876	8,184	32.6	28.7	40.7	50.8	26.7	20.5
Armenia	..	171	..	62.0	..	15.8	..	22.8
Australia	13,388	17,874	33.9	33.8	31.5	33.6	34.7	32.6
Austria	14,104	28,274	8.4	10.2	54.9	32.2	36.7	57.6
Azerbaijan	..	*692*	..	*28.0*	..	*24.6*	..	*47.4*
Bangladesh	554	1,305	71.1	70.7	14.1	16.2	15.0	13.2
Belarus	..	*431*	..	*30.9*	..	*28.8*	..	*40.4*
Belgium[a]	24,298	36,584	24.9	19.8	22.5	28.7	52.6	51.6
Benin	113	*170*	46.9	*65.9*	13.3	*9.4*	40.7	*24.7*
Bolivia	291	435	61.5	60.7	20.6	17.2	17.5	22.3
Bosnia and Herzegovina	..	..	..	..	..	..	..	..
Botswana	371	*517*	57.4	*42.2*	15.1	*24.4*	27.5	*33.5*
Brazil	6,733	11,924	44.4	37.4	22.4	24.1	33.2	38.5
Bulgaria	600	1,465	40.5	44.0	31.5	35.8	28.0	20.3
Burkina Faso	196	..	64.8	..	16.3	..	18.9	..
Burundi	59	*33*	62.7	*60.6*	28.8	*30.3*	8.5	*9.1*
Cambodia	*64*	190	*25.0*	57.4	..	4.2	*75.0*	38.4
Cameroon	1,018	..	45.3	..	27.4	..	27.3	..
Canada	27,479	38,469	21.1	21.1	39.8	29.5	39.2	49.4
Central African Republic	166	..	50.0	..	30.7	..	19.9	..
Chad	223	..	45.3	..	31.4	..	23.8	..
Chile	1,983	3,959	47.4	51.3	21.5	24.2	31.1	24.4
China[†]	4,113	30,666	78.9	25.8	11.4	35.4	9.7	38.8
Hong Kong, China	11,018	24,456	30.0	24.9	43.2	53.9	26.7	21.3
Colombia	1,683	3,221	34.9	34.8	27.0	33.4	38.1	31.8
Congo, Dem. Rep.	689	..	44.3	..	2.3	..	53.4	..
Congo, Rep.	748	*553*	18.4	*43.0*	15.1	*9.2*	66.4	*47.7*
Costa Rica	540	*1,168*	41.3	*42.6*	28.9	*35.0*	30.0	*22.3*
Côte d'Ivoire	1,518	1,307	32.1	42.2	11.1	17.9	56.8	39.9
Croatia	..	2,029	..	18.9	..	35.1	..	46.0
Cuba	*375*	*536*	..	..	..	..	..	..
Czech Republic	..	5,624	..	14.2	..	26.2	..	59.6
Denmark	10,106	15,201	38.3	42.4	36.5	32.1	25.2	25.5
Dominican Republic	435	1,351	40.0	59.2	33.1	20.0	26.9	20.8
Ecuador	755	1,264	41.6	32.4	23.2	21.4	35.2	46.2
Egypt, Arab Rep.	3,327	5,959	44.0	35.7	3.9	18.1	52.1	46.2
El Salvador	296	630	45.9	47.8	20.6	17.3	33.4	34.9
Eritrea	..	..	..	..	..	..	..	..
Estonia	*123*	829	76.4	44.6	*15.4*	25.9	*8.1*	29.4
Ethiopia	348	*405*	76.4	*56.0*	3.2	*11.4*	20.1	*32.6*
Finland	7,432	7,556	26.1	26.4	37.2	26.9	36.6	46.7
France	50,455	63,118	34.7	31.1	24.4	29.5	40.8	39.5
Gabon	984	..	23.3	..	13.9	..	62.9	..
Gambia, The	36	*58*	63.9	*55.2*	22.2	*27.6*	11.1	*17.2*
Georgia	..	*335*	..	*23.9*	..	*67.5*	..	*8.7*
Germany	79,214	132,816	21.6	18.4	42.8	36.4	35.6	45.2
Ghana	226	*433*	54.9	*61.9*	5.8	*5.5*	38.9	*32.6*
Greece	2,756	4,976	34.0	*29.1*	39.5	*31.6*	26.5	*39.3*
Guatemala	363	760	41.0	47.9	27.5	24.1	31.7	28.0
Guinea	243	236	57.2	42.4	12.3	10.2	30.5	47.5
Guinea-Bissau	17	*24*	52.9	*66.7*	17.6	*4.2*	23.5	*29.2*
Haiti	71	*370*	47.9	*88.1*	52.1	*10.0*	0.0	*2.2*
Honduras	213	427	45.5	63.5	17.8	14.1	37.1	22.7
† Data for Taiwan, China	13,923	23,475	27.0	24.8	35.8	31.5	37.2	43.7

Structure of service imports | 4.8

	Commercial service imports $ millions		Transport % of total		Travel % of total		Other % of total	
	1990	1999	1990	1999	1990	1999	1990	1999
Hungary	2,264	4,188	8.8	10.7	25.8	28.5	65.3	60.8
India	5,943	17,185	57.5	40.6	6.6	11.7	35.9	47.7
Indonesia	5,898	11,303	47.4	27.0	14.2	20.4	38.4	52.6
Iran, Islamic Rep.	3,703	2,392	47.3	54.5	9.2	6.4	43.5	39.1
Iraq	..	..	..	..	..	..	..	..
Ireland	5,145	26,069	24.3	9.1	22.6	9.4	53.1	81.4
Israel	4,825	10,812	39.6	39.5	29.7	24.1	30.7	36.4
Italy	46,602	58,376	23.7	21.7	22.1	28.0	54.2	50.4
Jamaica	667	1,264	47.8	42.6	17.1	16.4	35.1	40.9
Japan	84,281	114,173	31.6	26.8	29.6	28.7	38.8	44.5
Jordan	1,118	1,584	52.1	35.8	30.1	28.7	17.9	35.5
Kazakhstan	..	1,104	..	36.2	..	35.7	..	28.2
Kenya	598	583	66.2	49.6	6.4	22.0	27.4	28.5
Korea, Dem. Rep.	..	..	..	..	..	..	..	..
Korea, Rep.	10,050	26,106	39.8	38.5	27.5	11.3	32.7	50.1
Kuwait	2,805	4,093	31.9	36.3	65.5	61.3	2.6	2.4
Kyrgyz Republic	..	177	..	52.5	..	1.7	..	45.8
Lao PDR	25	92	72.0	41.3	12.8	25.0	28.0	33.7
Latvia	120	628	82.5	30.4	10.8	42.7	6.7	26.9
Lebanon	..	..	..	..	..	..	..	..
Lesotho	48	50	68.8	74.0	25.0	26.0	8.3	2.0
Libya	926	915	41.9	50.2	45.8	36.5	12.4	13.3
Lithuania	..	747	..	28.5	..	45.6	..	25.8
Macedonia, FYR	..	297	..	48.5	..	10.4	..	41.1
Madagascar	172	326	43.6	43.6	23.3	36.5	33.1	20.2
Malawi	268	..	82.1	..	6.0	..	12.3	..
Malaysia	5,394	14,322	46.9	32.3	26.9	13.7	26.2	54.1
Mali	352	324	57.4	69.1	15.6	13.0	26.7	17.9
Mauritania	126	130	77.0	36.9	18.3	32.3	4.8	30.0
Mauritius	407	651	51.6	34.9	23.1	28.7	25.3	36.4
Mexico	10,063	13,796	25.0	41.6	54.9	32.9	20.2	25.5
Moldova	..	156	..	32.7	..	37.2	..	30.1
Mongolia	155	140	56.1	61.4	0.6	29.3	43.2	9.3
Morocco	940	1,541	58.3	40.6	19.9	29.8	21.9	29.7
Mozambique	206	396	57.8	27.0	0.0	..	42.2	73.0
Myanmar	72	429	36.1	32.9	22.2	6.3	41.7	61.1
Namibia	341	449	46.9	33.4	17.9	19.6	35.2	47.0
Nepal	159	202	40.9	36.6	28.3	35.1	30.8	28.2
Netherlands	28,995	47,465	37.7	30.1	25.4	23.6	36.9	46.2
New Zealand	3,251	4,496	40.6	32.9	29.5	33.2	30.0	33.8
Nicaragua	73	319	71.2	45.8	20.5	24.5	9.6	29.8
Niger	209	..	67.9	..	10.5	..	21.5	..
Nigeria	1,901	3,311	33.6	19.8	30.3	18.7	36.1	61.4
Norway	12,247	17,683	44.6	37.0	30.0	29.5	25.3	33.5
Oman	719	1,303	36.6	42.1	6.5	3.6	56.9	54.3
Pakistan	1,863	2,424	67.0	71.6	23.1	14.6	9.9	13.8
Panama	666	1,057	66.7	59.5	14.9	17.4	18.6	23.1
Papua New Guinea	393	728	35.6	24.9	12.7	7.3	51.7	67.9
Paraguay	361	464	61.5	59.3	19.7	27.2	18.6	13.6
Peru	1,071	2,050	43.4	39.1	27.6	23.9	28.9	37.0
Philippines	1,721	7,492	56.9	25.9	6.4	17.5	36.6	56.6
Poland	2,847	7,622	52.4	22.2	14.9	10.8	32.8	67.0
Portugal	3,772	6,519	48.5	31.3	23.0	34.7	28.6	34.0
Puerto Rico	..	..	..	..	..	..	..	..
Romania	787	1,753	65.6	32.6	13.1	22.5	21.3	44.9
Russian Federation	..	12,427	..	14.5	..	55.0	..	30.5

	Commercial service imports		Transport		Travel		Other	
	$ millions		% of total		% of total		% of total	
	1990	1999	1990	1999	1990	1999	1990	1999
Rwanda	96	115	68.8	60.9	24.0	14.8	7.3	25.2
Saudi Arabia	12,694	9,452	..	..	..	..	..	..
Senegal	368	389	60.1	60.9	12.5	13.6	27.4	25.7
Sierra Leone	67	..	29.9	..	32.8	..	37.3	..
Singapore	8,575	18,768	41.0	33.5	21.0	24.7	38.0	41.8
Slovak Republic	..	1,797	..	21.4	..	18.9	..	59.7
Slovenia	..	1,569	..	24.0	..	37.8	..	38.2
South Africa	4,096	5,207	49.3	43.8	27.6	34.7	23.1	21.5
Spain	15,197	29,998	30.8	25.8	28.0	18.4	41.2	55.8
Sri Lanka	620	1,325	64.2	59.8	11.9	15.2	23.9	24.9
Sudan	202	270	31.7	81.9	25.2	13.0	42.6	5.2
Sweden	16,959	23,006	23.2	17.4	37.1	35.3	39.7	47.3
Switzerland	11,086	16,106	24.1	21.9	53.0	44.7	22.9	33.4
Syrian Arab Republic	702	1,297	54.4	48.7	35.5	44.7	10.1	6.6
Tajikistan	..	..	..	..	..	..	..	..
Tanzania	288	773	58.0	23.7	8.0	55.8	34.0	20.6
Thailand	6,160	13,970	58.1	38.3	23.2	21.3	18.7	40.4
Togo	217	149	56.7	71.8	18.4	2.0	24.4	26.2
Trinidad and Tobago	460	235	51.7	52.8	26.5	28.5	21.7	18.7
Tunisia	682	1,106	51.5	52.0	26.2	21.6	22.4	26.4
Turkey	2,794	8,376	32.2	27.5	18.6	17.6	49.2	55.0
Turkmenistan	..	669	..	22.9	..	18.7	..	58.3
Uganda	195	693	58.5	34.9	..	19.8	41.5	45.3
Ukraine	..	2,292	..	15.6	..	15.9	..	68.5
United Arab Emirates	..	..	..	..	..	..	..	..
United Kingdom	44,608	81,376	33.2	28.9	41.1	44.7	25.6	26.4
United States	97,940	180,415	36.2	30.9	38.9	34.3	24.9	34.7
Uruguay	363	850	48.2	41.5	30.6	32.9	21.2	25.6
Uzbekistan	..	..	..	..	..	..	..	..
Venezuela, RB	2,390	4,824	33.5	30.9	42.8	50.8	23.7	18.3
Vietnam	126	3,156	..	..	..	..	..	..
West Bank and Gaza					..	..	..	..
Yemen, Rep.	493	510	45.2	55.7	13.0	16.3	42.0	28.0
Yugoslavia, FR (Serb./Mont.)	..	..	..	..	..	..	..	..
Zambia	370	..	76.8	..	14.6	..	8.6	..
Zimbabwe	460	754	51.7	54.1	14.3	18.4	33.7	27.5
World	**791,441 s**	**1,259,694 s**	**32.5 w**	**27.7 w**	**32.1 w**	**31.0 w**	**35.4 w**	**41.3 w**
Low income	28,554	41,532	50.6	34.2	14.0	16.2	35.3	49.6
Middle income	119,543	220,835	40.3	32.3	22.4	26.3	37.2	41.4
Lower middle income	51,693	101,734	42.0	30.1	12.6	30.1	45.3	39.8
Upper middle income	67,850	119,101	38.7	34.3	31.6	22.8	29.6	42.9
Low & middle income	148,097	262,367	42.6	32.6	20.6	24.6	36.8	42.8
East Asia & Pacific	34,483	104,917	51.2	31.7	20.9	21.4	27.8	46.8
Europe & Central Asia	24,333	51,360	24.8	21.0	8.6	30.1	66.6	48.9
Latin America & Carib.	33,099	51,955	37.3	40.1	35.7	30.7	27.0	29.2
Middle East & N. Africa	27,432	21,146	48.5	40.0	15.8	22.2	35.7	37.8
South Asia	9,176	18,692	60.7	42.7	11.2	12.3	28.2	45.1
Sub-Saharan Africa	19,574	14,297	45.8	33.9	18.0	25.1	36.1	41.0
High income	643,344	997,327	30.3	26.4	34.7	32.8	35.1	40.9
Europe EMU	275,214	400,191	26.9	22.1	32.2	29.1	40.9	48.8

Note: Shares may not sum to 100 percent because of rounding.

a. Includes Luxembourg.

About the data

Trade in services differs from trade in goods because services are produced and consumed at the same time. Thus services to a traveler may be consumed in the producing country (for example, use of a hotel room) but are classified as imports of the traveler's country. In other cases services may be supplied from a remote location; for example, insurance services may be supplied from one location and consumed in another. For further discussion of the problems of measuring trade in services see *About the data* for table 4.7.

The data on exports of services in table 4.7 and on imports of services in this table, unlike those in editions before 2000, include only commercial services and exclude the category "government services not included elsewhere." The data are compiled by the World Trade Organization (WTO) from balance of payments statistics provided by the International Monetary Fund (IMF) and from national statistics. Estimates of missing data provided by the WTO are used to compute regional and income group aggregates but are not shown in the tables.

Definitions

• **Commercial service imports** are total service imports minus imports of government services not included elsewhere. International transactions in services are defined by the IMF's *Balance of Payments Manual* (1993) as the economic output of intangible commodities that may be produced, transferred, and consumed at the same time. Definitions may vary among reporting economies. • **Transport** covers all transport services (sea, air, land, internal waterway, space, and pipeline) performed by residents of one economy for those of another and involving the carriage of passengers, movement of goods (freight), rental of carriers with crew, and related support and auxiliary services. Excluded are freight insurance, which is included in insurance services; goods procured in ports by nonresident carriers and repairs of transport equipment, which are included in goods; repairs of railway facilities, harbors, and airfield facilities, which are included in construction services; and rental of carriers without crew, which is included in other services. • **Travel** covers goods and services acquired from an economy by travelers in that economy for their own use during visits of less than one year for business or personal purposes. Travel services include the goods and services consumed by travelers, such as meals, lodging, and transport (within the economy visited). • **Other commercial services** include such activities as insurance and financial services, international telecommunications, and postal and courier services; computer data; news-related service transactions between residents and nonresidents; construction services; royalties and license fees; miscellaneous business, professional, and technical services; and personal, cultural, and recreational services.

Data sources

The data on imports of commercial services come from the WTO. Selected data appear in the WTO's *Annual Report*. The IMF publishes balance of payments data in its *International Financial Statistics* and *Balance of Payments Statistics Yearbook*.

4.9 | Structure of demand

	Household final consumption expenditure		General government final consumption expenditure		Gross capital formation		Exports of goods and services		Imports of goods and services		Gross domestic savings	
	% of GDP		% of GDP		% of GDP		% of GDP		% of GDP		% of GDP	
	1990	1999	1990	1999	1990	1999	1990	1999	1990	1999	1990	1999
Albania	61	90	19	11	29	17	15	11	23	30	21	−2
Algeria	56	51	16	17	29	27	23	28	25	23	27	32
Angola	36	36	34	32	12	24	39	57	21	48	30	32
Argentina	77	70	3	13	14	19	10	10	5	11	20	17
Armenia	46	99	18	11	47	19	35	21	46	50	36	−9
Australia	59	60	19	18	22	25	17	19	17	21	22	22
Austria	54	56	19	20	24	24	40	45	39	46	27	24
Azerbaijan	..	66	..	12	..	40	..	34	..	51	..	23
Bangladesh	86	79	4	5	17	22	6	13	14	19	10	17
Belarus	45	59	26	20	27	24	46	62	44	65	29	21
Belgium	54	54	21	21	23	21	71	76	70	72	25	25
Benin	87	84	11	10	14	18	14	17	26	28	2	6
Bolivia	77	76	12	15	13	19	23	17	24	27	11	9
Bosnia and Herzegovina	..	100	..	..[a]	..	35	..	27	..	62	..	0
Botswana	39	58	24	28	32	20	55	28	50	33	37	14
Brazil	59	62	19	19	20	20	8	11	7	12	21	19
Bulgaria	60	73	18	16	26	19	33	44	37	52	22	11
Burkina Faso	77	77	15	14	21	28	13	11	26	29	8	10
Burundi	95	84	11	16	15	9	8	9	28	18	−5	0
Cambodia	91	86	7	9	8	15	6	34	13	44	2	5
Cameroon	67	71	13	10	18	19	20	24	17	25	21	19
Canada	57	58	23	19	21	20	26	44	26	41	21	23
Central African Republic	86	81	15	12	12	14	15	17	28	24	−1	7
Chad	89	95	10	8	16	10	13	17	29	30	0	−3
Chile	62	65	10	12	25	21	35	29	31	27	28	23
China	50	47	12	13	35	37	18	22	14	19	38	40
Hong Kong, China	57	60	7	10	27	25	134	133	126	128	36	31
Colombia	66	68	9	21	19	13	21	18	15	19	24	11
Congo, Dem. Rep.	79	83	12	8	9	8	30	24	29	22	9	9
Congo, Rep.	62	60	14	11	16	22	54	78	46	70	24	30
Costa Rica	61	63	18	13	27	17	35	54	41	47	21	24
Côte d'Ivoire	72	66	17	11	7	16	32	44	27	38	11	23
Croatia	74	57	24	27	10	23	78	41	86	48	2	16
Cuba	..	..	..	..	..	..	..	..	..	..	..	..
Czech Republic	49	53	23	20	25	28	45	64	43	65	28	27
Denmark	49	50	26	26	20	20	36	37	31	33	25	24
Dominican Republic	80	75	5	8	25	25	34	30	44	39	15	17
Ecuador	69	65	9	10	17	13	33	37	27	26	23	24
Egypt, Arab Rep.	73	76	11	10	29	23	20	16	33	24	16	14
El Salvador	89	86	10	10	14	16	19	25	31	37	1	4
Eritrea	98	71	33	51	5	47	20	10	57	79	−31	−21
Estonia	62	58	16	24	30	25	60	77	54	83	22	19
Ethiopia	74	81	19	16	12	18	8	14	12	29	7	3
Finland	51	51	22	21	29	20	23	37	24	29	27	28
France	55	55	22	24	23	19	21	26	22	24	22	22
Gabon	50	49	13	17	22	28	46	45	31	38	37	35
Gambia, The	76	84	14	14	22	18	60	51	72	67	11	2
Georgia	..	90	..	12	..	17	..	27	..	46	..	−2
Germany	57	58	19	19	24	22	26	29	26	28	24	23
Ghana	85	83	9	11	14	23	17	34	26	50	5	6
Greece	71	69	16	15	23	23	19	19	28	25	13	16
Guatemala	84	85	7	6	14	17	21	19	25	27	10	9
Guinea	70	77	12	7	18	17	31	21	31	23	18	15
Guinea-Bissau	87	91	10	11	30	16	10	26	37	44	3	−2
Haiti	93	98	8	6	12	11	16	12	29	28	−1	−4
Honduras	66	69	14	11	23	33	36	43	40	57	20	19

Structure of demand 4.9

	Household final consumption expenditure		General government final consumption expenditure		Gross capital formation		Exports of goods and services		Imports of goods and services		Gross domestic savings	
	% of GDP		% of GDP		% of GDP		% of GDP		% of GDP		% of GDP	
	1990	1999	1990	1999	1990	1999	1990	1999	1990	1999	1990	1999
Hungary	61	63	11	10	25	29	31	53	29	55	28	26
India	66	68	12	12	25	23	7	12	10	15	23	20
Indonesia	59	62	9	6	31	24	25	35	24	27	32	32
Iran, Islamic Rep.	62	64	11	14	29	18	22	21	24	16	27	23
Iraq	..	..	..	..	..	..	..	..	..	..	..	..
Ireland	58	49	16	14	21	23	57	88	52	74	26	37
Israel	56	60	30	29	25	21	35	36	45	45	14	11
Italy	58	60	20	18	22	20	20	26	20	24	22	22
Jamaica	62	65	14	18	28	26	52	49	56	59	24	17
Japan	58	61	9	10	32	26	11	10	10	9	33	28
Jordan	74	73	25	25	32	21	62	44	93	62	1	3
Kazakhstan	52	72	18	6	32	18	74	45	75	40	30	23
Kenya	67	76	19	17	20	14	26	24	31	31	14	7
Korea, Dem. Rep.	..	..	..	..	..	..	..	..	..	..	..	..
Korea, Rep.	53	56	10	10	38	27	29	42	30	35	37	34
Kuwait	57	50	39	27	18	12	45	47	58	37	4	22
Kyrgyz Republic	71	78	25	19	24	18	29	42	50	57	4	3
Lao PDR	..	81	..	5	..	25	..	37	..	49	..	13
Latvia	53	66	9	19	40	26	48	47	49	58	39	15
Lebanon	140	98	25	15	18	28	18	11	100	51	−64	−13
Lesotho	137	115	14	20	53	47	17	27	121	109	−51	−35
Libya	..	..	..	..	..	..	..	..	..	..	..	..
Lithuania	57	65	19	22	33	23	52	40	61	50	24	13
Macedonia, FYR	72	74	19	19	19	21	26	41	36	56	9	7
Madagascar	86	87	8	8	17	13	17	25	27	33	6	5
Malawi	75	88	16	12	20	15	25	27	35	43	10	−1
Malaysia	52	42	14	11	32	22	75	122	72	97	34	47
Mali	80	77	14	13	23	21	17	25	34	36	6	10
Mauritania	69	78	26	15	20	18	46	39	61	49	5	7
Mauritius	65	66	12	11	31	28	65	64	72	69	24	23
Mexico	70	68	8	10	23	23	19	31	20	32	22	22
Moldova	58	74	15	19	29	22	32	50	34	65	27	7
Mongolia	57	63	30	18	34	26	21	50	42	55	13	21
Morocco	65	61	15	19	25	24	26	30	32	34	19	20
Mozambique	101	82	12	11	16	33	8	12	36	38	−12	7
Myanmar	89	90	..[a]	..[a]	13	11	3	0	5	1	11	10
Namibia	51	64	31	26	34	20	52	53	68	64	18	9
Nepal	83	77	9	10	18	20	11	23	21	30	8	13
Netherlands	47	50	24	23	24	22	58	61	55	56	28	27
New Zealand	63	65	17	15	19	19	28	31	27	30	20	20
Nicaragua	59	94	43	18	19	43	25	34	46	89	−2	−12
Niger	84	81	15	15	8	10	15	16	22	22	1	4
Nigeria	56	67	15	15	15	24	43	37	29	42	29	18
Norway	49	48	21	21	23	24	41	39	34	33	30	30
Oman	27	..	38	..	13	..	53	..	31	..	35	..
Pakistan	74	78	15	12	19	15	16	15	23	20	11	10
Panama	60	60	18	16	17	33	38	33	34	41	21	24
Papua New Guinea	59	66	25	13	24	18	41	45	49	42	16	21
Paraguay	77	82	6	9	23	23	33	23	39	37	17	9
Peru	74	69	8	11	16	22	16	15	14	17	18	20
Philippines	72	67	10	13	24	19	28	51	33	50	18	20
Poland	50	65	19	15	25	26	28	26	21	32	32	20
Portugal	63	64	16	19	27	25	33	31	40	40	20	16
Puerto Rico	..	..	..	..	..	..	..	..	..	..	..	..
Romania	66	70	13	15	30	20	17	30	26	34	21	16
Russian Federation	49	53	21	14	30	15	18	46	18	28	30	33

	Household final consumption expenditure		General government final consumption expenditure		Gross capital formation		Exports of goods and services		Imports of goods and services		Gross domestic savings	
	% of GDP		% of GDP		% of GDP		% of GDP		% of GDP		% of GDP	
	1990	1999	1990	1999	1990	1999	1990	1999	1990	1999	1990	1999
Rwanda	84	89	10	13	15	14	6	6	14	21	6	−1
Saudi Arabia	40	39	31	30	20	19	46	40	36	28	30	31
Senegal	76	76	15	11	14	19	25	33	30	39	9	13
Sierra Leone	82	95	10	11	9	0	24	14	25	20	8	−6
Singapore	46	39	10	10	37	33	202	..	195	..	44	52
Slovak Republic	54	54	22	19	33	32	27	62	36	67	24	27
Slovenia	55	55	19	21	17	28	84	53	74	57	26	24
South Africa	63	63	20	19	12	16	24	25	19	23	18	18
Spain	..	59	..	17	..	24	..	28	..	28	..	23
Sri Lanka	76	71	10	9	22	27	30	35	38	43	14	20
Sudan	..	..	..	..	..	..	..	..	..	..	..	..
Sweden	48	51	28	27	24	17	30	44	29	38	24	22
Switzerland	57	61	14	14	28	21	36	40	36	36	29	25
Syrian Arab Republic	70	70	14	11	15	29	28	29	27	40	16	18
Tajikistan	..	77	..	10	..	9	..	68	..	63	..	13
Tanzania[b]	81	86	18	12	26	17	13	13	37	28	1	2
Thailand	57	56	9	11	41	21	34	57	42	45	34	33
Togo	71	85	14	11	27	13	33	30	45	40	15	4
Trinidad and Tobago	59	62	12	11	13	21	45	50	29	44	29	27
Tunisia	58	60	16	16	32	27	44	42	51	44	25	24
Turkey	69	65	11	15	24	23	13	23	18	27	20	20
Turkmenistan	49	62	23	12	40	46	..	42	..	62	28	26
Uganda	92	85	8	10	13	16	7	11	19	23	1	5
Ukraine	57	57	17	22	27	20	28	53	29	52	26	21
United Arab Emirates	39	..	16	..	20	..	65	..	40	..	45	..
United Kingdom	63	66	20	18	20	18	24	26	27	27	18	16
United States	67	67	17	14	18	20	10	11	11	13	16	18
Uruguay	70	73	12	14	12	15	24	18	18	20	18	14
Uzbekistan	61	75	25	10	32	15	29	19	48	19	13	16
Venezuela, RB	62	70	8	8	10	16	39	22	20	15	29	22
Vietnam	86	70	8	7	13	25	26	44	33	52	6	23
West Bank and Gaza	..	93	..	26	..	39	..	17	..	75	..	−19
Yemen, Rep.	77	71	19	17	15	19	16	39	27	45	4	12
Yugoslavia, FR (Serb./Mont.)	..	..	..	..	..	..	..	..	..	..	..	..
Zambia	64	92	19	10	17	17	36	22	37	41	17	−1
Zimbabwe	63	74	19	15	17	11	23	45	23	46	17	11
World	**60 w**	**62 w**	**16 w**	**15 w**	**24 w**	**23 w**	**19 w**	**27 w**	**19 w**	**25 w**	**24 w**	**25 w**
Low income	66	68	12	11	24	22	17	24	20	26	21	20
Middle income	59	59	14	15	26	24	21	29	20	26	27	26
Lower middle income	57	57	13	14	30	26	21	31	22	28	30	30
Upper middle income	60	61	15	15	23	22	21	27	19	25	25	23
Low & middle income	60	61	14	14	26	23	21	28	20	26	26	25
East Asia & Pacific	54	53	11	11	35	30	26	38	26	32	35	36
Europe & Central Asia	55	60	18	16	28	21	23	40	24	37	26	25
Latin America & Carib.	65	66	13	15	19	20	14	17	12	18	21	19
Middle East & N. Africa	57	56	20	20	24	22	34	30	35	27	23	24
South Asia	69	70	12	11	24	22	9	13	13	17	20	18
Sub-Saharan Africa	66	68	18	17	15	18	27	29	26	31	16	15
High income	60	62	16	16	23	22	19	22	19	21	23	23
Europe EMU	56	57	20	20	23	21	28	33	28	31	24	23

a. Data on general government final consumption expenditure are not available separately; they are included in household final consumption expenditure. b. Data cover mainland Tanzania only.

Gross domestic product (GDP) from the expenditure side is made up of household final consumption expenditure, general government final consumption expenditure, gross capital formation (private and public investment in fixed assets and changes in inventories), and net exports (exports minus imports) of goods and services. Such expenditures are recorded in purchaser prices and so include net taxes on products.

Because policymakers have tended to focus on fostering the growth of output, and because data on production are easier to collect than data on spending, many countries generate their primary estimate of GDP using the production approach. Moreover, many countries do not estimate all the separate components of national expenditures but instead derive some of the main aggregates indirectly using GDP (based on the production approach) as the control total.

Household final consumption expenditure (private consumption in previous editions) is often estimated as a residual, by subtracting from GDP all other known expenditures. The resulting aggregate may incorporate fairly large discrepancies. When household consumption is calculated separately, the household surveys on which many of the estimates are based tend to be one-year studies with limited coverage. Thus the estimates quickly become outdated and must be supplemented by price- and quantity-based statistical estimating procedures. Complicating the issue, in many developing countries the distinction between cash outlays for personal business and those for household use may be blurred. The *World Development Indicators* includes in household consumption the expenditures of nonprofit institutions serving households.

General government final consumption expenditure (general government consumption in previous editions) includes expenditures on goods and services for individual consumption as well as those on services for collective consumption. Defense expenditures, including those on capital outlays—with certain exceptions—are treated as current spending.

Gross capital formation (gross domestic investment in previous editions) consists of outlays on additions to the economy's fixed assets plus net changes in the level of inventories. The 1993 United Nations System of National Accounts (1993 SNA) recognizes a third category of capital formation: net acquisition of valuables. Included in gross capital formation under the 1993 SNA guidelines are capital outlays on defense establishments that may be used by the general public, such as schools, airfields, and hospitals. These expenses were treated as consumption in the earlier version of the SNA. Data on capital formation may be estimated from

direct surveys of enterprises and administrative records or based on the commodity flow method using data from trade and construction activities. While the quality of data on fixed capital formation by government depends on the quality of government accounting systems (which tend to be weak in developing countries), measures of fixed capital formation by households and corporations—particularly capital outlays by small, unincorporated enterprises—are usually very unreliable.

Estimates of changes in inventories are rarely complete but usually include the most important activities or commodities. In some countries these estimates are derived as a composite residual along with household final consumption expenditure. According to national accounts conventions, adjustments should be made for appreciation of the value of inventory holdings due to price changes, but this is not always done. In highly inflationary economies this element can be substantial.

Data on exports and imports are compiled from customs reports and balance of payments data. Although the data on exports and imports from the payments side provide reasonably reliable records of cross-border transactions, they may not adhere strictly to the appropriate definitions of valuation and timing used in the balance of payments or, more important, correspond with the change-of-ownership criterion. This issue has assumed greater significance with the increasing globalization of international business. Neither customs nor balance of payments data usually capture the illegal transactions that occur in many countries. Goods carried by travelers across borders in legal but unreported shuttle trade may further distort trade statistics.

Domestic savings, a concept used by the World Bank, represent the difference between GDP and total consumption. Domestic savings also satisfy this fundamental identity: exports minus imports equal domestic savings minus capital formation. Domestic savings differ from savings as defined in the national accounts; this SNA concept represents the difference between disposable income and consumption.

For further discussion of the problems in building and maintaining national accounts see Srinivasan (1994), Heston (1994), and Ruggles (1994). For a classic analysis of the reliability of foreign trade and national income statistics see Morgenstern (1963).

- **Household final consumption expenditure** is the market value of all goods and services, including durable products (such as cars, washing machines, and home computers), purchased by households. It excludes purchases of dwellings but includes imputed rent for owner-occupied dwellings. It also includes payments and fees to governments to obtain permits and licenses. Here, household consumption expenditure includes the expenditures of nonprofit institutions serving households, even when reported separately by the country. In practice, household consumption expenditure may include any statistical discrepancy in the use of resources relative to the supply of resources. • **General government final consumption expenditure** includes all government current expenditures for purchases of goods and services (including compensation of employees). It also includes most expenditures on national defense and security, but excludes government military expenditures that are part of government capital formation. • **Gross capital formation** consists of outlays on additions to the fixed assets of the economy plus net changes in the level of inventories. Fixed assets include land improvements (fences, ditches, drains, and so on); plant, machinery, and equipment purchases; and the construction of roads, railways, and the like, including schools, offices, hospitals, private residential dwellings, and commercial and industrial buildings. Inventories are stocks of goods held by firms to meet temporary or unexpected fluctuations in production or sales, and "work in progress." According to the 1993 SNA, net acquisitions of valuables are also considered capital formation. • **Exports and imports of goods and services** represent the value of all goods and other market services provided to or received from the rest of the world. They include the value of merchandise, freight, insurance, transport, travel, royalties, license fees, and other services, such as communication, construction, financial, information, business, personal, and government services. They exclude labor and property income (formerly called factor services) as well as transfer payments. • **Gross domestic savings** are calculated as GDP less total consumption.

The national accounts indicators for most developing countries are collected from national statistical organizations and central banks by visiting and resident World Bank missions. The data for high-income economies come from OECD data files (see the OECD's *National Accounts, 1988–1998*, volumes 1 and 2). The United Nations Statistics Division publishes detailed national accounts for United Nations member countries in *National Accounts Statistics: Main Aggregates and Detailed Tables* and updates in the *Monthly Bulletin of Statistics*.

4.10 | Growth of consumption and investment

	Household final consumption expenditure				Household final consumption expenditure per capita		General government final consumption expenditure		Gross capital formation	
	$ millions		average annual % growth		average annual % growth		average annual % growth		average annual % growth	
	1990	1999	1980–90	1990–99	1980–90	1990–99	1980–90	1990–99	1980–90	1990–99
Albania	1,271	3,320	..	4.4	..	4.0	..	–1.9	–0.3	23.7
Algeria	34,865	24,559	1.9	1.0	–1.1	–1.0	4.7	3.6	–2.3	–0.8
Angola	3,674	2,296	–0.1	–3.8	..	..	6.7	–2.0	–5.1	10.8
Argentina	109,038	198,055	..	3.3	..	2.0	..	1.3	–5.2	9.4
Armenia	2,005	1,761	..	–5.6	..	–6.3	..	–2.9	..	–1.8
Australia	181,286	221,538	3.1	3.6	1.5	2.4	3.6	2.5	3.6	6.6
Austria	90,609	117,288	2.4	2.0	2.2	1.5	1.4	1.4	2.2	2.6
Azerbaijan	..	2,623	..	–0.7	..	–1.7	..	4.9	..	..
Bangladesh	24,988	35,601	4.7	2.7	2.2	1.1	5.0	4.9	1.4	9.1
Belarus	15,537	15,640	..	–2.2	..	–2.0	..	–3.1	..	–9.0
Belgium	107,952	133,147	1.7	1.4	1.6	1.1	0.5	1.3	3.2	1.6
Benin	1,602	1,979	1.9	4.2	–1.3	1.3	0.5	3.6	–5.3	5.0
Bolivia	3,741	6,344	1.2	3.6	–0.9	1.2	–3.8	3.6	1.0	9.5
Bosnia and Herzegovina	..	..	..	..	..	..	..	..	..	..
Botswana	1,473	3,489	5.9	5.4	2.4	2.8	13.6	5.2	13.8	–1.3
Brazil	275,761	464,530	1.6	4.3	–0.4	2.9	7.3	–1.5	0.2	3.2
Bulgaria	11,566	9,277	4.4	–1.5	4.4	–0.9	6.7	–11.0	2.4	0.2
Burkina Faso	2,141	1,978	2.6	3.4	0.1	1.0	6.2	2.6	8.6	5.9
Burundi	1,070	603	3.4	–1.8	0.5	–4.0	3.2	–1.6	6.9	–1.8
Cambodia	1,016	2,463	..	..	..	..	..	..	..	..
Cameroon	7,432	6,525	3.5	2.3	0.6	–0.4	6.8	–0.3	–2.6	0.0
Canada	323,850	366,938	3.3	2.5	2.1	1.5	2.5	–0.1	5.2	4.7
Central African Republic	1,274	855	1.5	..	..	..	–1.7	..	10.0	..
Chad	1,482	1,454	5.3	0.7	..	..	14.5	–2.5	..	4.9
Chile	18,759	43,992	2.0	7.3	0.3	5.7	0.4	4.4	6.4	11.1
China	174,249	475,926	8.8	8.8	7.2	7.7	9.8	9.4	10.8	12.4
Hong Kong, China	42,422	94,703	6.7	4.2	5.3	2.1	5.0	4.1	4.0	6.3
Colombia	26,357	58,812	2.6	3.0	0.5	1.0	4.2	9.7	1.4	5.2
Congo, Dem. Rep.	7,398	4,784	3.4	–6.6	0.0	–9.6	0.0	–16.1	–5.1	–2.6
Congo, Rep.	1,746	1,320	3.3	–0.5	0.4	–3.3	2.5	–6.5	–12.6	4.7
Costa Rica	4,406	9,613	2.9	5.0	0.1	2.9	1.1	1.8	5.3	4.9
Côte d'Ivoire	7,766	7,394	1.5	1.5	–2.0	–1.3	–0.1	1.9	–10.4	16.2
Croatia	13,527	11,715	..	..	..	..	..	..	..	..
Cuba	..	..	..	..	..	..	..	..	..	..
Czech Republic	17,195	28,349	..	2.5	..	2.6	..	–1.1	..	5.1
Denmark	65,430	87,937	1.4	2.7	1.4	2.2	0.9	2.4	4.7	4.4
Dominican Republic	5,633	13,070	1.6	5.0	–0.6	3.1	1.9	15.1	3.5	5.0
Ecuador	7,323	12,414	1.9	1.6	–0.7	–0.5	–1.4	–1.4	–3.8	–0.1
Egypt, Arab Rep.	30,933	66,131	4.6	4.1	2.0	2.1	3.1	2.5	0.0	6.4
El Salvador	4,273	10,704	0.8	5.5	–0.2	3.4	0.1	3.1	2.2	7.6
Eritrea	430	457	..	..	..	..	..	..	..	..
Estonia	4,074	3,028	..	–0.4	..	0.7	..	4.8	..	–2.3
Ethiopia	5,081	5,208	0.2	3.3	–2.8	1.1	4.5	5.5	2.1	12.1
Finland	68,939	65,275	3.9	1.7	3.4	1.3	3.2	0.8	3.4	0.7
France	672,960	784,791	2.2	1.1	1.7	0.7	2.6	1.9	3.3	0.4
Gabon	2,961	2,115	1.5	0.6	–1.8	–1.9	–0.6	6.2	–5.7	4.2
Gambia, The	240	330	–2.9	3.3	–6.3	–0.1	1.7	–4.0	0.0	5.8
Georgia	..	2,458	..	..	..	..	..	..	..	..
Germany	1,003,506	1,219,501	..	1.3	..	1.0	..	1.2	..	1.1
Ghana	5,016	6,446	2.8	4.0	–0.6	1.3	2.4	4.0	3.3	3.7
Greece	60,704	86,666	2.4	1.8	1.9	1.4	2.7	1.6	–0.8	5.5
Guatemala	6,398	15,423	1.2	4.3	–1.3	1.6	2.6	4.7	–1.8	6.0
Guinea	1,982	2,897	..	3.5	..	0.9	..	2.2	..	2.4
Guinea-Bissau	212	199	0.8	3.1	–1.3	0.8	7.2	0.9	12.9	–13.0
Haiti	2,785	4,210	0.9	–0.9	–1.0	–3.0	–4.4	–1.7	–0.6	2.2
Honduras	2,026	3,741	2.7	3.1	–0.5	0.2	3.3	0.5	2.9	7.5

Growth of consumption and investment | 4.10

	Household final consumption expenditure				Household final consumption expenditure per capita		General government final consumption expenditure		Gross capital formation	
	$ millions		average annual % growth		average annual % growth		average annual % growth		average annual % growth	
	1990	1999	1980–90	1990–99	1980–90	1990–99	1980–90	1990–99	1980–90	1990–99
Hungary	20,290	30,642	1.3	–0.8	1.7	–0.5	1.9	1.0	–0.9	9.4
India	214,302	304,210	4.6	4.9	2.5	3.0	7.7	6.4	6.5	6.5
Indonesia	67,388	88,166	5.6	6.2	3.7	4.4	4.6	0.2	6.7	3.7
Iran, Islamic Rep.	74,476	70,406	2.8	2.9	–0.6	1.2	–5.0	6.2	–2.5	2.3
Iraq	..	..	..	..	..	..	..	..	..	..
Ireland	27,957	45,806	2.2	5.1	1.9	4.3	–0.3	3.6		8.9
Israel	32,112	60,054	5.4	6.5	3.6	3.6	0.5	3.1	2.2	5.7
Italy	634,194	697,264	2.9	1.2	2.8	1.0	2.9	0.0	2.1	0.9
Jamaica	2,637	4,493	4.5	–0.9	3.3	–1.8	6.2	2.8	–0.1	4.7
Japan	1,721,698	2,328,143	3.7	1.7	3.2	1.4	2.4	2.2	5.3	0.5
Jordan	2,978	5,880	2.3	5.4	–1.5	1.2	2.3	6.5	–1.5	2.0
Kazakhstan	..	11,179	..	..	..	..	..	–2.3	..	–11.9
Kenya	5,309	7,619	4.6	2.3	1.0	–0.1	2.6	9.8	0.4	3.7
Korea, Dem. Rep.	..	..	..	..	..	..	..	..	..	..
Korea, Rep.	132,113	226,822	8.0	5.1	6.8	4.0	5.2	3.2	11.9	1.6
Kuwait	10,459	14,878	–1.4	..	..	..	2.2	..	–4.5	..
Kyrgyz Republic	..	971	..	–7.5	..	–8.4	..	–10.5	..	–5.8
Lao PDR	..	882	..	..	..	..	..	..	..	..
Latvia	6,578	4,103	..	..	..	..	5.0	8.4	3.4	–2.9
Lebanon	3,961	16,930	..	7.2	..	5.3	..	7.4	..	18.4
Lesotho	855	1,007	3.6	3.4	1.1	1.1	3.2	6.6	5.3	2.3
Libya	..	..	..	..	..	..	..	..	..	..
Lithuania	7,527	6,905	..	..	..	..	..	..	..	..
Macedonia, FYR	..	2,549	..	1.6	..	0.9	..	–0.3	..	2.2
Madagascar	2,649	3,275	–0.6	2.0	–3.3	–0.9	0.5	0.8	4.9	1.1
Malawi	1,345	1,597	1.5	5.8	–1.7	3.0	6.3	–4.8	–2.8	–8.9
Malaysia	22,806	32,832	3.3	5.4	0.4	2.8	2.7	5.2	3.1	6.2
Mali	1,933	1,987	1.0	2.6	–1.5	0.1	7.9	3.8	3.6	–1.4
Mauritania	705	744	1.4	3.7	–1.2	0.9	–3.8	–1.3	6.9	7.8
Mauritius	1,707	2,802	6.7	4.5	5.8	3.3	3.3	3.8	9.0	2.6
Mexico	182,791	329,144	1.1	1.9	–1.0	0.2	2.4	1.5	–3.3	3.8
Moldova	2,328	855	..	7.1	..	7.3	..	–2.7	..	–17.6
Mongolia	..	578	..	..	..	..	..	..	..	..
Morocco	16,833	21,206	4.3	2.8	2.0	1.0	2.1	3.2	1.2	1.8
Mozambique	2,530	3,273	–3.2	5.0	–4.6	2.6	–2.1	–6.2	3.8	10.7
Myanmar	..	..	0.6	3.9	..	..	..	..	–4.1	14.7
Namibia	1,188	1,981	1.3	2.8	–1.4	0.2	3.7	2.6	–2.9	2.5
Nepal	3,028	3,851	4.5	4.0	1.8	1.5	7.2	6.0	6.0	7.2
Netherlands	142,467	197,322	1.7	2.7	1.2	2.1	2.1	1.4	3.1	2.2
New Zealand	27,300	34,403	2.0	3.0	1.1	1.8	1.5	2.0	0.8	7.1
Nicaragua	592	2,127	–3.6	5.1	–6.2	2.2	3.4	–2.9	–4.8	11.9
Niger	2,079	1,641	0.0	2.3	–3.2	–1.1	4.4	0.8	–7.1	3.9
Nigeria	15,816	23,379	–2.6	1.7	–5.5	–1.1	–3.5	0.9	–8.5	4.0
Norway	57,047	74,153	2.2	3.3	1.9	2.8	2.3	2.4	0.7	5.5
Oman	2,810	..	..	..	..	..	..	..	25.5	..
Pakistan	28,561	44,577	4.5	5.1	1.8	2.5	10.3	0.7	5.8	1.8
Panama	3,022	5,449	4.2	4.9	2.1	3.1	1.2	2.3	–8.9	12.3
Papua New Guinea	1,902	2,358	0.4	4.9	–1.7	2.5	–0.1	2.2	–0.9	1.3
Paraguay	4,063	6,336	2.4	4.0	–0.7	1.3	1.5	6.2	–0.8	0.7
Peru	19,376	36,059	1.4	4.2	–0.8	2.4	–0.9	5.4	–3.8	9.6
Philippines	31,566	55,313	2.6	3.7	0.4	1.4	0.6	3.6	–2.1	3.6
Poland	28,281	100,282	..	5.2	..	5.1	..	3.5	..	10.6
Portugal	44,562	71,340	2.5	2.9	2.4	2.8	5.0	2.1	..	4.1
Puerto Rico	19,827	..	3.5	..	..	..	5.1	..	6.9	..
Romania	25,232	23,693	..	0.4	..	0.8	..	1.3	..	–6.4
Russian Federation	282,978	211,044	..	1.5	..	1.7	..	–0.9	..	–20.7

4.10 Growth of consumption and investment

	Household final consumption expenditure				Household final consumption expenditure per capita		General government final consumption expenditure		Gross capital formation	
	$ millions		average annual % growth		average annual % growth		average annual % growth		average annual % growth	
	1990	1999	1980–90	1990–99	1980–90	1990–99	1980–90	1990–99	1980–90	1990–99
Rwanda	2,162	1,734	1.4	1.1	–1.6	–0.4	5.2	–4.2	4.3	2.1
Saudi Arabia	41,621	54,115	..	..	..	..	..	..	..	..
Senegal	4,353	3,634	2.1	3.3	–0.8	0.6	3.3	–0.4	5.2	4.5
Sierra Leone	734	633	0.1	–2.3	..	..	0.0	–4.5	–1.1	–9.1
Singapore	16,972	34,336	5.8	5.9	4.1	3.9	6.6	8.3	3.1	8.5
Slovak Republic	8,350	10,647	3.8	0.3	3.5	0.1	4.8	1.4	0.3	8.2
Slovenia	6,917	11,094	..	3.9	..	4.0	..	3.1	..	10.8
South Africa	70,283	82,058	2.4	2.6	–0.2	0.6	3.5	0.7	–5.3	3.0
Spain	308,673	354,308	2.5	1.8	2.2	1.6	5.4	2.0	5.7	1.7
Sri Lanka	6,098	11,356	3.8	6.0	2.4	4.7	7.3	7.4	0.6	6.3
Sudan	..	..	0.0	..	..	..	–0.5	..	–1.8	..
Sweden	116,747	120,642	1.8	0.7	1.5	0.3	1.5	0.3	4.4	0.6
Switzerland	130,900	160,623	1.6	0.7	1.1	0.0	3.1	0.7	3.9	0.3
Syrian Arab Republic	8,607	9,815	3.6	2.3	0.2	–0.6	–3.6	5.4	–5.3	7.9
Tajikistan	..	1,433	..	..	..	..	..	..	..	..
Tanzania[a]	3,526	7,395	..	2.8	..	–0.1	..	–8.6	..	–2.8
Thailand	48,270	68,814	5.9	4.2	4.1	3.3	4.2	5.2	9.5	–2.9
Togo	1,158	1,201	4.7	4.3	1.6	1.3	–1.2	–1.0	2.7	–2.0
Trinidad and Tobago	2,975	4,267	–1.3	–1.6	–2.5	–2.3	–1.7	1.3	–10.1	18.2
Tunisia	7,152	12,578	2.9	4.1	0.3	2.5	3.8	4.0	–1.8	3.3
Turkey	103,378	125,395	..	3.7	..	2.2	..	4.0	..	4.2
Turkmenistan	..	..	..	..	..	..	..	..	..	..
Uganda	4,002	5,463	2.6	6.5	0.0	3.3	2.0	8.0	8.0	9.0
Ukraine	52,131	22,182	..	–8.0	..	–7.6	..	–4.7	..	–22.4
United Arab Emirates	12,726	..	4.6	..	..	..	–3.9	..	–8.7	..
United Kingdom	617,733	949,599	4.0	2.6	3.8	2.3	0.8	0.9	6.4	4.0
United States	3,842,000	5,836,800	3.8	3.2	2.9	1.9	2.8	0.5	4.4	6.7
Uruguay	6,525	15,096	0.7	5.6	0.1	4.8	1.8	2.6	–6.6	8.0
Uzbekistan	13,321	13,196	..	..	..	..	..	..	..	..
Venezuela, RB	30,171	71,750	1.3	0.4	–1.2	–1.8	2.0	–0.8	–5.3	3.5
Vietnam	5,597	19,690	..	9.1	..	7.2	..	10.9	..	22.4
West Bank and Gaza	..	4,155	..	3.3	..	–0.6	..	13.2	..	5.0
Yemen, Rep.	3,582	4,869	..	0.6	..	–2.8	..	–1.3	..	7.7
Yugoslavia, FR (Serb./Mont.)	..	..	..	..	..	..	..	..	..	..
Zambia	2,078	2,763	1.8	–3.5	–1.3	–6.0	–3.4	–6.9	–4.3	4.5
Zimbabwe	5,543	4,153	3.7	1.1	0.3	–1.1	4.7	–4.0	3.5	–1.6
World	**13,027,651 t**	**18,183,811 t**	**3.4 w**	**2.6 w**	**1.7 w**	**1.2 w**	**2.9 w**	**1.4 w**	**4.0 w**	**2.7 w**
Low income	570,956	714,434	3.8	3.7	1.4	1.6	5.6	2.5	4.3	1.3
Middle income	2,008,046	3,267,270	3.3	4.1	1.6	2.8	5.0	2.2	1.9	1.9
Lower middle income	993,993	1,484,153	..	4.6	..	3.4	3.9	4.9	4.4	0.0
Upper middle income	1,024,681	1,785,270	2.5	3.7	0.7	2.2	5.5	0.4	0.4	4.4
Low & middle income	2,572,276	3,979,084	3.4	4.0	1.4	2.4	5.1	2.2	2.2	1.8
East Asia & Pacific	496,295	994,893	6.9	6.5	5.2	5.1	5.9	6.0	9.3	6.8
Europe & Central Asia	657,914	659,861	..	2.2	..	2.0	..	0.3	..	–10.0
Latin America & Carib.	739,621	1,349,097	1.5	3.4	–0.5	1.8	5.6	0.1	–1.5	4.8
Middle East & N. Africa	221,775	355,041	..	..	..	..	..	..	..	..
South Asia	281,154	405,563	4.6	4.7	2.4	2.8	8.0	5.6	5.8	6.2
Sub-Saharan Africa	192,251	218,866	1.6	2.5	–1.3	–0.2	2.7	0.8	–3.9	3.1
High income	10,444,693	14,384,915	3.4	2.3	2.8	1.6	2.5	1.2	4.5	2.9
Europe EMU	3,200,493	3,704,215	..	1.5	..	1.2	..	1.3	..	1.7

a. Data cover mainland Tanzania only.

Growth of consumption and investment | 4.10

About the data

Measures of growth in consumption and capital formation are subject to two kinds of inaccuracy. The first stems from the difficulty of measuring expenditures at current price levels, as described in *About the data* for table 4.9. The second arises in deflating current price data to measure volume growth, where results depend on the relevance and reliability of the price indexes and weights used. Measuring price changes is more difficult for investment goods than for consumption goods because of the one-time nature of many investments and because the rate of technological progress in capital goods makes capturing change in quality difficult. (An example is computers—prices have fallen as quality has improved.) Many countries estimate capital formation from the supply side, identifying capital goods entering an economy directly from detailed production and international trade statistics. This means that the price indexes used in deflating production and international trade, reflecting delivered or offered prices, will determine the deflator for capital formation expenditures on the demand side.

The data in the table on household final consumption expenditure (private consumption in previous editions) in current U.S. dollars are converted from national currencies using official exchange rates or an alternative conversion factor as noted in *Primary data documentation*. (For a discussion of alternative conversion factors see *Statistical methods*.) Growth rates of household final consumption expenditure, household final consumption expenditure per capita, general government final consumption expenditure, and gross capital formation are estimated using constant price data. (Consumption and capital formation as shares of GDP are shown in table 4.9.)

To obtain government consumption in constant prices, countries may deflate current values by applying a wage (price) index or extrapolate from the change in government employment. Neither technique captures improvements in productivity or changes in the quality of government services. Deflators for household consumption are usually calculated on the basis of the consumer price index. Many countries estimate household consumption as a residual that includes statistical discrepancies accumulated from other domestic sources; thus these estimates lack detailed breakdowns of expenditures.

Definitions

- **Household final consumption expenditure** is the market value of all goods and services, including durable products (such as cars, washing machines, and home computers), purchased by households. It excludes purchases of dwellings but includes imputed rent for owner-occupied dwellings. It also includes payments and fees to governments to obtain permits and licenses. The *World Development Indicators* includes in household consumption expenditure the expenditures of nonprofit institutions serving households, even when reported separately by the country. In practice, household consumption expenditure may include any statistical discrepancy in the use of resources relative to the supply of resources. • **General government final consumption expenditure** includes all government current expenditures for purchases of goods and services (including compensation of employees). It also includes most expenditures on national defense and security, but excludes government military expenditures that are part of government capital formation. • **Gross capital formation** consists of outlays on additions to the fixed assets of the economy plus net changes in the level of inventories. Fixed assets include land improvements (fences, ditches, drains, and so on); plant, machinery, and equipment purchases; and the construction of roads, railways, and the like, including schools, offices, hospitals, private residential dwellings, and commercial and industrial buildings. Inventories are stocks of goods held by firms to meet temporary or unexpected fluctuations in production or sales, and "work in progress." According to the 1993 SNA, net acquisitions of valuables are also considered capital formation.

Data sources

The national accounts indicators for most developing countries are collected from national statistical organizations and central banks by visiting and resident World Bank missions. Data for high-income economies come from Organisation for Economic Co-operation and Development (OECD) data files (see the OECD's *National Accounts, 1988–1998,* volumes 1 and 2). The United Nations Statistics Division publishes detailed national accounts for United Nations member countries in *National Accounts Statistics: Main Aggregates and Detailed Tables* and publishes updates in the *Monthly Bulletin of Statistics.*

Figure 4.10

Rising investment in Asia

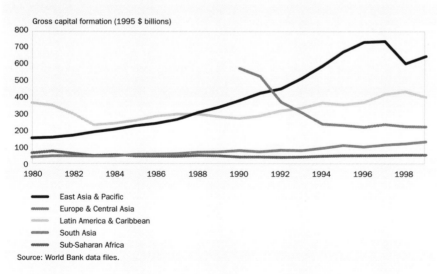

Gross capital formation (1995 $ billions)

- — East Asia & Pacific
- — Europe & Central Asia
- — Latin America & Caribbean
- — South Asia
- — Sub-Saharan Africa

Source: World Bank data files.

In 1999 gross capital formation in East Asia recovered from the financial crisis and continued to increase in South Asia.

4.11 Central government finances

	Current revenue[a]		Total expenditure		Overall budget deficit (including grants)		Financing from abroad		Domestic financing		Debt and interest payments	
	% of GDP		% of GDP		% of GDP		% of GDP		% of GDP		Total debt % of GDP	Interest % of current revenue
	1990	1998	1990	1998	1990	1998	1990	1998	1990	1998	1998	1998
Albania	..	19.3	..	29.8	..	−8.5	..	2.5	..	6.0	46.4	40.3
Algeria	..	27.8	..	31.5	..	−3.6	..	3.9	..	−6.9	67.3	14.3
Angola	..	..	..	..	..	..	..	..	..	..	..	..
Argentina	10.4	13.8	10.6	15.4	−0.4	−1.5	0.2	3.0	0.2	−1.5	..	16.2
Armenia	..	..	..	..	..	..	..	..	..	..	..	..
Australia	24.9	23.6	23.3	23.7	2.0	2.8	0.2	−0.9	−2.2	−1.9	16.8	6.1
Austria	33.9	37.5	37.5	40.5	−4.4	..	0.5	..	3.9	..	60.2	9.3
Azerbaijan	..	19.0	..	24.7	..	−3.9	..	..	..	..	..	2.0
Bangladesh	..	..	..	..	..	..	..	..	..	..	..	..
Belarus	31.2	30.9	37.6	32.2	−5.1	−0.9	2.7	−0.3	2.4	1.2	21.4	2.5
Belgium	42.6	43.7	47.7	45.7	−5.5	−1.8	−0.3	−0.9	5.8	2.7	114.6	16.7
Benin	..	..	..	..	..	..	..	..	..	..	..	..
Bolivia	13.7	17.5	16.4	22.0	−1.7	−2.3	0.7	1.9	1.0	0.5	46.9	9.4
Bosnia and Herzegovina	..	..	..	..	..	..	..	..	..	..	..	..
Botswana	51.1	44.3	33.8	35.3	11.3	8.4	0.0	0.5	−11.4	−8.9	11.0	1.3
Brazil	22.8	23.9	34.9	24.6	−5.8	−7.3	..	..	..	..	..	14.4
Bulgaria	47.1	33.9	55.1	33.5	−8.3	2.8	−0.8	−0.7	9.1	−2.1	..	13.0
Burkina Faso	11.0	..	15.0	..	−1.3	..	..	..	..	..	..	..
Burundi	18.2	16.9	28.7	24.9	−3.3	−4.6	4.9	4.1	−1.6	0.5	154.0	10.3
Cambodia	..	..	..	..	..	..	..	..	..	..	..	..
Cameroon	15.4	16.5	21.2	14.8	−5.9	1.6	5.2	−1.2	1.2	−0.4	107.9	19.5
Canada	21.6	21.8	26.2	21.5	−4.8	0.4	0.2	−0.5	4.6	0.0	75.1	16.7
Central African Republic	..	..	..	..	..	..	..	..	..	..	..	..
Chad	6.7	..	21.8	..	−4.7	..	5.0	..	−0.3	..	..	..
Chile	20.6	23.0	20.4	22.6	0.8	0.4	..	−0.2	..	−0.2	13.9	3.0
China	6.3	6.3	10.1	9.3	−1.9	−2.2	0.8	0.1	1.1	2.2	..	..
Hong Kong, China	..	..	..	..	..	..	..	..	..	..	..	..
Colombia	12.6	11.7	11.6	16.6	3.9	−5.1	..	1.9	..	3.2	22.0	24.7
Congo, Dem. Rep.	10.1	5.3	18.8	10.4	−6.5	−0.8	0.0	0.0	6.5	0.8	160.4	0.2
Congo, Rep.	22.5	25.1	35.6	46.2	−14.1	−20.8	..	..	..	..	282.4	58.3
Costa Rica	18.3	20.7	20.4	21.9	−2.5	−1.2	0.2	..	2.3	..	..	15.7
Côte d'Ivoire	22.0	20.9	24.5	23.5	−2.9	−1.3	4.0	0.7	0.4	0.5	115.0	20.6
Croatia	33.0	45.4	37.6	45.6	−4.6	0.6	0.0	0.1	4.7	−0.7	..	3.2
Cuba	..	..	..	..	..	..	..	..	..	..	..	..
Czech Republic	..	33.1	..	35.5	..	−1.6	..	−0.8	..	2.4	11.7	3.2
Denmark	37.8	38.5	39.0	37.3	−0.7	1.7	..	..	..	..	64.0	12.2
Dominican Republic	12.0	16.7	11.7	16.3	0.6	0.6	0.0	−0.6	−0.6	0.0	22.9	3.2
Ecuador	18.2	..	14.5	..	3.7	..	..	..	..	..	..	..
Egypt, Arab Rep.	23.0	26.3	27.8	30.6	−5.7	−2.0	−0.7	−0.6	6.4	2.6	..	23.0
El Salvador	..	15.0	..	15.1	..	−1.4	..	1.0	..	0.4	25.7	8.1
Eritrea	..	..	..	..	..	..	..	..	..	..	..	..
Estonia	26.2	31.7	23.7	32.9	0.4	−0.1	0.0	−0.7	−0.4	0.7	4.3	1.0
Ethiopia	17.4	..	27.2	..	−9.8	..	2.8	..	7.0	..	..	..
Finland	30.6	31.9	30.3	33.4	0.2	−0.3	0.7	−1.1	−0.8	1.4	61.0	14.3
France	39.7	41.4	41.8	46.2	−2.1	−3.5	1.1	0.2	1.0	5.0	..	7.4
Gabon	20.6	..	20.2	..	3.2	..	2.7	..	−5.8	..	..	..
Gambia, The	19.4	..	23.6	..	−0.8	..	..	..	..	..	..	..
Georgia	..	11.2	..	15.7	..	−3.6	..	0.6	..	3.0	59.0	23.5
Germany	27.5	31.3	29.3	32.6	−2.1	−0.9	1.5	2.1	0.6	−1.2	38.6	7.3
Ghana	12.5	..	13.2	..	0.2	..	1.3	..	−1.5	..	..	..
Greece	27.5	23.5	51.7	30.8	−22.7	−4.4	1.5	2.4	21.1	2.0	113.1	38.4
Guatemala	..	..	..	..	..	..	..	..	..	..	..	..
Guinea	16.0	10.3	22.9	16.9	−3.3	−4.1	4.1	4.2	−0.8	−0.1	..	30.8
Guinea-Bissau	..	..	..	..	..	..	..	..	..	..	..	..
Haiti	..	..	..	..	..	..	..	..	..	..	..	..
Honduras	..	..	..	..	..	..	..	..	..	..	..	..

Central government finances | 4.11

	Current revenue[a]		Total expenditure		Overall budget deficit (including grants)		Financing from abroad		Domestic financing		Debt and interest payments	
											Total debt % of GDP	Interest % of current revenue
	% of GDP		% of GDP		% of GDP		% of GDP		% of GDP			
	1990	1998	1990	1998	1990	1998	1990	1998	1990	1998	1998	1998
Hungary	52.9	36.2	52.1	44.1	0.8	−6.2	−0.5	0.4	−0.3	5.8	61.5	21.3
India	12.6	11.5	16.3	14.9	−7.7	−4.8	0.6	0.1	7.1	4.8	49.7	35.7
Indonesia	18.8	16.0	18.4	17.6	0.4	−2.7	0.7	5.0	−1.1	−2.3	53.3	19.7
Iran, Islamic Rep.	18.1	18.7	19.9	24.5	−1.8	−5.7	0.0	0.0	1.8	5.7	..	0.8
Iraq	..	..	..	..	..	..	..	..	..	..	..	..
Ireland	33.6	*31.9*	37.7	*33.0*	−2.4	*0.7*	..	..	..	..	..	*13.3*
Israel	39.4	41.8	50.7	47.8	−5.3	−1.4	0.8	1.3	4.6	0.1	113.0	14.3
Italy	38.2	40.6	47.4	43.8	−10.2	−3.1	*0.0*	..	*9.9*	..	..	18.1
Jamaica	..	..	..	..	..	..	..	..	..	..	..	..
Japan	14.4	..	15.7	..	−1.6	..	*0.0*	..	−1.7	..	..	..
Jordan	26.1	25.2	35.8	33.2	−3.5	−5.8	3.0	2.4	0.5	3.4	100.5	13.3
Kazakhstan	..	11.3	..	18.5	..	−4.2	..	3.0	..	1.2	16.9	7.1
Kenya	22.4	*27.2*	27.5	*29.1*	−3.8	*−0.9*	1.3	*−0.2*	4.5	*1.1*	..	*28.0*
Korea, Dem. Rep.	..	..	..	..	..	..	..	..	..	..	..	..
Korea, Rep.	17.5	*20.0*	16.2	*17.4*	−0.7	*−1.3*	−0.2	*1.5*	0.9	*−0.3*	*10.4*	*2.5*
Kuwait	58.7	44.8	55.3	50.6	..	−5.7	..	..	..	..	..	3.9
Kyrgyz Republic	..	17.6	..	22.0	..	−3.0	..	*1.1*	..	4.3	..	..
Lao PDR	..	..	..	..	..	..	..	..	..	..	..	..
Latvia	..	33.5	..	34.7	..	0.2	..	0.4	..	−0.5	10.4	2.2
Lebanon	..	17.0	..	32.1	..	−15.1	..	9.4	..	5.6	107.1	75.3
Lesotho	39.0	44.6	51.2	50.3	−1.0	−3.7	7.9	0.7	−6.9	3.0	68.7	4.9
Libya	..	..	..	..	..	..	..	..	..	..	..	..
Lithuania	*31.9*	26.7	*28.9*	30.3	*1.4*	−0.4	..	1.8	..	−1.3	15.6	4.2
Macedonia, FYR	..	..	..	..	..	..	..	..	..	..	..	..
Madagascar	11.6	*8.7*	16.0	*17.3*	−0.9	*−1.3*	2.1	*1.4*	−1.2	*−0.1*	..	*54.1*
Malawi	20.7	..	26.5	..	−1.7	..	..	..	..	..	..	..
Malaysia	26.4	*23.1*	29.3	*19.7*	−2.0	*2.9*	−0.7	*−0.1*	2.8	*−1.2*	..	*10.2*
Mali	..	..	..	..	..	..	..	..	..	..	..	..
Mauritania	..	..	..	..	..	..	..	..	..	..	..	..
Mauritius	22.6	20.8	22.6	21.9	−0.4	0.9	−0.4	−1.0	0.8	0.1	34.4	13.3
Mexico	15.3	13.0	17.9	14.7	−2.5	−1.4	0.3	0.5	2.3	0.9	27.8	16.5
Moldova	..	30.2	..	35.9	..	−3.2	..	−3.3	..	6.5	82.8	15.5
Mongolia	*18.3*	19.5	*21.5*	23.0	*−6.0*	−10.8	*7.0*	7.8	*−1.0*	3.0	78.9	6.5
Morocco	26.4	..	28.8	..	−2.2	..	3.9	..	−1.6	..	..	..
Mozambique	..	..	..	..	..	..	..	..	..	..	..	..
Myanmar	10.5	7.2	16.0	7.7	−5.1	−0.4	0.0	0.1	5.1	0.4	..	..
Namibia	31.4	..	33.5	..	−1.2	..	1.8	..	−0.6	..	..	..
Nepal	8.4	10.4	17.2	17.3	−6.8	−4.6	5.4	2.8	1.4	1.9	66.4	12.0
Netherlands	45.0	*44.1*	49.5	*45.9*	−4.3	*−1.6*	−0.3	1.9	4.6	−1.9	55.6	*9.5*
New Zealand	42.6	34.1	44.0	33.4	4.0	0.5	..	..	..	..	38.7	7.1
Nicaragua	33.5	..	72.0	..	−35.6	..	12.7	..	22.9	..	..	..
Niger	..	..	..	..	..	..	..	..	..	..	..	..
Nigeria	..	..	..	..	..	..	..	..	..	..	..	..
Norway	42.4	41.8	41.3	37.2	0.5	−1.6	−0.6	−0.6	0.0	2.2	19.9	4.1
Oman	38.9	24.6	39.5	31.6	−0.8	−6.6	−3.9	6.4	4.7	0.2	27.1	7.2
Pakistan	19.1	16.2	22.4	21.8	−5.4	−6.4	2.3	1.6	3.1	4.8	79.1	42.2
Panama	25.6	24.9	23.7	27.9	3.0	−0.7	−3.4	3.9	0.4	−3.2	..	17.2
Papua New Guinea	25.2	*24.2*	34.7	27.1	−3.5	−1.7	0.4	−1.5	3.0	3.2	65.8	*17.1*
Paraguay	12.3	..	9.4	..	2.9	..	−0.9	..	−2.1	..	..	..
Peru	12.5	17.6	20.6	18.0	−8.1	−0.1	5.4	0.3	2.7	−0.2	..	10.6
Philippines	16.2	17.2	19.6	19.1	−3.5	−1.9	0.4	0.5	3.1	1.4	66.2	21.7
Poland	..	35.4	..	37.5	..	−1.0	..	0.2	..	0.8	42.9	9.1
Portugal	31.6	34.7	37.9	39.0	−4.4	−1.2	−1.3	−2.1	5.8	3.4	0.8	8.4
Puerto Rico	..	..	..	..	..	..	..	..	..	..	..	..
Romania	34.4	*26.5*	33.8	*31.9*	0.9	*−3.9*	0.0	*0.9*	−0.9	3.0	..	*13.9*
Russian Federation	..	18.6	..	25.5	..	−5.3	..	3.2	..	2.1	140.4	29.7

| | Current revenue[a] (% of GDP) | | Total expenditure (% of GDP) | | Overall budget deficit (including grants) (% of GDP) | | Financing from abroad (% of GDP) | | Domestic financing (% of GDP) | | Debt and interest payments | |
| | | | | | | | | | | | Total debt % of GDP | Interest % of current revenue |
	1990	1998	1990	1998	1990	1998	1990	1998	1990	1998	1998	1998
Rwanda	10.8	..	18.9	..	–5.3	..	2.5	..	2.8	..	..	..
Saudi Arabia	..	..	..	..	..	..	..	..	..	..	..	..
Senegal	..	..	..	..	..	..	..	..	..	..	..	..
Sierra Leone	4.0	10.2	6.0	17.2	–1.8	–5.8	0.4	4.9	1.4	0.9	116.2	21.0
Singapore	26.9	24.9	21.4	19.8	10.8	3.4	–0.1	0.0	–10.7	–3.4	85.5	2.9
Slovak Republic	..	34.6	..	39.1	..	–4.2	..	4.6	..	–0.5	29.2	7.9
Slovenia	39.8	39.1	38.6	40.1	0.3	–0.8	0.1	0.3	–0.4	0.4	23.9	3.2
South Africa	26.3	27.3	30.1	30.4	–4.1	–2.6	–0.1	0.0	4.1	2.6	51.0	21.3
Spain	29.1	28.8	32.4	32.9	–3.1	–2.9	0.7	1.7	2.4	1.2	55.6	14.1
Sri Lanka	21.0	17.2	28.4	24.9	–7.8	–8.0	3.6	1.0	4.2	7.0	89.2	31.4
Sudan	..	..	..	..	..	..	..	..	..	..	..	..
Sweden	42.6	38.2	39.4	41.6	1.0	–0.5	–0.3	0.7	–0.7	–0.2	..	15.7
Switzerland	20.8	24.8	23.3	28.3	–0.9	0.5	0.0	0.0	0.9	–0.5	28.8	3.7
Syrian Arab Republic	21.9	21.7	21.8	22.4	0.3	–0.7	..	1.6	..	–0.9	..	..
Tajikistan	..	9.3	..	12.7	..	–2.5	..	2.3	..	0.3	99.0	14.8
Tanzania	..	..	..	..	..	..	..	..	..	..	..	..
Thailand	18.5	16.2	14.1	22.7	4.6	–7.7	–1.5	1.3	–3.1	6.4	10.8	1.2
Togo	..	..	..	..	..	..	..	..	..	..	..	..
Trinidad and Tobago	..	..	..	..	..	..	..	..	..	..	..	..
Tunisia	30.7	29.3	34.6	31.7	–5.4	–0.4	1.8	0.0	3.6	0.4	59.6	11.6
Turkey	13.7	23.7	17.4	32.1	–3.0	–8.4	0.0	–1.5	3.0	9.9	41.4	49.9
Turkmenistan	..	..	..	..	..	..	..	..	..	..	..	..
Uganda	..	..	..	..	..	..	..	..	..	..	..	..
Ukraine	..	..	..	..	..	..	..	..	..	..	..	..
United Arab Emirates	1.6	3.4	11.5	11.1	0.4	–0.3	0.0	0.0	–0.4	0.3	..	0.0
United Kingdom	36.1	37.2	37.5	36.9	0.6	0.6	0.2	–0.2	–0.8	–0.3	49.8	8.8
United States	18.9	20.7	22.7	19.9	–3.8	0.8	0.2	–0.1	3.6	–0.7	42.8	14.1
Uruguay	23.8	29.9	23.3	30.9	0.3	–0.8	1.4	..	–1.7	..	..	4.9
Uzbekistan	..	..	..	..	..	..	..	..	..	..	..	..
Venezuela, RB	23.7	17.2	20.7	20.7	0.0	–3.7	1.0	0.2	–1.0	3.5	..	13.8
Vietnam	..	18.2	..	20.1	..	–1.1	..	0.9	..	0.2	..	3.0
West Bank and Gaza	..	..	..	..	..	..	..	..	..	..	..	..
Yemen, Rep.	19.6	35.0	28.8	36.2	–9.1	–2.2	3.3	3.1	5.9	–0.9	..	9.7
Yugoslavia, FR (Serb./Mont.)	..	..	..	..	..	..	..	..	..	..	..	..
Zambia	..	..	..	..	..	..	..	..	..	..	..	..
Zimbabwe	24.1	29.1	27.3	35.3	–5.3	–4.9	0.9	–0.1	4.4	5.0	57.5	24.2
World	22.5 w	26.4 w	25.8 w	27.9 w	–3.0 w	–1.5 w	0.6 m	.. m	1.1 m	.. m	.. m	11.8 m
Low income	15.5	13.9	18.3	17.0	–4.8	–4.0	..	..	..	..	..	..
Middle income	17.4	19.1	21.5	20.5	–2.5	–3.0	0.3	0.4	0.4	0.9	..	11.6
Lower middle income	12.7	14.2	15.3	18.8	–1.5	–4.0	0.5	0.7	1.6	1.3	..	11.6
Upper middle income	20.4	22.2	25.5	22.8	–3.1	–3.5	–0.1	0.2	0.6	0.7	26.5	11.2
Low & middle income	17.1	18.6	21.1	20.1	–2.8	–3.1	..	..	..	..	..	..
East Asia & Pacific	13.2	10.1	14.4	13.2	–0.8	–3.0	0.2	0.7	2.0	1.8	..	9.9
Europe & Central Asia	..	25.0	..	30.8	..	–4.7	..	0.4	..	1.2	41.4	8.5
Latin America & Carib.	18.8	20.1	25.5	21.0	–3.5	–4.2	..	..	..	..	..	12.2
Middle East & N. Africa	..	..	..	..	..	..	1.8	2.4	3.6	0.4	..	11.6
South Asia	13.8	12.4	17.6	16.3	–7.3	–5.1	3.0	1.3	3.6	4.8	72.7	33.5
Sub-Saharan Africa	24.0	..	27.7	..	–3.5	..	..	..	..	..	..	..
High income	23.9	28.7	27.0	30.2	–3.0	–1.1	0.2	0.0	2.4	0.0	49.2	8.6
Europe EMU	34.7	37.1	38.6	40.0	–3.7	–2.3	0.5	–0.9	3.9	1.4	57.9	11.8

a. Excluding grants.

About the data

Tables 4.11–4.13 present an overview of the size and role of central governments relative to national economies. The International Monetary Fund's (IMF) *Manual on Government Finance Statistics* describes the government as the sector of the economy responsible for "implementation of public policy through the provision of primarily nonmarket services and the transfer of income, supported mainly by compulsory levies on other sectors" (1986, p. 3). The definition of government generally excludes nonfinancial public enterprises and public financial institutions (such as the central bank).

Units of government meeting this definition exist at many levels, from local administrative units to the highest level of national government. Inadequate statistical coverage precludes the presentation of subnational data, however, making cross-country comparisons potentially misleading.

Central government can refer to one of two accounting concepts: consolidated or budgetary. For most countries central government finance data have been consolidated into one account, but for others only budgetary central government accounts are available. Countries reporting budgetary data are noted in *Primary data documentation*. Because budgetary accounts do not necessarily include all central government units, the picture they provide of central government activities is usually incomplete. A key issue is the failure to include the quasi-fiscal operations of the central bank. Central bank losses arising from monetary operations and subsidized financing can result in sizable quasi-fiscal deficits. Such deficits may also result from the operations of other financial intermediaries, such as public development finance institutions. Also missing from the data are governments' contingent liabilities for unfunded pension and insurance plans.

Data on government revenues and expenditures are collected by the IMF through questionnaires distributed to member governments and by the Organisation for Economic Co-operation and Development. Despite the IMF's efforts to systematize and standardize the collection of public finance data, statistics on public finance are often incomplete, untimely, and noncomparable.

Government finance statistics are reported in local currency. The indicators here are shown as percentages of GDP. Many countries report government finance data according to fiscal years; see *Primary data documentation* for the timing of these years. For further discussion of government finance statistics see *About the data* for tables 4.12 and 4.13.

Figure 4.11

Developing economies tend to run larger fiscal deficits

Largest overall budget deficits as % of GDP, 1997–99

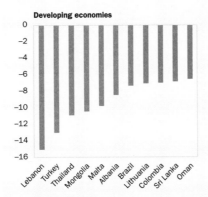

Developing economies

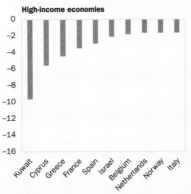

High-income economies

Note: Data refer to the most recent year available in 1997–99.
Source: International Monetary Fund, Government Finance Statistics data files.

Definitions

- **Current revenue** includes all revenue from taxes and current nontax revenues (other than grants) such as fines, fees, recoveries, and income from property or sales. • **Total expenditure** includes nonrepayable current and capital expenditures. It does not include government lending or repayments to the government or government acquisition of equity for public policy purposes. • **Overall budget deficit** is current and capital revenue and official grants received, less total expenditure and lending minus repayments. • **Financing from abroad** (obtained from nonresidents) and **domestic financing** (obtained from residents) refer to the means by which a government provides financial resources to cover a budget deficit or allocates financial resources arising from a budget surplus. The data include all government liabilities—other than those for currency issues or demand, time, or savings deposits with government—or claims on others held by government and changes in government holdings of cash and deposits. They exclude government guarantees of the debt of others. • **Debt** is the entire stock of direct government fixed term contractual obligations to others outstanding on a particular date. It includes domestic debt (such as debt held by monetary authorities, deposit money banks, nonfinancial public enterprises, and households) and foreign debt (such as debt to international development institutions and foreign governments). It is the gross amount of government liabilities not reduced by the amount of government claims against others. Because debt is a stock rather than a flow, it is measured as of a given date, usually the last day of the fiscal year. • **Interest payments** include interest payments on government debt—including long-term bonds, long-term loans, and other debt instruments—to both domestic and foreign residents.

Data sources

The data on central government finances are from the IMF's *Government Finance Statistics Yearbook, 2000* and IMF data files. Each country's accounts are reported using the system of common definitions and classifications in the IMF's *Manual on Government Finance Statistics* (1986). See these sources for complete and authoritative explanations of concepts, definitions, and data sources.

4.12 | Central government expenditures

	Goods and services		Wages and salaries[a]		Interest payments		Subsidies and other current transfers		Capital expenditure	
	% of total expenditure		% of total expenditure		% of total expenditure		% of total expenditure		% of total expenditure	
	1990	1998	1990	1998	1990	1998	1990	1998	1990	1998
Albania	..	17	..	9	..	26	..	42	..	16
Algeria	..	36	..	25	..	13	..	27	..	24
Angola	..	..	..	..	..	..	..	..	..	..
Argentina	30	20	23	15	8	14	57	58	5	8
Armenia	..	..	..	..	..	..	..	..	..	..
Australia	27	27	2	3	8	6	56	61	9	5
Austria	25	25	10	10	9	9	57	61	9	6
Azerbaijan	..	33	..	10	..	2	..	45	..	21
Bangladesh	..	..	..	..	..	..	..	..	..	..
Belarus	37	25	2	9	2	2	46	58	16	15
Belgium	19	19	14	13	21	16	56	60	5	5
Benin	..	..	..	..	..	..	..	..	..	..
Bolivia	63	39	36	24	6	8	16	38	15	16
Bosnia and Herzegovina	..	..	..	..	..	..	..	..	..	..
Botswana	51	48	23	25	2	2	25	31	21	19
Brazil	16	20	9	10	78	14	39	64	2	3
Bulgaria	35	32	3	8	10	13	52	46	3	9
Burkina Faso	60	..	51	..	6	..	11	..	23	..
Burundi	34	50	22	28	5	7	10	12	51	24
Cambodia	..	..	..	..	..	..	..	..	..	..
Cameroon	51	52	39	33	5	22	13	13	26	13
Canada	21	17	11	10	20	17	57	65	2	2
Central African Republic	..	..	..	..	..	..	..	..	..	..
Chad	41	..	28	..	2	..	3	..	56	..
Chile	28	29	18	20	10	3	51	52	11	17
China	..	..	..	..	..	..	..	..	..	..
Hong Kong, China	..	..	..	..	..	..	..	..	..	..
Colombia	26	21	18	15	10	17	42	42	22	20
Congo, Dem. Rep.	73	94	23	54	7	0	4	1	16	5
Congo, Rep.	56	53	49	19	22	32	20	6	2	10
Costa Rica	57	48	43	36	12	15	20	27	11	10
Côte d'Ivoire	69	43	38	25	1	18	30	9	0	29
Croatia	54	48	22	24	0	3	42	38	3	10
Cuba	..	..	..	..	..	..	..	..	..	..
Czech Republic	..	14	..	8	..	3	..	74	..	9
Denmark	20	21	12	12	15	13	61	64	3	3
Dominican Republic	39	50	29	38	4	3	13	20	44	24
Ecuador	42	..	38	..	23	..	16	..	18	..
Egypt, Arab Rep.	42	41	23	20	14	20	26	15	17	24
El Salvador	..	71	..	42	..	8	..	8	..	23
Eritrea	..	..	..	..	..	..	..	..	..	..
Estonia	25	43	8	11	0	1	73	47	8	9
Ethiopia	77	..	40	..	5	..	9	..	16	..
Finland	20	18	10	7	3	14	70	63	7	5
France	26	24	17	16	5	7	63	65	6	4
Gabon	63	..	37	..	0	..	6	..	32	..
Gambia, The	41	..	21	..	16	..	9	..	34	..
Georgia	..	42	..	9	..	17	..	36	..	5
Germany	32	31	8	8	5	7	58	57	5	4
Ghana	50	..	32	..	11	..	20	..	19	..
Greece	31	34	21	28	20	29	41	20	8	17
Guatemala	..	..	..	..	..	..	..	..	..	..
Guinea	37	34	18	23	7	19	4	6	53	33
Guinea-Bissau	..	..	..	..	..	..	..	..	..	..
Haiti	..	..	..	..	..	..	..	..	..	..
Honduras	..	..	..	..	..	..	..	..	..	..

Central government expenditures | 4.12

	Goods and services		Wages and salaries[a]		Interest payments		Subsidies and other current transfers		Capital expenditure	
	% of total expenditure		% of total expenditure		% of total expenditure		% of total expenditure		% of total expenditure	
	1990	1998	1990	1998	1990	1998	1990	1998	1990	1998
Hungary	27	16	6	8	6	18	64	51	4	9
India	24	24	11	12	22	27	43	38	11	11
Indonesia	23	19	16	9	13	18	21	28	43	34
Iran, Islamic Rep.	53	65	40	51	0	1	22	12	25	22
Iraq	..	..	..	..	..	..	..	..	..	..
Ireland	19	18	14	13	21	13	54	61	7	9
Israel	38	33	14	15	18	12	37	48	6	6
Italy	17	19	13	16	21	17	54	58	8	5
Jamaica	..	..	..	..	..	..	..	..	..	..
Japan	14	..	..	..	19	..	54	..	13	..
Jordan	55	62	44	44	18	10	11	9	16	19
Kazakhstan	..	28	..	13	..	4	..	58	..	9
Kenya	51	45	31	28	19	26	10	18	20	12
Korea, Dem. Rep.	..	..	..	..	..	..	..	..	..	..
Korea, Rep.	35	27	13	13	4	3	46	49	15	22
Kuwait	62	60	31	33	0	3	20	23	18	13
Kyrgyz Republic	..	..	..	..	..	..	..	..	..	..
Lao PDR	..	..	..	..	..	..	..	..	..	..
Latvia	..	30	..	12	..	2	..	60	..	7
Lebanon	..	30	..	21	..	40	..	11	..	18
Lesotho	40	76	22	35	11	4	5	0	45	19
Libya	..	..	..	..	..	..	..	..	..	..
Lithuania	12	49	6	15	..	4	67	37	20	10
Macedonia, FYR	..	..	..	..	..	..	..	..	..	..
Madagascar	37	25	25	18	9	27	9	8	43	39
Malawi	54	..	23	..	14	..	8	..	24	..
Malaysia	41	42	26	26	20	12	16	24	24	23
Mali	..	..	..	..	..	..	..	..	..	..
Mauritania	..	..	..	..	..	..	..	..	..	..
Mauritius	47	46	37	35	15	13	22	29	17	12
Mexico	25	24	18	17	45	15	17	49	14	12
Moldova	..	15	..	6	..	13	..	61	..	11
Mongolia	30	34	7	10	1	6	56	47	13	14
Morocco	48	..	35	..	16	..	8	..	28	..
Mozambique	..	..	..	..	..	..	..	..	..	..
Myanmar	..	..	..	..	..	..	..	..	29	49
Namibia	73	..	..	..	1	..	10	..	15	..
Nepal	..	..	..	..	..	7	..	..	..	..
Netherlands	15	15	9	9	9	9	70	72	6	3
New Zealand	19	51	12	..	15	7	64	39	2	3
Nicaragua	43	..	23	..	0	..	14	..	4	..
Niger	..	..	..	..	..	..	..	..	..	..
Nigeria	..	..	..	..	..	..	..	..	..	..
Norway	19	21	8	8	6	5	69	70	5	5
Oman	76	76	22	28	6	6	7	5	11	14
Pakistan	44	48	..	..	25	31	20	9	12	12
Panama	64	49	49	35	8	15	26	25	2	11
Papua New Guinea	61	56	34	27	11	15	18	22	11	7
Paraguay	54	..	36	..	10	..	19	..	17	..
Peru	30	42	17	20	37	10	25	32	8	16
Philippines	44	57	29	34	34	20	7	15	16	9
Poland	..	25	..	12	..	9	..	61	..	5
Portugal	38	41	27	32	18	7	33	38	12	13
Puerto Rico	..	..	..	..	..	..	..	..	..	..
Romania	26	29	12	13	0	12	57	50	17	9
Russian Federation	..	28	..	11	..	22	..	47	..	4

4.12 Central government expenditures

	Goods and services		Wages and salaries[a]		Interest payments		Subsidies and other current transfers		Capital expenditure	
	% of total expenditure		% of total expenditure		% of total expenditure		% of total expenditure		% of total expenditure	
	1990	1998	1990	1998	1990	1998	1990	1998	1990	1998
Rwanda	53	..	29	..	5	..	16	..	33	..
Saudi Arabia	..	..	..	..	..	..	..	..	..	..
Senegal	..	..	..	..	..	..	..	..	..	..
Sierra Leone	77	39	35	20	18	13	1	24	8	24
Singapore	51	52	27	25	14	4	12	7	24	37
Slovak Republic	..	24	..	13	..	7	..	57	..	13
Slovenia	40	39	20	22	1	3	52	50	7	8
South Africa	53	27	23	17	14	19	23	49	10	5
Spain	19	16	13	11	9	12	63	68	9	4
Sri Lanka	33	39	17	21	23	22	23	19	21	21
Sudan	..	..	..	..	..	..	..	..	..	..
Sweden	15	17	6	6	11	14	72	67	2	2
Switzerland	31	30	5	5	3	3	61	63	5	4
Syrian Arab Republic	..	..	..	..	..	..	..	..	27	38
Tajikistan	..	49	..	6	..	11	..	30	..	11
Tanzania	..	..	..	..	..	..	..	..	..	..
Thailand	60	44	35	27	13	1	9	6	18	49
Togo	..	..	..	..	..	..	..	..	..	..
Trinidad and Tobago	..	..	..	..	..	..	..	..	..	..
Tunisia	34	41	28	35	10	11	35	28	22	20
Turkey	52	32	38	24	18	37	16	24	13	7
Turkmenistan	..	..	..	..	..	..	..	..	..	..
Uganda	..	..	..	..	..	..	..	..	..	..
Ukraine	..	..	..	..	..	..	..	..	..	..
United Arab Emirates	88	81	33	35	0	0	10	14	1	5
United Kingdom	30	29	13	6	9	9	52	58	10	4
United States	27	21	10	8	15	15	50	60	8	4
Uruguay	35	32	20	15	8	5	50	59	7	5
Uzbekistan	..	..	..	..	..	..	..	..	..	..
Venezuela, RB	31	24	23	16	16	12	37	45	16	20
Vietnam	..	..	..	..	..	3	..	..	..	29
West Bank and Gaza	..	..	..	..	..	..	..	..	..	..
Yemen, Rep.	64	46	55	32	8	9	6	26	33	18
Yugoslavia, FR (Serb./Mont.)	..	..	..	..	..	..	..	..	..	..
Zambia	..	..	..	..	..	..	..	..	..	..
Zimbabwe	56	48	37	36	16	20	18	26	10	6
World	**39 m**	**32 m**	**23 m**	**17 m**	**10 m**	**10 m**	**23 m**	**43 m**	**13 m**	**10 m**
Low income	..	..	..	..	..	..	..	..	..	..
Middle income	42	32	23	20	10	10	23	38	16	12
Lower middle income	46	39	29	24	12	10	18	28	17	16
Upper middle income	35	28	22	16	9	10	32	49	11	11
Low & middle income	..	..	..	..	..	..	..	..	..	..
East Asia & Pacific	41	45	27	24	10	10	19	24	18	25
Europe & Central Asia	..	29	..	10	..	8	..	47	..	9
Latin America & Carib.	35	35	23	20	10	11	25	40	11	16
Middle East & N. Africa	53	46	35	32	10	10	11	12	23	20
South Asia	33	39	..	..	23	25	23	19	12	12
Sub-Saharan Africa	53	..	..	..	7	..	10	..	20	..
High income	23	29	13	12	13	8	56	58	7	5
Europe EMU	20	22	13	12	9	11	57	59	7	5

Note: Components include expenditures financed by grants in kind and other cash adjustments to total expenditure.

a. Part of goods and services.

Central government expenditures | 4.12

About the data

Government expenditures include all nonrepayable payments, whether current or capital, requited or unrequited. Total central government expenditure as presented in the International Monetary Fund's (IMF) *Government Finance Statistics Yearbook* is a more limited measure of general government consumption than that shown in the national accounts (see table 4.10) because it excludes consumption expenditures by state and local governments. At the same time, the IMF's concept of central government expenditure is broader than the national accounts definition because it includes government gross capital formation and transfer payments.

Expenditures can be measured either by function (education, health, defense) or by economic type (wages and salaries, interest payments, purchases of goods and services). Functional data are often incomplete, and coverage varies by country because functional responsibilities stretch across levels of government for which no data are available. Defense expenditures, which are usually the central government's responsibility, are shown in table 5.7. For more information on education expenditures see table 2.11; for more on health expenditures see table 2.15.

The classification of expenditures by economic type can also be problematic. For example, the distinction between current and capital expenditure may be arbitrary, and subsidies to state-owned enterprises or banks may be disguised as capital financing. Subsidies may also be hidden in special contractual pricing for goods and services.

Expenditure shares may not sum to 100 percent because expenditures financed by grants in kind and other cash adjustments (which may be positive or negative) are not shown.

For further discussion of government finance statistics see *About the data* for tables 4.11 and 4.13.

Figure 4.12

Public interest payments are a big burden on developing economies
Highest central government interest payments as % of total expenditure

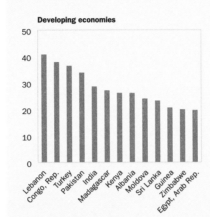

Developing economies

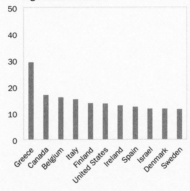

High-income economies

Note: For developing economies data refer to the most recent year available in 1996–99, and for high-income economies, in 1997–99.
Source: International Monetary Fund, Government Finance Statistics data files.

High public interest payments continue to strain the national budgets of developing countries. For about 13 developing economies interest payments are equal to 20 percent or more of their central government's total expenditure.

Definitions

- **Total expenditure of the central government** includes both current and capital (development) expenditures and excludes lending minus repayments. • **Goods and services** include all government payments in exchange for goods and services, whether in the form of wages and salaries to employees or other purchases of goods and services. • **Wages and salaries** consist of all payments in cash, but not in kind, to employees in return for services rendered, before deduction of withholding taxes and employee contributions to social security and pension funds. • **Interest payments** are payments made to domestic sectors and to nonresidents for the use of borrowed money. (Repayment of principal is shown as a financing item, and commission charges are shown as purchases of services.) Interest payments do not include payments by government as guarantor or surety of interest on the defaulted debts of others, which are classified as government lending. • **Subsidies and other current transfers** include all unrequited, nonrepayable transfers on current account to private and public enterprises, and the cost to the public of covering the cash operating deficits on sales to the public by departmental enterprises. • **Capital expenditure** is spending to acquire fixed capital assets, land, intangible assets, government stocks, and nonmilitary, nonfinancial assets. Also included are capital grants.

Data sources

The data on central government expenditures are from the IMF's *Government Finance Statistics Yearbook, 2000* and IMF data files. Each country's accounts are reported using the system of common definitions and classifications in the IMF's *Manual on Government Finance Statistics* (1986). See these sources for complete and authoritative explanations of concepts, definitions, and data sources.

4.13 Central government revenues

	Taxes on income, profits, and capital gains		Social security taxes		Taxes on goods and services		Taxes on international trade		Other taxes		Nontax revenue	
	% of total current revenue		% of total current revenue		% of total current revenue		% of total current revenue		% of total current revenue		% of total current revenue	
	1990	1998	1990	1998	1990	1998	1990	1998	1990	1998	1990	1998
Albania	..	7	..	14	..	40	..	15	..	1	..	23
Algeria	..	60	..	0	..	13	..	16	..	1	..	9
Angola	..	..	..	..	..	..	..	..	..	..	..	..
Argentina	2	16	44	26	20	42	14	7	10	1	10	8
Armenia	..	..	..	..	..	..	..	..	..	..	..	..
Australia	65	68	0	0	21	21	4	3	2	2	8	6
Austria	19	26	37	40	25	25	1	0	9	4	9	6
Azerbaijan	..	21	..	24	..	38	..	9	..	3	..	6
Bangladesh	..	..	..	..	..	..	..	..	..	..	..	..
Belarus	12	10	32	34	40	38	5	7	9	3	2	7
Belgium	35	37	35	33	24	25	0	0	3	3	3	2
Benin	..	..	..	..	..	..	..	..	..	..	..	..
Bolivia	5	8	9	9	31	50	7	7	11	13	38	14
Bosnia and Herzegovina	..	..	..	..	..	..	..	..	..	..	..	..
Botswana	39	17	0	0	2	4	13	12	0	0	46	67
Brazil	20	17	31	36	24	23	2	2	6	5	16	17
Bulgaria	30	15	23	23	18	35	2	6	1	1	27	20
Burkina Faso	23	..	0	..	30	..	33	..	7	..	8	..
Burundi	21	20	6	6	37	38	24	28	1	0	10	7
Cambodia	..	..	..	..	..	..	..	..	..	..	..	..
Cameroon	18	19	6	0	21	21	14	30	4	4	28	25
Canada	51	54	16	19	17	17	3	1	0	0	13	9
Central African Republic	..	..	..	..	..	..	..	..	..	..	..	..
Chad	19	..	0	..	39	..	24	..	10	..	8	..
Chile	12	19	8	6	43	46	12	8	3	4	21	16
China	31	7	0	0	18	79	14	6	0	4	37	4
Hong Kong, China	..	..	..	..	..	..	..	..	..	..	..	..
Colombia	29	37	0	0	30	42	20	10	1	0	19	11
Congo, Dem. Rep.	27	25	1	0	18	18	46	28	1	9	7	20
Congo, Rep.	..	..	0	0	16	30	21	12	2	0	35	58
Costa Rica	10	12	29	28	27	40	23	8	1	0	14	11
Côte d'Ivoire	16	21	7	6	27	17	29	51	11	3	9	3
Croatia	17	12	52	31	24	45	3	7	0	1	3	5
Cuba	..	..	..	..	..	..	..	..	..	..	..	..
Czech Republic	..	15	..	45	..	33	..	2	..	1	..	3
Denmark	37	36	4	4	41	42	0	0	3	4	15	13
Dominican Republic	21	17	4	4	23	35	40	37	1	1	10	6
Ecuador	62	..	0	..	22	..	13	..	1	..	2	..
Egypt, Arab Rep.	19	22	15	0	14	17	14	13	11	12	27	37
El Salvador	..	19	..	12	..	40	..	8	..	1	..	19
Eritrea	..	..	..	..	..	..	..	..	..	..	..	..
Estonia	27	20	28	34	41	40	1	0	1	0	2	6
Ethiopia	29	..	0	..	25	..	15	..	2	..	30	..
Finland	31	29	9	10	47	44	1	0	3	2	9	13
France	17	20	44	42	28	29	0	0	3	4	7	6
Gabon	24	..	1	..	23	..	18	..	2	..	32	..
Gambia, The	13	..	0	..	37	..	43	..	1	..	6	..
Georgia	..	11	..	20	..	45	..	11	..	0	..	13
Germany	16	15	53	48	24	20	0	0	0	0	6	16
Ghana	23	..	0	..	30	..	39	..	0	..	8	..
Greece	22	39	29	2	43	55	0	0	8	8	8	7
Guatemala	..	..	..	..	..	..	..	..	..	..	..	..
Guinea	9	10	0	1	15	6	47	77	0	3	28	3
Guinea-Bissau	..	..	..	..	..	..	..	..	..	..	..	..
Haiti	..	..	..	..	..	..	..	..	..	..	..	..
Honduras	..	..	..	..	..	..	..	..	..	..	..	..

Central government revenues | 4.13

	Taxes on income, profits, and capital gains		Social security taxes		Taxes on goods and services		Taxes on international trade		Other taxes		Nontax revenue	
	% of total current revenue		% of total current revenue		% of total current revenue		% of total current revenue		% of total current revenue		% of total current revenue	
	1990	1998	1990	1998	1990	1998	1990	1998	1990	1998	1990	1998
Hungary	18	19	29	29	31	34	6	4	0	2	16	12
India	15	24	0	0	36	28	29	21	0	0	20	27
Indonesia	62	61	0	3	24	24	6	4	3	0	5	6
Iran, Islamic Rep.	10	20	8	11	4	11	13	18	4	1	60	39
Iraq	..	..	..	..	..	..	..	..	..	..	..	..
Ireland	37	*42*	15	*13*	38	*37*	0	*0*	3	*4*	7	*4*
Israel	36	36	9	14	33	31	2	1	4	4	14	15
Italy	37	33	29	31	29	26	0	0	2	3	3	6
Jamaica	..	..	..	..	..	..	..	..	..	..	..	..
Japan	69	..	0	..	17	..	1	..	7	..	5	..
Jordan	16	10	0	0	21	31	27	23	7	8	29	28
Kazakhstan	..	12	..	32	..	44	..	5	..	4	..	3
Kenya	30	*34*	0	*0*	43	*37*	16	*15*	1	*1*	10	*14*
Korea, Dem. Rep.	..	..	..	..	..	..	..	..	..	..	..	..
Korea, Rep.	34	*27*	5	*9*	35	*34*	12	*6*	5	*10*	9	*14*
Kuwait	1	1	0	0	0	0	2	2	0	0	97	97
Kyrgyz Republic	..	15	..	0	..	55	..	6	..	5	..	19
Lao PDR	..	..	..	..	..	..	..	..	..	..	..	..
Latvia	..	13	..	32	..	41	..	2	..	0	..	12
Lebanon	..	9	..	0	..	8	..	44	..	13	..	26
Lesotho	11	18	0	0	21	12	57	48	0	0	11	22
Libya	..	..	..	..	..	..	..	..	..	..	..	..
Lithuania	*20*	14	*28*	30	*40*	49	*1*	2	*3*	0	*8*	5
Macedonia, FYR	..	..	..	..	..	..	..	..	..	..	..	..
Madagascar	13	*18*	0	*0*	19	*24*	48	*53*	2	*2*	18	*2*
Malawi	37	..	0	..	33	..	16	..	1	..	13	..
Malaysia	31	*36*	1	*1*	20	*26*	18	*13*	3	*5*	28	*18*
Mali	..	..	..	..	..	..	..	..	..	..	..	..
Mauritania	..	..	..	..	..	..	..	..	..	..	..	..
Mauritius	14	12	4	5	21	31	46	30	6	6	9	16
Mexico	31	36	13	12	56	59	6	4	2	2	11	10
Moldova	..	4	..	28	..	52	..	4	..	0	..	12
Mongolia	*24*	11	*14*	23	*31*	33	*17*	1	*0*	1	*15*	31
Morocco	24	..	4	..	38	..	18	..	4	..	13	..
Mozambique	..	..	..	..	..	..	..	..	..	..	..	..
Myanmar	18	18	0	0	28	27	14	5	0	0	41	51
Namibia	34	..	0	..	25	..	27	..	1	..	13	..
Nepal	11	15	0	0	36	37	31	27	5	5	17	17
Netherlands	31	*25*	35	*41*	22	*23*	0	*0*	3	*5*	9	*7*
New Zealand	53	62	0	0	27	28	2	2	3	2	15	6
Nicaragua	17	..	9	..	35	..	19	..	8	..	13	..
Niger	..	..	..	..	..	..	..	..	..	..	..	..
Nigeria	..	..	..	..	..	..	..	..	..	..	..	..
Norway	16	21	24	23	34	38	1	1	1	1	24	17
Oman	23	17	0	0	1	2	2	4	1	3	73	74
Pakistan	9	23	0	0	30	27	31	17	0	13	30	21
Panama	17	17	20	19	17	..	12	..	3	4	31	31
Papua New Guinea	37	*56*	0	*0*	14	*10*	25	*30*	3	*2*	20	*1*
Paraguay	9	..	0	..	21	..	20	..	24	..	25	..
Peru	5	21	7	8	50	49	17	10	19	4	7	15
Philippines	28	40	0	0	31	28	25	17	3	6	13	10
Poland	..	25	..	30	..	33	..	3	..	1	..	8
Portugal	23	27	25	25	34	36	2	0	4	2	12	10
Puerto Rico	..	..	..	..	..	..	..	..	..	..	..	..
Romania	19	*30*	23	*27*	33	*27*	1	*6*	15	*2*	10	*8*
Russian Federation	..	7	..	39	..	39	..	7	..	1	..	6

	Taxes on income, profits, and capital gains		Social security taxes		Taxes on goods and services		Taxes on international trade		Other taxes		Nontax revenue	
	% of total current revenue		% of total current revenue		% of total current revenue		% of total current revenue		% of total current revenue		% of total current revenue	
	1990	1998	1990	1998	1990	1998	1990	1998	1990	1998	1990	1998
Rwanda	18	..	7	..	34	..	26	..	4	..	12	..
Saudi Arabia	..	..	..	..	..	..	..	..	..	..	..	..
Senegal	..	..	..	..	..	..	..	..	..	..	..	..
Sierra Leone	31	17	0	0	23	33	40	46	0	0	5	3
Singapore	26	28	0	0	16	17	2	1	14	13	43	42
Slovak Republic	..	24	..	33	..	31	..	5	..	0	..	7
Slovenia	12	15	47	36	27	37	8	4	0	4	5	5
South Africa	51	53	2	2	34	33	4	3	2	2	8	7
Spain	32	30	38	39	22	25	2	0	0	0	5	6
Sri Lanka	11	12	0	0	46	52	29	16	5	4	10	16
Sudan	..	..	..	..	..	..	..	..	..	..	..	..
Sweden	18	14	31	34	29	28	1	0	9	12	13	11
Switzerland	15	15	51	51	23	23	1	1	3	3	7	7
Syrian Arab Republic	31	34	0	0	31	17	7	12	7	6	24	31
Tajikistan	..	6	..	14	..	64	..	12	..	0	..	3
Tanzania	..	..	..	..	..	..	..	..	..	..	..	..
Thailand	24	29	0	1	41	49	22	9	4	1	8	12
Togo	..	..	..	..	..	..	..	..	..	..	..	..
Trinidad and Tobago	..	..	..	..	..	..	..	..	..	..	..	..
Tunisia	13	19	13	17	19	35	28	14	5	4	22	11
Turkey	43	40	0	0	32	37	6	2	3	7	15	15
Turkmenistan	..	..	..	..	..	..	..	..	..	..	..	..
Uganda	..	..	..	..	..	..	..	..	..	..	..	..
Ukraine	..	..	..	..	..	..	..	..	..	..	..	..
United Arab Emirates	0	0	2	2	36	51	0	0	0	0	62	47
United Kingdom	39	39	17	17	28	31	0	0	7	7	9	5
United States	52	57	35	32	3	3	2	1	1	1	8	6
Uruguay	7	13	27	29	36	40	10	4	12	12	5	7
Uzbekistan	..	..	..	..	..	..	..	..	..	..	..	..
Venezuela, RB	64	18	4	4	3	31	7	11	0	3	22	33
Vietnam	..	20	..	0	..	33	..	25	..	9	..	13
West Bank and Gaza	..	..	..	..	..	..	..	..	..	..	..	..
Yemen, Rep.	26	18	0	0	10	7	17	10	5	2	43	63
Yugoslavia, FR (Serb./Mont.)	..	..	..	..	..	..	..	..	..	..	..	..
Zambia	..	..	..	..	..	..	..	..	..	..	..	..
Zimbabwe	45	43	0	0	26	24	17	20	1	2	10	10
World	23 m	19 m	4 m	10 m	27 m	34 m	13 m	6 m	3 m	2 m	13 m	12 m
Low income	..	..	..	..	..	..	..	..	..	..	..	..
Middle income	22	17	4	12	25	38	14	7	3	2	16	11
Lower middle income	23	15	0	9	26	40	19	9	4	1	17	12
Upper middle income	22	17	7	22	23	33	11	4	3	2	16	9
Low & middle income	21	..	1	..	26	..	17	..	3	..	15	13
East Asia & Pacific	31	20	0	0	26	33	16	6	3	1	17	12
Europe & Central Asia	..	14	..	30	..	40	..	5	..	1	..	7
Latin America & Carib.	17	17	9	10	27	..	13	..	3	2	14	12
Middle East & N. Africa	21	19	2	0	17	12	15	15	5	4	28	30
South Asia	11	19	0	0	36	32	30	19	3	4	18	19
Sub-Saharan Africa	23	..	0	..	25	..	27	..	1	..	10	..
High income	32	28	20	20	27	28	1	0	3	3	9	8
Europe EMU	31	28	35	32	27	26	0	0	3	3	7	8

Note: Components may not sum to 100 percent as a result of adjustments to tax revenue.

Central government revenues | 4.13

The International Monetary Fund (IMF) classifies government transactions as receipts or payments and according to whether they are repayable or nonrepayable. If nonrepayable, they are classified as capital (meant to be used in production for more than a year) or current, and as requited (involving payment in return for a benefit or service) or unrequited. Revenues include all nonrepayable receipts (other than grants), the most important of which are taxes. Grants are unrequited, nonrepayable, noncompulsory receipts from other governments or from international organizations. Transactions are generally recorded on a cash rather than an accrual basis. Measuring the accumulation of arrears on revenues or payments on an accrual basis would typically result in a higher deficit. Transactions within a level of government are not included, but transactions between levels are included. In some instances the government budget may include transfers used to finance the deficits of autonomous, extrabudgetary agencies.

The IMF's *Manual on Government Finance Statistics* (1986) describes taxes as compulsory, unrequited payments made to governments by individuals, businesses, or institutions. Taxes traditionally have been classified as either direct (those levied directly on the income or profits of individuals and corporations) or indirect (sales and excise taxes and duties levied on goods and services). This distinction may be a useful simplification, but it has no particular analytical significance.

Social security taxes do not reflect compulsory payments made by employers to provident funds or other agencies with a similar purpose. Similarly, expenditures from such funds are not reflected in government expenditure (see table 4.12). The revenue shares shown in this table may not sum to 100 percent because adjustments to tax revenues are not shown.

For further discussion of taxes and tax policies see *About the data* for table 5.5. For further discussion of government revenues and expenditures see *About the data* for tables 4.11 and 4.12.

Figure 4.13

Some developing economies tax as much as high-income economies
Highest tax revenue as % of GDP

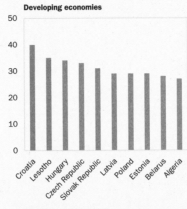

Developing economies

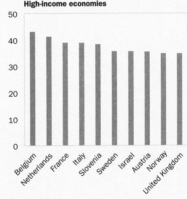

High-income economies

Note: For developing economies data refer to the most recent year available in 1998–99, and for high-income economies, in 1997–99.
Source: World Bank staff estimates.

• **Taxes on income, profits, and capital gains** are levied on the actual or presumptive net income of individuals, on the profits of enterprises, and on capital gains, whether realized on land, securities, or other assets. Intragovernmental payments are eliminated in consolidation. • **Social security taxes** include employer and employee social security contributions and those of self-employed and unemployed people. • **Taxes on goods and services** include general sales and turnover or value added taxes, selective excises on goods, selective taxes on services, taxes on the use of goods or property, and profits of fiscal monopolies. • **Taxes on international trade** include import duties, export duties, profits of export or import monopolies, exchange profits, and exchange taxes. • **Other taxes** include employer payroll or labor taxes, taxes on property, and taxes not allocable to other categories. They may include negative values that are adjustments (for example, for taxes collected on behalf of state and local governments and not allocable to individual tax categories). • **Nontax revenue** includes requited, nonrepayable receipts for public purposes, such as fines, administrative fees, or entrepreneurial income from government ownership of property, and voluntary, unrequited, nonrepayable receipts other than from government sources. It does not include proceeds of grants and borrowing, funds arising from the repayment of previous lending by governments, incurrence of liabilities, and proceeds from the sale of capital assets.

The data on central government revenues are from the IMF's *Government Finance Statistics Yearbook, 2000* and IMF data files. Each country's accounts are reported using the system of common definitions and classifications in the IMF's *Manual on Government Finance Statistics* (1986). The IMF receives additional information from the Organisation for Economic Cooperation and Development on the tax revenues of some of its members. See the IMF sources for complete and authoritative explanations of concepts, definitions, and data sources.

4.14 | Monetary indicators and prices

	Money and quasi money		Claims on private sector		Claims on governments and other public entities		GDP implicit deflator		Consumer price index		Food price index	
	annual % growth of M2		annual growth as % of M2		annual growth as % of M2		average annual % growth		average annual % growth		average annual % growth	
	1990	1999	1990	1999	1990	1999	1980–90	1990–99	1980–90	1990–99	1980–90	1990–99
Albania	..	22.3	..	1.4	..	6.1	–0.4	44.7	..	*32.1*	..	35.7
Algeria	11.4	13.7	12.2	3.5	3.2	21.3	8.1	19.1	9.1	19.5	6.9	20.8
Angola	..	526.0	..	56.3	..	70.3	*5.9*	813.7	..	787.0	..	..
Argentina	1,113.3	4.1	1,444.7	–2.4	1,573.2	9.6	391.1	6.2	390.6	10.6	279.2	9.9
Armenia	..	14.0	..	8.8	..	–4.8	..	269.2	..	*97.8*	..	..
Australia	12.8	11.7	15.3	14.3	–2.2	–1.7	7.2	1.3	7.9	2.0	7.3	2.5
Austria[a]	..	..	..	..	..	..	3.3	2.2	3.2	2.4	2.6	1.7
Azerbaijan	..	21.5	..	1.6	..	–12.8	..	250.0	..	*224.9*	*1.5*	257.5
Bangladesh	10.4	15.5	9.2	8.7	–0.2	6.7	9.5	4.1	..	5.5	10.4	*4.3*
Belarus	..	132.7	..	77.5	..	85.3	..	393.5	..	*383.7*	*2.4*	420.6
Belgium[a]	..	..	..	..	..	..	4.4	2.2	4.2	2.0	4.0	1.0
Benin	28.6	34.8	–1.3	21.8	12.4	–19.9	1.7	9.4	..	*9.9*	..	*8.2*
Bolivia	52.8	5.7	40.8	5.4	18.0	0.6	327.2	9.1	322.5	9.3	322.0	9.4
Bosnia and Herzegovina	..	..	..	..	..	..	..	*–1.0*	..	..	..	..
Botswana	–14.0	26.3	12.6	18.5	–52.4	–15.9	13.6	10.0	10.0	10.7	10.7	11.4
Brazil	1,289.2	7.4	1,566.4	4.8	3,093.6	–4.5	284.0	263.9	285.6	253.5	238.2	247.8
Bulgaria	*53.8*	11.9	*1.9*	9.5	*84.5*	–9.5	1.8	112.0	*6.3*	129.3	..	137.0
Burkina Faso	–0.5	2.6	3.6	1.5	–1.5	1.5	3.3	6.0	*1.0*	6.1	–0.5	*5.4*
Burundi	9.6	47.3	15.4	24.6	–6.9	23.7	4.4	11.8	7.1	15.8	6.1	6.7
Cambodia	..	17.3	..	8.8	..	–5.8	..	28.7	..	*7.1*	..	..
Cameroon	–1.7	13.3	0.9	7.1	–3.0	4.7	5.6	5.5	8.7	7.3	3.9	*3.6*
Canada	7.8	5.0	9.2	4.7	0.6	–0.2	4.5	1.4	5.3	1.7	4.6	1.5
Central African Republic	–3.7	11.1	–1.6	–0.2	2.3	9.0	7.9	4.9	3.2	*6.7*	*2.0*	7.8
Chad	–2.4	–2.6	1.3	–0.1	–17.3	6.5	1.4	7.6	*0.6*	8.7	..	*9.2*
Chile	23.5	14.8	21.4	12.5	16.4	0.5	20.7	8.0	20.6	9.7	20.8	9.2
China	28.9	14.7	26.5	9.8	1.5	1.1	5.9	8.2	..	9.9	8.8	..
Hong Kong, China	*8.5*	8.3	*7.9*	–8.6	*–1.0*	2.4	7.7	5.2	..	6.8	6.8	5.5
Colombia	33.0	13.8	*8.7*	–4.9	*–5.1*	4.5	24.8	22.7	22.7	21.7	24.5	19.3
Congo, Dem. Rep.	195.4	..	18.0	..	429.7	..	62.9	*1,423.1*	57.1	*2,089.0*	..	..
Congo, Rep.	18.5	19.9	5.1	26.1	–12.6	–12.9	0.5	8.6	0.9	10.0	4.1	*10.2*
Costa Rica	27.5	21.7	7.3	11.3	8.2	–10.5	23.6	16.7	23.0	16.2	23.0	14.4
Côte d'Ivoire	–2.6	–1.7	–3.9	–5.1	–3.0	3.8	2.8	8.2	5.4	7.8	6.0	..
Croatia	..	–1.8	..	–6.4	..	5.1	..	104.9	*304.1*	105.4	*246.3*	103.0
Cuba	..	..	..	..	..	..	..	..	..	..	..	..
Czech Republic	..	2.6	..	–3.8	..	0.8	..	12.4	..	8.5	..	..
Denmark	6.5	–0.9	3.0	2.3	–3.1	–1.0	5.8	2.0	5.5	2.0	4.8	1.9
Dominican Republic	42.5	23.7	19.1	19.4	0.7	4.3	21.6	9.9	22.4	9.0	25.2	*11.7*
Ecuador	101.6	99.2	46.7	117.7	–22.4	61.3	36.4	33.8	35.8	34.5	*43.0*	*33.7*
Egypt, Arab Rep.	28.7	5.7	6.3	11.8	25.3	2.2	13.7	8.8	17.4	9.6	19.0	*8.3*
El Salvador	32.4	9.1	8.8	9.2	9.6	0.7	16.3	8.1	19.6	9.4	21.4	11.1
Eritrea	..	..	..	..	..	..	..	9.4	..	..	..	..
Estonia	*71.1*	24.7	*27.6*	6.2	*–11.3*	4.7	2.3	62.7	..	*25.3*	..	58.5
Ethiopia	18.5	6.8	–1.0	9.9	23.0	18.1	*4.6*	7.6	4.0	*6.0*	3.7	*12.4*
Finland[a]	..	..	..	..	..	..	6.7	2.0	6.2	1.5	5.8	–0.8
France[a]	..	..	..	..	..	..	5.8	1.6	5.8	1.7	5.7	*1.1*
Gabon	3.3	–3.0	0.7	0.2	–20.6	–6.7	1.8	5.8	5.1	*5.7*	*2.8*	5.1
Gambia, The	8.4	12.1	7.8	7.8	–35.4	3.8	17.9	3.7	20.0	4.3	20.4	*4.9*
Georgia	..	21.1	..	30.9	..	54.6	..	..	..	1.0	..	5.4
Germany[a,b]	..	..	..	..	..	..	..	*1.9*	2.2	2.4	..	1.5
Ghana	13.3	16.2	4.9	25.0	–0.8	53.5	42.1	27.4	39.1	29.2	33.1	*28.5*
Greece	14.3	16.7	4.6	19.7	16.3	11.6	18.0	9.5	18.7	9.8	18.0	9.2
Guatemala	25.8	12.5	15.0	10.7	0.5	2.2	14.6	10.9	14.0	10.7	14.6	10.6
Guinea	–17.4	–77.5	*13.1*	–6.1	*2.9*	*1.1*	..	5.6	..	..	..	*9.1*
Guinea-Bissau	574.6	21.5	90.5	–8.0	460.7	25.3	57.4	37.6	..	37.6	..	..
Haiti	2.5	23.0	–0.6	5.2	0.4	10.1	7.3	22.2	5.2	23.2	4.1	*19.2*
Honduras	21.4	24.7	13.0	17.1	–10.5	–14.6	5.7	19.7	6.3	19.5	5.1	20.4

Monetary indicators and prices 4.14

	Money and quasi money		Claims on private sector		Claims on governments and other public entities		GDP implicit deflator		Consumer price index		Food price index	
	annual % growth of M2		annual growth as % of M2		annual growth as % of M2		average annual % growth		average annual % growth		average annual % growth	
	1990	1999	1990	1999	1990	1999	1980–90	1990–99	1980–90	1990–99	1980–90	1990–99
Hungary	29.2	15.5	22.8	11.5	69.9	−19.2	8.9	20.6	9.6	21.5	9.5	21.0
India	15.1	17.2	5.9	9.9	10.5	6.8	8.0	8.5	8.6	9.5	8.4	9.8
Indonesia	44.6	12.5	66.9	−49.8	−6.7	74.7	8.6	14.7	8.3	13.1	8.6	9.6
Iran, Islamic Rep.	18.0	21.5	14.7	15.5	5.8	9.5	14.4	27.0	18.2	27.1	16.3	28.7
Iraq	..	..	..	..	..	..	10.3	..	..	..	14.3	..
Ireland[a]	..	..	..	..	..	..	6.6	3.4	6.8	2.1	10.5	2.3
Israel	19.4	15.5	18.5	12.5	4.9	1.6	101.1	10.7	101.7	10.5	102.4	9.1
Italy[a]	..	..	..	..	..	..	10.0	4.0	9.1	3.9	8.2	3.3
Jamaica	21.5	12.2	12.5	8.4	−16.0	7.2	18.6	27.6	15.1	26.1	16.2	32.9
Japan	8.2	3.4	9.7	−2.1	1.5	6.0	1.7	0.1	1.7	0.9	1.6	0.7
Jordan	8.3	15.5	4.7	4.6	1.0	−1.0	4.3	3.5	5.7	3.9	4.7	4.3
Kazakhstan	..	84.4	..	44.0	..	−2.3	..	255.9	..	87.2	..	..
Kenya	20.1	6.0	8.0	6.4	21.5	−2.4	9.1	14.9	11.1	16.7	..	19.4
Korea, Dem. Rep.	..	..	..	..	..	..	..	..	..	..	..	..
Korea, Rep.	17.2	27.4	36.1	24.4	−1.2	−0.9	6.1	5.8	4.9	5.3	5.0	5.6
Kuwait	0.7	1.6	3.3	2.8	−3.1	−1.8	−2.8	..	2.9	2.0	1.2	2.7
Kyrgyz Republic	..	33.7	..	13.3	..	−6.4	..	129.7	..	..	..	..
Lao PDR	7.8	78.4	3.6	39.2	7.0	..	37.6	22.9	..	24.1	..	..
Latvia	..	8.3	..	8.2	..	2.8	−0.2	57.9	..	34.6	..	..
Lebanon	55.1	11.7	27.6	5.6	18.5	5.2	..	24.0	..	..	..	..
Lesotho	8.4	−5.1	6.8	1.5	−17.4	54.2	12.1	9.6	13.6	10.5	13.2	13.0
Libya	19.0	7.4	2.0	10.1	15.0	−9.0	0.2	..	..	..	..	..
Lithuania	..	7.7	..	8.1	..	5.1	..	90.7	..	40.2	..	..
Macedonia, FYR	..	32.0	..	31.8	..	−16.3	..	94.2	..	91.4	242.1	89.1
Madagascar	4.5	19.2	23.8	3.2	−14.8	5.2	17.1	20.6	16.6	19.8	15.7	20.1
Malawi	11.1	26.5	15.8	1.9	−12.8	24.7	14.6	34.3	16.9	33.8	16.3	37.4
Malaysia	10.6	16.9	20.8	5.5	−1.2	0.3	1.7	3.9	2.6	4.0	1.3	5.1
Mali	−4.9	1.0	0.1	10.2	−13.4	1.1	4.5	8.1	..	5.8	..	..
Mauritania	11.5	2.1	20.2	26.5	1.5	−40.3	8.4	6.0	7.1	6.3	..	..
Mauritius	21.2	15.2	10.8	7.8	0.8	−0.6	9.5	6.3	6.9	7.0	7.4	7.1
Mexico	81.9	11.8	48.5	−0.5	13.6	−2.1	71.5	19.5	73.8	19.9	73.1	19.9
Moldova	358.0	42.9	53.3	10.7	469.1	18.2	..	142.1	..	16.0	..	..
Mongolia	31.6	31.6	40.2	−3.5	38.5	−2.3	−1.6	66.6	..	53.7	..	..
Morocco	21.5	10.2	12.4	6.2	−4.9	−4.0	7.1	3.2	7.0	4.2	6.7	5.5
Mozambique	37.2	31.8	22.0	21.2	−5.1	2.2	38.3	36.4	..	34.9	..	..
Myanmar	37.7	29.7	12.8	7.6	24.2	22.9	12.2	26.1	11.5	27.1	11.9	28.4
Namibia	30.3	18.4	15.4	4.3	−4.2	4.7	13.7	9.8	12.6	9.9	14.9	9.8
Nepal	18.5	21.6	5.7	9.3	7.3	3.4	11.1	8.6	10.2	9.0	10.1	9.6
Netherlands[a]	..	..	..	..	..	..	1.6	1.9	2.0	2.4	1.2	1.5
New Zealand	12.5	5.0	4.2	10.7	−1.7	1.2	10.8	1.4	11.0	1.9	9.9	1.2
Nicaragua	7,677.8	18.8	4,932.9	23.6	12,679.2	15.7	422.3	38.6	535.7	35.1	..	..
Niger	−4.1	15.4	−5.1	−2.1	1.4	6.5	1.9	6.4	0.7	6.6	−1.5	..
Nigeria	32.7	31.7	7.8	21.7	27.1	55.0	16.7	31.6	21.5	36.2	21.6	37.6
Norway	5.6	1.7	5.0	7.3	−0.6	−14.7	5.6	2.1	7.4	2.1	7.8	1.8
Oman	10.0	6.4	9.6	10.4	−10.9	−10.5	−3.6	−2.9	..	0.2	..	0.4
Pakistan	11.6	4.3	5.9	6.7	7.7	−1.9	6.7	10.6	6.3	10.3	6.6	10.8
Panama	36.6	8.5	0.8	23.3	−25.7	−0.7	1.9	2.1	1.4	1.1	1.9	1.0
Papua New Guinea	4.3	9.2	−0.9	−0.8	9.9	2.7	5.3	7.1	5.6	8.7	4.6	6.8
Paraguay	52.5	11.7	33.1	7.6	−9.5	−6.2	24.4	13.4	21.9	13.8	24.9	13.5
Peru	6,384.9	14.5	2,123.7	6.4	2,129.5	9.0	220.2	31.0	246.1	31.6	..	29.2
Philippines	22.5	16.8	15.7	−1.8	3.4	3.8	14.9	8.6	13.4	8.5	14.1	7.9
Poland	160.1	19.4	20.8	16.9	75.6	1.8	..	25.0	50.9	27.8	52.4	24.2
Portugal[a]	..	..	..	..	..	..	18.0	5.6	17.1	4.8	16.9	3.7
Puerto Rico	..	..	..	..	..	..	3.5	3.7	..	..	2.8	8.4
Romania	26.4	44.9	..	0.9	0.0	16.3	2.5	105.6	..	108.9	1.8	104.9
Russian Federation	..	56.7	..	27.9	..	31.3	..	190.4	..	116.1	..	..

	Money and quasi money		Claims on private sector		Claims on governments and other public entities		GDP implicit deflator		Consumer price index		Food price index	
	annual % growth of M2		annual growth as % of M2		annual growth as % of M2		average annual % growth		average annual % growth		average annual % growth	
	1990	1999	1990	1999	1990	1999	1980–90	1990–99	1980–90	1990–99	1980–90	1990–99
Rwanda	5.6	7.9	–10.0	6.4	26.8	5.1	4.0	16.3	3.9	18.0	6.6	..
Saudi Arabia	4.6	6.8	–4.5	0.5	4.2	–17.7	–4.9	1.2	–0.8	1.2	–0.4	1.2
Senegal	–4.8	13.1	–8.4	7.3	–5.3	1.1	6.5	5.0	6.2	6.0	5.3	8.3
Sierra Leone	74.0	37.8	4.9	–2.6	228.7	50.2	62.8	31.1	72.4	31.4	71.0	..
Singapore	20.0	8.5	13.7	–2.9	–4.9	1.9	1.9	1.6	1.6	1.8	0.9	1.9
Slovak Republic	..	11.6	..	–5.5	..	7.0	1.8	11.4	..	13.0	1.6	9.0
Slovenia	123.0	15.1	96.1	16.7	–10.4	1.0	..	23.3	..	28.0	252.3	30.3
South Africa	11.4	10.9	13.7	12.0	1.8	–0.5	15.5	10.2	14.8	9.1	15.1	11.3
Spain[a]	..	..	..	..	..	..	9.3	4.1	9.0	3.9	9.3	3.1
Sri Lanka	21.1	12.4	16.2	9.9	6.8	7.8	11.0	9.4	10.9	10.3	10.9	10.9
Sudan	48.8	23.5	12.6	–0.4	29.4	16.6	41.0	66.6	37.6	81.1	38.0	..
Sweden	..	..	..	..	..	..	7.4	2.2	7.0	2.1	8.2	–0.6
Switzerland	0.8	13.3	11.7	7.2	1.0	0.0	3.4	1.4	2.9	1.7	3.1	0.4
Syrian Arab Republic	26.1	13.4	3.4	0.7	11.4	–3.5	15.3	8.7	23.2	7.8	24.5	6.4
Tajikistan	..	..	..	..	..	..	..	..	..	..	..	..
Tanzania	41.9	18.6	22.6	6.1	80.6	8.9	..	23.1	31.0	22.6	30.2	24.2
Thailand	26.7	5.4	30.0	–6.0	–4.0	1.7	3.9	4.6	3.5	5.1	2.7	6.4
Togo	9.5	8.4	1.8	–7.9	6.9	0.0	4.8	7.6	2.5	9.3	1.2	..
Trinidad and Tobago	6.2	4.2	2.7	5.2	–1.9	–3.8	2.4	5.8	10.7	5.9	14.6	13.3
Tunisia	7.6	18.9	5.9	10.2	1.8	5.0	7.4	4.6	7.4	4.6	8.3	4.6
Turkey	53.2	98.3	42.9	24.8	2.2	54.8	45.2	78.3	44.9	81.5	..	83.4
Turkmenistan	..	22.6	..	0.3	..	82.3	..	516.9	..	..	..	..
Uganda	60.2	13.6	..	8.4	–0.9	–0.5	113.8	13.8	102.5	11.6	..	13.4
Ukraine	..	41.3	..	20.7	..	29.4	..	339.1	..	413.4	..	..
United Arab Emirates	–8.2	11.5	1.3	8.0	–4.8	1.8	0.8	2.4	..	..	..	..
United Kingdom	..	..	..	..	..	..	5.7	3.0	5.8	2.9	4.6	2.0
United States	4.9	8.2	1.1	10.2	0.6	0.6	3.8	2.1	4.2	2.7	3.8	3.8
Uruguay	118.5	13.1	56.2	10.5	25.8	4.0	62.7	35.2	61.1	38.2	62.0	34.7
Uzbekistan	..	..	..	..	..	..	..	293.0	..	..	..	..
Venezuela, RB	71.2	20.9	17.0	6.4	43.7	–1.7	19.3	47.5	20.9	51.8	29.7	50.1
Vietnam	..	48.8	..	59.2	..	–47.4	210.8	16.8	..	..	..	..
West Bank and Gaza	..	..	..	..	..	..	..	9.1	..	..	..	..
Yemen, Rep.	11.3	13.8	1.4	5.0	10.2	–14.2	..	26.1	..	32.6	..	..
Yugoslavia, FR (Serb./Mont.)	..	..	..	..	..	..	..	..	..	..	..	..
Zambia	47.9	27.7	22.8	11.9	195.2	36.3	42.2	57.4	72.5	80.8	42.8	73.0
Zimbabwe	15.1	35.9	13.5	14.0	5.0	–4.3	11.6	23.6	13.8	25.4	14.6	32.5

a. As members of the European Monetary Union, these countries share a single currency, the euro. b. Data prior to 1990 refer to the Federal Republic of Germany before unification.

Money and the financial accounts that record the supply of money lie at the heart of a country's financial system. There are several commonly used definitions of the money supply. The narrowest, M1, encompasses currency held by the public and demand deposits with banks. M2 includes M1 plus time and savings deposits with banks that require a notice for withdrawal. M3 includes M2 as well as various money market instruments, such as certificates of deposit issued by banks, bank deposits denominated in foreign currency, and deposits with financial institutions other than banks. However defined, money is a liability of the banking system, distinguished from other bank liabilities by the special role it plays as a medium of exchange, a unit of account, and a store of value.

The banking system's assets include its net foreign assets and net domestic credit. Net domestic credit includes credit to the private sector and general government, and credit extended to the nonfinancial public sector in the form of investments in short- and long-term government securities and loans to state enterprises; liabilities to the public and private sectors in the form of deposits with the banking system are netted out. Net domestic credit also includes credit to banking and nonbank financial institutions.

Domestic credit is the main vehicle through which changes in the money supply are regulated, with central bank lending to the government often playing the most important role. The central bank can regulate lending to the private sector in several ways—for example, by adjusting the cost of the refinancing facilities it provides to banks, by changing market interest rates through open market operations, or by controlling the availability of credit through changes in the reserve requirements imposed on banks and ceilings on the credit provided by banks to the private sector.

Monetary accounts are derived from the balance sheets of financial institutions—the central bank, commercial banks, and nonbank financial intermediaries. Although these balance sheets are usually reliable, they are subject to errors of classification, valuation, and timing and to differences in accounting practices. For example, whether interest income is recorded on an accrual or a cash basis can make a substantial difference, as can the treatment of nonperforming assets. Valuation errors typically arise with respect to foreign exchange transactions, particularly in countries with flexible exchange rates or in those that have undergone a currency devaluation during the reporting period. The valuation of financial derivatives and the net liabilities of the banking system can also be difficult.

The quality of commercial bank reporting also may be adversely affected by delays in reports from bank branches, especially in countries where branch accounts are not computerized. Thus the data in the balance sheets of commercial banks may be based on preliminary estimates subject to constant revision. This problem is likely to be even more serious for nonbank financial intermediaries.

Controlling inflation is one of the primary goals of monetary policy and is intimately linked to the growth in money supply. Inflation is measured by the rate of increase in a price index, but actual price change can also be negative. Which index is used depends on which set of prices in the economy is being examined. The GDP deflator reflects changes in prices for total gross domestic product. The most general measure of the overall price level, it takes into account changes in government consumption, capital formation (including inventory appreciation), international trade, and the main component, household final consumption expenditure. The GDP deflator is usually derived implicitly as the ratio of current to constant price GDP, resulting in a Paasche index. It is defective as a general measure of inflation for use in policy because of the long lags in deriving estimates and because it is often only an annual measure.

Consumer price indexes are more current and produced more frequently. They are also constructed explicitly, based on surveys of the cost of a defined basket of consumer goods and services. Nevertheless, consumer price indexes should be interpreted with caution. The definition of a household and the geographic (urban or rural) and income group coverage of consumer price surveys can vary widely across countries, as can the basket of goods chosen. In addition, the weights are derived from household expenditure surveys, which, for budgetary reasons, tend to be conducted infrequently in developing countries, leading to poor comparability over time. Although a useful indicator for measuring consumer price inflation within a country, consumer price indexes are of less value in making comparisons across countries. Like consumer price indexes, food price indexes too should be interpreted with caution because of the high variability across countries in the items covered.

The least-squares method is used to calculate the growth rates of the GDP implicit deflator, consumer price index, and food price index.

- **Money and quasi money** comprise the sum of currency outside banks, demand deposits other than those of the central government, and the time, savings, and foreign currency deposits of resident sectors other than the central government. This definition of the money supply is frequently called M2; it corresponds to lines 34 and 35 in the International Monetary Fund's (IMF) *International Financial Statistics* (IFS). The change in money supply is measured as the difference in end-of-year totals relative to M2 in the preceding year. • **Claims on private sector** (IFS line 32d) include gross credit from the financial system to individuals, enterprises, nonfinancial public entities not included under net domestic credit, and financial institutions not included elsewhere. • **Claims on governments and other public entities** (IFS line 32an + 32b + 32bx + 32c) usually comprise direct credit for specific purposes such as financing the government budget deficit, loans to state enterprises, advances against future credit authorizations, and purchases of treasury bills and bonds, net of deposits by the public sector. Public sector deposits with the banking system also include sinking funds for the service of debt and temporary deposits of government revenues. • **GDP implicit deflator** measures the average annual rate of price change in the economy as a whole for the periods shown. • **Consumer price index** reflects changes in the cost to the average consumer of acquiring a basket of goods and services that may be fixed or change at specified intervals, such as yearly. The Laspeyres formula is generally used. • **Food price index** is a subindex of the consumer price index.

The monetary and financial data in this table are published by the IMF in its monthly *International Financial Statistics* and annual *International Financial Statistics Yearbook*. The IMF collects data on the financial systems of its member countries. The World Bank receives data from the IMF in electronic files that may contain more recent revisions than the published sources. The GDP deflator data are from the World Bank's national accounts files. The food price index data are from the United Nations Statistics Division's *Statistical Yearbook* and *Monthly Bulletin of Statistics*. The discussion of monetary indicators draws from an IMF publication by Marcello Caiola, *A Manual for Country Economists* (1995).

4.15 Balance of payments current account

	Goods and services				Net income		Net current transfers		Current account balance		Gross international reserves	
	Exports $ millions		Imports $ millions		$ millions		$ millions		$ millions		$ millions	
	1990	1999	1990	1999	1990	1999	1990	1999	1990	1999	1990	1999
Albania	354	544	485	1,101	−2	75	15	326	−118	−155	..	404
Algeria	13,462	12,943	10,106	9,602	−2,268	−2,668	333	..	1,420	..	2,703	6,146
Angola	3,992	5,370	3,385	6,947	−765	−491	−77	1,820	−236	−249	..	496
Argentina	14,800	27,747	6,846	32,589	−4,400	−7,739	998	268	4,552	−12,312	6,222	26,350
Armenia	..	383	..	919	..	55	..	174	..	−307	1	331
Australia	49,843	73,353	53,056	84,220	−13,176	−12,272	358	−36	−16,031	−23,175	19,319	21,956
Austria	63,694	94,920	61,580	95,901	−942	−2,723	−6	−2,043	1,166	−5,747	17,228	18,923
Azerbaijan	392	1,282	348	1,919	0	−526	106	56	150	−1,106	0	673
Bangladesh	1,903	6,031	4,156	8,527	−122	−135	802	2,237	−1,573	−394	660	1,634
Belarus	3,661	6,683	3,557	6,984	−1	−65	79	108	182	−257	..	294
Belgium[a]	138,605	194,213	135,098	183,454	2,316	5,486	−2,197	−4,560	3,627	11,685	23,789	13,346
Benin	364	537	454	792	−25	−14	97	111	−18	−157	69	403
Bolivia	977	1,311	1,086	1,989	−249	−201	159	324	−199	−556	511	1,190
Bosnia and Herzegovina	..	..	..	..	..	..	..	..	..	..	..	..
Botswana	2,005	3,044	1,987	2,512	−106	−266	69	252	−19	517	3,331	6,299
Brazil	35,170	55,746	28,184	63,648	−11,608	−18,859	799	1,688	−3,823	−25,073	9,200	35,717
Bulgaria	6,950	5,793	8,027	6,558	−758	−219	125	300	−1,710	−685	670	3,383
Burkina Faso	349	320	758	741	0	−35	332	143	−77	−312	305	298
Burundi	89	61	318	130	−15	−9	174	51	−69	−27	112	53
Cambodia	314	1,131	507	1,407	−21	−25	120	235	−93	−66	..	393
Cameroon	2,251	2,244	1,931	2,289	−478	−468	−39	117	−196	−396	37	4
Canada	149,538	277,674	149,118	258,985	−19,388	−21,640	−796	677	−19,764	−2,273	23,530	28,650
Central African Republic	220	156	410	244	−22	−20	123	66	−89	−42	123	136
Chad	271	311	488	494	−21	−15	192	36	−46	−161	132	95
Chile	10,221	19,406	9,166	18,058	−1,737	−1,880	198	452	−485	−80	6,784	14,761
China[†]	57,374	218,494	46,706	189,797	1,055	−17,974	274	4,944	11,997	15,667	34,476	161,414
Hong Kong, China	100,413	211,825	94,084	203,330	0	3,516	..	−1,470	6,329	10,541	24,656	96,255
Colombia	8,679	13,865	6,858	13,351	−2,305	−1,421	1,026	846	542	−61	4,869	8,198
Congo, Dem. Rep.	2,557	1,446	2,497	1,385	−770	−752	−27	33	−738	−658	261	..
Congo, Rep.	1,488	1,792	1,282	1,916	−460	−92	3	−20	−251	−252	10	39
Costa Rica	1,963	8,193	2,346	7,182	−233	−1,763	192	102	−424	−649	525	1,461
Côte d'Ivoire	3,503	5,346	3,445	4,137	−1,091	−692	−181	−479	−1,214	38	21	643
Croatia	..	8,118	..	9,791	..	−350	..	500	..	−1,522	167	3,025
Cuba	..	..	..	..	..	..	..	..	..	..	..	..
Czech Republic	..	33,188	..	33,989	..	−739	..	509	..	−1,032	..	12,936
Denmark	48,902	65,687	41,415	58,080	−5,708	−2,904	−408	−2,123	1,372	2,580	11,226	22,908
Dominican Republic	1,832	7,987	2,233	9,289	−249	−975	371	1,848	−280	−429	69	695
Ecuador	3,262	5,263	2,519	4,090	−1,210	−1,319	107	1,101	−360	955	1,009	1,763
Egypt, Arab Rep.	9,151	13,537	13,710	21,109	−912	995	4,836	4,869	−634	−1,708	3,620	15,190
El Salvador	973	3,135	1,624	4,651	−132	−283	631	1,557	−152	−242	595	2,140
Eritrea	88	66	278	597	0	6	171	244	−19	−282	..	..
Estonia	664	3,943	711	4,248	−13	−102	97	112	36	−295	198	856
Ethiopia	672	914	1,069	1,873	−67	−52	220	302	−244	−709	55	467
Finland	31,180	48,507	33,456	37,850	−3,735	−2,482	−952	−1,033	−6,962	7,141	10,415	8,665
France	285,389	382,008	283,238	343,498	−3,896	11,045	−8,199	−12,976	−9,944	36,579	68,291	67,925
Gabon	2,730	2,775	1,812	1,936	−617	−609	−134	−164	168	−256	279	18
Gambia, The	168	246	192	312	−11	−6	59	25	23	−46	55	111
Georgia	..	739	..	1,260	..	119	..	182	..	−220	..	133
Germany	474,713	626,025	423,497	605,298	20,832	−12,708	−23,745	−27,332	48,303	−19,313	104,547	93,407
Ghana	983	2,584	1,506	3,839	−111	−132	411	620	−223	−766	309	535
Greece	13,018	14,863	19,564	25,601	−1,709	−1,632	4,718	7,510	−3,537	−4,860	4,721	19,352
Guatemala	1,568	3,480	1,812	5,016	−196	−205	227	715	−213	−1,026	362	1,252
Guinea	829	791	953	926	−149	−82	70	78	−203	−138	80	200
Guinea-Bissau	26	56	88	80	−22	−14	39	40	−45	−6	18	35
Haiti	318	580	515	1,261	−18	−22	193	223	−22	−38	10	83
Honduras	1,032	2,280	1,127	3,056	−237	−126	280	690	−51	−211	47	1,264
† Data for Taiwan, China	74,175	135,774	67,015	130,396	4,361	2,671	−601	−2,188	10,920	5,861	77,653	110,139

Balance of payments current account | 4.15

	Goods and services				Net income		Net current transfers		Current account balance		Gross international reserves	
	Exports $ millions		Imports $ millions		$ millions		$ millions		$ millions		$ millions	
	1990	1999	1990	1999	1990	1999	1990	1999	1990	1999	1990	1999
Hungary	12,035	27,496	11,017	28,302	−1,427	−1,642	787	347	379	−2,101	1,185	10,983
India	23,028	54,047	31,485	67,250	−1,757	−3,133	2,069	12,638	−8,145	−3,699	5,637	36,005
Indonesia	29,295	55,821	27,511	42,151	−5,190	−9,799	418	1,914	−2,988	5,785	8,657	27,345
Iran, Islamic Rep.	19,741	20,208	22,292	15,593	378	−282	2,500	497	327	−1,897	..	..
Iraq	..	..	..	..	..	..	..	..	..	..	..	..
Ireland	26,786	81,672	24,576	68,887	−4,955	−13,476	2,384	1,285	−361	595	5,362	5,333
Israel	17,312	35,891	20,228	40,816	−1,975	−3,281	5,060	6,324	170	−1,881	6,598	22,605
Italy	219,971	292,293	218,573	269,117	−14,712	−11,492	−3,164	−5,379	−16,479	6,304	88,595	45,301
Jamaica	2,217	3,356	2,390	3,928	−430	−333	291	649	−312	−256	168	555
Japan	323,692	464,692	297,306	395,527	22,492	49,839	−4,800	−12,139	44,078	106,865	87,828	293,948
Jordan	2,511	3,520	3,754	4,979	−215	−155	1,046	2,004	−411	390	1,139	2,770
Kazakhstan	5,758	6,921	5,862	6,749	−175	−500	168	157	−111	−171	..	2,001
Kenya	2,228	2,653	2,705	3,153	−418	−163	368	674	−527	11	236	792
Korea, Dem. Rep.	..	..	..	..	..	..	..	..	..	..	..	..
Korea, Rep.	73,295	171,692	76,360	143,972	−87	−5,159	1,149	1,916	−2,003	24,476	14,916	74,114
Kuwait	8,268	13,964	7,169	12,079	7,738	5,180	−4,951	−2,004	3,886	5,062	2,929	5,561
Kyrgyz Republic	..	528	..	705	..	−75	..	68	..	−185	..	254
Lao PDR	102	468	212	580	−1	−39	56	240	−55	90	8	135
Latvia	1,090	2,913	997	3,605	2	−48	96	93	191	−647	..	913
Lebanon	511	1,817	2,836	8,717	622	323	1,818	2,689	115	−3,888	4,210	10,452
Lesotho	100	216	754	829	433	244	286	148	65	−221	72	500
Libya	11,469	7,335	8,960	5,291	174	311	−481	−219	2,201	2,136	7,225	8,622
Lithuania	..	4,238	..	5,337	..	−258	..	163	..	−1,194	107	1,249
Macedonia, FYR	..	1,441	..	1,926	..	−44	..	420	..	−109	..	460
Madagascar	471	940	809	1,238	−161	−45	234	88	−265	−289	92	227
Malawi	443	592	549	998	−80	−83	99	..	−86	..	142	254
Malaysia	32,665	95,971	31,765	76,140	−1,872	−5,497	102	−1,728	−870	12,606	10,659	30,931
Mali	420	647	830	936	−37	−42	225	126	−221	−178	198	355
Mauritania	471	364	520	413	−46	−26	86	215	−10	140	59	228
Mauritius	1,722	2,660	1,916	2,797	−23	−19	97	104	−119	−52	761	749
Mexico	48,805	148,125	51,915	156,268	−8,316	−12,337	3,975	6,314	−7,451	−14,166	10,217	31,828
Moldova	..	580	..	752	..	39	..	110	..	−23	0	186
Mongolia	493	530	1,096	656	−44	0	7	74	−640	−52	23	137
Morocco	6,239	10,624	7,783	11,960	−988	−986	2,336	2,154	−196	−167	2,338	5,894
Mozambique	229	586	996	1,638	−97	−144	448	313	−415	−429	232	669
Myanmar	641	1,182	1,182	1,798	−61	−18	77	402	−526	−232	410	333
Namibia	1,220	1,601	1,584	1,930	37	62	354	398	28	130	50	305
Nepal	379	1,150	761	1,496	71	25	60	153	−251	−168	354	887
Netherlands	159,304	248,744	147,652	226,669	−620	1,378	−2,943	−6,178	8,089	17,275	34,401	19,262
New Zealand	11,683	16,926	11,699	17,596	−1,576	−3,912	138	242	−1,453	−4,341	4,129	4,455
Nicaragua	392	839	682	2,011	−217	−200	202	785	−305	−587	166	514
Niger	533	288	728	424	−54	−24	14	−14	−236	−174	226	42
Nigeria	14,550	13,855	6,909	12,063	−2,738	−2,578	85	1,292	4,988	506	4,129	6,485
Norway	47,078	54,768	38,911	54,440	−2,700	−898	−1,476	−1,591	3,992	−2,161	15,788	20,744
Oman	5,577	7,236	3,342	5,361	−254	−666	−874	−1,402	1,106	−192	1,784	2,852
Pakistan	6,217	8,838	9,351	11,688	−966	−1,808	2,748	2,471	−1,352	−2,187	1,046	2,117
Panama	4,438	6,888	4,193	7,700	−255	−684	219	164	209	−1,333	344	823
Papua New Guinea	1,381	2,175	1,509	1,800	−103	−273	156	17	−76	120	427	223
Paraguay	2,514	3,261	2,169	3,544	2	1	43	47	390	−235	675	988
Peru	4,120	7,636	4,087	8,853	−1,733	−1,548	281	943	−1,419	−1,822	1,891	9,050
Philippines	11,430	39,012	13,967	36,767	−872	5,171	714	494	−2,695	7,910	2,036	15,029
Poland	19,037	38,522	15,095	52,213	−3,386	−1,010	2,511	2,214	3,067	−12,487	4,674	25,494
Portugal	21,554	34,046	27,146	46,612	−96	−1,540	5,507	3,937	−181	−10,169	20,579	14,510
Puerto Rico	..	..	..	..	..	..	..	..	..	..	..	..
Romania	6,380	9,868	9,901	11,380	161	−411	106	626	−3,254	−1,297	1,374	3,651
Russian Federation	53,883	84,889	48,915	52,571	−4,500	−11,900	..	542	468	20,960	..	12,325

	Goods and services				Net income		Net current transfers		Current account balance		Gross international reserves	
	Exports $ millions		Imports $ millions		$ millions		$ millions		$ millions		$ millions	
	1990	1999	1990	1999	1990	1999	1990	1999	1990	1999	1990	1999
Rwanda	145	98	359	297	–17	–11	145	207	–86	–2	44	174
Saudi Arabia	47,445	56,136	43,939	44,574	7,979	2,925	–15,637	–14,076	–4,152	412	13,437	18,331
Senegal	1,453	1,316	1,840	1,626	–129	–70	153	75	–363	–304	22	411
Sierra Leone	210	73	215	151	–71	–15	7	..	–69	..	5	39
Singapore	67,489	139,333	64,953	123,216	1,006	6,300	–421	–1,163	3,122	21,254	27,748	76,843
Slovak Republic	..	12,101	..	13,154	..	–299	..	198	..	–1,155	..	3,745
Slovenia	7,900	10,522	6,930	11,403	–38	–25	46	123	978	–782	112	3,168
South Africa	27,119	33,320	21,017	30,005	–4,096	–2,852	60	–926	2,065	–464	2,583	7,497
Spain	83,595	164,355	100,870	170,640	–3,533	–9,508	2,799	3,172	–18,009	–12,621	57,238	37,999
Sri Lanka	2,293	5,566	2,965	6,717	–167	–253	541	911	–298	–493	447	1,654
Sudan	532	832	1,453	1,551	–784	–1,292	407	457	–1,299	–1,555	11	189
Sweden	70,560	107,472	70,490	94,471	–4,473	–3,420	–1,936	–3,599	–6,339	5,982	20,324	16,749
Switzerland	96,928	118,985	96,388	106,867	8,746	21,151	–2,329	–4,150	6,957	29,119	61,284	60,492
Syrian Arab Republic	5,030	5,457	2,955	5,202	–401	–543	88	489	1,762	201	..	..
Tajikistan	185	741	238	688	0	–16	..	31	–53	67	..	56
Tanzania	538	1,190	1,474	2,241	–185	–76	562	534	–559	–593	193	775
Thailand	29,229	71,410	35,870	56,345	–853	–2,991	213	353	–7,281	12,428	14,258	34,781
Togo	663	476	847	688	–32	–27	132	90	–84	–140	358	126
Trinidad and Tobago	2,289	3,394	1,427	3,006	–397	–409	–6	22	459	–644	513	962
Tunisia	5,203	8,793	6,039	9,249	–455	–889	828	902	–463	–443	867	2,325
Turkey	21,042	45,724	25,652	48,726	–2,508	–3,537	4,493	5,175	–2,625	–1,364	7,626	24,427
Turkmenistan	1,238	1,376	857	2,046	0	–6	66	105	447	–571	..	1,513
Uganda	246	726	676	1,834	–77	–14	78	375	–429	–746	44	763
Ukraine	..	17,058	..	15,237	..	–869	..	706	..	1,658	469	1,094
United Arab Emirates	..	..	..	..	..	..	..	..	..	..	4,891	10,790
United Kingdom	238,568	373,777	263,985	397,874	–818	14,814	–7,624	–6,697	–33,859	–15,981	43,146	41,834
United States	537,143	956,244	615,990	1,221,218	28,552	–18,483	–26,653	–48,024	–76,948	–331,481	173,094	136,450
Uruguay	2,158	3,586	1,659	4,069	–321	–192	8	70	186	–605	1,446	2,604
Uzbekistan	..	3,170	..	3,059	..	–174	..	49	..	–14	..	..
Venezuela, RB	18,806	22,122	9,451	16,985	–774	–1,518	–302	70	8,279	3,689	12,733	15,110
Vietnam	1,913	14,229	1,901	14,354	–412	–890	49	951	–351	–64	429	2,002
West Bank and Gaza	..	..	..	..	..	..	..	..	..	..	..	..
Yemen, Rep.	1,490	2,615	2,170	3,018	–454	–721	1,872	1,256	739	–228	441	1,486
Yugoslavia, FR (Serb./Mont.)	..	..	..	..	..	..	..	..	..	..	..	..
Zambia	1,360	904	1,897	1,036	–437	–418	380	..	–594	..	201	45
Zimbabwe	2,012	2,525	2,001	2,315	–263	–357	112	..	–140	..	295	480

World	4,252,962 t	7,004,744 t	4,257,720 t	6,999,639 t
Low income	131,098	225,966	148,122	246,302
Middle income	700,877	1,456,089	666,531	1,375,775
Lower middle income	290,650	651,755	302,997	597,718
Upper middle income	409,318	804,287	365,839	777,390
Low & middle income	830,849	1,682,063	813,885	1,622,056
East Asia & Pacific	239,776	675,446	240,892	569,915
Europe & Central Asia	188,457	322,305	186,906	321,322
Latin America & Carib.	169,993	352,429	147,151	378,818
Middle East & N. Africa	134,094	158,908	134,829	152,767
South Asia	34,113	76,267	49,041	96,465
Sub-Saharan Africa	80,657	96,584	74,680	102,676
High income	3,418,188	5,321,940	3,429,968	5,378,612
Europe EMU	1,518,518	2,188,671	1,476,235	2,078,619

a. Includes Luxembourg.

Balance of payments current account | 4.15

The balance of payments records an economy's transactions with the rest of the world. Balance of payments
accounts are divided into two groups: the current
account, which records transactions in goods, services, income, and current transfers; and the capital and
financial account, which records capital transfers, acquisition or disposal of nonproduced, nonfinancial assets,
and transactions in financial assets and liabilities. This
table presents data from the current account with the
addition of gross international reserves.

The balance of payments is a double-entry accounting system that shows all flows of goods and services
into and out of a country; all transfers that are the counterpart of real resources or financial claims provided to
or by the rest of the world without a quid pro quo, such
as donations and grants; and all changes in residents'
claims on, and liabilities to, nonresidents that arise from
economic transactions. All transactions are recorded
twice—once as a credit and once as a debit. In principle the net balance should be zero, but in practice the
accounts often do not balance. In these cases a bal-
ancing item, net errors and omissions, is included.

Discrepancies may arise in the balance of payments because there is no single source for balance
of payments data and therefore no way to ensure
that the data are fully consistent. Sources include customs data, monetary accounts of the banking system,
external debt records, information provided by enterprises, surveys to estimate service transactions, and
foreign exchange records. Differences in collection
methods—such as in timing, definitions of residence
and ownership, and the exchange rate used to value
transactions—contribute to net errors and omissions.
In addition, smuggling and other illegal or quasi-legal
transactions may be unrecorded or misrecorded. For
further discussion of issues relating to the recording
of data on trade in goods and services see *About the
data* for tables 4.4–4.8.

The concepts and definitions underlying the data here
are based on the fifth edition of the International Monetary Fund's (IMF) *Balance of Payments Manual* (1993).
The fifth edition redefined as capital transfers some
transactions previously included in the current account,
such as debt forgiveness, migrants' capital transfers,
and foreign aid to acquire capital goods. Thus the current account balance now reflects more accurately net
current transfer receipts in addition to transactions in
goods, services (previously nonfactor services), and
income (previously factor income). Many countries maintain their data collection systems according to the
fourth edition. Where necessary, the IMF converts data
reported in such systems to conform with the fifth edi-

tion (see *Primary data documentation*). Values are in
U.S. dollars converted at market exchange rates.

The data in this table come from the IMF's Balance
of Payments and International Financial Statistics databases, supplemented by estimates by World Bank staff
for countries whose national accounts are recorded in
fiscal years (see *Primary data documentation*) and
countries for which the IMF does not collect balance of
payments statistics. In addition, World Bank staff make
estimates of missing data for the most recent year.

Figure 4.15

The top recipients of workers' remittances
$ billions

India
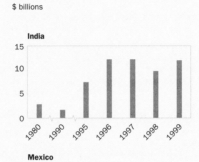

Mexico

Turkey
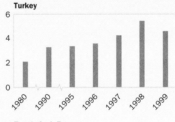

Egypt, Arab Rep.
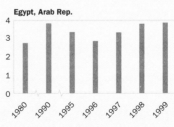

Source: International Monetary Fund, Balance of
Payments data files.

**Workers' remittances to India and Mexico have
increased dramatically over the past 20 years.**

• **Exports and imports of goods and services** comprise all transactions between residents of a country
and the rest of the world involving a change in ownership of general merchandise, goods sent for processing
and repairs, nonmonetary gold, and services. • **Net
income** refers to receipts and payments of employee
compensation to nonresident workers, and investment
income (receipts and payments on direct investment,
portfolio investment, and other investments, and
receipts on reserve assets). Income derived from the
use of intangible assets is recorded under business
services. • **Net current transfers** are recorded in
the balance of payments whenever an economy provides or receives goods, services, income, or financial
items without a quid pro quo. All transfers not considered to be capital are current. • **Current account
balance** is the sum of net exports of goods and services, net income, and net current transfers. • **Gross
international reserves** comprise holdings of monetary
gold, special drawing rights, reserves of IMF members held by the IMF, and holdings of foreign exchange
under the control of monetary authorities. The gold component of these reserves is valued at year-end (31
December) London prices ($385 an ounce in 1990 and
$290.25 an ounce in 1999).

More information about the design and compilation of
the balance of payments can be found in the IMF's *Balance of Payments Manual,* fifth edition (1993), *Balance
of Payments Textbook* (1996a), and *Balance of Payments Compilation Guide* (1995). The balance of payments data are published in the IMF's *Balance of
Payments Statistics Yearbook* and *International Financial Statistics.* The World Bank exchanges data with
the IMF through electronic files that in most cases are
more timely and cover a longer period than the published sources. The IMF's International Financial Statistics and Balance of Payments databases are
available on CD-ROM.

	Total external debt		Long-term debt		Public and publicly guaranteed debt				Private nonguaranteed external debt		Use of IMF credit	
					Total		IBRD loans and IDA credits					
	$ millions		$ millions		$ millions		$ millions		$ millions		$ millions	
	1990	1999	1990	1999	1990	1999	1990	1999	1990	1999	1990	1999
Albania	349	975	36	865	36	849	0	296	0	16	0	81
Algeria	27,877	28,015	26,416	25,913	26,416	25,913	1,208	1,540	0	0	670	1,906
Angola	8,594	10,871	7,605	9,248	7,605	9,248	0	214	0	0	0	0
Argentina	62,232	147,880	48,676	111,887	46,876	84,568	2,609	8,314	1,800	27,320	3,083	4,478
Armenia	..	932	..	682	..	682	..	361	..	0	..	201
Australia	..	..	..	..	..	..	..	..	..	..	..	..
Austria	..	..	..	..	..	..	..	..	..	..	..	..
Azerbaijan	..	1,036	..	600	..	493	..	199	..	107	..	407
Bangladesh	12,768	17,534	11,987	16,962	11,987	16,962	4,159	6,459	0	0	626	318
Belarus	..	1,136	..	864	..	851	..	122	..	13	..	178
Belgium	..	..	..	..	..	..	..	..	..	..	..	..
Benin	1,292	1,686	1,218	1,472	1,218	1,472	326	574	0	0	18	92
Bolivia	4,275	6,157	3,864	4,508	3,687	3,864	587	1,110	177	643	257	247
Bosnia and Herzegovina	..	1,962	..	1,828	..	1,826	..	951	..	3	..	94
Botswana	561	462	556	442	556	442	169	34	0	0	0	0
Brazil	119,877	244,673	94,340	206,326	87,669	95,233	8,427	6,822	6,671	111,093	1,821	8,827
Bulgaria	10,865	9,872	9,809	8,246	9,809	7,602	0	829	0	644	0	1,250
Burkina Faso	834	1,518	750	1,295	750	1,295	282	753	0	0	0	121
Burundi	907	1,131	851	1,050	851	1,050	398	599	0	0	43	12
Cambodia	1,854	2,262	1,688	2,136	1,688	2,136	0	180	0	0	27	73
Cameroon	6,676	9,443	5,595	7,969	5,365	7,614	889	1,025	230	355	121	196
Canada	..	..	..	..	..	..	..	..	..	..	..	..
Central African Republic	698	913	624	830	624	830	265	403	0	0	37	24
Chad	524	1,142	464	1,045	464	1,045	186	527	0	0	31	69
Chile	19,226	37,762	14,687	32,269	10,425	5,655	1,874	885	4,263	26,614	1,156	0
China	55,301	154,223	45,515	136,541	45,515	108,163	5,881	19,308	0	28,378	469	0
Hong Kong, China	..	..	..	..	..	..	..	..	..	..	..	..
Colombia	17,222	34,538	15,784	30,572	14,671	19,434	3,874	1,968	1,113	11,139	0	0
Congo, Dem. Rep.	10,274	11,906	9,010	8,188	9,010	8,188	1,161	1,318	0	0	521	412
Congo, Rep.	4,947	5,031	4,200	3,932	4,200	3,932	239	232	0	0	11	29
Costa Rica	3,756	4,182	3,367	3,402	3,063	3,186	412	151	304	216	11	0
Côte d'Ivoire	17,251	13,170	13,223	11,295	10,665	9,699	1,920	2,068	2,558	1,596	431	620
Croatia	..	9,443	..	8,555	..	5,443	..	387	..	3,112	..	197
Cuba	..	..	..	..	..	..	..	..	..	..	..	..
Czech Republic	6,383	22,582	3,983	15,317	3,983	13,440	0	324	0	1,878	0	0
Denmark	..	..	..	..	..	..	..	..	..	..	..	..
Dominican Republic	4,372	4,771	3,518	3,665	3,420	3,665	258	290	99	0	72	55
Ecuador	12,108	14,506	10,029	13,259	9,866	12,756	848	883	164	503	265	0
Egypt, Arab Rep.	32,949	30,404	28,372	26,110	27,372	25,998	2,401	2,034	1,000	112	125	0
El Salvador	2,149	4,014	1,938	2,961	1,913	2,649	164	310	26	312	0	0
Eritrea	..	254	..	254	..	254	..	56	..	0	..	0
Estonia	..	2,879	..	1,612	..	206	..	88	..	1,407	..	25
Ethiopia	8,630	5,551	8,479	5,360	8,479	5,360	851	1,739	0	0	6	95
Finland	..	..	..	..	..	..	..	..	..	..	..	..
France	..	..	..	..	..	..	..	..	..	..	..	..
Gabon	3,983	3,978	3,150	3,290	3,150	3,290	69	68	0	0	140	86
Gambia, The	369	459	308	425	308	425	102	173	0	0	45	11
Georgia	..	1,652	..	1,325	..	1,308	..	346	..	17	..	320
Germany	..	..	..	..	..	..	..	..	..	..	..	..
Ghana	3,881	6,928	2,816	5,907	2,783	5,647	1,423	3,117	33	260	745	310
Greece	..	..	..	..	..	..	..	..	..	..	..	..
Guatemala	3,080	4,660	2,605	3,290	2,478	3,129	293	258	127	162	67	0
Guinea	2,476	3,518	2,253	3,057	2,253	3,057	420	1,014	0	0	52	127
Guinea-Bissau	692	931	630	832	630	832	146	228	0	0	5	17
Haiti	935	1,190	797	1,049	797	1,049	324	504	0	0	38	45
Honduras	3,718	5,333	3,487	4,670	3,420	4,231	635	1,027	66	439	32	210

	Total external debt		Long-term debt		Public and publicly guaranteed debt				Private nonguaranteed external debt		Use of IMF credit	
					Total $ millions		IBRD loans and IDA credits $ millions					
	$ millions		$ millions						$ millions		$ millions	
	1990	1999	1990	1999	1990	1999	1990	1999	1990	1999	1990	1999
Hungary	21,202	29,042	17,931	25,499	17,931	16,064	1,512	654	0	9,436	330	0
India	83,717	94,393	72,550	90,324	71,062	82,380	20,996	26,746	1,488	7,944	2,623	26
Indonesia	69,872	150,096	58,242	119,819	47,982	72,554	10,385	12,106	10,261	47,265	494	10,248
Iran, Islamic Rep.	9,020	10,357	1,797	6,739	1,797	6,183	86	437	0	556	0	0
Iraq	..	..	..	..	..	..	..	..	..	..	..	..
Ireland	..	..	..	..	..	..	..	..	..	..	..	..
Israel	..	..	..	..	..	..	..	..	..	..	..	..
Italy	..	..	..	..	..	..	..	..	..	..	..	..
Jamaica	4,674	3,913	3,970	3,071	3,937	2,905	672	393	34	166	357	83
Japan	..	..	..	..	..	..	..	..	..	..	..	..
Jordan	8,177	8,947	7,043	7,574	7,043	7,546	593	895	0	28	94	498
Kazakhstan	..	6,182	..	5,248	..	3,413	..	1,082	..	1,835	..	460
Kenya	7,058	6,562	5,642	5,604	4,762	5,385	2,056	2,311	880	220	482	132
Korea, Dem. Rep.	..	..	..	..	..	..	..	..	..	..	..	..
Korea, Rep.	34,968	129,784	24,168	88,916	18,768	57,231	3,337	8,358	5,400	31,685	0	6,125
Kuwait	..	..	..	..	..	..	..	..	..	..	..	..
Kyrgyz Republic	..	1,699	..	1,449	..	1,130	..	342	..	319	..	190
Lao PDR	1,768	2,527	1,758	2,471	1,758	2,471	131	405	0	0	8	53
Latvia	..	2,657	..	1,499	..	865	..	200	..	634	..	47
Lebanon	1,779	8,441	358	6,239	358	5,568	34	234	0	671	0	0
Lesotho	396	686	378	662	378	662	112	241	0	0	15	17
Libya	..	..	..	..	..	..	..	..	..	..	..	..
Lithuania	..	3,584	..	2,806	..	1,892	..	200	..	915	..	230
Macedonia, FYR	..	1,433	..	1,264	..	1,135	..	333	..	129	..	102
Madagascar	3,704	4,409	3,335	4,023	3,335	4,023	797	1,361	0	0	144	63
Malawi	1,558	2,751	1,385	2,596	1,382	2,596	854	1,603	3	0	115	88
Malaysia	15,328	45,939	13,422	38,390	11,592	18,929	1,102	900	1,830	19,460	0	0
Mali	2,467	3,183	2,336	2,798	2,336	2,798	498	1,035	0	0	69	193
Mauritania	2,096	2,528	1,789	2,138	1,789	2,138	264	417	0	0	70	107
Mauritius	984	2,464	910	1,891	762	1,155	195	122	148	736	22	0
Mexico	104,442	166,960	81,809	138,424	75,974	87,531	11,030	11,027	5,835	50,893	6,551	4,473
Moldova	..	943	..	736	..	722	..	276	..	14	..	175
Mongolia	..	891	..	816	..	816	0	130	..	0	0	51
Morocco	24,458	19,060	23,301	18,877	23,101	17,284	3,138	3,221	200	1,593	750	0
Mozambique	4,650	6,959	4,231	6,372	4,211	4,625	268	702	19	1,747	74	200
Myanmar	4,695	5,999	4,466	5,333	4,466	5,333	716	722	0	0	0	0
Namibia	..	..	..	..	..	..	..	..	..	..	..	..
Nepal	1,640	2,970	1,572	2,910	1,572	2,910	668	1,147	0	0	44	18
Netherlands	..	..	..	..	..	..	..	..	..	..	..	..
New Zealand	..	..	..	..	..	..	..	..	..	..	..	..
Nicaragua	10,707	6,986	8,281	5,905	8,281	5,799	299	607	0	106	0	155
Niger	1,726	1,621	1,487	1,473	1,226	1,424	461	694	261	49	85	68
Nigeria	33,439	29,358	31,935	22,673	31,545	22,423	3,321	2,613	391	250	0	0
Norway	..	..	..	..	..	..	..	..	..	..	..	..
Oman	2,736	3,603	2,400	1,768	2,400	1,768	52	6	0	0	0	0
Pakistan	20,663	34,423	16,643	30,816	16,506	28,594	3,922	7,220	138	2,221	836	1,704
Panama	6,678	7,313	3,988	6,415	3,988	5,783	462	288	0	632	272	149
Papua New Guinea	2,594	2,847	2,461	2,576	1,523	1,517	349	337	938	1,059	61	22
Paraguay	2,105	2,514	1,732	1,762	1,713	1,672	320	212	19	91	0	0
Peru	20,064	32,284	13,959	25,194	13,629	20,709	1,188	2,417	330	4,485	755	735
Philippines	30,580	52,022	25,241	44,454	24,040	33,568	4,044	4,246	1,201	10,886	912	1,822
Poland	49,366	54,268	39,263	48,325	39,263	33,151	55	2,185	0	15,174	509	0
Portugal	..	..	..	..	..	..	..	..	..	..	..	..
Puerto Rico	..	..	..	..	..	..	..	..	..	..	..	..
Romania	1,140	9,367	230	7,968	223	5,985	0	1,662	7	1,984	0	458
Russian Federation	59,340	173,940	47,540	142,958	47,540	120,375	0	6,707	0	22,583	0	15,238

	Total external debt		Long-term debt		Public and publicly guaranteed debt				Private nonguaranteed external debt		Use of IMF credit	
					Total $ millions		IBRD loans and IDA credits $ millions					
	$ millions		$ millions		$ millions		$ millions		$ millions		$ millions	
	1990	1999	1990	1999	1990	1999	1990	1999	1990	1999	1990	1999
Rwanda	712	1,292	664	1,162	664	1,162	340	692	0	0	0	76
Saudi Arabia	..	..	..	..	..	..	..	..	..	..	..	..
Senegal	3,736	3,705	3,000	3,125	2,940	3,111	835	1,314	60	14	314	272
Sierra Leone	1,151	1,249	604	938	604	938	92	300	0	0	108	195
Singapore	..	..	..	..	..	..	..	..	..	..	..	..
Slovak Republic	2,008	9,150	1,505	7,440	1,505	4,457	0	219	0	2,983	0	133
Slovenia	..	..	..	..	..	..	..	..	..	..	..	..
South Africa	..	24,158	..	10,378	..	9,148	0	1	..	1,230	0	0
Spain	..	..	..	..	..	..	..	..	..	..	..	..
Sri Lanka	5,863	9,472	5,048	8,268	4,947	8,182	946	1,671	102	86	410	258
Sudan	14,762	16,132	9,651	9,348	9,155	8,852	1,048	1,211	496	496	956	715
Sweden	..	..	..	..	..	..	..	..	..	..	..	..
Switzerland	..	..	..	..	..	..	..	..	..	..	..	..
Syrian Arab Republic	17,053	22,369	14,902	16,142	14,902	16,142	523	73	0	0	0	0
Tajikistan	..	889	..	697	..	595	..	126	..	102	..	101
Tanzania	6,451	7,967	5,793	6,628	5,781	6,595	1,493	2,610	12	32	140	312
Thailand	28,165	96,335	19,842	69,486	12,531	31,011	2,530	2,816	7,311	38,475	1	3,431
Togo	1,275	1,500	1,075	1,262	1,075	1,262	398	622	0	0	87	83
Trinidad and Tobago	2,512	2,462	2,055	1,629	1,782	1,485	41	85	273	144	329	0
Tunisia	7,690	11,872	6,880	10,259	6,662	9,487	1,406	1,364	218	772	176	76
Turkey	49,424	101,796	39,924	77,433	38,870	50,095	6,429	3,009	1,054	27,338	0	890
Turkmenistan	..	2,015	..	1,692	..	1,678	..	9	..	14	..	0
Uganda	2,583	4,077	2,161	3,564	2,161	3,564	969	2,043	0	0	282	372
Ukraine	..	14,136	..	11,015	..	10,027	..	1,954	..	988	..	2,806
United Arab Emirates	..	..	..	..	..	..	..	..	..	..	..	..
United Kingdom	..	..	..	..	..	..	..	..	..	..	..	..
United States	..	..	..	..	..	..	..	..	..	..	..	..
Uruguay	4,415	7,447	3,114	5,496	3,045	5,108	359	476	69	389	101	157
Uzbekistan	..	4,163	..	3,245	..	2,920	..	203	..	325	..	202
Venezuela, RB	33,170	35,852	28,159	32,842	24,509	25,216	974	1,130	3,650	7,627	3,012	741
Vietnam	23,270	23,260	21,378	20,529	21,378	20,529	59	989	0	0	112	355
West Bank and Gaza	..	..	..	..	..	..	..	..	..	..	..	..
Yemen, Rep.	6,352	4,610	5,160	3,729	5,160	3,729	602	1,216	0	0	0	409
Yugoslavia, FR (Serb./Mont.)[a]	17,792	12,949	16,802	10,175	12,942	7,416	2,433	468	3,860	2,759	467	76
Zambia	6,916	5,853	4,554	4,571	4,552	4,498	813	1,736	2	73	949	1,171
Zimbabwe	3,247	4,566	2,649	3,451	2,464	3,211	449	925	185	240	7	369
World	.. s	.. s	.. s	.. s	.. s	.. s	.. s	.. s	.. s	.. s	.. s	.. s
Low income	418,954	572,468	357,695	485,071	340,680	420,284	66,693	99,552	17,015	64,787	11,251	24,887
Middle income[b]	1,040,914	2,000,146	822,425	1,601,283	773,927	1,129,627	73,982	106,080	48,498	471,656	23,401	54,001
Lower middle income	525,037	978,009	430,190	787,459	411,736	627,249	41,634	63,424	18,454	160,210	6,074	28,609
Upper middle income[b]	515,877	1,022,137	392,236	813,825	362,191	502,378	32,348	42,656	30,044	311,446	17,327	25,392
Low & middle income[b]	1,459,868	2,572,614	1,180,120	2,086,354	1,114,607	1,549,912	140,675	205,632	65,513	536,443	34,652	78,887
East Asia & Pacific	274,028	674,846	222,768	539,148	195,732	361,881	28,644	50,628	27,035	177,267	2,085	22,180
Europe & Central Asia	219,826	486,066	177,663	390,845	172,742	296,118	10,429	23,531	4,921	94,727	1,305	23,862
Latin America & Carib.	475,443	813,828	379,716	672,956	354,665	429,982	35,877	39,467	25,051	242,974	18,298	20,498
Middle East & N. Africa	183,798	216,763	137,615	163,240	136,113	159,314	10,074	11,068	1,502	3,926	1,815	2,901
South Asia	129,899	164,753	112,990	155,211	111,263	144,959	30,716	43,312	1,727	10,251	4,537	2,323
Sub-Saharan Africa	176,874	216,359	149,368	164,954	144,092	157,658	24,935	37,626	5,276	7,297	6,612	7,124
High income												
Europe EMU												

a. Data for 1990 refer to the former Socialist Federal Republic of Yugoslavia. Data for 1999 are estimates and reflect borrowings by the former Socialist Federal Republic of Yugoslavia that are not yet allocated to the successor republics. b. Includes data for Gibraltar not included in other tables.

About the data

Data on the external debt of low- and middle-income economies are gathered by the World Bank through its Debtor Reporting System. World Bank staff calculate the indebtedness of developing countries using loan-by-loan reports submitted by these countries on long-term public and publicly guaranteed borrowing, along with information on short-term debt collected by the countries or collected from creditors through the reporting systems of the Bank for International Settlements and the Organisation for Economic Co-operation and Development. These data are supplemented by information on loans and credits from major multilateral banks, loan statements from official lending agencies in major creditor countries, and estimates from World Bank and International Monetary Fund (IMF) staff. In addition, data on private nonguaranteed debt for 76 countries either reported to the World Bank or estimated by Bank staff are included.

The coverage, quality, and timeliness of debt data vary across countries. Coverage varies for both debt instruments and borrowers. With the widening spectrum of debt instruments and investors and the expansion of private nonguaranteed borrowing, comprehensive coverage of long-term external debt becomes more complex. Reporting countries differ in their capacity to monitor debt, especially private nonguaranteed debt. Even data on public and publicly guaranteed debt are affected by coverage and accuracy in reporting—again because of monitoring capacity and sometimes because of unwillingness to provide information. A key part often underreported is military debt.

Because debt data are normally reported in the currency of repayment, they have to be converted into U.S. dollars to produce summary tables. Stock figures (amount of debt outstanding) are converted using end-period exchange rates, as published in the IMF's *International Financial Statistics* (line ae). Flow figures are converted at annual average exchange rates (line rf). Projected debt service is converted using end-period exchange rates. Debt repayable in multiple currencies, goods, or services and debt with a provision for maintenance of value of the currency of repayment are shown at book value.

Because flow data are converted at annual average exchange rates and stock data at year-end exchange rates, year-to-year changes in debt outstanding and disbursed are sometimes not equal to net flows (disbursements less principal repayments); similarly, changes in debt outstanding including undisbursed debt differ from commitments less repayments. Discrepancies are particularly significant when exchange rates have moved sharply during the year. Cancellations and reschedulings of other lia-

bilities into long-term public debt also contribute to the differences.

Variations in reporting rescheduled debt also affect cross-country comparability. For example, rescheduling under the auspices of the Paris Club of official creditors may be subject to lags between the completion of the general rescheduling agreement and the completion of the specific, bilateral agreements that define the terms of the rescheduled debt. Other areas of inconsistency include country treatment of arrears and of nonresident national deposits denominated in foreign currency.

Figure 4.16

Liabilities to the World Bank increased in most regions in 1999
Debt outstanding and disbursed ($ billions)

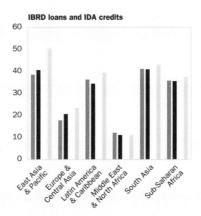

IBRD loans and IDA credits

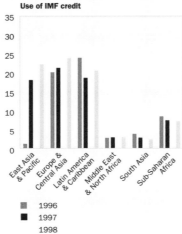

Use of IMF credit

- 1996
- 1997
- 1998
- 1999

Source: World Bank data files.

The World Bank's liabilities in Sub-Saharan Africa declined slightly in 1999 (by 1 percent). The International Monetary Fund's liabilities remained concentrated in three regions.

Definitions

- **Total external debt** is debt owed to nonresidents repayable in foreign currency, goods, or services. It is the sum of public, publicly guaranteed, and private nonguaranteed long-term debt, use of IMF credit, and short-term debt. Short-term debt includes all debt having an original maturity of one year or less and interest in arrears on long-term debt. • **Long-term debt** is debt that has an original or extended maturity of more than one year. It has three components: public, publicly guaranteed, and private nonguaranteed debt. • **Public and publicly guaranteed debt** comprises long-term external obligations of public debtors, including the national government and political subdivisions (or an agency of either) and autonomous public bodies, and external obligations of private debtors that are guaranteed for repayment by a public entity. • **IBRD loans and IDA credits** are extended by the World Bank Group. The International Bank for Reconstruction and Development (IBRD) lends at market rates. Credits from the International Development Association (IDA) are at concessional rates. • **Private nonguaranteed external debt** comprises long-term external obligations of private debtors that are not guaranteed for repayment by a public entity. • **Use of IMF credit** denotes repurchase obligations to the IMF for all uses of IMF resources (excluding those resulting from drawings on the reserve tranche). These obligations, shown for the end of the year specified, comprise purchases outstanding under the credit tranches, including enlarged access resources, and all special facilities (the buffer stock, compensatory financing, extended fund, and oil facilities), trust fund loans, and operations under the structural adjustment and enhanced structural adjustment facilities.

Data sources

The main sources of external debt information are reports to the World Bank through its Debtor Reporting System from member countries that have received IBRD loans or IDA credits. Additional information has been drawn from the files of the World Bank and the IMF. Summary tables of the external debt of developing countries are published annually in the World Bank's *Global Development Finance* and on its *Global Development Finance* CD-ROM.

4.17 External debt management

	Indebtedness classification[a]	Present value of debt		Total debt service				Public and publicly guaranteed debt service		Short-term debt	
		% of GNI	% of exports of goods and services	% of GNI		% of exports of goods and services		% of central government current revenue		% of total debt	
	1999	1999	1999	1990	1999	1990	1999	1990	1999	1990	1999
Albania	L	18	67	0.1	1.0	0.9	3.7	..	..	89.8	3.0
Algeria	M	63	206	14.7	11.7	63.4	37.8	..	34.2	2.8	0.7
Angola	S	286	156	4.0	38.6	8.1	21.1	..	..	11.5	14.9
Argentina	S	56	456	4.6	9.3	37.0	75.9	32.5	..	16.8	21.3
Armenia	M	..	135	..	..	..	11.9	..	..	..	5.3
Australia	..	..	..	..	..	..	..	..	..	..	..
Austria	..	..	..	..	..	..	..	..	..	..	..
Azerbaijan	L	22	57	..	2.5	..	6.5	..	6.2	..	2.8
Bangladesh	M	23	140	2.5	1.7	28.4	10.1	..	..	1.2	1.5
Belarus	L	4	16	..	0.8	..	3.2	..	..	..	8.2
Belgium	..	..	..	..	..	..	..	..	..	..	..
Benin	S	40[b]	148[b]	2.1	2.9	8.2	10.9	..	..	4.3	7.2
Bolivia	S	37[b]	193[b]	8.3	6.1	38.6	32.0	41.3	17.3	3.6	22.8
Bosnia and Herzegovina	S	35	..	..	8.9	..	..	..	..	..	2.0
Botswana	L	7	11	2.9	1.5	4.4	2.4	5.5	..	1.0	4.3
Brazil	S	33	399	1.8	9.2	22.2	110.9	3.9	..	19.8	12.1
Bulgaria	S	77	157	7.2	9.3	19.4	19.1	12.9	14.5	9.7	3.8
Burkina Faso	M	25[b]	158[b]	1.2	2.5	6.8	15.7	9.1	..	10.1	6.8
Burundi	S	96	1,072	3.8	4.1	43.4	45.6	..	15.7	1.5	6.2
Cambodia	M	61	161	2.7	1.1	..	2.9	..	..	7.5	2.3
Cameroon	S	76	292	4.9	6.3	22.5	24.3	16.8	24.6	14.4	13.5
Canada	..	..	..	..	..	..	..	..	..	..	..
Central African Republic	S	54	365	2.0	1.8	13.2	12.1	..	..	5.4	6.5
Chad	M	43	208	0.7	2.1	4.4	10.3	5.6	..	5.7	2.5
Chile	M	53	175	9.7	7.7	25.9	25.4	25.6	5.1	17.6	14.5
China	L	14	59	2.0	2.1	11.7	9.0	23.9	..	16.8	11.5
Hong Kong, China	..	..	..	..	..	..	..	..	..	..	..
Colombia	M	40	218	10.2	7.9	40.9	42.9	61.2	44.7	8.4	11.5
Congo, Dem. Rep.	S	..	..	4.1	..	13.5	..	0.0	..	7.2	27.8
Congo, Rep.	S	287	265	22.9	1.5	35.3	1.4	..	0.0	14.9	21.3
Costa Rica	L	30	48	7.2	4.0	23.9	6.4	32.8	15.1	10.0	18.7
Côte d'Ivoire	S	117[b]	220[b]	13.7	13.9	35.4	26.2	22.1	42.3	20.8	9.5
Croatia	L	47	106	..	8.5	..	19.4	..	7.7	..	7.3
Cuba	..	..	..	..	..	..	..	..	..	..	..
Czech Republic	L	43	64	..	6.9	..	10.3	..	13.7	37.6	32.2
Denmark	..	..	..	..	..	..	..	..	..	..	..
Dominican Republic	L	28	47	3.4	2.3	10.4	3.9	16.1	..	17.9	22.0
Ecuador	M	76	211	11.1	9.2	32.5	25.7	45.1	..	15.0	8.6
Egypt, Arab Rep.	L	27	127	7.3	1.9	22.3	9.0	16.3	..	13.5	14.1
El Salvador	L	31	81	4.4	2.9	15.3	7.6	..	13.6	9.8	26.2
Eritrea	L	19	71	..	0.5	..	1.9	..	..	..	0.1
Estonia	M	54	68	..	10.5	..	13.2	..	3.8	..	43.1
Ethiopia	S	55	374	3.5	2.5	34.9	16.8	13.4	..	1.7	1.7
Finland	..	..	..	..	..	..	..	..	..	..	..
France	..	..	..	..	..	..	..	..	..	..	..
Gabon	S	108	148	3.3	14.1	6.4	19.3	7.6	..	17.4	15.1
Gambia, The	M	67	103	12.9	5.5	22.2	8.5	49.1	..	4.3	4.8
Georgia	M	45	136	..	3.8	..	11.4	..	23.9	..	0.4
Germany	..	..	..	..	..	..	..	..	..	..	..
Ghana	M	66[b]	190[b]	6.4	6.9	36.9	19.9	26.2	..	8.2	10.3
Greece	..	..	..	..	..	..	..	..	..	..	..
Guatemala	L	24	109	2.9	2.3	12.6	10.3	..	..	13.3	29.4
Guinea	S	67	294	6.3	3.7	20.0	16.1	33.0	27.5	6.9	9.5
Guinea-Bissau	S	347	1,222	3.6	4.7	31.0	16.4	..	..	8.2	8.8
Haiti	M	17	124	1.1	1.4	10.1	10.0	..	..	10.8	8.1
Honduras	M	63	122	13.7	7.0	35.3	13.5	..	..	5.4	8.5

External debt management | 4.17

	Indebtedness classification[a]	Present value of debt		Total debt service				Public and publicly guaranteed debt service		Short-term debt	
		% of GNI	% of exports of goods and services	% of GNI		% of exports of goods and services		% of central government current revenue		% of total debt	
	1999	1999	1999	1990	1999	1990	1999	1990	1999	1990	1999
Hungary	M	60	99	13.4	16.1	34.3	26.6	21.4	17.7	13.9	12.2
India	L	16	104	2.6	2.3	32.7	15.0	14.5	14.2	10.2	4.3
Indonesia	S	113	255	9.1	13.5	33.3	30.3	34.4	35.3	15.9	13.3
Iran, Islamic Rep.	L	8	42	0.5	4.2	3.2	22.6	0.3	5.1	80.1	34.9
Iraq	..	..	..	..	..	..	..	..	..	..	..
Ireland	..	..	..	..	..	..	..	..	..	..	..
Israel	..	..	..	..	..	..	..	..	..	..	..
Italy	..	..	..	..	..	..	..	..	..	..	..
Jamaica	M	59	92	17.4	11.2	26.9	17.4	..	..	7.4	19.4
Japan	..	..	..	..	..	..	..	..	..	..	..
Jordan	S	104	150	16.4	8.2	20.3	11.8	52.1	25.9	12.7	9.8
Kazakhstan	L	41	89	..	8.9	..	19.4	..	36.2	..	7.7
Kenya	M	49	193	9.8	6.8	35.4	26.7	26.6	..	13.2	12.6
Korea, Dem. Rep.	..	..	..	..	..	..	..	..	..	..	..
Korea, Rep.	L	31	71	3.3	10.7	10.8	24.6	10.5	..	30.9	26.8
Kuwait	..	..	..	..	..	..	..	..	..	..	..
Kyrgyz Republic	S	104	228	..	10.0	..	21.8	..	10.0	..	3.5
Lao PDR	S	100	290	1.1	2.6	8.7	7.7	..	..	0.1	0.1
Latvia	L	39	79	..	7.4	..	15.0	..	2.0	..	41.8
Lebanon	M	52	..	2.9	6.1	3.3	..	..	20.2	79.9	26.1
Lesotho	L	45	92	2.3	4.6	4.2	9.4	9.4	..	0.7	1.1
Libya	..	..	..	..	..	..	..	..	..	..	..
Lithuania	L	34	80	..	2.7	..	6.3	..	6.0	..	15.3
Macedonia, FYR	L	37	83	..	13.4	..	29.9	..	..	..	4.7
Madagascar	S	80	304	7.6	4.5	45.5	17.1	42.9	..	6.1	7.3
Malawi	S	84[b]	246[b]	7.5	3.9	29.3	11.4	27.2	..	3.7	2.5
Malaysia	M	64	48	10.3	6.4	12.6	4.8	31.4	..	12.4	16.4
Mali	S	56[b]	193[b]	2.8	4.1	12.3	14.3	..	..	2.5	6.0
Mauritania	S	169	422	13.6	11.4	29.9	28.4	..	..	11.3	11.2
Mauritius	M	61	95	5.9	6.3	8.8	9.7	13.5	17.5	5.3	23.3
Mexico	L	37	108	4.5	8.5	20.7	25.1	19.5	..	15.4	14.4
Moldova	M	74	126	..	14.6	..	24.9	..	33.6	..	3.4
Mongolia	M	59	93	..	3.0	..	4.8	..	10.3	..	2.6
Morocco	M	51	135	7.2	9.1	21.5	24.4	21.3	..	1.7	1.0
Mozambique	M	28[b]	167[b]	3.3	3.3	26.2	20.0	..	..	7.4	5.6
Myanmar	S	..	369	..	..	9.0	7.9	2.2	..	4.9	11.1
Namibia	..	..	..	..	..	..	..	..	..	..	..
Nepal	L	32	122	1.9	2.1	13.4	7.9	18.2	19.4	1.5	1.4
Netherlands	..	..	..	..	..	..	..	..	..	..	..
New Zealand	..	..	..	..	..	..	..	..	..	..	..
Nicaragua	S	271[b]	475[b]	1.6	9.1	3.9	16.1	2.6	..	22.7	13.3
Niger	S	55[b]	362[b]	4.1	2.5	17.4	16.8	..	..	8.9	4.9
Nigeria	S	91	185	13.0	2.9	22.6	6.0	..	..	4.5	22.8
Norway	..	..	..	..	..	..	..	..	..	..	..
Oman	L	..	48	7.8	..	12.3	9.7	17.4	19.5	12.3	50.9
Pakistan	S	43	252	4.6	5.2	23.0	30.5	18.1	17.8	15.4	5.5
Panama	S	77	81	6.9	8.4	6.2	8.8	10.4	..	36.2	10.2
Papua New Guinea	S	77	120	17.9	6.2	37.2	9.6	33.2	23.5	2.8	8.8
Paraguay	L	31	68	6.0	3.0	12.2	6.6	46.8	..	17.7	29.9
Peru	S	63	354	1.9	5.8	10.8	32.7	4.9	22.9	26.7	19.7
Philippines	M	65	110	8.1	8.4	27.0	14.3	39.5	41.7	14.5	11.0
Poland	L	33	125	1.7	5.4	4.9	20.4	..	4.3	19.4	11.0
Portugal	..	..	..	..	..	..	..	..	..	..	..
Puerto Rico	..	..	..	..	..	..	..	..	..	..	..
Romania	L	27	90	0.0	9.3	0.3	31.3	0.0	..	79.8	10.0
Russian Federation	M	35	153	2.0	3.1	..	13.5	..	10.7	19.9	9.1

	Indebtedness classification[a]	Present value of debt		Total debt service				Public and publicly guaranteed debt service		Short-term debt	
		% of GNI	% of exports of goods and services	% of GNI		% of exports of goods and services		% of central government current revenue		% of total debt	
	1999	1999	1999	1990	1999	1990	1999	1990	1999	1990	1999
Rwanda	S	36	655	0.8	1.6	14.0	29.6	5.4	..	6.6	4.2
Saudi Arabia	..	..	..	..	..	..	..	..	..	..	..
Senegal	M	53	169	5.9	5.0	20.0	16.1	..	..	11.3	8.3
Sierra Leone	S	136	1,234	2.7	3.3	10.1	29.9	30.6	..	38.1	9.3
Singapore	..	..	..	..	..	..	..	..	..	..	..
Slovak Republic	L	44	69	2.1	8.8	..	13.9	..	8.8	25.0	17.2
Slovenia	..	..	..	..	..	..	..	..	..	..	..
South Africa	L	19	70	..	3.8	..	13.9	..	8.5	..	57.0
Spain	..	..	..	..	..	..	..	..	..	..	..
Sri Lanka	L	45	104	4.8	3.4	13.7	7.9	16.7	14.4	6.9	10.0
Sudan	S	172	1,717	0.4	0.6	7.5	6.5	..	..	28.1	37.6
Sweden	..	..	..	..	..	..	..	..	..	..	..
Switzerland	..	..	..	..	..	..	..	..	..	..	..
Syrian Arab Republic	S	146	377	11.0	2.5	23.2	6.4	22.7	..	12.6	27.8
Tajikistan	L	37	92	..	2.6	..	6.5	..	..	..	10.2
Tanzania[c]	S	53	370	4.4	2.2	32.9	15.6	..	..	8.0	12.9
Thailand	M	78	127	6.3	13.6	16.9	22.0	20.7	21.7	29.5	24.3
Togo	M	82	216	5.4	2.9	11.9	7.7	..	..	8.9	10.3
Trinidad and Tobago	L	39	72	9.7	7.0	19.3	13.1	..	..	5.1	33.8
Tunisia	M	59	122	12.0	7.6	24.5	15.9	32.2	22.4	8.2	13.0
Turkey	M	52	185	4.9	7.4	29.4	26.2	30.9	18.1	19.2	23.1
Turkmenistan	M	54	116	..	14.5	..	31.1	..	..	..	16.0
Uganda	S	27 [b]	225 [b]	3.4	2.9	58.9	23.7	..	..	5.4	3.5
Ukraine	L	34	75	..	7.5	..	16.3	..	..	..	2.2
United Arab Emirates	..	..	..	..	..	..	..	..	..	..	..
United Kingdom	..	..	..	..	..	..	..	..	..	..	..
United States	..	..	..	..	..	..	..	..	..	..	..
Uruguay	M	37	177	11.0	5.2	40.8	25.0	32.0	15.5	27.2	24.1
Uzbekistan	L	24	131	..	3.2	..	17.6	..	..	..	17.2
Venezuela, RB	M	38	156	10.6	5.6	23.2	23.2	36.2	23.3	6.0	6.3
Vietnam	M	76	151	..	4.9	8.9	9.8	..	27.7	7.7	10.2
West Bank and Gaza	..	..	..	..	..	..	..	..	..	..	..
Yemen, Rep.	M	58	91	3.8	2.5	5.6	4.0	..	5.6	18.8	10.2
Yugoslavia, FR (Serb/Mont.)	..	..	..	..	..	..	..	..	..	..	..
Zambia	S	172	548	6.7	14.6	14.9	46.6	..	..	20.4	1.9
Zimbabwe	M	78	159	5.5	12.3	23.1	25.3	17.4	..	18.2	16.3

World	.. w			.. w	.. w	.. w	.. w			.. w	.. w
Low income				4.8	4.9	22.9	18.8			11.9	10.9
Middle income				3.9	6.5	17.2 [d]	21.8 [d]			18.7 [d]	17.2 [d]
Lower middle income				3.9	4.5	21.2	15.7			16.9	16.6
Upper middle income				3.8	8.3	13.9 [d]	27.3 [d]			20.6 [d]	17.9 [d]
Low & middle income				4.0	6.2	18.1 [d]	21.4 [d]			16.8 [d]	15.8 [d]
East Asia & Pacific				4.4	6.1	15.7	15.8			17.9	16.8
Europe & Central Asia				2.9	5.8	..	18.0			18.6	14.7
Latin America & Carib.				4.2	8.4	24.4	41.6			16.3	14.8
Middle East & N. Africa				5.1	4.6	14.9	13.7			24.1	23.4
South Asia				2.9	2.6	28.9	15.7			9.5	4.4
Sub-Saharan Africa				..	4.8	12.9	13.9			11.8	20.5
High income											
Europe EMU											

a. S = severely indebted, M = moderately indebted, L = less indebted. b. Data are from debt sustainability analyses undertaken as part of the Heavily Indebted Poor Countries (HIPC) Initiative. Present value estimates for these countries are for public and publicly guaranteed debt only, and export figures exclude workers' remittances. c. Data refer to mainland Tanzania only. d. Includes data for Gibraltar not included in other tables.

About the data

The indicators in the table measure the relative burden on developing countries of servicing external debt. The present value of external debt provides a measure of future debt service obligations that can be compared with the current value of such indicators as gross national income, or GNI (gross national product, or GNP, in previous editions), and exports of goods and services. This table shows the present value of total debt service both as a percentage of GNI in 1999 and as a percentage of exports in 1999. The ratios compare total debt service obligations with the size of the economy and its ability to obtain foreign exchange through exports. Because workers' remittances are an important source of foreign exchange for many countries, they are included in the value of exports used to calculate debt indicators. Public and publicly guaranteed debt service is compared with the size of the central government budget. The ratios shown here may differ from those published elsewhere because estimates of exports and GNI have been revised to incorporate data available as of 1 February 2001.

The present value of external debt is calculated by discounting the debt service (interest plus amortization) due on long-term external debt over the life of existing loans. Short-term debt is included at its face value. The data on debt are in U.S. dollars converted at official exchange rates (see *About the data* for table 4.16). The discount rate applied to long-term debt is determined by the currency of repayment of the loan and is based on reference rates for commercial interest established by the Organisation for Economic Co-operation and Development. Loans from the International Bank for Reconstruction and Development (IBRD) and credits from the International Development Association (IDA) are discounted using an SDR (special drawing rights) reference rate, as are obligations to the International Monetary Fund (IMF). When the discount rate is greater than the interest rate of the loan, the present value is less than the nominal sum of future debt service obligations.

The ratios in the table are used to assess the sustainability of a country's debt service obligations, but there are no absolute rules that determine what values are too high. Empirical analysis of the experience of developing countries and their debt service performance has shown that debt service difficulties become increasingly likely when the ratio of the present value of debt to exports reaches 200 percent and the ratio of debt service to GNI exceeds 40 percent. Still, what constitutes a sustainable debt burden varies from one country to another. Countries with fast-growing economies and exports are likely to be able to sustain higher debt levels.

The World Bank classifies countries by their level of indebtedness for the purpose of developing debt management strategies. The most severely indebted countries may be eligible for debt relief under special programs such as the Heavily Indebted Poor Countries (HIPC) Initiative. Indebted countries may also apply to the Paris and London Clubs for renegotiation of obligations to public and private creditors. In 1999 countries with a present value of debt service greater than 220 percent of exports or 80 percent of GNI were classified as severely indebted; countries that were not severely indebted but whose present value of debt service exceeded 132 percent of exports or 48 percent of GNI were classified as moderately indebted; and countries that did not fall into the above two groups were classified as less indebted.

Figure 4.17

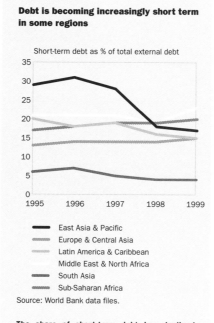

Debt is becoming increasingly short term in some regions

Short-term debt as % of total external debt

- East Asia & Pacific
- Europe & Central Asia
- Latin America & Caribbean
- Middle East & North Africa
- South Asia
- Sub-Saharan Africa

Source: World Bank data files.

The share of short-term debt has declined significantly in East Asia and Pacific. But it has been increasing in Europe and Central Asia, the Middle East and North Africa, and Sub-Saharan Africa.

Definitions

- **Indebtedness** is assessed on a three-point scale: severely indebted (S), moderately indebted (M), and less indebted (L). • **Present value of debt** is the sum of short-term external debt plus the discounted sum of total debt service payments due on public, publicly guaranteed, and private nonguaranteed long-term external debt over the life of existing loans. • **Total debt service** is the sum of principal repayments and interest actually paid in foreign currency, goods, or services on long-term debt, interest paid on short-term debt, and repayments (repurchases and charges) to the IMF. • **Public and publicly guaranteed debt service** is the sum of principal repayments and interest actually paid on long-term obligations of public debtors and long-term private obligations guaranteed by a public entity. • **Short-term debt** includes all debt having an original maturity of one year or less and interest in arrears on long-term debt.

Data sources

The main sources of external debt information are reports to the World Bank through its Debtor Reporting System from member countries that have received IBRD loans or IDA credits. Additional information has been drawn from the files of the World Bank and the IMF. The data on GNI and exports of goods and services are from the World Bank's national accounts files. Summary tables of the external debt of developing countries are published annually in the World Bank's *Global Development Finance* and on its *Global Development Finance* CD-ROM.

STATES AND MARKETS

Global digital opportunities

Rich countries spend more on information and communications technology (ICT) per capita than poor countries. But for a given income, some countries outpace others by a wide margin.

ICT spending refers to spending on information technology plus telecommunications equipment and services. More broadly, ICT is the set of activities that facilitate the processing, transmission, and display of information by electronic means.

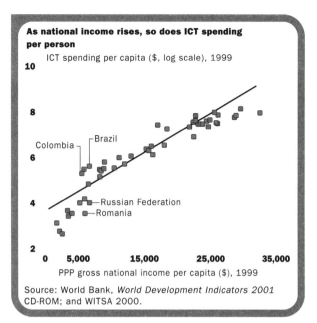

As national income rises, so does ICT spending per person

ICT spending per capita ($, log scale), 1999

Colombia

Brazil

Russian Federation

Romania

PPP gross national income per capita ($), 1999

Source: World Bank, *World Development Indicators 2001* CD-ROM; and WITSA 2000.

Colombia spends about seven times as much on ICT per person as Romania—and Brazil about five times as much as Russia—although all four countries have similar per capita incomes (about $6,000 in purchasing power parity terms).

Do you want bread—or computers? You want both.

The digital and information revolution has changed the way the world learns, communicates, does business, and cures illnesses. New information and communications technologies (ICT) offer vast opportunities for progress in all walks of life in all countries—opportunities for economic growth, improved health, better service delivery, learning through distance education, and social and cultural advances.

None of these benefits will come automatically. Technology may continue to develop at breakneck speed. But it has to be matched with market-oriented reforms that promote competition and entrepreneurial freedom and place a high priority on universal education. And all that requires support from the international community. Otherwise, many countries will be unable to compete and grow in this networked and globalized world.

Where do countries stand in the ICT race? What are the opportunities and benefits from a networked society? What are the risks of staying on the sidelines of the ICT revolution? What can developing country governments do to improve their development prospects? What can the international community, including the private sector and nongovernmental organizations, do to help bridge the digital divide and ensure a better world for all? Here are some of the answers.

What is the digital divide?

It's the gap between those with access to ICT and those without.

And the gap is big: rich countries have about 15 percent of the world's population but about 80 percent of the world's personal computers (PCs) and almost 90 percent of its Internet users. On average, a high-income country has 40 times as many computers per capita as a Sub-Saharan African country.

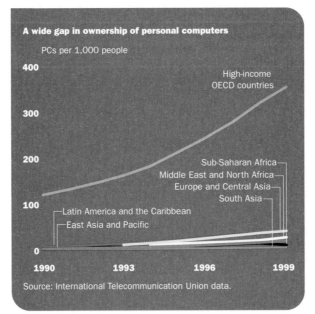

A wide gap in ownership of personal computers

PCs per 1,000 people

High-income OECD countries

Sub-Saharan Africa
Middle East and North Africa
Europe and Central Asia
South Asia

Latin America and the Caribbean
East Asia and Pacific

1990 1993 1996 1999

Source: International Telecommunication Union data.

Although the gap in PC ownership between high-income OECD countries and developing regions is wide, PC ownership is growing twice as fast in developing as in rich countries.

You have to look at ICT in the context of the development framework

---

Where do countries stand in the ICT race?

Five years ago only five Sub-Saharan African countries had Internet access. This year all the region's countries are connected. And at 36 percent, its annual growth in Internet hosts is almost twice the world's average.

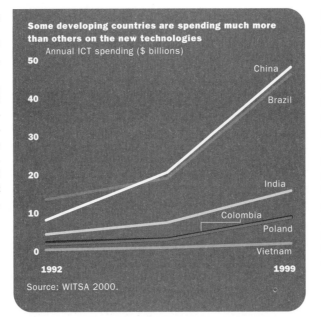

Some developing countries are spending much more than others on the new technologies

Annual ICT spending ($ billions)

China
Brazil
India
Colombia
Poland
Vietnam

1992 1999

Source: WITSA 2000.

Many countries are bridging the divide. Since 1992 China has increased its ICT spending about 30 percent a year, going from 0.6 percent of worldwide ICT spending to 2.2 percent in 1999. During the same period the number of PCs in China grew more than 40 percent a year.

Only 0.6 percent of the people in developing countries have access to the Internet, compared with 30 percent in the United States. But in 1992–99 such developing countries as China, Brazil, and India had some of the world's fastest growing ICT markets.

Opportunities in the digital society . . .

The advances in ICT can benefit high-income and developing countries through:
• Lower transaction and distribution costs.
• Broader markets and more effective marketing.
• Greater competition.
• Job creation and social stability.
• Social applications (such as distance education).
• New ways of forming social relationships, fostering human interaction, and bringing the poor and isolated into the global economy.

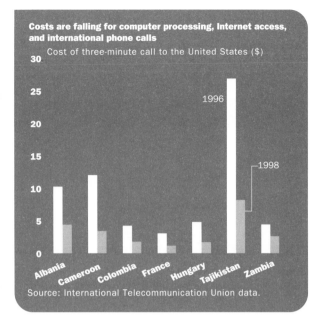

Costs are falling for computer processing, Internet access, and international phone calls

Cost of three-minute call to the United States ($)

1996
1998

Albania Cameroon Colombia France Hungary Tajikistan Zambia

Source: International Telecommunication Union data.

Consumers everywhere can benefit from the lower costs. And workers with ICT skills in developing countries may have opportunities for employment in firms that export software, accounting services, and insurance claims processing.

You cannot forget about growth, or about legal and financial systems

---

. . . and the risks of being excluded

There are wide disparities in access to information and telecommunications not only between countries, but also within countries. In Panama the wealthiest fifth of the population are 43 times as likely to have private telephones as the poorest fifth. And in South Africa households in cities are 10 times as likely to have access to private telephones as those in rural areas.

And without access to telephones and a reliable electricity supply, access to the Internet is not possible.

Teledensity differs by income and between rural and urban areas

% with telephones

	Poorest 20%	Richest 20%
Nepal, 1996	0.0	11.0
Panama, 1997	1.7	73.8
South Africa, 1993	0.6	75.0

	Rural households	Urban households
Nepal, 1996	0.1	10.4
Panama, 1997	9.2	57.5
South Africa, 1993	4.7	45.7

Source: World Bank, Living Standards Measurement Study survey data.

The Kisiizi Hospital in Uganda, far from a paved road and without phone service for many years, finally got wireless phone service through a satellite connection. But at a high cost. Phone calls to the capital cost $2.50 a minute, equal to five days' work for a wage earner.

The alternatives can be costly in both time and money. A nurse at the hospital found that a five-minute call to Kampala to find out about a training course cost the same as bus fare for the 800-kilometer round-trip to the capital. The rural poor often have little choice but to pay the high costs of phone service. The alternative may be to remain cut off from access to information and basic services.

Computers matter in the digital society . . .

Many developing countries increased the number of PCs in classrooms remarkably between 1992 and 1999. While Brazil had about 35,000 PCs in schools in 1992, it had 400,000 in 1999, about a 12-fold increase. In high-income economies the number of PCs in schools has grown about 3 percent a year on average.

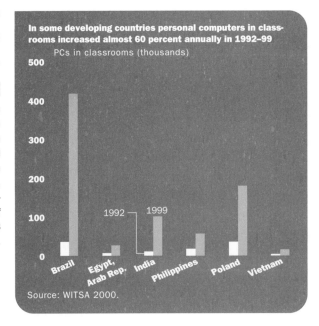

In some developing countries personal computers in classrooms increased almost 60 percent annually in 1992–99

PCs in classrooms (thousands)

1992 — 1999

Brazil   Egypt, Arab Rep.   India   Philippines   Poland   Vietnam

Source: WITSA 2000.

Internet capacity is not a simple linear function of economic and political development. Instead, it is driven by complex interactions that could be termed "post-industrialization." That is what Ohio State University researchers Kristopher Robison and Edward Crenshaw (2000) found after analyzing economic, political, and ICT development in 75 high-income and developing countries.

Nor can you forget about health, education, and good government

---

. . . but literacy is also important

Technology alone is not the answer. It also takes an educated workforce to create opportunities for more equitable development. So it may be just as important for a government to promote universal literacy as it is to promote PC ownership and universal Internet access.

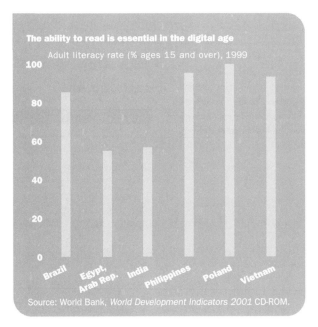

The ability to read is essential in the digital age

Adult literacy rate (% ages 15 and over), 1999

Brazil   Egypt, Arab Rep.   India   Philippines   Poland   Vietnam

Source: World Bank, *World Development Indicators 2001* CD-ROM.

Closing the digital divide requires a literate, educated population and open and transparent government and business.

Governments face many challenges in creating an environment for growth in ICT

Connecting computers to the Internet is only the first step toward a fully networked society. Much more is needed.

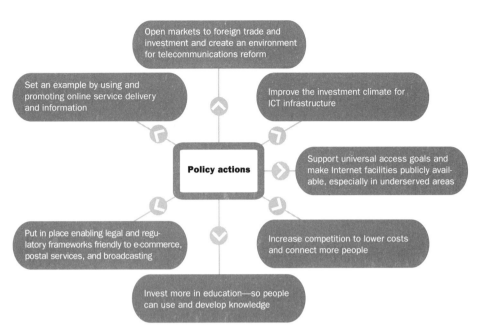

Open markets to foreign trade and investment and create an environment for telecommunications reform

Set an example by using and promoting online service delivery and information

Improve the investment climate for ICT infrastructure

Policy actions

Support universal access goals and make Internet facilities publicly available, especially in underserved areas

Put in place enabling legal and regulatory frameworks friendly to e-commerce, postal services, and broadcasting

Increase competition to lower costs and connect more people

Invest more in education—so people can use and develop knowledge

Indeed, to meet the challenge of poverty . . .

E-government puts government at the service of the public . . .

In 1997 an interagency committee in Chile created a communications and information technology unit to promote e-government. The aim is to improve coordination within and among agencies and thus increase the efficiency and effectiveness of public services. New management tools will improve transparency and probity, reducing opportunities for corruption. They will also support democracy, by empowering people and allowing wider participation.

With the new e-system, companies register in the area in which they want to do business with the public sector. When a government agency wants to purchase goods or services, the e-system automatically emails contract specifications to the registered companies, giving all of them an equal opportunity. The system also provides information on the procurement process, such as the winning bid.

From Chile's e-government procurement program alone, savings of at least $200 million a year are projected—about 1.4 percent of central government spending in 1997. Every dollar saved is an extra dollar for health care, social security, or public housing.

. . . saving citizens time and improving their satisfaction

In the Indian state of Andhra Pradesh innovative e-government activities have increased the efficiency of public administration in delivering public services. These innovations include Computer-Aided Registration of Deeds (CARD) and the Twin Cities Network Services (TWINS).

Operating in 214 locations, CARD completes all transactions through a computerized process. It has reduced the time required for an encumbrance certificate or property valuation from about a week to only 15 minutes. People can now register their property in a day rather than the several days it used to take.

TWINS, operating in the cities of Hyderabad and Secunderabad, provides an interface for issuing citizens certificates, permits, and licenses and for citizens to pay utility bills. A pilot center at Banjara Hills handles about 3,000 transactions a day, serving citizens in minutes. TWINS will expand to 18 other locations in the two cities through a public-private partnership, handling about a million transactions a year.

The international community has a role

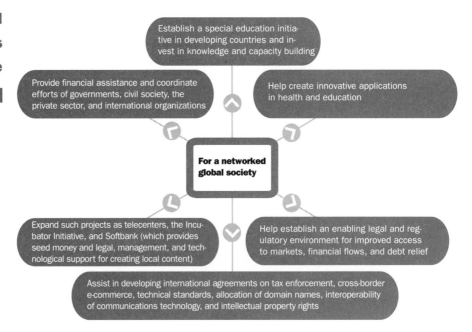

Establish a special education initiative in developing countries and invest in knowledge and capacity building

Provide financial assistance and coordinate efforts of governments, civil society, the private sector, and international organizations

Help create innovative applications in health and education

For a networked global society

Expand such projects as telecenters, the Incubator Initiative, and Softbank (which provides seed money and legal, management, and technological support for creating local content)

Help establish an enabling legal and regulatory environment for improved access to markets, financial flows, and debt relief

Assist in developing international agreements on tax enforcement, cross-border e-commerce, technical standards, allocation of domain names, interoperability of communications technology, and intellectual property rights

---

. . . you cannot forget anything in the development paradigm

Incubator Initiative

Efforts to start up ICT-related companies in developing countries face many constraints: no seed capital, too few experienced managers, and more. That is why the Incubator Initiative will provide technical assistance to a network of incubators to help create dynamic, results-oriented ICT enterprises. The initiative is an effort of the Information for Development Program (InfoDev), which is supported by the World Bank and other donors.

www.infodev.org

African Virtual University

The African Virtual University, sponsored by the World Bank Group, offers technology-based distance education to all of Africa. It delivers quality education to students and professionals, mainly in business, engineering, and information technology—areas where Africa lacks the skills needed to participate in the knowledge economy. The university has 25 learning centers in 15 Sub-Saharan countries, reaching more than 12,000 students.

www.avu.org

Bridges.org

Bridges.org's mission is to empower people in developing and emerging economies to use technology for themselves by:

• Providing public education about technology use.

• Promoting policies that remove barriers to the use of technology.

• Creating a body of knowledge about digital divide issues.

www.bridges.org

"DOT Force"— Digital Opportunity Taskforce

The DOT Force, drawn from the Group of Eight (G-8)—Canada, France, Germany, Italy, Japan, the Russian Federation, the United Kingdom, and the United States—held its inaugural meeting in Japan on 27–28 November 2000. In opening the meeting, Japanese Prime Minister Yoshiro Mori expressed hope for broad cooperation and concrete proposals to bridge the digital divide. The DOT Force will investigate promising solutions and report its findings at the G-8 summit in July 2001.

www.dotforce.org

ICT touches every facet of development and poverty reduction

New information and communications technologies can break down some of the physical barriers to participation in society faced by many poor people.

OPPORTUNITY—sustainable growth and employment

Distance education promises to broaden knowledge and information—some of the most important factors in economic growth.

EMPOWERMENT—more transparent governance, better service delivery, greater ability to communicate and participate

"Voice" and participation lead to better accountability and better governance.

SECURITY—reduced vulnerability to ill health and natural disasters

ICT can overcome the "information poverty" that health workers in developing countries face every day.

> ICT gives us the opportunity to leverage the transfer of knowledge, empowering people in ways that were previously not possible. It is not just a technical tool; it may be the answer to greater equity in the world, and a greater peace.
>
> James D. Wolfensohn
> President
> World Bank

	Private fixed investment		Domestic credit to private sector		Proceeds from privat- ization[a]	Investment in infrastructure projects with private participation[a]							
	% of gross domestic fixed investment		% of GDP		$ millions	Telecommunications $ millions		Energy $ millions		Transport $ millions		Water and sanitation $ millions	
	1990	1998	1990	1999	1990–99	1990–94	1995–99	1990–94	1995–99	1990–94	1995–99	1990–94	1995–99
Albania	..	..	..	3.6	28.5	..	..	..	..	..	..	..	..
Algeria	..	..	44.5	5.5	55.1	..	..	..	..	..	..	..	..
Angola	..	..	..	2.8	6.2	..	..	..	..	..	..	..	..
Argentina	67.4	91.5	15.6	25.0	44,588.0	9,262.0	10,838.3	9,806.5	10,394.3	5,293.4	8,863.1	4,075.0	3,182.6
Armenia	..	..	40.4	9.2	219.2	..	442.0	..	..	..	..	..	..
Australia	88.4[b]	90.4[b]	64.4	85.0	..	..	..	..	..	..	..	..	..
Austria	..	92.3	91.2	102.9	..	..	..	..	..	..	..	..	..
Azerbaijan	..	..	9.4	3.4	15.8	14.0	82.0	..	..	..	..	..	..
Bangladesh	61.4	68.0	16.7	23.5	59.6	116.0	472.4	..	538.7	..	..	..	..
Belarus	..	..	..	9.7	10.8	10.0	15.0	..	500.0	..	..	..	..
Belgium	92.8[b]	92.6[b]	35.5	77.5	..	..	..	..	..	..	..	..	..
Benin	44.8	66.1	20.3	10.9	39.0	..	..	..	..	..	..	..	..
Bolivia	39.3	66.5	24.0	63.9	1,045.4	20.0	729.4	..	1,112.0	..	146.6	..	682.0
Bosnia and Herzegovina	..	..	..	..	..	..	..	..	..	..	..	..	..
Botswana	..	..	9.4	15.2	..	..	40.0	..	..	..	..	..	..
Brazil	76.7[b]	82.9[b]	38.9	34.5	69,607.7	..	44,283.0	303.0	32,135.1	328.1	17,646.8	2.5	1,503.0
Bulgaria	3.6	55.2	7.2	14.6	3,199.0	21.0	219.0	..	..	..	..	..	..
Burkina Faso	..	..	19.0	11.8	7.5	..	..	..	..	..	..	..	..
Burundi	..	..	13.7	20.9	4.2	0.5	..	..	..	..	..	..	..
Cambodia	86.0	76.3	..	6.4	..	30.1	101.7	..	89.0	..	220.0	..	..
Cameroon	..	..	26.7	8.9	133.1	..	12.7	..	..	30.8	95.0	..	..
Canada	86.3[b]	89.4[b]	79.4	87.1	..	..	..	..	..	..	..	..	..
Central African Republic	..	..	7.2	4.3	..	..	1.1	..	..	..	..	..	..
Chad	..	62.5	7.3	3.6	..	..	2.0	..	..	..	..	..	..
Chile	79.5	87.7	47.2	67.5	2,138.4	2,354.9	3,364.4	1,755.8	3,525.3	112.4	3,090.3	127.6	3,162.3
China	33.8[c]	45.8[c]	87.7	122.0	20,593.2	..	5,970.0	5,447.8	12,922.1	5,820.5	9,650.8	42.8	605.4
Hong Kong, China	..	..	165.1	159.4	..	..	..	..	..	..	..	..	..
Colombia	61.6	59.9	30.8	33.4	5,979.5	1,355.4	1,302.8	540.0	6,831.4	783.0	856.2	..	210.0
Congo, Dem. Rep.	..	..	1.8	..	..	..	50.0	..	..	..	..	..	..
Congo, Rep.	..	..	15.7	11.6	..	..	47.1	..	325.0	..	..	..	..
Costa Rica	78.9	76.8	12.6	18.5	50.8	..	..	18.3	271.5	..	24.0	..	..
Côte d'Ivoire	57.9	72.0	36.5	16.0	597.4	..	742.3	109.6	260.6	..	178.0	..	..
Croatia	..	..	..	36.9	1,318.3	..	978.0	..	368.5	..	672.2	..	..
Cuba	..	..	..	..	706.0	371.0	..	..	..	..	..	..	..
Czech Republic	86.4	82.0	..	56.8	5,633.1	41.0	6,274.0	356.0	944.1	..	283.7	16.0	48.8
Denmark	91.8[b]	91.7[b]	52.2	34.8	..	..	..	..	..	..	..	..	..
Dominican Republic	73.1	77.8	27.5	32.7	643.4	5.0	163.0	87.5	928.5	..	..	..	..
Ecuador	67.2	70.2	15.1	61.0	169.3	13.8	684.6	..	30.0	12.5	686.8	..	..
Egypt, Arab Rep.	62.3	67.5	30.6	59.7	2,905.4	..	2,045.7	..	700.0	..	197.7	..	..
El Salvador	81.5	76.7	20.1	43.9	1,070.1	..	610.5	..	892.2	..	..	..	..
Eritrea	..	..	..	..	2.0	..	..	..	..	..	..	..	..
Estonia	..	..	20.2	26.4	778.2	136.1	628.2	..	26.5	..	1.0	..	..
Ethiopia	..	..	24.0	30.8	172.0	..	..	..	..	..	..	..	..
Finland	86.9[b]	84.5[b]	86.7	52.1	..	..	..	..	..	..	..	..	..
France	..	..	96.1	82.9	..	..	..	..	..	..	..	..	..
Gabon	..	..	13.0	10.7	..	..	..	..	624.8	..	..	..	624.8
Gambia, The	66.8	67.9	11.0	13.7	..	..	..	..	..	..	..	..	..
Georgia	..	..	..	7.7	..	..	31.0	..	25.0	..	..	..	..
Germany	..	..	91.9	118.2	..	..	..	..	..	..	..	..	..
Ghana	..	..	4.9	12.0	888.4	20.0	441.1	..	..	..	..	..	..
Greece	..	83.9[b]	35.9	46.5	..	..	..	..	..	..	..	..	..
Guatemala	80.0	86.6	14.2	20.7	1,351.2	..	1,366.3	100.0	1,202.9	..	10.0	..	..
Guinea	..	..	3.5	3.5	45.0	..	120.3	..	36.4	..	..	..	..
Guinea-Bissau	28.1	46.0	22.0	7.3	0.5	..	..	23.2	..	..	..	23.2	..
Haiti	..	..	12.1	14.3	16.5	..	1.5	..	..	..	..	..	..
Honduras	..	..	31.1	41.8	74.1	..	38.1	70.0	85.0	..	..	..	..

Private sector development 5.1

	Private fixed investment		Domestic credit to private sector		Proceeds from privat-ization[a]	Investment in infrastructure projects with private participation[a]							
	% of gross domestic fixed investment		% of GDP		$ millions	Telecommunications $ millions		Energy $ millions		Transport $ millions		Water and sanitation $ millions	
	1990	1998	1990	1999	1990–99	1990–94	1995–99	1990–94	1995–99	1990–94	1995–99	1990–94	1995–99
Hungary	..	..	46.3	25.4	13,998.9	1,610.2	6,300.2	..	3,812.1	1,086.0	135.0	..	180.3
India	59.1	69.6	25.3	26.1	8,983.4	93.0	9,176.7	2,865.8	7,910.8	126.9	915.3	..	..
Indonesia	67.5	77.0	46.9	20.1	6,134.8	1,119.0	7,245.5	352.5	9,747.1	709.8	2,223.1	3.8	872.2
Iran, Islamic Rep.	53.6	55.5	32.5	29.4	..	..	28.0	..	..	..	..	..	..
Iraq	..	..	..	..	..	..	..	..	..	..	..	..	..
Ireland	88.8[b]	87.7[b]	47.6	89.2	..	..	..	..	..	..	..	..	..
Israel	..	..	57.6	83.2	..	..	..	..	..	..	..	..	..
Italy	..	..	56.5	59.2	..	..	..	..	..	..	..	..	..
Jamaica	..	..	39.0	32.3	385.5	169.0	235.5	246.0	43.0	30.0	..	..	..
Japan	84.2[b]	78.5[b]	200.6	115.3	..	..	..	..	..	..	..	..	..
Jordan	..	..	72.3	71.0	63.8	43.0	17.8	..	..	..	182.0	..	55.0
Kazakhstan	..	..	..	9.5	6,375.9	30.0	1,548.0	..	2,084.5	..	..	..	..
Kenya	42.0	58.8	32.8	30.5	318.3	..	55.0	..	154.0	..	53.4	..	..
Korea, Dem. Rep.	..	..	..	..	..	..	..	..	..	..	..	..	..
Korea, Rep.	88.1	79.5	65.5	93.4	..	2,379.0	10,940.5	300.0	3,173.2	..	2,634.0	..	..
Kuwait	..	..	52.1	66.1	..	..	..	..	..	..	..	..	..
Kyrgyz Republic	..	..	..	5.1	139.5	..	94.0	..	..	..	..	..	..
Lao PDR	..	..	1.0	8.6	32.0	..	152.9	..	535.5	..	..	..	..
Latvia	..	..	..	16.7	490.9	180.0	446.7	..	106.0	..	75.0	..	..
Lebanon	..	..	79.4	74.0	..	50.0	323.0	..	..	..	..	..	..
Lesotho	..	..	15.6	15.2	16.2	..	10.0	..	..	..	..	..	..
Libya	..	..	..	..	..	..	..	..	..	..	..	..	..
Lithuania	..	..	..	13.0	1,535.6	30.0	809.5	..	20.0	..	..	..	..
Macedonia, FYR	..	..	..	22.3	679.3	..	..	..	..	..	..	..	..
Madagascar	46.5	47.1	16.9	8.4	9.0	5.0	10.0	..	..	..	..	..	..
Malawi	51.5	21.0	12.8	5.6	18.9	..	18.4	..	..	..	6.0	..	..
Malaysia	64.5	65.4	69.4	144.0	10,159.6	2,010.5	4,380.0	5,663.8	1,610.5	2,769.3	8,196.4	3,976.7	1,056.0
Mali	..	..	12.8	18.2	21.9	..	..	..	..	..	..	..	..
Mauritania	69.3	36.5	43.5	23.9	1.1	..	..	..	..	..	..	..	..
Mauritius	63.2	69.5	33.2	58.5	..	..	..	..	109.3	..	..	..	..
Mexico	..	92.1	17.5	16.2	28,593.0	15,795.0	11,736.9	..	2,250.3	7,430.9	5,151.9	516.7	199.7
Moldova	..	..	5.9	12.0	26.6	..	59.6	..	60.0	..	..	..	..
Mongolia	..	..	17.7	8.6	..	..	24.1	..	..	..	..	..	..
Morocco	65.6	70.7	34.0	54.6	3,102.2	..	1,240.0	2,300.0	4,819.9	..	..	..	4,050.9
Mozambique	..	..	17.6	17.0	138.2	..	29.0	..	..	..	..	..	200.0
Myanmar	..	..	4.7	8.6	..	..	4.0	..	..	..	50.0	..	..
Namibia	61.5	48.6	22.7	49.2	..	..	22.0	..	..	..	..	..	..
Nepal	..	..	12.8	29.1	15.1	..	..	125.7	98.2	..	..	..	..
Netherlands	..	..	79.6	107.3	..	..	..	..	..	..	..	..	..
New Zealand	..	..	76.9	118.0	..	..	..	..	..	..	..	..	..
Nicaragua	54.7	67.9	112.6	54.1	130.3	6.6	24.5	..	232.4	..	..	..	..
Niger	..	..	12.3	3.8	..	..	..	..	..	..	..	..	..
Nigeria	..	..	9.4	13.8	730.2	..	63.5	..	..	..	..	..	..
Norway	83.5[b]	85.1[b]	82.2	83.6	..	..	..	..	..	..	..	..	..
Oman	..	..	22.9	44.6	60.1	..	..	204.5	..	..	77.5	..	..
Pakistan	51.7	66.2	27.7	28.2	1,992.3	581.5	107.5	1,638.7	5,023.8	..	418.3	..	..
Panama	87.1	82.9	46.7	116.4	1,427.3	..	1,429.2	..	669.2	169.9	994.6	..	25.0
Papua New Guinea	79.7	76.9	28.6	17.1	223.6	..	..	..	50.0	..	..	..	..
Paraguay	86.7	65.1	15.8	29.3	42.0	33.2	199.3	..	..	..	58.0	..	..
Peru	83.1	84.2	11.8	28.4	8,134.4	1,645.0	4,868.5	431.2	2,671.8	..	86.8	..	..
Philippines	81.8	76.0	22.3	46.6	3,960.0	591.8	5,137.9	4,502.1	6,998.0	..	3,005.6	..	5,820.0
Poland	84.9	82.1	3.0	23.6	12,171.9	273.0	4,893.0	..	624.8	3.1	2.3	..	..
Portugal	..	..	49.5	98.9	..	..	..	..	..	..	..	..	..
Puerto Rico	..	..	..	..	..	..	..	..	..	..	..	..	..
Romania	9.8	43.3	..	8.4	1,865.7	5.0	1,879.3	..	100.0	..	23.4	..	..
Russian Federation	..	..	..	11.5	2,671.6	223.1	4,695.0	1,100.0	2,281.3	..	400.0	..	108.0

	Private fixed investment		Domestic credit to private sector		Proceeds from privat- ization[a]	Investment in infrastructure projects with private participation[a]							
	% of gross domestic fixed investment		% of GDP		$ millions	Telecommunications $ millions		Energy $ millions		Transport $ millions		Water and sanitation $ millions	
	1990	1998	1990	1999	1990–99	1990–94	1995–99	1990–94	1995–99	1990–94	1995–99	1990–94	1995–99
Rwanda	..	..	6.9	9.2	..	..	15.0	..	..	..	..	..	..
Saudi Arabia	..	..	61.0	69.3	..	..	..	..	..	..	..	..	..
Senegal	..	..	26.5	16.4	410.7	..	267.3	..	159.0	..	..	..	..
Sierra Leone	..	..	2.4	2.1	1.6	..	..	..	..	..	..	..	..
Singapore	..	..	97.5	115.3	..	..	..	..	..	..	..	..	..
Slovak Republic	..	..	..	37.3	1,979.4	109.2	488.5	..	..	..	..	..	..
Slovenia	..	..	34.9	36.1	521.1	..	..	..	..	..	..	..	..
South Africa	65.6	67.3	81.0	136.3	2,964.2	542.2	4,501.6	..	3.0	..	1,386.4	..	170.0
Spain	..	..	79.6	87.9	..	..	..	..	..	..	..	..	..
Sri Lanka	..	..	19.6	24.6	804.5	43.6	905.0	..	83.7	..	240.0	..	..
Sudan	..	..	4.8	1.8	..	..	6.0	..	..	..	..	..	..
Sweden	..	..	128.6	106.4	..	..	..	..	..	..	..	..	..
Switzerland	..	..	167.9	174.2	..	..	..	..	..	..	..	..	..
Syrian Arab Republic	..	..	7.5	8.2	..	..	..	..	..	..	..	..	..
Tajikistan	..	..	..	..	..	..	0.2	..	..	..	..	..	..
Tanzania	..	..	13.9	4.6	272.3	1.8	66.9	6.0	150.0	..	16.5	..	..
Thailand	84.8	66.2	83.4	130.1	2,985.8	3,664.0	4,034.9	674.8	4,944.9	695.9	1,700.0	..	239.3
Togo	..	..	22.6	17.0	38.1	..	..	..	..	..	..	..	..
Trinidad and Tobago	..	..	44.7	45.1	276.2	47.0	146.7	..	207.0	..	..	..	..
Tunisia	50.5	51.2	55.1	65.1	523.0	..	..	627.0	265.0	..	..	..	..
Turkey	..	..	16.7	22.5	4,654.4	74.0	3,269.7	718.0	2,992.2	..	505.0	..	1,202.0
Turkmenistan	..	..	..	1.5	..	..	..	..	..	..	..	..	..
Uganda	..	..	4.0	5.9	174.4	16.0	118.0	..	..	..	..	..	..
Ukraine	..	..	2.6	8.8	31.5	90.0	723.0	..	..	..	..	..	..
United Arab Emirates	..	..	37.4	59.1	..	..	..	..	..	..	..	..	..
United Kingdom	87.3[b]	92.9[b]	116.0	123.4	..	..	..	..	..	..	..	..	..
United States	..	..	93.1	145.3	..	..	..	..	..	..	..	..	..
Uruguay	71.5	70.5	32.4	49.6	17.0	13.0	63.7	..	..	96.0	138.0	10.0	..
Uzbekistan	..	..	..	..	212.0	2.5	357.4	..	..	..	..	..	..
Venezuela, RB	34.8	53.8	25.4	13.0	6,072.0	4,185.7	4,850.9	..	63.0	100.0	268.0	..	..
Vietnam	..	..	2.5	21.7	7.6	..	..	..	286.0	10.0	70.0	..	38.2
West Bank and Gaza	..	..	..	..	..	..	155.0	..	..	..	..	..	..
Yemen, Rep.	..	..	6.3	5.9	..	25.0	..	..	..	..	190.0	..	..
Yugoslavia, FR (Serb./Mont.)	..	..	..	..	921.7	..	1,275.0	..	..	..	..	..	..
Zambia	..	..	8.9	7.4	826.0	..	39.2	..	274.0	..	..	..	..
Zimbabwe	..	..	23.0	27.2	217.8	..	46.0	..	1,180.0	18.0	70.0	..	..
World	78.1w	76.0w	98.6w	109.0w	..s	..s	..s	..s	..s	..s	..s	..s	..s
Low income	48.1	53.7	26.6	21.7	2,121.0	21,270.9	5,121.5	26,560.0	895.5	4,505.6	27.0	1,121.3	
Middle income	72.2	74.8	45.4	58.3	47,786.1	160,933.7	35,082.7	113,412.8	24,700.0	67,400.1	8,767.3	23,173.0	
Lower middle income	..	..	55.6	69.0	8,673.8	44,182.3	16,687.5	52,873.7	7,331.9	17,858.9	42.8	13,020.5	
Upper middle income	73.8	77.9	41.1	48.6	39,112.3	116,751.4	18,395.2	60,539.1	17,368.1	49,541.2	8,724.5	10,152.5	
Low & middle income	64.5	66.9	41.7	52.7	49,907.1	182,204.6	40,204.2	139,972.8	25,595.5	71,905.7	8,794.3	24,294.3	
East Asia & Pacific	63.3	50.2	71.4	104.1	9,826.4	38,129.5	16,733.9	39,728.8	10,005.5	27,749.9	4,023.3	8,631.1	
Europe & Central Asia	..	..	..	18.9	2,849.1	35,506.3	2,174.0	13,945.0	1,089.1	2,097.6	16.0	1,539.1	
Latin America & Carib.	74.3	79.8	28.5	29.4	35,694.0	87,286.0	13,395.8	63,524.0	14,325.2	38,032.1	4,731.8	8,964.6	
Middle East & N. Africa	..	..	41.9	47.2	118.0	3,809.5	3,131.5	5,784.9	..	647.2	..	4,105.9	
South Asia	55.9	71.8	24.6	26.1	834.1	10,693.8	4,630.2	13,655.2	126.9	1,573.6			
Sub-Saharan Africa	..	..	42.7	66.2	585.5	6,779.5	138.8	3,334.9	48.8	1,805.3	23.2	1,053.6	
High income	81.9	79.2	110.9	129.3	..	..	..	..	..	..	..	..	..
Europe EMU	..	..	75.7	89.2	..	..	..	..	..	..	..	..	..

a. Data refer to total for the period shown. For differences in concepts and definitions between proceeds from privatization and investment in infrastructure projects with private participation see *About the data.* b. Data refer to investment by both private and public corporations. c. Data refer to investment by individuals, shareholding units, jointly owned units, collectively owned units, foreign-funded units, and units in Hong Kong, China; Macao, China; and Taiwan, China.

Private sector development | 5.1

This new table includes some private sector indicators from previous editions, such as private fixed investment, domestic credit to the private sector, and proceeds from privatization. In addition, it includes data from the World Bank's Private Participation in Infrastructure (PPI) Project Database.

Private fixed investment consists of outlays on additions to fixed assets—improvements to land, construction of infrastructure and buildings, and purchases of plant, machinery, and equipment—by the private sector. When direct estimates of private investment are unavailable, private fixed investment is estimated as the difference between total gross fixed investment and consolidated public investment. Total investment may be estimated directly from surveys of enterprises and administrative records or indirectly using the commodity flow method. Consolidated measures of public investment may omit important subnational units of government and in some cases may include financial as well as physical capital investment. As the difference between two estimated quantities, private fixed investment may be undervalued or overvalued and subject to large errors over time. When private domestic investment accounts for a large share of total investment, it may reflect a highly competitive and efficient private sector—or one that is subsidized and protected.

This concept of private investment is the one used by the International Finance Corporation (IFC) in its *Trends in Private Investment in Developing Countries 2000,* the source for most countries in the table. But for other countries, most notably members of the Organisation for Economic Co-operation and Development (OECD), the concepts and definitions of the 1993 System of National Accounts (SNA) are used. Since IFC data conform to the concepts and definitions of the 1968 SNA, the data are not strictly comparable. While the IFC data on private investment represent only the capital expenditure decisions of the private sector, in the 1993 SNA the term *fixed capital formation by households and corporations* includes capital expenditures by both private and public corporations. Countries reporting on this basis are footnoted in the table. (For further discussion on measuring gross capital formation see *About the data* for table 4.9.)

The data on domestic credit to the private sector are taken from the banking survey of the International Monetary Fund's (IMF) *International Financial Statistics* or, when data are unavailable, from its monetary survey. The monetary survey includes monetary authorities (the central bank) and deposit money banks. In addition to these, the banking survey includes other banking institutions, such as savings and loan institutions,

finance companies, and development banks. In some cases credit to the private sector may include credit to state-owned or partially state-owned enterprises.

Privatization—the transfer of productive assets from the public to the private sector—has been one of the defining economic changes of the past two decades. The data on proceeds from privatization measure the proceeds from the divestiture, or sale (direct or through share issues), of state-owned enterprises.

Direct sales are the most common method of privatization, accounting for more than half of privatization revenues in 1999. Direct sales enable governments to attract strategic investors who can transfer capital, technology, and managerial know-how to newly privatized enterprises. Share issues in domestic and international capital markets are the second most common method, accounting for most of the remaining sales. The estimates of privatization proceeds include data on the largest transactions, but in some cases total privatization revenues may be underreported because of lack of data.

Large sales proceeds do not necessarily imply major changes in the control of stock of state-owned enterprises. For example, selling equity may not change effective control. It may only generate revenue, with no gains in efficiency. A preliminary analysis suggests that the increase in proceeds from privatization in recent years is due to a larger number of countries privatizing a few firms rather than to a radical restructuring of ownership in many countries (Haggarty and Shirley 1997).

Private participation in infrastructure has made important contributions in improving the efficiency of infrastructure services and in extending their delivery to poor people. The privatization trend in infrastructure that began in the 1970s and 1980s took off in the 1990s. Developing countries have been at the head of this wave, pioneering better approaches to providing infrastructure services and reaping the benefits of increased competition and customer focus.

The data on investment in infrastructure projects with private participation refer to all investment (public and private) in projects in which a private company assumes operating risk during the operating period or assumes development and operating risk during the contract period. Foreign state-owned companies are considered private entities for the purposes of this measure. The data are from the World Bank's PPI Project Database, which tracks about 1,700 projects, newly owned or managed by private companies, that reached financial closure in low- and middle-income economies in 1990–99. For more information go to www.world bank.org/html/fpd/privatesector/PPIDBweb/Intro.htm.

• **Private fixed investment** covers gross outlays by the private sector (including private nonprofit agencies) on additions to its fixed domestic assets. Gross domestic fixed investment includes similar outlays by the public sector. No allowance is made for the depreciation of assets. • **Domestic credit to private sector** refers to financial resources provided to the private sector—such as through loans, purchases of nonequity securities, and trade credits and other accounts receivable—that establish a claim for repayment. For some countries these claims include credit to public enterprises. • **Proceeds from privatization** cover all sales of public assets to private entities through public offers, direct sales, management and employee buyouts, concessions or licensing agreements, and joint ventures. • **Investment in infrastructure projects with private participation** covers infrastructure projects in telecommunications, energy (electricity and natural gas transmission and distribution), transport, and water and sanitation that have reached financial closure and directly or indirectly serve the public. Movable assets, incinerators, stand-alone solid waste projects, and small projects such as windmills are excluded. The types of projects included are operations and management contracts, operations and management contracts with major capital expenditure, greenfield projects (in which a private entity or a public-private joint venture builds and operates a new facility), and divestiture.

The data on private investment are from the International Finance Corporation's *Trends in Private Investment in Developing Countries 2000,* OECD data files (see OECD, *National Accounts, 1960–97,* volumes 1 and 2), and World Bank estimates. The data on domestic credit are from the IMF's *International Financial Statistics.* The data on privatization proceeds are from various sources, including reports from official privatization agencies and World Bank estimates, supplemented by such publications as the *Financial Times, Privatization International, Institutional Investor, International Financing Review, Latin Finance, Project Finance,* the *Middle East Economic Digest,* and *Euromoney.* The data on investment in infrastructure projects with private participation are from the World Bank's Private Participation in Infrastructure (PPI) Project Database (www.worldbank.org/html/fpd/ privatesector/PPIDBweb/Intro.htm).

	Foreign direct investment % of gross capital formation		Entry and exit regulations[a]			Composite ICRG risk rating[b]	Institutional Investor credit rating[b]	Euromoney country credit-worthiness rating[b]	Moody's sovereign long-term debt rating[b]		Standard & Poor's sovereign long-term debt rating[b]	
				Repatriation of income	Repatriation of capital				Foreign currency	Domestic currency	Foreign currency	Domestic currency
	1990	1999	Entry 1999	1999	1999	December 2000	September 2000	September 2000	January 2001	January 2001	January 2001	January 2001
Albania	0.0	6.6	..	..	..	62.3	15.8	30.4	..	..	..	..
Algeria	0.0	0.1	..	..	..	59.0	33.1	37.7	..	..	..	..
Angola	–27.9	23.2	..	..	..	49.3	11.9	21.0	..	..	..	..
Argentina	9.3	44.2	F	F	F	68.8	45.8	55.0	B1	B1	BB–	BB
Armenia	..	34.0	..	..	..	56.8	..	30.5	..	..	..	..
Australia	10.9	6.7	..	..	..	80.5	82.1	88.0	Aa2	Aaa	AA+	AAA
Austria	1.7	5.6	..	..	..	81.5	87.1	92.3	Aaa	Aaa	AAA	AAA
Azerbaijan	..	32.2	..	..	..	59.5	..	36.9	..	..	..	..
Bangladesh	0.1	1.8	F	F	F	62.5	28.4	35.0	..	..	..	..
Belarus	..	3.5	..	..	..	59.8	14.4	30.7	..	..	..	..
Belgium	..	..	..	..	..	79.3	83.1	89.6	Aa1	Aa1	AA+	AA+
Benin	0.4	7.9	..	..	..	..	17.0	28.5	..	..	..	..
Bolivia	4.4	64.7	..	..	..	69.5	28.6	42.5	B1	B1	B+	BB
Bosnia and Herzegovina	..	0.0	..	..	..	..	..	15.8	..	..	..	..
Botswana	8.0	3.1	F	F	F	79.3	56.1	51.8	..	..	..	..
Brazil	1.1	21.3	F	F	F	64.5	45.0	51.3	B1	B1	BB–	BB+
Bulgaria	0.1	34.1	F	F	F	67.3	37.1	42.5	B2	B1	B+	BB–
Burkina Faso	0.0	0.1	..	..	..	61.8	19.8	32.0	..	..	..	..
Burundi	0.6	0.3	..	..	..	..	15.6	20.8	..	..	..	..
Cambodia	0.0	28.1	..	..	..	..	..	32.9	..	..	..	..
Cameroon	–5.7	3.0	..	..	..	59.0	16.3	29.7	..	..	..	..
Canada	6.4	19.6	..	..	..	84.8	89.6	89.1	Aa1	Aa1	AA+	AAA
Central African Republic	0.5	3.3	..	..	..	..	..	25.1	..	..	..	..
Chad	0.0	9.5	..	..	..	..	13.9	27.8	..	..	..	..
Chile	7.8	62.8	R	F	D	74.8	67.2	65.8	Baa1	A1	A–	AA
China	2.8	10.5	S	F	F	73.8	60.6	59.8	A3	..	BBB	..
Hong Kong, China	..	..	..	..	..	79.3	68.3	77.4	A3	Aa3	A	A+
Colombia	6.7	10.1	A	F	F	60.3	44.0	48.9	Ba2	Baa2	BB	BBB
Congo, Dem. Rep.	–1.4	0.2	..	..	..	45.0	8.8	22.9	..	..	..	..
Congo, Rep.	0.0	0.0	..	..	..	55.3	11.1	22.0	..	..	..	..
Costa Rica	8.3	25.5	..	..	..	75.8	47.5	53.4	Ba1	Ba1	BB	BB+
Côte d'Ivoire	6.6	19.2	F	F	F	54.0	24.1	32.5	..	..	..	..
Croatia	..	29.2	F	F	F	70.3	45.8	49.7	Baa3	Baa1	BBB–	BBB+
Cuba	..	..	..	..	..	62.3	14.1	10.7	Caa1	..	..	..
Czech Republic	2.4	33.7	F	F	F	73.3	60.9	63.1	Baa1	A1	A–	AA–
Denmark	4.2	24.8	..	..	..	86.3	88.9	92.8	Aaa	Aaa	AA+	AAA
Dominican Republic	7.5	30.6	..	..	..	73.8	37.0	43.8	B1	B1	B+	SD
Ecuador	6.7	28.1	F	F	F	52.8	18.3	29.3	Caa2	Caa1	B–	B–
Egypt, Arab Rep.	5.9	5.2	F	F	F	69.3	51.0	56.4	Ba1	Baa1	BBB–	A–
El Salvador	0.3	11.4	..	..	..	75.0	46.3	51.1	Baa3	Baa2	BB+	BB+
Eritrea	..	0.0	..	..	..	..	..	22.2	..	..	..	..
Estonia	7.2	23.8	F	F	F	73.8	55.1	55.7	Baa1	A1	BBB+	A–
Ethiopia	1.5	0.2	..	..	..	61.3	15.9	31.0	..	..	..	..
Finland	2.0	18.7	..	..	..	87.0	89.1	91.4	Aaa	Aaa	AA+	AA+
France	4.6	14.3	..	..	..	78.5	93.6	92.3	Aaa	Aaa	AAA	AAA
Gabon	5.7	–2.8	..	..	..	64.5	23.8	33.0	..	..	..	..
Gambia, The	0.0	20.0	..	..	..	66.3	..	37.6	..	..	..	..
Georgia	0.0	17.8	..	..	..	..	21.0	30.4	..	..	..	..
Germany	1.0	11.2	..	..	..	84.3	94.6	92.8	Aaa	Aaa	AAA	AAA
Ghana	1.8	0.9	F	F	F	53.8	29.5	37.6	..	..	..	..
Greece	5.2	3.9	F	F	F	73.3	70.0	78.7	A2	A2	A–	A–
Guatemala	4.6	4.9	..	..	..	69.8	37.1	47.3	Ba2	Ba1	..	..
Guinea	3.6	10.0	..	..	..	57.0	15.1	26.8	..	..	..	..
Guinea-Bissau	2.7	1.1	..	..	..	45.0	..	20.0	..	..	..	..
Haiti	2.2	1.9	..	..	..	55.5	12.5	24.3	..	..	..	..
Honduras	6.3	13.0	..	..	..	63.0	26.6	38.8	B2	B2	..	..

Investment climate | 5.2

	Foreign direct investment		Entry and exit regulations[a]			Composite ICRG risk rating[b]	Institutional investor credit rating[b]	Euro-money country credit-worthiness rating[b]	Moody's sovereign long-term debt rating[b]		Standard & Poor's sovereign long-term debt rating[b]	
	% of gross capital formation		Entry	Repatriation of income	Repatriation of capital				Foreign currency	Domestic currency	Foreign currency	Domestic currency
	1990	1999	1999	1999	1999	December 2000	September 2000	September 2000	January 2001	January 2001	January 2001	January 2001
Hungary	0.0	14.0	F	F	F	72.0	64.9	65.2	A3	A1	A–	A+
India	0.2	2.1	A	F	F	61.8	51.5	53.8	Ba2	Ba2	BB	BBB
Indonesia	3.1	–9.2	R	RS	RS	54.8	27.4	38.5	B3	B3	B–	B
Iran, Islamic Rep.	–1.1	0.3	..	..	..	68.5	27.0	40.6	B2	Ba2	..	..
Iraq	..	..	..	..	..	47.0	12.6	9.0	..	..	..	..
Ireland	6.3	87.6	..	..	..	84.3	88.5	89.7	Aaa	Aaa	AA+	AA+
Israel	1.1	11.1	F	F	F	68.0	64.4	72.7	A2	A2	A–	AA–
Italy	2.6	2.8	..	..	..	78.8	84.2	87.1	Aa3	Aa3	AA	AA
Jamaica	11.7	28.9	R	F	F	67.8	33.8	42.7	Ba3	Baa3	B	B+
Japan	0.2	1.1	..	..	..	83.8	87.7	90.7	Aa1	Aa2	AAA	AAA
Jordan	3.0	9.4	F	F	F	70.8	41.9	46.3	Ba3	Ba3	BB–	BBB–
Kazakhstan	1.2	56.8	..	..	..	66.0	34.4	42.5	B1	B1	BB–	BB
Kenya	3.4	1.0	R	F	F	60.3	25.0	37.6	..	..	..	..
Korea, Dem. Rep.	..	..	..	..	..	48.5	6.2	4.7	..	..	..	..
Korea, Rep.	0.8	8.5	R	F	F	78.0	63.3	66.3	Baa2	Baa1	BBB	A
Kuwait	..	2.0	..	..	..	80.3	64.4	73.3	Baa1	..	A	A+
Kyrgyz Republic	..	15.7	..	..	..	..	23.8	35.8	..	..	..	..
Lao PDR	..	14.7	..	..	..	..	..	24.0	..	..	..	..
Latvia	1.1	21.1	F	F	F	71.0	47.9	53.1	Baa2	A2	BBB	A–
Lebanon	1.2	4.2	F	F	F	61.0	36.8	46.8	B1	B1	B+	BB–
Lesotho	5.1	64.1	..	..	..	..	25.0	36.4	..	..	..	..
Libya	..	..	..	..	..	66.8	31.5	19.3	..	..	..	..
Lithuania	0.0	20.0	F	F	F	71.8	43.7	50.8	Ba1	Baa1	BBB–	BBB+
Macedonia, FYR	..	4.1	..	..	..	..	..	37.4	..	..	..	..
Madagascar	4.2	2.9	..	..	..	63.0	..	29.5	..	..	..	..
Malawi	0.0	22.4	..	..	..	56.8	19.6	31.7	..	..	..	..
Malaysia	16.4	8.8	R	F	F	75.8	59.5	61.1	Baa2	A3	BBB	A
Mali	–1.3	3.4	..	..	..	61.8	14.0	28.1	..	..	..	..
Mauritania	3.4	3.5	..	..	..	..	..	27.2	..	..	..	..
Mauritius	5.0	4.2	R	F	F	..	54.6	59.1	Baa2	A2	..	..
Mexico	4.3	10.5	F	F	F	73.0	56.7	59.7	Baa3	Baa1	BB+	BBB+
Moldova	0.0	13.1	..	..	..	49.5	15.8	29.2	B3	Caa1	..	..
Mongolia	..	12.7	..	..	..	67.0	..	30.6	..	..	B	B
Morocco	2.5	0.0	F	F	F	67.8	47.3	55.1	Ba1	..	BB	BBB
Mozambique	2.3	29.6	..	..	..	55.3	19.8	28.6	..	..	..	..
Myanmar	..	..	..	..	..	58.3	17.4	26.4	..	..	..	..
Namibia	..	..	F	F	F	75.8	37.4	23.6	..	..	..	..
Nepal	0.9	0.4	..	..	..	..	26.9	32.7	..	..	..	..
Netherlands	14.8	39.4	..	..	..	88.0	94.5	92.9	Aaa	Aaa	AAA	AAA
New Zealand	21.3	7.4	..	..	..	77.5	78.7	85.3	Aa2	Aaa	AA+	AAA
Nicaragua	0.0	30.7	..	..	..	52.3	21.8	26.3	B2	B2	..	..
Niger	–0.5	7.3	..	..	..	61.8	10.9	25.4	..	..	..	..
Nigeria	14.0	11.8	R	F	F	59.3	18.1	32.1	..	..	..	..
Norway	3.7	8.6	..	..	..	90.5	90.1	94.2	Aaa	Aaa	AAA	AAA
Oman	10.2	..	F	F	F	78.3	54.9	66.2	Baa2	Baa2	BBB–	BBB
Pakistan	3.2	4.3	F	F	F	53.8	19.2	32.0	Caa1	Caa1	B–	B+
Panama	14.8	0.7	..	..	..	72.8	46.7	52.2	Ba1	..	BB+	BB+
Papua New Guinea	19.7	46.3	..	..	..	66.5	30.2	37.2	B1	B1	B+	BB
Paraguay	6.3	5.3	..	..	..	65.8	32.5	41.3	B2	B1	B	BB–
Peru	0.9	17.3	F	F	F	69.5	42.3	39.1	Ba3	Baa3	BB–	BB+
Philippines	5.0	4.0	S	F	F	65.0	49.4	52.8	Ba1	Baa3	BB+	BBB+
Poland	0.6	17.8	F	F	F	73.8	62.2	63.6	Baa1	A2	BBB+	A+
Portugal	13.6	10.8	..	..	..	78.8	83.6	83.3	Aa2	Aa2	AA	AA
Puerto Rico	..	..	..	..	..	..	..	..	..	..	..	..
Romania	0.0	15.4	F	F	F	58.5	30.3	36.6	B3	Caa1	B–	B
Russian Federation	0.0	5.3	F	F	F	66.3	26.7	37.9	B3	B3	B–	B–

5.2 | Investment climate

	Foreign direct investment		Entry and exit regulations[a]			Composite ICRG risk rating[b]	Institutional Investor credit rating[b]	Euromoney country creditworthiness rating[b]	Moody's sovereign long-term debt rating[b]		Standard & Poor's sovereign long-term debt rating[b]	
	% of gross capital formation		Entry	Repatriation of income	Repatriation of capital				Foreign currency	Domestic currency	Foreign currency	Domestic currency
	1990	1999	1999	1999	1999	December 2000	September 2000	September 2000	January 2001	January 2001	January 2001	January 2001
Rwanda	2.1	0.6	..	..	..			21.1	..	..	..	..
Saudi Arabia[c]	..	..	C	RS	RS	76.0	57.0	68.3	Baa3	Ba1	..	..
Senegal	7.2	4.0	..	..	..	62.5	23.5	34.3	..	..	B+	B+
Sierra Leone	37.9	51.0	..	..	..	37.8	6.4	23.1	..	..	..	..
Singapore	41.5	25.1	..	..	..	90.5	87.8	90.0	Aa1	Aaa	AAA	AAA
Slovak Republic	0.0	5.6	F	F	F	71.5	49.1	53.0	Ba1	Baa2	BB+	BBB+
Slovenia	5.0	3.2	C	RS	RS	75.8	67.0	68.9	A2	Aa3	A	AA
South Africa	..	6.7	F	F	F	68.0	55.1	57.7	Baa3	Baa1	BBB–	A–
Spain	10.1	6.5	..	..	..	79.0	86.1	87.3	Aa2	Aa2	AA+	AA+
Sri Lanka	2.4	4.1	R	RS	RS	59.0	33.3	39.8	..	..	..	..
Sudan	..	..	..	..	..	49.5	8.7	25.5	..	..	..	..
Sweden	3.5	147.2	..	..	..	84.0	87.2	91.1	Aa1	Aaa	AA+	AAA
Switzerland	9.3	14.2	..	..	..	89.5	95.6	96.9	Aaa	Aaa	AAA	AAA
Syrian Arab Republic	3.7	1.6	..	..	..	69.3	23.1	39.0	..	..	..	..
Tajikistan	..	14.9	..	..	..		14.2	17.8	..	..	..	..
Tanzania	0.0	12.1	..	..	..	59.5	20.3	28.9	..	..	..	..
Thailand	6.9	23.8	R	F	F	75.3	53.2	59.5	Baa3	Baa1	BBB–	A–
Togo	0.0	15.9	..	..	..	58.5	17.6	28.6	..	..	..	..
Trinidad and Tobago	17.1	44.0	R	F	F	72.8	52.0	54.2	Baa3	Baa1	BBB–	BBB+
Tunisia	1.9	6.3	F	F	F	72.5	54.5	57.5	..	Baa2	BBB	A
Turkey	1.9	1.8	F	F	F	55.5	46.8	52.7	B1	..	B+	B+
Turkmenistan	..	5.4	..	..	..		21.9	31.8	B2	..	..	..
Uganda	0.0	21.1	..	..	..	64.3	22.7	33.7	..	..	..	..
Ukraine	0.0	6.5	F	F	F	61.8	17.7	33.1	Caa1	Caa3	..	..
United Arab Emirates	..	..	..	..	..	80.0	66.3	75.6	..	..	..	..
United Kingdom	16.3	33.4	..	..	..	83.5	89.1	91.5	Aaa	Aaa	AAA	AAA
United States	4.8	10.6	..	..	..	82.0	91.6	94.3	Aaa	Aaa	AAA	AAA
Uruguay	0.0	7.2	..	..	..	73.3	53.5	56.8	Baa3	Baa3	BBB–	BBB+
Uzbekistan	1.5	7.1	..	..	..		20.2	34.2	..	..	..	..
Venezuela, RB	9.1	20.0	F	F	F	70.0	37.9	43.8	B2	B3	B	..
Vietnam	1.9	22.1	..	..	..	70.0	28.0	38.4	B1	..	..	..
West Bank and Gaza	..	..	..	..	..		..	..	..	..	..	..
Yemen, Rep.	–19.2	–11.8	..	..	..	63.5	..	34.0	..	..	..	..
Yugoslavia, FR (Serb./Mont.)	..	..	..	..	..	45.5	12.7	14.8	..	..	..	..
Zambia	35.7	29.6	..	..	..	57.3	15.5	27.0	..	..	..	..
Zimbabwe	–0.8	9.2	R	F	F	40.3	17.1	33.4	..	..	..	..
World	**4.2 w**	**10.2 w**				**67.8 m**	**37.0 m**	**38.8 m**				
Low income	1.1	3.0				57.3	17.6	29.4				
Middle income	2.3	14.0				69.5	44.0	48.9				
Lower middle income	1.8	9.7				67.6	34.4	41.3				
Upper middle income	3.0	17.8				72.8	54.8	56.2				
Low & middle income	2.1	12.4				63.3	28.0	35.0				
East Asia & Pacific	3.5	9.6				67.0	39.8	38.4				
Europe & Central Asia	0.3	11.6				66.2	30.3	36.9				
Latin America & Carib.	3.8	22.3				69.5	40.1	45.6				
Middle East & N. Africa	2.3	0.8				68.5	39.3	46.3				
South Asia	0.5	2.4				60.4	27.6	33.8				
Sub-Saharan Africa	2.0	9.3				58.8	17.9	28.6				
High income	4.8	9.6				81.8	87.2	89.7				
Europe EMU	5.8	12.9				80.4	87.8	90.5				

a. Entry and exit regulations are classified as free (F), relatively free (R), delayed (D), special classes of shares (S), authorized investors only (A), restricted (RS), and closed (C). For explanations of these terms see *About the data*. b. This copyrighted material is reprinted with permission from the following data providers: PRS Group, 6320 Fly Road, Suite 102, PO Box 248, East Syracuse, NY 13057; Institutional Investor Inc., 488 Madison Avenue, New York, NY 10022; Euromoney Publications PLC, Nestor House, Playhouse Yard, London EC4V 5EX, UK; Moody's Investor Service, 99 Church Street, New York, NY 10007; and Standard & Poor's Rating Services, The McGraw-Hill Companies, Inc., 1221 Avenue of the Americas, New York, NY 10020. Prior written consent from the original data providers cited must be obtained for third-party use of these data. c. Foreigners are barred from investing directly in the Saudi stock market, but they may invest indirectly through mutual funds.

Investment climate | 5.2

As investment portfolios become increasingly global, investors as well as governments seeking to attract investment must have a good understanding of trends in foreign direct investment and country risk. This table presents information on foreign direct investment, country risk and creditworthiness ratings from several major international rating services, and information on the regulation of entry to and exit from emerging stock markets reported by Standard & Poor's.

The statistics on foreign direct investment are based on balance of payments data reported by the International Monetary Fund (IMF), supplemented by data on net foreign direct investment reported by the Organisation for Economic Co-operation and Development and official national sources. (For a detailed discussion of data on foreign direct investment see *About the data* for table 6.7.)

Entry and exit restrictions on investments are among the mechanisms by which countries attempt to reduce the risk to their economies associated with foreign investment. Yet such restrictions may increase the risk or uncertainty perceived by investors. Many countries close industries considered strategic to foreign or non-resident investors. And national law or corporate policy may limit foreign investment in a company or in certain classes of stocks.

The entry and exit regulations summarized in the table refer to "new money" investment by foreign institutions; other regulations may apply to capital invested through debt conversion schemes or to capital from other sources. The regulations reflected here are formal ones. But even formal regulations may have very different effects in different countries because of differences in the bureaucratic culture, the speed with which applications are processed, and the extent of red tape. The regulations on entry are evaluated using the terms *free* (no significant restrictions), *relatively free* (some registration procedures required to ensure repatriation rights), *special classes* (foreigners restricted to certain classes of stocks designated for foreign investors), *authorized investors only* (only approved foreign investors may buy stocks), and *closed* (closed or access severely restricted, as for nonresident nationals only). Regulations on repatriation of income and capital are evaluated as *free* (repatriation done routinely) or *restricted* (repatriation requires registration with or permission of a government agency that may restrict the timing of exchange release).

Most risk ratings are numerical or alphabetical indexes, with a higher number or a letter closer to the beginning of the alphabet meaning lower risk (a good prospect). (For more on the rating processes of the rat-

ing agencies see the data sources.) Risk ratings may be highly subjective, reflecting external perceptions that do not always capture the actual situation in a country. But these subjective perceptions are the reality that policymakers face. Countries not rated by credit risk rating agencies typically do not attract registered flows of private capital. The risk ratings presented here are included for their analytical usefulness and are not endorsed by the World Bank.

The PRS Group's *International Country Risk Guide* (ICRG) collects information on 22 components of risk, groups it into three major categories (political, financial, and economic), and converts it into a single numerical risk assessment ranging from 0 to 100. Ratings below 50 indicate very high risk, and those above 80 very low risk. Ratings are updated monthly.

Institutional Investor country credit ratings are based on information provided by leading international banks. Responses are weighted using a formula that gives more importance to responses from banks with greater worldwide exposure and more sophisticated country analysis systems. Countries are rated on a scale of 0 to 100 (highest risk to lowest), and ratings are updated every six months.

Euromoney country creditworthiness ratings are based on nine weighted categories (covering debt, economic performance, political risk, and access to financial and capital markets) that assess country risk. The ratings, also on a scale of 0 to 100 (highest risk to lowest), are based on polls of economists and political analysts supplemented by quantitative data such as debt ratios and access to capital markets.

Moody's sovereign long-term debt ratings are opinions of the ability of entities to honor senior unsecured financial obligations and contracts denominated in foreign currency (foreign currency issuer ratings) or in their domestic currency (domestic currency issuer ratings).

Standard & Poor's ratings of sovereign long-term foreign and domestic currency debt are based on current information furnished by obligors or obtained by Standard & Poor's from other sources it considers reliable. A Standard & Poor's issuer credit rating (one form of which is a sovereign credit rating) is a current opinion of an obligor's capacity and willingness to pay its financial obligations as they come due (its creditworthiness). This opinion does not apply to any specific financial obligation, as it does not take into account the nature and provisions of obligations, their standing in bankruptcy or liquidation, statutory preferences, or the legality and enforceability of obligations.

• **Foreign direct investment** is net inflows of investment to acquire a lasting management interest (10 percent or more of voting stock) in an enterprise operating in an economy other than that of the investor. It is the sum of equity capital, reinvestment of earnings, other long-term capital, and short-term capital as shown in the balance of payments. Gross capital formation (gross domestic investment in previous editions) is the sum of gross fixed capital formation, changes in inventories, and acquisitions less disposals of valuables. • **Regulations on entry to emerging stock markets** are assessed on a scale from free to closed (see *About the data*). • **Regulations on repatriation of income** (dividends, interest, and realized capital gains) and **repatriation of capital from emerging stock markets** are evaluated as free or restricted (see *About the data*). • **Composite International Country Risk Guide (ICRG) risk rating** is an overall index, ranging from 0 to 100, based on 22 components of risk. • **Institutional Investor credit rating** ranks, from 0 to 100, the chances of a country's default. • **Euromoney country creditworthiness rating** ranks, from 0 to 100, the risk of investing in an economy. • **Moody's sovereign foreign and domestic currency long-term debt ratings** assess the risk of lending to governments. An entity's ability to meet its senior financial obligations is rated from Aaa (offering exceptional financial security) to C (usually in default, with potential recovery values low). Modifiers 1–3 are applied to ratings from Aa to B, with 1 indicating a high ranking in the rating category. • **Standard & Poor's sovereign foreign and domestic currency long-term debt ratings** range from AAA (extremely strong capacity to meet financial commitments) through CC (currently highly vulnerable). Ratings from AA to CCC may be modified by a plus or minus sign to show relative standing in the category. An obligor rated SD (selective default) has failed to pay one or more financial obligations when due.

The data on foreign direct investment are based on estimates compiled by the IMF in its *Balance of Payments Statistics Yearbook,* supplemented by World Bank staff estimates. The data on entry and exit regulations are from Standard & Poor's *Emerging Stock Markets Factbook 2000.* The country risk and creditworthiness ratings are from the PRS Group's monthly *International Country Risk Guide,* the monthly *Institutional Investor,* the monthly *Euromoney,* Moody's Investors Service's *Sovereign, Subnational and Sovereign-Guaranteed Issuers,* and Standard & Poor's Sovereign List in *Credit Week.*

5.3 | Stock markets

	Market capitalization				Value traded		Turnover ratio		Listed domestic companies		S&P/IFC investable index	
	$ millions		% of GDP		% of GDP		value of shares traded as % of capitalization				% change in price index	
	1990	**2000**	**1990**	**1999**	**1990**	**1999**	**1990**	**2000**	**1990**	**2000**	**1999**	**2000**
Albania	..	..	..	..	..	..	..	..	..	..	..	..
Algeria	..	..	..	..	..	..	..	..	..	..	..	..
Angola	..	..	..	..	..	..	..	..	..	..	..	..
Argentina	3,268	166,068	2.3	29.6	0.6	2.7	33.6	4.8	179	127	33.4	−25.1
Armenia	..	25	..	1.4	..	0.1	..	4.6	..	95	..	..
Australia	108,879	427,683	35.1	105.9	12.9	26.2	31.6	28.0	1,089	1,217	..	..
Austria	11,476	33,025	7.1	15.9	11.5	6.1	110.3	37.9	97	97	..	..
Azerbaijan	..	4	..	0.1	..	..	..	..	..	2	..	..
Bangladesh	321	1,186	1.1	1.9	0.0	1.7	1.5	74.4	134	221	−17.5 [a]	28.5 [a]
Belarus	..	..	..	..	..	..	..	..	..	..	..	..
Belgium	65,449	184,942	33.1	74.5	3.2	23.8	..	27.5	182	172	..	..
Benin	..	..	..	..	..	..	..	..	..	..	..	..
Bolivia	..	116	..	1.4	..	0.0	..	1.0	..	18	..	..
Bosnia and Herzegovina	..	..	..	..	..	..	..	..	..	..	..	..
Botswana	261	978	6.7	17.5	0.2	0.6	6.1	4.8	9	16	45.6 [a]	−6.9 [a]
Brazil	16,354	226,152	3.5	30.3	1.2	11.6	23.6	43.5	581	459	66.9	−10.3
Bulgaria	..	617	..	5.7	..	0.4	..	9.2	..	503	−23.7 [a]	−30.0 [a]
Burkina Faso	..	..	..	..	..	..	..	..	..	..	..	..
Burundi	..	..	..	..	..	..	..	..	..	..	..	..
Cambodia	..	..	..	..	..	..	..	..	..	..	..	..
Cameroon	..	..	..	..	..	..	..	..	..	..	..	..
Canada	241,920	800,914	42.2	126.1	12.4	57.4	26.7	54.2	1,144	3,767	..	..
Central African Republic	..	..	..	..	..	..	..	..	..	..	..	..
Chad	..	..	..	..	..	..	..	..	..	..	..	..
Chile	13,645	60,401	45.0	101.1	2.6	10.2	6.3	9.4	215	258	35.8	−15.2
China	2,028	580,991	0.5	33.4	0.2	38.1	158.9	158.3	14	1,086	102.2	−9.8
Hong Kong, China	83,397	609,090	111.5	383.2	46.3	154.1	43.1	51.4	284	695	..	..
Colombia	1,416	9,560	3.5	13.4	0.2	0.8	5.6	3.8	80	126	−19.7	−43.8
Congo, Dem. Rep.	..	..	..	..	..	..	..	..	..	..	..	..
Congo, Rep.	..	..	..	..	..	..	..	..	..	..	..	..
Costa Rica	475	2,303	5.6	15.2	0.1	1.4	5.8	12.0	82	22	..	..
Côte d'Ivoire	549	1,185	5.1	13.5	0.2	0.8	3.4	2.6	23	41	−12.1 [a]	−25.6 [a]
Croatia	..	2,742	..	12.7	..	0.4	..	7.4	2	64	−18.1 [a]	10.9 [a]
Cuba	..	..	..	..	..	..	..	..	..	..	..	..
Czech Republic	..	11,002	..	22.2	..	7.8	..	60.3	..	131	4.2	−0.6
Denmark	39,063	105,293	29.3	60.4	8.3	35.2	28.0	60.0	258	233	..	..
Dominican Republic	..	141	..	0.8	..	..	..	..	..	6	..	..
Ecuador	69	704	0.5	2.2	..	0.1	0.0	5.5	65	30	−77.1 [a]	33.0 [a]
Egypt, Arab Rep.	1,765	28,741	4.1	36.8	0.3	10.1	..	34.7	573	1,076	24.2	−45.6
El Salvador	..	2,141	..	17.2	..	0.4	..	2.7	..	40	..	..
Eritrea	..	..	..	..	..	..	..	..	..	..	..	..
Estonia	..	1,846	..	34.2	..	5.4	..	18.9	..	23	42.3 [a]	4.5 [a]
Ethiopia	..	..	..	..	..	..	..	..	..	..	..	..
Finland	22,721	349,409	16.6	269.5	2.9	86.1	..	44.3	73	147	..	..
France	314,384	1,475,457	25.9	103.0	9.6	53.8	..	62.4	578	968	..	..
Gabon	..	..	..	..	..	..	..	..	..	..	..	..
Gambia, The	..	..	..	..	..	..	..	..	..	..	..	..
Georgia	..	..	..	..	..	..	..	..	..	..	..	..
Germany	355,073	1,432,190	22.2	67.8	21.4	64.3	139.3	107.5	413	933	..	..
Ghana	76	502	1.2	11.8	..	0.3	0.0	1.5	13	22	−33.5 [a]	−50.9 [a]
Greece	15,228	110,839	17.9	163.3	4.6	150.9	36.3	63.7	145	329	64.4	−44.6
Guatemala	..	215	..	1.2	..	0.0	..	2.9	..	5	..	..
Guinea	..	..	..	..	..	..	..	..	..	..	..	..
Guinea-Bissau	..	..	..	..	..	..	..	..	..	..	..	..
Haiti	..	..	..	..	..	..	..	..	..	..	..	..
Honduras	40	458	1.3	8.7	0.0	..	0.0	..	26	71	..	..

Stock markets | 5.3

	Market capitalization				Value traded		Turnover ratio		Listed domestic companies		S&P/IFC investable index	
	$ millions		% of GDP		% of GDP		value of shares traded as % of capitalization				% change in price index	
	1990	2000	1990	1999	1990	1999	1990	2000	1990	2000	1999	2000
Hungary	505	12,021	1.5	33.7	0.3	29.7	6.3	90.7	21	60	15.2	−28.2
India	38,567	148,064	12.2	41.3	6.9	27.3	65.9	133.6	2,435	5,937	81.0	−31.1
Indonesia	8,081	26,834	7.1	45.0	3.5	14.0	75.8	32.9	125	290	95.1	−61.0
Iran, Islamic Rep.	34,282	21,830	..	19.7	..	2.0	30.4	12.4	97	295	..	..
Iraq	..	..	..	..	..	..	..	..	..	..	..	..
Ireland	..	42,458	..	45.5	..	54.1	..	90.9	..	84	..	..
Israel	3,324	64,081	6.3	63.3	10.5	15.3	95.8	36.3	216	654	54.9	14.7
Italy	148,766	728,273	13.5	62.2	3.9	45.8	26.8	82.7	220	241	..	..
Jamaica	911	3,582	21.5	36.7	0.8	0.6	3.4	2.5	44	46	1.9 a	45.6 a
Japan	2,917,679	4,546,937	98.2	104.6	54.0	42.5	43.8	52.5	2,071	2,470	36.8 b	−27.2 b
Jordan	2,001	4,943	49.8	72.2	10.1	6.8	20.0	7.7	105	163	−3.6	−24.5
Kazakhstan	..	2,260	..	14.3	..	0.2	..	1.2	..	17	..	..
Kenya	453	1,283	5.3	13.2	0.1	0.7	2.2	3.6	54	57	−27.5 a	−8.1 a
Korea, Dem. Rep.	..	..	..	..	..	..	..	..	..	..	..	..
Korea, Rep.	110,594	148,649	43.8	75.8	30.1	180.3	61.3	233.2	669	704	106.5	−57.1
Kuwait	..	18,814	..	63.6	..	20.5	..	32.8	..	76	..	..
Kyrgyz Republic	..	..	..	..	..	0.0	..	..	..	..	..	..
Lao PDR	..	..	..	..	..	..	..	..	..	..	..	..
Latvia	..	563	..	6.2	..	0.7	..	48.6	..	64	−9.5 a	41.2 a
Lebanon	..	1,583	..	13.8	..	1.9	..	6.7	..	12	−18.0	−18.6 a
Lesotho	..	..	..	..	..	..	..	..	..	..	..	..
Libya	..	..	..	..	..	..	..	..	..	..	..	..
Lithuania	..	1,588	..	10.7	..	2.7	..	14.8	..	54	9.5 a	4.9 a
Macedonia, FYR	..	8	..	0.2	..	0.8	..	348.3	..	2	..	..
Madagascar	..	..	..	..	..	..	..	..	..	..	..	..
Malawi	..	..	..	..	..	..	..	..	..	..	..	..
Malaysia	48,611	116,935	110.4	184.0	24.7	61.4	24.6	44.6	282	795	44.5	−23.3
Mali	..	..	..	..	..	..	..	..	..	..	..	..
Mauritania	..	..	..	..	..	..	..	..	..	..	..	..
Mauritius	268	1,331	10.1	38.7	0.2	1.8	1.9	5.0	13	40	−6.7 a	−19.4 a
Mexico	32,725	125,204	12.5	31.8	4.6	7.5	44.0	32.3	199	179	78.5	−21.5
Moldova	..	38	..	3.3	..	0.3	..	97.9	..	58	..	..
Mongolia	..	32	..	3.5	..	3.8	..	7.3	..	418	..	..
Morocco	966	10,899	3.7	39.1	0.2	7.2	..	9.2	71	53	−7.8	−19.1
Mozambique	..	..	..	..	..	..	..	..	..	..	..	..
Myanmar	..	..	..	..	..	..	..	..	..	..	..	..
Namibia	21	311	0.7	22.5	..	0.7	0.0	4.5	3	13	5.3	−37.8 a
Nepal	..	418	..	8.4	..	0.5	..	6.9	..	108	..	..
Netherlands	119,825	695,209	40.5	176.6	13.6	239.2	29.0	145.1	260	344	..	..
New Zealand	8,835	28,352	20.5	51.9	4.5	21.9	17.3	45.0	171	114	..	..
Nicaragua	..	..	..	..	..	..	..	..	..	..	..	..
Niger	..	..	..	..	..	..	..	..	..	..	..	..
Nigeria	1,372	4,237	4.8	8.4	0.0	0.4	0.9	7.3	131	195	−10.3 a	−10.3 a
Norway	26,130	63,696	22.6	41.6	12.1	35.4	54.4	90.2	112	195	..	..
Oman	1,061	3,463	9.4	29.4	0.9	13.0	12.3	14.2	55	131	7.2	7.2 a
Pakistan	2,850	6,581	7.1	12.0	0.6	36.2	8.7	475.5	487	762	37.5	−16.7
Panama	226	3,584	3.4	37.5	0.0	0.5	0.9	1.5	13	31	..	..
Papua New Guinea	..	..	..	..	..	..	..	..	..	..	..	..
Paraguay	..	423	..	5.5	..	0.2	..	3.5	..	55	..	..
Peru	812	10,562	3.1	25.8	0.4	4.4	19.3	12.6	294	230	19.7	−28.1
Philippines	5,927	51,554	13.4	62.8	2.7	25.7	13.6	15.8	153	230	0.9	−43.6
Poland	144	31,279	0.2	19.1	0.0	7.2	89.7	49.9	9	225	22.3	−3.5
Portugal	9,201	66,488	13.0	58.5	2.4	35.9	16.9	63.0	181	125	38.4	38.4
Puerto Rico	..	..	..	..	..	..	..	..	..	..	..	..
Romania	..	1,069	..	2.6	..	0.9	..	23.1	..	5,555	−36.5 a	−25.3 a
Russian Federation	244	38,922	0.0	18.0	..	0.7	..	36.9	13	249	284.0	−32.2

5.3 | Stock markets

	Market capitalization				Value traded		Turnover ratio		Listed domestic companies		S&P/IFC investable index	
	$ millions		% of GDP		% of GDP		value of shares traded as % of capitalization				% change in price index	
	1990	2000	1990	1999	1990	1999	1990	2000	1990	2000	1999	2000
Rwanda	..	..	..	..	..	..	..	..	..	..	..	..
Saudi Arabia	48,213	67,171	40.8	43.4	1.9	10.6	..	27.1	59	75	42.3 [a]	42.3 [a]
Senegal	..	..	..	..	..	..	..	..	..	..	..	..
Sierra Leone	..	..	..	..	..	..	..	..	..	..	..	..
Singapore	34,308	198,407	93.6	233.6	55.4	115.4	..	66.9	150	355	..	..
Slovak Republic	..	742	..	3.7	..	2.4	..	129.8	..	838	-24.2	0.4
Slovenia	..	2,547	..	10.9	..	3.7	..	20.7	24	38	-3.7 [a]	-9.5 [a]
South Africa	137,540	204,952	122.8	200.2	7.3	55.6	..	33.9	732	616	56.1	-17.3
Spain	111,404	431,668	21.7	72.4	8.0	124.9	..	178.5	427	718	..	..
Sri Lanka	917	1,074	11.4	9.9	0.5	1.3	5.8	11.0	175	239	-6.0	-41.7
Sudan	..	..	..	..	..	..	..	..	..	..	..	..
Sweden	97,929	373,278	41.2	156.4	7.4	99.8	14.9	73.1	258	277	..	..
Switzerland	160,044	693,127	70.1	268.1	29.6	208.5	..	78.0	182	239	..	..
Syrian Arab Republic	..	..	..	..	..	..	..	..	..	..	..	..
Tajikistan	..	..	..	..	..	..	..	..	..	..	..	..
Tanzania	..	181	..	2.1	..	0.1	..	3.4	..	4	..	..
Thailand	23,896	29,489	28.0	46.9	26.8	33.5	92.6	53.2	214	381	42.3	-54.1
Togo	..	..	..	..	..	..	..	..	..	..	..	..
Trinidad and Tobago	696	4,330	13.7	63.6	1.1	1.4	10.0	3.1	30	27	-3.3 [a]	8.5 [a]
Tunisia	533	2,828	4.3	12.9	0.2	2.0	3.3	23.3	13	44	16.8 [a]	9.0 [a]
Turkey	19,065	69,659	12.6	60.7	3.9	43.8	42.5	206.2	110	315	254.5	-51.2
Turkmenistan	..	..	..	..	..	..	..	..	..	..	..	..
Uganda	..	..	..	..	..	..	..	..	..	..	..	..
Ukraine	..	1,881	..	2.9	..	0.3	..	19.6	..	139	20.2 [a]	75.2 [a]
United Arab Emirates	..	28,211	..	70.5	..	..	..	..	..	53	..	..
United Kingdom	848,866	2,933,280	85.9	203.4	28.2	95.6	33.4	51.9	1,701	1,945	14.5 [c]	-10.2 [c]
United States	3,059,434	16,635,114	53.2	181.8	30.5	202.9	53.4	123.5	6,599	7,651	19.5 [d]	-10.1 [d]
Uruguay	..	168	..	0.8	..	0.0	..	0.9	36	17	..	..
Uzbekistan	..	119	..	0.7	..	0.2	..	..	..	4	..	..
Venezuela, RB	8,361	8,128	17.2	7.3	4.6	0.8	43.0	8.9	76	85	-12.4	18.7
Vietnam	..	..	..	..	..	..	..	..	..	..	..	..
West Bank and Gaza	..	848	..	20.1	..	3.6	..	20.9	..	22	..	..
Yemen, Rep.	..	..	..	..	..	..	..	..	..	..	..	..
Yugoslavia, FR (Serb./Mont.)	..	10,817	..	..	..	..	..	0.0	..	16	..	..
Zambia	..	291	..	9.2	..	0.4	..	4.7	..	8	..	..
Zimbabwe	2,395	2,432	27.3	44.8	0.6	4.0	2.9	10.8	57	69	140.6	-24.6

World	9,399,659 s	36,030,812 s	50.7 w	119.0 w	28.5 w	102.6 w	48.3 w	87.6 w	25,424 s	49,612 s		
Low income	54,588	268,082	9.8	31.7	4.7	19.6	53.8	114.4	3,446	8,332		
Middle income	430,570	2,159,585	21.2	41.1	8.0	30.6	78.3	81.6	4,914	16,539		
Lower middle income	58,226	751,776	5.9	31.0	..	22.8	..	130.8	1,833	11,420		
Upper middle income	372,344	1,407,809	27.3	49.8	8.5	37.2	36.5	28.3	3,081	5,119		
Low & middle income	485,158	2,427,667	19.9	39.8	7.6	29.0	70.8	87.6	8,360	24,871		
East Asia & Pacific	197,109	955,379	21.3	52.4	13.2	68.1	117.2	139.4	1,443	3,754		
Europe & Central Asia	19,065	265,209	2.1	24.6	..	11.2	..	46.6	110	8,968		
Latin America & Carib.	78,470	584,986	7.7	29.7	2.1	7.3	29.7	20.8	1,748	1,938		
Middle East & N. Africa	5,265	151,895	27.8	33.9	1.5	7.3	..	32.5	817	1,874		
South Asia	42,655	194,475	10.8	34.0	5.6	25.2	53.9	128.3	3,231	7,199		
Sub-Saharan Africa	142,594	275,724	52.0	121.0	..	32.3	..	23.0	1,011	1,138		
High income	8,914,501	33,603,143	55.3	138.7	31.8	120.9	49.3	94.0	17,064	24,741		
Europe EMU	1,168,755	5,475,059	22.0	84.0	7.3	71.0	..	94.0	2,485	3,880		

Note: Because aggregates for market capitalization are unavailable for 2000, those shown refer to 1999.

a. Data refer to the S&P/IFC Global index. b. Data refer to the Nikkei 225 index. c. Data refer to the FT 100 index. d. Data refer to the S&P 500 index.

Stock markets | 5.3

Financial market development is closely related to an economy's overall development. At low levels of economic development, commercial banks tend to dominate the financial system. As economies grow, specialized financial intermediaries and equity markets develop.

The stock market indicators presented in the table include measures of size (market capitalization and number of listed domestic companies) and liquidity (value traded as a percentage of GDP, and turnover ratio). The comparability of such indicators between countries may be limited by conceptual and statistical weaknesses, such as inaccurate reporting and differences in accounting standards. The percentage change in stock market prices in U.S. dollars, from the Standard & Poor's Investable (S&P/IFCI) and Global (S&P/IFCG) country indexes, is an important measure of overall performance. Regulatory and institutional factors that can affect investor confidence, such as the existence of a securities and exchange commission and the quality of investor protection laws, may influence the functioning of stock markets but are not included in this table.

Stock market size can be measured in a number of ways, each of which may produce a different ranking among countries. Market capitalization shows the overall size of the stock market in U.S. dollars and as a percentage of GDP. The number of listed domestic companies is another measure of market size. Market size is positively correlated with the ability to mobilize capital and diversify risk.

Market liquidity, the ability to easily buy and sell securities, is measured by dividing the total value traded by GDP. This indicator complements the market capitalization ratio by showing whether market size is matched by trading. The turnover ratio—the value of shares traded as a percentage of market capitalization—is also a measure of liquidity, as well as of transactions costs. (High turnover indicates low transactions costs.) The turnover ratio complements the ratio of value traded to GDP, because the turnover ratio is related to the size of the market and the value traded ratio to the size of the economy. A small, liquid market will have a high turnover ratio but a low value traded ratio. Liquidity is an important attribute of stock markets because, in theory, liquid markets improve the allocation of capital and enhance prospects for long-term economic growth. A more comprehensive measure of liquidity would include trading costs and the time and uncertainty in finding a counterpart in settling trades.

Standard & Poor's maintains a series of indexes for investors interested in investing in stock markets in developing countries. (Standard & Poor's acquired these indexes of emerging market performance from the International Finance Corporation on 30 December 1999, as well as rights to publish the *Emerging Stock Markets Factbook*.) At the core of the Standard & Poor's family of emerging market indexes, the S&P/IFCG indexes are intended to represent the most active stocks in the markets they cover and to be the broadest possible indicator of market movements. The S&P/IFCI indexes apply the same calculation methodology as the S&P/IFCG indexes but include only the subset of S&P/IFCG markets that Standard & Poor's has determined to be "investable." The indexes are designed to measure returns on emerging market stocks that are legally and practically open to foreign portfolio investment. The S&P/IFCG indexes cover 54 markets, providing regular updates on 2,228 stocks; the S&P/IFCI indexes cover 30 markets and 1,197 stocks. They are widely used benchmarks for international portfolio management. See Standard & Poor's (2000b) for further information on the indexes.

Because markets included in Standard & Poor's emerging markets category vary widely in level of development, it is best to look at the entire category to identify the most significant market trends. And it is useful to remember that stock market trends may be distorted by currency conversions, especially when a currency has registered a significant devaluation.

Figure 5.3

Trends in market capitalization are mixed in developing regions

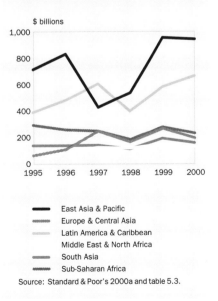

Source: Standard & Poor's 2000a and table 5.3.

• **Market capitalization** (also known as market value) is the share price times the number of shares outstanding. • **Value traded** refers to the total value of shares traded during the period. • **Turnover ratio** is the total value of shares traded during the period divided by the average market capitalization for the period. Average market capitalization is calculated as the average of the end-of-period values for the current period and the previous period. • **Listed domestic companies** are the domestically incorporated companies listed on the country's stock exchanges at the end of the year. This indicator does not include investment companies, mutual funds, or other collective investment vehicles. • **S&P/IFC Investable index price change** is the U.S. dollar price change in the stock markets covered by the S&P/IFCI country index, supplemented by the S&P/IFCG country index.

The data on stock markets are from Standard & Poor's *Emerging Stock Markets Factbook 2000*, supplemented by other data from Standard & Poor's. The firm collects data through an annual survey of the world's stock exchanges, supplemented by information provided by its network of correspondents and by Reuters. Standard & Poor's acquired the International Finance Corporation's Emerging Market Database on 30 December 1999. The GDP data are from the World Bank's national accounts data files. *About the data* is based on Demirgüç-Kunt and Levine (1996a).

5.4 | Financial depth and efficiency

	Domestic credit provided by banking sector		Liquid liabilities		Quasi-liquid liabilities		Ratio of bank liquid reserves to bank assets		Interest rate spread		Spread over LIBOR	
									Lending minus deposit rate percentage points		Lending rate minus LIBOR percentage points	
	% of GDP		% of GDP		% of GDP		%					
	1990	1999	1990	1999	1990	1999	1990	1999	1990	1999	1990	1999
Albania	..	46.5	..	57.9	..	37.5	..	11.0	2.1	8.7	16.7	16.2
Algeria	74.7	50.0	73.6	46.1	24.9	18.2	1.3	2.5	..	2.5	..	5.6
Angola	..	7.9	..	25.0	..	17.4	..	14.9	..	43.7	..	74.9
Argentina	32.4	35.6	11.5	31.6	7.1	23.9	7.4	2.6	..	3.0	..	5.6
Armenia	58.7	10.7	79.9	11.1	42.9	5.8	13.6	10.5	..	11.5	..	33.4
Australia	71.6	90.4	55.1	68.8	43.3	48.7	1.5	1.7	4.5	4.0	9.9	2.1
Austria	120.9	131.7					2.1	2.3	..	3.4	..	0.2
Azerbaijan	57.2	11.8	33.5	13.2	11.6	5.0	4.5	7.9	..	..	..	..
Bangladesh	23.9	33.5	23.4	31.4	16.8	23.0	12.8	9.7	4.0	5.4	7.7	8.7
Belarus	..	20.8	..	17.5	..	9.4	..	12.4	..	27.2	..	45.6
Belgium	70.3	147.3	..	..	..	..	0.2	1.0	6.9	4.3	4.7	1.3
Benin	22.4	6.7	26.7	25.6	5.9	7.4	29.3	10.7	9.0	..	7.7	..
Bolivia	30.7	67.7	24.5	56.1	18.0	48.5	18.8	4.8	18.0	23.1	33.5	30.0
Bosnia and Herzegovina	..	..	..	..	..	..	..	..	..	..	..	..
Botswana	−46.4	−69.7	22.1	31.2	13.7	23.6	11.0	7.5	1.8	5.2	−0.4	9.2
Brazil	89.8	51.8	26.4	31.8	18.5	25.3	6.7	8.4	..	..	..	..
Bulgaria	118.5	18.6	71.9	30.4	53.6	17.2	10.2	7.9	8.9	9.6	42.4	7.4
Burkina Faso	13.7	13.5	21.3	23.0	7.5	6.8	12.7	4.7	9.0	..	7.7	..
Burundi	23.2	34.0	18.2	23.6	6.5	6.4	2.8	4.1	..	..	4.0	9.8
Cambodia	..	7.4	..	12.2	..	7.7	..	29.1	..	10.2	..	12.1
Cameroon	31.2	17.7	22.6	15.6	10.1	5.7	3.4	4.7	11.0	17.0	10.2	16.6
Canada	85.8	96.5	74.5	76.0	59.9	54.8	1.6	0.6	1.3	1.5	5.7	1.0
Central African Republic	12.9	11.7	15.3	16.7	1.8	1.5	2.8	2.0	11.0	17.0	10.2	16.6
Chad	11.5	10.6	14.6	10.9	0.6	0.6	3.6	2.7	11.0	17.0	10.2	16.6
Chile	73.0	72.5	40.7	52.2	32.8	41.4	3.8	4.0	8.6	4.1	40.5	7.2
China	90.0	130.6	79.2	147.8	41.4	90.4	15.7	15.0	0.7	3.6	1.0	0.4
Hong Kong, China	156.3	140.8	181.7	222.6	166.8	207.7	0.1	0.2	3.3	4.0	1.7	3.1
Colombia	35.9	41.4	29.8	35.1	19.3	26.3	26.3	29.6	8.8	9.1	36.9	25.0
Congo, Dem. Rep.	25.3	..	12.9	..	2.1	..	2.0	..	..	..	..	119.2
Congo, Rep.	29.1	20.2	22.0	15.4	6.1	2.0	2.0	3.7	11.0	17.0	10.2	16.6
Costa Rica	23.8	28.0	33.9	33.1	23.8	21.3	68.5	17.8	11.4	11.4	24.2	20.3
Côte d'Ivoire	44.5	25.9	28.8	24.3	10.9	6.9	2.1	5.3	9.0	..	7.7	..
Croatia	..	45.9	..	39.2	..	29.5	..	6.8	..	10.6	..	9.5
Cuba	..	..	..	..	..	..	..	..	..	..	..	..
Czech Republic	..	62.7	..	67.9	..	43.5	..	18.0	..	4.2	..	3.3
Denmark	63.0	57.4	59.0	55.9	29.4	24.5	1.1	6.2	6.2	4.7	5.8	1.7
Dominican Republic	31.5	36.6	28.4	34.4	13.1	21.6	31.1	54.9	15.2	9.0	29.3	19.6
Ecuador	17.2	80.9	23.3	46.1	12.9	34.2	23.1	7.7	−6.0	15.1	29.2	58.6
Egypt, Arab Rep.	106.8	99.8	87.9	83.3	60.7	63.8	17.1	13.3	7.0	3.7	10.7	7.6
El Salvador	32.0	43.5	30.6	48.4	19.6	38.5	33.4	26.4	3.2	4.7	12.9	10.1
Eritrea	..	..	..	..	..	..	..	..	..	..	..	..
Estonia	65.0	34.6	136.2	34.4	93.5	12.0	43.1	19.4	..	4.5	26.6	3.3
Ethiopia	67.6	58.5	42.2	40.7	12.6	19.0	23.3	12.6	3.6	4.3	−2.3	5.2
Finland	83.1	55.8	54.4	48.0	..	..	4.1	4.1	4.1	3.5	3.3	−0.7
France	104.4	102.2	..	..	..	..	1.0	0.4	6.1	3.7	2.3	0.9
Gabon	20.0	22.5	17.8	16.6	6.6	6.6	2.0	5.7	11.0	17.0	10.2	16.6
Gambia, The	3.4	14.2	20.7	32.8	8.8	16.8	8.8	16.5	15.2	11.5	18.2	18.6
Georgia	..	20.2	..	8.2	..	3.1	..	14.1	..	18.8	..	28.0
Germany	105.4	145.2	67.9	78.1	..	..	3.2	6.6	4.5	6.4	3.3	3.4
Ghana	13.2	36.2	14.1	19.0	3.4	8.6	20.2	8.0	..	..	..	..
Greece	101.3	94.8	70.6	63.8	56.9	45.0	22.4	31.4	8.1	6.3	19.3	9.6
Guatemala	17.4	16.9	21.2	23.0	11.8	11.6	31.8	10.7	5.1	11.6	15.0	14.1
Guinea	6.0	6.1	0.8	2.0	0.8	2.0	6.2	14.7	0.2	..	12.9	..
Guinea-Bissau	77.5	14.1	68.9	30.3	4.4	1.0	10.8	13.2	13.1	..	37.4	..
Haiti	32.9	25.7	31.4	33.6	15.9	22.0	74.9	30.4	..	15.5	..	17.5
Honduras	40.9	28.3	33.6	49.3	18.8	34.9	6.6	22.0	8.3	10.2	8.7	24.7

Financial depth and efficiency | 5.4

	Domestic credit provided by banking sector		Liquid liabilities		Quasi-liquid liabilities		Ratio of bank liquid reserves to bank assets		Interest rate spread		Spread over LIBOR	
									Lending minus deposit rate percentage points		Lending rate minus LIBOR percentage points	
	% of GDP		% of GDP		% of GDP		%		points		points	
	1990	1999	1990	1999	1990	1999	1990	1999	1990	1999	1990	1999
Hungary	105.5	52.1	43.8	46.2	19.0	27.7	28.5	..	4.1	3.1	20.5	10.9
India	51.6	49.6	43.2	52.4	28.2	36.1	14.8	12.2	..	..	8.2	7.1
Indonesia	45.5	60.5	40.4	57.2	29.1	46.9	4.5	5.3	3.3	1.9	12.5	22.3
Iran, Islamic Rep.	70.8	54.1	57.6	45.0	31.1	25.3	66.0	54.8	..	..	..	..
Iraq	..	..	..	..	..	..	..	..	..	..	..	..
Ireland	55.2	93.8	44.5	..	..	..	4.8	1.8	5.0	3.2	3.0	–2.1
Israel	106.2	87.5	70.2	97.3	63.6	90.1	11.9	13.0	12.0	5.0	18.1	11.0
Italy	89.4	92.0	70.5	..	..	..	12.0	1.2	7.3	4.0	5.8	0.2
Jamaica	34.8	45.4	51.0	48.9	37.8	32.6	37.4	8.3	6.6	13.5	22.2	21.6
Japan	266.8	144.0	187.5	125.8	159.6	77.4	1.5	1.8	3.4	2.0	–1.4	–3.3
Jordan	117.9	86.1	131.2	109.6	77.8	78.7	20.5	28.0	2.2	3.2	2.0	6.5
Kazakhstan	..	11.3	..	14.5	..	3.6	..	7.7	..	..	..	..
Kenya	52.9	49.8	43.3	46.6	29.3	31.9	9.9	22.7	5.1	12.8	10.4	17.0
Korea, Dem. Rep.	..	..	..	..	..	..	..	..	..	..	..	..
Korea, Rep.	65.7	96.6	54.6	93.8	45.7	84.6	6.3	17.2	0.0	1.4	1.7	4.0
Kuwait	243.0	110.7	192.2	85.3	153.9	70.1	1.2	1.1	0.4	2.8	4.1	3.1
Kyrgyz Republic	..	14.6	..	13.6	..	5.0	..	16.7	..	25.3	..	55.4
Lao PDR	5.1	10.3	7.2	15.2	3.1	13.0	3.4	20.0	2.5	18.6	20.0	26.6
Latvia	..	20.0	..	28.3	..	10.9	..	8.9	..	9.2	..	8.8
Lebanon	132.6	134.9	193.7	153.7	170.9	145.8	3.9	8.1	23.1	7.0	31.6	14.1
Lesotho	30.1	–0.2	38.8	31.1	22.4	13.4	24.1	38.3	7.4	11.6	12.1	13.7
Libya	..	..	..	..	..	..	40.8	26.3	1.5	3.8	–1.3	1.6
Lithuania	..	15.5	..	21.1	..	8.7	..	17.2	..	8.1	..	7.7
Macedonia, FYR	..	21.0	..	19.3	..	9.1	..	4.4	..	9.1	..	15.0
Madagascar	26.2	15.2	17.8	26.0	5.3	10.8	8.5	22.5	5.3	12.7	17.5	22.6
Malawi	20.5	8.9	22.0	16.4	12.2	8.1	32.8	10.3	8.9	20.4	12.7	48.2
Malaysia	75.7	151.6	64.4	136.0	43.0	110.8	5.9	8.3	1.3	3.2	–1.1	1.9
Mali	13.7	16.8	20.5	22.6	5.5	6.0	50.8	4.9	9.0	..	7.7	..
Mauritania	54.7	0.3	28.5	14.1	7.0	4.3	6.1	2.1	5.0	..	1.7	..
Mauritius	45.1	76.7	63.3	81.8	49.1	70.6	8.8	15.0	5.4	10.7	9.7	16.2
Mexico	36.6	28.8	22.8	28.9	16.4	20.4	4.2	6.4	..	16.3	..	20.5
Moldova	62.8	28.9	70.3	20.6	35.4	8.5	8.3	14.3	..	8.0	..	30.1
Mongolia	68.5	11.4	52.4	23.5	13.8	11.3	2.0	11.8	..	17.9	..	32.2
Morocco	60.1	85.5	61.0	78.7	18.4	20.2	11.3	5.1	0.5	7.1	0.7	8.1
Mozambique	15.6	6.6	26.5	25.5	5.2	11.7	61.5	14.3	..	11.8	..	14.2
Myanmar	32.8	26.8	27.9	25.7	7.8	9.9	271.8	25.1	2.1	5.1	–0.3	10.7
Namibia	20.4	53.7	24.4	51.1	14.3	27.2	4.4	3.1	10.6	7.7	17.4	13.1
Nepal	28.9	41.3	32.2	48.5	18.5	32.3	12.7	20.2	2.5	4.0	6.1	5.9
Netherlands	103.0	126.8	..	..	..	..	0.3	0.8	8.4	0.7	3.4	–2.0
New Zealand	81.6	119.0	77.9	92.7	64.8	78.5	0.8	10.3	4.4	3.9	7.7	3.1
Nicaragua	206.6	147.0	56.9	66.6	23.1	54.8	20.2	15.4	12.5	11.9	13.7	16.7
Niger	16.2	9.4	19.8	7.4	8.3	1.9	42.9	1.6	9.0	..	7.7	..
Nigeria	23.7	19.5	23.6	21.6	10.3	9.2	11.6	5.4	5.5	7.5	17.0	14.9
Norway	89.5	74.1	59.9	55.9	27.0	11.8	0.5	3.4	4.6	2.8	5.9	2.7
Oman	16.6	44.7	28.9	37.0	19.3	28.3	6.9	4.1	1.4	2.2	1.4	4.9
Pakistan	50.9	48.4	39.8	46.2	10.0	18.9	8.9	7.2	..	..	..	..
Panama	52.7	106.6	41.1	81.8	33.0	69.8	..	..	3.6	3.1	3.7	4.6
Papua New Guinea	35.8	31.2	35.2	32.8	24.0	18.0	3.2	3.6	6.9	3.4	7.2	13.5
Paraguay	14.9	28.2	21.4	33.3	12.8	24.1	31.0	51.2	8.1	10.5	22.7	24.8
Peru	20.2	27.6	24.8	34.2	11.8	21.5	22.0	16.4	2,335.0	14.5	4,766.2	25.4
Philippines	26.9	68.8	36.8	67.5	28.2	54.3	11.7	8.3	4.6	3.6	15.8	6.4
Poland	18.8	39.3	32.8	42.8	16.6	28.5	20.6	10.7	462.5	5.8	495.9	11.6
Portugal	69.9	103.9	..	..	..	..	29.0	11.4	7.8	2.8	13.5	–0.2
Puerto Rico	..	..	..	..	..	..	..	..	..	..	..	..
Romania	79.7	18.7	60.4	25.7	32.7	20.3	1.2	14.3	..	..	..	..
Russian Federation	..	32.7	..	21.7	..	10.1	..	7.8	..	26.0	..	34.3

	Domestic credit provided by banking sector		Liquid liabilities		Quasi-liquid liabilities		Ratio of bank liquid reserves to bank assets		Interest rate spread		Spread over LIBOR	
	% of GDP		% of GDP		% of GDP		%		Lending minus deposit rate percentage points		Lending rate minus LIBOR percentage points	
	1990	1999	1990	1999	1990	1999	1990	1999	1990	1999	1990	1999
Rwanda	17.1	13.7	14.9	15.1	7.0	6.1	4.3	26.0	6.3	..	4.9	..
Saudi Arabia	58.7	38.2	47.9	57.8	21.9	27.7	5.6	10.0	..	..	..	..
Senegal	33.8	22.8	22.9	24.1	9.7	9.2	14.1	1.8	9.0	..	7.7	..
Sierra Leone	26.3	50.1	13.1	16.1	2.6	5.0	64.1	13.7	12.0	17.3	44.2	21.4
Singapore	75.7	94.7	123.6	121.2	100.6	99.6	3.7	4.2	2.7	4.1	–1.0	0.4
Slovak Republic	..	60.2	..	64.1	..	45.3	..	6.5	..	6.7	..	15.7
Slovenia	..	43.4	..	46.7	..	36.7	..	4.1	..	5.1	..	7.0
South Africa	97.8	155.0	44.6	45.1	27.2	12.7	3.3	7.0	2.1	5.8	12.7	12.6
Spain	106.1	108.9	..	..	..	..	8.7	2.6	5.4	2.1	7.7	–1.5
Sri Lanka	43.1	34.5	35.2	40.4	22.9	30.6	9.9	11.5	–6.4	–4.8	4.7	1.6
Sudan	20.4	7.3	20.1	10.3	2.9	3.6	79.5	5.3	..	..	..	..
Sweden	140.5	115.8	52.4	45.3	..	..	1.9	1.3	6.8	3.9	8.4	0.1
Switzerland	179.0	185.4	145.2	164.3	118.6	122.2	1.1	0.9	–0.9	2.7	–0.9	–1.5
Syrian Arab Republic	56.6	27.5	54.7	51.2	10.5	17.3	46.0	28.4	..	..	..	..
Tajikistan	..	..	..	..	..	..	..	..	..	..	..	..
Tanzania	34.6	12.8	19.9	19.0	6.3	9.3	5.3	3.9	..	14.1	..	16.5
Thailand	91.1	141.9	74.9	114.4	66.0	98.7	3.1	6.5	2.2	4.3	6.1	3.6
Togo	21.3	23.8	36.1	24.7	19.1	8.0	59.0	1.6	9.0	..	7.7	..
Trinidad and Tobago	58.5	52.8	54.6	59.0	42.7	46.6	13.5	8.8	6.9	8.5	4.6	11.6
Tunisia	62.5	69.6	51.5	55.2	26.7	31.9	1.6	3.7	..	..	..	..
Turkey	19.4	49.8	24.1	51.8	16.4	46.2	16.3	19.9	..	..	..	..
Turkmenistan	..	30.5	..	14.9	..	5.1	..	35.7	..	..	..	..
Uganda	17.8	7.5	7.6	14.6	1.4	6.7	17.9	8.3	7.4	12.8	30.4	16.1
Ukraine	83.2	25.9	50.1	17.0	9.0	6.0	49.0	10.4	..	34.3	..	49.5
United Arab Emirates	34.7	58.9	46.3	56.9	37.7	40.9	4.4	6.1	..	..	..	..
United Kingdom	121.4	127.0	..	..	..	..	0.5	0.4	2.2	2.7	6.4	–0.1
United States	110.9	164.2	65.5	62.4	49.4	46.4	2.3	6.6	1.9	2.7	1.7	2.6
Uruguay	46.7	54.4	58.1	48.7	51.5	42.4	31.1	18.4	76.6	39.0	166.1	47.9
Uzbekistan	..	..	..	..	..	..	..	..	..	..	..	..
Venezuela, RB	37.4	16.7	41.1	21.3	29.4	10.8	21.9	18.8	7.7	10.8	27.2	26.7
Vietnam	15.9	22.4	22.7	32.5	9.3	15.9	13.3	12.1	..	5.3	..	9.4
West Bank and Gaza	..	..	..	..	..	..	..	..	..	..	..	..
Yemen, Rep.	62.8	21.8	57.1	35.3	10.8	15.8	121.2	24.2	..	3.8	..	16.6
Yugoslavia, FR (Serb./Mont.)	..	..	..	..	..	..	..	..	..	..	..	..
Zambia	67.8	59.7	21.8	18.6	10.6	11.8	33.7	12.9	9.5	20.3	26.8	35.1
Zimbabwe	41.7	45.4	41.8	32.8	30.3	16.3	12.2	11.5	2.9	16.9	3.4	50.0
World	**123.5 w**	**126.2 w**	**85.7 w**	**78.0 w**	**69.1 w**	**54.0 w**	**9.9 m**	**9.2 m**				
Low income	44.7	43.3	37.2	43.7	22.3	29.1	12.8	12.1				
Middle income	65.0	70.5	44.8	66.1	28.1	45.2	12.6	8.5				
Lower middle income	65.5	84.1	63.7	86.9	37.9	55.6	17.9	10.6				
Upper middle income	64.4	58.4	34.9	47.6	24.2	35.9	7.1	7.2				
Low & middle income	61.0	66.4	43.3	62.7	26.9	42.7	12.8	10.8				
East Asia & Pacific	73.4	114.5	63.9	120.4	42.7	84.1	5.9	8.7				
Europe & Central Asia	..	38.0	..	34.7	..	22.9	..	10.4				
Latin America & Carib.	59.2	41.6	25.5	32.5	17.8	24.4	22.5	18.1				
Middle East & N. Africa	69.5	59.6	62.1	60.1	30.7	32.8	14.2	10.0				
South Asia	49.0	47.6	41.1	49.7	25.3	33.1	12.7	11.5				
Sub-Saharan Africa	56.8	77.9	32.5	33.4	17.3	12.7	11.6	5.3				
High income	137.2	147.8	97.8	79.8	82.8	..	1.9	1.4				
Europe EMU	97.0	115.9	..	..	..	..	3.7	2.2				

Financial depth and efficiency | 5.4

Households and institutions save and invest independently. The financial system's role is to intermediate between them and to cycle available funds to where they are needed. Savers accumulate claims on financial institutions, which pass these funds to their final users. As an economy develops, this indirect lending by savers to investors becomes more efficient and gradually increases financial assets relative to GDP. This wealth allows increased saving and investment, facilitating and enhancing economic growth. As more specialized savings and financial institutions emerge, more financing instruments become available, spreading risks and reducing costs to liability holders. As securities markets mature, savers can invest their resources directly in financial assets issued by firms.

The ratio of domestic credit provided by the banking sector to GDP is used to measure the growth of the banking system because it reflects the extent to which savings are financial. In a few countries governments may hold international reserves as deposits in the banking system rather than in the central bank. Since the claims on the central government are a net item (claims of central government minus central government deposits), this net figure may be negative, resulting in a negative figure for domestic credit provided by the banking sector.

Liquid liabilities include bank deposits of generally less than one year plus currency. Their ratio to GDP indicates the relative size of these readily available forms of money—money that the owners can use to buy goods and services without incurring any cost. Quasi-liquid liabilities are long-term deposits and assets—such as certificates of deposit, commercial paper, and bonds—that can be converted into currency or demand deposits, but at a cost. The ratio of bank liquid reserves to bank assets captures the banking system's liquidity. In countries whose banking system is liquid, adverse macroeconomic conditions should be less likely to lead to banking and financial crises. Data on domestic credit and liquid and quasi-liquid liabilities are cited on an end-of-year basis.

No less important than the size and structure of the financial sector is its efficiency, as indicated by the margin between the cost of mobilizing liabilities and the earnings on assets—or the interest rate spread. A narrowing of the interest rate spread reduces transactions costs, which lowers the overall cost of investment and is therefore crucial to economic growth. Interest rates reflect the responsiveness of financial institutions to competition and price incentives. The interest rate spread, also known as the intermediation margin, is a summary measure of a banking sys-

tem's efficiency. To the extent that information about interest rates is inaccurate, banks do not monitor all bank managers, or the government sets deposit and lending rates, the interest rate spread may not be a reliable measure of efficiency. The spread over LIBOR reflects the differential between a country's lending rate and the London interbank offered rate (ignoring expected changes in the exchange rate). Interest rates are expressed as annual averages.

In some countries financial markets are distorted by restrictions on foreign investment, selective credit controls, and controls on deposit and lending rates. Interest rates may reflect the diversion of resources to finance the public sector deficit through statutory reserve requirements and direct borrowing from the banking system. And where state-owned banks dominate the financial sector, noncommercial considerations may unduly influence credit allocation. The indicators in the table provide quantitative assessments of each country's financial sector, but qualitative assessments of policies, laws, and regulations are needed to analyze overall financial conditions. Recent events in East Asia highlight the risks of weak financial intermediation, poor corporate governance, and deficient government policies, including procyclical macroeconomic policy responses to large capital inflows.

The accuracy of financial data depends on the quality of accounting systems, which are weak in some developing economies. Some of the indicators in the table are highly correlated, particularly the ratios of domestic credit, liquid liabilities, and quasi-liquid liabilities to GDP, because changes in liquid and quasi-liquid liabilities flow directly from changes in domestic credit. Moreover, the precise definition of the financial aggregates presented varies by country.

The indicators reported here do not capture the activities of the informal sector, which remains an important source of finance in developing economies. Personal credit or credit extended through community-based pooling of assets may be the only source of credit available to small farmers, small businesses, or home-based producers. And in financially repressed economies the rationing of formal credit forces many borrowers and lenders to turn to the informal market, which is very expensive, or to self-financing and family savings.

- **Domestic credit provided by banking sector** includes all credit to various sectors on a gross basis, with the exception of credit to the central government, which is net. The banking sector includes monetary authorities, deposit money banks, and other banking institutions for which data are available (including institutions that do not accept transferable deposits but do incur such liabilities as time and savings deposits). Examples of other banking institutions include savings and mortgage loan institutions and building and loan associations.
- **Liquid liabilities** are also known as broad money, or M3. They are the sum of currency and deposits in the central bank (M0), plus transferable deposits and electronic currency (M1), plus time and savings deposits, foreign currency transferable deposits, certificates of deposit, and securities repurchase agreements (M2), plus travelers checks, foreign currency time deposits, commercial paper, and shares of mutual funds or market funds held by residents. • **Quasi-liquid liabilities** are the M3 money supply less M1. • **Ratio of bank liquid reserves to bank assets** is the ratio of domestic currency holdings and deposits with the monetary authorities to claims on other governments, nonfinancial public enterprises, the private sector, and other banking institutions. • **Interest rate spread** is the interest rate charged by banks on loans to prime customers minus the interest rate paid by commercial or similar banks for demand, time, or savings deposits. • **Spread over LIBOR** (London interbank offered rate) is the interest rate charged by banks on short-term loans in local currency to prime customers minus LIBOR. LIBOR is the most commonly recognized international interest rate and is quoted in several currencies. The average three-month LIBOR on U.S. dollar deposits is used here.

Data sources

The data on credit, liabilities, bank reserves, and interest rates are collected from central banks and finance ministries and reported in the print and electronic editions of the International Monetary Fund's *International Financial Statistics*.

	Tax revenue	Taxes on income, profits, and capital gains		Domestic taxes on goods and services		Export duties		Import duties		Highest marginal tax rate[a]		
	% of GDP	% of total taxes		% of value added in industry and services		% of tax revenue		% of tax revenue		Individual rate %	on income over $	Corporate rate %
	1999	1990	1999	1990	1999	1990	1999	1990	1999	1999	1999	1999
Albania	14.8	..	9.4	..	16.9	..	0.0	..	19.3	..	..	..
Algeria	27.5	..	72.3	..	3.7	..	0.0	..	15.3	..	..	..
Angola	..	..	..	..	..	..	..	..	..	..	..	..
Argentina	12.6	2.7	17.3	2.2	6.5	9.3	0.1	2.6	6.9	35	200,000	35
Armenia	..	..	..	..	..	..	..	..	..	..	..	..
Australia	22.1	70.9	72.8	5.4	5.0	0.1	0.0	4.4	2.8	47	30,579	36
Austria	35.3	20.8	27.0	9.0	9.5	0.0	..	1.6	..	50	59,590	34
Azerbaijan	19.0	..	23.1	..	11.3	..	..	..	9.0	40	3,704	30
Bangladesh	..	..	..	..	..	..	..	..	..	..	..	..
Belarus	27.8	12.1	11.6	17.3	15.2	3.6	0.0	0.4	8.0	..	..	..
Belgium	43.0	36.1	37.5	10.4	11.2	0.0	0.0	0.0	0.0	55	69,993	39
Benin	..	..	..	..	..	..	..	..	..	..	..	..
Bolivia	13.9	7.9	10.5	5.8	9.0	0.0	0.0	11.1	7.1	13	..	25
Bosnia and Herzegovina	..	..	..	..	..	..	..	..	..	..	..	..
Botswana	..	71.7	..	1.0	..	0.0	..	24.7	..	30	17,960	15
Brazil	19.9	24.5	19.9	7.1	6.9	0.0	0.0	2.5	3.0	28	17,881	15
Bulgaria	26.8	40.6	16.9	10.4	13.6	0.0	0.0	2.5	4.2	40	9,403	27
Burkina Faso	..	24.7	..	4.9	..	1.1	..	33.1	..	..	..	..
Burundi	16.7	23.4	22.5	16.6	17.0	3.1	0.0	23.2	16.4	..	..	..
Cambodia	..	..	..	..	..	..	..	..	..	20	39,915	20
Cameroon	12.8	25.1	26.0	4.3	6.9	1.7	3.9	18.9	31.6	60	13,321	39
Canada	19.8	59.3	59.3	3.7	..	0.0	0.0	3.2	1.6	29	38,604	38
Central African Republic	..	..	..	..	..	..	..	..	..	..	..	..
Chad	..	20.3	..	3.9	..	..	..	..	..	..	..	..
Chile	18.4	15.8	20.7	10.4	13.3	..	..	..	..	45	6,526	15
China	6.1	49.8	7.8	1.5	6.1	0.0	0.0	22.1	6.6	45	12,079	30
Hong Kong, China	..	..	..	..	..	..	..	..	..	17	13,583	16
Colombia	10.6	36.4	39.9	4.8	6.0	2.0	0.0	22.5	8.5	35	32,221	35
Congo, Dem. Rep.	4.3	28.5	31.0	2.6	2.3	4.1	1.9	45.1	32.5	50	13,167	..
Congo, Rep.	8.7	..	4.1	..	6.7	0.0	0.0	32.3	30.6	50	14,210	45
Costa Rica	18.4	11.5	16.5	6.9	9.8	8.0	0.7	18.2	4.9	25	14,185	30
Côte d'Ivoire	19.9	18.1	22.7	8.9	5.1	3.7	13.1	28.4	34.7	10	4,263	35
Croatia	40.4	..	12.0	..	24.6	..	0.0	..	7.6	35	5,556	..
Cuba	..	..	..	..	..	..	..	..	..	..	..	..
Czech Republic	33.0	..	14.3	..	14.4	..	0.0	..	2.0	40	36,979	35
Denmark	33.3	43.5	43.1	16.1	16.4	0.0	0.0	0.1	0.0	59	..	32
Dominican Republic	15.7	23.8	17.6	3.1	6.7	0.1	0.0	41.4	36.8	25	14,309	25
Ecuador	..	62.9	..	4.5	..	0.3	..	12.1	..	0	66,226	0
Egypt, Arab Rep.	16.6	26.4	34.4	4.1	5.9	0.0	0.0	18.9	19.9	32	14,706	40
El Salvador	12.6	..	26.1	..	6.5	..	0.0	..	9.5	30	22,857	25
Eritrea	..	..	..	..	..	..	..	..	..	..	..	..
Estonia	28.5	27.5	21.0	14.8	14.5	0.0	0.0	0.8	0.0	26	..	26
Ethiopia	..	40.9	..	9.1	..	2.8	..	18.0	..	..	..	..
Finland	27.7	34.5	33.6	15.2	14.5	0.0	0.0	1.0	0.0	38	61,164	28
France	38.9	18.7	20.9	11.7	12.2	0.0	0.0	0.0	0.0	..	..	33
Gabon	..	35.9	..	5.0	..	2.8	..	23.4	..	55	2,290	40
Gambia, The	..	13.7	..	12.2	..	0.2	..	45.6	..	..	..	..
Georgia	10.2	..	11.6	..	11.0	..	0.0	..	4.7	..	..	..
Germany	26.3	17.5	17.3	7.4	6.4	0.0	0.0	0.0	0.0	53	66,690	30
Ghana	..	25.1	..	6.8	..	12.4	..	28.7	..	35	7,102	35
Greece	21.9	23.3	41.6	13.0	14.0	0.0	0.0	0.1	0.1	45	56,271	35
Guatemala	..	..	..	..	..	..	..	..	..	25	26,740	28
Guinea	10.7	12.6	10.1	3.2	0.8	51.7	0.2	11.2	42.9	..	..	..
Guinea-Bissau	..	..	..	..	..	..	..	..	..	..	..	..
Haiti	..	..	..	..	..	..	..	..	..	..	..	..
Honduras	..	..	..	..	..	..	..	..	..	30	75,758	15

Tax policies | 5.5

	Tax revenue	Taxes on income, profits, and capital gains		Domestic taxes on goods and services		Export duties		Import duties		Highest marginal tax rate[a]		
	% of GDP	% of total taxes		% of value added in industry and services		% of tax revenue		% of tax revenue		Individual rate %	on income over $	Corporate rate %
	1999	1990	1999	1990	1999	1990	1999	1990	1999	1999	1999	1999
Hungary	33.6	21.2	21.8	22.6	15.3	1.3	0.0	5.6	3.7	40	4,566	18
India	9.1	18.6	33.2	7.4	5.2	0.1	0.1	35.8	28.4	30	3,538	35
Indonesia	15.7	65.4	67.4	5.5	5.2	0.1	0.5	6.6	2.4	30	6,623	30
Iran, Islamic Rep.	16.0	24.7	25.2	1.0	5.4	0.0	0.0	18.6	11.5	54	174,171	54
Iraq	..	..	..	..	..	..	..	..	..	..	..	..
Ireland	30.6	39.7	43.5	13.4	12.5	0.0	0.0	0.0	0.0	46	14,799	32
Israel	35.5	42.4	41.9	..	..	0.0	0.0	1.4	0.8	50	57,789	36
Italy	38.8	37.7	38.8	11.5	10.4	0.0	0.0	0.0	0.0	46	81,665	37
Jamaica	..	..	..	..	..	..	..	..	..	25	2,712	33
Japan	..	73.0	..	2.5	..	0.0	..	1.4	..	50	259,291	35
Jordan	18.3	22.9	14.6	6.8	9.8	0.0	0.0	34.7	27.8	..	..	..
Kazakhstan	8.3	..	18.0	..	6.8	..	0.0	..	4.4	30	..	30
Kenya	..	32.9	..	15.9	..	0.0	..	17.8	..	33	382	33
Korea, Dem. Rep.	..	..	..	..	..	..	..	..	..	..	..	..
Korea, Rep.	17.3	37.5	31.0	6.7	7.1	0.0	0.0	13.0	7.4	40	66,236	28
Kuwait	3.4	19.5	8.2	0.0	..	0.0	0.0	76.8	0.0	0	..	0
Kyrgyz Republic	10.1	..	14.5	..	11.7	..	0.0	..	6.3	..	..	30
Lao PDR	..	..	..	..	..	..	..	..	..	40	1,064	..
Latvia	29.3	..	15.0	..	15.6	..	0.0	..	1.5	25	..	25
Lebanon	12.6	..	15.1	..	1.9	..	..	..	39.0	..	..	..
Lesotho	34.9	12.7	23.2	12.8	7.5	0.2	..	63.6	..	..	..	..
Libya	..	..	..	..	..	..	..	..	..	..	..	..
Lithuania	24.6	22.2	13.4	16.4	16.0	..	0.0	..	1.8	33	..	29
Macedonia, FYR	..	..	..	..	..	..	..	..	..	..	..	..
Madagascar	..	15.7	..	3.6	..	8.5	..	50.1	..	..	..	..
Malawi	..	42.5	..	13.9	..	0.0	..	18.7	..	38	948	38
Malaysia	18.9	42.5	44.4	6.3	6.9	9.7	2.2	15.1	13.3	30	39,474	28
Mali	..	..	..	..	..	..	..	..	..	..	..	..
Mauritania	..	..	..	..	..	..	..	..	..	..	..	..
Mauritius	18.2	15.2	13.9	6.4	9.6	4.6	0.0	45.7	30.7	30	2,220	15
Mexico	11.7	34.2	40.4	10.2	8.8	0.1	0.0	6.9	4.8	40	200,000	35
Moldova	21.6	..	5.1	..	18.3	..	0.0	..	3.8	..	..	..
Mongolia	15.3	28.2	9.6	8.1	12.6	0.0	2.2	19.6	4.1	..	..	..
Morocco	24.6	27.3	24.1	12.1	..	0.3	0.0	20.3	19.5	44	6,445	35
Mozambique	..	..	..	..	..	..	..	..	..	20	792	35
Myanmar	3.5	29.8	36.2	6.8	4.7	0.0	0.0	23.3	9.5	30	..	30
Namibia	..	39.4	..	10.0	..	3.6	..	26.9	..	40	16,129	40
Nepal	8.5	13.0	20.0	6.6	6.5	0.4	1.3	37.0	31.7	..	..	..
Netherlands	41.1	33.6	26.6	10.5	10.3	0.0	0.0	0.0	0.0	60	56,075	35
New Zealand	29.8	62.2	65.5	12.4	..	0.0	0.0	2.5	1.7	33	18,134	33
Nicaragua	..	20.0	..	16.9	..	..	..	21.3	..	30	18,083	30
Niger	..	..	..	..	..	..	..	..	..	..	..	..
Nigeria	..	..	..	..	..	..	..	..	..	25	1,395	28
Norway	34.7	21.7	25.6	14.9	16.1	0.1	0.0	0.6	0.7	28	6,835	28
Oman	6.4	87.6	63.6	0.3	..	0.0	0.0	7.8	19.9	0	..	12
Pakistan	13.1	12.8	27.8	8.6	6.8	0.0	0.0	44.4	16.1	..	..	..
Panama	17.2	24.4	24.8	4.8	..	1.3	0.2	15.8	..	30	200,000	30
Papua New Guinea	18.4	47.0	51.3	5.0	2.9	2.1	5.1	29.3	27.8	47	48,251	25
Paraguay	..	12.4	..	3.6	..	0.0	..	18.8	..	0	..	30
Peru	13.8	5.7	25.0	6.8	8.9	7.7	0.0	9.9	11.6	30	47,985	30
Philippines	14.4	32.5	42.6	6.4	5.6	0.0	0.0	28.4	20.0	33	12,773	33
Poland	29.0	..	21.6	..	12.6	..	0.0	..	3.1	40	15,192	34
Portugal	31.3	25.7	29.6	11.5	13.0	0.0	0.0	2.6	0.0	40	36,478	34
Puerto Rico	..	..	..	..	..	..	..	..	..	33	50,000	20
Romania	24.4	21.0	32.8	15.3	9.6	..	0.0	0.6	6.3	45	4,080	38
Russian Federation	19.0	..	11.9	..	8.9	..	4.5	..	5.5	35	6,036	35

	Tax revenue	Taxes on income, profits, and capital gains		Domestic taxes on goods and services		Export duties		Import duties		Highest marginal tax rate[a]		
	% of GDP	% of total taxes		% of value added in industry and services		% of tax revenue		% of tax revenue		Individual		Corporate rate %
										rate %	on income over $	
	1999	1990	1999	1990	1999	1990	1999	1990	1999	1999	1999	1999
Rwanda	..	20.0	..	5.6	..	7.4	..	20.7	..	..	..	..
Saudi Arabia	..	..	..	..	..	..	..	..	..	0	..	45
Senegal	..	..	..	..	..	..	..	..	..	50	22,469	35
Sierra Leone	9.9	33.0	17.6	1.9	6.8	0.4	0.0	41.3	47.4	..	..	..
Singapore	14.5	44.6	47.3	4.3	4.2	0.0	0.0	3.5	2.2	28	240,964	26
Slovak Republic	30.7	..	24.8	..	10.9	..	0.0	..	4.9	42	29,258	40
Slovenia	38.3	..	13.8	..	19.9	..	0.0	..	3.3	..	..	..
South Africa	25.6	55.0	58.0	10.3	10.7	0.0	0.0	3.9	3.0	45	20,391	30
Spain	26.9	34.0	31.8	..	7.4	0.0	0.0	1.7	0.0	40	77,139	35
Sri Lanka	14.9	12.0	17.0	14.7	13.0	4.2	0.0	27.4	16.7	35	4,405	35
Sudan	..	..	..	..	..	..	..	..	..	..	..	..
Sweden	35.5	20.6	15.5	12.6	..	0.0	0.0	0.6	0.0	31	27,198	28
Switzerland	23.1	17.0	16.6	..	..	0.0	0.0	6.9	1.0	..	..	45
Syrian Arab Republic	15.0	40.2	48.9	9.6	..	1.3	3.1	8.2	13.6	..	..	..
Tajikistan	9.0	..	6.3	..	8.5	..	0.0	..	12.6	..	..	..
Tanzania	..	..	..	..	..	..	..	..	..	35	12,335	30
Thailand	13.7	26.2	33.3	8.8	8.0	0.2	0.3	23.7	10.3	37	108,430	30
Togo	..	..	..	..	..	..	..	..	..	..	..	..
Trinidad and Tobago	..	..	..	..	..	..	..	..	..	35	7,937	35
Tunisia	25.9	16.0	21.5	7.1	12.2	0.4	0.1	35.1	13.4	..	..	..
Turkey	21.3	51.2	45.2	5.9	12.9	0.0	0.0	7.3	1.7	40	159,898	30
Turkmenistan	..	..	..	..	..	..	..	..	..	..	..	..
Uganda	..	..	..	..	..	..	..	..	..	30	3,578	30
Ukraine	..	..	..	..	..	..	..	..	..	40	5,953	30
United Arab Emirates	1.8	0.0	0.0	0.6	..	..	..	..	..	0	..	20
United Kingdom	34.6	43.2	41.8	10.2	11.4	0.0	0.0	0.0	0.0	40	46,589	31
United States	19.5	56.1	59.5	..	..	0.0	0.0	1.7	1.0	40	283,150	35
Uruguay	24.7	7.1	16.8	9.4	12.4	0.6	0.1	8.1	3.5	..	..	30
Uzbekistan	..	..	..	..	..	..	..	..	..	45	2,400	33
Venezuela, RB	12.8	82.2	29.2	0.8	5.9	0.0	0.0	7.1	13.2	34	78,500	34
Vietnam	14.7	..	23.5	..	7.6	..	0.0	..	26.4	50	5,695	32
West Bank and Gaza	..	..	..	..	..	..	..	..	..	..	..	..
Yemen, Rep.	10.3	44.9	45.9	2.7	2.9	0.0	0.0	29.2	25.9	..	..	..
Yugoslavia, FR (Serb./Mont.)	..	..	..	..	..	..	..	..	..	..	..	..
Zambia	..	..	..	..	..	..	..	..	..	30	742	35
Zimbabwe	26.1	49.7	48.2	8.4	9.7	0.0	0.0	18.8	19.0	50	20,455	35

a. These data are from PricewaterhouseCoopers's *Individual Taxes: Worldwide Summaries 1999–2000* and *Corporate Taxes: Worldwide Summaries 1999–2000*, copyright 1999 by PricewaterhouseCoopers by permission of John Wiley and Sons, Inc.

Tax policies | 5.5

Taxes are compulsory, unrequited payments made to governments by individuals, businesses, or institutions. They are considered unrequited because governments provide nothing specifically in return for them, although taxes typically are used to provide goods or services to individuals or communities on a collective basis. The sources of the revenue received by governments and the relative contributions of these sources are determined by policy choices about where and how to impose taxes and by changes in the structure of the economy. Tax policy may reflect concerns about distributional effects, economic efficiency (including corrections for externalities), and the practical problems of administering a tax system. There is no ideal level of taxation. But taxes influence incentives and thus the behavior of economic actors and the country's competitiveness.

The level of taxation is typically measured by tax revenue as a share of GDP. Comparing levels of taxation across countries provides a quick overview of the fiscal obligations and incentives facing the private sector. In this table tax data measured in local currencies are normalized by scaling variables in the same units to ease cross-country comparisons. The table refers only to central government data, which may significantly understate the total tax burden, particularly in countries where provincial and municipal governments are large or have considerable tax authority.

Low ratios of tax collections to GDP may reflect weak administration and large-scale tax avoidance or evasion. They may also reflect the presence of a sizable parallel economy with unrecorded and undisclosed incomes. Tax collection ratios tend to rise with income, with higher-income countries relying on taxes to finance a much broader range of social services and social security than lower-income countries are able to provide.

As countries develop, their capacity to tax residents directly typically expands and indirect taxes become less important as a source of revenue. Thus the share of taxes on income, profits, and capital gains is one measure of an economy's (and tax system's) level of development. In the early stages of development governments tend to rely on indirect taxes because the administrative costs of collecting them are relatively low. The two main indirect taxes are international trade taxes (including customs revenues) and domestic taxes on goods and services. The table shows these domestic taxes as a percentage of value added in industry and services. Agriculture and mining are excluded from the denominator because indirect taxes on goods originating from these sectors are usually negligible. What is missing here is a measure of the uniformity of these taxes across industries and along the value added chain of production. Without such data no clear inferences can be drawn about how neutral a tax system is between subsectors. "Surplus" revenues raised by some governments by charging higher prices for goods produced under monopoly by state-owned enterprises are not counted as tax revenues. Similarly, losses from charging below-market prices for products are rarely identified as subsidies.

Export and import duties are shown separately because the burden they impose on the economy (and thus growth) is likely to be large. Export duties, typically levied on primary (particularly agricultural) products, often take the place of direct taxes on income and profits, but they reduce the incentive to export and encourage a shift to other products. High import duties penalize consumers, create protective barriers—which promote higher-priced output and inefficient production—and implicitly tax exports. By contrast, lower trade taxes enhance openness—to foreign competition, knowledge, technologies, and resources—energizing development in many ways. The economies growing fastest over the past 15 years have not relied on tax revenues from imports. Seeing this pattern, many developing countries have lowered tariffs over the past decade, a trend that is expected to continue. In some countries, such as members of the European Union, most customs duties are collected by a supranational authority; these revenues are not reported in the individual countries' accounts.

The tax revenues collected by governments are the outcomes of systems that are often complex, containing many exceptions, exemptions, penalties, and other inducements that affect the incidence of taxes and thus influence the decisions of workers, managers, and entrepreneurs. A potentially important influence on both domestic and international investors is a tax system's progressivity, as reflected in the highest marginal tax rate on individual and corporate income. Figures for individual marginal tax rates generally refer to employment income. For some countries the highest marginal tax rate is also the basic or flat rate, and other surtaxes, deductions, and the like may apply.

• **Tax revenue** comprises compulsory, unrequited, nonrepayable receipts collected by central governments for public purposes. It includes interest collected on tax arrears and penalties collected on nonpayment or late payment of taxes and is shown net of refunds and other corrective transactions. • **Taxes on income, profits, and capital gains** include taxes levied by central governments on the actual or presumptive net income of individuals and profits of enterprises. Also included are taxes on capital gains, whether realized or not, on the sale of land, securities, and other assets. Social security contributions based on gross pay, payroll, or number of employees are not included, but those based on personal income after deductions and personal exemptions are included. • **Domestic taxes on goods and services** include all taxes and duties levied by central governments on the production, extraction, sale, transfer, leasing, or delivery of goods and rendering of services, or on the use of goods or permission to use goods or perform activities. These include general sales taxes, turnover or value added taxes, excise taxes, and motor vehicle taxes. • **Export duties** include all levies collected on goods at the point of export. Rebates on exported goods—that is, repayments of previously paid general consumption taxes, excise taxes, or import duties—should be deducted from the gross receipts of the appropriate taxes, not from export duty receipts. • **Import duties** comprise all levies collected on goods at the point of entry into the country. They include levies for revenue purposes or import protection, whether on a specific or ad valorem basis, as long as they are restricted to imported products. • **Highest marginal tax rate** is the highest rate shown on the schedule of tax rates applied to the taxable income of individuals and corporations. Also presented are the income levels above which the highest marginal tax rates for individuals apply.

The definitions used here are from the International Monetary Fund's (IMF) *Manual on Government Finance Statistics* (1986). The data on tax revenues are from print and electronic editions of the IMF's *Government Finance Statistics Yearbook*. The data on individual and corporate tax rates are from PricewaterhouseCoopers's *Individual Taxes: Worldwide Summaries 1999–2000* and *Corporate Taxes: Worldwide Summaries 1999–2000*.

5.6 Relative prices and exchange rates

	Exchange rate arrangements[a]		Official exchange rate	Ratio of official to parallel exchange rate	Real effective exchange rate	Purchasing power parity conversion factor		Interest rate			Key agricultural producer prices	
			local currency units to $		1995 = 100	local currency units to international $		Deposit %	Lending %	Real %	Wheat $ per metric ton	Maize $ per metric ton
	Classification 1999	Structure 1999	1999	1999	1999	1990	1999	1999	1999	1999	1998	1998
Albania	IF	U	137.7	1.0	..	1.9	47.0	12.9	21.6	18.7	..	..
Algeria	MF	U	66.6	0.9	110.2	4.9	21.0	8.5	11.0	0.1	337	..
Angola	IF	U	2.8	..	..	0.0	0.4	36.6	80.3	−70.0	..	..
Argentina	CB	U	1.0	1.0	..	0.3	0.6	8.0	11.0	13.3	85	56
Armenia	IF	U	535.1	1.0	115.3	..	..	27.3	38.8	38.8	127	128
Australia	IF	U	1.5	1.0	98.6	1.4	1.3	3.5	7.5	6.3	127	128
Austria	Euro	U	12.9 [b]	1.0	92.8	12.6	13.2	2.2	5.6	4.7	117	117
Azerbaijan	MF	U	4,120.2	1.0	..	..	724.8	..	..	..	204	132
Bangladesh	P	U	49.1	1.0	..	9.1	11.6	8.7	14.1	9.1	172	173
Belarus	MF	M	..	..	..	..	41.9	23.8	51.0	−64.2	..	..
Belgium	Euro	U	37.9 [b]	1.0	91.8	33.5	36.2	2.4	6.7	5.7	..	..
Benin	FF	U	615.7	1.0	..	150.7	260.5	3.5	..	..	..	285
Bolivia	P	U	5.8	1.0	118.4	1.3	2.5	12.3	35.4	31.7	195	119
Bosnia and Herzegovina	CB	U	..	..	..	..	..	..	..	..	..	..
Botswana	P	D	4.6	1.0	..	1.1	2.1	9.5	14.6	6.6	90	110
Brazil	IF	U	1.8	..	..	0.0	0.8	26.0	..	..	164	125
Bulgaria	CB	U	1.8	1.0	118.7	0.0	0.6	3.2	12.8	9.4	..	..
Burkina Faso	FF	U	615.7	1.0	..	119.8	149.8	3.5	..	..	..	148
Burundi	MF	D	563.6	1.0	90.0	49.0	104.3	..	15.2	11.4	156	357
Cambodia	MF	D	3,807.8	1.0	..	66.5	741.9	7.3	17.6	11.3	..	158
Cameroon	FF	U	615.7	1.0	109.7	184.8	233.9	5.0	22.0	23.5	79	120
Canada	IF	U	1.5	1.0	97.6	1.2	1.2	4.9	6.4	4.7	87	78
Central African Republic	FF	U	615.7	1.0	91.3	129.9	157.1	5.0	22.0	20.4	..	469
Chad	FF	U	615.7	1.0	..	107.2	148.0	5.0	22.0	27.6	258	271
Chile	IF	U	508.8	1.0	105.4	141.6	264.2	8.5	12.6	8.8	201	148
China	P	U	8.3	0.9	106.9	1.2	1.8	2.3	5.8	8.4	134	109
Hong Kong, China	CB	U	7.8	1.0	..	6.1	8.3	4.5	8.5	14.3	..	..
Colombia	IF	U	1,756.2	..	102.7	120.4	637.2	21.3	30.4	15.9	261	156
Congo, Dem. Rep.	IF	U	4.0	0.7	340.7	0.0	..	..	124.6	..	..	..
Congo, Rep.	FF	U	615.7	1.0	..	456.5	657.4	5.0	22.0	−0.3	..	238
Costa Rica	P	U	285.7	1.0	103.5	41.3	136.6	14.3	25.7	11.6	..	179
Côte d'Ivoire	FF	U	615.7	1.0	103.5	160.4	268.3	3.5	..	..	..	133
Croatia	MF	U	7.1	1.0	97.1	..	4.3	4.3	14.9	10.5	..	..
Cuba	..	..	..	..	..	..	..	..	..	..	..	..
Czech Republic	MF	U	34.6	1.0	114.8	..	13.7	4.5	8.7	6.2	..	..
Denmark	P	U	7.0	1.0	97.5	8.2	8.8	2.4	7.1	4.3	127	..
Dominican Republic	MF	D	16.0	1.0	105.6	2.5	6.0	16.1	25.0	17.5	..	226
Ecuador	Other	U	11,786.8	1.0	80.3	286.2	4,341.3	48.9	64.0	1.3	185	203
Egypt, Arab Rep.	P	M	3.4	..	..	0.7	1.4	9.2	13.0	11.0	189	158
El Salvador	P	U	8.8	1.0	..	2.4	4.1	10.7	15.5	14.9	..	220
Eritrea	IF	U	..	..	..	..	1.6	..	..	..	..	..
Estonia	CB	U	14.7	..	..	0.1	6.3	4.2	8.7	4.6	..	..
Ethiopia	MF	U	7.9	1.0	..	0.7	1.2	6.3	10.6	9.1	205	128
Finland	Euro	U	5.6 [b]	1.0	90.7	5.9	6.1	1.2	4.7	4.0	159	..
France	Euro	U	6.2 [b]	1.0	93.2	6.5	6.6	2.7	6.4	6.0	126	119
Gabon	FF	U	615.7	1.0	96.2	324.2	368.1	5.0	22.0	16.4	..	163
Gambia, The	IF	U	11.4	1.0	99.8	1.8	2.3	12.5	24.0	30.6	..	241
Georgia	IF	U	2.0	1.0	..	..	0.4	14.6	33.4	21.9	796	724
Germany	Euro	U	1.8 [b]	1.0	89.5	..	2.0	2.4	8.8	7.8	124	139
Ghana	IF	U	2,647.3	1.0	125.7	91.7	582.5	23.6	..	..	..	242
Greece	P	U	305.6	1.0	102.5	117.8	235.4	8.7	15.0	11.6	220	171
Guatemala	MF	U	7.4	1.0	..	1.4	3.3	8.0	19.5	13.6	234	156
Guinea	IF	U	1,387.4	1.0	..	211.8	357.6	..	..	..	..	226
Guinea-Bissau	FF	U	615.7	1.0	..	12.0	167.3	3.5	..	..	..	..
Haiti	IF	U	16.9	1.0	..	1.3	6.4	7.4	22.9	12.1	..	271
Honduras	P	U	14.2	1.0	..	1.2	5.2	20.0	30.2	17.1	51	264

Relative prices and exchange rates | 5.6

	Exchange rate arrangements[a]		Official exchange rate	Ratio of official to parallel exchange rate	Real effective exchange rate	Purchasing power parity conversion factor		Interest rate			Key agricultural producer prices	
			local currency units to $		1995 = 100	local currency units to international $		Deposit %	Lending %	Real %	Wheat $ per metric ton	Maize $ per metric ton
	Classification 1999	Structure 1999	1999	1999	1999	1990	1999	1999	1999	1999	1998	1998
Hungary	P	U	237.1	1.0	109.0	21.3	99.8	13.3	16.3	6.8	113	92
India	IF	U	43.1	1.0	..	4.7	8.6	..	12.5	9.0	142	95
Indonesia	IF	U	7,855.1	1.0	..	603.8	1,892.4	25.7	27.7	13.2	..	87
Iran, Islamic Rep.	P	D	1,752.9	0.2	245.4	172.9	1,192.8	..	..	..	289	297
Iraq	P	U	0.3	..	..	..	..	..	..	..	..	..
Ireland	Euro	U	0.7 [b]	1.0	94.2	0.6	0.7	0.1	3.3	–0.4	114	..
Israel	P	U	4.1	1.0	105.7	1.7	3.7	11.3	16.4	9.1	149	1,330
Italy	Euro	U	1,817.4 [b]	1.0	110.9	1,330.1	1,665.1	1.6	5.6	4.0	181	161
Jamaica	MF	U	39.0	1.0	..	3.7	29.9	13.5	27.0	17.7	..	1,182
Japan	IF	U	113.9	1.0	89.3	174.8	157.1	0.1	2.2	3.1	1,295	1,082
Jordan	P	U	0.7	1.0	..	0.3	0.3	..	..	..	379	223
Kazakhstan	IF	U	119.5	0.5	..	..	25.6	..	..	..	72	85
Kenya	MF	U	70.3	1.0	..	8.5	24.9	9.6	22.4	14.6	262	151
Korea, Dem. Rep.	..	..	..	..	..	..	..	..	..	..	..	..
Korea, Rep.	IF	U	1,188.8	1.0	..	467.4	657.1	7.9	9.4	11.2	494	398
Kuwait	P	U	0.3	1.0	..	..	..	5.8	8.6	..	..	..
Kyrgyz Republic	MF	U	39.0	1.0	..	..	3.9	35.6	60.9	16.9	18	34
Lao PDR	MF	D	7,102.0	1.0	..	168.0	1,357.3	13.4	32.0	–42.1	..	96
Latvia	P	U	0.6	1.0	..	..	0.2	5.0	14.2	12.0	..	..
Lebanon	P	U	1,507.8	1.0	..	294.5	..	12.5	19.5	..	283	304
Lesotho	P	U	6.1	1.0	83.3	0.9	1.4	7.5	19.1	11.0	195	137
Libya	P	D	0.5	0.2	..	..	..	3.2	7.0	..	897	889
Lithuania	CB	U	4.0	1.0	..	..	1.7	4.9	13.1	9.5	..	..
Macedonia, FYR	P	U	56.9	1.0	73.7	0.1	20.8	11.4	20.4	20.8	..	..
Madagascar	IF	U	6,283.8	1.0	..	482.2	1,945.3	15.3	28.0	16.6	102	192
Malawi	MF	U	44.1	1.0	111.7	1.2	12.6	33.2	53.6	8.0	64	35
Malaysia	P	U	3.8	..	83.3	1.4	1.6	4.1	7.3	7.5	..	98
Mali	FF	U	615.7	1.0	..	133.2	201.4	3.5	..	..	124	128
Mauritania	MF	U	209.5	1.0	..	34.8	48.0	..	..	..	197	223
Mauritius	IF	U	25.2	1.0	..	6.6	10.0	10.9	21.6	15.0	..	208
Mexico	IF	U	9.6	1.0	..	1.4	5.8	9.6	25.9	8.4	141	146
Moldova	IF	U	10.5	1.0	100.1	..	1.4	27.5	35.5	–3.1	..	..
Mongolia	IF	U	1,021.9	1.0	..	2.7	230.0	19.8	37.7	32.7	..	..
Morocco	P	U	9.8	1.0	106.1	3.0	3.5	6.4	13.5	12.5	255	205
Mozambique	IF	U	12,775.1	1.0	..	303.4	3,410.9	7.9	19.6	17.3	21	13
Myanmar	P	D	6.3	..	..	..	..	11.0	16.1	..	..	..
Namibia	P	U	6.1	1.0	..	1.0	2.0	10.8	18.5	9.4	183	154
Nepal	P	U	68.2	1.0	..	6.2	11.7	7.3	11.3	2.6	119	106
Netherlands	Euro	U	2.1 [b]	1.0	93.7	2.1	2.1	2.7	3.5	2.1	118	232
New Zealand	IF	U	1.9	1.0	90.9	1.5	1.4	4.6	8.5	8.5	153	131
Nicaragua	P	U	11.8	1.0	104.3	0.0	2.4	10.3	22.1	9.8	..	189
Niger	FF	U	615.7	1.0	..	117.8	157.3	3.5	..	..	372	582
Nigeria	MF	U	92.3	0.9	78.9	3.5	30.6	12.8	20.3	6.5	1,555	1,201
Norway	MF	U	7.8	1.0	97.8	8.7	9.4	5.4	8.2	1.5	528	..
Oman	..	U	0.4	1.0	..	..	..	8.1	10.3	..	..	..
Pakistan	P	U	49.1	0.8	92.4	5.8	11.8	..	..	..	125	19
Panama	Other	U	1.0	1.0	..	0.6	0.6	6.9	10.1	10.8	..	265
Papua New Guinea	IF	U	2.6	1.0	84.2	0.5	0.8	15.5	18.9	5.9	..	84
Paraguay	MF	U	3,119.1	1.0	97.7	389.4	1,027.7	19.7	30.2	25.4	109	121
Peru	IF	U	3.4	1.0	..	0.1	1.5	16.3	30.8	25.9	242	246
Philippines	IF	U	39.1	1.0	96.4	5.3	10.6	8.2	11.8	3.3	..	138
Poland	IF	U	4.0	1.0	112.3	0.3	1.9	11.2	17.0	9.5	164	146
Portugal	Euro	U	188.2 [b]	1.0	98.9	90.9	133.4	2.4	5.2	1.1	147	144
Puerto Rico	..	..	..	..	..	..	..	..	..	..	..	..
Romania	MF	U	15,332.8	1.0	97.8	5.9	3,845.4	..	..	..	93	73
Russian Federation	IF	M	24.6	1.0	80.9	..	4.2	13.7	39.7	–14.5	..	..

	Exchange rate arrangements[a]		Official exchange rate	Ratio of official to parallel exchange rate	Real effective exchange rate	Purchasing power parity conversion factor		Interest rate			Key agricultural producer prices	
			local currency units to $		1995 = 100	local currency units to international $		Deposit %	Lending %	Real %	Wheat $ per metric ton	Maize $ per metric ton
	Classification 1999	Structure 1999	1999	1999	1999	1990	1999	1999	1999	1999	1998	1998
Rwanda	IF	U	333.9	1.0	..	32.1	88.9	7.9	..	..	226	145
Saudi Arabia	P	U	3.7	1.0	106.4	2.6	2.4	6.1	..	..	387	478
Senegal	FF	U	615.7	1.0	..	175.3	224.4	3.5	..	..	..	135
Sierra Leone	IF	D	1,804.2	..	116.2	38.0	544.4	9.5	26.8	1.5	..	33
Singapore	MF	U	1.7	1.0	95.4	1.7	1.8	1.7	5.8	7.2	101	89
Slovak Republic	MF	U	41.4	1.0	100.0	5.8	14.3	14.4	21.1	13.6	101	89
Slovenia	MF	U	181.8	1.0	..	..	114.7	7.2	12.4	5.4	198	160
South Africa	IF	U	6.1	1.0	84.6	1.0	2.1	12.2	18.0	10.4	146	103
Spain	Euro	U	156.2 [b]	1.0	96.7	104.4	130.6	1.8	3.9	0.8	157	154
Sri Lanka	P	U	70.4	1.0	..	9.3	17.8	11.8	7.0	2.3	..	218
Sudan	IF	U	252.6	1.0	..	..	..	..	..	..	248	117
Sweden	IF	U	8.3	1.0	97.0	9.0	9.8	1.6	5.5	5.0	126	..
Switzerland	IF	U	1.5	1.0	89.4	2.0	2.0	1.2	3.9	3.2	525	368
Syrian Arab Republic	P	M	11.2	1.0	..	8.9	13.2	..	..	..	884	823
Tajikistan	MF	U	..	..	..	..	..	..	..	..	..	..
Tanzania	IF	U	744.8	1.0	..	71.6	395.8	7.8	21.9	11.7	348	348
Thailand	IF	U	37.8	1.0	..	10.2	12.7	4.7	9.0	11.9	..	87
Togo	FF	U	615.7	1.0	105.3	90.5	134.4	3.5	..	..	..	216
Trinidad and Tobago	P	U	6.3	1.0	110.2	2.9	4.1	8.5	17.0	11.3	..	349
Tunisia	P	U	1.2	1.0	101.5	0.3	0.4	..	..	..	284	..
Turkey	P	U	418,782.9	1.0	..	1,445.1	188,358.0	78.4	..	..	133	126
Turkmenistan	P	D	5,200.0	0.7	..	..	1,074.3	..	..	..	..	..
Uganda	IF	U	1,454.8	1.0	88.0	112.8	348.4	8.7	21.5	16.5	681	358
Ukraine	MF	U	4.1	0.9	126.6	..	0.7	20.7	55.0	24.5	..	..
United Arab Emirates	P	U	3.7	1.0	..	3.3	..	..	..	..	..	..
United Kingdom	IF	U	0.6	1.0	127.6	0.6	0.7	..	5.3	2.8	118	176
United States	IF	U	1.0	1.0	119.4	1.0	1.0	..	8.0	6.3	99	61
Uruguay	P	U	11.3	1.0	112.5	0.6	8.1	14.2	53.3	46.2	119	125
Uzbekistan	MF	U	..	..	..	..	37.3	..	..	..	..	..
Venezuela, RB	P	U	605.7	1.0	152.3	23.0	475.3	21.3	32.1	3.5	123	445
Vietnam	P	U	13,916.2	1.0	..	..	2,773.9	..	..	..	..	..
West Bank and Gaza	..	..	..	..	..	..	..	..	..	..	..	..
Yemen, Rep.	IF	U	155.7	1.0	..	14.0	77.4	18.3	22.0	1.9	191	221
Yugoslavia, FR (Serb./Mont.)	..	..	..	..	..	..	..	..	..	..	..	..
Zambia	IF	U	2,388.0	1.0	111.9	17.3	1,006.7	20.3	40.5	15.4	211	79
Zimbabwe	P	U	38.3	1.0	..	0.9	6.3	38.5	55.4	4.9	211	122

a. Exchange rate arrangements are given for the end of the year in 1999. Exchange rate classifications include independent floating (IF), managed floating (MF), pegged (P), currency board (CB), and several exchange arrangements (euro means that the euro is used, FF that the currency is pegged to the French franc, and other that the currency of another country is used as legal tender). Exchange rate structures include dual exchange rates (D), multiple exchange rates (M), and unitary rate (U). b. Data refer to the exchange rate for the national currency relative to the U.S. dollar. On 1 January 1999 irrevocably fixed factors for converting national currencies of European Monetary Union members to the euro were established. The average annual euro–U.S. dollar exchange rate in 1999 was 0.94.

Relative prices and exchange rates | 5.6

In a market-based economy the choices households, producers, and governments make about the allocation of resources are influenced by relative prices, including the real exchange rate, real wages, real interest rates, and commodity prices. Relative prices also reflect, to a large extent, the choices of these agents. Thus relative prices convey vital information about the interaction of economic agents in an economy and with the rest of the world.

The exchange rate is the price of one currency in terms of another. Official exchange rates and exchange rate arrangements are established by governments (other exchange rates fully recognized by governments include market rates, which are determined largely by legal market forces, and, for countries maintaining multiple exchange arrangements, principal rates, secondary rates, and tertiary rates). Parallel, or black market, exchange rates reflect unofficial rates negotiated by traders and are by nature difficult to measure. Parallel exchange rate markets often account for only a small share of transactions and so may be both thin and volatile. But in countries with weak policies and financial systems they often represent the "going" rate. The parallel rates used here are collected by the Monetary Research Institute and published in its *MRI Bankers' Guide to Foreign Currency.*

Real effective exchange rates are derived by deflating a trade-weighted average of the nominal exchange rates that apply between trading partners. For most high-income countries the weights are based on trade in manufactured goods with other high-income countries during 1989–91, and an index of relative, normalized unit labor costs is used as the deflator. (Normalization smooths a time series by removing short-term fluctuations while retaining changes of a large amplitude over the longer economic cycle.) For other countries the weights prior to 1990 take into account trade in manufactured and primary products during 1980–82, and the weights from January 1990 onward this trade during 1988–90, and an index of relative changes in consumer prices is used as the deflator. An increase in the real effective exchange rate represents an appreciation of the local currency. Because of conceptual and data limitations, changes in real effective exchange rates should be interpreted with caution.

The official or market exchange rate is often used to compare prices in different currencies. But because market imperfections are extensive and exchange rates reflect at best the relative prices of tradable goods, the volume of goods and services that a U.S. dollar buys in the United States may not correspond to what a U.S. dollar converted to another country's

currency at the official exchange rate would buy in that country. The alternative approach is to convert national currency estimates of gross national income to a common currency by using conversion factors that reflect equivalent purchasing power. Purchasing power parity (PPP) conversion factors are based on price and expenditure surveys conducted by the International Comparison Programme (ICP) and represent the conversion factors applied to equalize price levels across countries. See *About the data* for table 1.1 for further discussion of the PPP conversion factor.

Many interest rates coexist in an economy, reflecting competitive conditions, the terms governing loans and deposits, and differences in the position and status of creditors and debtors. In some economies interest rates are set by regulation or administrative fiat. In economies with imperfect markets or where reported nominal rates are not indicative of effective rates, it may be difficult to obtain data on interest rates that reflect actual market transactions. Deposit and lending rates are collected by the International Monetary Fund (IMF) as representative interest rates offered by banks to resident customers. The terms and conditions attached to these rates differ by country, however, limiting their comparability. Real interest rates are calculated by adjusting nominal rates by an estimate of the inflation rate in the economy. A negative real interest rate indicates a loss in the purchasing power of the principal. The real interest rates in the table are calculated as $(i-P)/(1+P)$, where i is the nominal interest rate and P is the inflation rate (as measured by the GDP deflator).

The table also shows prices for two key agricultural commodities, wheat and maize. The prices received by farmers, used here, are important determinants of the type and volume of agricultural production. In theory these prices should refer to national average farmgate, or first-point-of-sale, transactions. But depending on the country's institutional arrangements—whether it relies on market wholesale prices, government fixed prices, or support prices—the data may not always refer to the same selling points. These data come from the Food and Agriculture Organization (FAO), with most originating from official national publications or FAO questionnaires. As the data show, the prices received by farmers are often not equalized across international markets (even after adjusting for freight, transport, and insurance costs and for differences in quality). Market imperfections such as taxes, subsidies, and trade barriers drive a wedge between domestic and international prices.

- **Exchange rate arrangement** describes the arrangement that an IMF member country has furnished to the IMF under article IV, section 2(a) of the IMF's Articles of Agreement. *Exchange rate classification* indicates how the exchange rate is determined in the main market when there is more than one market: floating (managed or independent), pegged (conventional, within horizontal bands, crawling peg, or crawling band), currency board (implicit legislative commitment to exchange domestic currency for a specified foreign currency at a fixed exchange rate), and exchange arrangement (country uses the euro, currency is pegged to the French franc, or another country's currency is used as legal tender). *Exchange rate structure* shows whether countries have a unitary exchange rate or dual or multiple rates. • **Official exchange rate** refers to the exchange rate determined by national authorities or to the rate determined in the legally sanctioned exchange market. It is calculated as an annual average based on monthly averages (local currency units relative to the U.S. dollar). • **Ratio of official to parallel exchange rate** measures the premium people must pay, relative to the official exchange rate, to exchange the domestic currency for U.S. dollars in the black market. • **Real effective exchange rate** is the nominal effective exchange rate (a measure of the value of a currency against a weighted average of several foreign currencies) divided by a price deflator or index of costs. • **Purchasing power parity conversion factor** is the number of units of a country's currency required to buy the same amount of goods and services in the domestic market as a U.S. dollar would buy in the United States. • **Deposit interest rate** is the rate paid by commercial or similar banks for demand, time, or savings deposits. • **Lending interest rate** is the rate charged by banks on loans to prime customers. • **Real interest rate** is the lending interest rate adjusted for inflation as measured by the GDP deflator. • **Key agricultural producer prices** are domestic producer prices converted to U.S. dollars using the official exchange rate.

The information on exchange rate arrangements is from the IMF's *Exchange Arrangements and Exchange Restrictions Annual Report, 1999*. The official and real effective exchange rates and deposit and lending rates are from the IMF's *International Financial Statistics*. The estimates of parallel market exchange rates are from the Monetary Research Institute's *MRI Bankers' Guide to Foreign Currency*. PPP conversion factors are from the World Bank. The agricultural price data are from the FAO's *Production Yearbook*. The real interest rates are calculated using World Bank data on the GDP deflator.

5.7 Defense expenditures and trade in arms

	Military expenditures				Armed forces personnel				Arms trade			
	% of GNI		% of central government expenditure		Total thousands		% of labor force		Exports % of total exports		Imports % of total imports	
	1992	1997	1992	1997	1992	1997	1992	1997	1992	1997	1992	1997
Albania	4.7	1.4	10.1	4.9	65	52	4.2	3.2	0.0	0.0	0.0	1.3
Algeria	1.8	3.9	5.9	12.0	126	124	1.7	1.3	0.0	0.0	0.1	5.6
Angola	24.2	20.5	24.6	36.3	128	95	2.8	1.8	0.0	0.0	1.5	3.5
Argentina	1.9	1.2	16.0	6.3	65	65	0.5	0.5	0.0	0.0	0.3	0.2
Armenia	3.5	3.5	..	..	20	60	1.1	3.2	0.0	2.1	0.0	0.0
Australia	2.5	2.2	9.2	8.6	68	65	0.8	0.7	0.1	0.0	2.1	1.4
Austria	1.0	0.9	2.4	1.9	52	48	1.4	1.3	0.2	0.0	0.1	0.3
Azerbaijan	2.9	1.9	9.0	10.8	43	75	1.4	2.2	0.0	1.3	0.0	0.0
Bangladesh	1.3	1.4	11.2	10.7	107	110	0.2	0.2	0.0	0.0	1.0	0.7
Belarus	1.9	1.7	4.9	4.8	102	65	1.9	1.2	0.0	6.7	0.0	0.0
Belgium	1.8	1.5	3.7	3.2	79	46	1.9	1.1	0.3	0.1	0.2	0.2
Benin	1.3	1.3	6.3	6.8	7	8	0.3	0.3	0.0	0.0	0.0	0.0
Bolivia	2.2	1.9	10.4	6.7	32	33	1.2	1.1	0.0	0.0	0.9	1.6
Bosnia and Herzegovina	16.5	5.9	..	14.1	60	40	3.2	2.4	0.0	0.0	0.0	6.5
Botswana	4.4	5.1	10.3	13.4	7	8	1.2	1.2	0.0	0.0	1.1	0.9
Brazil	1.1	1.8	3.5	3.9	296	296	0.4	0.4	0.5	0.1	0.9	0.7
Bulgaria	3.3	3.0	7.9	9.2	99	80	2.3	1.9	3.1	2.4	0.0	0.2
Burkina Faso	2.4	2.8	11.5	12.3	9	9	0.2	0.2	0.0	0.0	1.1	0.0
Burundi	2.7	6.1	7.8	25.8	13	35	0.4	1.0	0.0	0.0	0.0	16.5
Cambodia	4.9	4.1	..	25.8	135	60	2.7	1.0	0.0	0.0	0.0	0.9
Cameroon	1.6	3.0	8.2	17.7	12	13	0.2	0.2	0.0	0.0	0.0	0.7
Canada	2.0	1.3	7.5	..	82	61	0.5	0.4	0.7	0.3	0.6	0.2
Central African Republic	2.0	3.9	8.3	27.7	4	5	..	..	0.0	0.0	0.0	0.0
Chad	4.0	2.7	17.3	12.6	38	35	1.3	1.0	0.0	0.0	4.1	2.1
Chile	2.5	3.9	11.7	17.8	92	102	1.7	1.7	0.0	0.0	1.0	0.3
China	2.8	2.2	19.8	17.6	3,160	2,600	0.5	0.4	1.3	0.6	1.6	0.4
Hong Kong, China	..	..	..	..	..	..	..	..	..	..	..	..
Colombia	2.4	3.7	14.7	19.9	139	149	0.9	0.9	0.0	0.0	1.7	0.8
Congo, Dem. Rep.	3.0	5.0	16.1	41.4	45	50	0.3	0.3	0.0	0.0	0.0	2.4
Congo, Rep.	5.7	4.1	13.5	12.3	10	10	1.0	0.9	0.0	0.0	0.0	1.1
Costa Rica	1.4	0.6	7.5	3.1	8	10	0.7	0.7	0.0	0.0	0.2	0.1
Côte d'Ivoire	1.5	1.1	4.3	4.0	15	15	0.3	0.3	0.0	0.0	0.0	0.0
Croatia	7.7	6.3	19.2	20.1	103	58	4.6	2.7	0.0	0.0	0.0	0.1
Cuba	2.4	2.3	..	..	175	55	3.5	1.0	0.0	0.0	4.5	0.0
Czech Republic	2.7	1.9	6.9	5.8	107	55	1.9	1.0	1.6	0.4	0.0	0.5
Denmark	2.0	1.7	4.8	3.9	28	29	1.0	1.0	0.0	0.0	0.5	0.5
Dominican Republic	0.9	1.1	6.8	7.3	22	22	0.7	0.6	0.0	0.0	0.2	0.1
Ecuador	3.5	4.0	25.4	20.3	57	58	1.5	1.3	0.0	0.0	1.2	3.2
Egypt, Arab Rep.	3.7	2.8	8.5	11.0	424	430	2.2	1.9	0.7	0.1	19.4	12.1
El Salvador	2.1	0.9	13.4	6.7	49	15	2.4	0.6	0.0	0.0	4.1	0.3
Eritrea	..	7.8	..	18.1	55	55	3.2	2.9	0.0	0.0	0.0	0.0
Estonia	0.5	1.5	2.2	4.5	3	7	0.4	0.9	0.0	0.0	1.2	0.2
Ethiopia	3.7	1.9	17.9	7.9	120	100	0.5	0.4	0.0	0.0	0.0	0.0
Finland	2.2	1.7	4.3	4.3	33	35	1.3	1.3	0.0	0.1	2.1	1.2
France	3.4	3.0	7.6	6.4	522	475	2.1	1.8	0.9	2.0	0.2	0.1
Gabon	3.1	2.0	10.1	7.0	7	10	1.4	1.8	0.0	0.0	0.0	0.0
Gambia, The	3.6	3.7	19.7	15.0	1	1	0.2	0.2	0.0	0.0	2.1	11.9
Georgia	2.4	1.4	..	9.6	25	11	0.9	0.4	0.0	0.0	0.0	1.1
Germany	2.1	1.6	6.3	4.7	442	335	1.1	0.8	0.3	0.1	0.6	0.2
Ghana	0.8	0.7	4.6	2.4	7	7	0.1	0.1	0.0	0.0	0.0	0.0
Greece	4.4	4.6	13.5	13.8	208	206	4.8	4.6	0.2	0.3	3.9	3.1
Guatemala	1.5	1.4	14.0	15.0	44	30	1.4	0.8	0.0	0.0	0.2	0.1
Guinea	1.4	1.5	7.0	8.0	15	12	0.5	0.4	0.0	0.0	0.0	3.7
Guinea-Bissau	3.2	3.2	7.6	13.0	11	7	2.3	1.3	0.0	0.0	0.0	0.0
Haiti	1.5	..	14.7	..	8	0	0.3	0.0	0.0	0.0	0.0	0.8
Honduras	1.4	1.3	5.5	5.6	17	10	0.9	0.5	0.0	0.0	2.9	0.5

Defense expenditures and trade in arms | 5.7

	Military expenditures				Armed forces personnel				Arms trade			
	% of GNI		% of central government expenditure		Total thousands		% of labor force		Exports % of total exports		Imports % of total imports	
	1992	1997	1992	1997	1992	1997	1992	1997	1992	1997	1992	1997
Hungary	2.1	1.9	3.8	4.3	78	50	1.6	1.0	0.4	0.0	0.0	0.5
India	2.5	2.8	12.4	14.3	1,260	1,260	0.3	0.3	0.0	0.3	2.9	1.0
Indonesia	1.4	2.3	7.2	13.1	283	280	0.3	0.3	0.1	0.0	0.4	1.0
Iran, Islamic Rep.	3.0	3.0	14.9	11.6	528	575	3.2	3.1	0.1	0.1	3.3	5.8
Iraq	9.7	4.9	..	..	407	400	8.2	6.8	0.0	0.0	0.0	0.0
Ireland	1.4	1.2	3.8	3.3	13	17	1.0	1.1	0.0	0.0	0.1	0.1
Israel	11.7	9.7	23.3	20.9	181	185	8.8	7.5	4.8	1.6	7.9	3.6
Italy	2.1	2.0	3.9	4.1	471	419	1.9	1.7	0.3	0.3	0.2	0.2
Jamaica	1.0	0.9	3.0	2.4	3	3	0.2	0.2	0.0	0.0	0.6	0.2
Japan	1.0	1.0	6.3	6.6	242	250	0.4	0.4	0.0	0.0	0.9	0.8
Jordan	8.8	9.0	27.3	25.0	100	102	9.9	7.9	0.0	0.0	1.2	3.2
Kazakhstan	2.9	1.3	14.2	4.4	15	34	0.2	0.4	0.0	0.0	0.0	3.3
Kenya	3.0	2.1	9.9	7.2	24	24	0.2	0.2	0.0	0.0	1.2	1.2
Korea, Dem. Rep.	25.0	27.5	28.5	..	1,200	1,100	10.9	9.1	13.1	8.1	7.9	2.1
Korea, Rep.	3.7	3.4	19.8	14.6	750	670	3.7	2.9	0.1	0.0	1.5	0.8
Kuwait	77.0	7.5	96.3	26.8	12	28	2.1	4.1	0.2	0.0	13.8	24.3
Kyrgyz Republic	0.7	1.6	7.1	..	12	14	0.6	0.7	0.0	0.0	0.0	0.0
Lao PDR	9.8	3.4	23.6	17.5	37	50	..	..	0.0	0.0	2.3	1.4
Latvia	1.6	0.9	2.5	..	5	5	0.4	0.4	0.0	0.0	0.0	0.0
Lebanon	4.0	3.0	18.5	8.4	37	57	3.1	4.0	0.0	0.0	0.0	0.5
Lesotho	3.6	2.5	10.5	6.1	2	2	0.3	0.2	0.0	0.0	0.0	0.0
Libya	7.6	6.1	16.4	19.7	85	70	6.3	4.7	0.1	0.0	1.7	0.1
Lithuania	0.7	0.8	2.5	2.8	10	12	0.5	0.6	0.0	0.0	0.0	0.1
Macedonia, FYR	2.2	2.5	..	10.2	10	15	1.2	1.6	0.0	0.0	0.0	0.0
Madagascar	1.1	1.5	5.4	8.5	21	21	0.4	0.3	0.0	0.0	0.0	0.0
Malawi	1.1	1.0	3.9	2.9	10	8	0.2	0.2	0.0	0.0	0.0	0.0
Malaysia	3.2	2.2	10.3	9.9	128	110	1.7	1.3	0.0	0.0	0.6	0.9
Mali	2.3	1.7	9.4	7.2	12	10	0.3	0.2	0.0	0.0	0.0	1.5
Mauritania	3.5	2.3	13.3	9.8	16	11	1.6	1.0	0.0	0.0	0.0	0.0
Mauritius	0.4	0.3	1.5	1.2	1	1	0.2	0.2	0.0	0.0	0.3	0.4
Mexico	0.5	1.1	3.7	6.2	175	250	0.5	0.7	0.0	0.0	0.5	0.1
Moldova	0.5	1.0	1.5	1.9	9	11	0.4	0.5	0.0	7.9	0.8	0.0
Mongolia	2.6	1.9	9.3	5.1	21	20	1.9	1.6	0.0	0.0	0.0	0.0
Morocco	4.5	4.3	14.3	12.9	195	195	2.1	1.8	0.0	0.0	1.4	1.9
Mozambique	7.6	2.8	17.0	9.2	50	14	0.6	0.2	0.0	0.0	0.6	0.0
Myanmar	8.3	7.6	74.3	75.5	286	322	1.3	1.4	0.0	0.0	23.0	13.6
Namibia	2.2	2.7	5.6	7.3	8	8	1.4	1.2	0.0	0.0	0.0	0.3
Nepal	1.0	0.8	6.0	5.1	35	35	0.4	0.3	0.0	0.0	0.0	0.0
Netherlands	2.5	1.9	6.9	6.4	90	57	1.3	0.8	0.1	0.3	0.4	0.3
New Zealand	1.6	1.3	4.0	3.9	11	10	0.6	0.5	0.0	0.0	1.2	0.7
Nicaragua	3.1	1.5	8.1	4.5	15	14	1.0	0.8	13.5	0.0	0.6	0.0
Niger	1.3	1.1	7.9	6.9	5	5	0.1	0.1	0.0	0.0	0.0	1.4
Nigeria	2.6	1.4	15.6	12.3	76	76	0.2	0.2	0.0	0.0	2.0	0.7
Norway	3.1	2.1	6.4	4.8	36	33	1.6	1.4	0.1	0.0	1.7	0.7
Oman	20.5	26.1	40.2	36.4	35	38	6.6	6.2	0.0	0.0	0.3	3.2
Pakistan	7.4	5.7	27.9	24.2	580	610	1.4	1.3	0.4	0.0	6.7	5.2
Panama	1.3	1.4	5.7	4.8	11	12	1.1	1.1	2.0	0.0	0.5	0.3
Papua New Guinea	1.5	1.3	4.2	4.1	4	5	0.2	0.2	0.0	0.0	4.0	0.0
Paraguay	1.8	1.3	13.2	10.5	16	16	1.0	0.9	0.0	0.0	0.7	0.1
Peru	1.8	2.1	11.1	13.4	112	115	1.4	1.3	0.0	0.0	1.4	3.0
Philippines	1.9	1.5	10.2	7.9	107	105	0.4	0.3	0.0	0.0	1.6	0.3
Poland	2.3	2.3	8.8	5.6	270	230	1.4	1.2	0.2	0.2	0.0	0.4
Portugal	2.7	2.4	6.4	5.9	80	72	1.7	1.4	0.1	0.0	0.6	0.3
Puerto Rico	..	..	..	..	..	..	..	..	..	..	..	..
Romania	3.3	2.4	7.9	6.9	172	200	1.6	1.9	0.5	0.1	0.6	2.2
Russian Federation	8.0	5.8	28.0	30.9	1,900	1,300	2.5	1.7	5.9	2.6	0.0	0.0

	Military expenditures				Armed forces personnel				Arms trade			
	% of GNI		% of central government expenditure		Total thousands		% of labor force		Exports % of total exports		Imports % of total imports	
	1992	1997	1992	1997	1992	1997	1992	1997	1992	1997	1992	1997
Rwanda	4.4	4.4	21.7	22.2	30	40	0.8	1.0	0.0	0.0	0.0	6.7
Saudi Arabia	26.8	14.5	72.5	35.8	172	180	3.1	2.7	0.0	0.0	25.2	40.4
Senegal	2.8	1.6	13.5	8.5	18	14	0.5	0.4	0.0	0.0	1.0	0.0
Sierra Leone	3.2	5.9	17.7	33.0	8	5	0.5	0.3	0.0	0.0	6.8	0.0
Singapore	5.2	5.7	27.2	19.4	56	55	3.9	3.5	0.0	0.1	0.4	0.3
Slovak Republic	2.2	2.1	5.1	8.0	33	44	1.2	1.5	0.7	0.5	3.5	0.1
Slovenia	2.1ᵃ	1.7ᵃ	..	..	15	10	1.5	1.0	0.0	0.0	0.0	0.2
South Africa	3.2	1.8	9.8	5.6	75	75	0.5	0.5	0.4	1.2	1.3	0.1
Spain	1.6	1.5	6.4	6.0	198	107	1.2	0.6	0.3	0.5	0.4	0.4
Sri Lanka	3.8	5.1	13.6	21.2	110	110	1.5	1.4	0.0	0.0	0.3	1.5
Sudan	7.8	4.6	60.0	53.8	82	105	0.9	1.0	0.0	0.0	13.4	1.3
Sweden	2.6	2.5	5.3	5.4	70	60	1.5	1.3	1.5	1.1	0.3	0.5
Switzerland	1.8	1.4	7.2	5.8	31	39	0.9	1.0	1.2	0.1	0.7	0.4
Syrian Arab Republic	9.7	5.6	39.0	26.2	408	320	10.9	6.9	0.6	0.0	11.2	1.7
Tajikistan	0.3	1.7	0.7	10.6	3	10	0.1	0.4	0.0	0.0	0.0	0.0
Tanzania	2.2	1.3	10.0	10.7	46	35	0.3	0.2	0.0	0.0	0.3	1.5
Thailand	2.6	2.3	16.6	12.1	283	288	0.9	0.8	0.0	0.0	1.2	1.5
Togo	2.9	2.0	13.2	11.6	8	12	0.5	0.7	0.0	0.0	0.0	1.3
Trinidad and Tobago	1.5	1.5	4.8	5.4	2	2	0.4	0.4	0.0	0.0	0.0	0.2
Tunisia	2.4	2.0	7.1	5.3	35	35	1.1	1.0	0.0	0.0	0.3	0.3
Turkey	3.8	4.0	18.8	14.7	704	820	2.8	2.8	0.1	0.0	6.6	3.3
Turkmenistan	..	4.6	..	15.6	28	21	1.7	1.1	1.4	0.0	0.0	0.0
Uganda	2.4	4.2	11.7	23.9	70	50	0.8	0.5	0.0	0.0	2.3	2.3
Ukraine	1.9	3.7	..	8.4	438	450	1.7	1.8	0.0	3.5	0.0	0.0
United Arab Emirates	5.7	6.9	49.5	46.5	55	60	5.3	4.7	0.0	0.1	4.2	4.7
United Kingdom	3.8	2.7	9.3	7.1	293	218	1.0	0.7	3.3	2.3	1.3	0.7
United States	4.8	3.3	21.1	16.3	1,920	1,530	1.5	1.1	5.6	4.6	0.3	0.2
Uruguay	2.3	1.4	8.0	4.4	25	25	1.8	1.7	0.0	0.0	0.5	0.3
Uzbekistan	2.7	2.5	6.0	6.1	40	65	0.5	0.7	0.0	1.7	0.0	0.1
Venezuela, RB	2.6	2.2	11.9	9.8	75	75	1.0	0.8	0.0	0.0	0.9	1.8
Vietnam	3.4	2.8	14.5	11.1	857	650	2.4	1.7	0.4	0.0	0.4	1.1
West Bank and Gaza	..	..	..	..	..	..	..	..	..	..	..	..
Yemen, Rep.	9.4	8.1	29.8	17.4	64	69	1.5	1.3	0.0	0.0	0.2	5.5
Yugoslavia, FR (Serb./Mont.)	..	4.9	..	..	137	115	2.8	2.3	0.0	1.6	0.0	0.4
Zambia	3.3	1.1	9.3	3.9	16	21	0.5	0.5	0.0	0.0	0.0	0.0
Zimbabwe	3.8	3.8	10.1	11.9	48	40	1.0	0.8	0.3	0.0	4.1	0.5
World	**3.2 w**	**2.5 w**	**12.7 w**	**10.8 w**	**24,539 t**	**22,157 t**	**1.0 w**	**0.8 w**	**1.2 w**	**1.0 w**	**1.1 w**	**1.0 w**
Low income	2.7	2.9	12.5	13.3	6,511	6,206	0.7	0.6	0.1	0.4	2.0	1.2
Middle income	4.0	2.9	19.0	15.3	12,357	11,074	1.0	0.8	0.8	0.4	2.7	1.8
Lower middle income	4.3	3.2	19.4	17.6	9,856	8,573	1.0	0.8	1.6	0.8	2.4	1.5
Upper middle income	3.8	2.7	18.6	12.7	2,501	2,501	1.2	1.0	0.1	0.1	3.0	2.1
Low & middle income	3.8	2.9	18.0	15.0	18,868	17,280	0.9	0.7	0.7	0.4	2.6	1.8
East Asia & Pacific	2.9	2.5	17.2	15.1	7,256	6,264	0.8	0.6	0.4	0.2	1.3	0.8
Europe & Central Asia	5.2	4.0	19.3	16.6	4,311	3,899	2.1	1.7	2.9	1.4	1.4	0.8
Latin America & Carib.	1.3	1.8	5.8	6.6	1,443	1,362	0.8	0.7	0.2	0.0	0.7	0.5
Middle East & N. Africa	14.2	6.9	48.2	22.9	2,633	2,614	3.3	2.9	0.1	0.0	10.5	13.1
South Asia	3.1	3.1	14.9	15.6	2,142	2,133	0.4	0.4	0.1	0.2	3.3	1.6
Sub-Saharan Africa	3.1	2.3	10.5	7.8	1,083	1,008	0.5	0.4	0.2	0.5	1.3	0.5
High income	3.1	2.4	11.8	9.9	5,671	4,877	1.3	1.1	1.4	1.2	0.7	0.7
Europe EMU	2.3	2.0	5.9	5.1	1,981	1,612	1.5	1.2	0.4	0.5	0.4	0.2

Note: Data for some countries are based on partial or uncertain data or rough estimates; see U.S. Department of State (1999).

a. Data provided by national authorities.

Defense expenditures and trade in arms | 5.7

Although national defense is an important function of government and security from external threats contributes to economic development, high levels of defense spending burden the economy and may impede growth. Comparisons of defense spending between countries should take into account the many factors that influence perceptions of vulnerability and risk, including historical and cultural traditions, the length of borders that need defending, the quality of relations with neighbors, and the role of the armed forces in the body politic.

Data on defense spending from governments are often incomplete and unreliable. Even in countries where parliaments vigilantly review government budgets and spending, defense spending and trade in arms often do not receive close scrutiny. For a detailed critique of the quality of such data see Ball (1984) and Happe and Wakeman-Linn (1994).

The International Monetary Fund's (IMF) *Government Finance Statistics Yearbook* is the primary source of data on defense spending. It uses a consistent definition of defense spending based on the United Nations' classification of the functions of government and the North Atlantic Treaty Organization (NATO) definition. The IMF checks data on defense spending for broad consistency with other macroeconomic data reported to it but is not always able to verify the accuracy and completeness of the data. Moreover, country coverage is affected by delays or failure to report data. Thus most researchers supplement the IMF's data with assessments by other organizations. However, these organizations rely heavily on reporting by governments, on confidential intelligence estimates of varying quality, on sources that they do not or cannot reveal, and on one another's publications. The data in this table are the latest available from the U.S. Department of State's Bureau of Verification and Compliance (formerly the Bureau of Arms Control).

Definitions of military spending differ depending on whether they cover civil defense, reserves and auxiliary forces, police and paramilitary forces, dual-purpose forces such as military and civilian police, military grants in kind, pensions for military personnel, and social security contributions paid by one part of government to another. Official government data may omit parts of military spending, disguise financing through extrabudgetary accounts or unrecorded use of foreign exchange receipts, or fail to include military assistance or secret military equipment imports. Current spending is more likely to be reported than capital spending. In some cases a more accurate estimate of military spending can be obtained by adding the value of estimated arms imports and nominal military expenditures. This method may understate or overstate spending

in a particular year, however, because payments for arms may not coincide with deliveries.

The data on armed forces refer to military personnel on active duty, including paramilitary forces. These data exclude civilians in the defense establishment and so are not consistent with the data on military spending on personnel. Moreover, because they exclude payments to personnel not on active duty, they underestimate the share of the labor force working for the defense establishment. Because governments rarely report the size of their armed forces, such data typically come from intelligence sources. The Bureau of Verification and Compliance attributes its data to unspecified U.S. government sources.

The Standard International Trade Classification does not clearly distinguish trade in military goods. For this and other reasons, customs-based data on trade in arms are of little use, so most compilers rely on trade publications, confidential government information on third-country trade, and other sources. The construction of defense production facilities and the licensing fees paid for the production of arms are included in trade data when they are specified in military transfer agreements. Grants in kind are usually included as well. Definitional issues include treatment of dual-use equipment such as aircraft, use of military establishments such as schools and hospitals by civilians, and purchases by nongovernment buyers. Bureau of Verification and Compliance data do not include arms supplied to subnational groups. Valuation problems arise when data are reported in volume terms and the purchase price must be estimated. Differences between sources may reflect reporting lags or differences in the period covered. Most compilers revise their time-series data regularly, so estimates for the same year may not be consistent between publication dates.

The data on U.S. arms exports were substantially revised upward in last year's edition of the *World Development Indicators*, based on data from the most recent edition of the Bureau of Verification and Compliance's *World Military Expenditures and Arms Transfers* (U.S. Department of State 1999). Revisions were made in commercial arms sales made directly by U.S. firms to foreign importers under authorization of the U.S. Department of State in accordance with U.S. regulations on international traffic in arms. Under the previous methodology the commercial arms component was represented by preliminary data on the deliveries made under approved export licenses. But because of weaknesses in data reporting, the extent to which authorized exports matched actual exports was uncertain. The new methodology assumes that deliveries constitute 50 percent of total authorizations by country. These deliveries are then distributed in a fixed pattern over the years of the license.

• **Military expenditures** for NATO countries are based on the NATO definition, which covers military-related expenditures of the defense ministry (including recruiting, training, construction, and the purchase of military supplies and equipment) and other ministries. Civilian-type expenditures of the defense ministry are excluded. Military assistance is included in the expenditures of the donor country, and purchases of military equipment on credit are included at the time the debt is incurred, not at the time of payment. Data for other countries generally cover expenditures of the ministry of defense (excluded are expenditures on public order and safety, which are classified separately).
• **Armed forces personnel** refer to active duty military personnel, including paramilitary forces if those forces resemble regular units in their organization, equipment, training, or mission. • **Arms trade** comprises exports and imports of military equipment usually referred to as "conventional," including weapons of war, parts thereof, ammunition, support equipment, and other commodities designed for military use. See *About the data* for more details.

The data on military expenditures, armed forces, and arms trade are from the Bureau of Verification and Compliance's *World Military Expenditures and Arms Transfers 1998* (U.S. Department of State 1999).

5.8 | Transport infrastructure

	Roads			Railways			Air		
	Total road network km 1995–99[a]	Paved roads % 1995–99[a]	Goods hauled million ton-km 1995–99[a]	Passenger-km per $ million of PPP GDP 1995–99[a]	Goods transported ton-km per $ million of PPP GDP 1995–99[a]	Diesel locomotives available % 1995–99[a]	Aircraft departures thousands 1999	Passengers carried thousands 1999	Air freight million ton-km 1999
Albania	18,000	39.0	1,830	9,196	1,941	..	1	20	0
Algeria	104,000	68.9	..	11,146	..	..	36	2,937	15
Angola	51,429	10.4	..	..	..	..	6	531	37
Argentina	215,471	29.4	..	28,665	..	..	184	9,192	241
Armenia	16,718	96.2	69	6,232	49,717	30	4	343	12
Australia	913,000	38.7	..	..	..	..	338	30,007	1,794
Austria	200,000	100.0	16,100	41,307	75,075	89	127	6,057	341
Azerbaijan	27,327	92.3	2,968	..	..	..	9	572	58
Bangladesh	201,182	9.5	..	22,570	4,706	81	6	1,215	143
Belarus	65,994	94.8	9,232	202,576	463,691	93	6	212	2
Belgium	145,850	80.7	35,000	32,012	30,522	86	233	9,965	535
Benin	6,787	20.0	..	..	..	..	2	84	14
Bolivia	49,400	5.5	..	6,460	..	..	24	1,873	20
Bosnia and Herzegovina	21,846	52.3	..	..	..	..	4	60	1
Botswana	10,217	55.0	..	..	..	..	6	144	0
Brazil	1,724,924	9.5	..	865	31,150	..	678	28,273	1,450
Bulgaria	37,284	92.0	168	96,104	122,533	37	15	735	12
Burkina Faso	12,100	16.0	..	..	..	..	3	147	14
Burundi	..	..	..	..	..	..	0	0	0
Cambodia	35,769	7.5	1,200	..	77,235	..	..	..	..
Cameroon	34,300	12.5	..	14,371	40,811	68	5	293	47
Canada	901,903	35.3	76,694	1,945	429,555	..	310	24,039	1,861
Central African Republic	24,307	2.7	60	..	..	..	2	84	14
Chad	33,400	0.8	..	..	..	..	2	84	14
Chile	79,353	18.9	..	4,907	7,802	65	93	5,188	1,139
China	1,526,389	..	572,430	82,693	260,427	82	548	55,853	3,295
Hong Kong, China	1,760	100.0	..	..	..	..	72	12,593	4,546
Colombia	112,988	14.4	31	62	1,948	..	209	8,665	635
Congo, Dem. Rep.	..	..	..	700	..	15	..	..	..
Congo, Rep.	12,800	9.7	..	36,264	..	35	6	132	14
Costa Rica	35,876	22.0	3,070	..	..	50	32	1,055	85
Côte d'Ivoire	50,400	9.7	..	6,512	21,081	53	6	260	14
Croatia	..	..	..	34,782	58,859	63	16	833	2
Cuba	60,858	49.0	..	..	..	53	16	1,259	63
Czech Republic	128,854	100.0	36,694	53,029	138,506	86	35	1,853	26
Denmark	71,462	100.0	15,300	40,275	11,786	..	110	5,971	198
Dominican Republic	4,969	78.7	..	..	..	..	0	10	0
Ecuador	43,197	18.9	3,959	..	..	..	20	1,387	33
Egypt, Arab Rep.	64,000	78.1	31,500	317,220	16,164	..	44	4,620	270
El Salvador	10,029	19.8	..	..	..	..	32	1,624	44
Eritrea	4,010	21.8	..	..	..	..	..	..	..
Estonia	50,436	21.3	3,929	19,842	486,631	80	11	302	1
Ethiopia	28,652	13.3	..	..	..	..	25	861	102
Finland	77,900	64.5	26,500	29,933	87,619	88	108	6,050	252
France	893,500	100.0	245,400	50,392	42,145	93	747	49,691	4,962
Gabon	7,670	8.2	..	11,254	65,276	89	7	423	54
Gambia, The	2,700	35.4	..	..	..	..	..	..	..
Georgia	20,215	93.5	420	44,361	200,857	34	3	159	3
Germany	656,140	99.1	315,900	31,471	38,962	92	710	54,550	6,620
Ghana	37,800	24.1	..	6,221	..	..	5	304	31
Greece	117,000	91.8	17,000	11,850	2,101	..	91	6,267	103
Guatemala	14,118	34.5	..	..	..	..	7	506	3
Guinea	30,500	16.5	..	..	..	..	1	59	1
Guinea-Bissau	4,400	10.3	..	..	..	..	0	0	0
Haiti	4,160	24.3	..	..	..	..	..	..	..
Honduras	13,603	20.4	..	..	..	..	..	..	..

Transport infrastructure | 5.8

	Roads			Railways			Air		
	Total road network km 1995–99[a]	Paved roads % 1995–99[a]	Goods hauled million ton-km 1995–99[a]	Passenger-km per $ million of PPP GDP 1995–99[a]	Goods transported ton-km per $ million of PPP GDP 1995–99[a]	Diesel locomotives available % 1995–99[a]	Aircraft departures thousands 1999	Passengers carried thousands 1999	Air freight million ton-km 1999
Hungary	188,203	43.4	14	94,085	72,243	64	30	1,944	40
India	2,465,877	56.5	958	195,355	136,165	90	181	16,005	531
Indonesia	346,863	46.3	..	28,490	8,725	83	135	8,047	362
Iran, Islamic Rep.	167,157	56.3	..	18,506	43,629	47	76	8,277	107
Iraq	47,400	86.0	..	..	..	..	0	0	0
Ireland	92,500	94.1	5,900	18,714	4,599	82	133	11,949	138
Israel	16,115	100.0	..	3,243	9,132	92	52	4,033	1,034
Italy	654,676	100.0	219,800	38,135	18,054	79	357	28,049	1,614
Jamaica	19,000	70.7	..	..	..	..	21	1,670	30
Japan	1,156,371	76.0	300,670	77,409	7,608	88	662	105,960	8,226
Jordan	6,640	100.0	..	..	35,549	90	16	1,252	194
Kazakhstan	109,445	89.7	4,506	177,393	1,474,814	..	18	667	23
Kenya	63,800	13.9	..	13,457	44,821	..	25	1,358	66
Korea, Dem. Rep.	31,200	6.4	..	..	..	..	1	59	2
Korea, Rep.	86,990	74.5	74,504	46,461	19,459	88	206	31,319	8,359
Kuwait	4,450	80.6	..	..	..	..	17	2,130	268
Kyrgyz Republic	18,500	91.1	1,220	..	..	..	8	312	6
Lao PDR	22,321	13.8	..	..	..	..	6	197	2
Latvia	73,227	38.6	4,161	108,396	770,302	88	9	199	0
Lebanon	6,350	95.0	..	..	..	..	10	719	103
Lesotho	5,940	18.3	..	..	..	..	0	1	0
Libya	24,484	57.1	..	..	..	..	6	571	0
Lithuania	73,650	91.3	7,740	28,379	328,042	88	11	250	2
Macedonia, FYR	8,684	63.8	1,210	15,959	40,430	40	6	488	2
Madagascar	30,623	11.6	..	..	..	..	19	635	32
Malawi	16,451	19.0	..	0	11,535	..	4	112	1
Malaysia	65,877	75.8	..	8,221	7,203	65	165	14,985	1,425
Mali	15,100	12.1	..	30,578	35,377	..	2	84	14
Mauritania	7,660	11.3	..	..	..	..	4	187	14
Mauritius	1,910	96.0	..	..	..	..	11	831	169
Mexico	318,952	34.3	179,085	2,578	61,435	68	310	19,263	309
Moldova	12,657	87.0	952	..	..	..	1	43	1
Mongolia	49,250	3.5	123	253,483	684,165	..	2	225	8
Morocco	57,646	56.3	2,557	18,176	52,224	75	44	3,392	58
Mozambique	30,400	18.7	110	..	..	..	6	235	7
Myanmar	28,200	12.2	..	..	..	..	11	537	6
Namibia	63,258	8.3	..	5,607	133,970	89	8	201	4
Nepal	7,700	41.5	..	..	..	..	12	583	16
Netherlands	116,500	90.0	46,500	41,134	9,712	88	224	19,741	4,053
New Zealand	92,075	61.9	..	..	51,030	..	228	8,892	842
Nicaragua	18,000	10.1	..	..	..	..	1	59	1
Niger	10,100	7.9	..	..	..	..	2	84	14
Nigeria	194,394	30.9	..	512	4,915	18	8	668	30
Norway	90,880	75.5	12,796	..	..	..	335	15,020	201
Oman	32,800	30.0	..	..	..	..	22	1,933	126
Pakistan	254,410	43.0	96,802	81,899	17,118	..	65	4,972	330
Panama	11,400	34.6	..	..	..	..	21	933	15
Papua New Guinea	19,600	3.5	..	..	..	..	24	1,102	18
Paraguay	29,500	9.5	..	..	..	..	8	232	0
Peru	77,999	12.9	..	1,397	4,640	..	37	1,900	6
Philippines	199,950	19.8	..	915	4	..	36	5,004	241
Poland	381,046	65.6	70,452	77,593	171,756	55	44	2,141	80
Portugal	68,732	..	14,200	30,125	13,406	88	104	7,325	223
Puerto Rico	14,400	100.0	..	..	..	..	..	..	..
Romania	198,589	69.3	13,457	97,692	135,241	78	18	980	12
Russian Federation	570,719	..	138	129,048	1,102,493	..	321	18,600	872

	Roads			Railways			Air		
	Total road network km 1995–99[a]	Paved roads % 1995–99[a]	Goods hauled million ton-km 1995–99[a]	Passenger-km per $ million of PPP GDP 1995–99[a]	Goods transported ton-km per $ million of PPP GDP 1995–99[a]	Diesel locomotives available % 1995–99[a]	Aircraft departures thousands 1999	Passengers carried thousands 1999	Air freight million ton-km 1999
Rwanda	14,900	9.1	..	..	..	..	..	..	..
Saudi Arabia	151,470	30.1	..	998	3,811	80	107	12,329	1,000
Senegal	14,576	29.3	..	6,609	37,365	79	2	103	14
Sierra Leone	11,300	8.0	..	..	..	..	0	19	0
Singapore	3,066	100.0	..	..	..	..	68	15,283	5,451
Slovak Republic	42,713	86.7	8,474	57,115	215,427	87	6	111	0
Slovenia	20,126	90.6	3,440	21,848	89,048	..	11	556	4
South Africa	534,131	11.8	..	25,701	283,106	96	101	7,374	686
Spain	663,795	99.0	103,000	26,047	16,714	87	414	33,559	816
Sri Lanka	11,285	95.0	30	59,310	1,865	..	10	1,422	179
Sudan	11,900	36.3	..	..	..	42	7	390	34
Sweden	210,907	77.5	32,700	36,988	96,543	..	237	12,933	290
Switzerland	71,115	..	14,500	..	..	..	276	16,209	1,879
Syrian Arab Republic	41,792	23.1	..	5,688	28,030	100	11	668	18
Tajikistan	13,700	82.7	..	..	..	..	4	156	2
Tanzania	88,200	4.2	..	73,054	73,054	66	5	190	2
Thailand	64,600	97.5	..	26,781	7,923	72	95	15,951	1,671
Togo	7,520	31.6	..	..	..	..	2	84	14
Trinidad and Tobago	8,320	51.1	..	..	..	..	21	1,112	55
Tunisia	23,100	78.9	..	21,247	41,961	71	20	1,923	19
Turkey	385,960	34.0	150,974	14,726	20,238	74	111	10,097	313
Turkmenistan	13,597	..	..	..	..	..	3	220	17
Uganda	..	..	..	1,366	4,924	..	3	179	22
Ukraine	168,674	96.6	18,206	296,128	941,037	87	28	891	13
United Arab Emirates	1,088	100.0	..	..	..	..	44	5,848	1,114
United Kingdom	371,603	100.0	159,500	..	..	..	906	68,235	4,926
United States	6,348,227	58.8	1,534,430	1,020	350,942	..	8,512[b]	634,365[b]	27,317[b]
Uruguay	8,983	90.0	..	6,931	6,126	..	11	728	18
Uzbekistan	43,463	87.3	..	42,559	304,816	..	31	1,658	69
Venezuela, RB	96,155	33.6	..	0	342	65	130	4,690	68
Vietnam	93,300	25.1	..	18,843	9,807	95	29	2,600	99
West Bank and Gaza	..	..	..	..	..	..	..	..	..
Yemen, Rep.	64,725	8.1	..	..	..	..	8	731	21
Yugoslavia, FR (Serb./Mont.)	48,603	59.3	1,244	..	..	..	..	..	..
Zambia	66,781	..	..	24,892	74,141	62	1	42	0
Zimbabwe	18,338	47.4	..	18,082	151,076	70	13	567	35

World	**55.3 m**						**20,645 s**	**1,558,788 s**	
Low income	18.7						730	47,663	
Middle income	47.8						4,345	308,412	
Lower middle income	45.7						2,001	156,778	
Upper middle income	49.7						2,344	151,634	
Low & middle income	29.9						5,074	356,075	
East Asia & Pacific	17.4						1,350	136,788	
Europe & Central Asia	87.0						754	43,845	
Latin America & Carib.	20.1						1,930	91,253	
Middle East & N. Africa	52.1						428	42,080	
South Asia	43.0						282	24,711	
Sub-Saharan Africa	15.3						331	17,397	
High income	90.3						15,571	1,202,713	
Europe EMU	99.0						3,186	227,778	

a. Data are for the latest year available in the period shown. b. Data cover only the carriers designated by the U.S. Department of Transportation as major and national air carriers.

Transport infrastructure | 5.8

About the data

Transport infrastructure—highways, railways, ports and waterways, and airports and air traffic control systems—and the services that flow from it are crucial to the activities of households, producers, and governments. Because performance indicators vary significantly by transport mode and by focus (whether physical infrastructure or the services flowing from that infrastructure), highly specialized and carefully specified indicators are required. The table provides selected indicators of the size and extent of roads, railways, and air transport systems and the volume of freight and passengers carried.

Data for most transport sectors are not internationally comparable. Unlike for demographic statistics, national income accounts, and international trade data, the collection of infrastructure data has not been "internationalized." Data on roads are collected by the International Road Federation (IRF), and data on air transport by the International Civil Aviation Organization (ICAO). National road associations are the primary source of IRF data; in countries where such an association is lacking or does not respond, other agencies are contacted, such as road directorates, ministries of transport or public works, or central statistical offices. As a result, the compiled data are of uneven quality.

Even when data are available, they are often of limited value because of incompatible definitions, inappropriate geographical units of observation, lack of timeliness, and variations in the nature of the terrain. Data on passengers carried, for example, may be distorted because of "ticketless" travel or breaks in journeys; in such cases the statistics may report the number of passenger-kilometers for two passengers rather than one. Measurement problems are compounded because the mix of transported commodities changes over time, and in some cases shorter-haul traffic has been excluded from intercity traffic. Finally, the quality of transport service (reliability, transit time, and condition of goods delivered) is rarely measured but may be as important as quantity in assessing an economy's transport system. Serious efforts are needed to create international databases whose comparability and accuracy can be gradually improved.

The air transport data represent the total (international and domestic) scheduled traffic carried by the air carriers registered in a country. Countries submit air transport data to ICAO on the basis of standard instructions and definitions issued by ICAO. In many cases, however, the data include estimates by ICAO for nonreporting carriers. Where possible, these estimates are based on previous submissions supplemented by information published by the air carriers, such as flight schedules.

The data represent the air traffic carried on scheduled services, but changes in air transport regulations in Europe have made it more difficult to classify traffic as scheduled or nonscheduled. Thus recent increases shown for some European countries may be due to changes in the classification of air traffic rather than actual growth. For countries with few air carriers or only one, the addition or discontinuation of a home-based air carrier may cause significant changes in air traffic.

Figure 5.8

More roads are paved in Europe and Central Asia

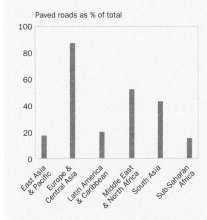

Paved roads as % of total

Note: Data are for the latest year available in 1995–99.
Source: Table 5.8.

In Europe and Central Asia almost 90 percent of the road network is paved, but in Sub-Saharan Africa only about 15 percent is. Roads carry 80–90 percent of Sub-Saharan Africa's freight and passengers and provide the only access to most rural communities. But about 50 percent of its rural road network and 30 percent of its urban network are in poor condition.

Definitions

• **Total road network** includes motorways, highways, and main or national roads, secondary or regional roads, and all other roads in a country. • **Paved roads** are those surfaced with crushed stone (macadam) and hydrocarbon binder or bituminized agents, with concrete, or with cobblestones, as a percentage of all the country's roads, measured in length. • **Goods hauled by road** are the volume of goods transported by road vehicles, measured in millions of metric tons times kilometers traveled. • **Railway passengers** refer to the total number of passengers transported times kilometers traveled per million dollars of GDP measured in purchasing power parity (PPP) terms (for a discussion of PPP see *About the data* for table 1.1). • **Goods transported by rail** are the tonnage of goods transported times kilometers traveled per million dollars of GDP measured in purchasing power parity (PPP) terms. • **Diesel locomotives available** are those in service as a percentage of all diesel locomotives. • **Aircraft departures** are the number of domestic and international takeoffs of air carriers registered in the country. • **Air passengers carried** include both domestic and international aircraft passengers of air carriers registered in the country. • **Air freight** is the sum of the metric tons of freight, express, and diplomatic bags carried on each flight stage (the operation of an aircraft from takeoff to its next landing) multiplied by the stage distance of air carriers registered in the country.

Data sources

The data on roads are from the International Road Federation's *World Road Statistics* and from Eurostat (europa.eu.int/eurostat.html). The railway data are from a database maintained by the World Bank's Transportation, Water, and Urban Development Department, Transport Division. The air transport data are from the International Civil Aviation Organization's *Civil Aviation Statistics of the World* and ICAO staff estimates.

5.9 Power and communications

	Electric power		Telephone mainlines[a]							Mobile phones[a]	International telecommunications[a]	
	Consumption per capita kwh **1998**	Transmission and distribution losses % of output **1998**	per 1,000 people **1999**	In largest city per 1,000 people **1999**	Waiting list thousands **1999**	Waiting time years **1999**	per employee **1999**	Revenue per line $ **1999**	Cost of local call $ per 3 minutes **1999**	per 1,000 people **1999**	Outgoing traffic minutes per subscriber **1999**	Cost of call to U.S. $ per 3 minutes **1999**
Albania	678	47	36	93	98.5	3.9	32	596	0.02	3	525	*4.37*
Algeria	563	19	52	*55*	640.0	6.0	90	166	*0.02*	2	90	*4.70*
Angola	60	28	8	21	21.1	1.5	46	1,195	0.06	2	363	5.13
Argentina	1,891	15	201	247	*58.2*	0.1	373	1,319	0.09	121	62	2.80
Armenia	930	28	155	*206*	71.0	..	61	153	0.11	2	62	..
Australia	8,717	6	520	516	0.0	0.0	116	1,422	0.16	343	112	0.60
Austria	6,175	6	472	*519*	0.0	0.0	160	1,249	0.17	514	305	*1.60*
Azerbaijan	1,584	16	95	*184*	88.4	3.1	65	84	*0.12*	23	44	9.80
Bangladesh	81	16	3	24	172.0	4.4	23	591	0.03	1	104	*6.00*
Belarus	2,761	7	257	*341*	440.0	2.6	99	61	0.01	2	60	6.10
Belgium	7,249	5	502	501	..	*0.0*	218	1,009	0.16	314	312	2.00
Benin	46	71	*7*	36	*13.7*	4.0	*30*	1,182	*0.11*	*1*	*297*	6.90
Bolivia	409	12	62	*107*	7.5	0.1	*103*	826	0.09	52	65	3.70
Bosnia and Herzegovina	539	22	96	*447*	70.0	2.2	201	431	*0.03*	14	264	*3.70*
Botswana	..	..	77	*179*	*11.8*	0.7	71	974	0.02	74	323	3.60
Brazil	1,793	17	149	165	*2,400.0*	0.7	176	728	0.03	89	24	1.80
Bulgaria	3,166	13	354	564	330.0	3.5	111	131	*0.00*	42	34	..
Burkina Faso	..	..	4	36	..	..	38	1,137	0.10	0	206	11.00
Burundi	..	..	3	*54*	*10.0*	7.9	33	480	0.03	0	129	7.30
Cambodia	..	..	3	16	..	..	38	771	0.03	8	263	..
Cameroon	185	20	6	38	*50.0*	6.2	43	725	*0.06*	0	291	*3.39*
Canada	15,071	7	655	..	0.0	0.0	238	969	..	226	266	*1.20*
Central African Republic	..	..	3	*15*	*1.8*	>10.0	25	1,181	0.49	1	439	8.00
Chad	..	..	1	*7*	*0.6*	0.5	*23*	2,199	*0.16*	0	292	14.07
Chile	2,082	8	207	*282*	*58.3*	0.1	223	766	0.12	151	80	2.90
China	746	7	86	294	*812.0*	0.0	159	259	0.01	34	18	*6.70*
Hong Kong, China	5,244	12	576	576	0.0	0.0	106	1,760	0.00	636	703	2.60
Colombia	866	21	160	*322*	1,155.0	1.7	151	413	0.04	75	33	2.20
Congo, Dem. Rep.	110	3	*0*	..	..	..	..	..	..	0	..	..
Congo, Rep.	83	38	..	..	..	..	..	..	..	1	..	..
Costa Rica	1,450	5	204	*478*	34.7	0.4	178	333	0.02	35	122	2.00
Côte d'Ivoire	..	..	15	57	*33.1*	1.1	59	1,911	0.07	18	325	*7.86*
Croatia	2,463	21	365	*324*	*72.0*	0.9	151	473	*0.03*	66	198	*5.66*
Cuba	954	18	39	86	..	..	29	1,396	0.09	0	75	7.30
Czech Republic	4,747	8	371	965	74.0	0.2	161	585	0.15	189	96	2.00
Denmark	6,033	6	685	..	0.0	0.0	193	1,140	0.12	495	180	*1.77*
Dominican Republic	627	28	98	129	..	..	*200*	..	..	50	234	*3.90*
Ecuador	625	21	91	339	..	..	164	400	0.01	31	91	4.90
Egypt, Arab Rep.	861	12	75	*151*	1,290.0	2.3	84	395	0.03	8	32	5.84
El Salvador	559	13	76	*198*	..	..	50	752	0.06	62	151	*2.40*
Eritrea	..	..	7	40	19.3	6.8	59	726	0.02	0	92	*8.24*
Estonia	3,531	18	357	*381*	39.3	1.5	191	514	0.07	268	143	*3.41*
Ethiopia	22	10	3	*42*	225.0	>10.0	30	404	0.03	0	64	7.37
Finland	14,129	4	557	*677*	0.0	0.0	132	1,417	0.13	651	151	*1.75*
France	6,287	6	582	..	0.0	0.0	200	836	0.12	366	129	1.00
Gabon	749	10	32	*89*	*10.0*	>10.0	36	2,064	*0.15*	7	491	..
Gambia, The	..	..	23	77	16.9	6.4	31	1,028	0.30	4	219	*6.18*
Georgia	1,257	14	123	289	105.0	3.0	69	47	..	19	70	..
Germany	5,681	4	590	591	0.0	0.0	215	1,070	0.11	286	152	0.80
Ghana	289	1	8	*54*	..	..	44	1,078	0.08	4	190	..
Greece	3,739	7	528	*727*	21.6	0.2	260	891	0.07	367	130	*2.59*
Guatemala	322	21	55	..	..	..	130	403	0.09	30	140	0.80
Guinea	..	..	6	19	*1.3*	0.2	56	373	0.10	3	266	*9.04*
Guinea-Bissau	..	..	*7*	*109*	*3.0*	>10.0	*34*	1,902	0.15	*0*	384	..
Haiti	33	54	9	..	*100.0*	>10.0	20	1,443	*0.00*	3	203	*7.10*
Honduras	446	21	44	99	170.0	5.7	72	697	0.06	12	54	4.20

Power and communications | 5.9

	Electric power		Telephone mainlines[a]							Mobile phones[a]	International telecommunications[a]	
	Consumption per capita kwh 1998	Transmission and distribution losses % of output 1998	per 1,000 people 1999	In largest city per 1,000 people 1999	Waiting list thousands 1999	Waiting time years 1999	per employee 1999	Revenue per line $ 1999	Cost of local call $ per 3 minutes 1999	per 1,000 people 1999	Outgoing traffic minutes per subscriber 1999	Cost of call to U.S. $ per 3 minutes 1999
Hungary	2,888	13	371	557	77.2	0.2	215	616	0.13	162	61	1.68
India	384	18	27	131	3,680.0	0.9	63	138	0.01	2	18	4.20
Indonesia	320	12	29	163	..	..	135	275	0.02	11	41	4.20
Iran, Islamic Rep.	1,343	15	125	286	1,200.0	1.4	177	218	0.01	7	24	7.71
Iraq	1,359	0	30	75	..	..	..	..	..	0	..	..
Ireland	4,760	9	478	..	..	..	118	1,637	0.17	447	573	1.54
Israel	5,475	6	471	..	22.0	0.2	422	1,089	0.05	472	279	3.30
Italy	4,431	7	462	..	0.0	0.0	335	1,129	0.13	528	117	1.40
Jamaica	2,252	10	199	..	183.0	3.5	160	908	0.06	56	137	5.20
Japan	7,322	3	558	814	0.0	0.0	392	1,613	0.09	449	28	2.10
Jordan	1,205	10	87	232	29.7	0.4	97	708	0.03	18	258	..
Kazakhstan	2,399	17	108	224	172.0	..	53	148	..	3	60	2.68
Kenya	129	25	10	71	121.0	9.6	20	1,044	0.05	1	85	11.17
Korea, Dem. Rep.	..	..	46	..	..	..	..	..	..	0	..	..
Korea, Rep.	4,497	7	438	521	0.0	0.0	298	768	0.04	500	43	1.80
Kuwait	13,800	..	240	47	0.0	0.0	60	841	0.00	158	394	5.41
Kyrgyz Republic	1,430	32	76	197	66.9	>10.0	50	64	..	1	65	15.48
Lao PDR	..	..	7	..	8.3	1.7	30	662	..	2	230	4.00
Latvia	1,879	20	300	403	19.7	..	161	452	0.12	112	77	3.00
Lebanon	1,820	16	201	96	..	..	124	580	0.07	194	124	4.45
Lesotho	..	..	10	58	20.0	>10.0	33	655	0.02	5	1,707	..
Libya	3,677	..	101	94	80.0	1.4	32	619	0.03	4	78	..
Lithuania	1,909	9	312	400	74.9	1.4	166	188	0.06	90	44	5.49
Macedonia, FYR	..	..	234	235	40.0	1.2	128	342	0.01	24	188	4.13
Madagascar	..	..	3	7	7.3	2.0	17	1,084	0.08	1	192	11.16
Malawi	..	..	4	37	31.6	>10.0	8	877	0.03	2	236	12.45
Malaysia	2,554	7	203	282	160.0	0.7	174	559	0.02	137	156	2.80
Mali	..	..	3	18	..	..	20	2,196	0.14	0	444	17.59
Mauritania	..	..	6	17	47.8	>10.0	34	1,730	0.09	0	489	..
Mauritius	..	..	224	306	29.1	1.2	145	473	0.04	89	122	4.60
Mexico	1,513	15	112	135	137.0	0.2	130	1,000	0.14	79	143	3.00
Moldova	688	26	127	304	118.0	..	73	71	0.02	4	70	3.53
Mongolia	..	..	39	101	39.6	6.0	22	329	0.08	13	40	5.65
Morocco	443	4	53	115	17.9	0.2	104	592	0.08	13	150	4.50
Mozambique	54	10	4	24	39.7	7.0	35	988	0.09	1	222	..
Myanmar	64	33	6	29	84.4	3.6	32	2,726	0.48	0	70	26.86
Namibia	..	..	64	317	5.4	0.7	63	840	0.05	18	572	..
Nepal	47	23	11	..	275.0	5.9	54	268	0.01	0	99	..
Netherlands	5,908	4	607	..	0.0	0.0	279	1,130	0.14	436	224	0.30
New Zealand	8,215	13	496	..	0.0	0.0	239	968	0.00	366	397	0.90
Nicaragua	281	29	30	74	108.0	8.4	65	635	0.09	9	339	3.20
Niger	..	..	2	18	..	..	19	1,243	..	0	340	..
Nigeria	85	32	4	11	42.0	..	35	3,738	..	0	141	..
Norway	24,607	8	709	823	0.0	0.0	133	1,546	0.08	613	178	1.05
Oman	2,828	16	90	165	3.9	0.5	106	1,552	0.07	49	463	..
Pakistan	337	25	22	62	298.0	1.5	51	349	0.02	2	29	..
Panama	1,211	23	164	250	..	..	106	764	..	86	116	4.40
Papua New Guinea	..	..	13	..	..	..	23	2,428	..	2	422	..
Paraguay	756	3	55	129	20.1	0.5	51	782	0.05	196	116	6.10
Peru	642	13	67	132	29.6	0.4	294	819	0.07	40	66	2.40
Philippines	451	16	39	146	..	..	219	690	0.00	38	45	4.80
Poland	2,458	11	263	199	1,800.0	1.5	121	578	0.07	102	61	3.65
Portugal	3,395	8	423	711	25.6	0.2	215	1,123	0.10	467	129	1.88
Puerto Rico	..	..	333	..	..	..	168	1,163	..	209	723	0.87
Romania	1,626	12	167	357	740.0	3.9	83	203	0.09	61	41	4.29
Russian Federation	3,937	11	210	448	6,530.0	3.9	71	133	0.02	9	33	6.12

	Electric power		Telephone mainlines[a]							Mobile phones[a]	International telecommunications[a]	
	Consumption per capita kwh 1998	Transmission and distribution losses % of output 1998	per 1,000 people 1999	In largest city per 1,000 people 1999	Waiting list thousands 1999	Waiting time years 1999	per employee 1999	Revenue per line $ 1999	Cost of local call $ per 3 minutes 1999	per 1,000 people 1999	Outgoing traffic minutes per subscriber 1999	Cost of call to U.S. $ per 3 minutes 1999
Rwanda	..	..	2	*40*	8.0	>10.0	45	1,424	0.04	2	376	..
Saudi Arabia	4,692	8	129	*253*	927.0	3.1	117	1,445	0.02	40	305	*6.41*
Senegal	111	11	18	*48*	24.0	1.0	118	1,099	0.12	9	220	*4.48*
Sierra Leone	..	..	*4*	*18*	25.0	>10.0	*18*	129	0.03	0	*236*	..
Singapore	6,771	4	482	482	0.0	0.0	222	1,313	0.02	419	719	1.70
Slovak Republic	3,899	8	307	670	69.3	0.5	116	269	0.12	170	98	2.10
Slovenia	5,096	6	378	661	5.7	0.2	225	481	*0.03*	309	197	*5.56*
South Africa	3,832	8	125	*415*	*116.0*	0.3	112	717	0.08	120	84	..
Spain	4,195	10	410	485	4.3	0.0	354	980	0.09	306	117	*1.88*
Sri Lanka	244	19	36	245	225.0	1.6	61	461	0.05	12	67	*4.49*
Sudan	47	31	9	45	355.0	7.0	95	403	0.02	0	99	*7.79*
Sweden	13,955	7	665	..	0.0	0.0	205	1,260	*0.13*	583	257	0.90
Switzerland	6,980	6	699	966	0.0	0.0	203	1,670	0.13	411	481	1.00
Syrian Arab Republic	838	..	99	*140*	2,820.0	>10.0	78	219	0.01	0	76	*26.71*
Tajikistan	2,045	14	35	*150*	*49.1*	..	52	*48*	*0.00*	0	42	*8.16*
Tanzania	53	22	5	*28*	29.6	1.6	40	840	0.08	2	77	*13.30*
Thailand	1,345	9	86	371	420.0	1.2	154	351	0.08	38	57	2.50
Togo	..	..	8	33	17.0	3.6	44	1,209	0.10	4	222	*11.44*
Trinidad and Tobago	3,478	8	216	200	10.0	0.5	100	814	0.04	30	243	3.30
Tunisia	824	11	90	*90*	83.7	0.9	129	445	0.03	6	165	*6.47*
Turkey	1,353	19	278	405	500.0	0.4	249	267	0.10	125	39	*3.31*
Turkmenistan	859	10	82	*155*	58.6	8.5	48	104	..	1	46	..
Uganda	..	..	3	*37*	9.2	3.0	34	1,522	0.15	3	179	*8.60*
Ukraine	2,350	17	199	418	2,650.0	9.6	80	82	0.00	4	38	..
United Arab Emirates	9,892	9	332	*373*	0.6	0.0	119	1,730	0.00	283	988	*3.77*
United Kingdom	5,327	8	567	..	0.0	0.0	167	1,505	0.19	457	180	1.10
United States	11,832	7	664	..	0.0	0.0	172	1,463	*0.09*	312	155	..
Uruguay	1,788	15	271	336	0.0	0.0	154	970	0.18	95	96	5.00
Uzbekistan	1,618	9	66	231	38.8	1.7	56	153	..	2	44	..
Venezuela, RB	2,566	23	109	*329*	392.0	..	218	1,461	0.09	143	63	*5.20*
Vietnam	232	16	27	133	..	..	*17*	304	0.08	4	22	..
West Bank and Gaza	..	..	72	..	38.8	0.8	..	302	0.05	14	165	*0.61*
Yemen, Rep.	96	26	17	77	131.0	4.5	61	273	0.02	2	109	..
Yugoslavia, FR (Serb./Mont.)	..	..	214	455	119.0	1.8	155	147	0.01	57	89	*12.08*
Zambia	539	11	9	24	12.3	7.2	24	1,346	0.05	3	170	*2.60*
Zimbabwe	896	17	21	*75*	*109.0*	5.1	*33*	640	*0.03*	15	275	*2.81*
World	**2,085 w**	**9 w**	**158 w**	***231**w*	**38,167.9 s**	**1.4 m**	**198 m**	**935 w**	**0.06 m**	**86 w**	**129 m**	***4.00**m*
Low income	362	17	26	125	*7,727.8*	5.9	72	212	0.06	3	141	..
Middle income	1,367	11	121	293	28,593.8	1.0	166	447	0.05	55	94	*4.36*
Lower middle income	1,064	10	102	293	21,600.7	1.5	148	267	0.05	33	77	*4.50*
Upper middle income	2,482	12	190	..	7,402.7	0.5	207	809	0.07	136	116	*3.73*
Low & middle income	913	12	79	211	37,980.2	2.0	152	412	0.05	32	105	*4.70*
East Asia & Pacific	787	8	82	*265*	1,901.9	1.2	178	355	0.03	42	51	*5.30*
Europe & Central Asia	2,652	13	213	*378*	12,480.6	2.0	126	256	0.07	47	61	*3.97*
Latin America & Carib.	1,452	16	130	..	*4,139.2*	0.5	189	847	0.09	82	116	3.20
Middle East & N. Africa	1,263	12	87	..	6,314.0	1.4	126	467	0.03	12	124	..
South Asia	341	19	23	118	4,352.5	1.6	61	174	0.02	2	67	*5.45*
Sub-Saharan Africa	454	10	*14*	*29*	*1,158.2*	6.0	99	793	0.07	5	222	..
High income	8,353	6	583	..	63.7	0.0	230	1,312	0.10	377	189	*1.78*
Europe EMU	5,504	6	526	..	29.9	0.0	248	1,037	0.13	383	152	*1.67*

a. Data are from the International Telecommunication Union's (ITU) *World Telecommunication Development Report 2000.* Please cite the ITU for third-party use of these data.

Power and communications | 5.9

An economy's production and consumption of electricity is a basic indicator of its size and level of development. Although a few countries export electric power, most production is for domestic consumption. Expanding the supply of electricity to meet the growing demand of increasingly urbanized and industrialized economies without incurring unacceptable social, economic, and environmental costs is one of the great challenges facing developing countries.

Data on electric power production and consumption are collected from national energy agencies by the International Energy Agency (IEA) and adjusted by the IEA to meet international definitions (for data on electricity production see table 3.9). Electricity consumption is equivalent to production less power plants' own use and transmission, distribution, and transformation losses. It includes consumption by auxiliary stations, losses in transformers that are considered integral parts of those stations, and electricity produced by pumping installations. It covers electricity generated by primary sources of energy—coal, oil, gas, nuclear, hydro, geothermal, wind, tide and wave, and combustible renewables—where data are available. Neither production nor consumption data capture the reliability of supplies, including breakdowns, load factors, and frequency of outages.

Over the past decade privatization and liberalization have spurred dramatic growth in telecommunications in many countries. The table presents some common performance indicators for telecommunications, including measures of supply and demand, service quality, productivity, economic and financial performance, and tariffs. The quality of data varies among reporting countries as a result of differences in regulatory obligations for the provision of data.

Demand for telecommunications is often measured by the sum of telephone mainlines and registered applicants for new connections. (A mainline is normally identified by a unique number that is the one billed.) In some countries the list of registered applicants does not reflect real current pending demand, which is often hidden or suppressed, reflecting an extremely short supply that has discouraged potential applicants from applying for telephone service. And in some cases waiting lists may overstate demand because applicants have placed their names on the list several times to improve their chances. Waiting time is calculated by dividing the number of applicants on the waiting list by the average number of mainlines added each year over the past three years. The number of mainlines no longer reflects a telephone system's full capacity because mobile telephones—whose use has been expanding rapidly in most countries, rich and poor—provide an alternative point of access.

The table includes four measures of efficiency in telecommunications: waiting list, waiting time, mainlines per employee, and revenue per mainline. Caution should be used in interpreting the estimates of mainlines per employee because firms often subcontract part of their work. The cross-country comparability of revenue per mainline may also be limited because, for example, some countries do not require telecommunications providers to submit financial information; the data usually do not include revenues from cellular and mobile phones or radio, paging, and data services; and there are definitional and accounting differences between countries.

Figure 5.9

Latin America leads developing regions in mobile phones

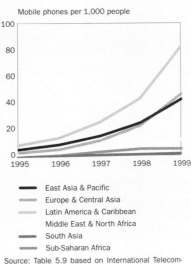

Mobile phones per 1,000 people

Source: Table 5.9 based on International Telecommunication Union data.

Mobile phone use is growing rapidly in all developing regions, but especially in Latin America, where there are about 60 percent as many mobile phones as there are fixed telephone lines. Mobile phones are even beginning to reach the poor and isolated, helping them participate in the global economy.

• **Electric power consumption** measures the production of power plants and combined heat and power plants less transmission, distribution, and transformation losses and own use by heat and power plants. • **Electric power transmission and distribution losses** are losses in transmission between sources of supply and points of distribution and in distribution to consumers, including pilferage. • **Telephone mainlines** are telephone lines connecting a customer's equipment to the public switched telephone network. Data are presented for the entire country and for the largest city. • **Waiting list** shows the number of applications for a connection to a mainline that have been held up by a lack of technical capacity. • **Waiting time** is the approximate number of years applicants must wait for a telephone line. • **Mainlines per employee** are calculated by dividing the number of mainlines by the number of telecommunications staff (with part-time staff converted to full-time equivalents) employed by telecommunications enterprises providing public telecommunications services. • **Revenue per line** is the revenue received by firms per mainline for providing telecommunications services. • **Cost of local call** is the cost of a three-minute call within the same exchange area using the subscriber's equipment (that is, not from a public phone). • **Mobile phones** refer to users of portable telephones subscribing to an automatic public mobile telephone service using cellular technology that provides access to the public switched telephone network, per 1,000 people. • **Outgoing traffic** is the telephone traffic, measured in minutes per subscriber, that originates in the country and has a destination outside the country. • **Cost of call to U.S.** is the cost of a three-minute peak rate call from the country to the United States.

The data on electricity consumption and losses are from the IEA's *Energy Statistics and Balances of Non-OECD Countries 1997–98,* the IEA's *Energy Statistics of OECD Countries 1997–98,* and the United Nations Statistics Division's *Energy Statistics Yearbook.* The telecommunications data are from the International Telecommunication Union's (ITU) *World Telecommunication Development Report 2000,* except for the data on telephone traffic, which are from *Direction of Traffic 1999,* published by TeleGeography and the ITU.

5.10 | The information age

	Daily newspapers	Radios	Television[a]		Fax machines[a]	Personal computers[a]	Internet						Infor-mation and com-munications technology expen-ditures
			Sets	Cable subscribers			Hosts			Monthly access charges[a]			
										Service provider charge	Telephone call charge		
	per 1,000 people 1996	per 1,000 people 1999	per 1,000 people 1999	per 1,000 people 1999	per 1,000 people 1999	per 1,000 people 1999	per 10,000 people[b] July 2000	Users thousands[a] 1999		$ 1998	$ 1998	Secure servers 2000	% of GDP 1999
Albania	36	217	113	..	4.8	5.2	0.35	3		..	..	1	..
Algeria	38	241	107	..	0.2	5.8	0.01	20		..	..		..
Angola	11	62	15	..	..	1.0	0.01	10		..	..		..
Argentina	123	681	293	163.2	2.4	49.2	47.34	900		30	11	219	3.41
Armenia	23	224	238	0.4	0.3	5.7	3.13	30		..	..	2	..
Australia	293	1,378	706	30.3	48.6	469.2	683.26	6,000		22	6	3,207	8.85
Austria	296	753	516	138.8	..	256.8	431.74	1,840		23	29	554	4.82
Azerbaijan	27	23	254	0.1	..	..	0.23	8		..	..	1	..
Bangladesh	9	50	7	..	..	1.0	0.00	50		..	..	1	..
Belarus	174	296	322	..	2.3	..	1.06	50		..	..	3	..
Belgium	160	792	523	369.6	..	315.2	352.15	1,400		23	20	310	5.88
Benin	2	110	11	..	..	1.5	0.04	10		..	..	1	..
Bolivia	55	676	118	5.1	..	12.3	1.73	78		..	..	5	..
Bosnia and Herzegovina	152	245	112	..	..	..	1.84	4		..	..	..	..
Botswana	27	156	20	..	2.3	31.0	14.63	12		..	..	..	..
Brazil	40	444	333	15.5	3.1	36.3	38.97	3,500		33	3	923	5.82
Bulgaria	257	543	408	28.8	..	26.6	18.80	235		..	..	19	1.76
Burkina Faso	1	33	11	..	..	1.0	0.19	4		..	..	..	..
Burundi	3	152	15	..	..	..	0.00	2		..	..	..	..
Cambodia	2	128	9	..	0.3	1.2	0.18	4		..	..	1	..
Cameroon	7	163	34	..	..	2.7	0.01	20		..	..	..	..
Canada	159	1,047	715	273.2	35.8	360.8	590.37	11,000		12	0	4,530	8.52
Central African Republic	2	83	6	..	0.1	1.4	0.02	1		..	..	..	..
Chad	0	242	1	..	0.0	1.3	0.01	1		..	..	..	..
Chile	98	355	240	44.9	2.7	66.6	33.78	700		32	7	112	5.74
China	..	334	292	47.2	1.6	12.2	0.69	8,900		39	26	171	4.86
Hong Kong, China	792	678	434	68.0	58.0	297.6	182.92	2,430		18	5	475	8.31
Colombia	46	560	199	15.5	4.6	33.7	10.15	664		..	..	54	8.85
Congo, Dem. Rep.	3	375	2	..	..	..	0.00	1		..	..	..	..
Congo, Rep.	8	124	13	..	..	3.5	0.01	1		..	..	..	..
Costa Rica	94	776	229	19.1	2.3	101.7	24.33	150		..	..	48	..
Côte d'Ivoire	17	164	70	0.0	..	5.5	0.37	20		..	..	1	..
Croatia	115	336	279	42.4	11.2	67.0	42.61	200		..	..	44	..
Cuba	118	355	246	..	..	9.9	0.33	35		..	..	2	..
Czech Republic	254	803	487	89.7	9.9	107.2	134.39	700		25	15	232	8.49
Denmark	309	1,318	621	250.5	..	414.0	692.29	1,500		23	27	332	6.94
Dominican Republic	52	178	96	..	..	..	9.16	25		..	..	8	..
Ecuador	70	420	205	16.4	..	20.1	1.67	35		..	..	9	..
Egypt, Arab Rep.	40	324	183	..	0.5	12.0	0.85	200		..	..	12	3.32
El Salvador	48	478	191	45.0	..	16.2	1.62	40		..	..	9	..
Eritrea	..	484	16	..	0.4	..	0.02	1		..	..	..	..
Estonia	174	966	555	75.9	..	135.2	249.29	200		..	..	70	..
Ethiopia	1	196	6	..	0.1	0.7	0.01	8		..	..	1	..
Finland	455	1,563	643	180.5	38.4	360.1	1,358.99	2,143		9	18	414	5.88
France	218	937	623	43.4	47.5	221.8	167.11	5,370		17	26	1,446	5.96
Gabon	29	500	251	8.3	0.4	8.4	0.21	3		..	..	..	..
Gambia, The	2	394	3	..	1.0	7.9	0.12	3		..	..	..	..
Georgia	..	555	474	2.8	..	..	2.06	20		..	..	13	..
Germany	311	948	580	226.3	79.1	297.0	233.29	14,400		23	28	4,441	5.27
Ghana	14	680	115	..	..	2.5	0.06	20		..	..	1	..
Greece	153	478	480	1.2	..	60.2	100.38	750		21	7	106	5.51
Guatemala	33	80	61	28.5	..	9.9	2.54	65		..	..	10	..
Guinea	..	49	44	0.0	0.4	3.4	0.00	5		..	..	..	..
Guinea-Bissau	5	44	..	..	0.4	..	0.11	2		..	..	..	..
Haiti	3	55	5	..	..	..	0.00	6		..	..	1	..
Honduras	55	395	95	8.1	..	9.5	0.19	20		..	..	4	..

	Daily newspapers	Radios	Television[a]		Fax machines[a]	Personal computers[a]	Internet					Information and communications technology expenditures
			Sets	Cable subscribers			Hosts		Monthly access charges[a]			
	per 1,000 people 1996	per 1,000 people 1999	per 1,000 people 1999	per 1,000 people 1999	per 1,000 people 1999	per 1,000 people 1999	per 10,000 people[b] July 2000	Users thousands[a] 1999	Service provider charge $ 1998	Telephone call charge $ 1998	Secure servers 2000	% of GDP 1999
Hungary	186	687	448	159.0	17.6	74.7	129.30	600	24	20	100	6.42
India	..	121	75	37.1	0.2	3.3	0.32	2,800	13	0	85	3.46
Indonesia	24	157	143	..	0.9	9.1	1.15	900	9	6	54	1.39
Iran, Islamic Rep.	28	264	157	..	..	52.4	0.11	100	..	..	..	..
Iraq	19	229	83	..	..	..	0.00	..	..	..	..	..
Ireland	150	699	406	171.2	27.4	404.9	227.43	679	21	14	290	6.48
Israel	290	519	328	188.5	..	245.7	260.58	800	25	3	..	7.36
Italy	104	880	488	2.8	31.4	191.8	272.96	7,000	23	14	940	4.72
Jamaica	62	795	189	98.8	..	43.0	2.26	60	..	..	5	..
Japan	578	960	719	125.4	127.0	286.9	269.25	27,060	41	14	4,139	7.06
Jordan	58	288	83	0.1	8.4	13.9	1.45	120	..	..	2	..
Kazakhstan	..	395	238	..	0.1	..	2.82	70	..	..	17	..
Kenya	9	104	22	..	..	4.2	0.32	35	..	..	..	..
Korea, Dem. Rep.	199	147	55	..	..	..	..	..	..	..	..	..
Korea, Rep.	393	1,033	361	150.1	..	181.8	100.65	10,860	12	14	313	4.42
Kuwait	374	632	480	..	31.6	121.3	23.15	100	..	..	3	..
Kyrgyz Republic	15	112	47	..	..	..	6.18	10	..	..	1	..
Lao PDR	4	143	10	..	..	2.3	0.00	2	..	..	..	..
Latvia	247	684	741	66.8	..	82.0	65.26	105	..	..	33	..
Lebanon	107	908	351	1.4	..	46.4	11.89	200	..	..	16	..
Lesotho	8	49	16	..	..	..	0.39	1	..	..	..	..
Libya	14	243	136	..	..	..	0.01	7	..	..	..	..
Lithuania	93	500	420	62.2	1.7	59.5	38.66	103	..	..	39	..
Macedonia, FYR	21	206	250	..	1.5	..	11.53	30	..	..	..	..
Madagascar	5	198	22	..	..	1.9	0.36	8	..	..	..	..
Malawi	3	250	3	..	0.1	0.9	0.00	10	..	..	..	..
Malaysia	158	419	174	5.2	8.1	68.7	27.55	1,500	1	8	128	5.20
Mali	1	54	12	0.0	..	1.0	0.05	10	..	..	1	..
Mauritania	0	151	96	..	1.3	27.2	0.20	13	..	..	..	..
Mauritius	75	368	230	..	26.1	95.7	27.70	55	..	..	11	..
Mexico	97	325	267	15.7	3.0	44.2	50.60	1,822	24	3	218	4.20
Moldova	60	742	297	17.8	0.2	8.0	4.02	25	..	..	3	..
Mongolia	27	151	61	10.8	3.0	9.2	0.70	6	..	..	1	..
Morocco	26	241	165	..	0.7	10.8	0.33	50	..	..	4	..
Mozambique	3	40	5	..	..	2.6	0.10	15	..	..	..	..
Myanmar	10	70	7	..	0.1	1.1	0.00	1	..	..	..	..
Namibia	19	144	38	..	..	29.5	19.76	6	..	..	2	..
Nepal	11	39	7	2.9	0.4	2.7	0.35	35	..	..	..	..
Netherlands	306	981	600	387.3	38.5	359.9	679.75	3,000	23	19	695	7.13
New Zealand	216	989	518	76.5	..	328.0	807.94	700	19	0	539	10.54
Nicaragua	30	277	69	40.5	..	8.1	2.18	20	..	..	4	..
Niger	0	66	27	..	..	0.4	0.12	3	..	..	..	..
Nigeria	24	224	68	..	..	6.4	0.01	100	..	..	1	..
Norway	588	916	648	184.3	50.0	446.6	1,121.12	2,000	13	22	318	6.93
Oman	29	598	575	..	2.7	26.4	3.00	50	..	..	1	..
Pakistan	23	104	119	0.1	2.0	4.3	0.40	80	..	..	6	..
Panama	62	300	192	..	..	32.0	10.21	45	..	..	25	..
Papua New Guinea	15	95	13	..	..	..	0.70	2	..	..	..	..
Paraguay	43	182	205	15.4	..	11.2	2.66	20	..	..	5	..
Peru	0	273	147	14.1	..	35.7	3.88	400	..	..	26	..
Philippines	79	159	110	9.4	..	16.9	2.21	500	31	0	59	2.71
Poland	113	522	387	122.0	..	62.0	67.14	2,100	14	20	251	4.90
Portugal	75	304	560	59.8	7.0	93.0	117.25	700	19	10	131	5.31
Puerto Rico	126	742	324	72.0	..	..	3.10	200	..	..	52	..
Romania	300	335	312	129.5	..	26.8	13.23	600	..	..	40	1.78
Russian Federation	105	418	421	23.1	0.4	37.4	19.50	2,700	20	0	242	1.55

	Daily newspapers	Radios	Television[a]		Fax machines[a]	Personal computers[a]	Internet						Information and communications technology expenditures
			Sets	Cable subscribers			Hosts		Monthly access charges[a]				
									Service provider charge	Telephone call charge			
	per 1,000 people 1996	per 1,000 people 1999	per 1,000 people 1999	per 1,000 people 1999	per 1,000 people 1999	per 1,000 people 1999	per 10,000 people[b] July 2000	Users thousands[a] 1999	$ 1998	$ 1998	Secure servers 2000	% of GDP 1999
---	---	---	---	---	---	---	---	---	---	---	---	---
Rwanda	0	102	0	..	0.1	..	0.42	5	..	..	1	..
Saudi Arabia	57	321	263	..	..	57.4	1.53	300	..	..	9	..
Senegal	5	142	41	..	..	15.1	0.51	30	..	..	..	..
Sierra Leone	4	274	13	0.0	0.5	..	0.16	2	..	..	1	..
Singapore	360	682	308	53.2	25.8	436.6	385.73	950	15	5	483	7.67
Slovak Republic	185	967	417	122.3	10.0	109.7	58.78	600	..	..	59	5.98
Slovenia	199	407	356	150.8	10.5	251.4	99.12	250	..	..	98	4.31
South Africa	32	333	129	..	3.6	54.7	43.12	1,820	17	8	470	7.17
Spain	100	333	547	13.3	..	119.4	136.51	4,652	23	16	857	4.03
Sri Lanka	29	209	102	0.0	..	5.6	0.91	65	..	..	5	..
Sudan	27	271	173	0.0	0.9	2.9	0.00	5	..	..	..	..
Sweden	445	932	531	221.5	..	451.4	703.91	3,666	22	17	934	9.28
Switzerland	337	1,000	518	357.1	..	461.9	582.23	1,427	18	14	..	7.48
Syrian Arab Republic	20	277	66	..	1.4	14.3	0.00	20	..	..	1	..
Tajikistan	20	142	328	..	0.3	..	0.36	2	..	..	..	..
Tanzania	4	279	21	..	..	2.4	0.16	25	..	..	..	..
Thailand	63	233	289	2.4	2.5	22.7	8.84	800	25	33	103	2.13
Togo	4	227	22	..	4.1	17.7	0.34	15	..	..	..	..
Trinidad and Tobago	123	535	337	3.9	..	54.2	41.88	30	..	..	8	..
Tunisia	31	158	190	0.0	3.4	15.3	0.10	30	..	..	4	..
Turkey	111	180	332	11.0	1.7	33.8	16.60	1,500	25	9	171	2.47
Turkmenistan	..	277	201	..	..	..	1.03	2	..	..	..	..
Uganda	2	127	28	..	..	2.5	0.07	25	..	..	1	..
Ukraine	54	884	413	15.7	0.0	15.8	6.57	200	..	..	37	..
United Arab Emirates	156	345	252	..	21.0	102.1	92.13	400	..	..	24	..
United Kingdom	329	1,435	652	52.6	..	302.5	348.34	12,500	20	29	5,374	9.35
United States	215	2,146	844	246.3	78.4	510.5	2,419.86	74,100	20	0	73,386	8.87
Uruguay	293	606	531	104.6	..	99.6	107.27	300	..	..	32	..
Uzbekistan	3	458	276	..	0.1	..	0.09	8	..	..	1	..
Venezuela, RB	206	470	185	25.9	3.0	42.2	6.48	525	28	39	86	3.44
Vietnam	4	107	184	..	0.4	8.9	0.01	100	..	..	4	7.40
West Bank and Gaza	..	..	..	..	..	..	..	..	..	..	..	..
Yemen, Rep.	15	64	286	..	..	1.7	0.06	10	..	..	..	..
Yugoslavia, FR (Serb./Mont.)	107	297	273	..	1.9	20.7	13.38	80	..	..	8	..
Zambia	12	160	145	..	0.1	7.2	0.86	15	..	..	..	..
Zimbabwe	19	390	180	..	..	13.0	2.61	20	..	..	..	..
World	.. w	420 w	268 w	58.5 w	12.3 w	68.4 w	152.47 w	241,864 s			110,498 s	
Low income	..	157	85		0.4	4.4	0.48	4,766			224	
Middle income	..	360	279	44.5	2.0	27.1	13.20	45,241			4,622	
Lower middle income	..	322	273	41.4	1.5	17.7	3.55	17,942			1,141	
Upper middle income	89	498	304	50.2	3.8	60.9	48.45	27,299			3,481	
Low & middle income	..	264	193	32.8	1.3	16.6	7.15	50,006			4,846	
East Asia & Pacific	..	302	252	46.8	1.5	17.0	3.98	23,593			844	
Europe & Central Asia	102	446	370	49.9	1.5	39.3	24.10	10,184			1,392	
Latin America & Carib.	71	419	272	29.6	3.1	37.7	29.62	9,687			1,946	
Middle East & N. Africa	33	272	175	..	..	25.4	0.67	1,153			72	
South Asia	..	113	71	36.3	0.3	3.2	0.31	3,034			97	
Sub-Saharan Africa	12	201	43	..	..	8.4	3.10	2,357			495	
High income	286	1,289	693	160.4	73.0	345.9	981.74	191,857			105,652	
Europe EMU	208	821	582	101.6	47.9	234.9	263.37	41,280			10,131	

a. Data are from the International Telecommunication Union's (ITU) *World Telecommunication Development Report 2000* and *Challenges to the Network: Internet for Development* (1999). Please cite the ITU for third-party use of these data. b. Data are from the Internet Software Consortium (www.isc.org).

The table includes indicators of the penetration of the information economy—newspapers, radios, television sets, fax machines, personal computers, and Internet hosts and users—as well as some of the economics of the information age—Internet access charges, the number of secure servers, and spending on information and communications technology. Other important indicators of information and communications technology—such as the use of teleconferencing or the use of the Internet in organizing conferences, distance education, and commercial transactions—are not collected systematically and so are not reported here. Important as all these indicators are, they fail to capture characteristics of the information disseminated, such as its quality.

The data on the number of daily newspapers in circulation and radio receivers in use are from statistical surveys carried out by the United Nations Educational, Scientific, and Cultural Organization (UNESCO). In some countries definitions, classifications, and methods of enumeration do not entirely conform to UNESCO standards. For example, newspaper circulation data should refer to the number of copies distributed, but in some cases the figures reported are the number of copies printed. In addition, many countries impose radio and television license fees to help pay for public broadcasting, discouraging radio and television owners from declaring ownership. Because of these and other data collection problems, estimates of the number of newspapers and radios vary widely in reliability and should be interpreted with caution.

The data for other electronic communications and information technology are from the International Telecommunication Union (ITU), the Internet Software Consortium, Netcraft, and the World Information Technology and Services Alliance. The ITU collects data on television sets and cable television subscribers through annual questionnaires sent to national broadcasting authorities and industry associations. Some countries require that television sets be registered. To the extent that households do not register their televisions or do not register all of their televisions, the data on licensed sets may understate the true number.

Because of different regulatory requirements for the provision of data, complete measurement of the telecommunications sector is not possible. Telecommunications data are compiled through annual questionnaires sent to telecommunications authorities and operating companies. The data are supplemented by annual reports and statistical yearbooks of telecommunications ministries, regulators, operators, and industry associations. In some cases estimates are derived from ITU documents or other references.

The data on fax machines exclude fax modems attached to computers. Some operators report only the equipment they sell, lease, or register, so the actual number is almost certainly much higher.

The estimates of personal computers are derived from an annual questionnaire, supplemented by other sources. In many countries mainframe computers are used extensively, and thousands of users can be connected to a single mainframe computer; thus the number of personal computers understates the total use of computers.

Internet hosts are computers connected directly to the worldwide network, each allowing many computer users to access the Internet. Hosts are assigned to countries on the basis of the host's country code, though this does not necessarily indicate that the host is physically located in that country. All hosts lacking a country code identification are assigned to the United States. The Internet Software Consortium changed the methods used in its Internet domain survey beginning in July 1998. The new survey is believed to be more reliable and to avoid the undercounting that occurs when organizations restrict download access to their domain data. Nevertheless, some measurement problems remain, so the number of Internet hosts shown for each country should be considered an approximation. In particular, most hosts are now under generic top-level domains (for example, .com, .net, and .org), which, unlike country code top-level domains (.de, .uk), have never had a geographic designation (see Zook 2000). For detailed analysis of Internet trends by country, it is best to use the original source data.

Data on Internet users are based on reported estimates, derived from reported counts of Internet service provider (ISP) subscribers, or calculated by multiplying the number of hosts by an estimated multiplier. The price of Internet access in many countries is a major constraint on universal access. The table shows both the ISP charge and the telephone call charge. ISP charges are similar across countries, but telephone call charges vary much more and are extremely high in some countries because of the monopolistic power of the telecommunications operator. As a result, the price of Internet access is much higher in developing than in high-income countries, especially relative to per capita income.

The number of secure servers, from the Netcraft Secure Server Survey, gives an indication of how many companies are conducting encrypted transactions over the Internet. The data on information and communications technology expenditures cover the world's 55 largest buyers of such technology among countries and regions, accounting for 98 percent of global spending.

• **Daily newspapers** refer to those published at least four times a week. • **Radios** refer to radio receivers in use for broadcasts to the general public. • **Television sets** refer to those in use. • **Cable television subscribers** are households that subscribe to a multichannel television service delivered by a fixed line connection. Some countries also report subscribers to pay television using wireless technology or those cabled to community antenna systems. • **Fax machines** are facsimile machines connected to the public switched telephone network. • **Personal computers** are self-contained computers designed to be used by a single individual. • **Internet hosts** are computers with active Internet Protocol (IP) addresses connected to the Internet. All hosts without a country code identification are assumed to be located in the United States. • **Internet users** are people with access to the worldwide network. • **Internet service provider charge** is the monthly dial-up access charge for 20 hours of use. It includes local telephone call charges and taxes but excludes the initial ISP connection charge. • **Internet telephone call charge** is the off-peak telephone call charge for 20 hours of Internet access. If a special Internet tariff exists, it is used instead. • **Secure servers** are servers using encryption technology in Internet transactions. • **Information and communications technology expenditures** include external spending on information technology ("tangible" spending on information technology products purchased by businesses, households, governments, and education institutions from vendors or organizations outside the purchasing entity), internal spending on information technology ("intangible" spending on internally customized software, capital depreciation, and the like), and spending on telecommunications and other office equipment.

The data on newspapers and radios are compiled by UNESCO. The data on television sets, cable television subscribers, fax machines, personal computers, Internet users, and Internet access charges are from the ITU. They are reported in the ITU's *World Telecommunication Development Report 2000, Challenges to the Network: Internet for Development* (1999), and the *World Telecommunications Indicators Database* (2000b). The data on Internet hosts are from the Internet Software Consortium (www.isc.org), and the data on secure servers from Netcraft (www.netcraft.com/). The data on information and communications technology expenditures are from *Digital Planet 2000: The Global Information Economy* by the World Information Technology and Services Alliance (WITSA), which uses data from the International Data Corporation.

5.11 | Science and technology

	Scientists and engineers in R&D	Technicians in R&D	Science and engineering students	Scientific and technical journal articles	Expenditures for R&D	High-technology exports		Royalty and license fees		Patent applications filed[a]	
			% of total tertiary students			$ millions	% of manufactured exports	Receipts $ millions	Payments $ millions		Non-residents
	per million people 1987–97[b]	per million people 1987–97[b]	1987–97[b]	1997	% of GNI 1987–97[b]	1999	1999	1999	1999	Residents 1998	1998
Albania	..	..	19	10	..	1	1	..	..	0	35,159
Algeria	..	..	58	139	..	16	4	..	..	42	264
Angola	..	..	24	2	..	..	..	..	..	..	..
Argentina	660	147	28	2,119	0.38	557	8	19	442	861	5,459
Armenia	1,485	177	29	178	..	3	2	..	..	77	33,822
Australia	3,357	797	24	11,793	1.80	1,624	11	344	1,124	9,097	48,609
Austria	1,627	812	33	3,432	1.53	6,384	13	120	623	3,023	144,017
Azerbaijan	2,791	188	37	71	0.21	..	..	..	..	0	33,507
Bangladesh	52	33	47	130	0.03	3	0	0	6	32	184
Belarus	2,248	266	48	548	1.07	157	4	1	1	919	34,350
Belgium	2,272	2,201	41	4,717	1.60	11,115	8	757	1,138	1,899	110,753
Benin	176	54	18	19	..	0	0	..	2	..	..
Bolivia	172	154	30	27	0.50	296	..	2	5	..	..
Bosnia and Herzegovina	..	..	..	8	..	..	..	..	..	0	34,441
Botswana	..	..	37	33	..	..	..	0	6	7	85
Brazil	168	59	27	3,908	0.81	3,453	13	133	1,283	2,535	48,331
Bulgaria	1,747	967	27	896	0.57	113	4	..	..	281	36,294
Burkina Faso	17	16	18	20	0.19	..	..	..	..	..	..
Burundi	33	32	20	11	0.31	..	..	0	0	1	4
Cambodia	..	..	13	3	..	..	..	..	..	..	..
Cameroon	..	..	45	73	..	..	..	..	..	..	..
Canada	2,719	1,070	16	19,910	1.66	23,935	15	1,178	2,602	4,841	60,841
Central African Republic	56	32	30	5	..	..	..	..	..	..	..
Chad	..	..	14	2	..	..	..	..	..	..	..
Chile	445	233	42	850	0.68	93	4	99	51	189	1,771
China	454	200	43	9,081	0.66	29,614	17	75	792	14,004	68,285
Hong Kong, China	..	..	36	2,080	..	4,398	21	..	..	128	14,539
Colombia	..	..	28	208	..	289	8	7	68	74	1,662
Congo, Dem. Rep.	..	..	..	15	..	..	..	..	..	2	27
Congo, Rep.	..	..	48	8	..	..	..	0	0	..	..
Costa Rica	532	..	20	73	0.21	2,707	..	1	31	..	..
Côte d'Ivoire	..	..	31	31	..	..	..	..	..	..	..
Croatia	1,916	714	30	544	1.03	270	8	..	..	273	12,633
Cuba	1,612	1,121	16	148	0.84	..	..	..	..	109	33,997
Czech Republic	1,222	693	28	2,024	1.20	2,141	9	43	137	641	38,555
Denmark	3,190	2,644	25	3,950	1.95	6,493	20	..	..	2,897	143,460
Dominican Republic	..	..	35	6	..	1	0	..	30	..	..
Ecuador	146	42	27	39	0.02	25	6	..	70	8	302
Egypt, Arab Rep.	459	341	12	1,108	0.22	3	0	47	329	494	1,139
El Salvador	20	356	59	3	..	38	7	1	20	..	..
Eritrea	..	..	30	0	..	..	..	..	..	..	..
Estonia	2,017	391	27	222	0.57	271	13	2	6	22	35,479
Ethiopia	..	..	26	103	..	..	..	..	0	4	0
Finland	2,799	1,966	39	3,897	2.78	8,547	24	648	375	4,796	142,088
France	2,659	2,873	37	26,509	2.25	55,834	23	1,983	2,297	20,298	109,717
Gabon	234	22	29	16	..	..	..	..	..	..	..
Gambia, The	..	..	..	25	..	0	19	..	..	5	60,267
Georgia	..	..	39	128	..	..	..	..	..	280	35,448
Germany	2,831	1,472	47	36,233	2.41	75,176	17	3,017	4,405	67,790	134,981
Ghana	..	..	32	78	..	32	14	..	0	6	66,167
Greece	773	314	26	2,123	0.47	484	10	0	58	68	111,271
Guatemala	104	112	..	15	0.16	71	9	..	..	11	196
Guinea	..	..	34	3	..	0	0	..	0	..	..
Guinea-Bissau	..	..	0	3	..	..	..	..	..	0	15,568
Haiti	..	..	..	2	..	2	4	..	..	3	6
Honduras	..	..	24	10	..	6	3	0	0	11	140

Science and technology | 5.11

	Scientists and engineers in R&D	Technicians in R&D	Science and engineering students	Scientific and technical journal articles	Expenditures for R&D	High-technology exports		Royalty and license fees		Patent applications filed[a]	
	per million people 1987–97[b]	per million people 1987–97[b]	% of total tertiary students 1987–97[b]	1997	% of GNI 1987–97[b]	$ millions 1999	% of manufactured exports 1999	Receipts $ millions 1999	Payments $ millions 1999	Residents 1998	Non-residents 1998
Hungary	1,099	510	32	1,717	0.68	4,839	23	62	307	751	37,956
India	149	108	25	8,439	0.73	1,415	6	23	315	2,111	7,997
Indonesia	182	..	39	123	0.07	2,731	10	..	..	0	32,910
Iran, Islamic Rep.	560	166	39	332	0.48	9	1	0	0	337	159
Iraq	..	..	41	35	..	..	..	..	..	68	18
Ireland	2,319	506	31	1,118	1.61	27,929	47	415	6,943	1,199	111,145
Israel	..	..	49	5,321	2.35	4,644	19	258	263	2,529	39,742
Italy	1,318	798	30	16,405	2.21	17,240	8	563	1,382	3,167	109,341
Jamaica	..	..	64	49	..	1	0	6	41	..	..
Japan	4,909	827	21	43,891	2.80	104,794	27	8,190	9,855	360,338	77,037
Jordan	94	10	26	177	0.26	15	2	..	..	0	0
Kazakhstan	..	..	20	119	0.32	130	8	..	..	1,245	34,093
Kenya	..	..	19	235	..	16	4	1	44	33	67,797
Korea, Dem. Rep.	..	..	..	0	..	..	..	..	..	0	33,918
Korea, Rep.	2,193	318	32	4,619	2.82	41,452	32	455	2,661	50,714	71,036
Kuwait	230	71	29	173	0.16	35	1	..	..	..	..
Kyrgyz Republic	584	50	14	9	0.20	5	6	..	..	111	33,794
Lao PDR	..	..	20	2	..	..	..	..	..	..	..
Latvia	1,049	351	23	141	0.43	40	4	10	10	195	35,768
Lebanon	..	..	30	81	..	..	..	..	..	..	..
Lesotho	..	..	19	2	..	..	..	13	0	6	67,485
Libya	..	..	..	12	..	..	..	..	..	12	23
Lithuania	2,028	631	31	198	0.70	377	12	0	14	135	35,703
Macedonia, FYR	1,335	546	47	49	..	18	2	2	3	84	35,049
Madagascar	12	37	25	..	0.18	4	3	1	10	0	34,941
Malawi	..	..	27	38	..	..	..	..	..	7	67,753
Malaysia	93	32	27	304	0.24	39,996	59	0	0	179	6,272
Mali	..	..	12	12	..	0	7	..	..	..	..
Mauritania	..	..	41	2	..	..	..	0	0	..	..
Mauritius	361	158	14	2	0.40	13	1	0	0	3	12
Mexico	214	74	32	1,915	0.33	24,070	21	42	554	472	44,249
Moldova	330	1,641	52	111	0.90	5	4	0	0	257	33,854
Mongolia	910	176	24	13	..	..	..	1	..	148	35,006
Morocco	..	..	41	271	..	10	0	6	201	90	237
Mozambique	..	..	42	9	..	..	..	..	..	..	..
Myanmar	..	..	56	3	..	..	..	2	0	..	..
Namibia	..	..	4	7	..	..	..	6	3	..	..
Nepal	..	..	13	35	..	..	..	0	0	..	..
Netherlands	2,219	1,358	39	11,008	2.08	39,917	33	2,388	3,426	5,751	109,325
New Zealand	1,663	809	20	2,308	1.04	611	15	49	317	1,353	38,381
Nicaragua	204	85	33	11	..	2	6	..	..	12	142
Niger	..	..	32	25	..	0	5	..	..	..	..
Nigeria	15	76	42	405	0.09	17	13	..	..	..	..
Norway	3,664	1,842	26	2,501	1.58	2,012	17	90	341	1,642	42,616
Oman	..	..	13	53	..	118	10	..	..	..	..
Pakistan	72	13	32	232	0.92	22	0	2	19	16	782
Panama	..	..	29	37	..	2	1	0	18	31	142
Papua New Guinea	..	..	10	31	..	34	..	..	..	..	..
Paraguay	..	..	20	4	..	3	3	189	3	..	..
Peru	233	10	34	63	..	46	5	4	60	48	756
Philippines	157	22	14	159	0.22	8,479	59	6	110	163	3,280
Poland	1,358	1,377	28	4,019	0.77	580	3	25	491	2,410	38,942
Portugal	1,182	167	36	1,085	0.62	1,062	5	27	292	119	145,023
Puerto Rico	..	..	..	..	..	..	..	..	..	..	..
Romania	1,387	581	21	751	0.72	237	4	4	38	1,308	36,518
Russian Federation	3,587	600	50	17,147	0.88	2,899	16	43	8	16,630	41,902

	Scientists and engineers in R&D	Technicians in R&D	Science and engineering students	Scientific and technical journal articles	Expenditures for R&D	High-technology exports		Royalty and license fees		Patent applications filed[a]	
	per million people 1987–97[b]	per million people 1987–97[b]	% of total tertiary students 1987–97[b]	1997	% of GNI 1987–97[b]	$ millions 1999	% of manufactured exports 1999	Receipts $ millions 1999	Payments $ millions 1999	Residents 1998	Non-residents 1998
Rwanda	35	8	28	5	0.04	..	..	0	1	..	..
Saudi Arabia	..	..	17	613		18	0	0	0	45	1,286
Senegal	3	4	21	58	0.01	34	13	0	2	..	..
Sierra Leone	..	..	17	8	..	..	..	..	..	0	33,154
Singapore	2,318	301	..	1,164	1.13	60,032	61	..	..	311	44,637
Slovak Republic	1,866	792	40	950	1.05	377	5	15	54	224	36,628
Slovenia	2,251	1,027	26	517	1.46	318	4	8	47	296	36,001
South Africa	1,031	315	29	1,927	0.70	1,055	8	71	162	..	8
Spain	1,305	343	31	11,210	0.90	6,945	8	344	1,831	3,119	144,770
Sri Lanka	191	47	34	61	..	109	3	..	..	81	34,974
Sudan	..	..	16	43	..	0	0	0	0	6	67,713
Sweden	3,826	3,166	38	8,219	3.76	15,100	22	1,386	1,147	8,599	140,894
Switzerland	3,006	1,374	34	6,935	2.60	16,283	22	..	..	6,026	141,553
Syrian Arab Republic	30	25	23	57	0.20	..	..	..	..	..	..
Tajikistan	666	..	17	29	..	..	..	..	..	37	33,742
Tanzania	..	..	37	89	..	6	6	0	4	..	..
Thailand	103	39	18	356	0.13	13,999	32	19	583	477	4,594
Togo	98	63	35	7	0.48	0	1	..	0	..	..
Trinidad and Tobago	..	..	58	41	..	21	2	0	0	17	34,969
Tunisia	125	57	33	188	0.30	125	3	10	3	46	128
Turkey	291	..	45	2,116	0.45	892	4	..	..	231	37,155
Turkmenistan	..	..	..	7	..	..	..	..	..	41	33,664
Uganda	21	14	17	46	0.57	2	11	0	0	7	67,603
Ukraine	2,171	575	42	2,163	..	..	..	..	..	5,327	36,623
United Arab Emirates	..	..	24	127	..	..	..	..	..	..	8
United Kingdom	2,448	1,017	34	38,530	1.95	66,942	30	7,942	6,301	28,889	147,298
United States	3,676	..	19	166,829	2.63	184,239	35	36,467	13,275	141,342	121,445
Uruguay	..	..	32	110	..	21	2	0	10	27	469
Uzbekistan	1,763	314	..	261	..	..	..	..	..	723	35,148
Venezuela, RB	209	32	26	429	0.49	71	3	0	0	201	2,323
Vietnam	..	..	..	106	..	..	..	..	..	30	35,748
West Bank and Gaza	..	..	..	..	..	..	..	..	..	..	..
Yemen, Rep.	..	..	5	10	..	..	..	..	..	..	..
Yugoslavia, FR (Serb./Mont.)	1,099	515	47	492	..	..	..	..	..	526	34,015
Zambia	..	..	16	23	..	..	..	..	..	7	86
Zimbabwe	..	..	24	100	..	10	2	..	..	8	66,264
World	.. w	.. w	35 w	512,637 s	2.18 w	959,990 s	21 w	67,641 s	66,837 s	785,229 s	5,034,563 s
Low income	..	..	28	13,572	0.47	2,890	6	41	370	9,241	1,105,167
Middle income	668	233	39	61,762	0.90	180,967	21	1,400	8,682	96,166	1,092,164
Lower middle income	763	255	41	35,148	0.58	60,566	18	434	2,461	37,173	613,666
Upper middle income	660	..	32	26,614	1.17	120,400	24	966	6,221	58,993	478,498
Low & middle income	..	..	35	75,334	0.85	183,857	20	1,440	9,052	105,407	2,197,331
East Asia & Pacific	492	193	43	14,817	1.32	136,271	31	558	4,147	65,506	284,777
Europe & Central Asia	2,533	..	44	34,905	0.77	13,222	11	207	1,068	32,728	940,242
Latin America & Carib.	..	..	30	10,093	0.62	31,706	16	503	2,710	4,003	241,989
Middle East & N. Africa	..	..	29	3,123	..	1,336	2	63	548	926	2,874
South Asia	137	98	24	8,896	0.66	131	4	23	321	2,143	43,155
Sub-Saharan Africa	..	..	29	3,499	..	1,190	9	86	258	101	684,294
High income	3,166	..	25	437,303	2.36	776,133	22	66,201	57,786	679,822	2,837,232
Europe EMU	2,127	1,510	38	115,641	2.15	239,033	19	10,379	22,820	111,399	1,405,323

a. Other patent applications filed in 1998 include those filed under the auspices of the African Intellectual Property Organization (25 by residents, 34,970 by nonresidents), African Regional Industrial Property Organization (24 by residents, 34,591 by nonresidents), European Patent Office (51,073 by residents, 62,335 by nonresidents), and Eurasian Patent Organization (293 by residents, 35,418 by nonresidents). The original information was provided by the World Intellectual Property Organization (WIPO). The International Bureau of WIPO assumes no liability or responsibility with respect to the transformation of these data. b. Data are for the latest year available; see *Primary data documentation* for the year.

Science and technology | 5.11

Science is advancing rapidly in virtually all fields, particularly biotechnology, and playing a growing economic role: countries unable to access, generate, and apply relevant scientific knowledge will fall even further behind. And there is greater appreciation of the need for high-quality scientific input into public policy issues such as regional and global environmental concerns.

Science and technology cover a range of issues too complex and too broad to be quantified by any single set of indicators, but those in the table shed light on countries' "technological base"—the availability of skilled human resources (students enrolled in science and engineering, and scientists, engineers, and technicians employed in research and development, or R&D), the number of scientific and technical articles published, the competitive edge countries enjoy in high-technology exports, sales and purchases of technology through royalties and licenses, and the number of patent applications filed.

The United Nations Educational, Scientific, and Cultural Organization (UNESCO) collects data on scientific and technical workers and R&D expenditures from member states, mainly through questionnaires and special surveys as well as from official reports and publications, supplemented by information from other national and international sources. UNESCO reports either the stock of scientists, engineers, and technicians or the number of economically active persons (people engaged in or actively seeking work in any branch of the economy on a given date) qualified to be scientists, engineers, or technicians. Stock data generally come from censuses and are less timely than measures of the economically active population. UNESCO supplements these data with estimates of the number of qualified scientists and engineers by counting the number of people who have completed education at ISCED (International Standard Classification of Education) levels 6 and 7; qualified technicians are estimated using the number of people who have completed education at ISCED level 5. The data on scientists, engineers, and technicians, normally calculated in terms of full-time-equivalent staff, cannot take into account the considerable variations in quality of training and education. Similarly, R&D expenditures are no guarantee of progress; governments need to pay close attention to the practices that make them effective.

The data on science and engineering students refer to those enrolled at the tertiary level, which normally requires as a minimum condition of admission the successful completion of education at the secondary level. These data are reported to UNESCO by national education authorities. (For further details on UNESCO education surveys see *About the data* for table 2.12.)

The methodology used for determining a country's high-technology exports was developed by the Organisation for Economic Co-operation and Development in collaboration with Eurostat. Termed the "product approach" to distinguish it from a "sectoral approach," the method is based on the calculation of R&D intensity (R&D expenditure divided by total sales) for groups of products from six countries (Germany, Italy, Japan, the Netherlands, Sweden, and the United States). Because industrial sectors characterized by a few high-technology products may also produce many low-technology products, the product approach is more appropriate for analyzing international trade than is the sectoral approach. To construct a list of high-technology manufactured products (services are excluded), the R&D intensity was calculated for products classified at the three-digit level of the Standard International Trade Classification revision 3. The final list was determined at the four- and five-digit level. At this level, since no R&D data were available, final selection was based on patent data and expert opinion. This methodology takes only R&D intensity into account. Other characteristics of high technology are also important, such as know-how, scientific and technical personnel, and technology embodied in patents; considering these characteristics would result in a different list. (See Hatzichronoglou 1997 for further details.)

The counts of scientific and technical journal articles include those published in a stable set of about 5,000 of the world's most influential scientific and technical journals, tracked since 1985 by the Institute of Scientific Information's Science Citation Index (SCI) and Social Science Citation Index (SSCI). (See *Definitions* for the fields covered.) The SCI and SSCI database covers the core set of scientific journals but may exclude some of regional or local importance. It may also reflect some bias toward English-language journals.

Most countries have adopted systems that protect patentable inventions. Under most patent legislation, to be protected by law (patentable), an idea must be new in the sense that it has not already been published or publicly used; it must be nonobvious (involve an inventive step) in the sense that it would not have occurred to any specialist in the industrial field, had such a specialist been asked to find a solution to the problem; and it must be capable of industrial application in the sense that it can be industrially manufactured or used. Information on patent applications filed is shown separately for residents and nonresidents of the country. The World Intellectual Property Organization estimates that at the end of 1998 about 4 million patents were in force in the world.

• **Scientists and engineers in R&D** are people trained at the tertiary level to work in any field of science who are engaged in professional R&D activity. • **Technicians in R&D** are people engaged in professional R&D activity who have received vocational or technical training in any branch of knowledge or technology. Most such jobs require three years beyond the first stage of secondary education. • **Science and engineering students** include students at the tertiary level in the following fields: engineering, natural science, mathematics and computers, and social and behavioral sciences. • **Scientific and technical journal articles** refer to scientific and engineering articles published in the following fields: physics, biology, chemistry, mathematics, clinical medicine, biomedical research, engineering and technology, and earth and space sciences. • **Expenditures for R&D** are current and capital expenditures on creative, systematic activity that increases the stock of knowledge. Included are fundamental and applied research and experimental development work leading to new devices, products, or processes. • **High-technology exports** are products with high R&D intensity. They include high-technology products such as in aerospace, computers, pharmaceuticals, scientific instruments, and electrical machinery. • **Royalty and license fees** are payments and receipts between residents and nonresidents for the authorized use of intangible, nonproduced, nonfinancial assets and proprietary rights (such as patents, copyrights, trademarks, industrial processes, and franchises) and for the use, through licensing agreements, of produced originals of prototypes (such as manuscripts and films). • **Patent applications filed** are applications filed with a national patent office for exclusive rights for an invention—a product or process that provides a new way of doing something or offers a new technical solution to a problem. A patent provides protection for the invention to the owner of the patent for a limited period, generally 20 years.

The data on technical personnel, science and engineering students, and R&D expenditures are from UNESCO's *Statistical Yearbook*. The data on scientific and technical journal articles are from the National Science Foundation's *Science and Engineering Indicators 2000*. The information on high-technology exports is from the United Nations' Commodity Trade (COMTRADE) database. The data on royalty and license fees are from the International Monetary Fund's *Balance of Payments Statistics Yearbook,* and the data on patents from the World Intellectual Property Organization's *Industrial Property Statistics*.

GLOBAL LINKS

Trade takes off

Growth in trade and growth in output tend to go hand in hand. Between 1990 and 1998 the 12 fastest growing developing countries saw their exports of goods and services grow by 14 percent and their output by 8 percent. The faster pace for exports implies a growing ratio of trade to GDP, one of the key indicators of globalization.

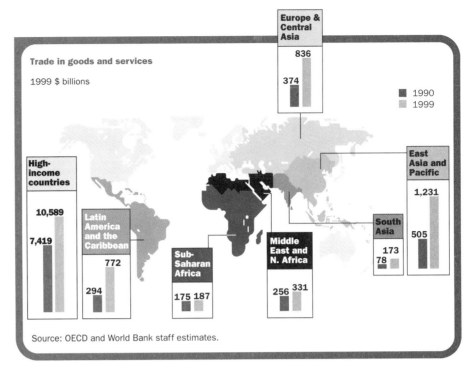

Trade in goods and services

1999 $ billions

- 1990
- 1999

Europe & Central Asia: 374 / 836

High-income countries: 7,419 / 10,589

Latin America and the Caribbean: 294 / 772

Sub-Saharan Africa: 175 / 187

Middle East and N. Africa: 256 / 331

South Asia: 78 / 173

East Asia and Pacific: 505 / 1,231

Source: OECD and World Bank staff estimates.

Evidence of globalization

At the opening of the 21st century the world's economies appear to be becoming more integrated: trade is expanding, capital markets have sprung up in developing and transition economies, tourism—and, in some places, migration—are rising, and new technologies have linked the farthest corners of the world. All these activities are evidence of a process that has come to be called globalization. By opening new markets, sharing knowledge, and increasing the efficiency of resources, globalization can expand opportunities for people and reduce poverty. But there are also risks. Globalization can increase vulnerability to external shocks. Increased competition creates losers as well as winners. And the rise of large, multinational corporations may contribute to a sense of helplessness and loss of control.

This is not the first time the world has experienced globalization. At the end of the 19th century massive migrations took place from Europe and Asia to Australia and North and South America. Between 1891 and 1900 more than 3.5 million immigrants landed in the United States, and 8.8 million more followed in the next decade. The 19th century also witnessed an enormous expansion in trade. In 1820 British trade stood at 3 percent of GDP. By 1870 it had reached 12 percent (Maddison 1995). The new technologies of steam power and telegraphs and telephones brought goods and people closer together. But globalization is not an inevitable process. In the 20th century wars, economic depression, protectionism, and restrictions on the movements of people interrupted the trend toward greater integration until the last two decades.

The growing impor-
tance of trade . . .

Trade in goods—primary commodities and manufactured articles—has
been the traditional basis of trade.
Although service trade has grown
quickly in the past two decades,
goods still account for 80 percent of
the value of world trade.

Growth in trade has been strongest
among upper-middle-income
economies, whose share of world
trade in goods (measured as the sum
of imports and exports) grew from 8
to 11 percent between 1990 and
1998. Their ratio of trade to GDP
measured in purchasing power parity

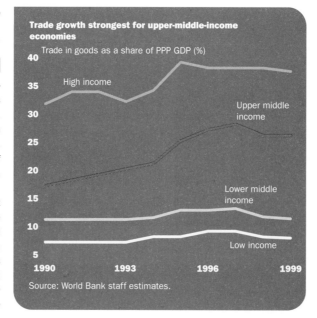

Trade growth strongest for upper-middle-income economies

Trade in goods as a share of PPP GDP (%)

High income

Upper middle income

Lower middle income

Low income

Source: World Bank staff estimates.

(PPP) terms now stands at more than
25 percent.

A few low-income economies have
also participated in the expansion of
world trade. Vietnam more than
tripled its share of world trade in
goods between 1990 and 1999. But
too many of the poorest countries
have been left out. The share of the
poorest 48 economies has
remained nearly constant at about 4
percent, and their ratio of trade to
PPP GDP remains below 10 percent.

Comparing trade with GDP measured
in PPP terms adjusts for the relative
size of domestic economies.
Nontraded goods and services produced in developing countries are
often undervalued relative to those in
high-income economies.

Trade

. . . and the high cost
of trade barriers

Border barriers—tariffs and quotas—
have begun to come down, but there
is still a long way to go. With the completion of the Uruguay Round of trade
negotiations in 1993, average import-
weighted tariffs in high-income countries fell to around 2.6 percent and in
developing countries to 13.3 percent.
Nontariff barriers have been reduced
or converted to tariffs, and foreign
exchange distortions reduced.

Lowering barriers reduces the cost
of trade and allows producers to
work more efficiently: inputs cost
less, and outputs can be sold where

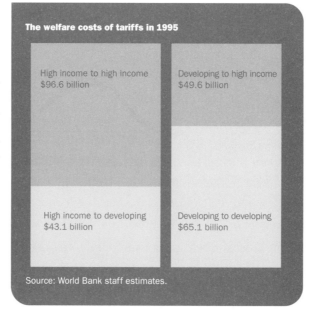

The welfare costs of tariffs in 1995

High income to high income
$96.6 billion

Developing to high income
$49.6 billion

High income to developing
$43.1 billion

Developing to developing
$65.1 billion

Source: World Bank staff estimates.

they obtain the best price. The result
is higher output and greater welfare.

The costs of tariffs can be measured
by the forgone gains from the trade
that is lost. Tariffs imposed by high-
income economies on trade with
developing economies cost an estimated $43.1 billion in 1995—three-
fourths as much as the OECD
countries provided in official development assistance in 1998. When
antidumping measures, protectionist
product standards, and barriers to
service trade are included, the
losses at least double.

Developing country tariff barriers
impose losses on high-income
economies—almost $50 billion. But
they cause even greater losses for other
developing countries—$65.1 billion.

Investment flows increase . . .

Foreign direct investment is now the largest form of private capital inflows to developing countries. World flows of foreign direct investment increased fourfold between 1990 and 1999, from $200 billion to $884 billion, and its ratio to GDP is generally rising in both high-income and developing countries.

But the surge in foreign direct investment began to slow after the financial crisis in 1997. In that year developing countries received 38 percent of world flows. By 1999 their share had fallen to 21 percent.

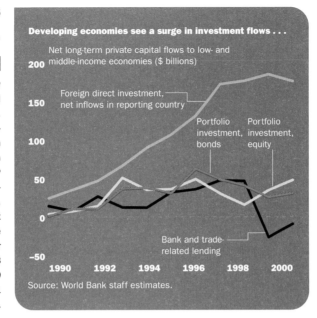

Developing economies see a surge in investment flows . . .

Net long-term private capital flows to low- and middle-income economies ($ billions)

Foreign direct investment, net inflows in reporting country

Portfolio investment, bonds

Portfolio investment, equity

Bank and trade-related lending

Source: World Bank staff estimates.

Foreign direct investment may have indirect benefits. It is often accompanied by transfers of skills and new technologies that increase its dynamic effects on growth. Portfolio investment is more volatile than foreign direct investment and requires careful management, but it can play an important role in deepening the domestic capital markets of more advanced developing countries. In general, the benefits of private capital flows will be greatest in countries with a well-educated workforce, good infrastructure, properly regulated capital markets, and a good business climate.

Capital flows

. . . but the distribution remains uneven

Private capital flows tend to go to countries with strong investment climates. Fifteen emerging market economies, mainly in East Asia, Latin America, and Europe, accounted for 83 percent of all net long-term private capital flows to developing countries in 1997. Most of these economies are middle income, so the increased capital flows in the past decade may have contributed to widening income differences across countries. Sub-Saharan Africa received only 5 percent of the total.

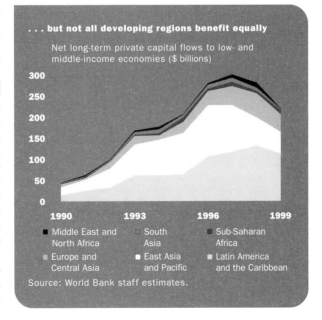

. . . but not all developing regions benefit equally

Net long-term private capital flows to low- and middle-income economies ($ billions)

- Middle East and North Africa
- South Asia
- Sub-Saharan Africa
- Europe and Central Asia
- East Asia and Pacific
- Latin America and the Caribbean

Source: World Bank staff estimates.

The capital markets of developing countries still are not globally integrated. The ratio of gross (two-way) capital flows to GDP measured in purchasing power parity terms has increased by about 250 percent since 1989 in developing as well as high-income economies. But the average for developing countries, 4.1 percent, is less than a ninth that for the highly integrated European Monetary Union (see table 6.1).

Foreign workers fill many jobs in high-income economies

Migration is perhaps the most tightly regulated form of international exchange. Migration policies differ widely, reflecting a complex mix of economic and political considerations. They are strongly influenced by the historical experience of the receiving country and its relationship with the supplying countries. Motivations for migration also differ. Some migrants seek only temporary opportunities or escape

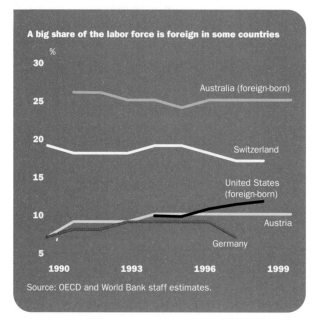

A big share of the labor force is foreign in some countries

%

Australia (foreign-born)

Switzerland

United States (foreign-born)

Austria

Germany

Source: OECD and World Bank staff estimates.

from conditions in their home country, while others relocate permanently.

Migration allows people to offer their skills where they are in short supply, which benefits both workers and the receiving economy. It may also benefit the workers' home country. Over the past five years receipts of workers' remittances in developing countries have averaged at least $50 billion a year.

Movement of people

Tourism is an important industry—it also brings people together

In 1999 world receipts from tourists were $455 billion. Developing economies received $132 billion, accounting for 7.7 percent of their exports of goods and services. Low-income economies receive the fewest international visitors, but are experiencing the fastest growth in tourism—an average of 10 percent a year.

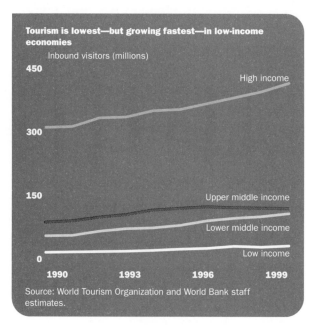

Tourism is lowest—but growing fastest—in low-income economies

Inbound visitors (millions)

High income

Upper middle income

Lower middle income

Low income

Source: World Tourism Organization and World Bank staff estimates.

For many countries tourism has been an attractive way to increase export earnings and to employ large numbers of relatively unskilled people. But successful tourism requires investment in hotels, transport facilities, and cultural attractions. And like all industries, tourism does best in a stable, secure environment.

Development assistance— important but declining

Over much of the past decade aid flows from the members of the OECD Development Assistance Committee have declined. They now represent less than 3 percent of gross national income for low-income economies and less than 0.5 percent for middle-income economies. In 1997 aid flows rose when a few countries increased their assistance to economies caught in the Asian financial crisis. More encouraging, some countries,

such as the Netherlands and the United Kingdom, have decided to maintain higher levels of assistance.

Foreign direct investment now exceeds official development assistance, but many of the poorest economies do not have access to international capital markets. Nor can they raise enough money out of domestic savings to finance their development programs. For countries capable of using aid effectively, aid can raise growth rates, improve the climate for investment, and create the conditions that allow all people, including the poor, to benefit from the global economy.

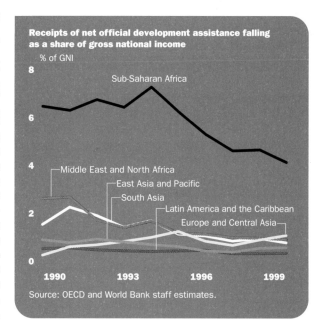

Receipts of net official development assistance falling as a share of gross national income

% of GNI

Sub-Saharan Africa

Middle East and North Africa
East Asia and Pacific
South Asia
Latin America and the Caribbean
Europe and Central Asia

Source: OECD and World Bank staff estimates.

Official development assistance

Slowing at the source

Many members of the Development Assistance Committee (DAC) have pledged to provide 0.7 percent of their gross national income (GNI) as aid, but only Denmark, the Netherlands, Norway, and Sweden have met this target. Some countries have curtailed their aid flows because of budget constraints. Others face skepticism from voters about the effectiveness of aid. But growing evidence of greater aid effectiveness strengthens the case for increasing the flow of official development assistance (ODA).

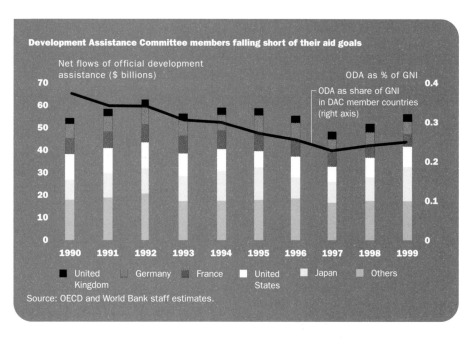

Development Assistance Committee members falling short of their aid goals

Net flows of official development assistance ($ billions)

ODA as % of GNI

ODA as share of GNI in DAC member countries (right axis)

■ United Kingdom   Germany   ■ France   ■ United States   ■ Japan   ■ Others

Source: OECD and World Bank staff estimates.

6.1 | Integration with the global economy

	Trade in goods				Growth in real trade less growth in real GDP	Gross private capital flows		Gross foreign direct investment	
	% of PPP GDP		% of goods GDP		percentage points	% of PPP GDP		% of PPP GDP	
	1989	1999	1989	1999	1989–99	1989	1999	1989	1999
Albania	6.7	13.9	34.5	51.9	10.5	4.2	1.9	0.0	0.4
Algeria	16.4	14.3	53.3	69.9	–1.3	0.9	..	0.0	..
Angola	31.1	16.0	80.6	87.5	..	1.4	..	1.4	..
Argentina	5.1	10.9	35.1	46.8	8.6	4.0	11.3	0.4	5.5
Armenia	..	13.3	..	..	–11.7	..	5.4	..	2.6
Australia	28.2	26.9	69.0	94.5	4.0	13.9	13.2	4.5	2.9
Austria	50.8	65.1	135.8	170.9	3.4	9.3	42.5	1.0	3.0
Azerbaijan	..	8.6	..	79.6	25.2	..	2.4	..	2.2
Bangladesh	4.1	6.8	30.0	53.0	7.1	0.2	0.8	0.0	0.1
Belarus	..	18.2	..	76.8	–5.1	..	0.8	..	0.3
Belgium	..	129.6	..	408.3	2.7	..	..	..	..
Benin	19.8	18.1	84.0	84.2	–1.7	7.3	3.7	2.0	1.0
Bolivia	12.5	14.6	..	..	1.9	2.5	7.3	0.7	5.3
Bosnia and Herzegovina	..	..	..	179.4	–0.7	..	..	..	..
Botswana	54.7	44.0	151.6	199.6	–3.6	5.4	3.2	0.7	0.6
Brazil	6.3	8.4	21.8	27.1	6.4	1.1	6.7	0.2	2.9
Bulgaria	59.7	22.9	205.3	150.3	–4.1	1.7	3.9	0.0	2.1
Burkina Faso	6.6	8.7	30.7	57.9	–1.6	0.8	..	0.1	..
Burundi	7.0	4.5	31.1	38.8	4.2	0.5	1.6	0.0	0.0
Cambodia	2.0	6.7	24.6	53.3	..	..	1.3	..	0.9
Cameroon	15.4	12.7	46.5	53.1	1.9	8.7	..	0.6	..
Canada	43.5	57.3	101.5	..	5.7	7.8	15.1	2.4	6.0
Central African Republic	8.9	12.0	33.2	61.0	..	0.9	..	0.2	..
Chad	8.7	9.8	54.3	69.4	–1.5	2.0	..	0.8	..
Chile	24.0	23.7	105.0	91.0	3.7	5.9	21.7	2.0	10.8
China	7.3	8.0	47.9	54.5	–6.5	0.6	2.7	0.3	1.0
Hong Kong, China	165.4	239.2	739.3	1,075.2	5.5	..	174.7	..	35.3
Colombia	6.7	9.3	48.1	59.1	6.0	0.9	3.3	0.4	0.8
Congo, Dem. Rep.	4.0	2.4	39.8	19.6	–4.9	..	..	..	..
Congo, Rep.	88.8	104.5	117.4	164.5	4.0	6.8	8.3	0.0	0.0
Costa Rica	19.9	40.6	107.4	165.9	5.1	1.3	5.0	0.7	2.1
Côte d'Ivoire	26.9	28.6	92.5	125.2	1.1	2.0	2.3	0.1	1.8
Croatia	..	36.5	..	113.9	..	..	12.0	..	4.8
Cuba	..	..	..	..	..	..	..	..	..
Czech Republic	..	41.6	..	..	9.9	..	11.2	..	4.0
Denmark	58.3	67.8	142.1	150.6	1.7	16.6	25.4	3.2	13.3
Dominican Republic	21.4	29.0	178.5	168.6	–1.0	1.9	4.6	0.4	2.9
Ecuador	15.5	20.1	83.4	79.5	1.2	3.3	4.4	0.6	1.9
Egypt, Arab Rep.	8.2	9.1	50.3	42.1	–1.4	2.4	2.7	1.0	0.5
El Salvador	11.4	16.1	84.6	86.7	7.9	0.5	3.8	0.3	0.9
Eritrea	..	..	..	..	3.9	..	..	..	..
Estonia	..	58.5	..	346.1	13.0	..	11.2	..	3.6
Ethiopia	..	5.5	..	..	0.9	0.6	0.7	0.0	..
Finland	54.2	61.3	87.1	131.1	5.2	16.7	49.4	4.2	14.3
France	37.6	44.0	102.3	120.6	3.8	19.1	29.2	3.0	10.9
Gabon	52.8	52.2	102.6	109.9	–1.5	12.1	..	2.5	..
Gambia, The	14.5	14.9	126.9	130.1	–3.0	1.4	1.5	1.1	0.7
Georgia	..	6.3	..	..	..	..	1.2	..	0.6
Germany	51.9	52.0	109.0	132.9	3.4	10.4	36.0	1.8	7.8
Ghana	11.3	15.1	66.4	112.6	6.0	0.5	0.8	0.1	0.0
Greece	20.7	25.5	87.1	88.7	3.0	1.9	7.7	0.7	0.7
Guatemala	11.5	16.6	..	..	3.9	1.2	10.9	0.3	4.2
Guinea	13.9	14.1	82.7	90.2	–1.8	1.2	1.0	0.1	0.5
Guinea-Bissau	12.8	17.9	57.2	88.5	–0.5	6.9	0.4	0.0	..
Haiti	4.0	10.7	30.0	..	0.0	0.5	1.4	0.1	0.1
Honduras	18.4	26.9	102.1	133.3	–0.6	1.4	2.9	0.5	1.6

Integration with the global economy | 6.1

	Trade in goods				Growth in real trade less growth in real GDP	Gross private capital flows		Gross foreign direct investment	
	% of PPP GDP		% of goods GDP		percentage points	% of PPP GDP		% of PPP GDP	
	1989	**1999**	**1989**	**1999**	**1989–99**	**1989**	**1999**	**1989**	**1999**
Hungary	18.2	46.1	98.8	*214.8*	7.6	0.4	10.3	0.0	2.0
India	3.2	3.6	19.8	*31.1*	4.5	0.3	0.6	0.0	0.1
Indonesia	12.1	12.3	63.4	81.3	2.5	0.4	2.4	0.2	0.9
Iran, Islamic Rep.	12.6	8.4	39.1	51.1	–9.0	1.5	*1.1*	0.0	*0.0*
Iraq	..	..	..	..	..	..	..	..	..
Ireland	92.1	120.1	196.6	*226.2*	7.0	24.3	179.3	0.2	25.2
Israel	43.3	52.4	..	..	3.5	5.7	12.1	0.4	3.1
Italy	30.5	35.0	86.0	104.5	4.0	9.0	27.0	0.4	1.2
Jamaica	36.5	40.2	159.8	140.3	0.5	3.3	13.7	0.9	6.7
Japan	20.8	23.2	43.5	*52.3*	3.0	11.1	30.9	2.0	1.2
Jordan	30.9	29.4	192.4	171.6	–1.1	1.2	2.6	0.2	0.9
Kazakhstan	..	12.5	..	125.8	8.5	..	4.3	..	2.2
Kenya	14.4	15.6	66.0	92.0	2.4	1.7	3.0	0.3	0.0
Korea, Dem. Rep.	..	..	..	..	..	..	..	..	..
Korea, Rep.	35.7	35.9	108.3	136.3	7.1	3.3	9.1	0.5	2.1
Kuwait	86.8	..	132.7	..	..	20.9	..	4.9	..
Kyrgyz Republic	..	8.4	..	125.8	–0.4	..	0.8	..	0.3
Lao PDR	7.3	11.2	47.0	*97.8*	..	2.0	1.7	0.1	1.1
Latvia	..	30.7	..	187.3	7.0	..	10.3	..	2.4
Lebanon	53.2	*39.0*	..	..	3.5	..	..	..	..
Lesotho	39.4	28.4	207.0	*198.2*	0.3	1.1	4.5	0.8	4.2
Libya	..	..	..	..	..	..	..	..	..
Lithuania	..	31.8	..	159.0	12.0	..	8.0	..	2.7
Macedonia, FYR	..	33.1	..	..	6.1	..	2.9	..	0.3
Madagascar	6.9	*6.7*	39.0	..	2.7	0.5	*0.5*	0.1	*0.1*
Malawi	20.1	16.6	64.6	98.0	–1.8	1.4	..	0.0	..
Malaysia	59.3	80.2	211.4	333.7	4.9	4.2	7.1	2.1	0.8
Mali	14.9	16.1	58.8	74.5	1.9	1.6	*1.7*	0.1	*0.5*
Mauritania	32.4	18.8	122.0	138.2	–3.0	2.0	*10.3*	0.1	*0.0*
Mauritius	41.8	34.4	199.0	185.2	0.3	3.0	2.1	0.7	0.5
Mexico	14.1	35.6	75.8	151.4	10.3	1.4	4.5	0.6	1.5
Moldova	..	11.9	..	170.5	14.6	..	5.8	..	0.6
Mongolia	43.4	18.7	..	135.9	..	33.8	2.2	0.0	0.7
Morocco	13.1	18.6	76.5	108.2	2.1	1.2	2.5	0.2	0.9
Mozambique	11.8	8.7	53.4	52.1	–2.0	0.0	*2.2*	0.0	*1.6*
Myanmar	..	..	..	..	..	..	..	..	..
Namibia	38.7	36.0	187.9	208.3	–0.2	*7.4*	*4.4*	*2.0*	*1.4*
Nepal	4.6	6.8	30.5	..	7.7	0.1	0.8	0.0	0.0
Netherlands	85.4	101.4	228.6	286.3	2.4	33.6	81.3	9.4	20.3
New Zealand	35.8	36.7	110.5	..	3.2	13.3	*19.8*	7.1	*12.0*
Nicaragua	11.0	21.3	168.3	194.0	6.0	0.6	4.1	0.0	2.7
Niger	10.9	8.5	51.2	55.5	–2.9	1.1	..	0.1	..
Nigeria	21.3	20.5	79.6	*83.2*	2.2	3.5	4.3	2.8	1.0
Norway	62.7	62.2	124.1	118.6	1.7	15.0	*32.8*	3.6	*7.3*
Oman	..	..	126.1	..	..	..	..	..	..
Pakistan	8.5	8.0	52.3	62.6	–0.9	0.6	*1.5*	0.2	*0.3*
Panama	15.1	26.3	103.8	191.5	–1.5	30.7	32.2	1.8	5.0
Papua New Guinea	46.1	27.5	148.0	113.0	–1.2	4.5	8.7	3.5	2.7
Paraguay	11.0	12.8	49.1	74.5	4.5	1.1	2.6	0.1	0.6
Peru	7.5	12.2	68.5	61.4	5.2	1.1	4.4	0.1	1.7
Philippines	9.5	24.5	77.1	*191.1*	5.6	1.2	5.3	0.3	0.3
Poland	*11.5*	22.4	*70.7*	99.5	11.3	*2.9*	5.4	*0.0*	2.6
Portugal	29.9	38.9	132.1	*144.4*	4.1	5.7	27.0	1.7	3.2
Puerto Rico	..	..	..	..	..	..	..	..	..
Romania	12.4	13.9	61.3	106.6	6.4	1.2	2.4	0.0	0.8
Russian Federation	..	10.6	..	58.6	1.5	..	1.1	..	0.5

	Trade in goods				Growth in real trade less growth in real GDP	Gross private capital flows		Gross foreign direct investment	
	% of PPP GDP		% of goods GDP		percentage points	% of PPP GDP		% of PPP GDP	
	1989	1999	1989	1999	1989–99	1989	1999	1989	1999
Rwanda	6.2	4.6	27.0	25.8	4.5	0.5	0.2	0.2	0.0
Saudi Arabia	37.5	36.0	109.6	..	..	9.9	11.6	0.5	0.4
Senegal	22.4	19.3	108.5	122.7	–1.7	3.6	3.5	0.4	1.5
Sierra Leone	10.6	4.3	38.9	22.1	–4.6	2.2	..	0.7	..
Singapore	264.5	275.1	854.6	739.5	..	56.8	54.2	10.6	13.3
Slovak Republic	..	37.6	..	240.0	11.1	..	11.2	..	1.3
Slovenia	..	58.5	..	183.0	0.0	..	6.8	..	0.7
South Africa	14.0ᵃ	14.2ᵃ	82.5ᵃ	99.9ᵃ	4.7	0.9	6.9	0.2	0.7
Spain	24.2	35.8	..	114.1	7.2	5.1	28.2	2.1	6.4
Sri Lanka	11.5	16.9	93.9	123.2	2.8	2.2	1.2	0.1	0.4
Sudan	..	..	..	..	..	..	..	..	..
Sweden	66.1	76.5	125.6	..	4.9	34.3	86.2	8.0	39.4
Switzerland	70.9	82.7	..	..	2.2	34.4	139.2	8.7	22.9
Syrian Arab Republic	18.1	10.4	108.5	..	–2.4	4.5	4.3	0.0	0.4
Tajikistan	..	..	..	147.9	..	..	..	..	..
Tanzania	12.6	14.3	50.8	41.8	–2.4	0.0	2.1	0.0	1.1
Thailand	23.9	29.4	123.6	171.2	3.0	3.8	4.9	1.0	1.8
Togo	14.5	13.0	93.7	97.7	–3.5	3.1	2.1	0.2	0.9
Trinidad and Tobago	38.4	44.6	145.8	165.5	1.0	6.2	8.6	2.0	7.5
Tunisia	24.5	25.5	164.1	126.7	–0.4	2.0	2.2	0.3	0.6
Turkey	11.1	16.2	47.4	76.3	7.4	1.6	3.7	0.3	0.3
Turkmenistan	..	16.8	..	100.4	–1.0	..	3.7	..	0.9
Uganda	4.8	7.4	15.1	44.2	4.5	0.7	1.3	0.0	0.9
Ukraine	..	13.6	..	104.8	7.8	..	2.9	..	0.3
United Arab Emirates	78.3	106.7	159.3	..	..	..	..	..	..
United Kingdom	36.8	44.8	102.2	118.4	3.6	37.6	66.1	7.4	23.0
United States	14.9	19.8	..	..	5.2	7.4	13.6	2.5	5.2
Uruguay	14.6	19.0	82.7	94.6	5.5	8.3	7.3	0.0	0.8
Uzbekistan	..	7.7	..	37.6	1.4	..	..	..	..
Venezuela, RB	22.6	26.6	92.1	80.1	3.9	7.1	11.7	0.4	2.8
Vietnam	7.4	16.0	112.4	..	22.3	..	..	..	..
West Bank and Gaza	..	..	..	..	0.2	..	..	..	..
Yemen, Rep.	25.9	34.1	90.2	116.2	1.3	6.2	3.9	1.5	1.7
Yugoslavia, FR (Serb./Mont.)	..	..	..	..	..	..	..	..	..
Zambia	33.9	18.7	75.9	81.7	2.2	36.9	..	2.5	..
Zimbabwe	14.9	14.4	69.2	171.3	8.0	0.3	..	0.0	..
World	**22.5 w**	**27.4 w**	**85.2 w**	**111.5 w**		**8.5 w**	**18.3 w**	**2.0 w**	**4.6 w**
Low income	7.2	7.8	41.3	60.0		0.8	1.2	0.2	0.3
Middle income	14.1	16.9	69.0	81.5		1.9	4.9	0.4	1.6
Lower middle income	11.5	11.7	65.3	65.3		1.2	2.9	0.3	1.0
Upper middle income	17.2	26.0	71.8	96.5		2.7	8.2	0.5	2.6
Low & middle income	12.3	14.7	63.3	81.3		1.6	4.1	0.3	1.3
East Asia & Pacific	14.5	15.3	82.7	91.1		1.3	3.8	0.4	1.1
Europe & Central Asia	..	17.7	..	83.4		..	3.7	..	1.1
Latin America & Carib.	10.2	18.2	49.8	74.6		2.2	7.3	0.4	3.0
Middle East & N. Africa	19.4	16.8	76.1	67.1		3.5	5.7	0.3	0.5
South Asia	4.0	4.6	25.6	38.1		0.3	0.6	0.0	0.1
Sub-Saharan Africa	15.9	16.3	78.1	95.6		2.1	4.9	0.6	0.7
High income	28.5	37.4	93.5	123.5		12.7	29.2	2.9	7.2
Europe EMU	38.7	52.7	112.0	153.3		14.0	37.1	2.4	7.8

a. Data refer to the South African Customs Union (Botswana, Lesotho, Namibia, South Africa, and Swaziland).

The growing importance of trade in the world economy is one indication of increasing global economic integration. Another is the increased size and importance of private capital flows to developing countries that have liberalized their financial markets. This table presents standardized measures of the size of trade and capital flows relative to gross domestic product. For three of the indicators GDP measured in purchasing power parity (PPP) terms, which adjust for differences in domestic prices, has been used in the denominator to better measure the relative size of the domestic economy. (No adjustment has been made to the numerators because goods and capital exchanged on international markets are assumed to be valued at international prices.)

The numerators are based on gross flows that capture the two-way flow of goods and capital. In conventional balance of payments accounting exports are recorded as a credit and imports as a debit. And in the financial account inward investment is a credit and outward investment a debit. Thus net flows, the sum of credits and debits, represent a balance in which many transactions are canceled out. Gross flows are a better measure of integration because they show the total value of financial transactions during a given period.

The growth of services has affected the historical record. Compared with the levels achieved at the end of the 19th century, trade in goods appears to have declined in importance relative to GDP, especially in economies with growing service sectors. Measuring merchandise trade relative to GDP after deducting value added by services thus provides a better measure of its relative size than does comparing it with total GDP, although this neglects the growing service component of most goods output.

Trade in services, traditionally called invisibles, is becoming an important element of global integration. The difference between the growth of real trade in goods and services and the growth of GDP helps to identify economies with dynamic trade regimes.

The investment indicators in the table were constructed from data recorded at the most detailed level available. Higher-level aggregates tend to be affected by the netting out of credits and debits and so produce a smaller total. The comparability of these indicators between countries and over time is affected by the accuracy and completeness of balance of payments records and by their level of detail.

• **Trade in goods as a share of PPP GDP** is the sum of merchandise exports and imports measured in current U.S. dollars divided by the value of GDP converted to international dollars using purchasing power parity rates. • **Trade in goods as a share of goods GDP** is the sum of merchandise exports and imports divided by the value of GDP after subtracting value added in services, all in current U.S. dollars. • **Growth in real trade less growth in real GDP** is the difference between annual growth in trade of goods and services and annual growth in GDP. Growth rates are calculated using constant price series taken from national accounts and are expressed as a percentage. • **Gross private capital flows** are the sum of the absolute values of direct, portfolio, and other investment inflows and outflows recorded in the balance of payments financial account, excluding changes in the assets and liabilities of monetary authorities and general government. The indicator is calculated as a ratio to GDP converted to international dollars using purchasing power parity rates. • **Gross foreign direct investment** is the sum of the absolute values of inflows and outflows of foreign direct investment recorded in the balance of payments financial account. It includes equity capital, reinvestment of earnings, other long-term capital, and short-term capital. This indicator differs from the standard measure of foreign direct investment, which captures only inward investment (see table 6.7). The indicator is calculated as a ratio to GDP converted to international dollars using purchasing power parity rates.

Data sources

The data on merchandise trade are from the World Trade Organization. The data on GDP in PPP terms come from the World Bank's International Comparison Programme database. The data on real trade and GDP growth come from the World Bank's national accounts files. Gross private capital flows and foreign direct investment were calculated using the International Monetary Fund's Balance of Payments database.

Gross private capital flows expand

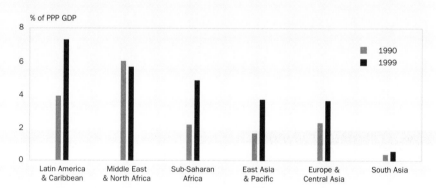

Note: Data for Europe and Central Asia refer to 1994 and 1999.
Source: International Monetary Fund, Balance of Payments database; and World Bank staff estimates.

Since 1990 gross private capital flows have increased in all regions except the Middle East and North Africa, with Sub-Saharan Africa and East Asia and Pacific experiencing the biggest increases.

6.2 Direction and growth of merchandise trade

High-income importers

Direction of trade % of world trade, 1999	European Union	Japan	United States	Other industrial	All industrial	Other high income	All high income
Source of exports							
High-income economies	31.8	2.9	11.8	6.6	53.0	5.1	58.1
Industrial economies	30.4	2.1	9.9	6.3	48.7	3.9	52.6
European Union	24.6	0.7	3.5	2.3	31.0	1.3	32.3
Japan	1.3		2.3	0.4	4.0	1.3	5.3
United States	2.7	1.0		3.3	7.1	1.1	8.1
Other industrial economies	1.8	0.4	4.1	0.3	6.6	0.3	6.9
Other high-income economies	1.4	0.8	1.9	0.3	4.4	1.2	5.5
Low- and middle-income economies	6.3	2.1	6.3	0.9	15.6	2.7	18.3
East Asia & Pacific	1.7	1.6	2.3	0.4	5.9	2.2	8.2
Europe & Central Asia	2.3	0.0	0.3	0.1	2.8	0.1	2.9
Latin America & Caribbean	0.7	0.1	3.1	0.1	4.1	0.1	4.2
Middle East & N. Africa	0.8	0.2	0.3	0.0	1.4	0.2	1.5
South Asia	0.3	0.0	0.2	0.0	0.6	0.1	0.7
Sub-Saharan Africa	0.5	0.0	0.2	0.0	0.8	0.1	0.9
World	38.1	5.0	18.1	7.4	68.6	7.8	76.4

Low- and middle-income importers

Direction of trade % of world trade, 1999	East Asia & Pacific	Europe & Central Asia	Latin America & Caribbean	Middle East & N. Africa	South Asia	Sub-Saharan Africa	All low & middle income	World
Source of exports								
High-income economies	6.3	3.3	4.1	1.5	0.7	0.8	16.7	74.8
Industrial economies	3.9	3.2	3.9	1.4	0.5	0.7	13.7	66.3
European Union	1.0	2.8	1.0	1.0	0.3	0.1	6.5	38.8
Japan	1.5	0.1	0.3	0.1	0.1	0.1	2.2	7.5
United States	1.0	0.2	2.5	0.3	0.1	0.1	4.2	12.3
Other industrial economies	0.4	0.1	0.1	0.1	0.1	0.0	0.8	7.7
Other high-income economies	2.3	0.1	0.2	0.1	0.2	0.1	3.0	8.5
Low- and middle-income economies	2.4	1.7	1.3	0.6	0.5	0.4	6.9	25.2
East Asia & Pacific	1.6	0.2	0.3	0.2	0.2	0.1	2.7	11.8
Europe & Central Asia	0.1	1.2	0.1	0.1	0.0	0.0	1.6	4.5
Latin America & Caribbean	0.1	0.1	0.9	0.1	0.0	0.0	1.2	5.3
Middle East & N. Africa	0.4	0.1	0.0	0.2	0.1	0.0	0.8	2.3
South Asia	0.1	0.0	0.0	0.0	0.0	0.0	0.2	1.0
Sub-Saharan Africa	0.1	0.0	0.0	0.0	0.0	0.2	0.4	1.3
World	8.7	4.9	5.4	2.2	1.1	1.3	23.6	100.0

Direction and growth of merchandise trade | 6.2

High-income importers

Nominal growth of trade annual % growth, 1989–99	European Union	Japan	United States	Other industrial	All industrial	Other high income	All high income
Source of exports							
High-income economies	5.3	3.1	6.8	5.2	5.5	7.7	5.7
Industrial economies	5.2	2.6	6.9	5.3	5.4	6.9	5.5
European Union	5.5	4.1	7.7	4.0	5.6	7.7	5.6
Japan	3.5		3.3	0.7	3.1	6.5	3.8
United States	5.2	2.6		7.1	5.6	7.1	5.8
Other industrial economies	3.6	0.3	9.0	4.3	6.3	4.9	6.2
Other high-income economies	7.5	4.7	6.2	4.2	6.1	11.1	7.0
Low- and middle-income economies	7.9	5.9	11.6	9.2	9.0	9.7	9.1
East Asia & Pacific	13.6	7.5	11.9	12.4	11.0	10.8	10.9
Europe & Central Asia	10.3	−2.6	14.0	10.1	10.1	12.6	10.2
Latin America & Caribbean	4.0	1.2	14.3	8.0	10.7	3.9	10.5
Middle East & N. Africa	2.2	6.0	2.0	−1.3	2.6	1.7	2.5
South Asia	8.3	−0.8	9.1	8.5	7.5	11.5	8.1
Sub-Saharan Africa	3.1	−0.7	2.2	5.4	2.7	12.3	3.2
World	5.7	4.2	8.2	5.6	6.2	8.4	6.4

Low- and middle-income importers

Nominal growth of trade annual % growth, 1989–99	East Asia & Pacific	Europe & Central Asia	Latin America & Caribbean	Middle East & N. Africa	South Asia	Sub-Saharan Africa	All low & middle income	World
Source of exports								
High-income economies	8.6	9.7	10.0	2.6	3.7	2.0	7.7	6.1
Industrial economies	7.3	9.6	10.0	2.6	2.3	1.8	7.2	5.9
European Union	8.0	11.5	9.0	2.6	2.7	1.7	7.4	5.9
Japan	7.2	−1.9	7.4	1.3	0.2	−0.8	5.7	4.3
United States	7.6	2.3	11.2	3.5	1.3	4.1	8.6	6.6
Other industrial economies	5.8	1.3	5.1	1.8	5.6	2.9	4.3	6.0
Other high-income economies	11.1	12.9	10.1	2.0	8.0	4.1	10.0	7.9
Low- and middle-income economies	15.2	5.0	11.5	3.6	10.9	11.1	9.4	9.2
East Asia & Pacific	18.4	8.8	20.8	9.6	13.2	13.6	15.9	11.9
Europe & Central Asia	3.0	5.6	12.4	−0.1	1.6	6.6	4.9	7.9
Latin America & Caribbean	6.1	−0.6	10.5	4.4	12.9	4.9	8.6	10.0
Middle East & N. Africa	16.1	−3.4	−1.9	1.0	9.5	12.9	5.6	3.5
South Asia	10.3	9.7	26.9	7.2	11.8	20.4	11.8	8.9
Sub-Saharan Africa	16.1	2.6	12.9	10.3	22.8	10.2	11.8	5.2
World	10.0	7.9	10.4	2.8	6.0	4.3	8.1	6.8

About the data

This table provides estimates of the flow of trade in goods between groups of economies. Twenty-three high-income countries and 23 developing countries report their trade data to the International Monetary Fund (IMF) each month. Together these countries account for about 80 percent of world exports. Trade by less timely reporters and by countries that do not report is estimated using reports of partner countries. Because the largest exporting and importing countries are reliable reporters, a large portion of the missing trade flows can be estimated from partner reports. Even so, a small amount of trade between developing countries, particularly in Africa, is not captured in partner data. In addition, estimates of intra-European trade have been significantly affected by changes in reporting methods following the creation of a customs union.

Most countries report their trade data in national currencies, which are converted using the IMF's published exchange rate series rf (official rate, period average) or rh (market rate, period average). Because imports are reported at c.i.f. (cost, insurance, and freight) valuations and exports at f.o.b. (free on board) valuations, the IMF divides partner country reports of import values by 1.10 to estimate equivalent export values. This approximation is more or less accurate, depending on the set of partners and the items traded. Other factors affecting the accuracy of trade data include lags in reporting, recording differences across countries, and whether the country reports trade according to the general or special system of trade. (For further discussion of the measurement of exports and imports see *About the data* for tables 4.5 and 4.6.)

The regional trade flows shown in this table were calculated from current price values. Growth rates therefore include the effects of changes in both volumes and prices.

Figure 6.2

More than three-quarters of world imports go to high-income economies

% of world imports of goods, 1999

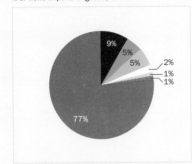

- East Asia & Pacific
- Europe & Central Asia
- Latin America & Caribbean
- Middle East & N. Africa
- South Asia
- Sub-Saharan Africa
- High-income economies

Source: International Monetary Fund, Direction of Trade database.

The high-income economies' share has remained much the same since the 1980s.

Definitions

- **Merchandise trade** includes all trade in goods. Trade in services is excluded. • **Regional groupings** are based on World Bank definitions and may differ from those used by other organizations. • **European Union** comprises Austria, Belgium, Denmark, Finland, France, Germany, Greece, Ireland, Italy, Luxembourg, the Netherlands, Portugal, Spain, Sweden, and the United Kingdom. • **Other industrial economies** include Australia, Canada, Iceland, New Zealand, Norway, and Switzerland. • **Other high-income economies** include Cyprus, Hong Kong (China), Israel, Kuwait, Macao (China), Malta, Qatar, Singapore, Taiwan (China), and the United Arab Emirates. Some small high-income economies such as Aruba, the Bahamas, and Bermuda have been included in the Latin America and Caribbean group.

Data sources

Intercountry trade flows are published in the IMF's *Direction of Trade Statistics Yearbook* and *Direction of Trade Statistics Quarterly;* the data in the table were calculated using the IMF's Direction of Trade database.

OECD trade with low- and middle-income economies | 6.3

	High-income OECD countries		European Union		Japan		United States	
Exports to low- and middle-income economies	**1990**	**1999**	**1990**	**1999**	**1990**	**1999**	**1990**	**1999**
$ billions								
Food	34.1	52.3	15.7	24.0	0.4	0.6	11.7	19.3
Cereals	14.0	13.8	4.1	4.6	0.1	0.1	6.2	7.1
Agricultural raw materials	11.6	14.3	3.1	4.6	0.8	1.2	5.6	5.0
Ores and nonferrous metals	9.3	14.9	2.9	5.2	0.9	2.6	2.9	3.4
Fuels	8.5	14.2	2.5	5.7	0.8	0.7	3.3	4.3
Crude petroleum	0.3	1.8	0.0	0.5	0.0	0.0	0.0	0.3
Petroleum products	5.6	8.9	2.4	4.7	0.8	0.6	2.2	3.1
Manufactured goods	303.4	593.4	147.5	282.2	68.7	111.7	72.6	178.8
Chemical products	45.4	81.2	23.4	42.5	5.9	10.8	12.2	22.2
Mach. & transport equip.	174.5	349.1	79.3	151.7	44.0	73.7	44.5	114.0
Other	83.6	163.1	44.8	88.0	18.8	27.2	15.9	42.5
Miscellaneous goods	11.6	23.8	3.7	10.5	0.8	3.0	4.7	9.4
Total	**378.5**	**712.9**	**175.5**	**332.1**	**72.5**	**119.8**	**100.7**	**220.2**
% of total exports								
Food	9.0	7.3	9.0	7.2	0.5	0.5	11.6	8.8
Cereals	3.7	1.9	2.3	1.4	0.1	0.1	6.2	3.2
Agricultural raw materials	3.1	2.0	1.8	1.4	1.1	1.0	5.5	2.3
Ores and nonferrous metals	2.5	2.1	1.7	1.6	1.2	2.2	2.9	1.6
Fuels	2.2	2.0	1.4	1.7	1.1	0.6	3.2	1.9
Crude petroleum	0.1	0.2	0.0	0.1	0.0	0.0	0.0	0.1
Petroleum products	1.5	1.2	1.3	1.4	1.0	0.5	2.2	1.4
Manufactured goods	80.2	83.2	84.0	85.0	94.9	93.2	72.1	81.2
Chemical products	12.0	11.4	13.3	12.8	8.2	9.0	12.1	10.1
Mach. & transport equip.	46.1	49.0	45.2	45.7	60.7	61.5	44.2	51.8
Other	22.1	22.9	25.5	26.5	26.0	22.7	15.8	19.3
Miscellaneous goods	3.1	3.3	2.1	3.2	1.1	2.5	4.7	4.3
Total	**100.0**	**100.0**	**100.0**	**100.0**	**100.0**	**100.0**	**100.0**	**100.0**

Imports from low- and middle-income economies	High-income OECD countries		European Union		Japan		United States	
	1990	1999	1990	1999	1990	1999	1990	1999
$ billions								
Food	64.2	94.1	34.6	44.5	10.7	19.4	15.5	24.7
Cereals	1.3	2.4	0.5	1.0	0.5	0.5	0.2	0.7
Agricultural raw materials	17.3	22.1	9.5	12.5	5.0	4.2	2.3	4.5
Ores and nonferrous metals	30.1	42.7	14.9	19.6	9.1	9.9	5.1	10.3
Fuels	144.2	139.8	58.3	56.4	33.5	26.4	48.8	50.8
Crude petroleum	107.5	100.7	46.6	40.0	20.8	14.5	37.3	40.8
Petroleum products	23.5	21.9	6.1	8.2	5.9	4.0	10.8	9.2
Manufactured goods	208.3	667.8	83.1	237.8	24.5	73.0	85.5	315.6
Chemical products	14.0	30.6	7.5	14.2	2.3	4.1	3.0	9.7
Mach. and transport equip.	59.1	292.0	18.2	92.9	3.7	27.0	32.1	153.7
Other	135.2	345.1	57.4	130.7	18.5	41.9	50.3	152.2
Miscellaneous goods	5.5	12.5	2.0	0.6	0.5	1.4	2.6	10.3
Total	**469.7**	**979.0**	**202.4**	**371.4**	**83.4**	**134.3**	**159.8**	**416.3**
% of total imports								
Food	13.7	9.6	17.1	12.0	12.8	14.4	9.7	5.9
Cereals	0.3	0.2	0.2	0.3	0.6	0.4	0.1	0.2
Agricultural raw materials	3.7	2.3	4.7	3.4	6.0	3.1	1.5	1.1
Ores and nonferrous metals	6.4	4.4	7.4	5.3	10.9	7.4	3.2	2.5
Fuels	30.7	14.3	28.8	15.2	40.2	19.6	30.5	12.2
Crude petroleum	22.9	10.3	23.0	10.8	24.9	10.8	23.3	9.8
Petroleum products	5.0	2.2	3.0	2.2	7.0	3.0	6.8	2.2
Manufactured goods	44.4	68.2	41.1	64.0	29.4	54.4	53.5	75.8
Chemical products	3.0	3.1	3.7	3.8	2.8	3.1	1.9	2.3
Mach. and transport equip.	12.6	29.8	9.0	25.0	4.5	20.1	20.1	36.9
Other	28.8	35.2	28.4	35.2	22.2	31.2	31.5	36.6
Miscellaneous goods	1.2	1.3	1.0	0.2	0.7	1.1	1.6	2.5
Total	**100.0**	**100.0**	**100.0**	**100.0**	**100.0**	**100.0**	**100.0**	**100.0**

OECD trade with low- and middle-income economies | 6.3

Trade flows between high-income members of the Organisation for Economic Co-operation and Development (OECD) and low- and middle-income economies reflect the changing mix of exports to and imports from developing economies. While food and primary commodities have continued to fall as a share of OECD imports, the share of manufactured goods supplied by developing countries has grown. At the same time, developing countries have increased their imports of manufactured goods from high-income countries— particularly capital-intensive goods such as machinery and transport equipment. Although trade between developing countries has grown substantially over the past decade (see table 6.5), high-income OECD countries remain the developing world's most important partners.

The aggregate flows in the table were compiled from intercountry flows recorded in the United Nations Statistics Division's Commodity Trade (COMTRADE) database. Partner country reports by high-income OECD countries were used for both exports and imports. Exports are recorded free on board (f.o.b.); imports include insurance and freight charges (c.i.f.). Revisions have been made to the time-series data as far back as 1990. Because of differences in sources of data, timing, and treatment of missing data, the data in this table may not be fully comparable with those used to calculate the direction of trade statistics in table 6.2 or the aggregate flows shown in tables 4.4–4.6.

For further discussion of merchandise trade statistics see *About the data* for tables 4.4–4.6 and 6.2.

Definitions

The product groups in the table are defined in accordance with the Standard International Trade Classification (SITC) revision 1: food (0, 1, 22, and 4) and cereals (04); agricultural raw materials (2 excluding 22, 27, and 28); ores and nonferrous metals (27, 28, and 68); fuels (3), crude petroleum (331), and petroleum products (332); manufactured goods (5–8 excluding 68), chemical products (5), machinery and transport equipment (7), and other manufactured goods (6 and 8 excluding 68); and miscellaneous goods (9). • **Exports** are all merchandise exports by high-income OECD countries to low- and middle-income economies as recorded in the United Nations Statistics Division's COMTRADE database. • **Imports** are all merchandise imports by high-income OECD countries from low- and middle-income economies as recorded in the United Nations Statistics Division's COMTRADE database. • **High-income OECD countries** in 1999 were Australia, Austria, Belgium, Canada, Denmark, Finland, France, Germany, Greece, Iceland, Ireland, Italy, Japan, Luxembourg, the Netherlands, New Zealand, Norway, Portugal, Spain, Sweden, Switzerland, the United Kingdom, and the United States. • **European Union** comprises Austria, Belgium, Denmark, Finland, France, Germany, Greece, Ireland, Italy, Luxembourg, the Netherlands, Portugal, Spain, Sweden, and the United Kingdom.

Data sources

COMTRADE data are available in machine-readable form from the United Nations Statistics Division. Although not as comprehensive as the underlying COMTRADE records, detailed statistics on international trade are published annually in the United Nations Conference on Trade and Development's (UNCTAD) *Handbook of International Trade and Development Statistics* and the United Nations Statistics Division's *International Trade Statistics Yearbook*.

6.4 | Primary commodity prices

	1960	1965	1970	1975	1980	1985	1990	1995	1998	1999	2000
World Bank commodity price index											
(1990 = 100)											
Non-energy commodities	187	187	175	166	174	133	100	103	93	85	86
Agriculture	208	193	182	179	192	146	100	110	102	90	87
Beverages	234	213	227	180	252	239	100	127	132	104	88
Food	184	197	186	223	193	126	100	98	99	85	84
Raw materials	220	174	145	121	145	103	100	114	82	86	91
Fertilizers	180	179	121	350	179	130	100	87	115	110	105
Metals and minerals	137	173	161	117	131	101	100	85	71	71	82
Petroleum	34	29	21	101	224	173	100	63	54	76	122
Steel products[a]	..	116	124	115	110	89	100	90	71	66	76
MUV G-5 index	21	22	25	45	72	69	100	119	106	103	101
Commodity prices											
(1990 $)											
Agricultural raw materials											
Cotton (cents/kg)	314	290	252	257	284	192	182	179	136	113	129
Logs, Cameroon ($/cu. m)[a]	168	183	171	280	349	253	343	285	270	261	273
Logs, Malaysian ($/cu. m)	154	162	172	149	272	177	177	215	153	181	188
Rubber (cents/kg)	377	234	162	124	198	111	86	133	68	61	68
Sawnwood, Malaysian											
($/cu. m)	721	726	699	494	550	448	533	622	456	582	591
Tobacco ($/mt)	8,390	5,858	4,287	4,075	3,161	3,807	3,392	2,223	3,143	2,944	2,960
Beverages (cents/kg)											
Cocoa	285	169	269	276	362	329	127	120	158	110	90
Coffee, robustas	270	323	369	298	450	386	118	233	172	144	90
Coffee, Arabica	446	464	457	319	481	471	197	280	281	222	190
Tea, avg., 3 auctions	497	463	333	253	230	255	206	125	193	178	186
Energy											
Coal, Australian ($/mt)	..	..	..	..	54.72	49.20	39.67	33.10	27.54	25.13	26.01
Coal, U.S. ($/mt)	..	..	..	..	59.86	67.93	41.67	32.94	32.40	32.11	32.76
Natural gas, Europe ($/mmbtu)	..	..	..	2.43	4.72	5.39	2.55	2.29	2.28	2.06	3.82
Natural gas, U.S. ($/mmbtu)	0.66	0.72	0.66	0.95	2.15	3.57	1.70	1.45	1.97	2.19	4.27
Petroleum ($/bbl)	7.87	6.57	4.82	23.07	51.21	39.62	22.88	14.45	12.31	17.49	27.97

About the data

Primary commodities are raw or partially processed materials that will be transformed into finished goods. They are often the most significant exports of developing countries, and revenues obtained from them have an important effect on living standards. Price data for primary commodities are collected from a variety of sources, including international study groups, trade journals, newspaper and wire service reports, government market surveys, and commodity exchange spot and near-term forward prices. This table is based on frequently updated price reports. When possible, the prices received by exporters are used; if export prices are

unavailable, the prices paid by importers are used. Annual price series are generally simple averages based on higher-frequency data. The constant price series in the table are deflated using the manufactures unit value (MUV) index for the G-5 countries (see below).

The commodity price indexes are calculated as Laspeyres index numbers in which the fixed weights are the 1987–89 export values for low- and middle-income economies, rebased to 1990. Each index represents a fixed basket of primary commodity exports. The non-energy commodity price index contains 37 price series for 31 non-energy commodities. Separate indexes are

compiled for petroleum and for steel products, which are not included in the non-energy commodity price index.

The MUV index is a composite index of prices for manufactured exports from the five major (G-5) industrial countries (France, Germany, Japan, the United Kingdom, and the United States) to low- and middle-income economies, valued in U.S. dollars. The index covers products in groups 5–8 of the Standard International Trade Classification (SITC) revision 1. To construct the MUV G-5 index, unit value indexes for each country are combined using weights determined by each country's export share.

Primary commodity prices | 6.4

	1960	1965	1970	1975	1980	1985	1990	1995	1998	1999	2000
Fertilizers ($/mt)											
Phosphate rock	65	60	44	148	65	49	40	29	41	43	43
TSP	256	250	169	448	250	177	132	126	163	150	136
Food											
Fats and oils ($/mt)											
Coconut oil	1,507	1,610	1,583	871	936	860	336	563	620	713	446
Groundnut oil	1,576	1,499	1,508	1,898	1,193	1,319	964	833	857	762	707
Palm oil	1,102	1,262	1,036	961	810	730	290	528	632	422	307
Soybeans	444	542	466	486	411	327	247	218	229	195	210
Soybean meal	377	435	410	343	364	229	200	166	160	147	187
Soybean oil	1,082	1,250	1,141	1,246	830	834	447	526	590	414	335
Grains ($/mt)											
Grain sorghum	182	219	206	248	179	150	104	100	92	82	87
Maize	209	255	233	265	174	164	109	104	96	87	88
Rice	519	550	503	755	570	287	271	270	287	240	201
Wheat	280	275	219	330	240	198	136	149	119	108	113
Other food											
Bananas ($/mt)	692	735	659	546	526	554	541	374	461	361	420
Beef (cents/kg)	356	408	519	294	383	314	256	160	163	178	192
Oranges ($/mt)	927	755	669	504	542	581	531	447	417	417	360
Sugar, EU domestic											
(cents/kg)	59	58	45	75	68	51	58	58	56	57	55
Sugar, U.S. domestic											
(cents/kg)	61	63	66	110	92	65	51	43	46	45	42
Sugar, world (cents/kg)	33	22	33	100	88	13	28	25	19	13	18
Metals and minerals											
Aluminum ($/mt)	2,430	2,194	2,215	1,763	2,022	1,517	1,639	1,518	1,279	1,318	1,535
Copper ($/mt)	3,270	5,972	5,629	2,737	3,030	2,066	2,661	2,468	1,558	1,523	1,797
Iron ore (cents/dmtu)	55	47	39	38	39	39	32	24	29	27	29
Lead (cents/kg)	96	147	121	92	126	57	81	53	50	49	45
Nickel ($/mt)	7,881	8,032	11,339	10,111	9,053	7,141	8,864	6,918	4,362	5,819	8,558
Tin (cents/kg)	1,061	1,801	1,463	1,521	2,330	1,682	609	522	522	523	539
Zinc (cents/kg)	119	144	118	164	106	114	151	87	97	104	112

a. Series not included in the non-energy index.

Definitions

• **Non-energy commodity price index** covers the 31 non-energy primary commodities that make up the agriculture, fertilizer, and metals and minerals indexes. • **Agriculture,** in addition to beverages, food, and agricultural raw materials, includes sugar, bananas, beef, and oranges. • **Beverages** include cocoa, coffee, and tea. • **Food** includes rice, wheat, maize, sorghum, soybeans, soybean oil, soybean meal, palm oil, coconut oil, and groundnut oil. • **Agricultural raw materials** include timber (logs and sawnwood), cotton, natural rubber, and tobacco. • **Fertilizers** include phosphate rock and triple superphosphate (TSP). • **Metals and minerals** include aluminum, copper, iron ore, lead, nickel, tin, and zinc. • **Petroleum price index**

refers to the average spot price of Brent, Dubai, and West Texas Intermediate crude oil, equally weighted. • **Steel products price index** is the composite price index for eight steel products based on quotations f.o.b. (free on board) Japan excluding shipments to China and the United States, weighted by product shares of apparent combined consumption (volume of deliveries) for Germany, Japan, and the United States. • **MUV G-5 index** is the manufactures unit value index for G-5 country exports to low- and middle-income economies. • **Commodity prices**—for definitions and sources see "Commodity Price Data" (also known as the "Pink Sheet") at the Global Prospects Web site (www.worldbank.org/prospects).

Data sources

Commodity price data and the G-5 MUV index are compiled by the World Bank's Development Prospects Group. Monthly updates of commodity prices are available on the Web at www.worldbank.org/prospects.

6.5 Regional trade blocs

Exports within bloc

$ millions	1970	1980	1985	1990	1995	1996	1997	1998	1999
High-income and low- and middle-income economies									
APEC[a]	58,633	357,697	494,464	901,560	1,688,182	1,754,745	1,868,642	1,733,713	1,904,911
CEFTA	1,157	7,766	6,302	4,235	12,118	12,874	13,169	14,223	13,135
European Union	76,451	456,857	419,134	981,260	1,259,699	1,273,430	1,162,419	1,226,988	1,376,314
NAFTA	22,078	102,218	143,191	226,273	394,472	437,804	496,423	521,649	581,162
Latin America and the Caribbean									
ACS	758	4,892	4,123	5,401	10,448	10,894	11,870	12,260	12,002
Andean Group	97	1,161	768	1,312	4,812	4,692	5,627	5,427	4,012
CACM	287	1,174	544	671	1,595	1,723	1,973	1,988	2,102
CARICOM	52	576	414	448	305	906	971	1,017	1,089
Central American Group of Four	176	692	310	399	1,026	1,106	1,299	1,171	1,237
Group of Three	59	706	534	1,046	3,460	3,130	4,022	3,918	3,009
LAIA	1,263	10,981	7,139	12,331	35,299	37,949	45,018	42,860	35,152
MERCOSUR	451	3,424	1,953	4,127	14,199	17,075	20,772	20,352	15,313
OECS	..	8	10	29	38	32	35	34	36
Africa									
CEMAC	22	75	84	139	120	164	161	153	121
CEPGL	3	2	9	7	8	9	6	8	9
COMESA	412	616	466	963	1,386	1,610	1,545	1,480	1,403
Cross-Border Initiative	209	447	294	613	1,002	1,191	1,144	1,105	994
ECCAS	162	89	131	163	163	212	211	198	167
ECOWAS	86	692	1,026	1,533	2,088	2,527	2,487	2,638	2,687
Indian Ocean Commission	5	10	4	24	64	69	74	94	29
MRU	1	7	4	0	1	4	7	8	8
SADC	483	617	843	1,630	3,373	3,963	4,471	3,789	3,880
UDEAC	22	75	84	139	120	163	160	152	120
UEMOA	52	460	397	614	555	707	733	779	832
Middle East and Asia									
Arab Common Market	110	671	529	911	1,368	1,149	465	516	443
ASEAN	1,456	13,350	14,343	28,648	81,911	86,923	88,770	71,669	81,929
Bangkok Agreement	132	1,464	1,953	4,476	12,070	13,128	13,647	13,259	15,390
EAEC	9,197	98,532	126,030	282,351	637,029	651,803	673,285	551,555	621,606
ECO	63	1,165	2,447	1,243	4,746	4,773	4,929	4,052	3,820
GCC	156	4,632	3,101	6,906	6,832	7,624	8,110	7,210	7,175
SAARC	99	613	601	863	2,024	2,147	2,007	2,861	2,680
UMA	60	109	274	958	1,109	1,115	927	918	918

Note: Regional bloc memberships are as follows: **Asia Pacific Economic Cooperation (APEC),** Australia, Brunei Darussalam, Canada, Chile, China, Hong Kong (China), Indonesia, Japan, the Republic of Korea, Malaysia, Mexico, New Zealand, Papua New Guinea, Peru, the Philippines, the Russian Federation, Singapore, Taiwan (China), Thailand, the United States, and Vietnam; **Central European Free Trade Area (CEFTA),** Bulgaria, the Czech Republic, Hungary, Poland, Romania, the Slovak Republic, and Slovenia; **European Union (EU; formerly European Economic Community and European Community),** Austria, Belgium, Denmark, Finland, France, Germany, Greece, Ireland, Italy, Luxembourg, the Netherlands, Portugal, Spain, Sweden, and the United Kingdom; **North American Free Trade Area (NAFTA),** Canada, Mexico, and the United States; **Association of Caribbean States (ACS),** Antigua and Barbuda, the Bahamas, Barbados, Belize, Colombia, Costa Rica, Cuba, Dominica, the Dominican Republic, El Salvador, Grenada, Guatemala, Guyana, Haiti, Honduras, Jamaica, Mexico, Nicaragua, Panama, St. Kitts and Nevis, St. Lucia, St. Vincent and the Grenadines, Suriname, Trinidad and Tobago, and República Bolivariana de Venezuela; **Andean Group,** Bolivia, Colombia, Ecuador, Peru, and República Bolivariana de Venezuela; **Central American Common Market (CACM),** Costa Rica, El Salvador, Guatemala, Honduras, and Nicaragua; **Caribbean Community and Common Market (CARICOM),** Antigua and Barbuda, the Bahamas (part of the Caribbean Community but not of the Common Market), Barbados, Belize, Dominica, Grenada, Guyana, Jamaica, Montserrat, St. Kitts and Nevis, St. Lucia, St. Vincent and the Grenadines, Suriname, and Trinidad and Tobago; **Central American Group of Four,** El Salvador, Guatemala, Honduras, and Nicaragua; **Group of Three,** Colombia, Mexico, and República Bolivariana de Venezuela; **Latin American Integration Association (LAIA; formerly Latin American Free Trade Area),** Argentina, Bolivia, Brazil, Chile, Colombia, Ecuador, Mexico, Paraguay, Peru, Uruguay, and República Bolivariana de Venezuela; **Southern Cone Common Market (MERCOSUR),** Argentina, Brazil, Paraguay, and Uruguay; **Organization of Eastern Caribbean States (OECS),** Antigua and Barbuda, Dominica, Grenada, Montserrat, St. Kitts and Nevis, St. Lucia, and St. Vincent and the Grenadines; **Economic and Monetary Community of Central Africa (CEMAC),** Cameroon, the Central African Republic, Chad, the Republic of Congo, Equatorial Guinea, Gabon, and São Tomé and Principe; **Economic Community of the Countries of the Great Lakes (CEPGL),** Burundi, the Democratic Republic of the Congo, and Rwanda; **Common Market for Eastern and Southern Africa (COMESA),** Angola, Burundi, Comoros, the Democratic Republic of the Congo, Djibouti, the Arab Republic of Egypt, Eritrea, Ethiopia, Kenya, Madagascar, Malawi, Mauritius, Rwanda, Seychelles, Sudan, Swaziland, Uganda, Tanzania, Zambia, and Zimbabwe; **Cross-Border Initiative,** Burundi, Comoros, Kenya, Madagascar, Malawi, Mauritius, Namibia, Rwanda, Seychelles, Swaziland, Tanzania, Uganda, Zambia, and Zimbabwe; **Economic Community of Central African States (ECCAS),** Angola, Burundi, Cameroon, the Central African Republic, Chad, the Democratic Republic of the Congo, the Republic of Congo, Equatorial Guinea, Gabon, Rwanda, and São Tomé and Principe; **Economic Community of West African States (ECOWAS),** Benin, Burkina Faso, Cape Verde, Côte d'Ivoire, the Gambia, Ghana, Guinea, Guinea-Bissau, Liberia, Mali, Mauritania, Niger, Nigeria, Senegal, Sierra Leone, and Togo; **Indian Ocean Commission,** Comoros, Madagascar, Mauritius, and Seychelles; **Mano River Union (MRU),** Guinea, Liberia, and Sierra Leone; **Southern African Development Community (SADC; formerly Southern African Development Coordination Conference),** Angola, Botswana, the Democratic Republic of the Congo, Lesotho, Malawi, Mauritius, Mozambique, Namibia, Seychelles, South Africa, Swaziland, Tanzania, Zambia, and Zimbabwe; **Central African Customs and Economic Union**

Regional trade blocs | 6.5

Exports within bloc

% of total exports	1970	1980	1985	1990	1995	1996	1997	1998	1999
High-income and low- and middle-income economies									
APEC[a]	57.8	57.9	67.7	68.3	71.9	72.1	71.8	69.7	71.9
CEFTA	12.9	14.8	13.8	9.9	14.6	14.4	13.4	13.0	11.9
European Union	59.5	60.8	59.2	65.9	62.4	61.4	55.5	57.0	62.6
NAFTA	36.0	33.6	43.9	41.4	46.2	47.6	49.1	51.7	54.6
Latin America and the Caribbean									
ACS	9.6	8.7	7.9	8.4	8.1	7.1	7.0	7.1	5.9
Andean Group	1.8	3.8	3.2	4.1	12.0	10.3	11.8	13.9	9.3
CACM	26.0	24.4	14.4	15.4	21.7	22.0	18.1	15.6	11.6
CARICOM	4.2	5.3	6.3	8.1	4.6	13.3	14.4	17.3	15.3
Central American Group of Four	20.1	18.1	10.9	13.7	22.0	22.0	19.9	16.3	11.3
Group of Three	1.1	1.8	1.3	2.0	3.2	2.4	2.8	2.7	1.8
LAIA	9.9	13.7	8.3	10.8	17.1	16.3	17.4	16.9	13.0
MERCOSUR	9.4	11.6	5.5	8.9	20.3	22.7	24.8	25.0	20.5
OECS	..	9.1	6.4	8.1	11.7	9.1	9.6	10.4	7.3
Africa									
CEMAC	4.8	1.6	1.9	2.3	2.2	2.3	2.1	2.3	1.6
CEPGL	0.4	0.1	0.8	0.5	0.5	0.5	0.4	0.6	0.7
COMESA	9.1	6.1	4.7	6.6	7.8	8.0	7.9	8.6	7.5
Cross-Border Initiative	9.3	8.8	6.9	10.3	11.9	12.4	12.7	13.5	11.8
ECCAS	9.8	1.4	1.7	1.4	1.6	1.6	1.6	1.8	1.3
ECOWAS	2.9	10.1	5.2	7.8	9.8	9.4	9.7	11.8	12.2
Indian Ocean Commission	1.1	0.5	0.2	0.7	1.8	1.7	2.0	2.5	0.8
MRU	0.2	0.8	0.4	0.0	0.1	0.3	0.5	0.5	0.6
SADC	8.0	2.0	3.8	4.8	8.7	9.4	10.4	10.2	9.3
UDEAC	4.9	1.6	1.9	2.3	2.2	2.3	2.1	2.3	1.6
UEMOA	6.5	9.6	8.7	12.9	10.0	9.8	11.8	11.6	12.7
Middle East and Asia									
Arab Common Market	2.3	2.4	2.0	2.7	6.7	4.4	1.7	2.6	1.6
ASEAN	22.9	18.7	19.8	19.8	25.4	25.4	25.0	21.7	22.2
Bangkok Agreement	2.7	3.7	3.7	3.7	5.1	5.3	5.2	5.1	5.3
EAEC	28.9	35.6	34.3	39.9	48.3	49.3	48.3	42.2	44.1
ECO	1.5	5.4	9.9	3.2	7.9	7.1	7.5	7.0	6.3
GCC	2.9	3.0	4.9	8.0	6.8	6.4	6.5	8.0	6.8
SAARC	3.2	4.8	4.5	3.2	4.4	4.3	4.0	5.3	4.7
UMA	1.4	0.3	1.0	2.9	3.8	3.4	2.7	3.3	2.5

(UDEAC; formerly Union Douanière et Economique de l'Afrique Centrale), Cameroon, the Central African Republic, Chad, the Republic of Congo, Equatorial Guinea, and Gabon; **West African Economic and Monetary Union (UEMOA),** Benin, Burkina Faso, Côte d'Ivoire, Guinea-Bissau, Mali, Niger, Senegal, and Togo; **Arab Common Market,** the Arab Republic of Egypt, Iraq, Jordan, Libya, Mauritania, the Syrian Arab Republic, and the Republic of Yemen; **Association of South-East Asian Nations (ASEAN),** Brunei, Cambodia, Indonesia, the Lao People's Democratic Republic, Malaysia, Myanmar, the Philippines, Singapore, Thailand, and Vietnam; **Bangkok Agreement,** Bangladesh, India, the Republic of Korea, the Lao People's Democratic Republic, the Philippines, Sri Lanka, and Thailand; **East Asian Economic Caucus (EAEC),** Brunei, China, Hong Kong (China), Indonesia, Japan, the Republic of Korea, Malaysia, the Philippines, Singapore, Taiwan (China), and Thailand; **Economic Cooperation Organization (ECO),** Afghanistan, Azerbaijan, the Islamic Republic of Iran, Kazakhstan, the Kyrgyz Republic, Pakistan, Tajikistan, Turkey, Turkmenistan, and Uzbekistan; **Gulf Cooperation Council (GCC),** Bahrain, Kuwait, Oman, Qatar, Saudi Arabia, and the United Arab Emirates; **South Asian Association for Regional Cooperation (SAARC),** Bangladesh, Bhutan, India, Maldives, Nepal, Pakistan, and Sri Lanka; and **Arab Maghreb Union (UMA),** Algeria, Libya, Mauritania, Morocco, and Tunisia.
a. No preferential trade agreement.

Total exports by bloc

% of world exports	1970	1980	1985	1990	1995	1996	1997	1998	1999
High-income and low- and middle-income economies									
APEC[a]	36.0	33.7	38.9	39.0	46.3	46.0	47.2	46.1	46.9
CEFTA	3.2	2.9	2.4	1.3	1.6	1.7	1.8	2.0	1.9
European Union	45.6	41.0	37.8	44.0	39.8	39.2	38.0	39.9	38.9
NAFTA	21.7	16.6	17.4	16.2	16.8	17.4	18.4	18.7	18.8
Latin America and the Caribbean									
ACS	2.8	3.1	2.8	1.9	2.6	2.9	3.1	3.2	3.6
Andean Group	1.9	1.7	1.3	0.9	0.8	0.9	0.9	0.7	0.8
CACM	0.4	0.3	0.2	0.1	0.1	0.1	0.2	0.2	0.3
CARICOM	0.4	0.6	0.3	0.2	0.1	0.1	0.1	0.1	0.1
Central American Group of Four	0.3	0.2	0.2	0.1	0.1	0.1	0.1	0.1	0.2
Group of Three	1.8	2.1	2.1	1.5	2.1	2.4	2.6	2.7	3.0
LAIA	4.5	4.4	4.6	3.4	4.1	4.4	4.7	4.7	4.8
MERCOSUR	1.7	1.6	1.9	1.4	1.4	1.4	1.5	1.5	1.3
OECS	..	0.0	0.0	0.0	0.0	0.0	0.0	0.0	0.0
Africa									
CEMAC	0.2	0.3	0.2	0.2	0.1	0.1	0.1	0.1	0.1
CEPGL	0.3	0.1	0.1	0.0	0.0	0.0	0.0	0.0	0.0
COMESA	1.6	0.6	0.5	0.4	0.4	0.4	0.4	0.3	0.3
Cross-Border Initiative	0.8	0.3	0.2	0.2	0.2	0.2	0.2	0.2	0.1
ECCAS	0.6	0.3	0.4	0.3	0.2	0.3	0.2	0.2	0.2
ECOWAS	1.1	0.4	1.0	0.6	0.4	0.5	0.5	0.4	0.4
Indian Ocean Commission	0.2	0.1	0.1	0.1	0.1	0.1	0.1	0.1	0.1
MRU	0.1	0.0	0.1	0.1	0.0	0.0	0.0	0.0	0.0
SADC	2.2	1.6	1.2	1.0	0.8	0.8	0.8	0.7	0.7
UDEAC	0.2	0.3	0.2	0.2	0.1	0.1	0.1	0.1	0.1
UEMOA	0.3	0.3	0.2	0.1	0.1	0.1	0.1	0.1	0.1
Middle East and Asia									
Arab Common Market	1.7	1.5	1.4	1.0	0.4	0.5	0.5	0.4	0.5
ASEAN	2.3	3.9	3.9	4.3	6.4	6.5	6.5	6.1	6.5
Bangkok Agreement	1.8	2.2	2.8	3.6	4.7	4.7	4.7	4.9	5.1
EAEC	11.3	15.1	19.6	20.9	26.0	25.0	25.3	24.2	24.9
ECO	1.5	1.2	1.3	1.1	1.2	1.3	1.2	1.1	1.1
GCC	1.9	8.5	3.4	2.5	2.0	2.2	2.3	1.7	1.9
SAARC	1.1	0.7	0.7	0.8	0.9	0.9	0.9	1.0	1.0
UMA	1.5	2.3	1.5	1.0	0.6	0.6	0.6	0.5	0.6

About the data

Trade blocs are groups of countries that have established special preferential arrangements governing trade between members. Although in some cases the preferences—such as lower tariff duties or exemptions from quantitative restrictions—may be no greater than those available to other trading partners, the general purpose of such arrangements is to encourage exports by bloc members to one another—sometimes called intratrade. Asia Pacific Economic Cooperation (APEC), which has no preferential arrangements, is included because of the volume of trade between its members. The table shows the value of merchandise intratrade for important regional trade blocs (service exports are excluded) as well as the size of intratrade relative to each bloc's total exports of goods and the share of the bloc's total exports in world exports.

The data on country exports are drawn from the International Monetary Fund's (IMF) Direction of Trade database and should be broadly consistent with those from other sources, such as the United Nations Statistics Division's Commodity Trade (COMTRADE) database. However, trade flows between many developing countries, particularly in Africa, are not well recorded. Thus the value of intratrade for certain groups may be understated. Data on trade between developing and high-income countries are generally complete.

Membership in the trade blocs shown is based on the most recent information available, from the World Bank Policy Research Report *Trade Blocs* (2000d) and from consultation with the World Bank's international trade unit. Although bloc exports have been calculated back to 1970 on the basis of current membership, most of the blocs came into existence in later years and their membership may have changed over time. For this reason, and because systems of preferences also change over time, intratrade in earlier years may not have been affected by the same preferences as in recent years. In addition, some countries belong to more than one trade bloc, so shares of world exports exceed 100 percent. Exports of blocs include all commodity trade, which may include items not specified in trade bloc agreements. Differences from previously published estimates may be due to changes in bloc membership or to revisions in the underlying data.

Definitions

• **Exports within bloc** are the sum of exports by members of a trade bloc to other members of the bloc. They are shown both in U.S. dollars and as a percentage of total exports by the bloc. • **Total exports by bloc** as a share of world exports are the ratio of the bloc's total exports (within the bloc and to the rest of the world) to total exports by all economies in the world.

Data sources

Data on merchandise trade flows are published in the IMF's *Direction of Trade Statistics Yearbook* and *Direction of Trade Statistics Quarterly;* the data in the table were calculated using the IMF's Direction of Trade database. The United Nations Conference on Trade and Development (UNCTAD) publishes data on intratrade in its *Handbook of International Trade and Development Statistics*. The information on trade bloc membership is from the World Bank Policy Research Report *Trade Blocs* (2000d) and the World Bank's international trade unit.

6.6 | Tariff barriers

	Year	All products					Primary products		Manufactured products	
		Simple mean tariff %	Standard deviation of tariff rates %	Weighted mean tariff %	Share of lines with international peaks %	Share of lines with specific tariffs %	Simple mean tariff %	Weighted mean tariff %	Simple mean tariff %	Weighted mean tariff %
Algeria	1993	21.7	16.9	14.1	44.2	0.0	19.3	8.5	21.9	17.1
	1998	25.0	16.4	17.4	52.2	0.0	20.8	14.9	25.4	18.7
Argentina	1992	12.2	7.7	12.7	31.0	0.0	10.0	5.8	12.5	13.8
	1999	11.0	8.3	10.7	39.5	0.0	10.4	5.9	11.0	11.2
Australia	1991	8.0	12.9	7.4	15.5	1.1	1.9	1.4	8.8	8.4
	1999	5.7	7.3	3.8	12.2	0.7	1.7	1.0	6.2	4.2
Bangladesh	1989	106.6	79.3	88.4	98.5	1.0	79.9	53.5	110.5	112.2
	1999	22.0	20.3	22.0	53.5	0.0	22.4	13.3	22.0	24.3
Belarus	1996	12.2	8.7	8.8	30.9	0.0	9.6	6.4	13.0	10.5
	1997	13.0	8.3	9.5	31.9	0.0	10.4	7.0	13.8	11.2
Bolivia	1993	9.7	1.1	9.4	0.0	0.0	10.0	10.0	9.7	9.3
	1999	9.0	2.0	8.9	0.0	0.0	10.0	10.0	8.9	8.7
Brazil	1989	42.2	17.2	32.0	92.4	0.2	38.2	18.4	42.5	37.9
	1999	13.6	7.8	12.6	54.0	0.0	10.8	5.3	13.9	14.5
Cameroon	1994	19.4	10.5	14.0	54.7	0.0	24.6	14.5	18.7	13.8
	1995	59.3	29.6	61.4	83.9	0.0	62.1	66.2	58.8	59.8
Canada	1989	7.7	7.0	6.5	14.2	2.5	4.5	2.4	8.4	7.2
	1999	4.4	22.3	3.2	5.3	2.5	12.2	5.9	2.7	2.9
Central African Republic	1995	7.1	9.5	7.1	16.4	0.4	8.1	8.2	6.7	6.6
	1997	7.0	9.5	6.7	16.3	0.4	7.6	9.3	6.8	5.6
Chad	1995	15.8	10.9	16.3	44.8	0.0	17.0	23.1	15.5	13.5
	1997	15.8	10.9	16.3	44.8	0.0	17.0	23.1	15.5	13.5
Chile	1992	11.0	0.5	11.0	0.0	0.0	11.0	11.0	11.0	10.9
	1999	10.0	0.5	9.9	0.0	0.0	10.0	10.0	10.0	9.9
China[†]	1992	41.3	30.8	32.6	78.4	0.0	36.7	14.3	42.4	36.5
	1998	16.8	11.1	15.7	43.4	0.4	16.5	21.1	16.9	14.7
Hong Kong, China	1988	0.0	0.0	0.0	0.0	0.0	0.0	0.0	0.0	0.0
	1998	0.0	0.0	0.0	0.0	0.0	0.0	0.0	0.0	0.0
Colombia	1991	5.7	8.2	6.4	1.6	0.0	7.8	7.9	5.6	6.2
	1999	11.8	6.2	10.7	22.9	0.0	13.1	13.5	11.6	10.1
Congo, Rep.	1994	20.6	9.3	16.3	62.1	0.0	22.1	20.5	20.3	14.8
	1997	17.6	8.6	16.7	36.0	0.0	18.0	15.2	17.5	17.0
Costa Rica	1995	8.1	8.5	7.8	24.3	0.0	12.0	9.8	7.7	7.4
	1999	3.3	7.8	3.3	9.9	0.0	6.4	5.5	3.0	2.9
Côte d'Ivoire	1993	25.3	12.1	22.0	75.6	0.0	26.8	21.3	25.1	22.6
	1996	19.2	10.7	14.2	53.2	0.0	21.2	14.6	18.8	14.1
Cuba	1993	13.2	8.0	10.1	25.9	0.0	13.8	8.5	13.1	11.6
	1997	11.4	6.6	8.2	9.6	0.0	11.3	5.3	11.4	9.8
Czech Republic	1996	7.0	6.4	5.8	5.7	0.0	8.2	4.1	6.7	6.3
	1999	6.8	20.2	5.8	5.4	0.0	12.3	5.3	5.4	5.9
Ecuador	1993	8.7	6.0	8.2	20.7	0.0	9.7	6.3	8.6	8.4
	1999	12.9	6.3	11.3	37.0	0.0	13.9	11.3	12.9	11.3
Egypt, Arab Rep.	1995	25.6	33.2	16.7	53.1	1.2	24.5	7.6	25.8	22.4
	1998	20.5	39.5	13.7	47.4	9.5	22.7	7.4	20.2	17.5
El Salvador	1995	10.3	7.8	9.3	27.8	0.0	13.3	10.9	9.9	8.8
	1998	6.7	8.2	6.5	18.6	1.2	11.5	8.8	6.0	5.6
European Union	1988	3.2	5.6	3.6	3.6	13.8	7.9	2.8	1.9	4.1
	1999	3.5	5.0	2.7	2.3	12.6	6.4	1.8	2.7	3.0
Gabon	1995	20.4	9.6	16.1	60.6	0.0	23.4	20.0	19.7	15.1
	1998	20.6	9.8	16.2	62.5	0.3	25.1	21.5	19.7	14.7
Guatemala	1995	10.0	7.5	8.6	25.7	0.0	13.1	10.6	9.6	8.1
	1998	8.0	8.8	6.9	25.1	0.0	10.4	10.4	7.6	6.0
Honduras	1995	9.7	7.5	8.4	25.3	0.0	13.2	12.3	9.3	7.6
	1999	8.1	7.8	7.6	25.5	0.0	12.2	12.5	7.5	6.3
† Data for Taiwan, China	1989	12.3	9.5	9.9	16.7	0.5	17.5	8.0	11.2	10.7
	1999	8.8	9.4	5.2	10.6	2.1	16.0	7.4	7.1	4.6

Tariff barriers | 6.6

		All products					Primary products		Manufactured products	
	Year	Simple mean tariff %	Standard deviation of tariff rates %	Weighted mean tariff %	Share of lines with international peaks %	Share of lines with specific tariffs %	Simple mean tariff %	Weighted mean tariff %	Simple mean tariff %	Weighted mean tariff %
India	1990	79.1	43.8	49.8	97.0	0.9	69.6	26.0	80.3	69.9
	1999	32.2	12.4	29.5	93.5	0.6	30.5	24.9	32.4	32.3
Indonesia	1989	21.9	19.7	13.2	50.5	0.3	20.4	5.8	22.3	15.6
	1999	10.9	14.1	6.2	26.9	0.1	11.9	2.8	10.7	7.4
Jamaica	1996	21.3	8.8	19.8	45.1	42.0	24.2	17.5	20.7	20.9
	1999	17.9	8.4	18.1	35.3	43.3	21.4	16.6	16.8	18.8
Japan	1988	5.9	8.0	3.3	8.6	11.5	10.4	3.7	4.7	2.9
	1999	4.8	7.3	2.3	7.6	2.6	8.6	4.0	3.4	1.4
Korea, Rep.	1988	18.8	8.1	13.8	72.8	10.2	19.8	8.1	18.7	16.9
	1999	8.7	5.9	5.9	4.8	0.8	12.8	5.5	7.8	6.1
Latvia	1996	4.3	7.5	2.2	2.2	0.0	8.3	1.5	3.2	2.6
	1997	5.6	9.2	3.2	3.0	0.0	10.0	4.0	4.2	2.9
Lithuania	1995	3.9	8.6	2.7	7.0	0.0	9.2	4.1	2.4	1.7
	1997	3.9	8.0	2.4	6.5	0.0	7.6	3.3	2.9	1.9
Malawi	1994	31.3	14.6	22.3	87.0	0.0	27.7	12.8	31.7	26.6
	1998	15.7	14.5	10.0	14.0	49.5	15.6	4.8	15.7	11.8
Malaysia	1988	20.6	19.9	13.8	54.0	6.2	21.9	12.1	20.3	14.3
	1997	7.1	31.0	4.9	15.9	0.4	6.0	7.3	7.5	4.6
Mauritius	1995	22.3	23.4	16.2	49.2	0.0	16.1	22.4	23.1	13.9
	1998	19.0	22.2	15.7	41.1	0.0	14.9	11.3	19.5	16.9
Mexico	1991	13.2	4.3	11.9	18.9	0.0	12.2	8.2	13.3	13.1
	1999	10.1	9.4	14.7	24.5	0.0	11.5	20.7	10.0	14.0
Morocco	1993	66.5	29.5	45.3	96.8	0.1	55.0	29.7	68.1	55.8
	1997	22.1	19.3	21.1	61.8	0.0	28.9	26.0	21.3	19.6
Mozambique	1994	5.0	0.0	5.0	0.0	0.0	5.0	5.0	5.0	5.0
	1997	16.9	14.3	17.5	38.1	0.0	21.1	22.2	16.2	15.6
Nepal	1993	24.0	24.4	19.4	50.6	6.6	15.7	14.2	26.0	21.9
	1999	17.7	21.0	18.0	18.7	7.3	12.9	14.5	18.9	19.7
New Zealand	1992	10.5	11.0	8.5	36.3	2.7	6.2	3.8	11.2	9.6
	1999	2.8	4.5	3.3	4.0	5.4	1.5	1.5	3.0	3.6
Nicaragua	1995	5.0	7.4	3.9	13.3	0.0	5.6	4.6	4.9	3.6
	1999	11.0	7.3	10.9	24.3	0.0	16.4	14.6	10.3	9.6
Nigeria	1988	26.0	16.7	23.8	62.9	0.4	33.3	32.3	25.2	21.4
	1995	21.8	15.7	20.0	9.7	80.5	29.5	20.8	20.2	19.9
Norway	1988	1.6	4.9	0.7	4.6	5.9	0.8	0.1	1.7	0.8
	1998	2.9	16.6	1.1	4.3	8.8	9.9	2.7	2.1	0.8
Oman	1992	5.5	8.2	7.4	1.5	0.0	7.2	14.1	5.2	5.5
	1997	4.8	0.9	4.7	0.0	0.0	4.0	3.0	4.9	5.0
Pakistan	1995	50.7	21.7	46.3	91.6	3.8	44.5	22.2	51.4	51.0
	1998	46.5	21.2	41.7	87.1	3.3	42.7	26.2	46.9	44.4
Paraguay	1991	15.7	11.4	12.6	42.3	0.0	15.4	4.5	15.8	14.5
	1999	9.0	7.4	6.9	28.4	0.0	10.2	9.1	9.0	6.2
Peru	1993	17.4	4.2	15.9	23.5	0.0	18.8	15.6	17.2	16.1
	1999	13.0	2.6	12.7	12.1	0.0	13.9	13.7	12.9	12.3
Philippines	1988	28.0	14.2	22.5	77.2	0.1	29.8	18.4	27.7	23.6
	1999	10.0	8.8	6.7	24.1	0.0	14.2	12.3	9.3	6.0
Poland	1991	15.3	10.6	14.9	37.8	0.0	16.1	19.2	15.2	13.5
	1996	13.1	23.8	8.6	21.6	5.6	26.5	10.9	9.4	7.9
Romania	1991	19.2	8.3	11.9	55.6	0.0	20.1	8.2	19.0	18.2
	1999	13.1	14.1	9.2	26.9	0.0	23.6	10.7	10.7	8.8
Russian Federation	1993	7.8	9.9	6.3	3.3	0.0	3.4	3.9	9.4	7.5
	1997	13.9	8.5	11.3	35.5	0.0	11.5	10.3	14.8	11.8
Saudi Arabia	1994	12.5	3.3	10.7	10.2	0.1	12.1	9.1	12.6	11.0
	1999	12.6	3.8	10.8	11.0	0.2	12.2	11.5	12.6	10.7
Singapore	1989	0.5	2.2	0.5	0.1	1.1	0.2	0.0	0.6	0.6
	1995	0.0	0.0	0.0	0.0	0.2	0.0	0.0	0.0	0.0

	Year	All products					Primary products		Manufactured products	
		Simple mean tariff %	Standard deviation of tariff rates %	Weighted mean tariff %	Share of lines with international peaks %	Share of lines with specific tariffs %	Simple mean tariff %	Weighted mean tariff %	Simple mean tariff %	Weighted mean tariff %
South Africa[a]	1988	12.7	11.8	12.0	32.3	18.8	6.3	4.3	12.9	12.4
	1999	8.5	10.2	4.4	22.0	20.5	8.0	1.6	8.6	5.1
Sri Lanka	1990	28.3	24.5	26.9	51.5	1.4	31.8	32.2	27.9	24.3
	1997	20.1	14.3	22.5	47.0	0.4	23.9	26.5	19.7	21.4
Switzerland	1990	0.0	0.0	0.0	0.0	40.1	0.0	0.0	0.0	0.0
	1999	0.0	0.0	0.0	0.0	25.5	0.0	0.0	0.0	0.0
Tanzania	1993	14.4	10.7	15.6	42.6	0.0	22.4	20.0	13.7	14.8
	1998	21.0	13.3	19.5	69.2	0.4	27.9	19.3	20.4	19.6
Thailand	1989	38.5	19.6	33.0	72.8	21.9	30.6	24.2	39.6	35.7
	1995	21.6	15.4	15.0	57.6	1.8	25.6	11.1	21.2	15.7
Trinidad and Tobago	1991	19.9	14.9	12.9	40.3	0.0	26.9	10.5	18.7	14.2
	1999	18.4	8.3	17.0	36.5	46.0	21.4	17.6	17.8	16.7
Tunisia	1990	28.4	10.0	26.6	97.3	0.0	25.1	18.6	28.7	28.6
	1998	30.1	13.1	28.9	90.5	0.0	29.6	21.4	30.2	30.2
Turkey	1993	7.0	4.9	6.0	5.4	0.0	6.0	8.1	7.2	5.2
	1997	8.2	12.8	5.7	8.3	0.3	23.2	5.2	5.9	5.8
Ukraine	1995	9.1	9.4	9.6	14.6	0.0	13.3	16.2	7.6	6.4
	1997	10.5	10.9	5.2	24.0	0.0	17.5	3.4	8.2	7.2
United States	1989	5.7	6.7	4.1	8.0	13.0	3.9	2.1	6.1	4.5
	1999	4.3	11.4	2.8	6.3	8.3	4.8	2.0	4.2	2.9
Uruguay	1992	7.5	5.8	5.8	0.0	0.0	8.4	5.9	7.5	5.8
	1999	4.6	4.3	4.7	0.0	0.0	4.2	3.2	4.7	5.2
Venezuela, RB	1992	15.7	11.4	16.4	47.4	1.0	17.7	14.5	15.5	16.7
	1999	12.6	5.9	13.1	25.4	0.0	13.4	14.3	12.6	12.8
Vietnam	1994	12.7	17.8	18.6	32.4	1.0	20.6	50.8	12.0	13.0
	1999	15.1	17.7	17.3	37.3	0.6	21.5	32.4	14.4	14.9
Zambia	1993	25.2	11.0	17.9	90.9	0.0	29.5	12.4	24.5	20.0
	1997	14.6	8.8	13.0	31.4	0.0	16.9	13.9	14.4	12.9
Zimbabwe	1996	40.8	15.0	38.1	94.4	1.5	34.2	32.1	41.4	38.9
	1998	22.2	17.9	17.5	46.5	0.0	27.0	23.8	21.7	16.7

a. Data refer to the South African Customs Union (Botswana, Lesotho, Namibia, South Africa, and Swaziland).

Tariff barriers | 6.6

Economies regulate their imports through a combination of tariff and nontariff measures. The most common form of tariff is an ad valorem duty, based on the value of the import, but tariffs may also be levied on a specific, or per unit, basis or may combine ad valorem and specific rates. Tariffs may be used to raise fiscal revenues or to protect domestic industries from foreign competition—or both. Nontariff barriers, which limit the quantity of imports of a particular good, take many forms. Some common ones are quotas, prohibitions, licensing schemes, export restraint arrangements, and health and quarantine measures.

Nontariff barriers are generally considered less desirable than tariffs because changes in an exporting country's efficiency and costs no longer result in changes in market share in the importing country. Further, the quotas or licenses that regulate trade become very valuable and resources are frequently wasted in attempts to acquire these assets. A high percentage of products subject to nontariff barriers suggests a protectionist trade regime, but the frequency of nontariff barriers does not measure how much they restrict trade. Moreover, a wide range of domestic policies and regulations (such as health regulations) may act as nontariff barriers. Because of the difficulty of combining nontariff barriers into an aggregate indicator, they are not included in this table.

The table shows new data on average tariffs, the dispersion of tariff rates, the proportion of tariff lines with duties exceeding 15 percent, and the proportion of lines subject to specific tariffs. The rates used in calculating the indicators here are effectively applied rates, which reflect the rates actually applied to partners in preferential trade agreements such as the North American Free Trade Agreement. Countries typically maintain a hierarchy of trade preferences applicable to specific trading partners. In previous years the indicators were based on most-favored-nation rates, which are equal to or higher than effectively applied rates.

Two measures of average tariffs are shown: the simple and the weighted mean tariff. Weighted mean tariffs are weighted by the value of the country's trade with each of its trading partners. Simple averages are frequently a better indicator of tariff protection than weighted averages, which are biased downward because higher tariffs discourage trade and reduce the weights applied to these tariffs. Specific duties—duties not expressed as a proportion of the declared value—have not been included in this year's table, but work is under way to estimate ad valorem equivalents.

Some countries set fairly uniform tariff rates across all imports. Others are more selective, setting high tariffs to protect favored domestic industries. The standard deviation of tariffs is a measure of the dispersion of tariff rates around their mean value. Highly dispersed rates increase the costs of protection substantially. But these nominal tariff rates tell only part of the story. The effective rate of protection—the degree to which the value added in an industry is protected—may exceed the nominal rate if the tariff system systematically differentiates among imports of raw materials, intermediate products, and finished goods.

Two other measures of tariff coverage are shown: the share of tariff lines with international peaks (those for which ad valorem tariff rates exceed 15 percent) and the share of tariff lines with specific duties (those not covered by ad valorem rates). Some countries—for example, Switzerland—apply only specific duties.

The indicators in this table were calculated from data supplied by the United Nations Conference on Trade and Development (UNCTAD). Data are classified using the Harmonized System of trade at the six- or eight-digit level. Tariff line data were matched to Standard International Trade Classification (SITC) revision 2 codes to define the commodity groups and import weights. Import weights were calculated for 1995 using the United Nations Statistics Division's Commodity Trade (COMTRADE) database. Data are shown only for the first and last year for which complete data are available. To conserve space, countries for which only a single year is available and countries that are members of the European Union have not been included. Data for the whole of the European Union are shown.

• **Primary products** are commodities classified in SITC revision 2 sections 0–4 plus division 68 (nonferrous metals). • **Manufactured products** are commodities classified in SITC revision 2 sections 5–9, excluding division 68. • **Simple mean tariff** is the unweighted average of the effectively applied rates for all products subject to tariffs. • **Standard deviation of tariff rates** measures the average dispersion of tariff rates around the simple mean. • **Weighted mean tariff** is the average of effectively applied rates weighted by the product import shares corresponding to each partner country. • **International peaks** are tariff rates that exceed 15 percent. • **Specific tariffs** are tariffs that are set on a per unit basis or that combine ad valorem and per unit rates.

All indicators in this table were calculated by World Bank staff using the World Integrated Trade Solution (WITS) system. Tariff data were provided by UNCTAD. Data on global imports come from the United Nations Statistics Division's COMTRADE database.

6.7 Global financial flows

	Net private capital flows		Foreign direct investment		Portfolio investment flows				Bank and trade-related lending	
					Bonds $ millions		Equity $ millions			
	$ millions		$ millions						$ millions	
	1990	1999	1990	1999	1990	1999	1990	1999	1990	1999
Albania	31	37	0	41	0	0	0	0	31	–4
Algeria	–424	–1,486	0	7	–16	0	0	3	–409	–1,496
Angola	235	2,373	–335	2,471	0	0	0	0	570	–98
Argentina	–203	32,296	1,836	23,929	–857	8,000	13	404	–1,195	–37
Armenia	..	122	0	122	..	0	..	0	..	0
Australia	..	..	7,465	5,655	..	..	..	..	..	..
Austria	..	..	653	2,834	..	..	..	..	..	..
Azerbaijan	..	596	0	510	..	0	..	0	..	86
Bangladesh	70	198	3	179	0	0	0	4	67	15
Belarus	..	394	0	225	..	0	..	0	..	169
Belgium[a]	..	..	8,047	38,392	..	..	..	..	..	..
Benin	1	31	1	31	0	0	0	0	0	0
Bolivia	3	1,016	27	1,016	0	0	0	0	–24	0
Bosnia and Herzegovina	..	0	..	0	..	0	..	0	..	0
Botswana	77	36	95	37	0	0	0	0	–19	–1
Brazil	563	22,793	989	32,659	129	2,683	0	1,961	–555	–14,510
Bulgaria	–67	1,112	4	806	65	18	0	102	–136	186
Burkina Faso	–1	10	0	10	0	0	0	0	–1	0
Burundi	–5	0	1	0	0	0	0	0	–6	0
Cambodia	0	122	0	126	0	0	0	0	0	–3
Cameroon	–125	–13	–113	40	0	0	0	0	–12	–53
Canada	..	..	7,581	25,129	..	..	..	..	..	..
Central African Republic	0	13	1	13	0	0	0	0	–1	0
Chad	–1	14	0	15	0	0	0	0	–1	–1
Chile	2,098	11,851	590	9,221	–7	862	320	18	1,194	1,750
China	8,107	40,632	3,487	38,753	–48	660	0	3,732	4,668	–2,514
Hong Kong, China	..	..	..	..	..	..	..	..	..	..
Colombia	345	3,635	500	1,109	–4	1,235	0	25	–151	1,267
Congo, Dem. Rep.	–24	1	–12	1	0	0	0	0	–12	0
Congo, Rep.	–100	5	0	5	0	0	0	0	–100	0
Costa Rica	23	924	163	669	–42	283	0	0	–99	–28
Côte d'Ivoire	57	74	48	350	–1	–46	0	8	10	–238
Croatia	..	2,392	0	1,408	..	539	..	0	..	444
Cuba	..	..	..	..	..	..	..	..	..	..
Czech Republic	876	4,837	207	5,093	0	175	0	500	669	–932
Denmark	..	..	1,132	8,482	..	..	..	..	..	..
Dominican Republic	130	1,404	133	1,338	0	–4	0	0	–3	70
Ecuador	183	944	126	690	0	–19	0	0	57	273
Egypt, Arab Rep.	682	1,558	734	1,065	–1	100	0	550	–51	–157
El Salvador	8	360	2	231	0	150	0	0	6	–21
Eritrea	..	0	0	0	..	0	..	0	..	0
Estonia	..	569	0	305	..	45	..	191	..	28
Ethiopia	–45	78	12	90	0	0	0	0	–57	–12
Finland	..	..	812	4,754	..	..	..	..	..	..
France	..	..	13,183	38,828	..	..	..	..	..	..
Gabon	103	209	74	200	0	0	0	0	29	9
Gambia, The	–8	14	0	14	0	0	0	0	–8	0
Georgia	..	86	0	82	..	0	..	0	..	4
Germany	..	..	2,532	52,232	..	..	..	..	..	..
Ghana	–5	–16	15	17	0	0	0	19	–20	–52
Greece	..	..	1,005	984	..	..	..	..	..	..
Guatemala	44	98	48	155	–11	–31	0	0	7	–26
Guinea	–1	63	18	63	0	0	0	0	–19	0
Guinea-Bissau	2	3	2	3	0	0	0	0	0	0
Haiti	8	30	8	30	0	0	0	0	0	0
Honduras	76	251	44	230	0	0	0	0	32	21

Global financial flows | 6.7

	Net private capital flows		Foreign direct investment		Portfolio investment flows				Bank and trade-related lending	
					Bonds $ millions		Equity $ millions			
	$ millions		$ millions						$ millions	
	1990	1999	1990	1999	1990	1999	1990	1999	1990	1999
Hungary	−308	4,961	0	1,950	921	605	150	592	−1,379	1,813
India	1,872	1,813	162	2,169	147	−1,126	105	1,302	1,458	−532
Indonesia	3,235	−8,416	1,093	−2,745	26	−1,458	312	1,273	1,804	−5,486
Iran, Islamic Rep.	−392	−1,385	−362	85	0	0	0	0	−30	−1,470
Iraq	..	..	..	..	..	..	..	..	..	..
Ireland	..	..	627	19,091	..	..	..	..	..	..
Israel	..	..	151	2,363	..	..	..	..	..	..
Italy	..	..	6,411	6,783	..	..	..	..	..	..
Jamaica	92	425	138	524	0	−65	0	0	−46	−33
Japan	..	..	1,777	12,308	..	..	..	..	..	..
Jordan	254	112	38	158	0	−9	0	11	216	−48
Kazakhstan	..	1,477	0	1,587	..	−200	..	0	..	90
Kenya	122	−51	57	14	0	0	0	5	65	−70
Korea, Dem. Rep.	..	..	..	..	..	..	..	..	..	..
Korea, Rep.	1,038	6,409	788	9,333	151	−1,414	518	12,426	−418	−13,935
Kuwait	..	..	..	72	..	..	..	..	..	..
Kyrgyz Republic	..	−16	0	36	..	0	..	0	..	−52
Lao PDR	6	79	6	79	0	0	0	0	0	0
Latvia	..	303	0	348	..	240	..	0	..	−285
Lebanon	12	1,771	6	250	0	−114	0	3	6	1,632
Lesotho	17	168	17	163	0	0	0	0	0	5
Libya	..	..	..	..	..	..	..	..	..	..
Lithuania	..	1,148	0	487	..	505	..	0	..	156
Macedonia, FYR	..	51	0	30	..	0	..	0	..	21
Madagascar	7	52	22	58	0	0	0	0	−15	−6
Malawi	2	60	0	60	0	0	0	0	2	0
Malaysia	770	3,247	2,333	1,553	−1,239	747	293	522	−617	426
Mali	−8	19	−7	19	0	0	0	0	−1	0
Mauritania	6	0	7	2	0	0	0	0	−1	−2
Mauritius	86	102	41	49	0	0	0	6	45	47
Mexico	8,253	26,780	2,634	11,786	661	5,621	563	1,129	4,396	8,244
Moldova	..	12	0	34	..	0	..	0	..	−22
Mongolia	..	28	0	30	..	0	..	0	..	−3
Morocco	341	−118	165	3	0	−35	0	91	176	−177
Mozambique	35	374	9	384	0	0	0	0	26	−10
Myanmar	153	203	161	216	0	0	0	0	−8	−14
Namibia	..	..	..	..	..	..	..	..	..	..
Nepal	−8	−8	6	4	0	0	0	0	−14	−13
Netherlands	..	..	10,676	34,154	..	..	..	..	..	..
New Zealand	..	..	1,735	745	..	..	..	..	..	..
Nicaragua	20	382	0	300	0	0	0	0	20	82
Niger	9	−8	−1	15	0	0	0	0	10	−23
Nigeria	467	860	588	1,005	0	0	0	2	−121	−146
Norway	..	..	1,003	3,597	..	..	..	..	..	..
Oman	−259	−413	141	60	0	0	0	11	−400	−484
Pakistan	181	−66	244	530	0	−75	0	0	−63	−521
Panama	127	685	132	22	−2	381	0	0	−4	282
Papua New Guinea	204	499	155	297	0	0	0	232	49	−30
Paraguay	67	109	76	72	0	0	0	0	−9	38
Peru	59	3,140	41	1,969	0	−255	0	289	18	1,138
Philippines	639	4,915	530	573	395	3,895	0	422	−286	25
Poland	71	10,452	89	7,270	0	1,096	0	721	−18	1,365
Portugal	..	..	2,610	1,112	..	..	..	..	..	..
Puerto Rico	..	..	..	..	..	..	..	..	..	..
Romania	4	714	0	1,041	0	−681	0	0	4	355
Russian Federation	5,556	3,780	0	3,309	310	0	0	644	5,246	−173

	Net private capital flows $ millions		Foreign direct investment $ millions		Portfolio investment flows				Bank and trade-related lending $ millions	
					Bonds $ millions		Equity $ millions			
	1990	1999	1990	1999	1990	1999	1990	1999	1990	1999
Rwanda	6	2	8	2	0	0	0	0	−2	0
Saudi Arabia	..	..	..	..	..	..	..	..	..	..
Senegal	42	54	57	60	0	0	0	0	−15	−6
Sierra Leone	36	1	32	1	0	0	0	0	4	0
Singapore	..	..	5,575	6,984	..	..	..	..	..	..
Slovak Republic	278	281	0	354	0	415	0	0	278	−488
Slovenia	..	..	..	181	..	..	..	..	..	..
South Africa	..	4,533	−89	1,376	..	234	..	3,855	..	−932
Spain	..	..	13,984	9,321	..	..	..	..	..	..
Sri Lanka	53	109	43	177	0	0	0	6	10	−74
Sudan	0	371	0	371	0	0	0	0	0	0
Sweden	..	..	1,982	59,386	..	..	..	..	..	..
Switzerland	..	..	5,987	9,944	..	..	..	..	..	..
Syrian Arab Republic	18	87	71	91	0	0	0	0	−53	−4
Tajikistan	..	10	0	24	..	0	..	0	..	−14
Tanzania	4	171	0	183	0	0	0	0	4	−13
Thailand	4,399	2,471	2,444	6,213	−87	−1,358	449	2,527	1,593	−4,911
Togo	0	30	0	30	0	0	0	0	0	0
Trinidad and Tobago	−69	713	109	633	−52	230	0	0	−126	−150
Tunisia	−121	739	76	350	−60	240	0	0	−137	149
Turkey	1,782	8,667	684	783	597	3,223	35	800	466	3,861
Turkmenistan	..	−54	0	80	..	0	..	0	..	−134
Uganda	16	221	0	222	0	0	0	0	16	−1
Ukraine	..	371	0	496	..	187	..	0	..	−311
United Arab Emirates	..	..	..	..	..	..	..	..	..	..
United Kingdom	..	..	32,518	84,812	..	..	..	..	..	..
United States	..	..	48,497	275,535	..	..	..	..	..	..
Uruguay	−192	65	0	229	−16	−137	0	0	−176	−26
Uzbekistan	..	658	0	113	..	0	..	0	..	545
Venezuela, RB	−126	3,130	451	3,187	345	134	0	67	−922	−258
Vietnam	16	828	16	1,609	0	0	0	0	0	−781
West Bank and Gaza	..	..	..	..	..	..	..	..	..	..
Yemen, Rep.	30	−150	−131	−150	0	0	0	0	161	0
Yugoslavia, FR (Serb./Mont.)	−837	0	67	0	0	0	0	0	−904	0
Zambia	194	151	203	163	0	0	0	0	−9	−12
Zimbabwe	85	70	−12	59	−30	−30	0	4	127	37
World	.. s	.. s	200,479 s	884,452 s	.. s	.. s	.. s	.. s	.. s	.. s
Low income	6,630	2,083	2,201	9,830	142	−2,548	417	2,616	3,870	−7,816
Middle income	36,030	216,992	22,064	175,577	1,018	27,993	2,341	31,839	10,606	−18,418
Lower middle income	20,673	83,086	9,584	66,214	1,099	8,126	484	13,289	9,506	−4,542
Upper middle income	15,357	133,906	12,480	109,364	−81	19,868	1,857	18,550	1,100	−13,875
Low & middle income	43,645	219,076	24,265	185,408	1,160	25,446	3,743	34,456	14,476	−26,233
East Asia & Pacific	19,405	51,062	11,135	56,041	−802	1,072	2,290	21,133	6,782	−27,184
Europe & Central Asia	7,667	43,164	1,051	26,534	1,893	6,167	235	3,550	4,488	6,914
Latin America & Carib.	12,626	111,367	8,188	90,352	101	19,067	1,111	3,893	3,226	−1,945
Middle East & N. Africa	399	979	2,504	1,461	−148	182	0	669	−1,957	−1,333
South Asia	2,173	2,054	464	3,070	147	−1,201	105	1,312	1,457	−1,127
Sub-Saharan Africa	1,374	10,449	923	7,949	−31	158	2	3,899	480	−1,558
High income	..	..	176,213	699,045	..	..	..	..	..	..
Europe EMU	..	..	59,535	207,501	..	..	..	..	..	..

a. Includes Luxembourg.

About the data

The data on foreign direct investment are based on balance of payments data reported by the International Monetary Fund (IMF), supplemented by data on net foreign direct investment reported by the Organisation for Economic Co-operation and Development (OECD) and official national sources. The internationally accepted definition of foreign direct investment is that provided in the fifth edition of the IMF's *Balance of Payments Manual* (1993).

Under this definition foreign direct investment has three components: equity investment, reinvested earnings, and short- and long-term intercompany loans between parent firms and foreign affiliates. However, many countries fail to report reinvested earnings, and the definition of long-term loans differs among countries. Foreign direct investment, as distinguished from other kinds of international investment, is made to establish a lasting interest in or effective management control over an enterprise in another country. As a guideline, the IMF suggests that investments should account for at least 10 percent of voting stock to be counted as foreign direct investment. In practice, many countries set a higher threshold.

The OECD has also published a definition, in consultation with the IMF, Eurostat, and the United Nations. Because of the multiplicity of sources and differences in definitions and reporting methods, there may be more than one estimate of foreign direct investment for a country and data may not be comparable across countries.

Foreign direct investment data do not give a complete picture of international investment in an economy. Balance of payments data on foreign direct investment do not include capital raised locally, which has become an important source of financing for investment projects in some developing countries. In addition, foreign direct investment data capture only cross-border investment flows involving equity participation and thus omit non-equity cross-border transactions such as intrafirm flows of goods and services. For a detailed discussion of the data issues see the World Bank's *World Debt Tables 1993–94* (volume 1, chapter 3).

Portfolio flow data are compiled from several official and market sources, including Euromoney databases and publications, Micropal, Lipper Analytical Services, published reports of private investment houses, central banks, national securities and exchange commissions, national stock exchanges, and the World Bank's Debtor Reporting System.

Gross statistics on international bond and equity issues are produced by aggregating individual trans-actions reported by market sources. Transactions of public and publicly guaranteed bonds are reported through the Debtor Reporting System by World Bank member economies that have received either loans from the International Bank for Reconstruction and Development or credits from the International Development Association. Information on private nonguaranteed bonds is collected from market sources, because official national sources reporting to the Debtor Reporting System are not asked to report the breakdown between private nonguaranteed bonds and private nonguaranteed loans. Information on transactions by nonresidents in local equity markets is gathered from national authorities, investment positions of mutual funds, and market sources.

The volume of portfolio investment reported by the World Bank generally differs from that reported by other sources because of differences in the classification of economies, in the sources, and in the method used to adjust and disaggregate reported information. Differences in reporting arise particularly for foreign investments in local equity markets because clarity, adequate disaggregation, and comprehensive and periodic reporting are lacking in many developing economies. By contrast, capital flows through international debt and equity instruments are well recorded, and for these the differences in reporting lie primarily in the classification of economies, the exchange rates used, whether particular tranches of the transactions are included, and the treatment of certain offshore issuances.

Definitions

• **Net private capital flows** consist of private debt and nondebt flows. Private debt flows include commercial bank lending, bonds, and other private credits; nondebt private flows are foreign direct investment and portfolio equity investment. • **Foreign direct investment** is net inflows of investment to acquire a lasting management interest (10 percent or more of voting stock) in an enterprise operating in an economy other than that of the investor. It is the sum of equity capital, reinvestment of earnings, other long-term capital, and short-term capital, as shown in the balance of payments. • **Portfolio investment flows** are net and include non-debt-creating portfolio equity flows (the sum of country funds, depository receipts, and direct purchases of shares by foreign investors) and portfolio debt flows (bond issues purchased by foreign investors). • **Bank and trade-related lending** covers commercial bank lending and other private credits.

Data sources

The data in this table are compiled from a variety of public and private sources, including the World Bank's Debtor Reporting System, the IMF's International Financial Statistics and Balance of Payments databases, and other sources mentioned in *About the data*. These data are also published in the World Bank's *Global Development Finance 2001*.

Net financial flows from Development
6.8 | Assistance Committee members

Net flows to part I countries

$ millions, 1999

	Official development assistance				Other official flows	Private flows					Net grants by NGOs	Total net flows
	Total	Bilateral grants	Bilateral loans	Contributions to multilateral institutions		Total	Foreign direct investment	Bilateral portfolio investment	Multilateral portfolio investment	Private export credits		
Australia	982	730	..	252	671		..	..	..	..	95	1,749
Austria	527	381	−37	183	23	1,334	831	..	..	503	80	1,963
Belgium	760	454	−17	323	−76	4,765	277	4,636	..	−148	78	5,528
Canada	1,699	1,195	−23	527	665	4,484	4,052	460	..	−29	137	6,984
Denmark	1,733	1,023	3	708	−189	410	344	..	..	67	37	1,992
Finland	416	286	−45	176	140	313	145	70	..	98	6	875
France	5,637	4,320	−195	1,512	−3	3,524	5,517	−1,388	..	−605	−32	9,125
Germany	5,515	3,236	42	2,238	−179	13,853	5,871	7,075	−229	1,136	992	20,181
Greece	194	77	2	115	1	..	..	..	..	..	..	195
Ireland	245	149	..	97	..	..	..	..	..	..	6	251
Italy	1,806	551	−100	1,355	19	9,484	1,655	8,335	..	−506	28	11,337
Japan	15,323	5,475	5,001	4,848	9,507	−4,297	5,277	−3,149	−4,070	−2,355	261	20,794
Luxembourg	119	89	..	30		..	..	..	..	..	6	124
Netherlands	3,134	2,359	−198	972	−8	4,581	4,103	−327	387	418	278	7,985
New Zealand	134	101	..	33	..	16	16	..	..	..	13	163
Norway	1,370	993	14	363	..	522	340	..	..	182	168	2,060
Portugal	276	273	−65	69	107	1,953	1,650	..	..	304	..	2,337
Spain	1,363	653	176	534	11	27,655	27,710	..	..	−55	..	29,029
Sweden	1,630	1,143	3	484	−1	1,192	665	..	..	527	71	2,892
Switzerland	969	719	..	250	21	2,236	1,834	..	..	402	..	3,226
United Kingdom	3,401	2,067	182	1,153	−24	6,160	6,361	−98	..	−104	480	10,017
United States	9,145	7,638	−790	2,297	4,793	32,218	22,724	9,319	−1,856	2,031	3,981	50,138
Total	**56,378**	**33,910**	**3,951**	**18,517**	**15,477**	**110,404**	**89,373**	**24,934**	**−5,768**	**1,866**	**6,684**	**188,943**

Net flows to part II countries

$ millions, 1999

	Official aid				Other official flows	Private flows				Net grants by NGOs	Total net flows
	Total	Bilateral grants	Bilateral loans	Contributions to multilateral institutions		Total	Foreign direct investment	Bilateral portfolio investment	Private export credits		
Australia	3	2	..	1	1	..	..	..	..	..	4
Austria	184	130	0	54	..	512	512	..	..	5	701
Belgium	82	6	..	75	−9	17,604	1,825	15,691	88	0	17,678
Canada	165	165	..	..	1,294	−21	..	..	−21	..	1,437
Denmark	128	99	29	..	25	..	..	..	..	2	155
Finland	74	41	−4	38	18	378	225	167	−14	..	470
France	550	148	−6	408	−11	8,229	3,953	4,058	218	..	8,767
Germany	729	366	−124	487	268	14,007	4,946	8,700	361	98	15,102
Greece	11	8	..	3	0	..	..	..	..	..	11
Ireland	..	..	..	..	..	..	..	..	..	..	0
Italy	92	8	−1	84	−1	6,137	−209	6,831	−486	..	6,228
Japan	67	72	−45	40	1,524	1,018	2,624	−1,656	50	..	2,609
Luxembourg	3	3	..	..	..	..	..	..	..	..	3
Netherlands	22	22	..	..	17	2,299	3,247	..	..	..	2,338
New Zealand	0	0	..	0	..	..	..	..	..	..	0
Norway	28	28	..	..	0	556	548	0	8	..	584
Portugal	28	0	..	28	..	2,782	2,779	..	3	..	2,809
Spain	13	13	..	..	−7	57	57	..	..	..	62
Sweden	99	94	0	6	−2	1,215	1,133	0	81	..	1,312
Switzerland	70	62	1	6	1	6,899	6,894	0	6	..	6,970
United Kingdom	326	98	0	228	..	−6,446	−1,734	−4,877	165	5	−6,115
United States	3,521	3,204	240	78	−96	16,221	15,693	3	526	2,121	21,767
Total	**6,193**	**4,568**	**89**	**1,535**	**3,021**	**71,446**	**42,490**	**28,917**	**986**	**2,232**	**82,892**

Net financial flows from Development Assistance Committee members | 6.8

The high-income members of the Organisation for Economic Co-operation and Development (OECD) are the main source of external finance for developing countries. This table shows the flow of financial resources from members of the OECD's Development Assistance Committee (DAC) to official and private recipients in developing and transition economies. DAC exists to help its members coordinate their development assistance and to encourage the expansion and improve the effectiveness of the aggregate resources flowing to developing and transition economies. In this capacity DAC monitors the flow of all financial resources, but its main concern is official development assistance (ODA). DAC has three criteria for ODA: It is undertaken by the official sector. It promotes economic development or welfare as a main objective. And it is provided on concessional terms, with a grant element of at least 25 percent on loans.

This definition excludes military aid and nonconcessional flows from official creditors, which are considered other official flows. (However, refinancing of military aid on concessional terms is included in ODA.) The definition includes capital projects, food aid, emergency relief, peacekeeping efforts, and technical cooperation. Also included are contributions to multilateral institutions, such as the United Nations and its specialized agencies, and concessional funding to the multilateral development banks. In 1999, to avoid double counting extrabudgetary expenditures reported by DAC countries and flows reported by the United Nations, all United Nations agencies revised their data to include only regular budgetary expenditures since 1990 (except for the World Food Programme and the United Nations High

Commissioner for Refugees, which revised their data from 1996 onward).

DAC maintains a list of countries and territories that are aid recipients. Part I of the list comprises those considered by DAC members to be eligible for ODA. Part II of the list, created after the collapse of the Soviet Union to monitor concessional flows to transition economies, consists of countries that are not considered eligible for ODA but nevertheless receive ODA-like flows. These flows are termed official aid.

The data in the table were compiled from replies by DAC member countries to questionnaires issued by the DAC Secretariat. Net flows of ODA, official aid, and other official resources are defined as gross disbursements of grants and loans minus repayments on earlier loans. Because the data are based on donor country reports, they do not provide a complete picture of the resources received by developing and transition economies, for three reasons. First, flows from DAC members are only part of the aggregate resource flows to these economies. Second, the data that record contributions to multilateral institutions measure the flow of resources made available to those institutions by DAC members, not the flow of resources from those institutions to developing and transition economies. Third, because some of the countries and territories on the DAC recipient list are normally classified as high income, the reported flows may overstate the resources available to low- and middle-income economies. High-income countries receive only a small fraction of all development assistance, however.

Net disbursements of ODA by some important donor countries that are not DAC members are shown in table 6.8a.

• **Official development assistance** comprises grants and loans, net of repayments, that meet the DAC definition of ODA and are made to countries and territories in part I of the DAC list of aid recipients. • **Official aid** comprises grants and ODA-like loans, net of repayments, to countries and territories in part II of the DAC list of aid recipients. • **Bilateral grants** are transfers in money or in kind for which no repayment is required. • **Bilateral loans** are loans extended by governments or official agencies that have a grant element of at least 25 percent and for which repayment is required in convertible currencies or in kind. • **Contributions to multilateral institutions** are concessional funding received by multilateral institutions from DAC members in the form of grants or capital subscriptions. • **Other official flows** are transactions by the official sector whose main objective is other than development or whose grant element is less than 25 percent. • **Private flows** consist of flows at market terms financed from private sector resources. They include changes in holdings of private long-term assets by residents of the reporting country. • **Foreign direct investment** is investment by residents of DAC member countries to acquire a lasting management interest (at least 10 percent of voting stock) in an enterprise operating in the recipient country. The data in the table reflect changes in the net worth of subsidiaries in recipient countries whose parent company is in the DAC source country. • **Bilateral portfolio investment** covers bank lending and the purchase of bonds, shares, and real estate by residents of DAC member countries in recipient countries. • **Multilateral portfolio investment** records the transactions of private banks and nonbanks in DAC member countries in the securities issued by multilateral institutions. • **Private export credits** are loans that are extended to recipient countries by the private sector in DAC member countries for the purpose of promoting trade and are supported by an official guarantee. • **Net grants by NGOs** are private grants by nongovernmental organizations, net of subsidies from the official sector. • **Total net flows** comprise ODA or official aid flows, other official flows, private flows, and net grants by NGOs.

The data on financial flows are compiled by DAC and published in its annual statistical report, *Geographical Distribution of Financial Flows to Aid Recipients,* and the DAC chairman's annual report, *Development Co-operation.* Data are available to registered users from the OECD Web site at www.oecd.org/dac/htm/online.htm.

Table 6.8a

Official development assistance from non-DAC donors, 1995–99

Net disbursements ($ millions)

	1995	1996	1997	1998	1999
OECD members (non-DAC)					
Czech Republic	..	..	..	16	15
Korea, Rep.	116	159	186	183	317
Poland	..	..	..	19	20
Turkey	107	88	77	102	..
Arab countries					
Kuwait	384	414	373	278	147
Saudi Arabia	192	327	251	288	185
United Arab Emirates	65	31	115	63	92
Other donors					
Estonia	..	..	..	0.2	0.4

Note: China also provides aid but does not disclose the amount.

Source: OECD data.

Net flows to part I countries	Net official development assistance							Aid appropriations		Untied aid[a]	
	$ millions		% of GNI		average annual % change in volume[b] 1993–94 to 1998–99	Per capita of donor country[b] $	$	% of central government budget		% of total ODA commitments	
	1994	1999	1994	1999		1994	1999	1994	1999	1994	1999
Australia	1,091	982	0.34	0.26	−0.2	55	50	1.2	1.0	44.4	86.7
Austria	655	527	0.33	0.26	−3.5	80	67	..	..	..	39.8
Belgium	727	760	0.32	0.30	1.6	70	77	..	..	..	39.0
Canada	2,250	1,699	0.43	0.28	−5.0	74	55	1.4	1.3	44.1	29.6
Denmark	1,446	1,733	1.03	1.01	3.5	285	331	2.3	2.9	..	70.8
Finland	290	416	0.31	0.33	2.5	61	84	1.0	1.1	47.0	84.7
France	8,466	5,637	0.64	0.39	−6.8	145	99	..	..	50.9	66.8[c]
Germany	6,818	5,515	0.33	0.26	−3.7	81	69	1.9	..	44.3	84.7
Greece		194		0.15			19		..		3.3
Ireland	109	245	0.25	0.31	15.9	34	66	0.6	..	..	..
Italy	2,705	1,806	0.27	0.15	−7.9	52	33	0.5	..	66.4	22.6
Japan	13,239	15,323	0.29	0.35	4.2	81	106	1.3	..	81.5	96.4
Luxembourg	59	119	0.40	0.66	15.9	143	281	1.2	..	..	96.1
Netherlands	2,517	3,134	0.76	0.79	4.3	161	203	3.1	3.2	94.8	94.1
New Zealand	110	134	0.24	0.27	4.8	29	36	..	0.6	..	..
Norway	1,137	1,370	1.05	0.91	3.6	269	298	1.9	..	85.0	99.1
Portugal	303	276	0.34	0.26	−1.6	33	28	..	0.3	93.2	96.6
Spain	1,305	1,363	0.28	0.23	0.9	34	35	0.9	0.9	..	26.1[c]
Sweden	1,819	1,630	0.96	0.70	−3.1	216	190	2.6	..	81.7	91.5
Switzerland	982	969	0.36	0.35	1.3	134	140	3.0	2.7	95.8	96.8
United Kingdom	3,197	3,401	0.31	0.23	−0.8	66	57	1.2	0.9	45.8	91.8
United States	9,927	9,145	0.14	0.10	−4.0	41	33	1.4	0.9	..	..
Total	**59,152**	**56,378**	**0.29**	**0.24**	**−1.4**	**71**	**66**	**1.4**	**1.2**	**66.1**	**83.8**

Net flows to part II countries	Net official aid						
	$ millions		% of GNI		average annual % change in volume[b] 1993–94 to 1998–99	Per capita of donor country[b] $	$
	1994	1999	1994	1999		1994	1999
Australia	4	3	0.00	0.00	−17.2	0	0
Austria	261	184	0.13	0.09	−10.2	32	24
Belgium	86	82	0.04	0.03	−1.9	8	8
Canada	73	165	0.01	0.03	17.4	2	5
Denmark	124	128	0.09	0.07	−4.6	24	24
Finland	51	74	0.05	0.06	9.6	11	15
France	650	550	0.05	0.04	1.9	11	10
Germany	2,527	729	0.12	0.03	−22.1	30	9
Greece		11		0.01			1
Ireland	16	0	0.04	0.00	−100.0	5	0
Italy	196	92	0.02	0.01	−6.6	4	2
Japan	247	67	0.01	0.00	−21.3	2	0
Luxembourg	7	3	0.05	0.01	−18.2	17	6
Netherlands	118	22	0.04	0.01	−17.3	8	1
New Zealand	1	0	0.00	0.00	−26.4	0	0
Norway	79	28	0.07	0.02	−12.9	19	6
Portugal	28	28	0.03	0.03	2.5	3	3
Spain	157	13	0.03	0.00	−40.8	4	0
Sweden	91	99	0.05	0.04	4.7	11	12
Switzerland	119	70	0.04	0.03	−6.9	16	10
United Kingdom	293	326	0.03	0.02	1.3	6	5
United States	2,422	3,521	0.03	0.04	7.0	10	13
Total	**7,550**	**6,193**	**0.04**	**0.03**	**−4.1**	**10**	**7**

a. Excluding administrative costs in 1994 and administrative costs and technical cooperation in 1999. b. At 1998 prices. c. Data refer to 1998.

Aid flows from Development Assistance Committee members | 6.9

About the data

As part of its work, the Development Assistance Committee (DAC) of the Organisation for Economic Co-operation and Development (OECD) assesses the aid performance of member countries relative to the size of their economies. As measured here, aid comprises bilateral disbursements of concessional financing to recipient countries plus the provision by donor governments of concessional financing to multilateral institutions. Volume amounts, at constant prices and exchange rates, are used to measure the change in real resources provided over time. Aid flows to part I recipients—official development assistance (ODA)—are tabulated separate from those to part II recipients—official aid (see *About the data* for table 6.8 for more information on the distinction between the two types of aid flows).

Measures of aid flows from the perspective of donors differ from aid receipts by recipient countries. This is because the concessional funding received by multilateral institutions from donor countries is recorded as an aid disbursement by the donor when the funds are deposited with a multilateral institution and recorded as a resource receipt by the recipient country when that institution makes a disbursement.

Ratios of aid to gross national income (GNI), aid per capita, and aid appropriations as a percentage of donor government budgets are calculated by the OECD. The denominators used in calculating these ratios may differ from corresponding values elsewhere in this book because of differences in timing or definitions.

For many European countries, adoption of the 1993 United Nations System of National Accounts has led to an apparent increase in their GNI. As a result, ratios of aid to GNI have fallen. DAC is reviewing the extent to which this phenomenon has affected measures of aid performance.

The proportion of untied aid is reported here because tying arrangements require recipients to purchase goods and services from the donor country or from a specified group of countries. Tying arrangements may be justified on the grounds that they prevent a recipient from misappropriating or mismanaging aid receipts, but they may also be motivated by a desire to benefit suppliers in the donor country. The same volume of aid may have different purchasing power depending on the relative costs of suppliers in countries to which the aid is tied and the degree to which each recipient's aid basket is untied. Thus tying arrangements may prevent recipients from obtaining the best value for their money and so reduce the value of the aid received.

Definitions

• **Net official development assistance** and **net official aid** record the actual international transfer by the donor of financial resources or of goods or services valued at the cost to the donor, less any repayments of loan principal during the same period. Data are shown at current prices and dollar exchange rates. • **Aid as a percentage of GNI** shows the donor's contributions of ODA or official aid as a share of its gross national income. • **Average annual percentage change in volume** and **aid per capita of donor country** are calculated using 1998 exchange rates and prices. • **Aid appropriations** are the share of ODA or official aid appropriations in the donor's national budget. • **Untied aid** is the share of ODA that is not subject to restrictions by donors on procurement sources.

Data sources

The data appear in the DAC chairman's annual report, *Development Co-operation*. The OECD also makes its data available on diskette, magnetic tape, and the Internet. Data are available to registered users from the OECD Web site at www.oecd.org/dac/htm/online.htm.

Figure 6.9

Declining efforts in aid

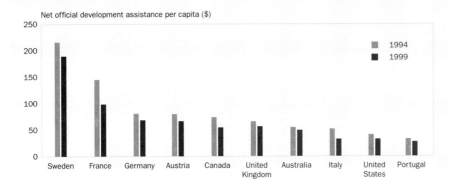

Net official development assistance per capita ($)

Source: OECD data.

Ten of 21 Development Assistance Committee (DAC) members (excluding Greece, which became a member in 1999) provided less aid per capita in 1999 than they did in 1994. Overall, aid per capita from DAC members fell from $71 to $66.

6.10 | Aid dependency

	Net official development assistance or official aid		Aid per capita		Aid dependency ratios							
					Aid as % of GNI		Aid as % of gross capital formation		Aid as % of imports of goods and services		Aid as % of central government expenditure	
	$ millions		$									
	1994	1999	1994	1999	1994	1999	1994	1999	1994	1999	1994	1999
Albania	165	480	52	142	8.4	12.8	46.4	77.7	21.3	43.2	..	..
Algeria	419	89	15	3	1.0	0.2	3.1	0.7	3.6	0.7	3.0	0.6
Angola	450	388	42	31	23.1	13.1	66.7	..	11.9	5.2	..	..
Argentina	147	91	4	2	0.1	0.0	0.3	0.2	0.4	0.2	0.4	..
Armenia	191	208	51	55	8.1	..	34.6	58.0	43.6	21.8	..	..
Australia												
Austria												
Azerbaijan	147	162	19	20	4.5	4.7	29.2	10.2	14.5	6.5	33.6	15.7
Bangladesh	1,752	1,203	15	9	5.0	2.5	28.2	11.8	36.0	13.7	..	..
Belarus	119	24	12	2	0.6	0.1	1.8	0.4	3.3	0.3	..	..
Belgium												
Benin	256	211	48	34	17.5	9.0	108.5	50.4	39.1	25.2	..	..
Bolivia	569	569	79	70	9.8	7.0	66.1	36.2	36.6	24.2	41.0	29.6
Bosnia and Herzegovina	391	1,063	107	274	..	22.8	..	69.4	..	..	..	..
Botswana	86	61	60	38	2.0	1.1	7.2	5.2	4.0	1.9	5.7	..
Brazil	253	184	2	1	0.0	0.0	0.2	0.1	0.5	0.2	0.1	..
Bulgaria	158	265	19	32	1.6	2.1	17.2	11.2	2.9	3.8	3.6	6.0
Burkina Faso	433	398	44	36	23.5	15.5	113.4	55.4	83.1	50.4	..	..
Burundi	312	74	52	11	34.2	10.5	319.3	114.0	109.1	52.5	119.0	39.8
Cambodia	327	279	32	24	13.6	9.0	73.3	..	35.0	19.2	..	..
Cameroon	730	434	57	30	10.0	5.0	60.6	24.3	35.0	15.7	83.9	31.1
Canada												
Central African Republic	165	117	51	33	19.9	11.3	165.6	77.9	61.9	44.7	..	..
Chad	213	188	33	25	18.5	12.4	108.5	118.7	50.3	36.6	..	..
Chile	151	69	11	5	0.3	0.1	1.2	0.5	0.9	0.3	1.4	0.4
China	3,225	2,324	3	2	0.6	0.2	1.4	0.6	2.7	1.1	6.3	..
Hong Kong, China	27	4	4	1	0.0	0.0	0.1	0.0	0.0	0.0	..	..
Colombia	77	301	2	7	0.1	0.4	0.4	2.7	0.5	1.9	0.8	1.9
Congo, Dem. Rep.	245	132	6	3	4.8	..	53.4	..	13.3	..	0.0	..
Congo, Rep.	362	140	145	49	23.9	8.4	37.6	28.4	19.1	7.0	55.8	20.1
Costa Rica	73	−10	22	−3	0.7	−0.1	3.3	−0.4	1.9	−0.1	2.9	−0.3
Côte d'Ivoire	1,594	447	118	29	23.1	4.3	165.7	24.5	46.4	8.9	75.9	18.0
Croatia	110	48	24	11	0.9	0.2	5.2	1.0	1.6	0.5	1.9	0.5
Cuba	47	58	4	5	..	..	..	..	..	..	..	..
Czech Republic	148	318	14	31	0.4	0.6	1.2	2.1	0.6	0.9	1.0	1.6
Denmark												
Dominican Republic	60	195	8	23	0.6	1.2	2.7	4.5	0.9	1.9	3.4	..
Ecuador	212	146	19	12	1.4	0.8	6.7	5.9	3.9	2.7	8.1	..
Egypt, Arab Rep.	2,690	1,579	47	25	5.2	1.8	31.3	7.8	17.2	7.2	13.9	..
El Salvador	305	183	55	30	3.8	1.5	19.0	9.0	10.2	3.6	..	9.0
Eritrea	157	148	45	37	23.7	19.5	173.5	48.6	..	24.5	..	..
Estonia	44	83	29	57	1.1	1.6	4.0	6.4	2.1	1.8	5.9	4.5
Ethiopia	1,071	633	20	10	22.2	9.9	144.4	54.4	90.2	32.4	..	..
Finland												
France												
Gabon	181	48	169	39	4.9	1.2	19.8	3.9	8.6	1.9	..	..
Gambia, The	70	33	65	26	19.5	8.6	105.9	47.3	27.4	10.3	..	..
Georgia	176	239	33	44	..	8.4	..	51.8	19.1	17.7	..	56.9
Germany												
Ghana	546	607	33	32	10.2	8.0	41.9	33.7	25.7	15.2	..	..
Greece												
Guatemala	217	293	22	26	1.7	1.6	10.7	9.2	6.4	5.5	..	..
Guinea	359	238	56	33	10.7	7.0	67.9	39.0	31.7	23.0	..	32.6
Guinea-Bissau	172	52	162	44	77.7	25.7	335.5	147.1	160.6	55.9	..	..
Haiti	601	263	86	34	30.7	6.1	899.8	55.5	243.8	20.4	..	..
Honduras	293	817	53	129	9.1	15.6	22.7	46.1	15.4	24.9	..	..

Aid dependency | 6.10

	Net official development assistance or official aid		Aid per capita		Aid dependency ratios							
					Aid as % of GNI		Aid as % of gross capital formation		Aid as % of imports of goods and services		Aid as % of central government expenditure	
	$ millions		$									
	1994	1999	1994	1999	1994	1999	1994	1999	1994	1999	1994	1999
Hungary	200	248	20	25	0.5	0.5	2.2	1.8	1.2	0.8	0.9	1.2
India	2,324	1,484	3	1	0.7	0.3	3.1	1.4	5.1	2.1	4.7	2.2
Indonesia	1,639	2,206	9	11	1.0	1.7	3.0	6.5	3.3	4.1	5.7	8.0
Iran, Islamic Rep.	130	161	2	3	0.2	0.1	0.8	0.8	0.8	1.0	0.7	0.3
Iraq	259	76	13	3	..	..	..	..	..	..	..	..
Ireland												
Israel	1,237	906	229	148	1.7	0.9	6.9	4.3	3.6	1.9	3.6	1.9
Italy												
Jamaica	109	−23	44	−9	2.7	−0.3	7.7	−1.2	3.1	−0.5	..	..
Japan												
Jordan	369	430	91	91	6.2	5.4	17.8	25.6	7.7	7.9	19.6	16.9
Kazakhstan	48	161	3	11	0.2	1.1	0.9	5.8	0.9	2.2	..	6.3
Kenya	675	308	26	10	10.0	2.9	57.5	21.4	23.7	9.2	29.2	..
Korea, Dem. Rep.	6	201	0	9	..	..	..	..	..	..	..	..
Korea, Rep.	−114	−55	−3	−1	0.0	0.0	−0.1	−0.1	−0.1	0.0	−0.2	..
Kuwait	3	7	2	4	0.0	0.0	0.1	0.2	0.0	0.1	0.0	0.1
Kyrgyz Republic	172	267	38	55	5.5	22.7	59.7	118.2	33.1	33.8	66.3	126.4
Lao PDR	216	294	48	58	14.0	21.1	..	..	31.7	46.7	..	..
Latvia	53	96	21	40	1.0	1.6	5.0	5.8	3.2	2.5	5.2	4.1
Lebanon	235	194	60	45	2.5	1.2	8.0	..	3.8	..	7.3	3.3
Lesotho	116	31	62	15	9.5	2.8	24.9	..	12.7	3.4	30.4	..
Libya	4	7	1	1	..	..	..	..	0.0	0.1	..	..
Lithuania	71	129	19	35	1.2	1.2	6.6	5.3	2.7	2.3	6.6	3.9
Macedonia, FYR	104	273	54	135	4.4	8.0	27.6	37.0	..	13.7	..	..
Madagascar	289	358	22	24	10.2	9.8	89.1	74.6	28.0	27.4	51.1	..
Malawi	467	446	49	41	41.0	25.1	135.7	166.2	48.6	40.8	..	..
Malaysia	66	143	3	6	0.1	0.2	0.2	0.8	0.1	0.2	0.4	..
Mali	441	354	47	33	25.3	14.0	91.5	65.0	54.0	35.8	..	..
Mauritania	267	219	118	84	27.4	23.6	125.5	128.2	46.0	49.2	..	..
Mauritius	14	42	13	35	0.4	1.0	1.3	3.5	0.6	1.5	1.8	4.1
Mexico	425	34	5	0	0.1	0.0	0.5	0.0	0.4	0.0	0.7	..
Moldova	54	102	12	24	2.0	8.5	6.8	39.9	6.9	12.3	..	29.3
Mongolia	182	219	81	92	27.6	25.4	107.2	91.3	40.8	33.0	117.9	99.0
Morocco	631	678	24	24	2.2	2.0	9.7	8.0	5.9	5.2	6.4	..
Mozambique	1,200	118	78	7	58.1	3.2	267.2	9.1	84.1	6.5	..	..
Myanmar	161	73	4	2	..	..	..	..	9.4	4.0	2.0	..
Namibia	137	178	91	104	4.5	5.7	19.3	28.7	6.7	8.4	..	..
Nepal	448	344	22	15	10.9	6.7	49.4	34.0	34.9	22.6	75.5	42.9
Netherlands												
New Zealand												
Nicaragua	597	675	139	137	46.5	33.0	143.9	69.1	42.1	30.1	100.9	..
Niger	377	187	43	18	24.6	9.4	231.9	90.7	80.9	41.3	..	..
Nigeria	190	152	2	1	0.9	0.5	4.1	1.8	1.5	1.0	..	..
Norway												
Oman	95	40	46	17	1.0	..	5.0	..	1.8	0.6	1.9	0.8
Pakistan	1,605	732	13	5	3.0	1.2	15.8	8.4	12.6	5.4	13.5	6.1
Panama	31	14	12	5	0.4	0.2	1.5	0.4	0.3	0.1	1.6	..
Papua New Guinea	322	216	77	46	6.3	6.3	28.1	33.7	13.7	10.3	20.5	22.6
Paraguay	93	78	20	14	1.2	1.0	5.1	4.3	2.2	2.1	..	..
Peru	336	452	15	18	0.8	0.9	3.4	4.0	3.6	4.1	4.0	4.5
Philippines	1,057	690	16	9	1.6	0.9	6.9	4.9	3.8	1.7	9.0	4.6
Poland	1,806	984	47	25	1.8	0.6	10.1	2.4	7.0	1.8	4.4	1.8
Portugal												
Puerto Rico												
Romania	144	373	6	17	0.5	1.1	2.1	5.5	1.8	3.1	1.5	..
Russian Federation	1,847	1,816	12	12	0.6	0.5	2.2	2.9	2.6	2.8	2.4	4.3

6.10 Aid dependency

| | Net official development assistance or official aid ($ millions) | | Aid per capita ($) | | Aid dependency ratios | | | | | | | |
| | | | | | Aid as % of GNI | | Aid as % of gross capital formation | | Aid as % of imports of goods and services | | Aid as % of central government expenditure | |
	1994	1999	1994	1999	1994	1999	1994	1999	1994	1999	1994	1999
Rwanda	714	373	115	45	95.3	19.2	809.8	133.8	149.7	118.0	..	..
Saudi Arabia	16	29	1	1	0.0	0.0	0.1	0.1	0.0	0.1	..	..
Senegal	640	534	79	58	18.3	11.4	94.7	59.1	38.1	30.3	..	..
Sierra Leone	275	74	63	15	33.9	11.3	348.6	3,751.0	78.0	44.5	154.4	..
Singapore	17	−1	5	0	0.0	0.0	0.1	0.0	0.0	0.0	0.2	..
Slovak Republic	78	318	15	59	0.5	1.6	2.4	5.1	0.9	2.3	..	4.3
Slovenia	32	31	16	16	0.2	0.2	1.1	0.6	0.4	0.3	0.6	0.4
South Africa	295	539	8	13	0.2	0.4	1.4	2.6	1.0	1.6	0.7	1.4
Spain												
Sri Lanka	595	251	33	13	5.2	1.6	18.8	5.8	10.5	3.5	18.7	6.6
Sudan	410	243	16	8	5.6	2.8	..	..	19.2	8.4	..	..
Sweden												
Switzerland												
Syrian Arab Republic	745	228	54	15	4.8	1.5	16.4	4.0	10.3	3.7	6.3	..
Tajikistan	67	122	12	20	..	6.6	..	75.9	8.8	17.3	..	..
Tanzania	965	990	34	30	22.2	11.3	86.8	66.4	49.1	41.7	..	..
Thailand	578	1,003	10	17	0.4	0.8	1.0	3.8	0.9	1.6	2.5	3.3
Togo	125	71	31	16	13.5	5.2	84.7	37.8	23.0	9.4	..	..
Trinidad and Tobago	21	26	17	20	0.5	0.4	2.1	1.8	1.1	0.8	1.6	..
Tunisia	106	244	12	26	0.7	1.2	2.8	4.4	1.3	2.4	2.1	3.7
Turkey	159	−10	3	0	0.1	0.0	0.6	0.0	0.5	0.0	0.5	0.0
Turkmenistan	25	21	6	4	0.8	0.7	..	1.4	1.1	1.0	..	..
Uganda	750	590	40	27	19.0	9.2	127.7	56.1	82.1	31.1	..	..
Ukraine	290	480	6	10	0.6	1.3	1.6	6.3	1.6	3.0	..	..
United Arab Emirates	−8	4	−4	1	0.0	0.0	−0.1	..	..	..	−0.2	0.1
United Kingdom												
United States												
Uruguay	74	22	23	7	0.5	0.1	2.9	0.7	1.9	0.4	1.3	0.3
Uzbekistan	28	134	1	5	0.1	0.8	0.7	5.0	..	..	..	..
Venezuela, RB	27	44	1	2	0.0	0.0	0.3	0.3	0.2	0.2	0.2	0.2
Vietnam	891	1,421	12	18	5.7	5.0	22.5	19.5	13.0	9.2	21.9	26.9
West Bank and Gaza	460	512	196	180	12.6	10.2	44.8	30.9	..	..	..	..
Yemen, Rep.	170	456	11	27	5.2	7.4	23.0	36.0	6.2	12.0	2.4	22.9
Yugoslavia, FR (Serb./Mont.)[a]	49	638	5	60	..	..	..	..	..	..	..	..
Zambia	718	623	82	63	23.1	20.8	260.4	113.3	41.6	41.8	..	..
Zimbabwe	560	244	52	21	8.5	4.7	27.9	38.0	19.8	9.0	31.4	..
World	67,506 s	59,125 s	12 w	10 w	0.3 w	0.2 w	1.1 w	0.9 w	1.1 w	0.7 w	..	..
Low income	29,422	22,399	13	9	2.9	2.2	13.8	10.0	13.7	7.9	..	..
Middle income	24,531	22,924	10	9	0.6	0.4	2.1	1.7	2.2	1.5	..	..
Lower middle income	18,315	17,816	9	9	0.9	0.7	3.3	2.6	3.4	2.5	..	..
Upper middle income	4,933	3,848	9	7	0.2	0.1	0.9	0.6	0.8	0.5	..	..
Low & middle income	58,475	48,473	12	10	1.1	0.8	4.3	3.8	4.3	2.6	..	..
East Asia & Pacific	9,431	9,811	5	5	0.7	0.5	1.7	1.7	2.0	1.5	..	..
Europe & Central Asia	9,728	10,878	21	23	1.0	1.0	4.6	4.7	3.9	3.1	..	..
Latin America & Carib.	5,684	5,856	12	12	0.4	0.3	1.6	1.4	1.8	1.3	..	..
Middle East & N. Africa	7,194	5,128	27	18	1.6	0.9	7.0	3.9	4.8	3.0	..	..
South Asia	7,057	4,254	6	3	1.6	0.7	7.2	3.3	10.0	4.1	..	..
Sub-Saharan Africa	19,381	12,546	34	20	7.2	4.1	39.4	22.0	20.2	10.6	..	..
High income	2,197	1,823	3	2	0.0	0.0	0.0	0.0	0.0	0.0	..	..
Europe EMU												

Note: Regional aggregates include data for economies that are not specified elsewhere. World and income group totals include aid not allocated by country or region.
a. Aid to the states of the former Socialist Federal Republic of Yugoslavia that is not otherwise specified is included in regional and income group aggregates.

Aid dependency | 6.10

Ratios of aid to gross national income (GNI), gross capital formation, imports, and public spending provide a measure of the recipient country's dependency on aid. But care must be taken in drawing policy conclusions. For foreign policy reasons some countries have traditionally received large amounts of aid. Thus aid dependency ratios may reveal as much about the donors' interests as they do about the recipients' needs. Ratios in Sub-Saharan Africa are generally much higher than those in other regions, and they increased in the 1980s. These high ratios are due only in part to aid flows. Many African countries saw severe erosion in their terms of trade in the 1980s, which, along with weak policies, contributed to falling incomes, imports, and investment. Thus the increase in aid dependency ratios reflects events affecting both the numerator and the denominator.

As defined here, aid includes official development assistance (ODA) and official aid. The data cover loans and grants from Development Assistance Committee (DAC) member countries, multilateral organizations, and certain Arab countries. They do not reflect aid given by recipient countries to other developing countries. As a result, some countries that are net donors (such as Saudi Arabia) are shown in the table as aid recipients (see table 6.8a).

The data in the table do not distinguish among different types of aid (program, project, or food aid; emergency assistance; peacekeeping assistance; or technical cooperation), each of which may have a very different effect on the economy. Technical cooperation expenditures do not always directly benefit the economy to the extent that they defray costs incurred outside the country on the salaries and benefits of technical experts and the overhead costs of firms supplying technical services.

In 1999, to avoid double counting extrabudgetary expenditures reported by DAC countries and flows reported by the United Nations, all United Nations agencies revised their data to include only regular budgetary expenditures since 1990 (except for the World Food Programme and the United Nations High Commissioner for Refugees, which revised their data from 1996 onward). These revisions have affected net official development assistance and official aid and, as a result, aid per capita and aid dependency ratios.

Because the table relies on information from donors, it is not consistent with information recorded by recipients in the balance of payments, which often excludes all or some technical assistance—particularly payments to expatriates made directly by the donor. Similarly, grant commodity aid may not always be recorded

in trade data or in the balance of payments. Moreover, although ODA estimates in balance of payments statistics are meant to exclude purely military aid, the distinction is sometimes blurred. Under DAC rules concessional refinancing of military aid may be counted as ODA; the definition used by the country of origin usually prevails.

The nominal values used here tend to overstate the amount of resources transferred. Changes in international prices and in exchange rates can reduce the purchasing power of aid. The practice of tying aid, still prevalent though declining in importance, also tends to reduce its purchasing power (see *About the data* for table 6.9).

The values for population, GNI, gross capital formation, imports of goods and services, and central government expenditure used in computing the ratios are taken from World Bank and International Monetary Fund databases. The ratios shown may therefore differ somewhat from those computed and published by the Organisation for Economic Co-operation and Development (OECD). Aid not allocated by country or region—including administrative costs, research into development issues, and aid to nongovernmental organizations—is included in the world total. Thus regional and income group totals do not sum to the world total.

DAC members distributed their aid in 1999 much as they had in the previous six years

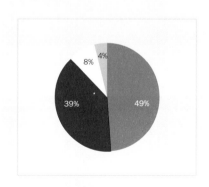

- Low income
- Lower middle income
  Upper middle income
- High income

Source: OECD data.

- **Net official development assistance** consists of disbursements of loans made on concessional terms (net of repayments of principal) and grants by official agencies of the members of DAC, by multilateral institutions, and by certain Arab countries to promote economic development and welfare in recipient economies listed as developing by DAC. Loans with a grant element of at least 25 percent are included in ODA, as are technical cooperation and assistance. • **Net official aid** refers to aid flows, net of repayments, from official donors to the transition economies of Eastern Europe and the former Soviet Union and to certain advanced developing countries and territories as determined by DAC. Official aid is provided under terms and conditions similar to those for ODA. • **Aid per capita** includes both ODA and official aid. • **Aid dependency ratios** are calculated using values in U.S. dollars converted at official exchange rates. For definitions of GNI, gross capital formation, imports of goods and services, and central government expenditure see *Definitions* for tables 1.1, 4.9, and 4.12.

Data on aid are compiled by DAC and published in its annual statistical report, *Geographical Distribution of Financial Flows to Aid Recipients,* and in the DAC chairman's annual report, *Development Co-operation.* The OECD also makes its data available on diskette, magnetic tape, and the Internet. Data are available to registered users from the OECD Web site at www.oecd.org/dac/htm/online.htm. The data on population, GNI, gross capital formation, imports of goods and services, and central government expenditure are from World Bank and International Monetary Fund databases.

	Total		Ten major DAC donors									Other DAC donors
$ millions, 1999		Japan	United States	Germany	France	United Kingdom	Netherlands	Canada	Sweden	Denmark	Norway	
Albania	253.0	15.2	23.8	22.2	2.6	16.3	2.0	0.2	5.3	2.2	6.3	156.9
Algeria	37.1	−5.0	0.1	2.2	74.3	0.1	0.4	2.0	2.0	0.0	1.0	−39.8
Angola	251.8	22.0	48.1	18.0	8.7	3.6	13.7	3.1	17.8	1.8	19.8	95.4
Argentina	31.3	37.0	0.4	11.6	7.6	0.3	0.1	0.9	0.1	−0.3	0.0	−26.4
Armenia	75.3	3.4	47.9	6.9	2.0	1.3	5.2	0.7	0.3	0.4	2.4	4.7
Australia												
Austria												
Azerbaijan	52.4	10.8	12.5	19.2	1.4	1.2	1.1	0.6	0.5	..	2.0	3.1
Bangladesh	607.3	123.7	113.6	46.6	14.1	114.9	36.1	29.0	25.2	42.0	34.2	27.9
Belarus	15.5	0.1	4.5	6.4	1.3	0.6	0.0	0.1	1.2	0.5	0.1	0.7
Belgium												
Benin	119.3	14.2	19.6	27.4	27.6	1.6	4.6	3.6	0.2	9.1	..	11.6
Bolivia	397.3	41.5	112.9	58.8	13.8	47.1	27.2	6.0	13.2	21.4	3.5	52.0
Bosnia and Herzegovina	734.5	36.4	218.9	65.0	115.7	6.9	77.0	14.3	30.4	2.1	31.6	136.2
Botswana	41.1	13.9	3.6	10.5	0.5	4.8	0.3	0.6	1.2	2.1	2.7	0.9
Brazil	98.4	149.4	−157.8	47.6	21.1	11.6	0.2	3.5	1.6	..	2.0	19.4
Bulgaria	137.1	30.6	37.2	29.6	10.0	4.9	1.9	0.2	0.0	2.3	..	20.5
Burkina Faso	232.0	28.2	11.2	36.5	55.5	0.5	21.1	6.9	0.6	34.6	1.5	35.4
Burundi	52.0	1.1	15.8	1.7	4.3	0.8	4.3	2.0	3.7	..	6.4	12.1
Cambodia	167.1	50.9	14.1	21.6	22.1	7.5	6.3	1.2	7.6	2.5	6.3	27.1
Cameroon	254.3	21.9	4.5	36.6	134.8	11.1	7.0	18.1	..	1.6	0.7	18.0
Canada												
Central African Republic	59.1	18.1	0.9	7.0	30.7	0.0	..	0.2	0.3	..	0.0	1.9
Chad	64.5	0.3	3.2	14.7	34.9	..	..	0.4	0.2	..	0.2	10.8
Chile	63.5	23.7	−5.2	29.6	9.1	1.2	0.7	0.8	1.7	0.2	0.6	1.1
China	1,821.6	1,226.0	38.3	304.6	46.2	59.3	34.2	31.4	3.8	0.1	14.1	63.8
Hong Kong, China	3.8	2.5	..	1.2	..	..	0.0	0.0	0.2	..	..	−0.2
Colombia	292.3	24.4	183.8	22.4	11.5	4.6	7.2	5.7	4.2	0.0	6.6	21.8
Congo, Dem. Rep.	87.0	0.1	11.2	12.2	9.6	2.4	2.5	2.6	9.3	..	2.2	35.0
Congo, Rep.	121.4	0.0	0.6	3.8	20.7	5.2	2.9	3.2	0.9	..	1.5	82.6
Costa Rica	−4.3	−5.7	−34.9	−2.3	3.0	8.3	6.2	4.0	1.7	1.2	0.3	14.0
Côte d'Ivoire	365.6	56.1	13.8	39.9	201.3	1.4	10.2	18.8	0.3	1.0	0.1	22.7
Croatia	27.8	0.5	10.6	1.8	2.0	1.7	0.2	0.3	1.9	..	3.8	4.9
Cuba	35.5	1.3	..	2.5	1.0	2.2	0.9	2.5	2.0	0.0	1.2	21.9
Czech Republic	29.8	1.7	0.5	13.1	7.0	1.4	0.4	0.7	..	1.0	..	4.1
Denmark												
Dominican Republic	151.9	29.7	22.5	10.8	1.7	47.1	0.1	1.2	0.1	0.2	0.4	38.1
Ecuador	128.9	25.3	14.0	14.5	8.0	11.6	8.7	6.7	1.4	2.9	2.0	33.7
Egypt, Arab Rep.	1,298.1	132.1	666.8	103.6	254.1	5.3	18.6	13.6	0.9	40.4	1.2	61.6
El Salvador	173.7	53.0	49.4	19.8	3.9	0.4	3.7	2.6	7.1	1.3	1.7	31.0
Eritrea	80.5	0.4	11.5	3.5	1.2	0.7	7.9	1.0	4.3	9.3	7.4	33.3
Estonia	28.6	0.3	0.9	4.2	0.8	0.4	0.1	0.4	4.1	11.8	0.8	5.0
Ethiopia	325.0	40.4	77.4	37.5	10.5	12.0	31.2	14.8	18.9	4.1	23.9	54.4
Finland												
France												
Gabon	34.5	0.3	1.7	1.2	29.0	..	..	1.3	..	..	..	1.0
Gambia, The	13.2	2.2	2.4	2.9	1.1	1.9	0.1	0.5	0.6	0.1	0.2	1.2
Georgia	77.7	10.2	30.2	17.0	1.2	1.5	5.2	0.5	2.0	..	2.3	7.4
Germany												
Ghana	355.6	101.8	40.9	37.6	3.8	91.8	11.8	12.8	0.6	38.0	0.8	15.9
Greece												
Guatemala	230.7	67.4	51.8	22.6	4.5	2.0	13.1	4.4	18.2	6.0	13.2	27.7
Guinea	111.1	16.5	21.8	25.9	37.4	0.9	..	5.6	0.6	..	0.1	2.4
Guinea-Bissau	32.1	1.5	3.9	0.7	1.9	0.0	2.2	0.3	5.5	0.3	0.0	15.9
Haiti	157.2	6.8	91.7	6.0	14.3	..	3.2	25.8	0.5	..	1.2	7.6
Honduras	355.1	66.3	86.1	43.1	18.6	3.3	14.4	23.8	29.0	4.9	3.7	61.8

Distribution of net aid by Development
Assistance Committee members | 6.11

$ millions, 1999	Total	Ten major DAC donors										Other DAC donors
		Japan	United States	Germany	France	United Kingdom	Netherlands	Canada	Sweden	Denmark	Norway	
Hungary	29.2	−33.6	5.1	28.8	8.8	3.5	0.1	0.9	0.0	9.5	..	6.1
India	838.3	634.0	8.1	29.6	−28.6	131.7	−4.9	4.8	13.3	25.1	9.9	15.3
Indonesia	2,169.4	1,605.8	207.3	−19.5	21.2	40.7	71.9	26.3	3.0	1.9	8.6	202.2
Iran, Islamic Rep.	138.4	48.0	..	59.3	9.0	0.8	0.0	..	0.0	0.1	1.8	19.4
Iraq	79.0	1.0	..	21.6	1.8	11.0	2.4	..	8.7	..	22.2	10.4
Ireland												
Israel	901.6	0.3	989.2	−90.9	..	..	..	0.1	0.5	..	..	2.4
Italy												
Jamaica	−22.7	−8.2	−15.3	−2.5	−0.8	4.2	−2.8	5.2	0.2	..	0.4	−3.0
Japan												
Jordan	325.3	60.8	170.2	58.9	9.9	7.5	0.8	2.3	1.9	0.0	3.0	10.0
Kazakhstan	133.6	67.5	44.6	13.0	1.9	2.0	0.1	1.5	0.1	0.1	1.1	1.7
Kenya	253.7	58.6	38.9	37.2	3.6	55.0	10.4	7.3	11.3	11.6	2.0	17.8
Korea, Dem. Rep.	165.1	..	146.3	2.9	0.1	0.5	0.3	0.2	4.4	..	3.9	6.6
Korea, Rep.	−53.8	−49.5	−44.0	25.8	9.7	..	..	..	0.7	..	..	3.5
Kuwait	5.6	0.1	..	0.8	4.5	..	..	..	0.2	..	..	..
Kyrgyz Republic	115.6	62.5	30.2	7.4	0.3	1.6	2.0	0.2	0.1	2.0	1.2	8.1
Lao PDR	210.5	132.5	6.0	21.7	10.7	0.9	0.5	0.7	11.6	2.8	6.6	16.6
Latvia	44.2	1.1	3.8	5.5	0.8	0.3	0.1	1.3	9.6	15.6	1.1	5.0
Lebanon	80.3	1.6	14.7	6.3	40.3	0.4	0.2	1.9	2.0	..	3.9	9.1
Lesotho	25.7	2.7	1.5	5.0	−0.3	4.4	..	0.3	0.1	2.2	0.2	9.7
Libya	3.3	0.0	..	2.1	0.9	..	..	..	..	..	..	0.4
Lithuania	61.3	1.7	7.8	7.4	1.9	0.5	..	0.8	10.0	26.8	1.8	2.6
Macedonia, FYR	136.5	25.9	32.1	13.9	8.1	12.2	12.4	0.7	8.0	0.3	7.0	16.0
Madagascar	192.5	49.1	29.1	16.9	79.3	0.9	1.4	0.3	..	0.2	3.4	12.0
Malawi	227.7	34.0	27.8	28.7	0.2	77.3	7.0	5.8	3.6	28.4	12.4	2.6
Malaysia	140.1	122.6	..	6.2	−2.0	1.0	0.2	1.5	0.1	11.4	0.2	−1.2
Mali	237.3	25.5	34.2	48.8	58.2	1.2	26.2	18.3	0.3	0.5	8.7	15.4
Mauritania	88.7	32.6	2.8	17.5	23.1	0.3	0.3	0.5	0.2	..	0.4	11.1
Mauritius	5.1	2.7	−0.7	−9.9	10.3	0.5	..	0.1	..	..	0.0	2.0
Mexico	21.9	−27.4	11.9	16.5	12.2	5.4	−0.3	2.1	0.2	−0.1	0.6	0.8
Moldova	51.2	3.5	36.9	2.7	0.8	1.1	1.7	0.2	2.2	0.1	..	2.0
Mongolia	138.2	94.0	12.5	19.6	1.9	0.7	2.8	0.3	0.8	1.9	1.7	2.1
Morocco	333.5	61.7	−16.8	30.9	223.7	0.4	2.6	4.3	0.6	−1.0	0.0	27.1
Mozambique	593.2	63.3	70.6	51.6	34.0	49.4	43.9	11.7	51.4	51.5	36.7	129.2
Myanmar	44.7	34.2	−0.4	1.6	1.6	1.2	1.2	0.2	0.2	0.4	1.9	2.5
Namibia	117.2	4.3	13.9	48.5	4.7	6.1	7.5	0.6	8.7	2.3	5.4	15.3
Nepal	204.8	65.6	16.7	22.1	2.0	26.4	6.1	4.7	1.0	23.8	7.4	29.0
Netherlands												
New Zealand												
Nicaragua	323.4	44.8	64.2	28.3	6.9	5.9	19.4	6.7	33.3	24.4	17.0	72.6
Niger	120.2	15.9	6.5	17.7	44.9	1.3	3.6	3.1	0.1	5.6	1.7	19.8
Nigeria	52.9	2.2	7.5	7.7	5.2	21.0	0.2	0.8	0.5	3.3	0.7	3.8
Norway												
Oman	8.8	9.0	−1.3	0.4	0.6	0.1	..	..	..	..	..	0.0
Pakistan	435.2	169.7	75.0	83.4	8.2	39.5	23.2	12.1	1.2	−1.5	4.7	19.8
Panama	15.2	4.1	−12.1	2.9	0.1	0.5	..	0.5	0.0	0.2	..	19.0
Papua New Guinea	212.2	37.1	6.2	3.0	0.1	..	0.4	..	0.1	..	0.2	165.0
Paraguay	65.5	32.8	4.9	10.6	0.6	0.1	..	0.2	1.1	..	1.4	13.8
Peru	407.3	189.1	124.0	11.3	7.5	6.9	12.4	7.7	3.4	4.2	1.7	39.2
Philippines	616.0	413.0	72.7	22.1	9.6	2.4	9.6	11.2	4.0	4.2	2.5	64.9
Poland	385.4	−2.6	37.7	62.2	13.9	9.5	0.4	125.3	8.5	18.0	0.1	112.6
Portugal												
Puerto Rico												
Romania	122.5	18.3	18.5	31.1	18.3	6.0	1.9	2.5	0.4	9.3	0.2	16.1
Russian Federation	1,599.9	0.5	1,350.7	81.2	15.9	46.2	11.7	13.5	18.9	9.4	18.3	33.8

$ millions, 1999	Total	Ten major DAC donors										Other DAC donors
		Japan	United States	Germany	France	United Kingdom	Netherlands	Canada	Sweden	Denmark	Norway	
Rwanda	180.5	8.0	39.8	18.8	5.4	26.5	20.3	6.2	13.1	1.4	4.8	36.1
Saudi Arabia	19.1	13.9	..	1.8	3.4	..	..	..	..	..	..	0.1
Senegal	416.2	59.1	23.2	26.4	226.4	0.7	5.8	17.5	0.2	−0.7	1.2	56.5
Sierra Leone	59.9	1.2	17.4	4.4	0.5	17.1	3.3	2.2	2.5	0.0	6.2	5.2
Singapore	−1.5	1.3	..	−5.9	2.1	0.1	0.1	0.8	..	..	..	0.1
Slovak Republic	35.0	2.5	12.3	6.4	3.3	3.7	..	0.7	0.1	0.5	..	5.5
Slovenia	1.3	−3.9	..	1.9	0.9	−0.7	..	0.0	..	0.1	..	3.0
South Africa	386.2	14.1	84.6	51.1	27.8	62.9	26.6	11.6	40.9	17.2	15.9	33.5
Spain												
Sri Lanka	207.7	136.0	5.2	10.8	−0.7	9.3	3.1	2.5	14.3	1.4	14.0	11.8
Sudan	158.5	0.6	71.5	12.7	2.7	13.2	15.3	4.1	6.8	0.6	14.2	16.9
Sweden												
Switzerland												
Syrian Arab Republic	172.3	136.2	..	15.3	12.8	0.2	0.1	0.1	0.1	..	1.3	6.4
Tajikistan	35.1	1.6	19.5	3.9	0.0	0.1	1.4	1.8	1.2	0.2	0.6	4.7
Tanzania	613.4	74.8	26.5	66.6	4.9	88.6	55.2	13.3	46.2	80.9	49.7	106.7
Thailand	994.8	880.3	2.0	57.5	−2.6	1.3	0.3	4.1	3.6	30.3	0.5	17.5
Togo	47.1	9.4	3.9	10.1	19.5	0.6	0.0	0.4	0.3	..	0.0	2.9
Trinidad and Tobago	0.2	2.3	..	−3.6	0.5	0.4	0.0	0.4	..	..	..	0.2
Tunisia	102.0	29.9	−20.7	−11.9	103.6	0.2	4.4	2.6	1.5	0.0	0.3	−7.9
Turkey	−66.4	−45.6	−72.8	5.6	23.4	2.3	3.3	0.1	1.5	−2.4	4.1	14.2
Turkmenistan	11.5	1.7	8.3	0.9	0.3	0.2	..	0.1	..	..	..	0.1
Uganda	357.5	28.2	47.4	28.6	1.6	96.4	26.5	2.6	20.3	58.9	25.5	21.6
Ukraine	401.7	0.9	319.7	29.5	5.3	13.4	0.6	14.1	3.5	4.5	0.1	10.3
United Arab Emirates	2.9	0.1	..	1.0	1.8	..	..	..	..	..	..	..
United Kingdom												
United States												
Uruguay	19.0	5.9	0.4	7.8	1.8	0.3	..	1.3	0.1	..	0.1	1.3
Uzbekistan	112.8	81.6	17.5	9.0	2.0	1.0	0.0	0.1	0.3	..	..	1.3
Venezuela, RB	34.2	5.8	1.1	3.4	3.7	0.3	0.1	0.9	0.1	1.5	..	17.4
Vietnam	1,017.7	680.0	−0.2	66.1	79.2	8.4	11.7	11.3	33.1	39.4	7.1	81.5
West Bank and Gaza	326.6	56.1	84.9	26.4	12.0	10.7	12.5	0.5	25.1	8.5	27.8	62.1
Yemen, Rep.	177.3	41.8	44.7	31.6	14.1	1.8	33.5	0.2	0.6	3.0	2.3	3.8
Yugoslavia, FR (Serb./Mont.)	635.2	0.1	36.9	119.1	3.0	1.1	63.1	30.3	19.6	0.0	96.2	265.8
Zambia	340.0	59.4	26.6	64.7	17.7	63.6	13.5	9.3	15.3	25.7	27.4	16.9
Zimbabwe	219.2	78.0	20.2	8.5	0.2	26.4	13.0	3.7	19.1	28.6	13.3	8.5
World	**42,519 s**	**10,503 s**	**10,291 s**	**3,520 s**	**4,266 s**	**2,346 s**	**2,184 s**	**1,337 s**	**1,240 s**	**1,153 s**	**1,034 s**	**4,645 s**
Low income	14,775	4,816	2,193	1,250	1,310	1,091	615	353	422	591	414	1,720
Middle income	16,037	4,411	3,989	1,710	1,503	661	442	387	345	273	344	1,972
Lower middle income	13,459	4,006	3,495	1,386	1,146	412	400	216	262	200	299	1,637
Upper middle income	1,756	342	−19	328	324	176	29	157	63	73	31	250
Low & middle income	40,790	10,502	9,301	3,602	3,582	2,344	2,064	1,335	1,239	1,153	1,034	4,633
East Asia & Pacific	8,322	5,337	638	536	257	147	140	94	79	95	54	945
Europe & Central Asia	6,249	330	3,000	592	321	179	197	216	174	133	201	907
Latin America & Carib.	4,131	814	1,238	398	167	316	149	137	165	76	70	602
Middle East & N. Africa	3,404	599	1,132	353	805	39	76	30	45	52	69	204
South Asia	2,476	1,166	251	210	−4	328	75	62	65	110	78	133
Sub-Saharan Africa	8,434	985	1,339	936	1,376	786	450	241	364	436	367	1,153
High income	1,729	1	990	−82	685	3	119	1	1	0	0	12
Europe EMU												

Note: Regional aggregates include data for economies that are not specified elsewhere. World and income group totals include aid not allocated by country or region.

Distribution of net aid by Development Assistance Committee members | 6.11

The data in the table show net bilateral aid to low- and middle-income economies from members of the Development Assistance Committee (DAC) of the Organisation for Economic Co-operation and Development (OECD). The DAC compilation includes aid to some countries and territories not shown in the table and small quantities to unspecified economies that are recorded only at the regional or global level. Aid to countries and territories not shown in the table has been assigned to regional totals based on the World Bank's regional classification system. Aid to unspecified economies has been included in regional totals and, when possible, in income group totals. Aid not allocated by country or region—including administrative costs, research on development issues, and aid to nongovernmental organizations—is included in the world total; thus regional and income group totals do not sum to the world total.

In 1999 all United Nations agencies revised their data to include only regular budgetary expenditures since 1990 (except for the World Food Programme and the United Nations High Commissioner for Refugees, which revised their data from 1996 onward). They did so to avoid double counting extrabudgetary expenditures reported by DAC countries and flows reported by the United Nations.

Because the data in the table are based on donor country reports of bilateral programs, they cannot be reconciled with recipient country reports. Nor do they reflect the full extent of aid flows from the reporting donor countries or those to recipient countries. A full accounting would include donor country contributions to multilateral institutions and the flow of resources from multilateral institutions to recipient countries as well as flows from countries that are not members of DAC. In addition, the expenditures countries report as official development assistance (ODA) have changed. For example, some DAC members providing aid to refugees within their own borders have reported these expenditures as ODA.

Some of the aid recipients shown in the table are themselves significant donors. See table 6.8a for a summary of ODA from non-DAC countries.

• **Net aid** comprises net bilateral official development assistance to part I recipients and net bilateral official aid to part II recipients (see *About the data* for table 6.8). • **Other DAC donors** are Australia, Austria, Belgium, Finland, Greece, Ireland, Italy, Luxembourg, New Zealand, Portugal, Spain, and Switzerland.

Data on aid are compiled by DAC and published in its annual statistical report, *Geographical Distribution of Financial Flows to Aid Recipients,* and in the DAC chairman's annual report, *Development Co-operation.* The OECD also makes its data available on diskette, magnetic tape, and the Internet. Data are available to registered users from the OECD Web site at www.oecd.org/dac/htm/online.htm.

Figure 6.11

The flow of aid from DAC members in 1999 reflected geopolitical interests and historical ties

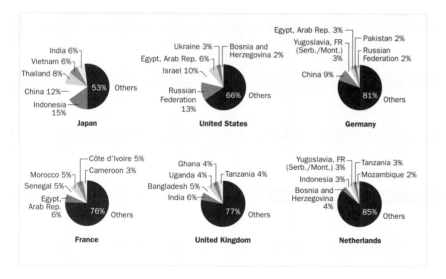

Source: OECD data.

The states and regions of the former Socialist Federal Republic of Yugoslavia drew a larger share of aid from DAC members in 1999. And aid from Japan to Asia increased after the East Asian financial crisis of 1997–98.

Net financial flows from multilateral institutions

| $ millions, 1999 | International financial institutions | | | | | | | United Nations | | | | | Total |
| | World Bank | | IMF | | Regional development banks | | | | | | | | |
	IDA	IBRD	Concessional	Non-concessional	Concessional	Non-concessional	Others	UNDP	UNFPA	UNICEF	WFP	Others	
Albania	80.6	0.0	17.7	0.0	0.0	−1.1	25.6	2.6	0.6	0.5	..	1.3	127.8
Algeria	0.0	−136.6	0.0	−53.7	0.0	35.6	132.0	1.2	0.8	1.0	1.2	5.0	−13.6
Angola	38.8	0.0	0.0	0.0	0.0	0.0	−1.3	7.5	2.1	4.7	27.6	3.2	82.7
Argentina	0.0	1,127.9	0.0	−823.8	−1.9	901.6	1.4	1.1	0.1	1.1	..	26.0	1,233.4
Armenia	65.7	−0.4	28.6	−13.1	0.0	−4.0	4.8	3.2	0.2	1.0	1.5	1.1	88.5
Australia													
Austria													
Azerbaijan	60.5	0.0	0.0	93.8	0.0	23.0	3.0	2.7	0.5	1.0	3.9	0.5	188.9
Bangladesh	339.6	−4.9	−93.0	0.0	190.6	20.4	9.6	13.9	6.0	13.9	23.1	7.1	526.2
Belarus	0.0	1.3	0.0	−58.1	0.0	−14.9	0.0	0.5	0.2	..	..	1.3	−69.8
Belgium													
Benin	42.9	0.0	0.0	1.0	14.4	0.0	−7.4	3.4	1.4	1.7	0.8	3.3	61.6
Bolivia	73.9	−13.5	−10.5	0.0	60.9	−28.5	29.1	2.2	1.0	1.3	4.5	3.2	123.5
Bosnia and Herzegovina	68.0	0.0	0.0	19.0	0.0	16.5	−209.9	5.3	0.3	1.6	..	0.6	−98.6
Botswana	−0.5	−11.2	0.0	0.0	−1.1	−15.2	−7.2	0.7	0.5	1.1	..	2.1	−31.0
Brazil	0.0	580.4	0.0	4,107.9	0.0	2,717.1	−1,299.7	0.3	1.3	1.7	..	65.1	6,174.1
Bulgaria	0.0	198.9	0.0	162.0	0.0	4.7	94.4	1.0	0.0	..	..	2.3	463.3
Burkina Faso	57.2	0.0	0.0	11.4	8.3	−1.9	−3.8	5.3	1.6	3.2	4.4	3.2	88.7
Burundi	7.6	0.0	−7.0	0.0	3.2	−2.0	−2.2	9.4	0.7	5.3	2.5	2.2	19.6
Cambodia	26.8	0.0	8.1	0.0	20.7	0.0	0.2	7.9	3.3	3.5	2.3	2.3	75.1
Cameroon	59.0	−63.3	0.0	43.4	34.0	−12.4	−14.6	1.7	0.7	1.3	5.3	4.0	59.0
Canada													
Central African Republic	−1.3	0.0	9.6	−3.3	0.0	−0.9	1.5	1.9	1.1	1.0	1.2	4.9	15.6
Chad	48.1	0.0	8.4	−1.8	23.6	0.0	−0.4	5.9	1.3	2.2	3.6	2.0	92.9
Chile	−0.7	−67.4	0.0	0.0	−0.9	28.9	0.8	0.2	0.1	0.9	..	1.9	−36.2
China	406.8	788.3	0.0	0.0	0.0	466.6	41.2	15.5	5.5	14.8	6.5	8.6	1,753.8
Hong Kong, China	..	..	..	..	..	..	..	−0.1	..	..	..	0.1	−0.1
Colombia	−0.7	218.0	0.0	0.0	−12.5	664.7	126.1	1.0	0.3	1.3	2.6	3.0	1,003.7
Congo, Dem. Rep.	0.0	0.0	0.0	−0.9	0.0	0.0	0.0	22.5	0.2	10.6	2.0	3.8	38.3
Congo, Rep.	0.0	0.0	0.0	−4.3	0.0	0.0	0.0	0.1	0.3	1.7	1.0	13.9	12.7
Costa Rica	−0.2	−22.6	0.0	0.0	−10.9	−21.3	31.3	0.3	0.2	0.6	..	2.6	−20.1
Côte d'Ivoire	52.8	−120.7	−8.1	0.0	6.4	−47.8	−17.5	2.0	1.5	2.7	0.9	11.7	−116.4
Croatia	0.0	66.6	0.0	−31.3	0.0	8.9	16.1	1.2	..	0.1	..	0.2	61.8
Cuba	..	..	..	..	..	..	..	1.0	0.4	1.0	3.2	3.3	8.7
Czech Republic	0.0	−35.7	0.0	0.0	0.0	0.0	44.7	0.4	..	..	..	3.5	12.9
Denmark													
Dominican Republic	−0.7	71.7	0.0	0.0	8.7	41.2	−5.1	0.7	0.7	0.9	1.2	7.1	126.4
Ecuador	−1.1	8.3	0.0	−67.6	−4.0	38.7	143.8	2.1	0.7	1.0	3.1	2.1	127.1
Egypt, Arab Rep.	13.7	−74.2	0.0	0.0	−1.4	−54.2	−33.6	3.9	3.2	3.3	3.6	9.4	−126.3
El Salvador	−0.8	6.3	0.0	0.0	−15.5	115.9	−44.0	1.3	0.5	0.8	2.8	1.1	68.3
Eritrea	19.5	0.0	0.0	0.0	5.1	0.0	18.5	3.9	1.6	2.3	0.4	2.2	53.5
Estonia	0.0	15.7	0.0	−4.0	0.0	−3.3	3.4	0.3	0.0	..	..	0.1	12.2
Ethiopia	136.8	0.0	−9.7	0.0	26.8	−11.1	16.9	7.7	0.9	14.4	19.9	19.9	222.4
Finland													
France													
Gabon	0.0	−5.9	0.0	−24.5	0.0	−15.5	−4.0	0.2	0.5	0.6	..	1.7	−47.0
Gambia, The	3.6	0.0	1.2	0.0	2.7	−1.0	2.2	2.6	0.7	0.8	1.3	1.8	15.9
Georgia	78.8	0.0	45.5	−21.5	0.0	−0.1	−9.8	3.5	0.2	0.9	1.7	0.6	99.8
Germany													
Ghana	198.6	−7.7	0.0	−15.1	14.6	−14.5	2.5	3.3	3.3	3.0	1.3	3.5	192.7
Greece													
Guatemala	0.0	55.3	0.0	0.0	6.8	116.5	14.8	0.5	0.7	0.9	4.5	1.4	201.4
Guinea	19.1	0.0	3.6	0.0	3.8	−5.5	19.5	3.7	0.7	1.9	0.5	27.7	75.0
Guinea-Bissau	−0.6	0.0	−0.6	2.9	0.0	0.0	−1.1	2.0	0.1	1.2	0.5	1.5	5.9
Haiti	8.2	0.0	−11.3	0.0	57.0	0.0	−0.7	4.0	2.0	1.6	3.4	1.2	65.3
Honduras	270.9	−46.9	0.0	100.2	57.4	−5.0	−46.8	3.9	0.9	1.4	2.8	2.3	341.0

Net financial flows from multilateral institutions 6.12

$ millions, 1999	International financial institutions							United Nations					Total
	World Bank		IMF		Regional development banks								
	IDA	IBRD	Conces-sional	Non-concessional	Conces-sional	Non-concessional	Others	UNDP	UNFPA	UNICEF	WFP	Others	
Hungary	0.0	−23.6	0.0	0.0	0.0	203.3	11.4	0.4	..	..	..	1.6	193.1
India	486.1	−254.3	−261.8	0.0	0.0	359.8	96.1	16.0	7.0	30.5	21.4	15.1	515.9
Indonesia	−12.4	732.7	0.0	1,382.4	0.5	737.4	26.7	4.1	4.1	7.7	0.1	7.6	2,890.8
Iran, Islamic Rep.	0.0	3.9	0.0	0.0	0.0	0.0	−16.9	0.3	1.2	1.7	0.7	16.8	7.6
Iraq	..	..	..	..	..	..	..	−19.4	0.3	1.5	1.2	4.9	−11.6
Ireland													
Israel	..	..	..	..	..	..	..	..	..	..	..	0.6	0.6
Italy													
Jamaica	0.0	−11.3	−19.0	0.0	−4.7	43.0	−6.6	−0.3	0.2	0.9	..	1.8	4.1
Japan													
Jordan	−2.6	95.1	0.0	40.3	0.0	0.0	110.1	0.7	0.6	0.9	2.2	86.5	333.7
Kazakhstan	0.0	189.8	0.0	−175.7	5.9	6.9	36.9	1.2	0.2	0.9	..	1.1	67.1
Kenya	55.1	−58.5	−59.8	0.0	3.5	−22.7	−10.4	5.4	3.0	4.9	8.5	21.0	−49.9
Korea, Dem. Rep.	..	..	..	..	..	..	..	1.3	0.5	4.9	0.4	2.7	9.8
Korea, Rep.	−3.5	843.6	0.0	−10,306.7	0.0	−15.6	0.0	0.6	..	..	..	1.5	−9,480.1
Kuwait	..	..	..	..	..	..	..	0.0	..	..	..	1.4	1.3
Kyrgyz Republic	21.6	0.0	0.0	19.5	62.4	25.0	15.2	2.4	0.6	0.9	..	0.7	148.2
Lao PDR	18.5	0.0	−8.0	0.0	37.7	0.0	18.3	4.4	1.0	2.9	..	1.8	76.5
Latvia	0.0	23.4	0.0	−15.1	0.0	7.7	252.3	0.4	0.0	..	..	0.7	269.4
Lebanon	0.0	36.4	0.0	0.0	0.0	0.0	56.3	1.3	0.5	0.9	..	48.2	143.5
Lesotho	8.1	−4.1	0.0	−5.9	−0.1	−2.4	−2.6	1.3	0.2	1.1	1.4	1.3	−1.9
Libya	..	..	..	..	..	..	..	−0.5	..	..	..	4.7	4.2
Lithuania	0.0	26.1	0.0	−16.5	0.0	8.1	59.1	0.5	0.0	..	..	0.5	77.8
Macedonia, FYR	45.0	8.8	0.0	1.9	0.0	−12.8	51.6	1.0	0.0	0.7	..	1.1	97.3
Madagascar	68.6	−0.7	6.3	0.0	16.9	−4.9	5.9	5.5	2.0	3.5	1.3	2.1	106.5
Malawi	74.6	−7.9	−3.3	−8.7	30.2	2.0	14.7	7.7	1.8	4.1	0.5	2.7	118.4
Malaysia	0.0	−68.7	0.0	0.0	0.0	−15.4	−1.1	0.3	0.3	0.7	..	2.0	−81.9
Mali	46.3	0.0	0.0	11.6	20.0	0.0	8.8	7.5	1.1	5.1	2.1	2.8	105.3
Mauritania	16.0	−2.0	−1.0	0.0	15.2	−8.1	−19.3	2.5	0.8	1.4	0.9	2.5	8.9
Mauritius	−0.6	−12.4	0.0	0.0	−0.4	−3.5	11.8	0.3	0.1	0.7	..	1.0	−3.1
Mexico	0.0	−484.4	−3,681.5	0.0	−2.3	119.2	5.5	0.8	1.4	1.0	..	11.4	−4,029.1
Moldova	39.3	30.7	0.0	2.9	0.0	4.1	−0.1	0.6	0.1	0.7	..	0.8	79.0
Mongolia	14.2	0.0	4.3	0.0	48.9	0.0	0.7	2.0	1.5	0.7	..	2.9	75.1
Morocco	−1.4	102.1	0.0	0.0	1.9	−16.3	49.2	2.2	2.0	1.8	2.6	3.2	147.2
Mozambique	78.0	0.0	0.0	−2.5	15.1	−5.1	20.2	8.8	3.3	6.6	2.2	2.7	129.3
Myanmar	0.0	0.0	0.0	0.0	0.0	0.0	−3.8	15.6	0.9	7.9	..	4.2	24.8
Namibia	..	..	..	..	..	..	..	1.6	0.7	1.7	0.3	4.7	8.9
Nepal	33.9	0.0	−5.9	0.0	52.6	0.0	0.7	8.9	3.9	4.5	10.0	9.9	118.4
Netherlands													
New Zealand													
Nicaragua	119.2	−5.7	104.4	0.0	75.2	−3.9	−5.3	3.3	1.8	0.9	16.5	1.6	308.0
Niger	19.3	0.0	−4.1	−1.9	11.7	0.0	0.3	5.4	1.4	5.5	2.0	2.6	42.2
Nigeria	72.2	−260.8	0.0	0.0	6.3	−44.4	−47.8	3.6	3.4	11.8	..	4.4	−251.3
Norway													
Oman	0.0	−3.5	0.0	0.0	0.0	0.0	12.3	..	..	0.8	..	1.5	11.1
Pakistan	134.9	209.1	413.6	−36.7	142.7	−99.5	121.7	7.4	5.5	9.4	9.6	17.8	935.5
Panama	0.0	9.2	0.0	−23.5	−9.4	35.3	5.1	−0.2	0.2	0.6	..	1.5	18.9
Papua New Guinea	−2.6	−23.0	0.0	−22.8	3.1	11.8	−4.2	−0.1	0.7	0.7	..	2.1	−34.3
Paraguay	−1.5	19.0	0.0	0.0	5.5	62.7	0.4	0.2	0.5	0.9	..	1.2	88.8
Peru	0.0	290.9	0.0	−146.5	−6.5	389.6	−33.2	−3.8	2.9	1.1	5.0	9.0	508.4
Philippines	3.2	−227.0	0.0	292.3	20.9	−96.6	−0.7	6.0	2.6	3.2	..	3.1	6.9
Poland	0.0	58.7	0.0	0.0	0.0	0.0	1.9	0.7	0.2	..	..	1.8	63.2
Portugal													
Puerto Rico													
Romania	0.0	232.6	0.0	−67.0	0.0	189.9	302.9	0.8	0.5	0.9	..	2.2	662.8
Russian Federation	0.0	388.3	0.0	−3,595.8	0.0	−13.3	−17.7	1.0	0.4	..	..	8.7	−3,228.5

| $ millions, 1999 | International financial institutions | | | | | | | United Nations | | | | | Total |
| | World Bank | | IMF | | Regional development banks | | | | | | | | |
	IDA	IBRD	Concessional	Non-concessional	Concessional	Non-concessional	Others	UNDP	UNFPA	UNICEF	WFP	Others	
Rwanda	63.5	0.0	0.0	20.8	9.2	0.0	-0.5	12.2	1.7	2.2	34.0	1.9	144.8
Saudi Arabia	..	..	..	..	..	..	..	0.0	..	0.0	..	9.1	9.2
Senegal	37.2	-4.5	-7.8	-5.3	5.3	-13.6	-34.1	2.3	1.8	1.5	3.9	4.9	-8.5
Sierra Leone	7.1	-0.6	-12.4	21.3	2.3	0.0	1.3	3.2	0.2	1.8	0.8	2.8	27.8
Singapore	..	..	..	..	..	..	..	..	..	..	..	0.4	0.4
Slovak Republic	0.0	-17.7	0.0	-52.1	0.0	-19.9	197.2	0.4	..	..	..	1.2	109.0
Slovenia	..	..	..	..	..	..	..	0.1	..	..	..	2.0	2.2
South Africa	0.0	0.7	0.0	0.0	0.0	0.0	0.0	3.0	0.9	5.2	..	6.7	16.4
Spain													
Sri Lanka	34.3	-6.1	-99.4	0.0	77.9	0.0	1.5	5.7	1.0	1.2	3.5	4.8	24.3
Sudan	0.0	-3.3	0.0	-37.8	0.0	-2.4	0.0	10.1	2.9	4.9	7.3	13.5	-4.9
Sweden													
Switzerland													
Syrian Arab Republic	-1.5	-21.2	0.0	0.0	0.0	0.0	-31.7	1.3	1.3	0.8	6.3	24.1	-20.6
Tajikistan	35.8	0.0	9.1	-5.1	0.0	0.0	0.0	3.2	0.7	1.1	4.3	0.7	49.7
Tanzania	174.9	-5.9	0.0	51.1	45.6	-1.2	-14.0	9.9	3.3	9.0	2.8	5.1	280.7
Thailand	-2.8	626.2	0.0	273.5	-1.8	195.9	-12.7	2.1	0.4	1.0	..	6.5	1,088.3
Togo	14.5	0.0	-9.5	0.0	3.2	0.0	-1.3	3.8	0.7	1.3	..	1.7	14.4
Trinidad and Tobago	0.0	2.4	0.0	0.0	-0.1	28.1	-1.5	0.0	..	..	..	0.9	29.9
Tunisia	-2.1	43.8	0.0	-50.1	0.0	28.0	82.4	0.5	0.6	0.9	..	2.0	106.0
Turkey	-5.9	-233.2	0.0	510.1	0.0	0.0	-253.4	0.6	0.8	1.2	..	6.5	26.7
Turkmenistan	0.0	-0.1	0.0	0.0	0.0	6.8	1.2	0.8	0.5	0.8	..	0.3	10.3
Uganda	121.7	0.0	0.0	-16.3	21.9	-13.9	3.5	6.2	4.1	6.2	3.0	19.8	156.0
Ukraine	0.0	420.2	0.0	81.5	0.0	2.4	0.0	0.9	0.2	..	..	3.1	508.2
United Arab Emirates	..	..	..	..	..	..	..	0.2	..	..	..	1.1	1.2
United Kingdom													
United States													
Uruguay	0.0	3.1	0.0	0.0	-1.5	296.6	-7.5	0.4	0.2	0.7	..	0.8	292.8
Uzbekistan	0.0	27.1	0.0	-25.1	0.1	86.9	0.1	1.5	0.9	1.2	..	0.7	93.3
Venezuela, RB	0.0	-89.6	0.0	-451.7	0.0	82.5	66.5	0.2	0.3	0.8	..	2.9	-388.2
Vietnam	156.1	0.0	0.0	-26.2	161.2	0.0	18.9	11.9	5.3	6.0	8.8	4.0	345.9
West Bank and Gaza	..	..	..	..	..	..	..	..	1.2	1.5	2.7	142.0	147.3
Yemen, Rep.	160.4	0.0	0.0	81.0	0.0	0.0	-9.4	8.1	1.8	2.5	3.9	7.4	255.7
Yugoslavia, FR (Serb./Mont.)	0.0	0.0	0.0	-0.6	0.0	0.0	0.0	1.4	..	..	1.5	0.0	2.3
Zambia	151.6	-8.7	13.7	0.0	18.8	-15.0	-3.1	5.5	1.0	2.8	4.3	6.4	177.4
Zimbabwe	19.8	-19.9	-27.8	0.0	16.5	-32.3	-3.8	4.6	1.7	1.2	..	3.0	-37.0
World	**4,508 s**	**5,089 s**	**-3,678 s**	**-8,886 s**	**1,463 s**	**7,400 s**	**290 s**	**516 s**	**187 s**	**569 s**	**356 s**	**1,161 s**	**8,975 s**
Low income	3,506	586	125	1,592	1,238	897	215	344	105	254	291	324	9,476
Middle income	1,002	4,503	-3,803	-10,478	225	6,503	75	66	45	77	65	632	-1,088
Lower middle income	1,006	2,577	-121	-2,869	244	2,173	899	53	37	59	65	407	4,531
Upper middle income	-4	1,926	-3,681	-7,610	-19	4,330	-825	13	7	17	0	201	-5,645
Low & middle income	4,508	5,089	-3,678	-8,886	1,463	7,400	290	411	162	336	356	1,058	8,510
East Asia & Pacific	612	2,669	4	-8,407	296	1,282	70	74	31	57	20	82	-3,210
Europe & Central Asia	489	1,378	101	-3,190	68	525	631	38	8	16	13	44	121
Latin America & Carib.	475	1,651	-3,628	2,695	230	5,619	-975	22	19	25	50	196	6,379
Middle East & N. Africa	167	46	0	22	2	-7	401	1	15	18	26	396	1,087
South Asia	1,032	-56	-47	-37	465	281	232	67	26	67	71	62	2,162
Sub-Saharan Africa	1,733	-598	-109	32	401	-299	-68	209	63	153	176	278	1,971
High income	..	..	..	..	..	..	..	..	..	..	..	7	7
Europe EMU													

Note: The aggregates for the regional development banks, the United Nations, and total net financial flows include amounts for economies that are not specified elsewhere.

Net financial flows from multilateral institutions 6.12

This table shows concessional and nonconcessional financial flows from the major multilateral institutions—the World Bank, the International Monetary Fund (IMF), regional development banks, United Nations agencies, and regional groups such as the Commission of the European Communities. Much of these data come from the World Bank's Debtor Reporting System.

The multilateral development banks fund their non-concessional lending operations primarily by selling low-interest, highly rated bonds (the World Bank, for example, has a AAA rating) backed by prudent lending and financial policies and the strong financial backing of their members. These funds are then on-lent at slightly higher interest rates, and with relatively long maturities (15–20 years), to developing countries. Lending terms vary with market conditions and the policies of the banks.

Concessional flows are defined by the Development Assistance Committee (DAC) as those containing a grant element of at least 25 percent. The grant element of loans is evaluated assuming a nominal, market interest rate of 10 percent. The grant element of a loan carrying a 10 percent interest rate is nil, and for a grant, which requires no repayment, it is 100 percent.

Concessional, or soft, lending by the World Bank Group is carried out through the International Development Association (IDA), although some loans by the International Bank for Reconstruction and Development (IBRD) are made on terms that may qualify as concessional under the DAC definition. Eligibility for IDA resources is based on gross national income (GNI) per capita; countries must also meet performance standards assessed by World Bank staff. Since 1 July 1999 the GNI per capita cutoff has been set at $885, measured in 1998 using the Atlas method (see *Users guide*). In exceptional circumstances IDA extends eligibility temporarily to countries that are above the cutoff and are undertaking major adjustment efforts but are not creditworthy for IBRD lending. An exception has also been made for small island economies. Lending by the International Finance Corporation is not included in this table.

The IMF makes concessional funds available through its Enhanced Structural Adjustment Facility (ESAF), the successor to the Structural Adjustment Facility, and through the IMF Trust Fund. Low-income countries facing protracted balance of payments problems are eligible for ESAF funds.

Regional development banks also maintain concessional windows for funds. In the *World Development Indicators* loans from the major regional development banks—the African Development Bank, Asian Development Bank, and Inter-American Development Bank—are recorded according to each institution's classification. In some cases nonconcessional loans by these institutions may be on terms that meet DAC's definition of concessional.

In 1999 all United Nations agencies revised their data to include only regular budgetary expenditures since 1990 (except for the World Food Programme and the United Nations High Commissioner for Refugees, which revised their data from 1996 onward). They did so to avoid double counting extrabudgetary expenditures reported by DAC countries and flows reported by the United Nations.

• **Net financial flows** recorded in this table are disbursements of public or publicly guaranteed loans and credits less repayments of principal. • **IDA** is the International Development Association, the soft loan window of the World Bank Group. • **IBRD** is the International Bank for Reconstruction and Development, the founding and largest member of the World Bank Group. • **IMF** is the International Monetary Fund. Its nonconcessional lending consists of the credit it provides to its members, principally to meet their balance of payments needs. It provides concessional assistance through the Enhanced Structural Adjustment Facility and the IMF Trust Fund. • **Regional development banks** include the African Development Bank, based in Abidjan, Côte d'Ivoire, which lends to all of Africa, including North Africa; the Asian Development Bank, based in Manila, Philippines, which serves countries in South Asia and East Asia and Pacific; the European Bank for Reconstruction and Development, based in London, England, which serves countries in Europe and Central Asia; the European Development Fund, based in Brussels, Belgium, which serves countries in Africa, the Caribbean, and the Pacific; and the Inter-American Development Bank, based in Washington, D.C., which is the principal development bank of the Americas. • **Others** is a residual category in the World Bank's Debtor Reporting System. It includes such institutions as the Caribbean Development Bank and European Investment Bank. • **United Nations** includes the United Nations Development Programme (UNDP), United Nations Population Fund (UNFPA), United Nations Children's Fund (UNICEF), World Food Programme (WFP), and other United Nations agencies, such as the United Nations High Commissioner for Refugees, United Nations Relief and Works Agency for Palestine Refugees in the Near East, and United Nations Regular Program for Technical Assistance. • **Concessional financial flows** cover disbursements made through concessional lending facilities. • **Nonconcessional financial flows** cover all other disbursements.

The data on net financial flows from international financial institutions come from the World Bank's Debtor Reporting System. These data are published in the World Bank's *Global Development Finance 2001*. The data on aid from United Nations agencies come from the DAC chairman's report, *Development Co-operation*. Data are available to registered users from the OECD Web site at www.oecd.org/dac/htm/online.htm.

	Foreign population[a]				Foreign labor force[b]		Inflows of foreign population					
	thousands		% of total population		% of total labor force		Total thousands[c]		Foreign workers thousands		Asylum seekers thousands	
	1990	1998	1990	1998	1990	1998	1990	1998	1990	1998	1990	1998
Austria	456	737	5.9	9.1	7.4	9.9	..	..	103	15	23	14
Belgium	905	892	9.1	8.7	7.1	8.8	50	51	..	7	13	22
Denmark	161	256	3.1	4.8	2.4	3.4	15	21	3	3	5	6
Finland	26	85	0.5	1.6	..	1.4	6	8	..	..	3	1
France	3,597	..	6.3	..	6.2	6.1	102[d]	138[d]	26	15	55	22
Germany	5,343	7,320	8.4	8.9	7.1[e]	..[e]	842[d]	606[d]	139	276	193	99
Ireland	80	111	2.3	3.0	2.6	3.2	..	21	1	6	0	5
Italy	781	1,250	1.4	2.2	..	..	..	111[d]	..	..	5	11
Japan	1,075	1,512	0.9	1.2	..	0.2	224	265	..	..	..	..
Luxembourg	113	153	29.4	35.6	45.3[e]	57.7[e]	9	11	17	..	0	2
Netherlands	692	662	4.6	4.2	3.1[e]	2.9[e]	81	82	..	..	21	45
Norway	143	165	3.4	3.7	2.3	3.0	16	27	..	..	4	9
Portugal	108	178	1.1	1.8	1.0	1.8	..	..	..	..	0	0
Spain	279	720	0.7	1.8	0.6	1.2	..	..	16	..	9	7
Sweden	484	500	5.6	5.6	5.4	5.1	53	36	..	..	29	13
Switzerland	1,100	1,348	16.3	19.0	18.9	17.3	101	75	47	26	36	41
United Kingdom	1,723	2,207	3.2	3.8	3.3	3.9	..	..	35	50	38	58

	Foreign-born population[a]				Foreign-born labor force[b]		Inflows of foreign population					
	thousands		% of total population		% of total labor force		Total thousands[c,d]		Foreign workers thousands		Asylum seekers thousands	
	1990	2000	1990	2000	1990	1998	1990	1998	1990	1998	1990	1998
Australia	..	..	..	..	..	24.8	121	77	43	26	4	8
Canada	..	..	..	..	..	..	214	174	230	..	37	23
United States	19,767[f]	28,379[g]	7.9[f]	10.4[g]	9.4	11.7	1,536	660	203	78	74	55

a. Data are from population registers or from registers of foreigners, except for France and the United States (censuses), Portugal (residence permits), and Ireland and the United Kingdom (labor force surveys), and refer to the population on 31 December of the year indicated. b. Data include the unemployed, except in Italy, Luxembourg, the Netherlands, Norway, and the United Kingdom. Cross-border workers and seasonal workers are excluded, unless otherwise noted. c. Inflow data are based on population registers and are not fully comparable because the criteria governing who gets registered differ from country to country. Counts for the Netherlands, Norway, and especially Germany include substantial numbers of asylum seekers. d. Data are based on residence permits or other sources. e. Includes cross-border workers. f. From the U.S. Census Bureau, *1990 Census of Population Listing*. g. From the U.S. Census Bureau, *Current Population Report* (March 2000).

Foreign labor and population in OECD countries | 6.13

About the data

The data in the table are based on national definitions and data collection practices and are not fully comparable across countries. Japan and the European members of the Organisation for Economic Co-operation and Development (OECD) have traditionally defined foreigners by nationality of descent. Australia, Canada, and the United States use place of birth, which is closer to the concept used in the United Nations' definition of the immigrant stock. Few countries, however, apply just one criterion in all circumstances. For this and other reasons, data based on the concept of foreign nationality and data based on the concept of foreign-born cannot be completely reconciled. See the notes to the table for other breaks in comparability between countries and over time.

Data on the size of the foreign labor force are also problematic. Countries use different permit systems to gather information on immigrants. Some countries issue a single permit for residence and work, while others issue separate residence and work permits. Differences in immigration laws across countries, particularly with respect to immigrants' access to the labor market, greatly affect the recording and measurement of migration and reduce the comparability of raw data at the international level. The data exclude temporary visitors and tourists (see table 6.14).

OECD countries are not the only ones that receive substantial migration flows. Migrant workers make up a significant share of the labor force in Gulf countries and in southern Africa, and people are displaced by wars and natural disasters throughout the world. Systematic recording of migration flows is difficult, however, especially in poor countries and those affected by civil disorder.

Definitions

• **Foreign (or foreign-born) population** is the number of foreign or foreign-born residents in a country. • **Foreign (or foreign-born) labor force as a percentage of total labor force** is the share of foreign or foreign-born workers in a country's workforce. • **Inflows of foreign population** are the gross arrivals of immigrants in the country shown. The total does not include asylum seekers, except as noted. • **Inflows of foreign workers** are the gross arrivals of foreign workers with legal employment status. The workers may be permanent or temporary. • **Asylum seekers** are those who apply for permission to remain in the country for humanitarian reasons.

Data sources

International migration data are collected by the OECD through information provided by national correspondents to the Continuous Reporting System on Migration (SOPEMI) network, which provides an annual overview of trends and policies. The data appear in the OECD's *Trends in International Migration 2000*.

Figure 6.13

OECD countries attracted immigrants from disparate locations in 1998

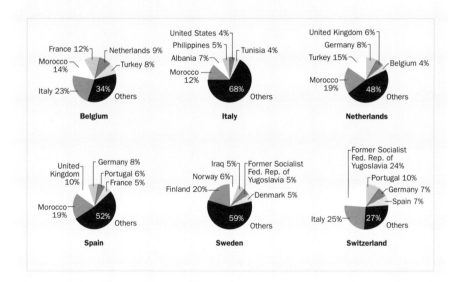

Source: OECD data.

A country's stock of immigrants reflects geographical, economic, and political connections with other countries. The largest share of immigrants in many OECD countries comes from other high-income economies.

6.14 Travel and tourism

	International tourism				International tourism receipts				International tourism expenditures			
	Inbound tourists thousands		Outbound tourists thousands		$ millions		% of exports		$ millions		% of imports	
	1990	1999	1990	1998	1990	1999	1990	1999	1990	1999	1990	1999
Albania	30	39	..	18	4	54	1.1	18.3	4	5	0.8	0.5
Algeria	1,137	755	3,828	1,377	64	24	0.5	0.2	149	40	1.5	0.5
Angola	46	45	..	3	13	13	0.3	0.2	38	70	1.1	1.3
Argentina	1,930	2,898	2,398	4,592	1,131	2,812	7.6	10.1	1,505	4,107	22.0	12.6
Armenia	15	41	..	..	..	27	..	7.0	..	34	..	3.7
Australia	2,215	4,459	2,170	3,161	4,088	7,525	8.2	10.3	4,535	5,792	8.5	6.9
Austria	19,011	17,467	8,527	13,263	13,417	11,088	21.1	11.7	7,748	9,195	12.6	9.6
Azerbaijan	77	63	..	343	42	125	3.3	12.4	..	170	..	7.0
Bangladesh	115	173	388	992	11	50	0.6	0.8	78	212	1.9	2.5
Belarus	..	355	..	969	..	22	..	0.3	..	124	..	1.4
Belgium	5,147	6,369	3,835	7,773	3,721	5,437	2.7	2.8	5,477	8,842	4.1	4.9
Benin	110	152	418	420	28	33	7.7	5.9	12	7	2.6	0.9
Bolivia	254	410	242	298	91	170	9.3	13.0	130	165	12.0	8.3
Bosnia and Herzegovina	1	89	..	..	..	13	..	..	..	..	..	..
Botswana	543	740	192	460	117	175	5.8	7.6	56	126	2.8	5.0
Brazil	1,091	5,107	1,188	4,598	1,444	3,994	4.1	7.2	1,559	3,059	5.5	4.8
Bulgaria	1,586	2,472	2,395	2,592	320	930	4.6	16.1	189	524	2.4	8.0
Burkina Faso	74	218	..	..	11	42	3.2	10.5	32	32	4.2	5.0
Burundi	109	15	24	16	4	1	4.5	1.4	17	12	5.3	6.9
Cambodia	17	368	..	41	50	190	15.9	16.8	..	13	..	1.0
Cameroon	89	135	..	..	53	40	2.4	1.7	279	107	14.5	5.2
Canada	15,209	19,557	20,415	17,640	6,339	10,025	4.2	3.6	10,931	11,302	7.3	4.4
Central African Republic	6	10	..	..	3	6	1.4	4.0	51	39	12.4	16.2
Chad	9	43	24	10	8	10	3.0	2.7	70	24	14.4	4.8
Chile	943	1,626	768	1,351	540	1,062	5.3	5.6	426	906	4.6	4.2
China	10,484	27,047	2,134	8,426	2,218	14,098	3.9	6.5	470	9,205	1.0	5.6
Hong Kong, China	6,581	11,328	2,043	4,197	5,032	7,210	5.0	3.4	..	..	..	..
Colombia	813	841	781	1,140	406	939	4.7	7.0	454	1,124	6.6	6.5
Congo, Dem. Rep.	55	53	..	..	7	2	0.3	0.1	16	7	0.6	0.5
Congo, Rep.	33	25	..	..	8	10	0.5	0.7	113	64	8.8	4.7
Costa Rica	435	1,027	191	330	275	1,002	14.0	12.2	148	428	6.3	6.0
Côte d'Ivoire	196	301	2	5	51	108	1.5	2.1	169	237	4.9	5.7
Croatia	7,049	3,443	..	..	1,704	2,502	..	30.8	729	712	..	7.3
Cuba	327	1,561	12	55	243	1,714	..	..	..	..	..	..
Czech Republic	7,278	16,031	3,510	..	419	3,035	..	9.1	455	1,474	..	4.3
Denmark	1,838	2,023	3,929	4,972	3,322	3,682	6.8	5.6	3,676	5,084	8.9	8.8
Dominican Republic	1,305	2,649	137	354	900	2,524	49.1	31.6	144	282	6.4	3.0
Ecuador	362	509	181	330	188	343	5.8	6.5	175	271	6.9	6.6
Egypt, Arab Rep.	2,411	4,489	2,012	2,854	1,100	3,903	12.0	28.8	129	1,153	0.9	5.3
El Salvador	194	658	525	868	18	211	1.8	6.7	61	81	3.8	1.8
Eritrea	169	57	..	..	..	28	..	42.7	..	..	..	..
Estonia	372	950	..	1,659	27	560	4.1	14.2	19	217	2.7	5.1
Ethiopia	79	91	89	140	25	11	3.7	1.1	11	46	1.0	2.8
Finland	1,572	2,700	1,169	4,743	1,167	1,460	3.7	3.0	2,791	1,944	8.3	5.1
France	52,497	73,042	19,430	18,077	20,184	31,699	7.1	8.3	12,423	17,732	4.4	5.2
Gabon	109	194	161	..	3	11	0.1	0.4	137	183	7.6	9.5
Gambia, The	100	91	..	..	26	33	15.5	12.5	8	16	4.2	5.7
Georgia	..	384	..	433	..	400	..	54.1	..	270	..	21.4
Germany	17,045	17,116	56,261	82,975	14,288	16,828	3.0	2.7	33,771	48,158	8.0	8.0
Ghana	146	373	..	..	81	284	8.2	11.2	13	24	0.9	0.7
Greece	8,873	12,000	1,651	1,935	2,587	8,765	19.9	34.7	1,090	3,989	5.6	5.2
Guatemala	509	823	289	391	185	570	11.8	16.4	100	157	5.5	3.1
Guinea	..	27	..	..	30	7	3.6	0.9	30	31	3.1	3.3
Guinea-Bissau	..	..	..	..	..	..	..	..	..	..	..	..
Haiti	144	147	..	..	46	57	14.5	11.9	37	37	7.2	3.6
Honduras	202	371	196	202	29	165	2.8	7.2	38	60	3.4	2.0

Travel and tourism | 6.14

	International tourism				International tourism receipts				International tourism expenditures			
	Inbound tourists thousands		Outbound tourists thousands		$ millions		% of exports		$ millions		% of imports	
	1990	1999	1990	1998	1990	1999	1990	1999	1990	1999	1990	1999
Hungary	20,510	12,930	13,596	12,317	824	3,394	6.8	12.3	477	1,191	4.3	4.2
India	1,707	2,482	2,281	3,811	1,513	3,036	6.6	5.6	393	1,713	1.2	2.9
Indonesia	2,178	4,700	688	2,076	2,105	4,045	7.2	7.4	836	2,102	3.0	4.8
Iran, Islamic Rep.	154	1,174	788	1,450	61	662	0.3	3.3	340	918	1.5	5.9
Iraq	748	51	239	..	55	13	..	..	..	..	..	..
Ireland	3,666	6,511	1,798	3,053	1,883	3,306	7.0	4.0	1,163	2,374	4.7	2.9
Israel	1,063	2,275	883	2,983	1,396	3,050	8.1	8.5	1,442	2,600	7.1	6.4
Italy	26,679	36,097	16,152	19,352	16,458	28,357	7.5	9.7	10,304	16,913	4.7	6.3
Jamaica	989	1,248	..	..	740	1,233	33.4	36.7	114	198	4.8	4.9
Japan	3,236	4,438	10,997	15,806	3,578	3,428	1.1	0.7	24,928	32,780	8.4	8.3
Jordan	572	1,358	1,143	1,347	512	795	20.4	22.6	336	355	9.0	7.1
Kazakhstan	..	..	..	..	..	289	..	4.3	..	143	..	1.9
Kenya	814	943	210	300	443	256	19.9	9.7	38	161	1.4	5.1
Korea, Dem. Rep.	115	130	..	..	..	..	..	..	..	..	..	..
Korea, Rep.	2,959	4,660	1,561	3,067	3,559	6,802	4.9	4.0	3,166	3,975	4.1	2.8
Kuwait	15	77	..	..	132	207	1.6	1.8	1,837	2,517	25.6	19.0
Kyrgyz Republic	..	69	..	32	2	7	..	1.2	..	4	..	0.5
Lao PDR	14	270	..	..	3	103	2.9	22.0	1	23	0.5	3.8
Latvia	..	489	..	1,961	7	111	0.6	3.8	13	265	1.3	7.4
Lebanon	210	673	..	1,650	..	807	..	67.2	..	..	..	..
Lesotho	171	186	..	..	17	19	17.0	8.8	12	12	1.6	1.4
Libya	96	40	425	650	6	28	0.1	0.4	424	150	4.7	2.8
Lithuania	780	1,422	..	3,241	..	550	..	13.0	..	341	..	6.4
Macedonia, FYR	..	..	..	..	..	..	..	..	..	..	..	..
Madagascar	53	138	34	35	40	100	8.5	10.6	40	111	4.9	9.0
Malawi	130	150	..	..	16	20	3.6	3.4	16	17	2.9	1.3
Malaysia	7,446	7,931	14,920	25,631	1,667	2,822	5.1	2.9	1,450	2,478	4.6	2.7
Mali	44	83	..	..	47	50	11.2	7.8	62	29	7.5	3.2
Mauritania	..	..	..	..	9	21	1.9	5.3	23	43	4.4	9.1
Mauritius	292	578	89	143	244	545	14.2	20.5	94	194	4.9	7.3
Mexico	17,176	19,043	7,357	9,637	5,467	7,223	11.2	4.9	5,519	4,541	10.6	2.9
Moldova	226	19	49	28	4	2	..	0.3	..	..	..	..
Mongolia	147	159	..	..	5	28	1.0	5.3	1	45	0.1	6.7
Morocco	4,024	3,824	1,202	1,480	1,259	1,960	20.2	18.4	184	460	2.4	3.8
Mozambique	..	..	..	..	..	..	..	..	..	..	..	..
Myanmar	21	198	..	..	9	35	1.4	3.0	16	27	1.4	1.0
Namibia	213	560	..	..	85	288	7.0	17.9	63	88	4.0	4.6
Nepal	255	492	82	122	64	168	16.9	14.6	45	78	5.9	4.7
Netherlands	5,795	9,881	9,000	13,560	4,155	7,092	2.6	2.9	7,376	11,366	5.0	5.0
New Zealand	976	1,607	717	1,166	1,030	2,083	8.8	12.3	958	1,405	8.2	8.9
Nicaragua	106	468	173	422	12	113	3.1	13.5	15	74	2.2	3.7
Niger	21	39	18	10	17	21	3.2	7.3	44	26	6.0	6.1
Nigeria	190	739	56	..	25	142	0.2	1.4	576	1,567	8.3	11.7
Norway	1,955	4,481	2,667	3,120	1,570	2,229	3.3	4.0	3,679	4,751	9.5	8.5
Oman	149	502	..	..	69	104	1.2	1.4	47	47	1.4	0.8
Pakistan	424	429	..	..	156	76	2.5	0.9	440	352	4.7	2.7
Panama	214	431	151	211	172	379	3.9	4.7	99	176	2.4	2.0
Papua New Guinea	41	70	66	63	41	104	3.0	4.8	50	73	3.3	4.1
Paraguay	280	272	264	318	128	595	5.1	13.7	103	142	4.7	3.1
Peru	317	944	329	616	217	913	5.3	12.2	295	466	7.2	4.4
Philippines	1,025	2,171	1,137	1,817	1,306	2,534	11.4	6.5	111	1,950	0.8	4.9
Poland	3,400	17,950	22,131	49,328	358	6,100	1.9	15.8	423	3,600	2.8	6.9
Portugal	8,020	11,600	2,268	2,425	3,555	5,169	16.5	15.2	867	2,291	3.2	4.9
Puerto Rico	2,560	3,024	996	1,250	1,366	2,138	..	..	630	815	..	..
Romania	3,009	3,209	11,247	6,893	106	254	1.7	2.6	103	395	1.0	3.5
Russian Federation	3,009	18,496	4,150	11,711	752	7,771	1.4	9.2	..	7,434	..	14.1

	International tourism				International tourism receipts				International tourism expenditures			
	Inbound tourists thousands		Outbound tourists thousands		$ millions		% of exports		$ millions		% of imports	
	1990	1999	1990	1998	1990	1999	1990	1999	1990	1999	1990	1999
Rwanda	16	2	..	..	10	19	6.9	18.3	23	17	6.4	4.8
Saudi Arabia	2,209	3,700	..	..	1,884	1,462	4.0	3.4	..	..	..	..
Senegal	246	369	..	..	167	166	11.5	12.6	105	53	5.7	3.4
Sierra Leone	98	..	..	..	19	..	9.1	..	4	..	1.9	..
Singapore	4,842	6,258	1,237	3,745	4,937	5,974	7.3	4.3	1,893	2,676	2.9	2.4
Slovak Republic	822	975	188	414	70	461	..	3.8	181	339	..	2.6
Slovenia	650	884	..	..	721	1,005	8.5	9.6	282	593	4.1	5.2
South Africa	1,029	6,253	616	3,080	992	2,738	3.7	7.9	1,117	1,842	5.3	5.6
Spain	34,085	51,772	10,698	13,203	18,593	32,913	22.2	20.0	4,254	5,624	4.2	3.3
Sri Lanka	298	436	297	518	132	275	5.8	4.9	74	224	2.5	3.3
Sudan	33	39	203	200	21	8	4.0	1.3	51	30	3.5	1.5
Sweden	1,900	2,595	6,232	11,422	2,906	3,894	4.1	3.6	6,286	7,557	8.9	8.0
Switzerland	13,200	10,800	9,627	12,213	7,411	7,355	7.6	6.2	5,873	6,963	6.1	6.5
Syrian Arab Republic	562	1,386	1,041	2,750	320	1,360	6.4	24.9	249	630	8.4	12.1
Tajikistan	..	511	..	..	..	..	..	..	..	..	..	..
Tanzania	153	450	301	150	65	733	12.1	61.6	23	550	1.6	24.5
Thailand	5,299	8,651	883	1,412	4,326	6,695	14.8	9.4	854	1,843	2.4	3.3
Togo	103	99	..	..	58	15	8.7	3.0	40	19	4.7	2.7
Trinidad and Tobago	195	336	254	250	95	201	4.2	6.9	122	67	8.6	2.1
Tunisia	3,204	4,832	1,727	1,526	948	1,560	18.2	17.7	179	168	3.0	1.8
Turkey	4,799	6,893	2,917	4,601	3,225	5,203	15.3	11.4	520	1,471	2.0	3.0
Turkmenistan	..	300	..	357	..	192	..	22.6	..	125	..	7.4
Uganda	69	238	..	..	10	142	4.1	22.4	8	137	1.2	8.3
Ukraine	..	7,500	..	8,241	..	5,407	..	30.7	..	4,482	..	23.8
United Arab Emirates	633	2,481	..	..	169	607	..	..	..	..	..	..
United Kingdom	18,013	25,740	31,150	50,872	13,762	20,972	5.8	5.6	17,560	32,267	6.7	8.3
United States	39,363	48,491	44,623	56,287	43,007	74,448	8.0	7.8	37,349	60,092	6.1	4.9
Uruguay	1,267	2,139	..	654	262	653	12.1	18.2	111	280	6.7	6.9
Uzbekistan	..	272	..	..	..	21	..	0.6	..	..	..	..
Venezuela, RB	525	587	309	524	496	656	2.6	3.0	1,023	1,646	10.8	9.7
Vietnam	250	1,782	..	168	85	86	4.4	0.7	..	..	..	..
West Bank and Gaza	..	..	..	..	..	..	..	..	..	..	..	..
Yemen, Rep.	52	88	..	..	20	84	1.3	4.9	64	83	2.9	3.0
Yugoslavia, FR (Serb./Mont.)	1,186	152	..	..	419	17	..	..	..	..	..	..
Zambia	141	456	..	..	41	85	3.0	9.4	54	59	2.8	4.4
Zimbabwe	605	2,328	200	213	60	145	3.0	5.7	66	110	3.3	4.8
World	461,483 t	668,484 t	458,115 t	670,815 t	265,000 t	454,724 t	6.0 w	6.3 w	268,275 t	416,224 t	6.3 w	6.0 w
Low income	12,966	29,365	..	33,972	7,927	16,773	4.9	8.0	..	14,758	3.8	6.4
Middle income	136,571	223,020	118,014	189,960	46,410	114,384	6.7	7.8	32,120	65,265	4.8	5.2
Lower middle income	52,227	104,775	47,372	64,372	21,207	61,351	8.5	9.5	..	31,689	2.4	5.6
Upper middle income	83,701	117,419	75,481	130,553	25,194	52,699	5.6	6.3	21,534	33,977	6.5	4.3
Low & middle income	150,018	252,568	145,818	239,501	54,338	131,631	6.4	7.7	38,507	84,903	4.7	5.3
East Asia & Pacific	30,457	58,837	23,210	46,785	15,682	38,396	6.5	5.6	7,146	20,058	2.9	4.1
Europe & Central Asia	59,439	99,660	87,991	126,738	9,734	37,023	7.4	11.3	..	23,952	2.6	7.6
Latin America & Carib.	33,354	48,755	17,289	28,720	15,651	31,477	8.3	7.4	13,049	18,893	9.0	5.0
Middle East & N. Africa	17,932	26,885	16,180	16,300	7,461	14,105	5.2	9.3	3,375	5,958	..	..
South Asia	3,004	4,481	3,503	6,258	1,968	3,949	5.8	5.2	1,048	2,591	2.1	2.9
Sub-Saharan Africa	7,052	17,850	..	..	3,080	6,355	3.8	7.2	3,683	6,267	5.5	6.4
High income	308,084	412,769	274,192	380,583	209,979	322,754	6.0	5.9	228,177	334,658	6.6	6.2
Europe EMU	175,237	235,026	129,933	180,133	98,292	144,878	6.5	7.0	86,625	125,177	6.0	6.3

The data in the table are from the World Tourism Organization. They are obtained primarily from questionnaires sent to government offices, supplemented with data published by official sources. Although the World Tourism Organization reports that progress has been made in harmonizing definitions and measurement units, differences in national practices still prevent full international comparability.

The data on international inbound and outbound tourists refer to the number of arrivals and departures of visitors within the reference period, not to the number of people traveling. Thus a person who makes several trips to a country during a given period is counted each time as a new arrival. International visitors include tourists (overnight visitors), same-day visitors, cruise passengers, and crew members.

Regional and income group aggregates are based on the World Bank's classification of countries and differ from those shown in the World Tourism Organization's *Yearbook of Tourism Statistics*. Countries not shown in the table but for which data are available are included in the regional and income group totals. World totals are no longer calculated by the World Tourism Organization. The aggregates in the table are calculated using the World Bank's weighted aggregation methodology (see *Statistical methods*) and differ from aggregates provided by the World Tourism Organization and published in previous editions of the *World Development Indicators*.

Figure 6.14

Rising tourism expenditures by people from low-income and lower-middle-income economies

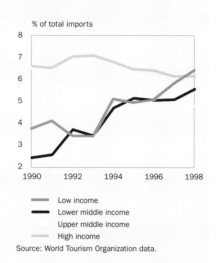

% of total imports

Source: World Tourism Organization data.

By 1998 tourism expenditures accounted for as large a share of imports for low-income and lower-middle-income economies as for high-income economies.

Definitions

• **International inbound tourists** are the number of visitors who travel to a country other than that in which they have their usual residence for a period not exceeding 12 months and whose main purpose in visiting is other than an activity remunerated from within the country visited. • **International outbound tourists** are the number of departures that people make from their country of usual residence to any other country for any purpose other than a remunerated activity in the country visited. • **International tourism receipts** are expenditures by international inbound visitors, including payments to national carriers for international transport. These receipts should include any other prepayment made for goods or services received in the destination country. They also may include receipts from same-day visitors, except in cases where these are important enough to justify a separate classification. Their share in exports is calculated as a ratio to exports of goods and services. • **International tourism expenditures** are expenditures of international outbound visitors in other countries, including payments to foreign carriers for international transport. These expenditures may include those by residents traveling abroad as same-day visitors, except in cases where these are so important as to justify a separate classification. Their share in imports is calculated as a ratio to imports of goods and services.

Data sources

The visitor and expenditure data are available in the World Tourism Organization's *Yearbook of Tourism Statistics* and *Compendium of Tourism Statistics, 1994–98*. The data in the table were updated from electronic files provided by the World Tourism Organization. The data on exports and imports are from the International Monetary Fund's *International Financial Statistics* and World Bank staff estimates.

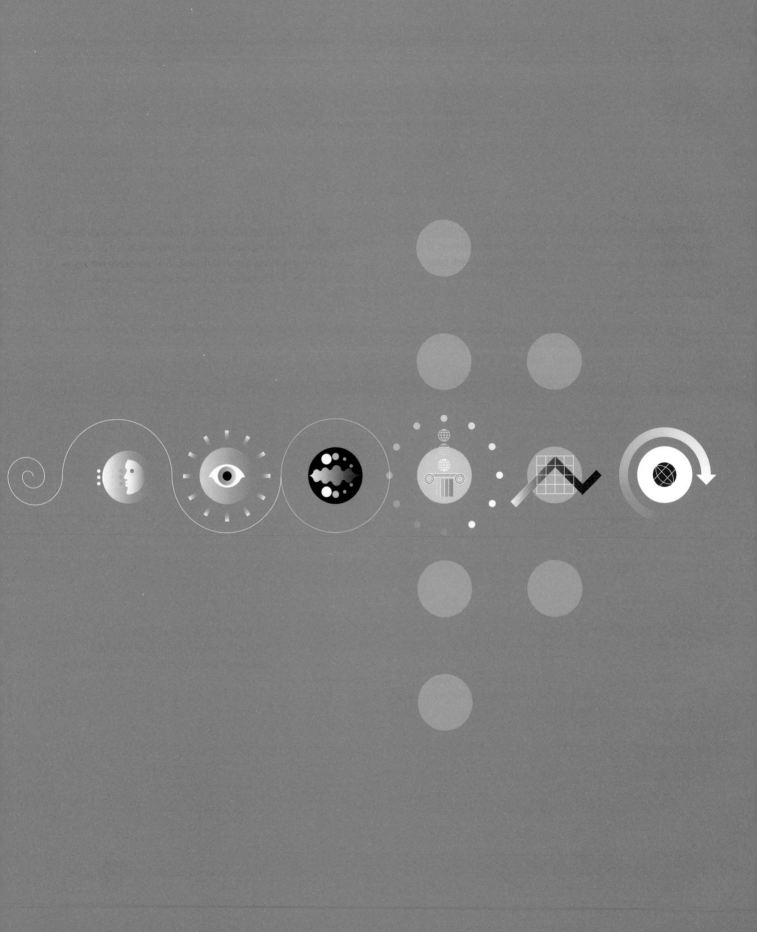

This section describes some of the statistical procedures used in preparing the *World Development Indicators.* It covers the methods employed for calculating regional and income group aggregates and for calculating growth rates, and it describes the World Bank's Atlas method for deriving the conversion factor used to estimate gross national income (GNI - formerly referred to as GNP) and GNI per capita in U.S. dollars. Other statistical procedures and calculations are described in the *About the data* sections that follow each table.

Aggregation rules

Aggregates based on the World Bank's regional and income classifications of economies appear at the end of most tables. These classifications are shown on the front and back cover flaps of the book. This year's edition of the *World Development Indicators,* like the two previous editions, includes aggregates for the member countries of the European Monetary Union (EMU). Members of the EMU on 1 January 2001 were Austria, Belgium, Finland, France, Germany, Ireland, Italy, Luxembourg, the Netherlands, Portugal, and Spain. Other classifications, such as the European Union and regional trade blocs, are documented in *About the data* for the tables in which they appear.

Because of missing data, aggregates for groups of economies should be treated as approximations of unknown totals or average values. Regional and income group aggregates are based on the largest available set of data, including values for the 148 economies shown in the main tables, other economies shown in table 1.6, and Taiwan, China. The aggregation rules are intended to yield estimates for a consistent set of economies from one period to the next and for all indicators. Small differences between sums of subgroup aggregates and overall totals and averages may occur because of the approximations used. In addition, compilation errors and data reporting practices may cause discrepancies in theoretically identical aggregates such as world exports and world imports.

Five methods of aggregation are used in the *World Development Indicators:*

- For group and world totals denoted in the tables by a *t*, missing data are imputed based on the relationship of the sum of available data to the total in the year of the previous estimate. The imputation process works forward and backward from 1995. Missing values in 1995 are imputed using one of several proxy variables for which complete data are available in that year. The imputed value is calculated so that it (or its proxy) bears the same relationship to the total of available data. Imputed values are usually not calculated if missing data account for more than a third of the total in the benchmark year. The variables used as proxies are GNI in U.S. dollars, total population, exports and imports of goods and services in U.S. dollars, and value added in agriculture, industry, manufacturing, and services in U.S. dollars.
- Aggregates marked by an *s* are sums of available data. Missing values are not imputed. Sums are not computed if more than a third of the observations in the series or a proxy for the series are missing in a given year.

- Aggregates of ratios are generally calculated as weighted averages of the ratios (indicated by *w*) using the value of the denominator or, in some cases, another indicator as a weight. The aggregate ratios are based on available data, including data for economies not shown in the main tables. Missing values are assumed to have the same average value as the available data. No aggregate is calculated if missing data account for more than a third of the value of weights in the benchmark year. In a few cases the aggregate ratio may be computed as the ratio of group totals after imputing values for missing data according to the above rules for computing totals.
- Aggregate growth rates are generally calculated as a weighted average of growth rates (and indicated by a *w*). In a few cases growth rates may be computed from time series of group totals. Growth rates are not calculated if more than half the observations in a period are missing. For further discussion of methods of computing growth rates see below.
- Aggregates denoted by an *m* are medians of the values shown in the table. No value is shown if more than half the observations for countries with a population of more than 1 million are missing.

Exceptions to the rules occur throughout the book. Depending on the judgment of World Bank analysts, the aggregates may be based on as little as 50 percent of the available data. In other cases, where missing or excluded values are judged to be small or irrelevant, aggregates are based only on the data shown in the tables.

Growth rates

Growth rates are calculated as annual averages and represented as percentages. Except where noted, growth rates of values are computed from constant price series. Three principal methods are used to calculate growth rates: least squares, exponential endpoint, and geometric endpoint. Rates of change from one period to the next are calculated as proportional changes from the earlier period.

Least-squares growth rate. Least-squares growth rates are used wherever there is a sufficiently long time series to permit a reliable calculation. No growth rate is calculated if more than half the observations in a period are missing.

The least-squares growth rate, *r,* is estimated by fitting a linear regression trend line to the logarithmic annual values of the variable in the relevant period. The regression equation takes the form

$$\ln X_t = a + bt,$$

which is equivalent to the logarithmic transformation of the compound growth equation,

$$X_t = X_0 (1 + r)^t.$$

In this equation X is the variable, t is time, and $a = \ln X_0$ and $b = \ln(1 + r)$ are parameters to be estimated. If b^* is the least-squares estimate of b, the average annual growth rate, r, is obtained as $[\exp(b^*) - 1]$ and is multiplied by 100 for expression as a percentage.

The calculated growth rate is an average rate that is representative of the available observations over the entire period. It does not necessarily match the actual growth rate between any two periods.

Exponential growth rate. The growth rate between two points in time for certain demographic indicators, notably labor force and population, is calculated from the equation

$$r = \ln(p_n/p_1)/n,$$

where p_n and p_1 are the last and first observations in the period, n is the number of years in the period, and ln is the natural logarithm operator. This growth rate is based on a model of continuous, exponential growth between two points in time. It does not take into account the intermediate values of the series. Nor does it correspond to the annual rate of change measured at a one-year interval, which is given by $(p_n - p_{n-1})/p_{n-1}$.

Geometric growth rate. The geometric growth rate is applicable to compound growth over discrete periods, such as the payment and reinvestment of interest or dividends. Although continuous growth, as modeled by the exponential growth rate, may be more realistic, most economic phenomena are measured only at intervals, in which case the compound growth model is appropriate. The average growth rate over n periods is calculated as

$$r = \exp[\ln(p_n/p_1)/n] - 1.$$

Like the exponential growth rate, it does not take into account intermediate values of the series.

World Bank Atlas method

In calculating GNI and GNI per capita in U.S. dollars for certain operational purposes, the World Bank uses the Atlas conversion factor. The purpose of the Atlas conversion factor is to reduce the impact of exchange rate fluctuations in the cross-country comparison of national incomes.

The Atlas conversion factor for any year is the average of a country's exchange rate (or alternative conversion factor) for that year and its exchange rates for the two preceding years, adjusted for the difference between the rate of inflation in the country and that in the G-5 countries (France, Germany, Japan, the United Kingdom, and the United States). A country's inflation rate is measured by the change in its GDP deflator.

The inflation rate for G-5 countries, representing international inflation, is measured by the change in the SDR deflator. (Special drawing rights, or SDRs, are the IMF's unit of account.) The SDR deflator is calculated as a weighted average of the G-5 countries' GDP deflators in SDR terms, the weights being the amount of each country's currency in one SDR unit. Weights vary over time because both the composition of the SDR and the relative exchange rates for each currency change. The SDR deflator is calculated in SDR terms first and then converted to U.S. dollars using the SDR to dollar Atlas conversion factor. The Atlas conversion factor is then applied to a country's GNI. The resulting GNI in U.S. dollars is divided by the midyear population to derive GNI per capita.

When official exchange rates are deemed to be unreliable or unrepresentative of the effective exchange rate during a period, an alternative estimate of the exchange rate is used in the Atlas formula (see below).

The following formulas describe the calculation of the Atlas conversion factor for year t:

$$e_t^* = \frac{1}{3}\left[e_{t-2}\left(\frac{p_t}{p_{t-2}} \Big/ \frac{p_t^{S\$}}{p_{t-2}^{S\$}} \right) + e_{t-1}\left(\frac{p_t}{p_{t-1}} \Big/ \frac{p_t^{S\$}}{p_{t-1}^{S\$}} \right) + e_t \right]$$

and the calculation of GNI per capita in U.S. dollars for year t:

$$Y_t^\$ = (Y_t/N_t)/e_t^*,$$

where e_t^* is the Atlas conversion factor (national currency to the U.S. dollar) for year t, e_t is the average annual exchange rate (national currency to the U.S. dollar) for year t, p_t is the GDP deflator for year t, $p_t^{S\$}$ is the SDR deflator in U.S. dollar terms for year t, $Y_t^\$$ is the Atlas GNI per capita in U.S. dollars in year t, Y_t is current GNI (local currency) for year t, and N_t is the midyear population for year t.

Alternative conversion factors

The World Bank systematically assesses the appropriateness of official exchange rates as conversion factors. An alternative conversion factor is used when the official exchange rate is judged to diverge by an exceptionally large margin from the rate effectively applied to domestic transactions of foreign currencies and traded products. This applies to only a small number of countries, as shown in *Primary data documentation*. Alternative conversion factors are used in the Atlas methodology and elsewhere in the *World Development Indicators* as single-year conversion factors.

Primary data documentation

The World Bank is not a primary data collection agency for most areas other than living standards surveys and debt. As a major user of socio-economic data, however, the World Bank places particular emphasis on data documentation to inform users of data in economic analysis and policymaking. The tables in this section provide information on the sources, treatment, and currentness of the principal demographic, economic, and environmental indicators in the *World Development Indicators*.

Differences in the methods and conventions used by the primary data collectors—usually national statistical agencies, central banks, and customs services—may give rise to significant discrepancies over time both among and within countries. Delays in reporting data and the use of old surveys as the base for current estimates may severely compromise the quality of national data.

Although data quality is improving in some countries, many developing countries lack the resources to train and maintain the skilled staff and obtain the equipment needed to measure and report demographic, economic, and environmental trends in an accurate and timely way. The World Bank recognizes the need for reliable data to measure living standards, track and evaluate economic trends, and plan and monitor development projects. Thus, working with bilateral and other multilateral agencies, it continues to fund and participate in technical assistance projects to improve statistical organization and basic data methods, collection, and dissemination.

The World Bank is working at several levels to meet the challenge of improving the quality of the data that it collates and disseminates. At the country level the Bank is carrying out technical assistance, training, and survey activities—with a view to strengthening national capacity—in the following areas:

- Poverty assessments in most borrower member countries.
- Living standards measurement and other household and farm surveys with partner national statistical agencies.
- National accounts and inflation.
- Price and expenditure surveys for the International Comparison Programme.
- Projects to improve statistics in the countries of the former Soviet Union.
- External debt management.
- Environmental and economic accounting.

Primary data documentation

	National currency	Fiscal year end	National accounts					Balance of payments and trade			Government finance	IMF special data dissemination
			Reporting period[a]	Base year	SNA price valuation	Alternative conversion factor	PPP survey year	Balance of Payments Manual in use	External debt	System of trade	Accounting concept	
Albania	Albanian lek	Dec. 31	CY	1995[b]	VAP		1996	BPM5	Actual	G		G
Algeria	Algerian dinar	Dec. 31	CY	1980	VAB			BPM5	Actual	S		
Angola	Angolan kwanza	Dec. 31	CY	1997	VAP	1991–96, 99		BPM4	Actual	S		
Argentina	Argentine peso	Dec. 31	CY	1993	VAB	1971–81, 84	1996	BPM5	Actual	S	C	S*
Armenia	Armenian dram	Dec. 31	CY	1994[b]	VAB	1993–95	1996	BPM5	Actual	G		
Australia	Australian dollar	Jun. 30	FY	1995[b]	VAP		1996	BPM5		G	C	S*
Austria	Austrian schilling[c]	Dec. 31	CY	1995[b]	VAP	1992–95	1996	BPM5		S	C	S*
Azerbaijan	Azeri manat	Dec. 31	CY	1999[b]	VAP		1996	BPM4	Actual	S		
Bangladesh	Bangladesh taka	Jun. 30	FY	1996[b]	VAP	1971–99	1993	BPM5	Preliminary	G		
Belarus	Belarussian rubel	Dec. 31	CY	1990[b]	VAB	1997	1996	BPM5	Actual	G	C	
Belgium	Belgian franc[c]	Dec. 31	CY	1995[b]	VAP		1996	BPM5		S	C	S*
Benin	CFA franc	Dec. 31	CY	1985	VAP	1992	1993	BPM4	Actual	S		
Bolivia	Boliviano	Dec. 31	CY	1990[b]	VAP	1960–85	1996	BPM5	Actual	S	C	G
Bosnia and Herzegovina	Convertible mark	Dec. 31	CY	1996	VAB			BPM5	Preliminary			
Botswana	Botswana pula	Jun. 30	FY	1986	VAP	1999	1993	BPM5	Actual	G	B	
Brazil	Brazilian real	Dec. 31	CY	1995	VAB	1999	1996	BPM5	Preliminary	S	C	S*
Bulgaria	Bulgarian leva	Dec. 31	CY	1990[b]	VAP	1985–92	1996	BPM5	Preliminary	G	C	G
Burkina Faso	CFA franc	Dec. 31	CY	1985	VAB	1992–93		BPM4	Actual	S	C	
Burundi	Burundi franc	Dec. 31	CY	1980	VAB			BPM5	Actual	S		
Cambodia	Cambodian riel	Dec. 31	CY	1989	VAP			BPM5	Actual	S		
Cameroon	CFA franc	Jun. 30	FY	1980	VAB	1970–99	1993	BPM5	Preliminary	S	C	G
Canada	Canadian dollar	Mar. 31	CY	1995[b]	VAP		1996	BPM5		G	C	S*
Central African Republic	CFA franc	Dec. 31	CY	1987	VAB			BPM4	Actual	S		
Chad	CFA franc	Dec. 31	CY	1995	VAB			BPM5	Estimate	S	C	
Chile	Chilean peso	Dec. 31	CY	1986	VAB		1996	BPM5	Actual	S	C	S*
China	Chinese yuan	Dec. 31	CY	1990	VAP	1987–93		BPM5	Estimate	S	B	
Hong Kong, China	Hong Kong dollar	Dec. 31	CY	1990	VAB		1993	BPM5		G		S*
Colombia	Colombian peso	Dec. 31	CY	1994	VAB	1992–94	1993	BPM5	Actual	S	C	S*
Congo, Dem. Rep.	Congo franc	Dec. 31	CY	1987	VAP	1993–99		BPM5	Actual	S	C	
Congo, Rep.	CFA franc	Dec. 31	CY	1978	VAP	1993	1993	BPM4	Estimate	S		
Costa Rica	Costa Rican colon	Dec. 31	CY	1991[b]	VAB			BPM5	Actual	S	C	
Côte d'Ivoire	CFA franc	Dec. 31	CY	1986	VAP		1993	BPM5	Estimate	S	C	G
Croatia	Croatian kuna	Dec. 31	CY	1997[b]	VAB	1993–99	1996	BPM5	Actual	G	C	S*
Cuba	Cuban peso	Dec. 31	CY	..	..					S		
Czech Republic	Czech koruna	Dec. 31	CY	1995[b]	VAB		1996	BPM5	Preliminary	G	C	S*
Denmark	Danish krone	Dec. 31	CY	1995[b]	VAP		1996	BPM5		G	C	S
Dominican Republic	Dominican peso	Dec. 31	CY	1990	VAP			BPM5	Actual	G	C	
Ecuador	Ecuadorian sucre	Dec. 31	CY	1975	VAP	1999	1996	BPM5	Estimate	S	B	S*
Egypt, Arab Rep.	Egyptian pound	Jun. 30	FY	1992	VAB	1965–91	1993	BPM5	Actual	S	C	
El Salvador	Salvadoran colone	Dec. 31	CY	1990	VAP	1982–90		BPM5	Actual	S	C	S*
Eritrea	Eritrean nakfa	Dec. 31	CY	1992	VAB			BPM4	Actual			
Estonia	Estonian kroon	Dec. 31	CY	1995[b]	VAB	1993–99	1996	BPM5	Actual	G	C	S*
Ethiopia	Ethiopian birr	Jul. 7	FY	1981	VAB	1989–99		BPM5	Actual	G	B	
Finland	Finnish markka[c]	Dec. 31	CY	1995[b]	VAP		1993	BPM5		G	C	S*
France	French franc[c]	Dec. 31	CY	1995[b]	VAP		1996	BPM5		S	C	S
Gabon	CFA franc	Dec. 31	CY	1991	VAP	1993	1993	BPM5	Estimate	S	B	
Gambia, The	Gambian dalasi	Jun. 30	FY	1987	VAB			BPM5	Actual	G	B	G
Georgia	Georgian lari	Dec. 31	CY	1994[b]	VAB		1996	BPM4	Actual	G		
Germany	Deutsche mark[c]	Dec. 31	CY	1995[b]	VAP		1996	BPM5		S	C	S*
Ghana	Ghanaian cedi	Dec. 31	CY	1975	VAP	1973–87		BPM5	Actual	S	C	
Greece	Greek drachma	Dec. 31	CY	1995[b]	VAP		1993	BPM4	Estimate	S	C	
Guatemala	Guatemalan quetzal	Dec. 31	CY	1958	VAP	1985–86		BPM5	Actual	S	B	
Guinea	Guinean franc	Dec. 31	CY	1994	VAB	1986	1993	BPM5	Estimate	S	C	
Guinea-Bissau	CFA franc	Dec. 31	CY	1986	VAB	1970–86		BPM5	Estimate	S		
Haiti	Haitian gourde	Sep. 30	FY	1976	VAP	1991–97		BPM5	Preliminary	G		
Honduras	Honduran lempira	Dec. 31	CY	1978	VAB	1988–89		BPM5	Actual	S		

	Latest population census	Latest household or demographic survey	Vital registration complete	Latest agricultural census	Latest industrial data	Latest water withdrawal data	Latest survey of scientists and engineers engaged in R&D	Latest survey of expenditure for R&D
Albania	1989	LSMS, 1996	Yes	1995	1990	1995		
Algeria	1998	PAPCHILD, 1992		1973	1996	1990		
Angola	1970			1964–65		1987		
Argentina	1991		Yes	1988	1995	1995	1995	1995
Armenia	1989		Yes		1991	1994		
Australia	1996		Yes	1990	1997	1995	1994	1994
Austria	1991		Yes	1990	1997	1995	1993	1995
Azerbaijan	1999		Yes			1995		
Bangladesh	1991	DHS, 1999–2000		1976	1997	1990	1993	1993
Belarus	1999		Yes	1994		1990	1995	1995
Belgium	1991		Yes	1990	1997	1980	1991	1991
Benin	1992	DHS, 1996		1992–93	1981	1994	1989	1989
Bolivia	1992	DHS, 1998			1995	1990	1991	1991
Bosnia and Herzegovina	1991		Yes		1991			
Botswana	1991	DHS, 1988		1993	1994	1992		
Brazil	2000	DHS, 1996		1996	1995	1992	1995	1995
Bulgaria	1992	LSMS, 1995	Yes		1997		1996	1996
Burkina Faso	1996	DHS, 1998–99		1993	1997	1992		
Burundi	1990				1991	1987	1989	1989
Cambodia	1998					1987		
Cameroon	1987	DHS, 1998		1972–73	1996	1987		
Canada	1996		Yes	1991	1997	1990	1993	1995
Central African Republic	1988	DHS, 1994–95			1993	1987	1990	1990
Chad	1993	DHS, 1996–97				1990		
Chile	1992		Yes	1997	1997	1990	1995	1995
China	1990	Population, 1995		1996	1996	1993	1995	1995
Hong Kong, China	1996		Yes		1997		1995	1995
Colombia	1993	DHS, 2000		1988	1996	1996	1982	1982
Congo, Dem. Rep.	1984			1990		1994		
Congo, Rep.	1984			1986	1988	1987	1984	1984
Costa Rica	1984	CDC, 1993	Yes	1973	1997	1997	1996	1991
Côte d'Ivoire	1998	DHS, 1999		1974–75	1997	1987		
Croatia	1991		Yes		1992	1996	1995	1995
Cuba	1981		Yes		1989	1995	1995	1995
Czech Republic	1991	CDC, 1993	Yes		1997	1995	1995	1995
Denmark	1991		Yes	1989	1997	1995	1993	1993
Dominican Republic	1993	DHS, 1996		1971	1984	1994		
Ecuador	1990	LSMS, 1995		1997	1997	1997	1990	1990
Egypt, Arab Rep.	1996	DHS, 1997	Yes	1989–90	1995	1993	1991	1995
El Salvador	1992	CDC, 1994		1970–71	1997	1992	1992	1992
Eritrea	1984	DHS, 1995			1997			
Estonia	2000		Yes	1994		1995	1995	1995
Ethiopia	1994	Family and fertility, 1990		1988–89	1991–92	1997	1987	
Finland	1990		Yes	1990	1997	1995	1995	1995
France	1999	Income, 1989	Yes	1988	1997	1995	1994	1994
Gabon	1993			1974–75	1982	1987	1987	1986
Gambia, The	1993				1982	1990		
Georgia	1989		Yes			1990	1991	1991
Germany			Yes	1993		1990	1993	1993
Ghana	2000	DHS, 1998		1984	1995	1970		
Greece	1991		Yes	1993	1996	1990	1993	1993
Guatemala	1994	DHS, 1998–99	Yes	1979	1988	1992	1988	1988
Guinea	1996	DHS, 1999		1996		1987	1984	1984
Guinea-Bissau	1991	SDA, 1991		1988		1991		
Haiti	1982	DHS, 1994–95		1971	1996	1991		
Honduras	1988	CDC, 1994		1993	1997	1992		

Primary data documentation

	National currency	Fiscal year end	National accounts					Balance of payments and trade			Government finance	IMF special data dissemination
			Reporting period[a]	Base year	SNA price valuation	Alternative conversion factor	PPP survey year	Balance of Payments Manual in use	External debt	System of trade	Accounting concept	
Hungary	Hungarian forint	Dec. 31	CY	1994[b]	VAB		1996	BPM5	Actual	S	C	S*
India	Indian rupee	Mar. 31	FY	1993	VAB			BPM5	Estimate	G	C	S
Indonesia	Indonesian rupiah	Mar. 31	CY	1993	VAP		1993	BPM5	Preliminary	S	C	S*
Iran, Islamic Rep.	Iranian rial	Mar. 20	FY	1982	VAB	1980–99	1993	BPM5	Preliminary	G	C	
Iraq	Iraqi dinar	Dec. 31	CY	1969	VAB					S		
Ireland	Irish pound[c]	Dec. 31	CY	1995[b]	VAP		1996	BPM5		G	C	S*
Israel	Israeli new shekel	Dec. 31	CY	1995[b]	VAB	1971–99	1996	BPM5		S	C	S*
Italy	Italian lira[c]	Dec. 31	CY	1995[b]	VAP		1996	BPM5		S	C	S*
Jamaica	Jamaica dollar	Dec. 31	CY	1986	VAP	1995–96, 99	1993	BPM5	Actual	G		
Japan	Japanese yen	Mar. 31	CY	1995	VAP		1996	BPM5		G	C	S*
Jordan	Jordan dinar	Dec. 31	CY	1994	VAB		1993	BPM5	Actual	G	B	G
Kazakhstan	Kazakh tenge	Dec. 31	CY	1993[b]	VAB	1994–95	1996	BPM5	Actual	G		G
Kenya	Kenya shilling	Jun. 30	CY	1982	VAB		1993	BPM5	Actual	G	B	
Korea, Dem. Rep.	Democratic Republic of Korea won	Dec. 31	CY	..	..			BPM5				
Korea, Rep.	Korean won	Dec. 31	CY	1995	VAP		1993	BPM5	Actual	S	C	S*
Kuwait	Kuwaiti dinar	Jun. 30	CY	1984	VAP			BPM5		S	C	G
Kyrgyz Republic	Kyrgyz som	Dec. 31	CY	1995[b]	VAB	1994–95	1996	BPM5	Actual	G		G
Lao PDR	Lao kip	Dec. 31	CY	1990	VAB	1960–89	1993	BPM5	Estimate			
Latvia	Latvian lat	Dec. 31	CY	1995[b]	VAP	1992–95	1996	BPM5	Actual	S	C	S*
Lebanon	Lebanese pound	Dec. 31	CY	1994	VAB		1993	BPM4	Estimate	G		
Lesotho	Lesotho loti	Mar. 31	CY	1995	VAB			BPM5	Actual	G	C	
Libya	Libyan dinar	Dec. 31	CY	1975	VAB	1986		BPM5		G		
Lithuania	Lithuanian litas	Dec. 31	CY	1995[b]	VAB	1992–95		BPM5	Actual	G	C	S*
Macedonia, FYR	Macedonian denar	Dec. 31	CY	1995[b]	VAB	1994–98	1996	BPM5	Actual	G		
Madagascar	Malagasy franc	Dec. 31	CY	1984	VAB		1993	BPM5	Estimate	S	C	
Malawi	Malawi kwacha	Mar. 31	CY	1994	VAB		1993	BPM5	Estimate	G	B	
Malaysia	Malaysian ringgit	Dec. 31	CY	1987	VAP		1993	BPM5	Preliminary	G		S*
Mali	CFA franc	Dec. 31	CY	1987	VAB		1993	BPM4	Actual	S		
Mauritania	Mauritanian ouguiya	Dec. 31	CY	1985	VAB			BPM4	Actual	S		
Mauritius	Mauritian rupee	Jun. 30	CY	1992	VAB		1993	BPM5	Actual	G	C	G
Mexico	Mexican new peso	Dec. 31	CY	1993[b]	VAB	1980–82	1996	BPM5	Actual	G	C	S*
Moldova	Moldovan leu	Dec. 31	CY	1996	VAB	1995	1996	BPM5	Actual	G		
Mongolia	Mongolian tugrik	Dec. 31	CY	1986	VAP		1996	BPM5	Estimate		C	G
Morocco	Moroccan dirham	Dec. 31	CY	1980	VAP		1993	BPM5	Actual	S	C	
Mozambique	Mozambican metical	Dec. 31	CY	1995	VAB	1992–95		BPM5	Actual	S		
Myanmar	Myanmar kyat	Mar. 31	FY	1985	VAP			BPM5	Actual	G	C	
Namibia	Namibia dollar	Mar. 31	CY	1990	VAB			BPM5	Estimate		C	
Nepal	Nepalese rupee	Jul. 14	FY	1985	VAB	1973–98	1993	BPM5	Actual	S	C	
Netherlands	Netherlands guilder[c]	Dec. 31	CY	1995[b]	VAP		1996	BPM5		S	C	S*
New Zealand	New Zealand dollar	Mar. 31	FY	1995	VAP		1996	BPM4		G	B	
Nicaragua	Nicaraguan gold cordoba	Dec. 31	CY	1980	VAP	1970–93		BPM5	Actual	S	C	
Niger	CFA franc	Dec. 31	CY	1987	VAP	1993		BPM5	Estimate	S		
Nigeria	Nigerian naira	Dec. 31	CY	1987	VAB	1971–98	1993	BPM5	Estimate	G		
Norway	Norwegian krone	Dec. 31	CY	1995[b]	VAP		1996	BPM5		G	C	S*
Oman	Rial Omani	Dec. 31	CY	1978	VAP		1993	BPM5	Estimate	G	B	
Pakistan	Pakistan rupee	Jun. 30	FY	1981	VAB	1972–99	1993	BPM5	Preliminary	G	C	
Panama	Panamanian balboa	Dec. 31	CY	1982[b]	VAP		1996	BPM5	Actual	S	C	G
Papua New Guinea	Papua New Guinea kina	Dec. 31	CY	1983	VAP			BPM5	Actual	G	B	
Paraguay	Paraguayan guarani	Dec. 31	CY	1982	VAP	1982–88		BPM5	Actual	S	C	
Peru	Peruvian new sol	Dec. 31	CY	1994	VAP	1985–91	1993	BPM5	Actual	G	C	S*
Philippines	Philippine peso	Dec. 31	CY	1985	VAP		1993	BPM5	Actual	G	B	S
Poland	Polish zloty	Dec. 31	CY	1990[b]	VAP	1994–98	1996	BPM5	Actual	S	C	S
Portugal	Portuguese escudo[c]	Dec. 31	CY	1995[b]	VAP		1996	BPM5		S	C	S*
Puerto Rico	U.S. dollar	Dec. 31	CY	1954	VAP							
Romania	Romanian leu	Dec. 31	CY	1993	VAB	1987–97	1996	BPM5	Preliminary	S	C	G
Russian Federation	Russian ruble	Dec. 31	CY	1997	VAB	1993–94, 99	1996	BPM5	Estimate	G	C	

	Latest population census	Latest household or demographic survey	Vital registration complete	Latest agricultural census	Latest industrial data	Latest water withdrawal data	Latest survey of scientists and engineers engaged in R&D	Latest survey of expenditure for R&D
Hungary	1990	Income, 1995	Yes	1994	1997	1995	1995	1995
India	1991	National family health, 1998–99		1986	1995	1990	1994	1994
Indonesia	1990	Socioeconomic, 1998		1993	1997	1990	1995	1995
Iran, Islamic Rep.	1991	Demographic, 1995		1988	1996	1993	1994	1994
Iraq	1997			1981	1997	1990	1993	1993
Ireland	1996		Yes	1991	1996	1995	1993	1993
Israel	1995		Yes	1983	1996	1997	1992	1992
Italy	1991		Yes	1990	1994	1993	1994	1994
Jamaica	1991	Reproductive Health, 1997	Yes	1979	1996	1993	1986	1986
Japan	2000		Yes	1990	1997	1992	1994	1994
Jordan	1994	Annual survey, 1999		1997	1997	1993	1986	1989
Kazakhstan	1999	DHS, 1999	Yes			1993		
Kenya	1999	DHS, 1998		1981	1995	1990		
Korea, Dem. Rep.	1993					1987		
Korea, Rep.	1995			1991	1997	1994	1994	1994
Kuwait	1995		Yes	1970	1996	1994	1984	1984
Kyrgyz Republic	1999	DHS, 1997	Yes			1994	1994	1995
Lao PDR	1995			1999		1987		
Latvia	2000		Yes	1994	1997	1994	1995	1995
Lebanon	1970			1999		1994	1980	1980
Lesotho	1996	DHS, 1991		1989–90	1985	1987		
Libya	1995	PAPCHILD, 1995		1987	1997	1995	1980	1980
Lithuania	1989		Yes	1994		1995	1996	1996
Macedonia, FYR	1994		Yes	1994	1996		1995	1995
Madagascar	1993	DHS, 1997		1984	1988	1990	1994	1995
Malawi	1998	DHS, 1996		1992–93	1997	1994		
Malaysia	1991		Yes		1996	1995	1992	1992
Mali	1998	DHS, 1995–96		1978		1987		
Mauritania	1988	PAPCHILD, 1990		1985		1990		
Mauritius	1990	CDC, 1991	Yes		1996	1974	1992	1992
Mexico	2000	Population, 1995		1991	1995	1998	1995	1995
Moldova	1989		Yes			1992	1995	1995
Mongolia	2000	Reproductive Health, 1998			1997	1993	1995	1995
Morocco	1994	DHS, 1995		1997	1997	1991		
Mozambique	1997	DHS, 1997				1992		
Myanmar	1983			1993	1997	1987		
Namibia	1991	DHS, 1992		1995	1994	1990		
Nepal	1991	DHS, 1996		1992	1996	1994	1980	1980
Netherlands	1991		Yes	1989	1997	1990	1991	1994
New Zealand	1996		Yes	1990	1997	1995	1993	1993
Nicaragua	1995	DHS, 1998		1963	1997	1998	1987	1987
Niger	1988	DHS, 1998		1980	1996	1990		
Nigeria	1991	Consumption and expenditure, 1992		1960	1994	1990	1987	1987
Norway	1990		Yes	1989	1997	1985	1995	1995
Oman	1993	Child health, 1989		1979	1997	1991		
Pakistan	1998	LSMS, 1991		1990	1997	1991	1990	1987
Panama	2000			1990	1997	1990	1986	1986
Papua New Guinea	1989	DHS, 1996				1987		
Paraguay	1992	DHS, 1990; CDC, 1998		1991	1997	1987		
Peru	1993	DHS, 1996		1994	1994	1992	1994	1995
Philippines	2000	DHS, 1998		1991	1997	1995	1992	1992
Poland	1988		Yes	1990	1997	1995	1995	1995
Portugal	1991		Yes	1989	1997	1990	1995	1995
Puerto Rico	1990		Yes	1987	1997			
Romania	1992	CDC, 1999	Yes		1996		1994	1995
Russian Federation	1989	LSMS, 1992	Yes	1994–95	1997	1994	1995	1995

	National currency	Fiscal year end	National accounts					Balance of payments and trade			Government finance	IMF special data dissemi- nation
			Reporting period[a]	Base year	SNA price valuation	Alternative conversion factor	PPP survey year	Balance of Payments Manual in use	External debt	System of trade	Accounting concept	
Rwanda	Rwanda franc	Dec. 31	CY	1985	VAB			BPM5	Estimate	G	C	
Saudi Arabia	Saudi Arabian riyal	Hijri year	Hijri year	1970	VAP		1993	BPM4	Estimate	G		
Senegal	CFA franc	Dec. 31	CY	1987	VAP		1993	BPM5	Preliminary	S		
Sierra Leone	Sierra Leonean leone	Jun. 30	CY	1990	VAB	1971–79, 87	1993	BPM5	Actual	G	B	
Singapore	Singapore dollar	Mar. 31	CY	1990	VAP		1993	BPM5		G	C	S*
Slovak Republic	Slovak koruna	Dec. 31	CY	1995[b]	VAP		1996	BPM5	Actual	G		S*
Slovenia	Slovenian tolar	Dec. 31	CY	1993[b]	VAB		1996	BPM5	Actual	S		S*
South Africa	South African rand	Mar. 31	CY	1995	VAB			BPM5	Estimate	G	C	S*
Spain	Spanish peseta[c]	Dec. 31	CY	1995[b]	VAP		1996	BPM5		S	C	S*
Sri Lanka	Sri Lankan rupee	Dec. 31	CY	1996	VAB	1995–99	1993	BPM5	Actual	G	C	G
Sudan	Sudanese dinar	Jun. 30	CY	1982	VAB	1980–91		BPM5	Estimate	G		
Sweden	Swedish krona	Jun. 30	CY	1995[b]	VAP		1996	BPM5		G	C	S*
Switzerland	Swiss franc	Dec. 31	CY	1995	VAP		1996	BPM5	Estimate	S	C	S*
Syrian Arab Republic	Syrian pound	Dec. 31	CY	1985	VAP	1970–99	1993	BPM5	Estimate	S	C	
Tajikistan	Tajik ruble	Dec. 31	CY	1985[b]	VAB	1992–98	1996	BPM4	Actual	G		
Tanzania	Tanzania shilling	Jun. 30	FY	1992	VAB		1993	BPM5	Actual	G		
Thailand	Thai baht	Sep. 30	CY	1988	VAP		1993	BPM5	Preliminary	G	C	S*
Togo	CFA franc	Dec. 31	CY	1978	VAP		1993	BPM5	Estimate	S		
Trinidad and Tobago	Trinidad and Tobago dollar	Dec. 31	CY	1985	VAP		1993	BPM5	Preliminary	S		
Tunisia	Tunisian dinar	Dec. 31	CY	1990	VAB		1993	BPM5	Actual	G	C	
Turkey	Turkish lira	Dec. 31	CY	1994	VAB		1996	BPM5	Actual	S	C	S*
Turkmenistan	Turkmen manat	Dec. 31	CY	1987[b]	VAB	1994–99	1996	BPM5	Estimate	G		
Uganda	Uganda shilling	Jun. 30	FY	1991	VAB	1980–99		BPM5	Actual	G		G
Ukraine	Ukraine hryvnia	Dec. 31	CY	1990[b]	VAB	1993–99	1996	BPM5	Actual	G		
United Arab Emirates	U.A.E. dirham	Dec. 31	CY	1985	VAB		1993	BPM4		GG	B	
United Kingdom	Pound sterling	Dec. 31	CY	1995[b]	VAP		1996	BPM5		G	C	S*
United States	U.S. dollar	Sep. 30	CY	1995[b]	VAP		1996	BPM5		G	C	S
Uruguay	Uruguayan peso	Dec. 31	CY	1983	VAP		1993	BPM5	Actual	S	C	
Uzbekistan	Uzbek sum	Dec. 31	CY	1997[b]	VAB	1991–99[d]	1996	BPM4	Actual	G		
Venezuela, RB	Venezuelan bolivar	Dec. 31	CY	1984	VAB		1993	BPM5	Preliminary	G		
Vietnam	Vietnamese dong	Dec. 31	CY	1989	VAP	1991	1993	BPM4	Estimate	G		
West Bank and Gaza	Israeli new shekel	Dec. 31	CY	1994	VAB		1993					
Yemen, Rep.	Yemen rial	Dec. 31	CY	1990	VAP	1991–96	1993	BPM5	Estimate	G	C	
Yugoslavia, FR (Serb./Mont.)	Yugoslav new dinar	Dec. 31	CY	..	VAP				Estimate	S		
Zambia	Zambian kwacha	Dec. 31	CY	1994	VAB	1990, 92	1993	BPM5	Actual	G	C	
Zimbabwe	Zimbabwe dollar	Jun. 30	CY	1990	VAB	1991, 98	1993	BPM5	Actual	G	C	

Note: For an explanation of the abbreviations used in the table see the notes.

a. Also applies to balance of payments reporting. b. Country uses the 1993 System of National Accounts methodology. c. European Monetary Union member currency linked to the euro. d. Official exchange rate does not take into account multiple exchange rate regime in Uzbekistan.

	Latest population census	Latest household or demographic survey	Vital registration complete	Latest agricultural census	Latest industrial data	Latest water withdrawal data	Latest survey of scientists and engineers engaged in R&D	Latest survey of expenditure for R&D
Rwanda	1991	DHS, 1992		1984	1986	1993	1995	1995
Saudi Arabia	1992	Demographic, 1999		1983		1992		
Senegal	1988	DHS, 1997		1960	1996	1990	1981	1981
Sierra Leone	1985	SHEHEA, 1989–90		1985	1986	1987		
Singapore	2000	General household, 1995	Yes		1997		1995	1995
Slovak Republic	1991		Yes		1997	1995	1995	1995
Slovenia	1991		Yes	1991	1997	1994	1995	1995
South Africa	1996	DHS, 1997			1996	1990	1993	1993
Spain	1991		Yes	1989	1997	1997	1994	1994
Sri Lanka	1981	DHS, 1993	Yes	1982	1995	1990	1985	1985
Sudan	1993	DHS, 1989–90			1997	1995		
Sweden	1990		Yes	1981	1997	1995	1993	1993
Switzerland	2000		Yes	1990	1996	1995	1990	1990
Syrian Arab Republic	1994	PAPCHILD, 1995		1981	1997	1993		
Tajikistan	1989		Yes	1994		1994	1992	1992
Tanzania	1988	DHS, 1999		1995	1997	1994		
Thailand	2000	DHS, 1987		1993	1997	1990	1995	1995
Togo	1981	DHS, 1998		1996	1997	1987		
Trinidad and Tobago	1990	DHS, 1987	Yes	1982	1997		1984	1984
Tunisia	1994	DHS, 1998		1961	1997	1996	1992	1992
Turkey	1997	DHS, 1993		1991	1997	1997	1995	1995
Turkmenistan	1995		Yes			1994		
Uganda	1991	DHS, 1995		1991	1997	1970		
Ukraine	1991		Yes			1992	1995	1995
United Arab Emirates	1995			1998	1981	1995		
United Kingdom	1991		Yes	1993	1997	1995	1993	1993
United States	2000	Current population, 1997	Yes	1997	1997	1995	1993	1995
Uruguay	1996		Yes	1990	1996	1990	1987	1987
Uzbekistan	1989	DHS, 1996	Yes			1994	1992	1992
Venezuela, RB	1990	LSMS, 1993	Yes	1997–98	1996	1970	1992	1992
Vietnam	1999	DHS, 1997		1994		1990	1985	1985
West Bank and Gaza	1997	Demographic, 1995		1971				
Yemen, Rep.	1994	DHS, 1997		1982–85		1990		
Yugoslavia, FR (Serb./Mont.)	1991		Yes	1981	1996		1995	1995
Zambia	1990	DHS, 1996		1990	1997	1994		
Zimbabwe	1992	DHS, 1999		1960	1997	1987		

- **Fiscal year end** is the date of the end of the fiscal year for the central government. Fiscal years for other levels of government and the reporting years for statistical surveys may differ, but if a country is designated as a fiscal year reporter in the following column, the date shown is the end of its national accounts reporting period. • **Reporting period** for national accounts and balance of payments data is designated as either calendar year (CY) or fiscal year (FY). Most economies report their national accounts and balance of payments data using calendar years, but some use fiscal years, which straddle two calendar years. In the *World Development Indicators* fiscal year data are assigned to the calendar year that contains the larger share of the fiscal year. If a country's fiscal year ends before June 30, the data are shown in the first year of the fiscal period; if the fiscal year ends on or after June 30, the data are shown in the second year of the period. Saudi Arabia follows a lunar year whose starting and ending dates change with respect to the solar year. Because the International Monetary Fund (IMF) reports most balance of payments data on a calendar year basis, balance of payments data for fiscal year reporters in the *World Development Indicators* are based on fiscal year estimates provided by World Bank staff. These estimates may differ from IMF data but allow consistent comparisons between national accounts and balance of payments data. • **Base year** is the year used as the base period for constant price calculations in the country's national accounts. Price indexes derived from national accounts aggregates, such as the GDP deflator, express the price level relative to prices in the base year. Constant price data reported in the *World Development Indicators* are rebased to a common 1995 reference year. See *About the data* for table 4.1 for further discussion. • **SNA price valuation** shows whether value added in the national accounts is reported at basic prices (VAB) or at producer prices (VAP). Producer prices include the value of taxes paid by producers and thus tend to overstate the actual value added in production. See *About the data* for table 4.2 for further discussion of national accounts valuation. • **Alternative conversion factor** identifies the countries and years for which a World Bank–estimated conversion factor has been used in place of the official (IFS line rf) exchange rate. See *Statistical methods* for further discussion of the use of alternative conversion factors. • **PPP survey year** refers to the latest available survey year for the International Comparison Programme's estimates of purchasing power parities (PPPs). • **Balance of Payments Manual in use** refers to the classification system used for compiling and

reporting data on balance of payments items in table 4.15. BPM4 refers to the fourth edition of the IMF's *Balance of Payments Manual* (1977), and BPM5 to the fifth edition (1993). Since 1995 the IMF has adjusted all balance of payments data to BPM5 conventions, but some countries continue to report using the older system. • **External debt** shows debt reporting status for 1999 data. *Actual* indicates that data are as reported, *preliminary* that data are preliminary and include an element of staff estimation, and *estimate* that data are staff estimates. • **System of trade** refers to the general trade system (G) or the special trade system (S). For imports under the general trade system, both goods entering directly for domestic consumption and goods entered into customs storage are recorded, at the time of their first arrival, as imports; under the special trade system goods are recorded as imports when declared for domestic consumption, whether at the time of entry or on withdrawal from customs storage. Exports under the general system comprise outward-moving goods: (a) national goods wholly or partly produced in the country; (b) foreign goods, neither transformed nor declared for domestic consumption in the country, that move outward from customs storage; and (c) nationalized goods that have been declared from domestic consumption and move outward without having been transformed. Under the special system of trade exports comprise categories (a) and (c). In some compilations categories (b) and (c) are classified as re-exports. Direct transit trade, consisting of goods entering or leaving for transport purposes only, is excluded from both import and export statistics. See *About the data* for tables 4.5 and 4.6 for further discussion. • **Government finance accounting concept** describes the accounting basis for reporting central government financial data. For most countries government finance data have been consolidated (C) into one set of accounts capturing all the central government's fiscal activities. Budgetary central government accounts (B) exclude central government units. See *About the data* for tables 4.11, 4.12, and 4.13 for further details. • **IMF special data dissemination** shows the countries that subscribe to the International Monetary Fund's (IMF) Special Data Dissemination Standard (SDDS) or the General Data Dissemination System (GDDS). *S* refers to countries that subscribe to the SDDS; *S** indicates subscribers that have posted data on the Dissemination Standards Bulletin Board web site; while *G* refers to countries that subscribe to the GDDS. (Posted data can be reached through the IMF Dissemination Standard Bulletin Board at dsbb.imf.org/.) The SDDS was established by the IMF to guide mem-

bers that have, or that might seek, access to international capital markets in the provision of their economic and financial data to the public. The GDDS helps guide member countries in the dissemination to the public of comprehensive, timely, accessible, and reliable economic, financial, and socio-demographic statistics. Member countries of the IMF voluntarily elect to participate in either the SDDS or the GDDS. Both the GDDS and the SDDS are expected to enhance the availability of timely and comprehensive statistics and therefore contribute to the pursuit of sound macroeconomic policies; the SDDS is also expected to contribute to the improved functioning of financial markets.

- **Latest population census** shows the most recent year in which a census was conducted and at least preliminary results have been released. • **Latest household or demographic survey** gives information on the surveys used in compiling household and demographic data presented in section 2. PAPCHILD is the Pan Arab Project for Child Development, DHS is Demographic and Health Survey, LSMS is Living Standards Measurement Study, SDA is Social Dimensions of Adjustment, CDC is Centers for Disease Control and Prevention, and SHEHEA is Survey of Household Expenditure and Household Economic Activities. • **Vital registration complete** identifies countries judged to have complete registries of vital (birth and death) statistics by the United Nations Statistics Division and reported in *Population and Vital Statistics Reports*. Countries with complete vital statistics registries may have more accurate and more timely demographic indicators. • **Latest agricultural census** shows the most recent year in which an agricultural census was conducted and reported to the Food and Agriculture Organization. • **Latest industrial data** refer to the most recent year for which manufacturing value added data at the three-digit level of the International Standard Industrial Classification (revision 2 or 3) are available in the UNIDO database. • **Latest water withdrawal data** refer to the most recent year for which data have been compiled from a variety of sources. See *About the data* for table 3.5 for more information. • **Latest surveys of scientists and engineers engaged in R&D and expenditure for R&D** refer to the most recent year for which data are available from a data collection effort by UNESCO in science and technology and research and development (R&D). See *About the data* for table 5.11 for more information.

Acronyms and abbreviations

Technical terms

AIDS	acquired immunodeficiency syndrome
BOD	biochemical oxygen demand
CFC	chlorofluorocarbon
c.i.f.	cost, insurance, and freight
CO₂	carbon dioxide
COMTRADE	United Nations Statistics Division's Commodity Trade database
CPI	consumer price index
cu. m	cubic meter
DHS	Demographic and Health Survey
DMTU	dry metric ton unit
DOTS	directly observed treatment, short-course (strategy)
DPT	diphtheria, pertussis, and tetanus
DRS	World Bank's Debtor Reporting System
ESAF	Enhanced Structural Adjustment Facility
f.o.b.	free on board
GDP	gross domestic product
GEMS	Global Environment Monitoring System
GIS	geographic information system
GNI	gross national income (formerly referred to as GNP)
GNP	gross national product (now referred to as GNI)
ha	hectare
HIPC	heavily indebted poor country
HIV	human immunodeficiency virus
ICD	International Classification of Diseases
ICRG	International Country Risk Guide
ICSE	International Classification of Status in Employment
ICT	information and communications technology
IP	Internet Protocol
ISCED	International Standard Classification of Education
ISIC	International Standard Industrial Classification
ISP	Internet service provider
kg	kilogram
km	kilometer
kwh	kilowatt-hour
LIBOR	London interbank offered rate
M0	currency and coins (monetary base)
M1	narrow money (currency and demand deposits)
M2	money plus quasi money
M3	broad money or liquid liabilities
mmbtu	millions of British thermal units
mt	metric ton
MUV	manufactures unit value
NEAP	national environmental action plan
NGO	nongovernmental organization
NO₂	nitrogen dioxide
ODA	official development assistance
PC	personal computer
PPI	private participation in infrastructure
PPP	purchasing power parity
R&D	research and development
S&P/IFCG	Standard & Poor's/International Finance Corporation Global (index)
S&P/IFCI	Standard & Poor's/International Finance Corporation Investable (index)
SDR	special drawing right
SITC	Standard International Trade Classification
SNA	System of National Accounts
SO₂	sulfur dioxide
SOPEMI	Continuous Reporting System on Migration
sq. km	square kilometer
STD	sexually transmitted disease
TB	tuberculosis
TFP	total factor productivity
ton-km	metric ton-kilometers
TSP	total suspended particulates

Organizations

ADB	Asian Development Bank
AfDB	African Development Bank
APEC	Asia Pacific Economic Cooperation
CDC	Centers for Disease Control and Prevention
CDIAC	Carbon Dioxide Information Analysis Center
CEC	Commission of the European Community
DAC	Development Assistance Committee of the OECD
EBRD	European Bank for Reconstruction and Development
EDF	European Development Fund
EFTA	European Free Trade Area
EIB	European Investment Bank
EMU	European Monetary Union
EU	European Union
Eurostat	Statistical Office of the European Communities
FAO	Food and Agriculture Organization
FYR	former Yugoslav Republic
G-5	France, Germany, Japan, United Kingdom, and United States
G-7	G-5 plus Canada and Italy
G-8	G-7 plus Russian Federation
GEF	Global Environment Facility
IBRD	International Bank for Reconstruction and Development
ICAO	International Civil Aviation Organization
ICP	International Comparison Programme
ICSID	International Centre for Settlement of Investment Disputes
IDA	International Development Association
IDB	Inter-American Development Bank
IDC	International Data Corporation
IEA	International Energy Agency
IFC	International Finance Corporation
ILO	International Labour Organization
IMF	International Monetary Fund
IRF	International Road Federation
ITU	International Telecommunication Union
IUCN	World Conservation Union
MIGA	Multilateral Investment Guarantee Agency
NAFTA	North American Free Trade Agreement
NATO	North Atlantic Treaty Organization
NSF	National Science Foundation
OECD	Organisation for Economic Co-operation and Development
PAHO	Pan American Health Organization
PARIS21	Partnership in Statistics for Development in the 21st Century
S&P	Standard & Poor's
UIP	Urban Indicators Programme
UN	United Nations
UNAIDS	Joint United Nations Programme on HIV/AIDS
UNCED	United Nations Conference on Environment and Development
UNCHS	United Nations Centre for Human Settlements (Habitat)
UNCTAD	United Nations Conference on Trade and Development
UNDP	United Nations Development Programme
UNECE	United Nations Economic Commission for Europe
UNEP	United Nations Environment Programme
UNESCO	United Nations Educational, Scientific, and Cultural Organization
UNFPA	United Nations Population Fund
UNHCR	United Nations High Commissioner for Refugees
UNICEF	United Nations Children's Fund
UNIDO	United Nations Industrial Development Organization
UNRISD	United Nations Research Institute for Social Development
UNSD	United Nations Statistics Division
USAID	U.S. Agency for International Development
WCMC	World Conservation Monitoring Centre
WFP	World Food Programme
WHO	World Health Organization
WIPO	World Intellectual Property Organization
WITSA	World Information Technology and Services Alliance
WTO	World Trade Organization
WWF	World Wide Fund for Nature

Credits

This book has drawn on a wide range of World Bank reports and numerous external sources, listed in the bibliography following this section. Many people inside and outside the World Bank helped in writing and producing the *World Development Indicators*. The team would like to particularly acknowledge the help and encouragement of Nick Stern and Jo Ritzen. It is also grateful to Jean Baneth, who provided valuable comments on the entire book. This note identifies those who made specific contributions. Numerous others, too many to acknowledge here, helped in many ways for which the team is extremely grateful.

1. World view

was prepared by Sulekha Patel and K. M. Vijayalakshmi. Eric Swanson wrote the introduction with contributions from Sulekha Patel and M. H. Saeed Ordoubadi. Masako Hiraga assisted in preparing tables and figures. The introduction drew heavily on *A Better World for All: Progress towards the International Development Goals* (IMF, OECD, United Nations, and World Bank 2000) and the articles in the "Development Spotlight" section of the October 2000 issue of the *OECD Observer*. Yonas Biru and William Prince provided substantial assistance with the data, preparing the estimates of gross national income in purchasing power parity terms.

2. People

was prepared by Sulekha Patel and Masako Hiraga in partnership with the World Bank's Human Development Network, Development Research Group, and Poverty Reduction and Economic Management Network's Gender Anchor. The Institute of Statistics of the United Nations Educational, Scientific, and Cultural Organization (UNESCO) provided substantial help in preparing the education data for this section. Sulekha Patel wrote the introduction, drawing on the World Bank's *Engendering Development through Gender Equality in Rights, Resources, and Voice* (2001) and with comments and guidance from Lucia Fork, Karen Mason, Elizabeth King, Susan Razzaz, and Waafas Ofosu-Amaah. Contributions to the section were provided by Eduard Bos and Davidson Gwatkin (demography, health, and nutrition); Martin Rama and Raquel Artecona (labor force and employment); Shaohua Chen and Martin Ravallion (poverty and income distribution); Christiaan Grootaert, Montserrat Pallares-Miralles, and Robert Palacios (vulnerability and security); and Lianqin Wang (education). Comments and suggestions at various stages of production also came from Jean Baneth, Eduard Bos, Eric Swanson, and Mead Over.

3. Environment

was prepared by M. H. Saeed Ordoubadi in partnership with the World Bank's Environmentally and Socially Sustainable Development Network and in collaboration with the World Bank's Development Research Group and Transportation, Water, and Urban Development Department. Important contributions were made by Robin White of the World Resources Institute, Orio Tampieri of the Food and Agriculture Organization, Laura Battlebury of the World Conservation Monitoring Centre, and Christine Auclair, Guenter Karl, Bildad Kagai, Markanley Rai, Pauline Maingi, and Moses Ayiemba of the Urban Indicators Programme, United Nations Centre for Human Settlements. Amy Heyman assisted with research and data preparation. John Dixon, Kirk Hamilton, and Nwanze Okidegbe provided invaluable comments and guidance. The World Bank's Environment Department and Rural Development Department devoted substantial staff resources to the book, for which the team is very grateful. Drawing on the discussion draft of the World Bank's environment strategy, M. H. Saeed Ordoubadi wrote the introduction to the section with valuable comments from John Dixon, Kirk Hamilton, Nwanze Okidegbe, and Bruce Ross-Larson. Other contributions were made by Susmita Dasgupta, Craig Meisner, and David Wheeler (water pollution); Jan Bojö, Katja Erickson, and Nanako Tsukahara (government commitment); and Kirk Hamilton and Katie Bolt (genuine savings). Valuable comments were also provided by Jean Baneth, the World Wide Fund for Nature (Madagascar branch), Elena Rabeson, Nadir Mohammad, and Leu Freinkman.

4. Economy

was prepared by K. M. Vijayalakshmi, Amy Heyman, and Eric Swanson in close collaboration with the Macroeconomic Data Team of the World Bank's Development Data Group, led by Soong Sup Lee. Eric Swanson wrote the introduction to this section, drawing valuable ideas from the work of Bill Easterly and Lant Pritchett. Substantial contributions to the section were provided by Barbro Hexeberg and Jong-goo Park (national accounts), Azita Amjadi and Amy Heyman (trade), Punam Chuhan and Ibrahim Levent (external debt), and K. M. Vijayalakshmi (balance of payments and OECD national accounts). The national accounts and balance of payments data for low- and middle-income economies were gathered from the World Bank's regional staff through the annual Unified Survey under the direction of Monica Singh and Mona Fetouh. Maja Bresslauer, Raquel Fok, and Soong Sup Lee worked on updating, estimating, and validating the databases for national accounts. The national accounts data for

OECD countries were processed by Mehdi Akhlaghi. The team is grateful to Guy Karsenty, Andreas Maurer, Vudda Meach, and Wladimir Tislenkoff at the World Trade Organization and Sanja Blazevic and Arunas Butkevicius at the United Nations Conference on Trade and Development (UNCTAD) for providing data on trade in goods and commercial services, to Tetsuo Yamada for help in obtaining the United Nations Industrial Development Organization (UNIDO) database, and to Jean Baneth for helpful comments.

5. States and markets

was prepared by David Cieslikowski in partnership with the World Bank's Private Sector and Infrastructure Network, its Poverty Reduction and Economic Management Network, the International Finance Corporation, and external partners. Amy Wilson gave invaluable assistance in preparing data, supported by Amy Heyman and Anat Lewin. David Cieslikowski wrote the introduction to the section, with substantial inputs from Carlos Braga, Charles Kenny, Robert Schware, Amy Wilson, and Christine Zhen-Wei Qiang. Other contributors include Carol Gabyzon, Ada Karina Izaguirre, Shokraneh Minovi, and Neil Roger (privatization and infrastructure projects); Bo Chung and Shannon Laughlin (Standard & Poor's emerging stock markets indexes); Yonas Biru (purchasing power parity conversion factors); Mariusz Sumlinksi (private investment); Maria Concetta Gasbarro and Michael Minges of the International Telecommunication Union (communications and information); Louis Thompson (transport); and Lise McLeod of the World Intellectual Property Organization (patents data).

6. Global links

was prepared by Amy Heyman and Eric Swanson. Eric Swanson wrote the introduction, drawing in part on ideas developed by Paul Collier, David Dollar, and Zmarak Shalizi. Valuable input was received from Will Martin and Ashoka Mody. Substantial help in preparing the data for this section came from Azita Amjadi and Francis Ng (trade); Betty Dow (commodity prices); Aki Kuwahara of UNCTAD and Jerzy Rozanski (tariffs); Shelly Fu, Ibrahim Levent, and Gloria Moreno (financial data); and Cecile Thoreau and Marc Kircher of the OECD (migration). The team wishes to acknowledge the considerable assistance of Yasmin Ahmad and Rudolphe Petras of the OECD, who provided data on aid flows, and Antonio Massieu and Rosa Songel of the World Tourism Organization.

Other parts

The maps on the inside covers were prepared by the World Bank's Map Design Unit. The *Users guide* was prepared by David Cieslikowski. *Statistical methods* was written by Eric Swanson. *Primary data documentation* was coordinated by K. M. Vijayalakshmi, who served as database administrator. *Acronyms and abbreviations* was prepared by Amy Heyman and Estela Zamora. The index was collated by Richard Fix.

Systems support

Mehdi Akhlaghi was responsible for database management and programming tables. Tariqul Khan was responsible for programming database updates and aggregations. Soong Sup Lee provided valuable systems support. And Estela Zamora provided assistance in updating the databases.

Administrative assistance and office technology support

Estela Zamora provided administrative assistance, and Nacer Megherbi and Shahin Outadi provided office technology support.

Design, production, and editing

Richard Fix coordinated all aspects of production with the Communications Development Incorporated team, led by Terry Fischer. Bruce Ross-Larson edited the section introductions and provided overall direction for design and planning. The design and production team was led by Garrett Cruce and included Megan Klose, and the editing team was led by Alison Strong and included Fiona Blackshaw, Daphne Levitas, Molly Lohman, Susan Quinn, and Stephanie Rostron. The team would also like to thank the design team of Peter Grundy and Tilly Northedge.

Client services

The Development Data Group's Client Services Team, led by Elizabeth Crayford, contributed to the design and planning of the *World Development Indicators* and the *Atlas* and helped coordinate work with the Office of the Publisher.

Publishing and dissemination

The Office of the Publisher, under the direction of Dirk Koehler, provided valuable assistance throughout the production process. Betty Sun, Jamila Abdelghani and Randi Park coordinated production, and Carlos Rossel supervised marketing and distribution. Lawrence MacDonald of Development Economics and Phillip Hay of External Affairs managed the communications strategy, and the regional operations group headed by Paul Mitchell helped coordinate the overseas release.

The Atlas

Production was managed by Richard Fix with guidance from David Cieslikowski and Elizabeth Crayford. The preparation of data benefited from the work on corresponding sections in the *World Development Indicators*. William Prince assisted with systems support and production of tables and graphs. Greg G. Prakas and Jeffrey Lecksell from the World Bank's Map Design Unit coordinated map production.

World Development Indicators CD-ROM

Design, programming, and testing were carried out by Reza Farivari and his team: Azita Amjadi, Ying Chi, Elizabeth Crayford, Sathyanarayanan Govindaraju, Amy Heyman, Tariqul Khan, Anat Lewin, and Nacer Megherbi. Yusri Harun prepared the text files. Masako Hiraga produced the social indicators tables and the education tables. William Prince coordinated production and provided quality assurance.

Client feedback

The team is also grateful to the many people who took the trouble to provide comments on its publications. Their feedback and suggestions have helped to improve this year's edition.

Bibliography

AbouZhar, Carla. 2000. "Maternal Mortality." *OECD Observer* (223): 29–30.

Ahmad, Sultan. 1992. "Regression Estimates of Per Capita GDP Based on Purchasing Power Parities." Policy Research Working Paper 956. World Bank, International Economics Department, Washington, D.C.

———. 1994. "Improving Inter-Spatial and Inter-Temporal Comparability of National Accounts." *Journal of Development Economics* 44: 53–75.

American Automobile Manufacturers Association. 1998. *World Motor Vehicle Data.* Detroit, Mich.

Analysys. 2000. "The Networking Revolution: Opportunities and Challenges for Developing Countries." Information for Development Program (InfoDev) working paper. World Bank, Washington, D.C.

Anker, Richard. 1998. *Gender and Jobs: Sex Segregation of Occupations in the World.* Geneva: International Labour Office.

Ball, Nicole. 1984. "Measuring Third World Security Expenditure: A Research Note." *World Development* 12(2): 157–64.

Barro, Robert J. 1991. "Economic Growth in a Cross-Section of Countries." *Quarterly Journal of Economics* 106(2): 407–44.

Behrman, Jere R., and Mark R. Rosenzweig. 1994. "Caveat Emptor: Cross-Country Data on Education and the Labor Force." *Journal of Development Economics* 44: 147–71.

Bloom, David E., and Jeffrey G. Williamson. 1998. "Demographic Transitions and Economic Miracles in Emerging Asia." *World Bank Economic Review* 12(3): 419–55.

Brown, Lester R., and others. 1999. *Vital Signs 1999: The Environmental Trends That Are Shaping Our Future.* New York and London: W. W. Norton for Worldwatch Institute.

Brown, Lester R., Christopher Flavin, Hilary F. French, and others. 1998. *State of the World 1998.* Washington, D.C.: Worldwatch Institute.

Brown, Lester R., Michael Renner, Christopher Flavin, and others. 1998. *Vital Signs 1998.* Washington, D.C.: Worldwatch Institute.

Bulatao, Rodolfo. 1998. *The Value of Family Planning Programs in Developing Countries.* Santa Monica, Calif.: Rand.

Caiola, Marcello. 1995. *A Manual for Country Economists.* Training Series 1, vol. 1. Washington, D.C.: International Monetary Fund.

Cassen, Robert, and associates. 1986. *Does Aid Work? Report to Intergovernmental Task Force on Concessional Flows.* Oxford: Clarendon Press.

Centro Latinoamericano de Demografía. Various years. *Boletín Demográfico.* Santiago.

Chen, Shaohua, and Martin Ravallion. 2000. "How Did the World's Poorest Fare in the 1990s?" Policy Research Working Paper 2409. World Bank, Development Research Group, Washington, D.C.

Collier, Paul, and David Dollar. 1999. "Aid Allocation and Poverty Reduction." Policy Research Working Paper 2041. World Bank, Development Research Group, Washington, D.C.

Collins, Wanda W., Emile A. Frison, and Suzanne L. Sharrock. 1997. "Global Programs: A New Vision in Agricultural Research." *Issues in Agriculture* (World Bank, Consultative Group on International Agricultural Research, Washington, D.C.) 12: 1–28.

Corrao, Marlo Ann, G. Emmanuel Guindon, Namita Sharma, and Donna Fakhrabadi Shokoohi, eds. 2000. *Tobacco Control Country Profiles.* Atlanta: American Cancer Society.

Demirgüç-Kunt, Asli, and Enrica Detragiache. 1997. "The Determinants of Banking Crises: Evidence from Developed and Developing Countries." Working paper. World Bank and International Monetary Fund, Washington, D.C.

Demirgüç-Kunt, Asli, and Ross Levine. 1996a. "Stock Market Development and Financial Intermediaries: Stylized Facts." *World Bank Economic Review* 10(2): 291–321.

———. 1996b. "Stock Markets, Corporate Finance, and Economic Growth: An Overview." *World Bank Economic Review* 10(2): 223–39.

Dixon, John, and Paul Sherman. 1990. *Economics of Protected Areas: A New Look at Benefits and Costs.* Washington, D.C.: Island Press.

DKT International. 1998. *1997 Contraceptive Social Marketing Statistics.* Washington, D.C.

Doyle, John J., and Gabrielle J. Persley, eds. 1996. *Enabling the Safe Use of Biotechnology: Principles and Practice.* Environmentally Sustainable Development Studies and Monographs Series, no. 10. Washington, D.C.: World Bank.

Drucker, Peter F. 1994. "The Age of Social Transformation." *Atlantic Monthly* 274 (November).

Dunn, David E. 2000. "The Knowledge Divide: Where Some Angels Dare." *OECD Observer,* 22 August. [www.oecdobserver.org/news/] (30 June 2000).

Easterly, William. 2000. "Growth Implosions, Debt Explosions, and My Aunt Marilyn: Do Growth Slowdowns Cause Public Debt Crises?" Policy Research Working Paper 2531. World Bank, Development Research Group, Washington, D.C.

Easterly, William, Michael Kremer, Lant Pritchett, and Lawrence H. Summers. 1993. "Good Policy or Good Luck?" *Journal of Monetary Economics* 32: 459–83.

Economist. 2000. "What the Internet Cannot Do." 19 August, p. 11.

Euromoney. 2000. September. London.

Eurostat (Statistical Office of the European Communities). Various years. *Demographic Statistics.* Luxembourg.

———. Various years. *Statistical Yearbook.* Luxembourg.

Evenson, Robert E., and Carl E. Pray. 1994. "Measuring Food Production (with Reference to South Asia)." *Journal of Development Economics* 44: 173–97.

Faiz, Asif, Christopher S. Weaver, and Michael P. Walsh. 1996. *Air Pollution from Motor Vehicles: Standards and Technologies for Controlling Emissions.* Washington, D.C.: World Bank.

Fallon, Peter, and Zafiris Tzannatos. 1998. *Child Labor: Issues and Directions for the World Bank.* Washington, D.C.: World Bank.

Fankhauser, Samuel. 1995. *Valuing Climate Change: The Economics of the Greenhouse.* London: Earthscan.

FAO (Food and Agriculture Organization). 1986. "Inter-Country Comparisons of Agricultural Production Aggregates." Economic and Social Development Paper 61. Rome.

———. 1996. *Food Aid in Figures 1994.* Vol. 12. Rome.

———. 2001. *State of the World's Forests 2001.* Rome.

———. Various years. *Fertilizer Yearbook.* FAO Statistics Series. Rome.

———. Various years. *Production Yearbook.* FAO Statistics Series. Rome.

———. Various years. *Trade Yearbook.* FAO Statistics Series. Rome.

Feldstein, Martin, and Charles Horioka. 1980. "Domestic Savings and International Capital Flows." *Economic Journal* 90(358): 314–29.

Filmer, Deon. 1999. "Educational Attainment and Enrollment around the World." World Bank, Development Research Group, Washington, D.C. [www.worldbank.org/research/projects/edattain/edpintro.htm].

Filmer, Deon, Elizabeth King, and Lant Pritchett. 1998. "Gender Disparity in South Asia." Policy

Research Working Paper 1867. World Bank, Development Research Group, Washington, D.C.

Frankel, Jeffrey. 1993. "Quantifying International Capital Mobility in the 1990s." In Jeffrey Frankel, ed., *On Exchange Rates.* Cambridge, Mass.: MIT Press.

Frankhauser, Pierre. 1994. "Fractales, tissus urbains et reseaux de transport." *Revue d'economie politique* 104: 435–55.

Fredricksen, Birger. 1993. *Statistics of Education in Developing Countries: An Introduction to Their Collection and Analysis.* Paris: UNESCO.

French, Kenneth, and James M. Poterba. 1991. "Investor Diversification and International Equity Markets." *American Economic Review* 81: 222–26.

Gallup, John L., and Jeffrey D. Sachs. 1998. "The Economic Burden of Malaria." Harvard Institute for International Development, Cambridge, Mass.

Gannon, Colin, and Zmarak Shalizi. 1995. "The Use of Sectoral and Project Performance Indicators in Bank-Financed Transport Operations." TWU Discussion Paper 21. World Bank, Transportation, Water, and Urban Development Department, Washington, D.C.

Gardner, Robert. 1998. "Education." Demographic and Health Surveys, Comparative Study 29. Macro International, Calverton, Md.

Gardner-Outlaw, Tom, and Robert Engelman. 1997. "Sustaining Water, Easing Scarcity: A Second Update." Population Action International, Washington, D.C.

GATT (General Agreement on Tariffs and Trade). 1966. *International Trade 1965.* Geneva.

———. 1989. *International Trade 1988–89.* Geneva.

Goldfinger, Charles. 1994. *L'utile et le futile: L'économie de l'immatériel.* Paris: Editions Odile Jacob.

Greaney, Vincent, and Thomas Kellaghan. 1996. *Monitoring the Learning Outcomes of Education Systems.* A Directions in Development book. Washington, D.C.: World Bank.

Group of Eight (G-8). 2000. *Okinawa Charter on Global Information Society.* Kyushu-Okinawa Summit Meeting 2000. [www.g8kyushu-okinawa.go.jp/e/documents/itl.html] (November 2000).

GTZ (German Agency for Technical Cooperation). 1999. *Fuel Prices and Taxation.* Eschborn, Germany.

Gupta, Sanjeev, Brian Hammond, and Eric Swanson. 2000. "Setting the Goals." *OECD Observer* (223): 15–17.

Haggarty, Luke, and Mary M. Shirley. 1997. "A New Database on State-Owned Enterprises." *World Bank Economic Review* 11(3): 491–513.

Happe, Nancy, and John Wakeman-Linn. 1994. "Military Expenditures and Arms Trade: Alternative Data Sources." IMF Working Paper 94/69. International Monetary Fund, Policy Development and Review Department, Washington, D.C.

Harrison, Ann. 1995. "Factor Markets and Trade Policy Reform." World Bank, Washington, D.C.

Hatter, Victoria L. 1985. *U.S. High-Technology Trade and Competitiveness.* Washington, D.C.: U.S. Department of Commerce.

Hatzichronoglou, Thomas. 1997. "Revision of the High-Technology Sector and Product Classification." STI Working Paper 1997/2. OECD Directorate for Science, Technology, and Industry, Paris.

Heck, W. W. 1989. "Assessment of Crop Losses from Air Pollutants in the U.S." In J. J. McKenzie and M. T. El Ashry, eds., *Air Pollution's Toll on Forests and Crops.* New Haven, Conn.: Yale University Press.

Heggie, Ian G. 1995. *Management and Financing of Roads: An Agenda for Reform.* World Bank Technical Paper 275. Washington, D.C.

Heston, Alan. 1994. "A Brief Review of Some Problems in Using National Accounts Data in Level of Output Comparisons and Growth Studies." *Journal of Development Economics* 44: 29–52.

Hettige, Hemamala, Muthukumara Mani, and David Wheeler. 1998. "Industrial Pollution in Economic Development: Kuznets Revisited." Policy Research Working Paper 1876. World Bank, Development Research Group, Washington, D.C.

IEA (International Energy Agency). Various years. *Energy Balances of OECD Countries.* Paris.

———. Various years. *Energy Statistics and Balances of Non-OECD Countries.* Paris.

———. Various years. *Energy Statistics of OECD Countries.* Paris.

IFC (International Finance Corporation). 2000. *Trends in Private Investment in Developing Countries 2000.* Washington, D.C.

IFPRI (International Food Policy Research Institute). 1999. *Soil Degradation: A Threat to Developing-Country Food Security by 2020.* Washington, D.C.

ILO (International Labour Organization). 1990. *ILO Manual on Concepts and Methods.* Geneva: International Labour Office.

———. 1999. *Key Indicators of the Labour Market.* Geneva: International Labour Office.

———. Various years. *Sources and Methods: Labour Statistics* (formerly *Statistical Sources and Methods*). Geneva: International Labour Office.

———. Various years. *Yearbook of Labour Statistics.* Geneva: International Labour Office.

IMF (International Monetary Fund). 1977. *Balance of Payments Manual.* 4th ed. Washington, D.C.

———. 1986. *A Manual on Government Finance Statistics.* Washington, D.C.

———. 1993. *Balance of Payments Manual.* 5th ed. Washington, D.C.

———. 1995. *Balance of Payments Compilation Guide.* Washington, D.C.

———. 1996a. *Balance of Payments Textbook.* Washington, D.C.

———. 1996b. *Manual on Monetary and Financial Statistics.* Washington, D.C.

———. 1999. *Exchange Arrangements and Exchange Restrictions Annual Report, 1999.* Washington, D.C.

———. Various years. *Balance of Payments Statistics Yearbook.* Parts 1 and 2. Washington, D.C.

———. Various issues. *Direction of Trade Statistics.* Quarterly. Washington, D.C.

———. Various years. *Direction of Trade Statistics Yearbook.* Washington, D.C.

———. Various years. *Government Finance Statistics Yearbook.* Washington, D.C.

———. Various issues. *International Financial Statistics.* Monthly. Washington, D.C.

———. Various years. *International Financial Statistics Yearbook.* Washington, D.C.

IMF (International Monetary Fund), OECD (Organisation for Economic Co-operation and Development), United Nations, and World Bank. 2000. *A Better World for All: Progress towards the International Development Goals.* Washington, D.C.

Institutional Investor. 2000. September. New York.

International Association for the Evaluation of Educational Achievement. 2000. *Third International Mathematics and Science Study, 1999.* Boston College, Lynch School of Education, International Study Center.

International Civil Aviation Organization. 1999. *Civil Aviation Statistics of the World, 1997.* Montreal.

Bibliography

International Helsinki Federation for Human Rights. 2000. *Women 2000: An Investigation into the Status of Women's Rights in Central and South-Eastern Europe and Newly Independent States.* Vienna.

International Road Federation. 2000. *World Road Statistics 2000.* Geneva.

International Telecommunication Union. 1999. *Challenges to the Network: Internet for Development.* Geneva.

———. 2000a. *World Telecommunication Development Report.* Geneva.

———. 2000b. *World Telecommunication Indicators Database.* 5th edition. Geneva.

International Working Group of External Debt Compilers (Bank for International Settlements, International Monetary Fund, Organisation for Economic Co-operation and Development, and World Bank). 1987. *External Debt Definitions.* Washington, D.C.

Inter-Secretariat Working Group on National Accounts (Commission of the European Communities, International Monetary Fund, Organisation for Economic Co-operation and Development, United Nations, and World Bank). 1993. *System of National Accounts.* Brussels, Luxembourg, New York, and Washington, D.C.

IPCC (Intergovernmental Panel on Climate Change). 2001. *Climate Change 2001.* Cambridge: Cambridge University Press.

Irwin, Douglas A. 1996. "The United States in a New Global Economy? A Century's Perspective." Papers and Proceedings of the 108th Annual Meeting of the American Economic Association. *American Economic Review* (May).

Isard, Peter. 1995. *Exchange Rate Economics.* Cambridge: Cambridge University Press.

IUCN (World Conservation Union). 1998. *1997 IUCN Red List of Threatened Plants.* Gland, Switzerland.

———. 2000. *2000 IUCN Red List of Threatened Animals.* Gland, Switzerland.

Journal of Development Economics. 1994. Special issue on Database for Development Analysis. Edited by T. N. Srinivasan. Vol. 44, no. 1.

Kaminsky, Graciela L., Saul Lizondo, and Carmen M. Reinhart. 1997. "Leading Indicators of Currency Crises." Policy Research Working Paper 1852. World Bank, Latin America and the Caribbean Region, Office of the Chief Economist, Washington, D.C.

Kaufmann, Daniel. 1998. "Challenges in the Next Stage of Anti-Corruption." In *New Perspectives on Combating Corruption.* Washington, D.C.: Transparency International and World Bank.

Khandker, Shahidur. 1998. *Fighting Poverty with Microcredit: Experience in Bangladesh.* Washington, D.C.: World Bank.

Knetter, Michael. 1994. "Why Are Retail Prices in Japan So High? Evidence from German Export Prices." NBER Working Paper 4894. National Bureau of Economic Research, Cambridge, Mass.

Kunte, Arundhati, Kirk Hamilton, John Dixon, and Michael Clemens. 1998. "Estimating National Wealth: Methodology and Results." Environmental Economics Series, no. 57. World Bank, Environment Department, Washington, D.C.

Leete, Richard. 2000. "Reproductive Health." *OECD Observer* (223): 31–32.

Lele, Uma, William Lesser, and Gesa Horstkotte-Wessler, eds. 2000. *Intellectual Property Rights in Agriculture: The World Bank's Role in Assisting Borrower and Member Countries.* Washington, D.C.: World Bank.

Lewis, Karen K. 1995. "Puzzles in International Financial Markets." In Gene Grossman and Kenneth Rogoff, eds., *Handbook of International Economics.* Vol. 3. Amsterdam: North Holland.

Lewis, Stephen R., Jr. 1989. "Primary Exporting Countries." In Hollis Chenery and T. N. Srinivasan, eds., *Handbook of Development Economics.* Vol. 2. Amsterdam: North Holland.

Litan, Robert E., and Alice M. Rivlin. 2000. "The Economy and the Internet: What Lies Ahead?" Conference Report 4. Brookings Institute, Washington, D.C.

Lovei, Magdolna. 1997. "Toward Effective Pollution Management." *Environment Matters* (fall): 52–53.

Lucas, R. E. 1988. "On the Mechanics of Economic Development." *Journal of Monetary Economics* 22: 3–22.

Maddison, Angus. 1995. *Monitoring the World Economy 1820–1992.* Paris: OECD.

Mani, Muthukumara, and David Wheeler. 1997. "In Search of Pollution Havens? Dirty Industry in the World Economy, 1960–95." World Bank, Policy Research Department, Washington, D.C.

Mani, Sunil. 1999. "Public Innovation Policies and Developing Countries in a Phase of Economic Liberalization." United Nations University Institute for New Technologies, Maastricht.

McConnell International. 2000. "Risk E-Business: Seizing the Opportunity of Global E-Readiness." In collaboration with World Information Technology and Services Alliance. [www.mcconnellinternational.com].

Midgley, Peter. 1994. *Urban Transport in Asia: An Operational Agenda for the 1990s.* World Bank Technical Paper 224. Washington, D.C.

Monetary Research Institute. 2000. *MRI Bankers' Guide to Foreign Currency* 3(36).

Moody's Investors Service. 2001. *Sovereign, Subnational and Sovereign-Guaranteed Issuers.* January. New York.

Moran, Katy. 1999. "Health: Indigenous Knowledge, Equitable Benefits." IK Notes, no. 15. World Bank, Africa Regional Office, Washington, D.C.

Morgenstern, Oskar. 1963. *On the Accuracy of Economic Observations.* Princeton, N.J.: Princeton University Press.

Murray, Christopher J. L., and Alan D. Lopez. 1996. *The Global Burden of Disease.* Cambridge, Mass.: Harvard University Press.

———, eds. 1998. *Health Dimensions of Sex and Reproduction: The Global Burden of Sexually Transmitted Diseases, HIV, Maternal Conditions, Perinatal Disorders, and Congenital Anomalies.* Cambridge, Mass.: Harvard University Press.

Narayan, Deepa, Robert Chambers, Meera K. Shah, and Patti Petesch. 2000. *Voices of the Poor: Crying Out for Change.* New York: Oxford University Press.

Narayan, Deepa, with Raj Patel, Kai Schafft, Anne Rademacher, and Sarah Koch-Schulte. 2000. *Voices of the Poor: Can Anyone Hear Us?* New York: Oxford University Press.

National Science Foundation. 2000. *Science and Engineering Indicators 2000.* Arlington, Va.

Netcraft. 2000. *Netcraft Secure Server Survey.* [www.netcraft.com].

Obstfeldt, Maurice. 1995. "International Capital Mobility in the 1990s." In P. B. Kenen, ed., *Understanding Interdependence: The Macroeconomics of the Open Economy.* Princeton, N.J.: Princeton University Press.

Obstfeldt, Maurice, and Kenneth Rogoff. 1996. *Foundations of International Macroeconomics.* Cambridge, Mass.: MIT Press.

OECD (Organisation for Economic Co-operation and Development). 1985. *Measuring Health Care 1960–1983: Expenditure, Costs, Performance.* Paris.

———. 1996. *Trade, Employment, and Labour Standards: A Study of Core Workers' Rights and International Trade*. Paris.

———. 1997. *Employment Outlook*. Paris.

———. 1999. *OECD Environmental Data: Compendium 1999*. Paris.

———. Various issues. *Main Economic Indicators*. Monthly. Paris.

———. Various years. *National Accounts*. Vol. 1, *Main Aggregates*. Paris.

———. Various years. *National Accounts*. Vol. 2, *Detailed Tables*. Paris.

———. Various years. *Trends in International Migration: Continuous Reporting System on Migration*. Paris.

OECD, Development Assistance Committee. Various years. *Development Co-operation*. Paris.

———. Various years. *Geographical Distribution of Financial Flows to Aid Recipients: Disbursements, Commitments, Country Indicators*. Paris.

Olson, Elizabeth. 1999. "Drug Groups and U.N. Offices Join to Develop Malaria Cures." *New York Times,* 31 October.

O'Meara, Molly. 1999. "Reinventing Cities for People and the Planet." Worldwatch Paper 147. Worldwatch Institute, Washington, D.C.

O'Rourke, Kevin, and Jeffrey Williamson. 2000. "When Did Globalization Begin?" NBER Working Paper 7632. National Bureau of Economic Research, Cambridge, Mass.

Orrego, Claudio, Carlos Osorio, and Rodrigo Mardones. 2001. "Technological Innovation in Public Sector Reform: Chile's Public Procurement E-System." PREM Note 50. World Bank, Poverty Reduction and Economic Management Network, Washington, D.C.

Palacios, Robert, and Montserrat Pallares-Miralles. 2000. "International Patterns of Pension Provision." Social Protection Discussion Paper 0009. World Bank, Human Development Network, Washington, D.C.

Pearce, David, and Giles Atkinson. 1993. "Capital Theory and the Measurement of Sustainable Development: An Indicator of Weak Sustainability." *Ecological Economics* 8: 103–08.

Pilling, David. 1999. "In Sickness and in Wealth." *Financial Times,* 22 October.

Plucknett, Donald L. 1991. "Saving Lives through Agricultural Research." *Issues in Agriculture* (World Bank, Consultative Group on International Agricultural Research, Washington, D.C.) 16.

Porter, Michael, Jeffrey Sachs, Andrew Warner, and Klaus Schwab. 2000. *The Global Competitiveness Report 2000*. New York: Oxford University Press.

PricewaterhouseCoopers. 1999a. *Corporate Taxes 1999–2000: Worldwide Summaries*. New York.

———. 1999b. *Individual Taxes 1999–2000: Worldwide Summaries*. New York.

Pritchett, Lant. 1996. "Measuring Outward Orientation in Developing Countries: Can It Be Done?" Policy Research Working Paper 566. World Bank, Country Economics Department, Washington, D.C.

———. 2000. "Understanding Patterns of Economic Growth: Searching for Hills among Plateaus, Mountains, and Plains." *World Bank Economic Review* 14(2): 221–50.

Pritchett, Lant, and Geeta Sethi. 1994. "Tariff Rates, Tariff Revenue, and Tariff Reform—Some New Facts." *World Bank Economic Review* 8(1): 1–16.

PRS Group. 2000. *International Country Risk Guide*. CD-ROM. December. East Syracuse, N.Y.

Rama, Martin, and Raquel Artecona. 1999. "A Database of Labor Market Indicators across Countries." World Bank, Development Research Group, Washington, D.C.

Ravallion, Martin. 1996. "Poverty and Growth: Lessons from 40 Years of Data on India's Poor." DECNote 20. World Bank, Development Economics Vice Presidency, Washington, D.C.

Ravallion, Martin, and Shaohua Chen. 1996. "What Can New Survey Data Tell Us about the Recent Changes in Living Standards in Developing and Transitional Economies?" World Bank, Policy Research Department, Washington, D.C.

———. 1997. "Can High-Inequality Developing Countries Escape Absolute Poverty?" *Economic Letters* 56: 51–57.

Robison, Kristopher Kyle, and Edward M. Crenshaw. 2000. "Cyber-Space and Post-Industrial Transformations: Cross-National Analysis of Internet Development." Paper presented at the Annual Meetings of the American Sociological Association, Washington, D.C., 14 August. Ohio State University, Department of Sociology, Columbus.

Rodriguez, Francisco, and Ernest J. Wilson III. 2000. "Are Poor Countries Losing the Information Revolution?" Information for Development Program (InfoDev) working paper. World Bank, Washington, D.C.

Rodrik, Dani. 1996. "Labor Standards in International Trade: Do They Matter and What Do We Do about Them?" Overseas Development Council, Washington, D.C.

Rogoff, Kenneth. 1996. "The Purchasing Power Parity Puzzle." *Journal of Economic Literature* 34: 647–68.

Romer, P. M. 1986. "Increasing Returns and Long-Run Growth." *Journal of Political Economy* 94: 1002–37.

Ruggles, Robert. 1994. "Issues Relating to the UN System of National Accounts and Developing Countries." *Journal of Development Economics* 44(1): 87–102.

Ryten, Jacob. 1998. "Fifty Years of ISIC: Historical Origins and Future Perspectives." ECA/STAT.AC. 63/22. United Nations Statistics Division, New York.

Sen, Amartya. 1988. "The Concept of Development." In Hollis Chenery and T. N. Srinivasan, eds., *Handbook of Development Economics*. Vol. 1. Amsterdam: North Holland.

Serageldin, Ismail. 1995. *Toward Sustainable Management of Water Resources*. A Directions in Development book. Washington, D.C.: World Bank.

Shiklovanov, Igor. 1993. "World Fresh Water Resources." In Peter H. Gleick, ed., *Water in Crisis: A Guide to Fresh Water Resources*. New York: Oxford University Press.

South Pacific Commission. 1999. *Pacific Island Populations Data Sheet 1999*. Noumea, New Caledonia.

Srinivasan, T. N. 1991. "Development Thought, Policy, and Strategy, Then and Now." Background paper to *World Development Report 1991*. World Bank, Washington, D.C.

———. 1994. "Database for Development Analysis: An Overview." *Journal of Development Economics* 44(1): 3–28.

Standard & Poor's. 2000a. *Emerging Stock Markets Factbook 2000*. New York.

———. 2000b. *The S&P Emerging Market Indices: Methodology, Definitions, and Practices*. New York.

———. 2001. *Credit Week*. January. New York.

Syrquin, Moshe. 1988. "Patterns of Structural Change." In Hollis Chenery and T. N. Srinivasan, eds., *Handbook of Development Economics*. Vol. 1. Amsterdam: North Holland.

Taylor, Alan M. 1996a. "International Capital Mobility in History: Purchasing Power Parity in the

Long Run." NBER Working Paper 5742. National Bureau of Economic Research, Cambridge, Mass.

———. 1996b. "International Capital Mobility in History: The Saving-Investment Relationship." NBER Working Paper 5743. National Bureau of Economic Research, Cambridge, Mass.

TeleGeography and International Telecommunication Union. 2000. *Direction of Traffic 1999.* Washington, D.C.

UNAIDS (Joint United Nations Programme on HIV/AIDS) and WHO (World Health Organization). 2000. *AIDS Epidemic Update.* December. [www.unaids.org].

UNCTAD (United Nations Conference on Trade and Development). Various years. *Handbook of International Trade and Development Statistics.* Geneva.

UNDP (United Nations Development Programme). 1995. *Human Development Report 1995.* New York: Oxford University Press.

UNECE (United Nations Economic Commission for Europe). 2000. *Women and Men in Europe and North America.* New York and Geneva.

UNEP (United Nations Environment Programme). 1991. *Urban Air Pollution.* Nairobi.

UNEP (United Nations Environment Programme) and WHO (World Health Organization). 1992. *Urban Air Pollution in Megacities of the World.* Cambridge, Mass.: Blackwell.

———. 1995. *City Air Quality Trends.* Nairobi.

UNESCO (United Nations Educational, Scientific, and Cultural Organization). 1998. *World Education Report 1998.* Paris: UNESCO Publishing and Bernan Press.

———. Various years. *Statistical Yearbook.* Paris: UNESCO Publishing and Bernan Press.

UNFPA (United Nations Population Fund). 2000. *Lives Together, Worlds Apart: The State of the World Population.* New York.

UNICEF (United Nations Children's Fund). 1999. *Women in Transition.* CEE/CIS/Baltics Regional Monitoring Report 6. MONEE Project. Florence.

———. Various years. *The State of the World's Children.* New York: Oxford University Press.

UNIDO (United Nations Industrial Development Organization). Various years. *International Yearbook of Industrial Statistics.* Vienna.

United Nations. 1947. *Measurement of National Income and the Construction of Social Accounts.* New York.

———. 1968. *A System of National Accounts: Studies and Methods.* Series F, no. 2, rev. 3. New York.

———. 1990. *International Standard Industrial Classification of All Economic Activities, Third Revision.* Statistical Papers Series M, no. 4, rev. 3. New York.

———. 1992. *Handbook of the International Comparison Programme.* Studies in Methods, Series F, no. 62. New York.

———. 1993. *System of National Accounts.* New York.

———. 2000a. *We the Peoples: The Role of the United Nations in the 21st Century.* New York.

———. 2000b. *The World's Women 2000: Trends and Statistics.* Department of Economic and Social Affairs. New York.

United Nations Administrative Committee on Coordination, Subcommittee on Nutrition. Various years. *Update on the Nutrition Situation.* Geneva.

United Nations Economic and Social Commission for Western Asia. 1997. *Purchasing Power Parities: Volume and Price Level Comparisons for the Middle East, 1993.* E/ESCWA/STAT/1997/2. Amman.

United Nations Population Division. 1996. *International Migration Policies 1995.* New York.

———. 1998. *World Population Prospects: The 1998 Revision.* New York.

———. 2000. *World Urbanization Prospects: The 1999 Revision.* New York.

———. Various years. *Levels and Trends of Contraceptive Use.* New York.

United Nations Statistics Division. 1985. *National Accounts Statistics: Compendium of Income Distribution Statistics.* New York.

———. 1993a. *Integrated Environmental and Economic Accounting.* New York.

———. 1993b. *International Trade Statistics Yearbook.* Vol. 1. New York.

———. Various years. *Energy Statistics Yearbook.* New York.

———. Various years. *International Trade Statistics Yearbook.* New York.

———. Various issues. *Monthly Bulletin of Statistics.* New York.

———. Various years. *National Accounts Statistics: Main Aggregates and Detailed Tables.* Parts 1 and 2. New York.

———. Various years. *National Income Accounts.* New York.

———. Various years. *Population and Vital Statistics Report.* New York.

———. Various years. *Statistical Yearbook.* New York.

U.S. Census Bureau. 1990. *1990 Census of Population Listing.* Washington, D.C.

———. 2000. *Current Population Report.* March. Washington, D.C.

U.S. Department of Health and Human Services. 1997. *Social Security Systems throughout the World.* Washington, D.C.

U.S. Department of State, Bureau of Verification and Compliance. 1999. *World Military Expenditures and Arms Transfers 1998.* Washington, D.C.

U.S. Environmental Protection Agency. 1995. *National Air Quality and Emissions Trends Report 1995.* Washington, D.C.

Walsh, Michael P. 1994. "Motor Vehicle Pollution Control: An Increasingly Critical Issue for Developing Countries." World Bank, Washington, D.C.

Watson, Robert, John A. Dixon, Steven P. Hamburg, Anthony C. Janetos, and Richard H. Moss. 1998. *Protecting Our Planet, Securing Our Future: Linkages among Global Environmental Issues and Human Needs.* A joint publication of the United Nations Environment Programme, U.S. National Aeronautics and Space Administration, and World Bank. Nairobi and Washington, D.C.

WCEFA Inter-Agency Commission (United Nations Development Programme, United Nations Educational, Scientific, and Cultural Organization, United Nations Children's Fund, and World Bank). 1990. *World Conference on Education for All: Meeting Basic Learning Needs—Final Report.* Jomtien, Thailand.

WCMC (World Conservation Monitoring Centre). 1992. *Global Biodiversity: Status of the Earth's Living Resources.* London: Chapman and Hall.

———. 1994. *Biodiversity Data Sourcebook.* Cambridge: World Conservation Press.

WHO (World Health Organization). 1977. *International Classification of Diseases.* 9th rev. Geneva.

———. 1997. *Coverage of Maternity Care.* Geneva.

———. 1999. *Global Tuberculosis Control Report 1999.* Geneva.

———. Various years. *World Health Report.* Geneva.

———. Various years. *World Health Statistics Annual.* Geneva.

WHO (World Health Organization) and UNICEF (United Nations Children's Fund). 1992. *Low*

Birth Weight: A Tabulation of Available Information. Geneva.

———. 2000. Global Water Supply and Sanitation Assessment 2000 Report. Geneva.

Windham, Douglas M. 1988. Indicators of Educational Effectiveness and Efficiency. Tallahassee, Fla.: Florida State University, Educational Efficiency Clearinghouse.

WITSA (World Information Technology and Services Alliance). 2000. Digital Planet 2000: The Global Information Economy. Based on research by International Data Corporation. Vienna, Va.

Wolf, Holger C. 1997. "Patterns of Intra- and Inter-State Trade." NBER Working Paper 5939. National Bureau of Economic Research, Cambridge, Mass.

Wolfensohn, James D. 2000a. "Development and International Cooperation in the Twenty-First Century: The Role of Information Technology in the Context of a Knowledge-Based Global Economy." Statement at the United Nations Economic and Social Council, New York, 5 July.

———. 2000b. "New Possibilities in Information Technology and Knowledge for Development in a Global Economy." Speech at Cambridge University, 24 June.

World Bank. 1990. World Development Report 1990: Poverty. New York: Oxford University Press.

———. 1991a. Developing the Private Sector: The World Bank's Experience and Approach. Washington, D.C.

———. 1991b. Gender and Poverty in India. A World Bank Country Study. Washington, D.C.

———. 1991c. World Development Report 1991: The Challenge of Development. New York: Oxford University Press.

———. 1992. World Development Report 1992: Development and the Environment. New York: Oxford University Press.

———. 1993a. The Environmental Data Book: A Guide to Statistics on the Environment and Development. Washington, D.C.

———. 1993b. Purchasing Power Parities: Comparing National Incomes Using ICP Data. Washington, D.C.

———. 1993c. World Development Report 1993: Investing in Health. New York: Oxford University Press.

———. 1995a. Private Sector Development in Low-Income Countries. Development in Practice series. Washington, D.C.

———. 1995b. Toward Gender Equality: The Role of Public Policy. Development in Practice series. Washington, D.C.

———. 1996a. Environment Matters (summer). Environment Department, Washington, D.C.

———. 1996b. Livable Cities for the 21st Century. A Directions in Development book. Washington, D.C.

———. 1996c. National Environmental Strategies: Learning from Experience. Environment Department, Washington, D.C.

———. 1997a. Can the Environment Wait? Priorities for East Asia. Washington, D.C.

———. 1997b. Expanding the Measure of Wealth: Indicators of Environmentally Sustainable Development. Environmentally Sustainable Development Studies and Monographs Series, no. 17. Washington, D.C.

———. 1997c. Private Capital Flows to Developing Countries: The Road to Financial Integration. A World Bank Policy Research Report. New York: Oxford University Press.

———. 1997d. Rural Development: From Vision to Action. Environmentally Sustainable Development Studies and Monographs Series, no. 12. Washington, D.C.

———. 1997e. Sector Strategy: Health, Nutrition, and Population. Human Development Network, Washington, D.C.

———. 1997f. World Development Report 1997: The State in a Changing World. New York: Oxford University Press.

———. 1998. 1998 Catalog: Operational Documents as of July 31, 1998. Washington, D.C.

———. 1999a. Fuel for Thought: Environmental Strategy for the Energy Sector. Environment Department, Energy, Mining, and Telecommunications Department, and International Finance Corporation, Washington, D.C.

———. 1999b. Greening Industry: New Roles for Communities, Markets, and Governments. A World Bank Policy Research Report. New York: Oxford University Press.

———. 1999c. Health, Nutrition, and Population Indicators: A Statistical Handbook. Human Development Network, Washington, D.C.

———. 1999d. World Development Report 1999/2000—Entering the 21st Century: The Changing Development Landscape. New York: Oxford University Press.

———. 2000a. Exploring the Implications of the HIV/AIDS Epidemic for Educational Planning in Selected African Countries: The Demographic Question. Washington, D.C.

———. 2000b. "Harness Global Knowledge, Information, and Connectivity for Development." News release 2001/016/S. Washington, D.C.

———. 2000c. Poverty Reduction and the World Bank: Progress in Fiscal 1999. Washington, D.C.

———. 2000d. Trade Blocs. A World Bank Policy Research Report. New York: Oxford University Press.

———. 2000e. World Development Report 2000/2001: Attacking Poverty. New York: Oxford University Press.

———. 2001. Engendering Development through Gender Equality in Rights, Resources, and Voice. A World Bank Policy Research Report. New York: Oxford University Press.

———. Forthcoming. Purchasing Power Parities: International Comparison of Volume and Price Levels. Washington, D.C.

———. Various issues. Global Commodity Markets. Quarterly. Washington, D.C.

———. Various years. Global Development Finance (formerly World Debt Tables). Washington, D.C. (Also available on CD-ROM.)

———. Various years. Global Economic Prospects and the Developing Countries. Washington, D.C.

———. Various years. World Development Indicators. Washington, D.C.

World Economic Forum. 2000. Global Digital Divide Initiative 2000: From the Global Digital Divide to the Global Digital Opportunity. Davos.

World Energy Council. 1995. Global Energy Perspectives to 2050 and Beyond. London.

World Intellectual Property Organization. 2000. Industrial Property Statistics. Publication A. Geneva.

World Resources Institute, International Institute for Environment and Development, and IUCN (World Conservation Union). Various years. World Directory of Country Environmental Studies. Washington, D.C.

World Resources Institute, UNEP (United Nations Environment Programme), and UNDP (United Nations Development Programme). 1994. World Resources 1994–95: A Guide to the Global Environment. New York: Oxford University Press.

World Resources Institute, UNEP (United Nations Environment Programme), UNDP (United Nations Development Programme), and World Bank. Various years. World Resources: A Guide *to the Global Environment.* New York: Oxford University Press.

World Tourism Organization. 2000a. *Compendium of Tourism Statistics 1994–98.* 20th ed. Madrid.

———. 2000b. *Yearbook of Tourism Statistics.* Vols. 1 and 2. Madrid.

WTO (World Trade Organization). Various years. *Annual Report.* Geneva.

Zimmermann, Klaus F. 1995. "European Migration: Push and Pull." In Michael Bruno and Boris Pleskovic, eds., *Proceedings of the World Bank Annual Conference on Development Economics 1994.* Washington, D.C.: World Bank.

Zook, Matthew. 2000. "Internet Metrics: Using Host and Domain Counts to Map the Internet." *International Journal on Knowledge Infrastructure Development, Management and Regulation* (University of California at Berkeley) 24(6/7).

Index of indicators

References are to table numbers.

Index of indicators

F

G

Index of indicators

Index of indicators

U

V

W